D1159626

Introduction to
Mathematical Programming

Introduction to
Mathematical Programming

Frederick S. Hillier
Professor of Operations Research
Stanford University

Gerald J. Lieberman
Professor of Operations Research and Statistics
Stanford University

McGraw-Hill Publishing Company

New York   St. Louis   San Francisco   Auckland   Bogotá
Caracas   Hamburg   Lisbon   London   Madrid   Mexico
Milan   Montreal   New Delhi   Oklahoma City   Paris
San Juan   São Paulo   Singapore   Sydney   Tokyo   Toronto

Introduction to Mathematical Programming

Copyright © 1990 by McGraw-Hill, Inc. All rights reserved. Printed in the United States of America. Except as permitted under the United States Copyright Act of 1976, no part of this publication may be reproduced or distributed in any form or by any means, or stored in a data base or retrieval system, without the prior written permission of the publisher.

1 2 3 4 5 6 7 8 9 0 DOH DOH 9 5 4 3 2 1 0

P/N   028939-5

This book was set in Times Roman by Waldman Graphics, Inc.
The editors were Eric M. Munson and Scott Amerman;
the production supervisor was Janelle S. Travers.
Design was done by Caliber Design Planning, Inc.
R. R. Donnelley & Sons Company was printer and binder.

Library of Congress Cataloging-in-Publication Data

Hillier, Frederick S.
    Introduction to mathematical programming / Frederick S. Hillier, Gerald J. Lieberman.
       p.   cm.
    P/N 028939-5
    1. Programming (Mathematics)   I. Lieberman, Gerald J.   II. Title.
    T57.7.H55   1990
    519.7—dc20   89-49756

About the Authors

Frederick S. Hillier was born and raised in Aberdeen, Washington, where he was an award winner in statewide high school contests in essay writing, mathematics, debate, and music. As an undergraduate at Stanford University, he ranked first in his engineering class, won the McKinsey Prize for technical writing, and won the Hamilton Award for combining excellence in engineering with notable achievements in the humanities and social sciences. Upon his graduation with a B.S. degree in Industrial Engineering, he was awarded three national fellowships (National Science Foundation, Tau Beta Pi, and Danforth) for graduate study at Stanford with specialization in operations research. After receiving his Ph.D. degree, he joined the faculty of Stanford University, where he is now Professor of Operations Research.

Dr. Hillier's research has extended into a variety of areas, including integer programming, queueing theory and its application, statistical quality control, and the application of operations research to the design of production systems and to capital budgeting. He has published widely, and his seminal papers have been selected for republication in books of selected readings at least ten times. He was the first-prize winner of a research contest on "Capital Budgeting of Interrelated Projects" sponsored by The Institute of Management Sciences and the U.S. Office of Naval Research. He also has served as Treasurer of the Operations Research Society of America, Vice President for Meetings of The Institute of Management Sciences (TIMS), and Co-General Chairman of the 1989 TIMS International Meeting in Osaka, Japan.

In addition to *Introduction to Mathematical Programming* and the two companion volumes, *Introduction to Operations Research* and *Introduction to Stochastic Models in Operations Research*, his books are *The Evaluation of Risky Interrelated Investments* (North-Holland, 1969) and *Queueing Tables and Graphs* (Elsevier North-Holland, 1981, co-authored by O. S. Yu, with D. M. Avis, L. D. Fossett, F. D. Lo, and M. I. Reiman).

Gerald J. Lieberman is Professor of Operations Research and Statistics at Stanford University. He has served as Vice Provost and Dean of Graduate Studies and Research at Stanford and was the founding chair of the Department of Operations Research. He is both an engineer (having received an undergraduate degree in mechanical engineering from Cooper Union) and an operations research statistician (with an A.M. from Columbia University in mathematical statistics and a Ph.D. from Stanford University in statistics).

His research interests have been in the stochastic areas of operations research, often at the interface of applied probability and statistics. He has extensive publications in the areas of reliability and quality control, and in the modeling of complex systems, including their optimal design, when resources are limited.

Dr. Lieberman's professional honors include being elected to the National Academy of Engineering, receiving the Shewhart Medal of the American Society for Quality Control, receiving the Cuthbertson Award for exceptional service to Stanford University, and serving as a fellow at the Center for Advanced Study in the Behavioral Sciences. He also served as President of The Institute of Management Sciences.

In addition to *Introduction to Mathematical Programming* and the two companion volumes, *Introduction to Operations Research* and *Introduction to Stochastic Models in Operations Research*, his books are *Handbook of Industrial Statistics* (Prentice-Hall, 1955, co-authored by A. H. Bowker), *Tables of the Non-Central t-Distribution* (Stanford University Press, 1957, co-authored by G. J. Resnikoff), *Tables of the Hypergeometric Probability Distribution* (Stanford University Press, 1961, co-authored by D. Owen), and *Engineering Statistics*, Second Edition (Prentice-Hall, 1972, co-authored by A. H. Bowker).

To Our Parents

Contents

Preface

During the 22 years since the initial publication of our textbook *Introduction to Operations Research* we have witnessed many new developments and considerable growth in the field. As subsequent editions expanded to keep up with this growth, many survey courses in operations research using this book were split into more specialized courses. Frequently, the new courses cover either mathematical programming (perhaps entitled "Deterministic Models in Operations Research") or stochastic models in operations research.

To accommodate this evolution, the current publication of the even bigger fifth edition of *Introduction to Operations Research* (nearly 1,000 pages) is being accompanied by the publication of two more compact split volumes, namely, *Introduction to Mathematical Programming* and *Introduction to Stochastic Models in Operations Research*. All of the material in the fifth edition has been duplicated in one or the other of the split volumes (and occasionally in both). In particular, this book, *Introduction to Mathematical Programming*, includes Part 1 (Introduction) with slight alterations, as well as Part 2 (Linear Programming), Part 3 (Mathematical Programming), and the first four appendices from the fifth edition. Because of its relevance for a course on deterministic models in operations research, the deterministic portion of Chap. 18 of *Introduction to Operations Research* (Inventory Theory) also has been included here as Chap. 15.

In addition, this book includes some topics from previous editions of *Introduction to Operations Research* that were subsequently deleted for space reasons. This material appears here as Secs. 7.6 (Multitime Period Problems), 7.7 (Multidivisional Multitime Period Problems), 7.8 (Stochastic Programming), 7.9 (Chance-Constrained Programming), and 9.5 (The Decomposition Principle for Multidivisional Problems), as well as a subsection (An Algorithm for the Assignment Problem) at the end of Sec. 7.4.

Those familiar with the fourth edition of *Introduction to Operations Research* will find that considerable new material has been added to the chapters appearing in both this book and the fifth edition. The major new topics are (1) a nontechnical introduction and evaluation of the new interior-point approach to solving linear programming problems proposed by N. Karmarkar (beginning of Sec. 4.9), (2) a technical introduction to this interior-point approach that highlights the most important concepts in an elementary way (Sec. 9.4), (3) the emerging new role for microcomputers in implementing some of the algorithms of operations research, including linear programming (end of Sec. 4.9, etc.), (4) the minimum cost flow problem (capacitated transshipment problem) and its special cases (Sec. 10.6), (5) the network simplex method for solving minimum cost flow problems (Sec. 10.7), and (6) an introduction to a recent algorithmic breakthrough for solving large binary integer programming problems by combining automatic problem preprocessing and the generation of cutting planes with clever branch-and-bound techniques (Sec. 13.6). In addition, several of the chapters here (Chaps. 3, 4, 5, 6, 7, 10, and 13) received a major revision, both to update the material and to increase the clarity of the presentation. Our goal has been to make this a real "student's book"—clear, interesting, and well organized, with lots of helpful examples and illustrations, good motivation and perspective, easy-to-find important material, enjoyable homework, etc.—without too much notation, terminology, and dense mathematics.

One notable feature in striving toward this goal is the incorporation of software that has been specifically designed to be a teaching supplement to this book. This software is packaged in the back of the book for use on either an IBM (or IBM-compatible) personal computer with a graphics card or a Macintosh. As described in Sec. 1.6, this is very different from the usual software now widely available for running the algorithms of operations research on microcomputers. Instead, it is true tutorial software for helping to learn the material in this book. The demonstration examples, the routines for interactive execution of algorithms, and the routines for post-optimality analysis should make the learning process far more efficient and effective, as well as more stimulating and enjoyable. We are convinced that this type of innovative software will usher in a new era for operations research education.

We feel very fortunate to have had the services of a talented and creative operations research student (and one with good family connections), Mark Hillier, to develop this software.

The prerequisites for a course using this book can be relatively modest because the mathematics has been kept at a relatively elementary level. Most of the material requires no mathematics beyond high school algebra. Calculus is used only in Chap. 14 (Nonlinear Programming) and in one example in Chap. 11 (Dynamic Programming). However, the mathematical maturity that a student achieves through taking an elementary calculus course is useful for the more advanced material throughout the book. Matrix notation is used in parts of Chap. 5 (The Theory of the Simplex Method), Chap. 6 (Duality Theory and Sensitivity Analysis), Chap. 9 (Other Algorithms for Linear Programming), and Chap. 14 (Nonlinear Programming), but the only background needed for this is presented in Appendix 3. An acquaintance with elementary probability theory is assumed in Sec. 7.8 (Stochastic Programming), Sec. 7.9 (Chance-Constrained Programming), and part of Sec. 10.8 (Project Planning and Control with PERT-CPM).

The content of the book is aimed largely at the upper division undergraduate level and at first-year (Master's level) graduate students. There is ample material for a one-quarter or one-semester course with some selectivity of topics, or for a longer course. There are several options for the scope of a one-term course. One is a general survey of mathematical programming (or deterministic models in operations research) that might, for example, cover Chaps. 1–4, a selection of topics from Chaps. 5–9, and most of Chaps. 10–14. A second option is a course just on linear programming that would cover most of Chaps. 1–10 and perhaps Chaps. 12 and 13. A third option is a course on other deterministic models, with linear programming as a prerequisite, where the focus is on Chaps. 10–15. A two-term course that combines the second and third options also is a ready possibility.

The book has been organized to provide great flexibility in the selection of topics. Chapters 1 and 2 provide a general introduction, and then Chap. 3 introduces linear programming. The remaining chapters can be covered nearly independently, except that they all use basic material presented in Chap. 3 and perhaps Chap. 4. There are only a few places where another prior chapter is used. In particular, Chap. 6 and Sec. 9.4 also draw upon Chap. 5. Sections 9.2 and 9.3 use parts of Chap. 6, and Sec. 9.5 deals with the problem structure formulated in Sec. 7.5. Section 10.6 assumes an acquaintance with the problem formulations in Secs. 7.1, 7.3, and 7.4, while prior exposure to Secs. 7.2 and 9.1 is helpful (but not essential) in Sec. 10.7.

We thank our wives, Ann and Helen, for their editorial and word processing assistance, as well as their encouragement and understanding when we devoted too many evenings and weekends to preparing this book and the companion volumes. Our children, David, John, and Mark Hillier, Janet Lieberman Argyres, and Joanne, Michael, and Diana Lieberman also have provided strong support in a variety of ways. Most of them have used *Introduction of Operations Research* as a text in their own college courses, given considerable advice, and even (in the case of Mark Hillier) become a full-fledged collaborator. It is a joy to have them on the team.

Acknowledgments

We are deeply indebted to many people for their part in making possible the preparation of this book and the two companion volumes. Those making helpful comments are literally too numerous to mention. However, we would like to acknowledge particularly the advice and support of Stephen Argyres, John Birge, Richard Cottle, Hans Daellenbach, George Dantzig, Cyrus Derman, B. Curtis Eaves, Philip Gill, Peter Glynn, David Hillier, Mark Hillier, Donald Iglehart, Carolyn Kinser, Steve Love, Tom Magnanti, Alan Manne, Douglas Montgomery, Walter Murray, Stephen Robinson, Matthew Rosenshine, Sheldon Ross, Michael Saunders, Siegfried Schaible, Gerald Shedler, N. D. Singpurwalla, John Tomlin, and Arthur Veinott, Jr. We also thank dozens of Stanford students who gave us helpful written suggestions at our request.

Ann Hillier devoted numerous long days and nights to sitting with a Macintosh, doing word processing, and constructing many figures and tables. She was a vital member of the team.

It was a real pleasure working with McGraw-Hill's thoroughly professional editorial and production staff, including Eric Munson, John Corrigan, and Scott Amerman.

FREDERICK S. HILLIER
GERALD J. LIEBERMAN

Introduction to
Mathematical Programming

1

Introduction

Mathematical programming is a relatively young field, having emerged during the middle decades of the twentieth century. However, its impact already has been quite remarkable. It has become a standard tool for improving the efficiency of business operations around the world. Indeed, a significant part of the continuing improvements in the economic productivity of the industrialized nations can be attributed to the use of mathematical programming.

The mathematical programming approach is to construct a *mathematical model* to represent the problem of interest. The model sometimes will include hundreds or thousands of variables to represent the *decisions* to be made, along with a similar number of inequalities (or equations) to represent the *constraints* on the decisions. An *algorithm* then is used to solve for an *optimal solution* with respect to the model.

Mathematical programming actually is part of a larger field called **operations research** that uses this same approach. Roughly speaking, mathematical programming is the part of operations research that focuses primarily on *deterministic* problems where probability theory is not needed. The other major area of operations research deals with systems that evolve *probabilistically* over time, as described in a companion volume, *Introduction to Stochastic Models in Operations Research*.

1

The overall operations research approach remains basically the same, regardless of whether the mathematical model eventually constructed falls into the category of mathematical programming or stochastic models. Therefore, before turning to the specific models of mathematical programming in Chap. 3 onward, we first give a general introduction to operations research in this chapter and then an overview of the operations research modeling approach in Chap. 2.

1.1 The Origins of Operations Research

Since the advent of the industrial revolution, the world has seen a remarkable growth in the size and complexity of organizations. The artisans' small shops of an earlier era have evolved into the billion-dollar corporations of today. An integral part of this revolutionary change has been a tremendous increase in the division of labor and segmentation of management responsibilities in these organizations. The results have been spectacular. However, along with its blessings, this increasing specialization has created new problems, problems that are still occurring in many organizations. One problem is a tendency for the many components of an organization to grow into relatively autonomous empires with their own goals and value systems, thereby losing sight of how their activities and objectives mesh with those of the overall organization. What is best for one component frequently is detrimental to another, so they may end up working at cross purposes. A related problem is that as the complexity and specialization in an organization increase, it becomes more and more difficult to allocate its available resources to its various activities in a way that is most effective for the organization as a whole. These kinds of problems and the need to find a better way to resolve them provided the environment for the emergence of operations research.

The roots of operations research can be traced back many decades, when early attempts were made to use a scientific approach in the management of organizations. However, the beginning of the activity called *operations research* has generally been attributed to the military services early in World War II. Because of the war effort, there was an urgent need to allocate scarce resources to the various military operations and to the activities within each operation in an effective manner. Therefore the British and then the American military management called upon a large number of scientists to apply a scientific approach to dealing with this and other strategic and tactical problems. In effect they were asked to do research on (military) operations. These teams of scientists were the first operations research teams. Their efforts allegedly were instrumental in winning the Air Battle of Britain, the Island Campaign in the Pacific, the Battle of the North Atlantic, and others.

Spurred on by the apparent success of operations research in the military, industry gradually became interested in this new field. As the industrial boom following the war was running its course, the problems caused by the increasing complexity and specialization in organizations were again coming to the forefront. It was becoming apparent to a growing number of people, including business consultants who had served on or with the operations research teams during the war, that these were basically the same problems that had been faced by the military but in a different context. In this way operations research began to creep into industry, business, and civil government. By 1951, it had already taken hold in Great Britain and was in the process of doing so in the United States. Since then the field has developed very rapidly, as will be described further in Sec. 1.3.

At least two other factors that played a key role in the rapid growth of operations research during this period can be identified. One was the substantial progress that was made early in improving the techniques available to operations research. After the war, many of the scientists who had participated on operations research teams or who had heard about this work were motivated to pursue research relevant to the field; important advancements in the state of the art resulted. A prime example is the simplex method for solving linear programming problems, developed by George Dantzig in 1947. Many of the standard tools of operations research, e.g., linear programming, dynamic programming, queueing theory, and inventory theory, were relatively well developed before the end of the 1950s. In addition to this rapid advancement in the theory of operations research, a second factor that gave great impetus to the growth of the field was the onslaught of the computer revolution. A large amount of computation is usually required to deal most effectively with the complex problems typically considered by operations research. Doing this by hand would often be out of the question. Therefore, the development of electronic digital computers, with their ability to perform arithmetic calculations thousands or even millions of times faster than a human being can, was a tremendous boon to operations research. Today, mainframe computers, minicomputers, or microcomputers are essential for solving real-world operations research problems.

1.2 The Nature of Operations Research

What is operations research? One way of trying to answer this question is to give a definition. For example, operations research may be described as a scientific approach to decision making that involves the operations of organizational systems. However, this description, like earlier attempts at a definition, is so general that it is equally applicable to many other fields as well. Therefore, perhaps the best way of grasping the unique nature of operations research is to examine its outstanding characteristics.

As its name implies, operations research involves "research on operations." This says something about both the approach and the area of application of the field. Thus operations research is applied to problems that concern how to conduct and coordinate the operations or activities within an organization. The nature of the organization is essentially immaterial, and, in fact, operations research has been applied extensively in business, industry, the military, civil government and agencies, hospitals, and so forth. Therefore, the breadth of application is unusually wide. The approach of operations research is that of the scientific method. In particular, the process begins by carefully observing and formulating the problem and then constructing a scientific (typically mathematical) model that attempts to abstract the essence of the real problem. It is then hypothesized that this model is a sufficiently precise representation of the essential features of the situation, so that the conclusions (solutions) obtained from the model are also valid for the real problem. This hypothesis is then modified and verified by suitable experimentation. Thus in a certain sense operations research involves creative scientific research into the fundamental properties of operations. However, there is more to it than this. Specifically, operations research is also concerned with the practical management of the organization. Therefore, to be successful it must also provide positive, understandable conclusions to the decision maker(s) when they are needed.

Still another characteristic of operations research is its broad viewpoint. As implied in the preceding section, operations research adopts an organizational point of view. Thus it attempts to resolve the conflicts of interest among the components of the organization in a way that is best for the organization as a whole. This does not imply that the study of each problem must give explicit consideration to all aspects of the organization; rather, the objectives being sought must be consistent with those of the overall organization. An additional characteristic that was mentioned in passing is that operations research attempts to find the best or optimal solution to the problem under consideration. Rather than being content with merely improving the status quo, the goal is to identify the best possible course of action. Although it must be interpreted carefully, this "search for optimality" is a very important theme in operations research.

All these characteristics lead quite naturally to still another one. It is evident that no single individual should be expected to be an expert on all the many aspects of operations research work or the problems typically considered; this would require a group of individuals having diverse backgrounds and skills. Therefore, when undertaking a full-fledged operations research study of a new problem, it is usually necessary to use a team approach. Such an operations research team typically needs to include individuals who collectively are highly trained in mathematics, statistics and probability theory, economics, business administration, electronic computing, engineering and the physical sciences, the behavioral sciences, and the special techniques of operations research. The team also needs to have the necessary experience and variety of skills to give appropriate consideration to the many ramifications of the problem throughout the organization and to execute effectively all the diverse phases of the operations research study.

In summary, operations research is concerned with optimal decision making in, and modeling of, deterministic and probabilistic systems that originate from real life. These applications, which occur in government, business, engineering, economics, and the natural and social sciences, are characterized largely by the need to allocate limited resources. In these situations, considerable insight can be obtained from scientific analysis such as that provided by operations research. The contribution from the operations research approach stems primarily from:

1. Structuring the real-life situation into a mathematical model, abstracting the essential elements so that a solution relevant to the decision maker's objectives can be sought. This involves looking at the problem in the context of the entire system.
2. Exploring the structure of such solutions and developing systematic procedures for obtaining them.
3. Developing a solution, including the mathematical theory, if necessary, that yields an optimal value of the system measure of desirability (or possibly comparing alternative courses of action by evaluating their measure of desirability).

1.3 The Impact of Operations Research

Operations research has had an increasingly great impact on the management of organizations in recent years. Both the number and the variety of its applications continue to grow rapidly, and no slowdown is in sight. In fact, with the exception of the advent of the electronic computer, the extent of this impact seems to be unrivaled by that of any other recent development.

After their success with operations research during World War II, the British and American military services continued to have active operations research groups, often at different levels of command. As a result, there now exists a large number of people called "military operations researchers" who are applying an operations research approach to problems of national defense. For example, they engage in tactical planning for requirements and use of weapon systems as well as consider the larger problems of the allocation and integration of effort. Some of their techniques involve quite sophisticated ideas in political science, mathematics, economics, probability theory, and statistics.

Operations research is also being used widely in other types of organizations, including business and industry. Sometimes the term "management science" is used as a designation for this activity. Almost all the dozen or so largest corporations in the world, and a sizable proportion of the small industrial organizations, either have well-established operations research groups or have integrated the activity into the regular components of the organization. Many industries, including aircraft and missile, automobile, communication, computer, electric power, electronics, food, metallurgy, mining, paper, petroleum, and transportation, have made widespread use of operations research. Financial institutions, governmental agencies, and hospitals are rapidly increasing their use of operations research.

As an example of the impact of operations research in government, the President's Commission on Aviation Safety issued its report in April 1988 on the strengths, weaknesses, and problems of the airspace system along with 15 recommendations for change. One of these recommendations stated that "Operations research (applied mathematics methods and models for solving complex operations problems) should be recognized as a standard approach for problem solving in the FAA."

To be more specific, consider some of the problems that have been solved by particular techniques of operations research. Linear programming has been used successfully in the solution of problems concerned with assignment of personnel, blending of materials, distribution and transportation, and investment portfolios. Dynamic programming has been applied successfully to such areas as planning advertising expenditures, distributing sales effort, and production scheduling. Queueing theory has had application in solving problems concerned with traffic congestion, servicing machines subject to breakdown, determining the level of a service force, air traffic scheduling, design of dams, production scheduling, and hospital operation. Other techniques of operations research, such as inventory theory, game theory, and simulation, also have been successfully applied to a variety of contexts.

The extent of operations research activities, and the profile of operations research practitioners, in U.S. corporations have been frequently surveyed. Among the surveys that emphasized the companies in *Fortune's* list of the top 500 corporations are the following:

1. In 1972, Turban[1] reported on a survey of operations research activities that provide a snapshot of activities in 1969. Mail questionnaires were sent to the directors of operations research/management science of 475 companies. These companies were selected from *Fortune's* list of the top 500, using the 300 largest industrial corporations, 50 industrial corporations drawn from the companies ranking between 300 and 500, and the 25 largest companies

[1] Turban, E.: "A Sample Survey of Operations Research Activities at the Corporate Level," *Operations Research,* **20**:708–721, 1972.

in each of the service categories, banks, utilities, merchandising, life insurance, and transportation. There were 107 questionnaires returned.

2. In 1977, Ledbetter and Cox[1] published the results of a survey of *Fortune's* 500 firms (1975 listing) concerning utilization of operations research techniques in their firms. There were 176 respondents.

3. In 1979, Thomas and DaCosta[2] reported on a survey of operations research activities in 1977. Mail questionnaires were sent to 420 individual corporations, including 260 firms from *Fortune's* 1975 list of the top 500, the largest 100 industrial firms in California, and the balance from California financial institutions. There were 150 questionnaires returned.

4. Finally, in 1983, Forgionne[3] published the results of a survey of corporate usage of operations research/management science. A questionnaire was mailed in 1982 to a random sample of 500 corporations drawn from the 1,500 largest U.S. corporations. There were 125 respondents.

Both Turban and Thomas and DaCosta indicated that nearly half of the companies reporting had a special department that was engaged mainly in operations research/management science (OR/MS) activities. While this ratio has remained virtually constant, and since all of the respondents used OR/MS techniques, Thomas and DaCosta concluded that "management science is becoming a part of the everyday activities of the modern firm and therefore is no longer a specialized function to be undertaken by a separate specialized department." An interesting finding of the Turban survey was that almost all of the specialized departments reported to the company president, vice-president, or controller.

All of the surveys attempted to find which techniques of OR/MS were most frequently used. Table 1.1 presents a ranking of the techniques of operations research that are being applied in U.S. corporations based upon the aforementioned surveys. It is interesting to note that all four surveys, representing a span of 14 years, are in basic agreement. Statistical analysis, simulation, and linear programming were the most widely used techniques, although PERT/CPM, inventory theory, and queueing theory were not far behind.

As noted earlier, OR/MS techniques are applied to a broad spectrum of corporate problem areas, and two of the surveys dealt with this issue. Table 1.2 presents a ranking of the application areas given in the Thomas and DaCosta and Forgionne surveys. The classification of application areas differs somewhat in these two surveys, but it is evident that capital budgeting, forecasting, inventory control, production planning, and project planning are the most popular application areas.

In 1976, Fabozzi and Valente[4] reported on the results of a questionnaire mailed to 1,000 firms in the United States in November 1974 concerning the use of mathematical programming (linear, nonlinear, and dynamic). There were 184 responses received. The authors found that the most important area of application of mathematical programming was production management (determination of product mix,

---

[1] Ledbetter, W. N., and J. F. Cox: "Are OR Techniques Being Used," *Industrial Engineering,* pp. 19–21, February 1977.

[2] Thomas, G., and J. DaCosta: "A Sample Survey of Corporate Operations Research," *Interfaces,* **9**:102–111, 1979.

[3] Forgionne, G. A.: "Corporate Management Science Activities: An Update," *Interfaces,* **13**:20–23, 1983.

[4] Fabozzi, F. J., and J. Valente: "Mathematical Programming in American Companies: A Sample Survey," *Interfaces,* **7**(1):93–98, November 1976.

Table 1.1 **Ranking of the Techniques of Operations Research/Management Science***

	Turban (1969)	Ledbetter and Cox (1975)	Thomas and DaCosta (1977)	Forgionne (1982)
Bayesian decision analysis	—	—	9	—
Delphi	—	—	13.5	—
Dynamic programming	6	6	10	7
Financial methods	—	—	13.5	—
Game theory	—	7	—	8
Heuristic programming	8.5	—	8	—
Integer and mixed programming	—	—	12	—
Inventory theory	4	—	5	—
Linear programming	3	2	3	4
Network models	—	4	—	—
Nonlinear programming	7	—	7	6
PERT/CPM	5	—	4	3
Risk analysis	—	—	11	—
Queueing theory	8.5	5	6	5
Simulation	2	3	2	2
Statistical analysis	1	1	1	1

* Rank 1 denotes the most frequently used technique.

allocation of resources, plant and machine scheduling, and work scheduling). The next largest area of application was financial and investment planning (capital budgeting, cash flow analysis, portfolio management for the employee pension fund, cash management, and merger and acquisitions analysis). The quality of results reported by these firms is given in Table 1.3.

Because of the great impact of operations research, professional societies devoted to this field and related activities have been founded in a number of countries throughout the world. In the United States, the Operations Research Society of America (ORSA), established in 1952, and The Institute of Management Sciences (TIMS),

Table 1.2 **Ranking of the Application Areas of Operations Research/Management Science***

	Thomas and DaCosta (1977)	Forgionne (1982)
Accounting	11	5
Advertising and sales research	8	—
Capital budgeting	4	2
Equipment replacement	9	—
Forecasting—market planning	1	6
Inventory control	2.5	4
Maintenance	10	9
Packaging	12	—
Personnel management	—	10
Plant location	6	8
Production planning and scheduling	2.5	3
Project planning	—	1
Quality control	7	7
Transportation	5	—

* Rank 1 denotes the most frequent application area.

Table 1.3 Quality of Results Reported by Firms Employing Mathematical Programming (Fabozzi and Valente Survey)

Results	Linear Programming		Nonlinear Programming		Dynamic Programming	
	No.	Percent	No.	Percent	No.	Percent
Good	102	76	38	57	27	53
Fair	21	16	19	28	15	29
Poor	6	3	6	9	3	6
Uncertain	7	5	4	6	6	12
Total	133	100	67	100	51	100

founded in 1953, each has over 6,000 members. ORSA publishes the journal *Operations Research* and TIMS *Management Science*. The two societies also jointly publish *Mathematics of Operations Research* and *Interfaces*. These four journals contain well over 3,000 pages per year reporting new research and applications in the field. In addition, there are many other similar journals published in such countries as the United States, England, France, India, Japan, Canada, and West Germany. Indeed, there are 32 member countries (including the United States) in the International Federation of Operational Research Societies (IFORS), with each country having a national operations research society.

Operations research also has had considerable impact in colleges and universities. Today most of the major American universities offer courses in this field, and many offer advanced degrees that are either in or with specialization in operations research. As a result, there are now thousands of students taking at least one course in operations research each year. Much of the basic research in the field is also being done in the universities.

1.4 Training for a Career in Operations Research

Because of the great growth of operations research, career opportunities in this field appear to be outstanding. The demand for trained people continues to far exceed the supply, and both attractive starting positions and rapid advancement are readily available. Because of the nature of their work, operations research groups tend to have a prominent staff position, with access to higher-level management in the organization. The problems they work on tend to be important, challenging, and interesting. Therefore, any individual with a mathematics and science orientation who is also interested in the practical management of organizations is likely to find a career in operations research very rewarding.

Three complementary types of academic training are particularly relevant for a career in operations research. The first is basic training in the fundamentals upon which operations research is based. This includes the basic methodology of mathematics and science as well as such topics as linear algebra and matrix theory, probability theory, statistical inference, stochastic processes, computer science, microeconomics, accounting and business administration, organization theory, and the behavioral sciences.

A second important type of training is in operations research per se, including special techniques of the field such as linear and nonlinear programming, dynamic

Major Field of Study	Percentage of Total at Degree Level			
	Bachelors	Masters	Doctorate	All Degree Levels
Operations research and management science	3	24	32	12
Mathematics and statistics	26	16	21	22
Business administration	20	27	2	22
Engineering	34	17	29	28
Other	17	16	16	16
Percentage of total	27	53	20	

programming, inventory theory, network flow theory, queueing models, reliability, game theory, and simulation. It should also include an introduction to the methodology of operations research, where the various techniques and their role in an operations research study involving specific problem areas would be placed in perspective. Often courses covering certain of these topics are offered in more than one department within a university, including departments of business, industrial engineering, mathematics, statistics, computer science, economics, and electrical engineering. This is a natural reflection of the broad scope of application of the field. Since it does spread across traditional disciplinary lines, separate programs or departments in operations research also are being established in some universities.

Finally, it is also good to have specialized training in some field other than operations research, for example, mathematics, statistics, industrial engineering, business, or economics. This additional training provides one with an area of special competence for applying operations research, and it should make that person a more valuable member of an operations research team.

The early operations researchers were people whose primary training and work had been in some traditional field, such as physics, chemistry, mathematics, engineering, or economics. They tended to have little or no formal education in operations research per se. However, as the body of special knowledge has expanded, it has become increasingly more difficult to enter the field without considerable prior education in this area. As a result, although it is still common for new operations researchers to have their college degree(s) in a traditional field, they generally have specialized too in operations research as part of their academic program. The traditional fields that have most commonly served as a vehicle into operations research are indicated in Table 1.4, which is based on the 1972 survey by Turban described in the preceding section. The Thomas and DaCosta survey confirms the diversity of the educational backgrounds of OR/MS practitioners. They noted that the percentage of Ph.D.'s has decreased from 20 percent to 13 percent during the time interval of the two surveys, which they speculate may be due to the "maturity" of the OR/MS techniques in industry and to the lack of need for separate specialized OR/MS departments.

Finally, in 1982[1] a survey of TIMS membership provided a profile of its membership, including information on educational training, job activity, and compensation

[1] Hall, J. R., Jr.: "Career Paths and Compensation in Management Science: Results of a TIMS Membership Survey," *Interfaces,* **14**(3):15–23, May–June 1984.

for professionals in industry, government, universities, and consulting. This report reinforced the point that, as the profession has matured, fewer people with formal training in non–operations research fields are entering the profession than occurred in the previous decades.

1.5 The Road Ahead

As an introduction to mathematical programming, this book is designed to acquaint students with the formulation, solution, and implementation of mathematical programming models for analyzing complex systems problems in industry or government. Chapter 2 provides an overview of the operations research modeling approach and describes the major phases of a typical operations research study. Chapters 3 to 9 present the topic of linear programming, a prominent area of mathematical programming concerned largely with how to allocate limited resources among the various activities of an organization. Chapters 10 to 15 deal with a variety of other mathematical programming models and their applications.

Many of these topics can be described in terms of typical examples of situations that are encountered in practice. Synopses of several such examples are presented here, with detailed solutions given in subsequent chapters.

The technique of *linear programming* is illustrated by a company that operates a reclamation center that collects several types of solid waste materials and then treats them so they can be amalgamated into a salable product. Different grades of this product can be made, depending upon the mix of the materials used. Although there is some flexibility in the mix for each grade, quality standards do specify a minimum or maximum percentage (by weight) of certain materials allowed in that product grade. Data are available on the cost of amalgamation and the selling price for each grade. The reclamation center collects its solid waste materials from some regular sources and so is normally able to maintain a steady production rate for treating these materials. Furthermore, the quantities available for collection and treatment each week, as well as the cost of treatment, for each type of material are known. Using the given information, the company is to determine just how much of each product grade to produce *and* the exact mix of materials to be used for each grade so as to maximize their total weekly profit (total sales income minus the total costs of *both* amalgamation and treatment).

Another example of linear programming concerns a steel producer who is facing an *air pollution problem* caused by pollutants emanating from the manufacturing plant. The three main types of pollutants in the airshed are particulate matter, sulfur oxides, and hydrocarbons. New standards require that the company reduce its annual emission of these pollutants. The steelworks has two primary sources of pollution, namely, the blast furnaces for making pig iron and the open-hearth furnaces for changing iron into steel. In both cases the engineers have decided that the most effective types of abatement methods are (1) increasing the height of the smokestacks, (2) using filter devices (including gas traps) in the smokestacks, and (3) including cleaner high-grade materials among the fuels for the furnaces. All these methods have known technological limits on how much emission they can eliminate. Fortunately, the methods can be used at any fraction of their abatement capacities. A cost analysis results in estimates of the total annual cost that is incurred by each abatement method when used by blast and open-hearth furnaces (cost of less-than-full-capacity use of a method is essentially proportional to its fractional capacity). With use of the aforementioned data, the optimal plan (minimum cost) for pollution abatement is to be determined. This plan would consist of specifying which types of abatement method would be used and at what fractions of their abatement capacities for (1) blast furnaces and (2) open-hearth furnaces.

One of the important special types of linear programming problems is called the *transportation problem*; a typical example deals with a company producing canned peas. The peas are prepared at several distantly located canneries and then shipped by truck to distributing warehouses throughout the western United States. Because the shipping costs are a major expense, management is initiating a study to reduce them as much as possible. For the upcoming season, an estimate has been made of what the output will be from each cannery, and each warehouse has been allocated a certain amount from the total supply of peas. This information (in units of truckloads), along with the shipping cost per truckload for each cannery-warehouse combination, is given. Using the data, the optimal plan for assigning these shipments to the various cannery-warehouse combinations that minimize total shipping costs is to be determined.

In addition to linear programming, there are a number of related mathematical programming techniques for dealing with similar kinds of problems. One of these is *dynamic programming,* which is concerned with making a sequence of interrelated decisions. It is illustrated by a job shop whose workload is subject to considerable seasonal fluctuation. However, machine operators are difficult to hire and costly to train, so the manager is reluctant to lay off workers during the slack seasons. The manager is likewise reluctant to maintain a peak payroll when it is not required. Furthermore, the manager is definitely opposed to overtime work on a regular basis. Because all work is done to custom orders, it is not possible to build up inventories during slack seasons. Therefore, the manager is in a dilemma as to what the policy should be regarding employment levels. Estimates are available for the personnel requirements during the four seasons of the year for the foreseeable future. Employment is not permitted to fall below these levels. Any employment above these levels is wasted. The salaries, hiring costs, and firing costs are known. Assuming that fractional levels of employment are possible because of a few part-time employees, the employment in each season that minimizes the total cost is to be determined.

Inventory theory is illustrated by a television manufacturing company that produces its own speakers, which are used in the production of its television sets. The television sets are assembled on a continuous production line at a known monthly rate. The speakers are produced in batches because they do not warrant setting up a continuous production line and because relatively large quantities can be produced in a short time. The company is interested in determining when and how many to produce. Several costs must be considered. (1) Each time a batch is produced, a setup cost is incurred. This cost includes the cost of "tooling up," administrative costs, record keeping, and so on. (2) The production of speakers in large batch sizes leads to a large inventory, resulting in a monthly cost for keeping a speaker in stock. This cost includes the cost of capital tied up, storage space, insurance, taxes, protection, and so forth. (3) A cost of producing a single speaker (excluding the setup cost) is incurred. (4) Company policy prohibits deliberately planning for shortages of any of its components. However, a shortage of speakers occasionally occurs, resulting in a monthly cost for each speaker unavailable when required. This cost includes the cost of installing speakers after the television set is fully assembled, storage space, delayed revenue, record keeping, and so on. Given data on these costs, the optimal batch size (and period between production) is to be determined.

1.6 Algorithms and OR COURSEWARE

An important part of this book is the presentation of the major **algorithms** (iterative solution procedures) of mathematical programming for solving the types of problems described in the previous section. Some of these algorithms are amazingly efficient and are routinely used on problems involving hundreds or thousands of variables. Outside the classroom, they normally are executed on computers because of the relatively extensive numerical calculations involved.

To aid the student in learning these algorithms, personal software (entitled OR COURSEWARE) is packaged in the back of the book. Separate diskettes are available for either an IBM (or IBM-compatible) personal computer with a graphics card or a Macintosh. (For an IBM personal computer that takes a $3\frac{1}{2}$-inch diskette, the *Instructor's Guide* which is available to the teacher includes equivalent diskettes of this size that can be copied.) Although use of this software will result in an enhancement of the textbook, it is not essential for the student to have access to a microcomputer to comprehend the material presented in the book.

Three types of routines are included in the software. One is *demonstration examples* that display and explain the algorithms in action. These "demos" supplement the examples in the book.

For doing homework problems, the second type of routine—*interactive execution of algorithms*—commonly will be used. The computer does all the routine calculations while the student focuses on learning and executing the logic of the algorithm.

A third type occasionally available is routines for *automatic execution of algorithms*. This type may be helpful for testing the validity of model formulations and for performing subsequent analysis, much as practitioners do with the output of production codes.

The OR COURSEWARE will be described at the end of Sec. 4.3, which is the first time it normally would be used. However, some students may wish to begin getting acquainted with it now, including reading the introduction that comes on the screen when a diskette is inserted into the disk drive. This introduction, and subsequent instructions, will guide the student through the complete use of the software.

2

Overview of Modeling

The bulk of this book is devoted to the mathematical methods of mathematical programming. This is quite appropriate because these quantitative techniques form the main part of what is known about mathematical programming. However, it does not imply that practical operations research studies are primarily mathematical exercises. As a matter of fact, the mathematical analysis often represents only a relatively small part of the total effort required. The purpose of this chapter is to place things into better perspective by describing all the major phases of a typical operations research study that uses mathematical programming.

One way of summarizing the usual phases of an operations research study is the following:[1]

1. Formulating the problem.
2. Constructing a mathematical model to represent the system under study.

[1] Ackoff, Russell L.: "The Development of Operations Research as a Science," *Operations Research,* **4**:265f, 1956.

13

3. Deriving a solution from the model.
4. Testing the model and the solution derived from it.
5. Establishing controls over the solution.
6. Putting the solution to work: implementation.

Each of these phases will be discussed in turn in the following sections.

2.1 Formulating the Problem

In contrast to textbook examples, most practical problems are initially communicated to an operations research team in a vague, imprecise way. Therefore, the first order of business is to study the relevant system and develop a well-defined statement of the problem to be considered. This includes determining such things as the appropriate objectives, the constraints on what can be done, interrelationships between the area to be studied and other areas of the organization, the possible alternative courses of action, time limits for making a decision, and so on. This process of problem formulation is a crucial one because it greatly affects how relevant the conclusions of the study will be. It is difficult to extract a ''right'' answer from the ''wrong'' problem! Consequently, this phase should be executed with considerable care, and the initial formulation should be continually reexamined in the light of new insights obtained during the later phases.

The first thing to recognize is that an operations research team is normally working in an *advisory capacity*. The team members are not just given a problem and told to solve it however they see fit. Instead, they are advising management (often one key decision maker). The team performs a detailed technical analysis of the problem and then presents its recommendations to management. Frequently, the report to management will identify a number of alternatives that are particularly attractive under different assumptions or over a different range of values of some policy parameter that can be evaluated only by management (e.g., the trade-off between *cost* and *benefits*). Management evaluates the study and its recommendations, takes into account a variety of intangible factors, and makes the final decision based on its best judgment. Consequently, it is vital for the operations research team to get on the same wavelength as management, including identifying the ''right'' problem from management's viewpoint, and to build the support of management for the course that the study is taking.

Determining the *appropriate objectives* is a very important aspect of problem formulation. To do this, it is necessary first to identify the member (or members) of management who actually will be making the decisions concerning the system under study and then to probe into this individual's thinking regarding the pertinent objectives. (Involving the decision maker from the outset also helps to build his or her support for the implementation of the study.) After the decision maker's objectives have been elicited, they should be analyzed and edited for the identification of the ultimate objectives that encompass the other objectives, the determination of the relative importance of these ultimate objectives, and the statement of them precisely in a way that does not eliminate worthwhile goals and alternatives.

By its nature, operations research is concerned with the welfare of the *entire organization* rather than that of only certain of its components. An operations research study seeks solutions that are optimal for the overall organization rather than suboptimal solutions that are best for only one component. Therefore, the objectives that are formulated should ideally be those of the entire organization. However, this is not always convenient to do. Many problems primarily concern only a portion of the organization, so the analysis would become unwieldy if the stated objectives were too general and if explicit consideration were given to all side effects on the rest of the organization. Granted that operations research takes the viewpoint of the overall organization, this does not imply that each problem should be broadened into a study of the entire organization. Instead, the objectives used in the study should be as specific as they can be while still encompassing the main goals of the decision maker and maintaining a reasonable degree of consistency with the higher-level objectives of the organization. Side effects on other segments of the organization must then be considered only to the extent that there are questions of consistency with these higher-level objectives.

For profit-making organizations, one possible approach to circumventing the problem of suboptimization is to use *long-run profit maximization* as the sole objective. The adjective *long-run* indicates that this objective provides the flexibility to consider activities that do not translate into profits *immediately* (e.g., research and development projects) but need to do so *eventually* in order to be worthwhile. At first glance, this approach appears to have considerable merit. In particular, this objective is specific enough to be used conveniently, and yet it seems to be broad enough to encompass the basic goal of profit-making organizations. In fact, some people believe that all other legitimate objectives can be translated into this one.

However, this is an oversimplification, and considerable caution is required! A number of studies of American corporations have found that the goal of *satisfactory profits,* combined with other objectives, is preferred over profit maximization. (In fact, inadequate consideration of long-run profits sometimes is cited as a major reason why American industry may be losing its competitive edge over that of other leading countries.) In particular, typical objectives might be to maintain stable profits, increase (or maintain) one's share of the market, provide for product diversification, maintain stable prices, improve worker morale, maintain family control of the business, and increase company prestige. Fulfilling these objectives might result in the achievement of long-run profit maximization, but the relationship is sufficiently obscure that it may not be convenient to incorporate them into this one objective.

Furthermore, there are additional considerations involving social responsibilities that are distinct from the profit motive. The five parties affected by a business firm located in a single country are: (1) the *owners* (stockholders), who desire profits (dividends, stock appreciation, etc.); (2) the *employees,* who desire steady employment at reasonable wages; (3) the *customers,* who desire a reliable product at a reasonable price; (4) the *vendors,* who desire integrity and a reasonable selling price for their goods; and (5) the *government* and, hence, the *nation,* which desires payment of fair taxes and consideration of the national interest. All five parties make essential contributions to the firm, and the firm should not be viewed as the exclusive servant of any one party for the exploitation of others. By the same token, international corporations acquire additional obligations to follow socially responsible practices. Therefore, although granting that management's prime responsibility is to make profits (which ultimately benefits all five parties), its broader social responsibilities also must be recognized.

After formulating the decision maker's problem, the next phase is to reformulate this problem into a form that is convenient for analysis. The conventional operations research approach for doing this is to construct a mathematical model that represents the essence of the problem. Before discussing how to formulate such a model, let us first explore the nature of models in general and of mathematical models in particular.

Models, or idealized representations, are an integral part of everyday life. Common examples of models include model airplanes, portraits, globes, and so on. Similarly, models play an important role in science and business, as illustrated by models of the atom, models of genetic structure, mathematical equations describing physical laws of motion or chemical reactions, graphs, organization charts, and industrial accounting systems. Such models are invaluable for abstracting the essence of the subject of inquiry, showing interrelationships, and facilitating analysis.

Mathematical models are also idealized representations, but they are expressed in terms of mathematical symbols and expressions. Such laws of physics as $F = ma$ and $E = mc^2$ are familiar examples. Similarly, the mathematical model of a business problem is the system of equations and related mathematical expressions that describe the essence of the problem. Thus, if there are n related quantifiable decisions to be made, they are represented as **decision variables** (say, $x_1, x_2, \ldots, x_n$) whose respective values are to be determined. The appropriate measure of performance (e.g., profit) is then expressed as a mathematical function of these decision variables (e.g., $P = 3x_1 + 2x_2 + \cdots + 5x_n$). This function is called the **objective function.** Any restrictions on the values that can be assigned to these decision variables are also expressed mathematically, typically by means of inequalities or equations (e.g., $x_1 + 3x_1x_2 + 2x_2 \leq 10$). Such mathematical expressions for the restrictions often are called **constraints.** The constants (coefficients or right-hand sides) in the constraints and the objective function are called the **parameters** of the model. The mathematical model might then say that the problem is to choose the values of the decision variables so as to maximize the objective function, subject to the specified constraints. Such a model, and minor variations of it, typify the models used in operations research.

You will see numerous examples of mathematical models throughout the remainder of this book. One particularly important type that is studied in Chaps. 3 to 9 is the **linear programming model,** where the mathematical functions appearing in both the objective function and the constraints are all linear functions. In the next chapter, specific linear programming models are constructed to fit such diverse problems as determining (1) the mix of products that maximizes profit, (2) the design of radiation therapy that effectively attacks a tumor while minimizing the damage to nearby healthy tissue, (3) the allocation of acreage to crops that maximizes total net return, and (4) the combination of pollution-abatement methods that achieves air quality standards at minimum cost.

Mathematical models have many advantages over a verbal description of the problem. One obvious advantage is that a mathematical model describes a problem much more concisely. This tends to make the overall structure of the problem more comprehensible, and it helps to reveal important cause-and-effect relationships. In this way, it indicates more clearly what additional data are relevant to the analysis. It also facilitates dealing with the problem in its entirety and considering all its interrelationships simultaneously. Finally, a mathematical model forms a bridge to the use of high-

16

powered mathematical techniques and computers to analyze the problem. Indeed, packaged software for both microcomputers and mainframe computers is becoming widely available for many mathematical models.

On the other hand, there are pitfalls to be avoided when using mathematical models. Such a model is necessarily an abstract idealization of the problem, so approximations and simplifying assumptions generally are required if the model is to be *tractable* (capable of being solved). Therefore, care must be taken to ensure that the model remains a valid representation of the problem. The proper criterion for judging the validity of a model is whether or not the model predicts the relative effects of the alternative courses of action with sufficient accuracy to permit a sound decision. Consequently, it is not necessary to include unimportant details or factors that have approximately the same effect for all the alternative courses of action considered. It is not even necessary that the absolute magnitude of the measure of performance be approximately correct for the various alternatives, provided that their relative values (i.e., the differences between their values) are sufficiently precise. Thus all that is required is that there be a high *correlation* between the prediction by the model and what would actually happen in the real world. To ascertain whether this requirement is satisfied, it is important to do considerable *testing* and consequent modifying of the model, which will be the subject of Sec. 2.4. Although this testing phase is placed later in the chapter, much of this *model validation* work actually is conducted during the model-building phase of the study to help guide the construction of the mathematical model.

When developing the model, a good approach is to begin with a very simple version and then move in evolutionary fashion toward more elaborate models that more nearly reflect the complexity of the real problem. This process of *model enrichment* continues only as long as the model remains tractable. The basic trade-off under consideration is between the *precision* and the *tractability* of the model. (See Selected Reference 6 for a detailed description of this process.)

A crucial step in formulating the mathematical model is constructing the objective function. This requires developing a quantitative measure of performance relative to each objective that has been formulated for the study. If there are multiple objectives, their respective measures commonly are then transformed and combined into a composite measure called the **overall measure of performance.** This overall measure might be something tangible (e.g., profit), corresponding to a higher goal of the organization, or it might be abstract (e.g., ''utility''). In the latter case, the task of developing this measure tends to be a complex one requiring a careful comparison of the objectives and their relative importance. After developing the overall measure of performance, the objective function is then obtained by expressing this measure as a mathematical function of the decision variables. Alternatively, there also are methods for explicitly considering multiple objectives simultaneously, and one of these (goal programming) is discussed in Chap. 8.

2.3 Deriving a Solution

After formulating a mathematical model for the problem under consideration, the next phase in an operations research study is to derive a solution from this model. You might think that this must be the major part of the study, but actually it is not in most cases. Sometimes, in fact, it is a relatively simple step, in which one of the standard **algorithms** (iterative solution procedures) of operations research is applied on a computer by using one of a number of readily available software packages. For experienced operations research practitioners, finding a solution is the "fun part," whereas the real work comes in the preceding and following steps, including the *post-optimality analysis* discussed later in this section.

Since much of this book is devoted to the subject of how to obtain solutions for various important types of mathematical models, little needs to be said about it here. However, we do need to discuss the nature of such solutions.

A common theme in operations research is the search for an **optimal,** or best, **solution.** Indeed, many procedures have been developed, and are presented in this book, for finding such solutions for certain kinds of problems. However, it needs to be recognized that these solutions are optimal only with respect to the model being used. Since the model necessarily is an idealized rather than an exact representation of the real problem, there cannot be any Utopian guarantee that the optimal solution for the model will prove to be the best possible solution that could have been implemented for the real problem. There just are too many imponderables and uncertainties associated with real problems. However, if the model is well formulated and tested, the resulting solution should tend to be a good approximation to the ideal course of action for the real problem. Therefore, rather than be deluded into demanding the impossible, the test of the practical success of an operations research study should be whether it provides a better guide for action than can be obtained by other means.

The eminent management scientist and Nobel Laureate in Economics, Herbert Simon, points out that **satisficing** is much more prevalent than optimizing in actual practice. In coining the term *satisficing* as a combination of the words *satisfactory* and *optimizing,* Simon is describing the tendency of managers to seek a solution that is "good enough" for the problem at hand. Rather than trying to develop an overall measure of performance to optimally reconcile conflicts between various desirable objectives (including well-established criteria for judging the performance of different segments of the organization), a more pragmatic approach may be used. Goals may be set to establish minimum satisfactory levels of performance in various areas, based perhaps on past levels of performance or on what the competition is achieving. If a solution is found that enables all of these goals to be met, it is likely to be adopted without further ado. Such is the nature of satisficing.

The distinction between optimizing and satisficing reflects the difference between theory and the realities frequently faced in trying to implement that theory in practice. In the words of one of England's OR leaders, Samuel Eilon, "optimizing is the science of the ultimate; satisficing is the art of the feasible."[1]

Operations research teams attempt to bring as much of the "science of the ultimate" as possible to the decision-making process. However, the successful team does so in full recognition of the overriding need of the decision maker to obtain a satisfactory guide for action in a reasonable period of time. Therefore, the goal of an operations research study should be to conduct the study in an optimal manner, regardless of whether this involves finding an optimal solution to the model or not. Thus, in addition to pursuing the "science of the ultimate," the team should also consider the cost of the study and the disadvantages of delaying its completion, and

---

[1] Eilon, Samuel: "Goals and Constraints in Decision-making," *Operational Research Quarterly,* **23**:3–15, 1972; address given at the 1971 Annual Conference of the Canadian Operational Research Society.

then attempt to maximize the net benefits resulting from the study. In recognition of this concept, operations research teams occasionally use only **heuristic procedures** (i.e., intuitively designed procedures that do not guarantee an optimal solution) to find a good **suboptimal solution.** This is most often the case when the time or cost required to find an optimal solution for an adequate model of the problem would be very large.

The discussion thus far has implied that an operations research study seeks to find only *one* solution, which may or may not be required to be optimal. In fact, this usually is not the case. An optimal solution for the original model may be far from ideal for the real problem. Therefore, **post-optimality analysis** is a very important part of most operations research studies.

In part, post-optimality analysis involves conducting **sensitivity analysis** to determine which parameters of the model are most critical (the ''sensitive parameters'') in determining the solution. Some or all of the parameters generally are an estimate of some quantity (e.g., unit profit) whose exact value will become known only after the solution has been implemented. Therefore, after identifying the sensitive parameters, special attention is given to estimating each one more closely, or at least its range of likely values. One then seeks a solution that remains a particularly good one for all of the various combinations of likely values of the sensitive parameters.

In some cases, certain parameters of the model represent policy decisions (e.g., resource allocations). If so, there frequently is some flexibility in the values assigned to these parameters. Perhaps some can be increased by decreasing others. Post-optimality analysis includes the investigation of such trade-offs.

In conjunction with the study phase discussed in the next section (testing the model and the solution), post-optimality analysis also involves obtaining a sequence of solutions that comprises a series of improving approximations to the ideal course of action. Thus the apparent weaknesses in the initial solution are used to suggest improvements in the model, its input data, and perhaps the solution procedure. A new solution is then obtained, and the cycle is repeated. This process continues until the improvements in the succeeding solutions become too small to warrant continuation. Even then, a number of alternative solutions (perhaps solutions that are optimal for one of several plausible versions of the model and its input data) may be presented to management for the final selection. As suggested in Sec. 2.1, this presentation of alternative solutions would normally be done whenever the final choice among these alternatives should be based on considerations that are best left to the judgment of management.

Ways in which the model and its solution are evaluated and improved will be discussed in the next section.

2.4 Testing the Model and the Solution

One of the first lessons of operations research is that it is generally not sufficient to rely solely on one's intuition. This caution applies not only in obtaining a solution to a problem but also in evaluating the model that has been formulated to represent this problem. As indicated in Sec. 2.2, the proper criterion for judging the validity of a model is whether or not it predicts the *relative effects* of the alternative courses of action with sufficient accuracy to permit a sound decision. No matter how plausible the model may appear to be, it should not be accepted on faith that this condition is satisfied. Given the difficulty of communicating and understanding all the aspects and subtleties of a complex operational problem, there is a distinct possibility that the operations research team either has not been given all the true facts of the situation or has not interpreted them properly. For example, an important factor or interrelationship may not have been incorporated into the model or perhaps certain parameters have not been estimated accurately.

Before undertaking more elaborate tests, check for obvious errors or oversights in the model. Reexamining the formulation of the problem and comparing it with the model may help to reveal any such mistakes. Another useful check is to make sure that all the mathematical expressions are *dimensionally consistent* in the units they use. Additional insight into the validity of the model can sometimes be obtained by varying the parameters and/or the decision variables and checking to see whether the output from the model behaves in a plausible manner. This is often especially revealing when the parameters or variables are assigned extreme values near their maxima or minima.

A more systematic approach to testing the model is to use a **retrospective test.** When it is applicable, this test involves using historical data to reconstruct the past and then determining how well the model and the resulting solution would have performed if it had been used. Comparing the effectiveness of this hypothetical performance with what actually happened then indicates whether using this model tends to yield a significant improvement over current practice. It may also indicate areas where the model has shortcomings and requires modifications. Furthermore, by using alternative solutions from the model and determining their hypothetical historical performances, considerable evidence can be gathered regarding how well the model predicts the relative effects of alternative courses of action.

On the other hand, a disadvantage of retrospective testing is that it uses the same data that guided the formulation of the model. The crucial question is whether or not the past is truly representative of the future. If it is not, then the model might perform quite differently in the future than it would have in the past.

To circumvent this disadvantage of retrospective testing, it is sometimes useful to continue the status quo temporarily. This provides new data that were not available when the model was constructed. These data are then used in the same ways as those described here to evaluate the model.

If the final model is to be used repeatedly, it is important to continue checking the model and its solution after the initial implementation to make sure that they remain valid as conditions evolve over time. The establishment of such controls is the subject of the next section.

2.5 Establishing Control Over the Solution

What happens after the testing phase has been completed and an acceptable model has been developed? If the model is to be used repeatedly, the next step is to install a well-documented *system* for applying the model. This system would include the model, the solution procedure (including post-optimality analysis), and operating procedures for implementation. Then, even as personnel changes, the system can be called on at regular intervals to provide a specific numerical solution.

It is evident that this solution remains valid for the real problem only as long as this specific model remains valid. However, conditions are constantly changing in the real world. Therefore, changes might well occur that would invalidate this model; e.g., the values of the parameters might change significantly. If these values should change, it is vital that the change be detected as soon as possible so that the model, its solution, and the resulting course of action can be modified accordingly. A plan for detecting such changes and making the needed modifications should be part of the system for applying the model.

This plan will include provision for maintaining a general surveillance of the situation. In addition, it is often worthwhile to establish *systematic procedures* for detecting change and controlling the solution. To do this, it is necessary to identify the *sensitive parameters* of the model by sensitivity analysis, as discussed in Sec. 2.3. Next, a procedure is established for detecting *statistically significant changes* in each of these sensitive parameters. This procedure can sometimes be established by the process control charts used in statistical quality control. Finally, provision is made for adjusting the solution and consequent course of action whenever such a change is detected.

2.6 Implementation

The last phase of an operations research study is to implement the final solution as approved by the decision maker. This phase is a critical one because it is here, and only here, that the benefits of the study are reaped. Therefore, it is important for the operations research team to participate in launching this phase, both to make sure that the solution is accurately translated into an operating procedure and to rectify any flaws in the solution that are then uncovered.

The success of the implementation phase depends a great deal upon the support of both top management and operating management. Consequently, as mentioned in Secs. 2.1 and 2.4, the operations research team should encourage the active participation of management in formulating the problem and evaluating the solution. Obtaining the guidance of management is valuable in its own right for identifying relevant special considerations and thereby avoiding potential pitfalls during these phases. However, making management a party to the study also serves to enlist their active support for its implementation.

The implementation phase involves several steps. First, the operations research team gives operating management a careful explanation of the solution to be adopted and how it relates to operating realities. Next, these two parties share the responsibility for developing the procedures required to put this solution into operation. Operating management then sees that a detailed indoctrination is given to the personnel involved, and the new course of action is initiated. If successful, the model and the solution procedure may be used periodically to provide guidance to management. With this in mind, the operations research team monitors the initial experience with the course of action taken and seeks to identify any modifications that should be made in the future.

Upon culminating a study, it is appropriate for the operations research team to document its methodology clearly and accurately enough so that the work is *reproducible*. *Replicability* should be part of the professional ethical code of the operations researcher. This condition is especially crucial when controversial public policy issues are being studied.

Although the remainder of this book focuses primarily on *constructing* and *solving* mathematical models, we have tried to emphasize in the present chapter that this constitutes only a portion of the overall process involved in conducting a typical operations research study. The other phases described here also are very important to the success of the study. Try to keep in perspective the role of the model and the solution procedure in the overall process as you move through the subsequent chapters. Then, after gaining a deeper understanding of mathematical models, we suggest that you plan to return to review this chapter again in order to further sharpen this perspective.

Many of the phases discussed in this chapter entail the use of *software tools,* including *decision support systems*. Operations research is closely intertwined with the use of computers. Until recently, these have been almost exclusively mainframe computers, but now microcomputers also are being widely used for dealing with smaller problems.

In concluding this discussion of the major phases of an operations research study, it should be emphasized that there are many exceptions to the "rules" prescribed in this chapter. By its very nature, operations research requires considerable ingenuity and innovation, so it is impossible to write down any standard procedure that should always be followed by operations research teams. Rather, the preceding description may be viewed as a model that roughly represents how successful operations research studies are conducted.

SELECTED REFERENCES

1. Ackoff, Russell L., and Patrick Rivett: *A Manager's Guide to Operations Research,* Wiley, New York, 1963.
2. Ackoff, Russell L., and Maurice W. Sasieni: *Fundamentals of Operations Research,* Wiley, New York, 1968.
3. Churchman, C. West, Russell L. Ackoff, and E. L. Arnoff: *Introduction to Operations Research,* Wiley, New York, 1957.
4. Huysmans, Jan H. B. M.: *The Implementation of Operations Research,* Wiley, New York, 1970.
5. Miller, David W., and Martin K. Starr: *Executive Decisions and Operations Research,* 2d ed., Prentice-Hall, Englewood Cliffs, N.J., 1969.
6. Morris, William T.: "On the Art of Modeling," *Management Science,* **13**:B707–717, 1967.
7. Williams, H. P.: *Model Building in Mathematical Programming,* 2d ed., Wiley, New York, 1985.

3

Introduction to Linear Programming

Many people rank the development of linear programming among the most important scientific advances of the mid-twentieth century, and we must agree with this assessment. Its impact since just 1950 has been extraordinary. Today it is a standard tool that has saved many thousands or millions of dollars for most companies or businesses of even moderate size in the various industrialized countries of the world, and its use in other sectors of society has been spreading rapidly. Dozens of textbooks have been written about the subject, and *published* articles describing important applications now number in the hundreds. A very major proportion of all scientific computation on computers is devoted to the use of linear programming.

What is the nature of this remarkable tool, and what kinds of problems does it address? You will gain insight into this as you work through subsequent examples. However, a verbal summary may help provide perspective. Briefly, the most common type of application involves the general problem of allocating *limited resources* among *competing activities* in the best possible (i.e., *optimal*) way. This problem of allocation can arise whenever one must select the level of certain activities that compete for

23

scarce resources necessary to perform those activities. The variety of situations to which this description applies is diverse indeed, ranging from the allocation of production facilities to products to the allocation of national resources to domestic needs, from portfolio selection to the selection of shipping patterns, from agricultural planning to the design of radiation therapy, and so on. However, the one common ingredient in each of these situations is the necessity for allocating resources to activities.

Linear programming uses a mathematical model to describe the problem of concern. The adjective *linear* means that all the mathematical functions in this model are required to be *linear functions*. The word *programming* does not refer here to computer programming; rather, it is essentially a synonym for planning. Thus linear programming involves the *planning of activities* to obtain an optimal result, i.e., a result that reaches the specified goal best (according to the mathematical model) among all feasible alternatives.

Although allocating resources to activities is the most common type of application, linear programming has numerous other important applications as well. In fact, *any* problem whose mathematical model fits the very general format for the linear programming model is a linear programming problem. Furthermore, a remarkably efficient solution procedure, called the **simplex method,** is available for solving linear programming problems of even enormous size. These are some of the reasons for the tremendous impact of linear programming in recent decades.

Because of its great importance, we devote this and the next six chapters specifically to linear programming. After this chapter introduces the general features of linear programming, Chaps. 4 and 5 focus on the simplex method. Chapter 6 discusses the further analysis of linear programming problems *after* the simplex method has been initially applied. Chapter 7 considers several *special types* of linear programming problems whose importance warrants individual study. Chapter 8 then concentrates on the formulation of linear programming models. Finally, Chap. 9 presents several widely used extensions of the simplex method, and then introduces a new *interior-point algorithm* that sometimes can solve even larger linear programming problems than the simplex method.

You also can look forward to seeing applications of linear programming to other areas of operations research in several later chapters.

We begin this chapter by developing a miniature prototype example of a linear programming problem. This example is small enough to be solved graphically in a straightforward way. We then present the general *linear programming model* and its basic assumptions. The chapter concludes with some additional examples of linear programming applications.

3.1 Prototype Example

The WYNDOR GLASS CO. is a producer of high-quality glass products, including windows and glass doors. It has three plants. Aluminum frames and hardware are made in Plant 1, wood frames are made in Plant 2, and Plant 3 is used to produce the glass and assemble the products.

Because of declining earnings, top management has decided to revamp the product line. Several unprofitable products are being discontinued, and this act will release production capacity to undertake one or both of two potential new products that have been in demand. One of these proposed products (product 1) is an 8-foot

Table 3.1 Data for Wyndor
Glass Co.

Plant	Product 1	Product 2	Capacity Available
1	1	0	4
2	0	2	12
3	3	2	18
Unit profit	$3	$5	

Capacity Used Per Unit Production Rate

glass door with aluminum framing. The other product (product 2) is a large (4 × 6 foot) double-hung wood-framed window. The Marketing Department has concluded that the company could sell as much of either product as could be produced with the available capacity. However, because both products would be competing for the same production capacity in Plant 3, it is not clear which *mix* between the two products would be *most profitable*. Therefore, management asked the Operations Research Department to study this question.

After some investigation, the OR Department determined (1) the percentage of each plant's production capacity that would be available for these products, (2) the percentages required by each product for each unit produced per minute, and (3) the unit profit for each product. This information is summarized in Table 3.1.

The OR Department immediately recognized that this was a linear programming problem of the classic **product mix** type, and it next undertook the formulation and solution of the problem.

FORMULATION AS A LINEAR PROGRAMMING PROBLEM: To formulate the mathematical (linear programming) model for this problem, let x_1 and x_2 represent the number of units produced per minute of products 1 and 2, respectively, and let Z be the resulting contribution to profit per minute. Thus x_1 and x_2 are the *decision variables* for the model. Using the bottom row of Table 3.1,

$$Z = 3x_1 + 5x_2.$$

The objective is to choose the values of x_1 and x_2 so as to *maximize $Z = 3x_1 + 5x_2$*, subject to the restrictions imposed on their values by the limited plant capacities available. Table 3.1 implies that each unit of product 1 produced per minute would use 1 percent of Plant 1 capacity, whereas only 4 percent is available. This restriction is expressed mathematically by the inequality $x_1 \leq 4$. Similarly, Plant 2 imposes the restriction that $2x_2 \leq 12$. The percentage of Plant 3 capacity consumed by choosing x_1 and x_2 as the new products' production rates would be $3x_1 + 2x_2$. Therefore, the mathematical statement of the Plant 3 restriction is $3x_1 + 2x_2 \leq 18$. Finally, since production rates cannot be negative, it is necessary to restrict the decision variables to be nonnegative: $x_1 \geq 0$ and $x_2 \geq 0$.

To summarize, in the mathematical language of linear programming, the problem is to choose the values of x_1 and x_2 so as to

Maximize $Z = 3x_1 + 5x_2,$

subject to the restrictions

$$x_1 \qquad\qquad \leq 4$$

$$2x_2 \leq 12$$

$$3x_1 + 2x_2 \leq 18$$

and

$$x_1 \geq 0, \qquad x_2 \geq 0.$$

(Notice how the layout of the coefficients of x_1 and x_2 in this linear programming model essentially duplicates the information summarized in Table 3.1.)

GRAPHICAL SOLUTION: This very small problem has only two decision variables, and therefore only two dimensions, so a graphical procedure can be used to solve it. This procedure involves constructing a two-dimensional graph with x_1 and x_2 as the axes. The first step is to identify the values of (x_1, x_2) that are permitted by the restrictions. This is done by drawing the lines that must border the range of permissible values. To begin, note that the nonnegativity restrictions, $x_1 \geq 0$ and $x_2 \geq 0$, require (x_1, x_2) to lie on the *positive* side of the axes (including actually on either axis). Next, observe that the restriction $x_1 \leq 4$ means that (x_1, x_2) cannot lie to the right of the line $x_1 = 4$. These results are shown in Fig. 3.1, where the shaded area contains the only values of (x_1, x_2) that are still allowed.

In a similar fashion, the restriction $2x_2 \leq 12$ implies that the line $2x_2 = 12$ should be added to the boundary of the permissible region. The final restriction, $3x_1 + 2x_2 \leq 18$, requires plotting the points (x_1, x_2) such that $3x_1 + 2x_2 = 18$ (another line) to complete the boundary. (Note that the points such that $3x_1 + 2x_2 \leq 18$ are those that lie either underneath or on the line $3x_1 + 2x_2 = 18$, so this is the limiting line beyond which the inequality ceases to hold.) The resulting region of permissible values of (x_1, x_2) is shown in Fig. 3.2.

The final step is to pick out the point in this region that maximizes the value of $Z = 3x_1 + 5x_2$. To discover how to perform this step efficiently, begin by trial and error. Try, for example, $Z = 10 = 3x_1 + 5x_2$ to see if there are in the permissible region any values of (x_1, x_2) that yield a value of Z as large as 10. By drawing the line $3x_1 + 5x_2 = 10$ (see Fig. 3.3), you can see that there are many points on this line that lie within the region. Therefore, try a larger value of Z, say, for example, $Z = 20 = 3x_1 + 5x_2$. Again, Fig. 3.3 reveals that a segment of the line $3x_1 +$

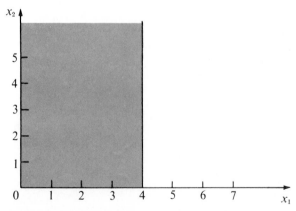

Figure 3.1   Shaded area shows values of (x_1, x_2) allowed by $x_1 \geq 0$, $x_2 \geq 0$, $x_1 \leq 4$.

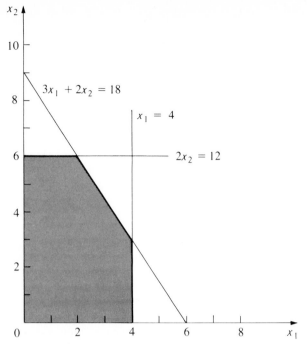

Figure 3.2 Shaded areas show permissible values of (x_1, x_2).

$5x_2 = 20$ lies within the region, so that the maximum permissible value of Z must be at least 20.

Now notice in Fig. 3.3 that the two lines just constructed are parallel, and that the line giving a larger value of Z ($Z = 20$) is farther up and away from the origin than the other line ($Z = 10$). Thus this trial-and-error procedure involves nothing

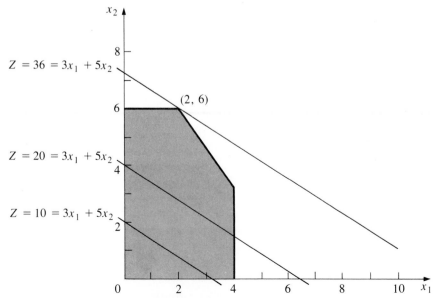

Figure 3.3 Value of (x_1, x_2) that maximizes $3x_1 + 5x_2$.

more than drawing a family of parallel lines containing at least one point in the permissible region and selecting the line that is the greatest distance from the origin (in the direction of increasing values of Z). This line passes through the point (2, 6) as indicated in Fig. 3.3, so that the equation is $3x_1 + 5x_2 = 3(2) + 5(6) = 36 = Z$. [Note that the point (2, 6) lies at the intersection of the two lines, $2x_2 = 12$ and $3x_1 + 2x_2 = 18$, shown in Fig. 3.2, so that this point can be calculated algebraically as the simultaneous solution of these two equations.]

Having seen the trial-and-error procedure for finding (2, 6), you now can streamline this approach for other problems. Rather than drawing several parallel lines, it is sufficient to form a single line with a ruler to establish the slope. Then move the ruler with fixed slope through the region of permissible values in the direction of improving Z. (When the objective is to *minimize Z,* move the ruler in the direction that *decreases* Z.) Stop moving the ruler at the last instant that it still passes through a point in this region. This point is the desired solution.

CONCLUSIONS: The OR Department used this approach to find that the desired solution is $x_1 = 2$, $x_2 = 6$, with $Z = 36$. This solution indicates that the Wyndor Glass Co. should produce products 1 and 2 at the rate of two per minute and six per minute, respectively, with a resulting profitability of $36/minute. No other mix of the two products would be so profitable—*according to the model.*

However, we emphasized in Chap. 2 that well-conducted operations research studies do not simply find *one* solution for the *initial* model formulated and stop. All six phases described in Chap. 2 are important, including thorough testing of the model (see Sec. 2.4) and post-optimality analysis (see Sec. 2.3).

In full recognition of these practical realities, the OR Department now is ready to evaluate the validity of the model more critically (to be continued in Sec. 3.3) and to perform sensitivity analysis on the effect of the estimates in Table 3.1 being inaccurate (to be continued in Sec. 6.7).

3.2 The Linear Programming Model

The Wyndor Glass Co. problem nicely illustrates a typical linear programming problem (miniature version). However, linear programming is too versatile to be completely described by any single example. In this section we discuss the general characteristics of linear programming problems, including the various legitimate forms of the mathematical model for linear programming.

Let us begin with some basic terminology and notation. The first column of Table 3.2 summarizes the components of the Wyndor Glass Co. problem. The second column then introduces more general terms for these same components that will fit

Table 3.2 **Common Terminology for Linear Programming**

Prototype Example	General Problem
Production capacities of plants 3 plants	Resources m resources
Production of products 2 products Production rate of product j (x_j)	Activities n activities Level of activity j (x_j)
Profit (Z)	Overall measure of performance (Z)

Table 3.3 **Data for Linear Programming Model**

	Resource Usage Per Unit of Activity				Amount of
	Activity				
Resource	1	2	$\cdots$	n	Resource Available
1	a_{11}	a_{12}	$\cdots$	a_{1n}	b_1
2	a_{21}	a_{22}	$\cdots$	a_{2n}	b_2
.			.		.
.			.		.
.			.		.
m	a_{m1}	a_{m2}	$\cdots$	a_{mn}	b_m
ΔZ/unit of activity	c_1	c_2	$\cdots$	c_n	
Level of activity	x_1	x_2	$\cdots$	x_n	

most linear programming problems. The key terms are *resources* and *activities,* where the number of each is denoted by m and n, respectively. As described in the introduction to the chapter, the resources are needed to perform these activities, but the amount available of each resource is limited, so a careful allocation of resources to activities must be made. Determining this allocation involves choosing the *levels* of the activities (the values of the *decision variables*) that achieve the best possible value of the *overall measure of performance Z.*

The standard notation of linear programming is summarized in Table 3.3. For activity j ($j = 1, 2, \ldots, n$), c_j is the increase in Z that would result from each unit of increase in x_j (the level of activity j). For resource i ($i = 1, 2, \ldots, m$), b_i is the amount available for allocation to the activities. Finally, a_{ij} is the amount of resource i consumed by each unit of activity j (for $i = 1, 2, \ldots, m$ and $j = 1, 2, \ldots, n$). This set of data (the a_{ij}, b_i, and c_j) constitutes the **parameters** (input constants) of the linear programming model.

Notice carefully the complete correspondence between Table 3.3 (except for the extra row added at the bottom) and Table 3.1.

A Standard Form of the Model

Proceeding just as for the example, we can now formulate the mathematical model for this general problem of allocating resources to activities. In particular, this model is to select the values for $x_1, x_2, \ldots, x_n$ (the **decision variables**) so as to

$$\text{Maximize} \quad Z = c_1x_1 + c_2x_2 + \cdots + c_nx_n,$$

subject to the restrictions

$$a_{11}x_1 + a_{12}x_2 + \cdots + a_{1n}x_n \leq b_1$$
$$a_{21}x_1 + a_{22}x_2 + \cdots + a_{2n}x_n \leq b_2$$
$$\vdots$$
$$a_{m1}x_1 + a_{m2}x_2 + \cdots + a_{mn}x_n \leq b_m,$$

and

$$x_1 \geq 0, \quad x_2 \geq 0, \quad \ldots, \quad x_n \geq 0.$$

We call this *our standard form*[1] for the linear programming problem. Any situation whose mathematical formulation fits this model is a linear programming problem.

[1] This is called *our* standard form rather than *the* standard form because some textbooks adopt other forms.

Notice that the model for the Wyndor Glass Co. problem fits our standard form, with $m = 3$ and $n = 2$.

Common terminology for the linear programming model can now be summarized. The function being maximized, $c_1x_1 + c_2x_2 + \cdots + c_nx_n$, is called the **objective function.** The restrictions normally are referred to as **constraints.** The first m constraints (those with a function $a_{i1}x_1 + a_{i2}x_2 + \cdots + a_{in}x_n$, representing the total usage of resource i, on the left) are sometimes called **functional constraints.** Similarly, the $x_j \geq 0$ restrictions are called **nonnegativity constraints.**

Other Forms

We now hasten to add that the preceding model does not actually fit the natural form of some linear programming problems. The other *legitimate forms* are the following:

1. Minimizing rather than maximizing the objective function:

$$\text{Minimize} \quad Z = c_1x_1 + c_2x_2 + \cdots + c_nx_n,$$

2. Some functional constraints with a greater-than-or-equal-to inequality:

$$a_{i1}x_1 + a_{i2}x_2 + \cdots + a_{in}x_n \geq b_i, \quad \text{for some values of } i,$$

3. Some functional constraints in equation form:

$$a_{i1}x_1 + a_{i2}x_2 + \cdots + a_{in}x_n = b_i, \quad \text{for some values of } i,$$

4. Deleting the nonnegativity constraints for some decision variables:

$$x_j \text{ unrestricted in sign}, \quad \text{for some values of } j.$$

Any problem that mixes some or all of these forms with the remaining parts of the preceding model is still a linear programming problem. Our interpretation of *allocating limited resources among competing activities* may no longer apply very well, if at all, but regardless of the interpretation or context, all that is required is that the mathematical statement of the problem fit the allowable forms.

In Sec. 4.6 you will see that all these other four legitimate forms can be rewritten in an equivalent way to fit the model just discussed. Thus every linear programming problem can be put into our standard form if so desired. We shall take advantage of this fact everywhere that procedures for solving linear programming problems are discussed (except Sec. 4.6) by assuming that the problems are in our standard form.

Terminology for Solutions of the Model

You may be used to having the term **solution** mean the final answer to a problem, but the convention in linear programming (and its extensions) is quite different. Here, *any* specification of values for the decision variables $(x_1, x_2, \ldots, x_n)$ is called a solution, regardless of whether it is a desirable or even an allowable choice. Different types of solutions are then identified by using an appropriate adjective.

A **feasible solution** is a solution for which *all* the constraints are satisfied.

In the example, (2, 3) and (4, 1) in Fig. 3.2 are feasible solutions, but $(-1, 3)$ and (4, 4) are infeasible solutions.

The **feasible region** is the collection of all feasible solutions.

The feasible region in the example is the entire shaded area in Fig. 3.2.

It is possible for a problem to have no feasible solutions. This would have happened in the example if the new products had been required to return a net profit of at least \$50/minute to justify discontinuing part of the current product line. The corresponding constraint, $3x_1 + 5x_2 \geq 50$, would eliminate the entire feasible region, so no mix of new products would be superior to the status quo.

Given that there are feasible solutions, the goal of linear programming is to find which one is best, as measured by the value of the objective function in the model.

An **optimal solution** is a feasible solution that has the *most favorable value* of the objective function.

Most favorable value means the largest or smallest value, depending upon whether the objective is maximization or minimization. Thus an optimal solution maximizes/minimizes the objective function over the entire feasible region.

Most problems will have just one optimal solution. However, it is possible to have more than one. This would occur in the example if the unit profitability of product 2 were changed to \$2, thereby changing the objective function to $Z = 3x_1 + 2x_2$, so that all the points on the line segment connecting (2, 6) and (4, 3) would be optimal. As in this case, any problem having multiple optimal solutions will have an infinite number of them.

Another possibility is that a problem has no optimal solutions. This occurs only if (1) it has no feasible solutions or (2) the constraints do not prevent increasing the value of the objective function (Z) indefinitely in the favorable direction (positive or negative). For example, the latter case would result if the last two functional constraints were mistakenly deleted in the example. A discussion of how the simplex method identifies these unusual cases is included in Secs. 4.5 (for case 2) and 4.6 (for case 1); we assume until then that they do not arise.

3.3 Assumptions of Linear Programming

All the assumptions of linear programming actually are implicit in the model formulation given in Sec. 3.2. However, it is good to highlight these assumptions so you can more easily evaluate how well linear programming applies to any given problem. Furthermore, we still need to see why the OR Department of the Wyndor Glass Co. concluded that a linear programming formulation provided a satisfactory representation of their problem.

Proportionality

Proportionality is an assumption about *individual* activities considered independently of the others (whereas the subsequent assumption of *additivity* concerns the effect of conducting activities *jointly*). Therefore, consider the case where only one of the n activities is undertaken. Call it activity k, so that $x_j = 0$ for all $j = 1, 2, \ldots, n$ except $j = k$.

The assumption is that (1) the overall measure of performance Z equals $c_k x_k$ and (2) the usage of each resource i equals $a_{ik} x_k$; that is, both quantities are directly *proportional* to the level of each activity k conducted by itself ($k = 1, 2, \ldots, n$).

Table 3.4 Examples of Satisfying or Violating Proportionality

Profit from Product 1

x_1	Proportionality Satisfied	Proportionality Violated		
		Case 1	Case 2	Case 3
0	0	0	0	0
1	3	2	3	3
2	6	5	7	5
3	9	8	12	6

This implies in particular that there is no extra startup cost associated with beginning the activity and that the proportionality holds over the entire range of levels of the activity.

To illustrate, consider the first term $(3x_1)$ in the objective function $(Z = 3x_1 + 5x_2)$ for the Wyndor Glass Co. problem. This term represents the profit generated per minute by producing product 1 at the rate of x_1 units per minute. The *proportionality satisfied* column of Table 3.4 shows the case that was assumed in Sec. 3.1, namely, that this profit is indeed proportional to x_1 so that $3x_1$ is the appropriate term for the objective function. By contrast, the next three columns show different hypothetical cases where the proportionality assumption would be violated.

Case 1 would arise if there were startup costs associated with initiating the production of product 1. For example, there might be costs involved with setting up the production facilities. There might also be costs associated with arranging the distribution of the new product. Because these are one-time costs, they would need to be amortized on a per-minute basis to be commensurable with Z (profit per minute). Suppose that this amortization were done and that the total startup cost amounted to $1/minute, but that the profit without considering the startup cost would be $3x_1$. This would mean that the contribution from product 1 to Z should be $(3x_1 - 1)$ for $x_1 > 0$. This function, which gives the numerical values shown for Case 1, certainly is *not* proportional to x_1.[1]

At first glance, it might appear that *Case 2* in Table 3.4 is quite similar to Case 1. However, Case 2 actually arises in a very different way. There no longer is a startup cost, and the profit from the first unit of product 1 per minute is indeed $3, as originally assumed. However, there now is an *increasing marginal return,* i.e., the increment in Z (ΔZ) due to incrementing x_1 by 1 keeps increasing as x_1 is increased, as summarized below:

$$\Delta Z = 3 \quad \text{when } x_1 = 0 \rightarrow x_1 = 1;$$
$$\Delta Z = 4 \quad \text{when } x_1 = 1 \rightarrow x_1 = 2;$$
$$\Delta Z = 5 \quad \text{when } x_1 = 2 \rightarrow x_1 = 3.$$

This violation of proportionality might occur because of economies that can sometimes be achieved at higher levels of production, e.g., through using more efficient high-volume machinery, longer production runs, and the learning-curve effect whereby workers become more efficient as they gain experience with a particular mode of

---

[1] If the contribution from product 1 to Z were $(3x_1 - 1)$ for *all* $x_1 \geq 0$, including $x_1 = 0$, then the fixed constant, -1, could be deleted from the objective function without changing the optimal solution, and proportionality would be restored. However, this "fix" does not work here because the -1 constant does not apply when $x_1 = 0$.

production. As the incremental cost goes down, the incremental profit will go up (assuming constant marginal revenue).

The reverse of Case 2 is *Case 3,* where there is a *decreasing marginal return,* as summarized below:

$$\Delta Z = 3 \quad \text{when } x_1 = 0 \rightarrow x_1 = 1;$$
$$\Delta Z = 2 \quad \text{when } x_1 = 1 \rightarrow x_1 = 2;$$
$$\Delta Z = 1 \quad \text{when } x_1 = 2 \rightarrow x_1 = 3.$$

This violation of proportionality might occur because the *marketing costs* need to go up more than proportionally to attain increases in the level of sales. For example, it might be possible to sell product 1 at the rate of one per minute ($x_1 = 1$) with no advertising, whereas attaining sales to sustain a production rate of $x_1 = 2$ might require a moderate amount of advertising, and $x_1 = 3$ might necessitate an extensive advertising campaign.

All three cases are hypothetical examples of ways in which the proportionality assumption could be violated. What is the actual situation? The actual profit from producing product 1 (or any other product) is derived from the sales revenue minus various direct and indirect costs. Inevitably, some of these cost components are not strictly proportional to the production rate, perhaps for one of the reasons illustrated above. However, the real question is whether, after cumulating all of the components of profit, proportionality is a reasonable approximation for practical modeling purposes. For the Wyndor Glass Co. problem the OR Department checked both the objective function and the functional constraints. The conclusion was that proportionality could indeed be assumed without serious distortion.

For other problems, what happens when the proportionality assumption does not hold even as a reasonable approximation? In most cases, this means you must use *nonlinear programming* instead (presented in Chap. 14). However, we do point out in Sec. 14.8 that a certain important kind of nonproportionality can still be handled by linear programming by reformulating the problem appropriately. Furthermore, if the assumption is violated only because of startup costs, there is an extension of linear programming (*mixed integer programming*) that can be used, as discussed in Sec. 13.2 (the fixed-charge problem).

Additivity

The proportionality assumption is not enough to guarantee that the objective function and constraint functions are linear. Cross-product terms will arise if there are interactions between some of the activities that would change the value of the overall measure of performance or the total usage of some resource. *Additivity* assumes that there are no such interactions between any of the activities, so that there are no cross-product terms in the model.

To be more specific, the additivity assumption (like proportionality) applies to both the objective function and the functions on the left-hand side of the functional constraints. The latter type of function represents the total usage of some resource. For both types of functions, the assumption concerns the comparison between the *total function value* from jointly conducting the activities at their respective levels $(x_1, x_2, \ldots, x_n)$ and the *individual contributions* to the function value from conducting each activity by itself (resetting all other variables to zero). For linear programming, these *individual contributions* are $c_j x_j$ for the objective function and $a_{ij} x_j$ for a constraint function.

The *additivity assumption* is that, for each function, the *total function value* can be obtained by *adding* the *individual contributions* from the respective activities.

To make this definition more concrete, and clarify why we need to worry about this assumption, let us look at some examples. Table 3.5 shows some possible cases for the objective function for the Wyndor Glass Co. problem. In each case, the *individual contributions* from the products are just as assumed in Sec. 3.1, namely, $3x_1$ for product 1 and $5x_2$ for product 2. The difference lies in the last row, which gives the *total function value* for Z when the two products are produced jointly. The *additivity satisfied* column shows the case where this *total function value* is obtained simply by adding the first two rows ($3 + 5 = 8$), so that $Z = 3x_1 + 5x_2$ as previously assumed. By contrast, the next two columns show hypothetical cases where the additivity assumption would be violated.

Case 1 corresponds to an objective function of $Z = 3x_1 + 5x_2 + x_1x_2$, so that $Z = 3 + 5 + 1 = 9$ for $(x_1, x_2) = (1, 1)$, thereby violating the additivity assumption that $Z = 3 + 5$. This case would arise if the two products were *complementary* in some way that *increases* profit. For example, suppose that a major advertising campaign would be required to market either new product produced by itself, but that the same single campaign can effectively promote both products if the decision is made to produce both of them. Because a major cost is saved for the second product, their joint profit is somewhat more than the *sum* of their individual profits when each is produced by itself.

Case 2 also violates the additivity assumption because of the extra term in its objective function, $Z = 3x_1 + 5x_2 - x_1x_2$, so that $Z = 3 + 5 - 1 = 7$ for $(x_1, x_2) = (1, 1)$. As the reverse of the first case, Case 2 would arise if the two products were *competitive* in some way that *decreases* their joint profit. For example, suppose that both products would need to use the same machinery and equipment. If either product were produced by itself, this machinery and equipment would be dedicated to this one use. However, producing both products would require switching the production processes back and forth, with substantial time and cost involved in temporarily shutting down the production of one product and setting up for the other. Because of this major extra cost, their joint profit is somewhat less than the *sum* of their individual profits when each is produced by itself.

The same kinds of interaction between activities can affect the additivity of the constraint functions. For example, consider the third functional constraint of the Wyndor Glass Co. problem, $3x_1 + 2x_2 \leq 18$. (This is the only constraint involving both products.) This constraint concerns the production capacity of Plant 3, where 18 percent is available for the two new products, and the function on the left-hand side ($3x_1 + 2x_2$) represents the percentage of the plant's capacity that would be used by

Table 3.5 Examples of Satisfying or Violating
Additivity for the Objective Function

	Value of Z		
	Additivity	Additivity Violated	
(x_1, x_2)	Satisfied	Case 1	Case 2
(1, 0)	3	3	3
(0, 1)	5	5	5
(1, 1)	8	9	7

Table 3.6 Examples of Satisfying or Violating
Additivity for a Functional Constraint

	Amount of Resource Used		
	Additivity	Additivity Violated	
(x_1, x_2)	Satisfied	Case 3	Case 4
(2, 0)	6	6	6
(0, 3)	6	6	6
(2, 3)	12	15	10.8

these products. The *additivity satisfied* column of Table 3.6 shows this case as is, whereas the next two columns display cases where the function has an extra cross-product term that violates additivity. For all three columns, the *individual contributions* from the products toward using the capacity of Plant 3 are just as assumed previously, namely, $3x_1$ for product 1 and $2x_2$ for product 2, or $3(2) = 6$ for $x_1 = 2$ and $2(3) = 6$ for $x_2 = 3$. As for Table 3.5, the difference lies in the last row, which now gives the *total function value* for capacity used when the two products are produced jointly.

The capacity used for Case 3 is given by the function, $3x_1 + 2x_2 + 0.5x_1x_2$, so the *total function value* is $6 + 6 + 3 = 15$ when $(x_1, x_2) = (2, 3)$, which violates the additivity assumption that the value is just $6 + 6 = 12$. This case can arise in exactly the same way as described for Case 2: namely, extra time is wasted switching the production processes back and forth between the two products. The extra cross-product term, $0.5x_1x_2$, would give the amount of capacity used in this way.

For Case 4, the function for capacity used is $3x_1 + 2x_2 - 0.1x_1^2x_2$, so the *total function value* for $(x_1, x_2) = (2, 3)$ is $6 + 6 - 1.2 = 10.8$. This case could arise in the following way. Similarly to Case 3, suppose that the two products require the same type of machinery and equipment, but suppose now that the time required to switch from one product to the other would be relatively small. Because each product goes through a sequence of production operations, individual production facilities normally dedicated to that product would incur occasional idle periods. During these otherwise idle periods, these facilities can be used by the other product, thereby saving production capacity. Consequently, the total production capacity used when the two products are produced jointly would be less than the *sum* of the capacities used by the individual products when each is produced by itself.

After analyzing the possible kinds of interaction between the two products illustrated by these four cases, the OR Department concluded that none played a major role in the actual Wyndor Glass Co. problem. Therefore, the additivity assumption was adopted as a reasonable approximation.

For other problems, if additivity is not a reasonable assumption, so that some or all of the mathematical functions of the model need to be *nonlinear* (because of the cross-product terms), you definitely enter the realm of nonlinear programming (Chap. 14).

Divisibility

Sometimes the decision variables have physical significance only if they have integer values. However, the optimal solution obtained by linear programming is often a noninteger one. Therefore, the *divisibility assumption* is that activity units can be *divided* into *any fractional levels,* so that noninteger values for the decision variables are permissible.

For the Wyndor Glass Co. problem, the decision variables represent production rates, which can have fractional values. Certain fractional values are more convenient than others because they correspond to integer numbers of people and machines working full time on the product. However, the OR Department concluded that these minor adjustments could be made easily after using the model to analyze the big picture and to identify approximately what the combination of production rates should be.

Frequently, linear programming is still applied even when an integer solution is required. If the solution obtained is a noninteger one, then the noninteger variables are merely rounded to integer values. This may be satisfactory, particularly if the decision variables are large, but it does have certain pitfalls (discussed in Sec. 13.3). If this approach cannot be used, then we are in the realm of *integer programming,* which is the topic of Chap. 13. However, it should be noted that linear programming *automatically* will obtain integer solutions to certain special types of problems, including some of those discussed in Chap. 7.

Certainty

The *certainty assumption* is that all the parameters of the model (the a_{ij}, b_i, and c_j values) are *known constants*. In real problems, this assumption is seldom satisfied precisely. Linear programming models usually are formulated to select some future course of action. Therefore, the parameters used would be based on a prediction of future conditions, which inevitably introduces some degree of uncertainty.

For this reason it is usually important to conduct a thorough **sensitivity analysis** after finding a solution that is optimal under the assumed parameter values. As discussed in Sec. 2.3, the general purpose is to identify the *sensitive* parameters (i.e., those that cannot be changed without changing the optimal solution), to try to estimate these more closely, and then to select a solution that remains a good one over the ranges of likely values of the sensitive parameters. This is what the OR Department will do for the Wyndor Glass Co. problem, as you will see in Sec. 6.7. However, it is necessary to acquire some more background before finishing that story.

Occasionally, the degree of uncertainty in the parameters is too great to be amenable to sensitivity analysis. In this case, it is necessary to treat the parameters explicitly as *random variables*. Formulations of this kind are discussed in Secs. 7.8 and 7.9.

The Assumptions in Perspective

We emphasized in Sec. 2.2 that a mathematical model is intended to be only an idealized representation of the real problem. Approximations and simplifying assumptions generally are required in order for the model to be tractable. Adding too much detail and precision can make the model too unwieldy for useful analysis of the problem. All that is really needed is that there be a reasonably high correlation between the prediction of the model and what would actually happen in the real problem.

This advice certainly is applicable to linear programming. It is very common in real applications of linear programming that almost *none* of the four assumptions hold completely. Except perhaps for the *divisibility assumption,* minor disparities are to be expected. This is especially true for the *certainty assumption,* so sensitivity analysis normally is a must to compensate for the violation of this assumption.

However, it is important for the operations research team to examine the four assumptions for the problem under study and analyze just how large the disparities are. If any of the assumptions are violated in a major way, then a number of useful alternative models are available, as presented in later chapters of the book. A disadvantage of these other models is that the algorithms available for solving them are not nearly as powerful as for linear programming, but this gap has been closing in some cases. For some applications, the powerful linear programming approach is used for the initial analysis and then a more complicated model is used to refine this analysis.

As you work through the examples in the next section, you will find it good practice to analyze how well each of the four assumptions of linear programming applies to these problems.

3.4 Additional Examples

The Wyndor Glass Co. problem is a prototype example of linear programming in several respects: It involves allocating limited resources among competing activities, its model fits our standard form, and its context is the traditional one of improved business planning. However, the applicability of linear programming is much wider. In this section we begin broadening our horizons. As you study the following examples, note that it is their underlying mathematical model rather than their context that characterizes them as linear programming problems. Then give some thought to how the same mathematical model could arise in many other contexts by merely changing the names of the activities and so forth.

These examples have been kept very small (by linear programming standards) for ease of reading. However, much larger versions of the problems, involving hundreds of constraints and variables, are readily solvable by linear programming.

Design of Radiation Therapy

MARY is a modern success story. She truly has it all—a very successful career, a leadership role in her community, many friends and admirers, as well as a loving husband and two fine children. But now tragedy has struck. Mary has just been diagnosed as having a cancer at a fairly advanced stage. Specifically, she has a large malignant tumor in the bladder area (a "whole bladder lesion").

Her family has arranged for Mary to receive the most advanced medical care available in the country in order to give her every possible chance for survival. Extensive *radiation therapy* (in combination with chemotherapy and surgery) is the only hope.

Radiation therapy involves using an external beam treatment machine to pass ionizing radiation through the patient's body, damaging both cancerous and healthy tissues. Normally, several beams are precisely administered from different angles in a two-dimensional plane. Due to attenuation, each beam delivers more radiation to the tissue near the entry point than to the tissue near the exit point. Scatter also causes some delivery of radiation to tissue outside the direct path of the beam. Because tumor cells are typically microscopically interspersed among healthy cells, the radiation dosage throughout the tumor region must be large enough to kill the malignant cells, which are slightly more radiosensitive, yet small enough to spare the healthy cells. At the same time, the aggregate dose to critical tissues must not exceed established tolerance levels in order to prevent complications that can be more serious than the disease itself. For the same reason, the total dose to the entire healthy anatomy also must be minimized.

Because of the need to carefully balance all of these factors, the design of radiation therapy is a very delicate process. The goal of the design is to select the combination of beams to be used, and the intensity of each one, in order to generate the best possible dose distribution. (The dose strength at any point in the body is measured in units called *kilorads*.) Once the treatment design has been developed, it then is administered in many installments, spread over several weeks.

In Mary's case, the size and location of her tumor make the design of her treatment an even more delicate process than usual. Figure 3.4 shows a diagram of a cross section of the tumor viewed from above, as well as nearby critical tissues to avoid. These tissues include critical organs (e.g., the rectum) as well as bony structures (e.g., the femurs and pelvis) that will attenuate the radiation. Also shown are the entry point and direction for the only two beams that can be used with any modicum of safety in this case. (Actually, we are simplifying the example at this point, because normally dozens of possible beams must be considered.)

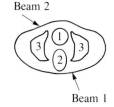

Beam 2

Beam 1

1. Bladder and tumor
2. Rectum, coccyx, etc.
3. Femur, part of pelvis, etc.

Figure 3.4 Cross section of Mary's tumor (viewed from above), nearby critical tissues, and the radiation beams being used.

For any proposed beam of given intensity, the analysis of what the resulting radiation absorption by various parts of the body would be requires a complicated process. In brief, based on careful anatomical analysis, the energy distribution within the two-dimensional cross section of the tissue can be plotted on an isodose map, where the contour lines represent dose strength as a percentage of the dose strength at the entry point. A fine grid then is placed over the isodose map. By summing the radiation absorbed in the squares containing each type of tissue, the average dose that is absorbed by the tumor, healthy anatomy, and critical tissues can be calculated. With more than one beam (administered sequentially), the radiation absorption is additive.

After thorough analysis of this type, the medical team has carefully estimated the data needed to design Mary's treatment, as summarized in Table 3.7. The first column lists the areas of the body that must be considered, and then the next two columns give the fraction of the radiation dose at the entry point for each beam that is absorbed by the respective areas on the average. For example, if the dose level at the entry point for beam 1 is 1 kilorad, then an average of 0.4 kilorad will be absorbed by the entire healthy anatomy in the two-dimensional plane, an average of 0.3 kilorad will be absorbed by nearby critical tissues, an average of 0.5 kilorad will be absorbed by the various parts of the tumor, and 0.6 kilorad will be absorbed by the center of

Table 3.7 **Data for Design of Mary's Radiation Therapy**

Area	Fraction of Entry Dose Absorbed by Area (Average)		Restriction on Total Average Dosage
	Beam 1	Beam 2	
Healthy anatomy	0.4	0.5	Minimize
Critical tissues	0.3	0.1	≤ 2.7
Tumor region	0.5	0.5	$= 6$
Center of tumor	0.6	0.4	ml 6

the tumor. The last column gives the restrictions on the total dosage from both beams that is absorbed on the average by the respective areas of the body. In particular, the average dosage absorption for the healthy anatomy must be *as small as possible,* the critical tissues must *not exceed* 2.7 kilorads, the average over the entire tumor must *equal* 6 kilorads, and the center of the tumor must be *at least* 6 kilorads.

Simultaneously satisfying all of these requirements will be very difficult. The medical team has confided to Mary's family that the disease has reached a critical stage where only a slim chance remains for successful treatment. The only chance for saving Mary's life lies with developing the best possible treatment design by using the most advanced optimization procedure available, namely (what else), *linear programming!*

FORMULATION AS A LINEAR PROGRAMMING PROBLEM: The two decision variables, x_1 and x_2, represent the dose (in kilorads) at the entry point for beam 1 and beam 2, respectively. Because the total dosage reaching the healthy anatomy is to be minimized, let Z denote this quantity. The data from Table 3.7 can then be used directly to formulate the following linear programming model.[1]

$$\text{Minimize} \quad Z = 0.4x_1 + 0.5x_2,$$

subject to

$$0.3x_1 + 0.1x_2 \leq 2.7$$

$$0.5x_1 + 0.5x_2 = 6$$

$$0.6x_1 + 0.4x_2 \geq 6$$

and

$$x_1 \geq 0, \quad x_2 \geq 0.$$

Notice the differences between this model and the one in Sec. 3.1 for the Wyndor Glass Co. problem. The latter model involved *maximizing Z,* and all of the functional constraints were in $\leq$ form. This new model does not fit this same standard form, but it does incorporate three other *legitimate* forms described in Sec. 3.2, namely, *minimizing Z,* functional constraints in $=$ form, and functional constraints in $\geq$ form.

However, both models have only two variables, so this new problem also can be solved by the graphical procedure illustrated in Sec. 3.1. Figure 3.5 shows the graphical solution. The *feasible region* consists of just the dark line segment between (6, 6) and (7.5, 4.5), because the points on this segment are the only ones that simultaneously satisfy all of the constraints. (Check this.) The dashed line is the objective function line that passes through the optimal solution, $(x_1, x_2) = (7.5, 4.5)$ with $Z = 5.25$. This solution is optimal rather than (6, 6) because *decreasing Z* (for positive values of Z) pushes the objective function line toward the origin (where $Z = 0$). $Z = 5.25$ for (7.5, 4.5) is less than $Z = 5.4$ for (6, 6).

But what about Mary? Over a gruelling treatment period of six weeks, the medical team implemented this optimal design of using a total dose at the entry point of 7.5 kilorads for beam 1 and 4.5 kilorads for beam 2. Mary's fighting spirit did the rest. As of this writing, she is alive and doing well!

---

[1] Actually, Table 3.7 simplifies the real situation, so the real model would be somewhat more complicated than this one, and would have dozens of variables and constraints. For details about the general situation, see Sonderman, David, and Philip G. Abrahamson: "Radiotherapy Treatment Design Using Mathematical Programming Models," *Operations Research,* **33**:705–725, 1985, and its ref. 1.

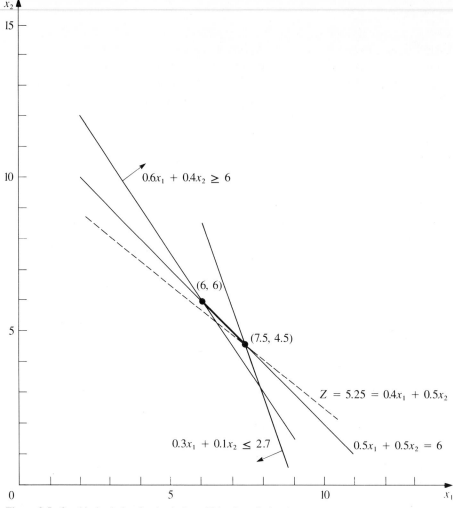

Figure 3.5 Graphical solution for the design of Mary's radiation therapy.

Regional Planning

One of the interesting social experiments in the Mediterranean region is the system of kibbutzim, or communal farming communities, in Israel. It is common for groups of kibbutzim to join together to share common technical services and to coordinate their production. Our next example concerns one such group of three kibbutzim, which we call the SOUTHERN CONFEDERATION OF KIBBUTZIM.

Overall planning for the Southern Confederation of Kibbutzim is done in its Coordinating Technical Office. This office currently is planning agricultural production for the coming year.

The agricultural output of each kibbutz is limited by both the amount of available irrigable land and by the quantity of water allocated for irrigation by the Water Commissioner (a national government official). These data are given in Table 3.8.

The crops suited for this region include sugar beets, cotton, and sorghum, and these are the three being considered for the upcoming season. These crops differ

Table 3.8 Resources Data for Southern
Confederation of Kibbutzim

Kibbutz	Usable Land (Acres)	Water Allocation (Acre Feet)
1	400	600
2	600	800
3	300	375

primarily in their expected net return per acre and their consumption of water. In addition, the Ministry of Agriculture has set a maximum quota for the total acreage that can be devoted to each of these crops by the Southern Confederation of Kibbutzim, as shown in Table 3.9.

The three kibbutzim belonging to the Southern Confederation have agreed that every kibbutz will plant the same proportion of its available irrigable land. For example, if kibbutz 1 plants 200 of its available 400 acres, then kibbutz 2 must plant 300 of its 600 acres, while kibbutz 3 plants 150 acres of its 300 acres. However, any combination of the crops may be grown at any of the kibbutzim. The job facing the Coordinating Technical Office is to plan how many acres to devote to each crop at the respective kibbutzim while satisfying the given restrictions. The objective is to maximize the total net return to the Southern Confederation as a whole.

FORMULATION AS A LINEAR PROGRAMMING PROBLEM: The quantities to be decided upon are the number of acres to devote to each of the three crops at each of the three kibbutzim. The decision variables, x_j ($j = 1, 2, \ldots, 9$), represent these nine quantities, as shown in Table 3.10. Since the measure of effectiveness Z is total net return, the resulting linear programming model for this problem is

Maximize $Z = 400(x_1 + x_2 + x_3) + 300(x_4 + x_5 + x_6) + 100(x_7 + x_8 + x_9)$,

subject to the following constraints:

   1. *Usable land for each kibbutz:*

$$x_1 + x_4 + x_7 \leq 400$$

$$x_2 + x_5 + x_8 \leq 600$$

$$x_3 + x_6 + x_9 \leq 300$$

   2. *Water allocation for each kibbutz:*

$$3x_1 + 2x_4 + x_7 \leq 600$$

Table 3.9 Crop Data for Southern Confederation of Kibbutzim

Crop	Maximum Quota (Acres)	Water Consumption (Acre Feet/Acre)	Net Return (Dollars/Acre)
Sugar beets	600	3	400
Cotton	500	2	300
Sorghum	325	1	100

Table 3.10 Decision Variables for Southern
Confederation of Kibbutzim Problem

	Allocation (Acres)		
	Kibbutz		
Crop	1	2	3
Sugar beets	x_1	x_2	x_3
Cotton	x_4	x_5	x_6
Sorghum	x_7	x_8	x_9

$$3x_2 + 2x_5 + x_8 \le 800$$

$$3x_3 + 2x_6 + x_9 \le 375$$

3. *Total acreage for each crop:*

$$x_1 + x_2 + x_3 \le 600$$

$$x_4 + x_5 + x_6 \le 500$$

$$x_7 + x_8 + x_9 \le 325$$

4. *Equal proportion of land planted:*

$$\frac{x_1 + x_4 + x_7}{400} = \frac{x_2 + x_5 + x_8}{600}$$

$$\frac{x_2 + x_5 + x_8}{600} = \frac{x_3 + x_6 + x_9}{300}$$

$$\frac{x_3 + x_6 + x_9}{300} = \frac{x_1 + x_4 + x_7}{400}$$

5. *Nonnegativity:*

$$x_j \ge 0, \quad \text{for } j = 1, 2, \ldots, 9.$$

This completes the model, except that the equality constraints are not yet in an appropriate form for a linear programming model because some of the variables are on the right-hand side. Hence their final form[1] is

$$3(x_1 + x_4 + x_7) - 2(x_2 + x_5 + x_8) = 0$$

$$x_2 + x_5 + x_8 - 2(x_3 + x_6 + x_9) = 0$$

$$4(x_3 + x_6 + x_9) - 3(x_1 + x_4 + x_7) = 0.$$

The Coordinating Technical Office formulated this model and then applied the simplex method (developed in the next chapter) to find the best solution. The solution they obtained is

$$(x_1, x_2, x_3, x_4, x_5, x_6, x_7, x_8, x_9) = (133\tfrac{1}{3}, 100, 25, 100, 250, 150, 0, 0, 0),$$

as shown in Table 3.11.

---

[1] Actually, any one of these equations is redundant and can be deleted if desired. Because of these equations, any two of the land constraints also could be deleted.

Table 3.11 Optimal Solution for Southern
Confederation of Kibbutzim Problem

| | Best Allocation (Acres) | | |
| Crop | Kibbutz | | |
	1	2	3
Sugar beets	$133\frac{1}{3}$	100	25
Cotton	100	250	150
Sorghum	0	0	0

Controlling Air Pollution

The NORI & LEETS CO., one of the major producers of steel in its part of the world, is located in the city of Steeltown and is the only large employer there. Steeltown has grown and prospered along with the company, which now employs nearly 50,000 residents. Therefore, the attitude of the townspeople always has been "What's good for Nori & Leets is good for the town." However, this attitude is now changing; uncontrolled air pollution from the company's furnaces is ruining the appearance of the city and endangering the health of its residents.

A recent stockholders' revolt resulted in the election of a new enlightened board of directors for the company. These directors are determined to follow socially responsible policies, and they have been discussing with Steeltown city officials and citizens' groups what to do about the air pollution problem. Together they have worked out stringent air quality standards for the Steeltown airshed.

The three main types of pollutants in this airshed are particulate matter, sulfur oxides, and hydrocarbons. The new standards require that the company reduce its annual emission of these pollutants by the amounts shown in Table 3.12. The board of directors has instructed management to have the engineering staff determine how to achieve these reductions in the most economical way.

The steelworks has two primary sources of pollution, namely, the blast furnaces for making pig iron and the open-hearth furnaces for changing iron into steel. In both cases the engineers have decided that the most effective types of abatement methods are (1) increasing the height of the smokestacks,[1] (2) using filter devices (including gas traps) in the smokestacks, and (3) including cleaner, high-grade materials among

Table 3.12 Clean Air Standards for
Nori & Leets Co.

Pollutant	Required Reduction in Annual Emission Rate (Million Pounds)
Particulates	60
Sulfur oxides	150
Hydrocarbons	125

[1] Subsequent to this study, this particular abatement method has become a controversial one. Because its effect is to reduce ground-level pollution by spreading emissions over a greater distance, environmental groups contend that this creates more acid rain by keeping sulfur oxides in the air longer. Consequently, the U.S. Environmental Protection Agency adopted new rules in 1985 to remove incentives for using tall smokestacks.

Table 3.13 **Reduction in Emission Rate from Maximum Feasible Use of Abatement Method for Nori & Leets Co.**

Pollutant	Taller Smokestacks		Filters		Better Fuels	
	Blast Furnaces	Open-hearth Furnaces	Blast Furnaces	Open-hearth Furnaces	Blast Furnaces	Open-hearth Furnaces
Particulates	12	9	25	20	17	13
Sulfur oxides	35	42	18	31	56	49
Hydrocarbons	37	53	28	24	29	20

the fuels for the furnaces. All these methods have technological limits on how much emission they can eliminate from each type of furnace, as shown (in millions of pounds per year) in Table 3.13.

However, the methods can be used at any fraction (including zero) of their abatement capacities shown in this table, and the fractions can be different for blast furnaces and open-hearth furnaces. For either type of furnace, the emission reduction achieved by each method is not substantially affected by whether or not the other methods also are used.

After these data were developed, it became clear that no single method by itself could achieve all the required reductions. On the other hand, combining all three methods at full capacity on both types of furnaces (which would be prohibitively expensive if the company's products are to remain competitively priced) is much more than adequate. Therefore, the engineers concluded that they would have to use some combination of the methods, perhaps with fractional capacities, based upon their relative costs. Furthermore, because of the differences between the blast and the open-hearth furnaces, the two types probably should not use the same combination.

An analysis was conducted to estimate the total annual cost that would be incurred by each abatement method. In addition to increased operating and maintenance expenses, consideration was given also to the initial costs (converted to an equivalent annual basis) of the method as well as any resulting loss in efficiency of the production process. This analysis led to the total cost estimates (in millions of dollars) given in Table 3.14 for using the methods at their full abatement capacities. It also was determined that the cost of a method being used at a lower level is essentially proportional to its fractional capacity. Thus, for any given fraction used, the total annual cost would be that fraction of the corresponding quantity in Table 3.14.

The stage now was set to develop the general framework of the company's plan for pollution abatement. This plan would consist of specifying which types of abate-

Table 3.14 **Total Annual Cost from Maximum Feasible Use of Abatement Method for Nori & Leets Co.**

Abatement Method	Blast Furnaces	Open-hearth Furnaces
Taller smokestacks	8	10
Filters	7	6
Better fuels	11	9

Table 3.15 Decision Variables (Fraction of Maximum Feasible Use of Abatement Method) for Nori & Leets Co.

Abatement Method	Blast Furnaces	Open-hearth Furnaces
Taller smokestacks	x_1	x_2
Filters	x_3	x_4
Better fuels	x_5	x_6

ment methods would be used and at what fractions of their abatement capacities for (1) the blast furnaces and (2) the open-hearth furnaces. Because of the combinatorial nature of the problem of finding a plan that satisfies the requirements with the smallest possible cost, an *operations research team* was formed to solve the problem. The team adopted a linear programming approach, formulating the model summarized next.

FORMULATION AS A LINEAR PROGRAMMING PROBLEM: This problem has six decision variables, x_j ($j = 1, 2, \ldots, 6$), each representing the usage of one of the three abatement methods for one of the two types of furnaces, expressed as a fraction of the abatement capacity. The ordering of these variables is shown in Table 3.15. Because the objective is to minimize total cost while satisfying the emission reduction requirements, the model is

$$\text{Minimize} \quad Z = 8x_1 + 10x_2 + 7x_3 + 6x_4 + 11x_5 + 9x_6,$$

subject to the following constraints:

1. *Emission reduction:*

$$12x_1 + 9x_2 + 25x_3 + 20x_4 + 17x_5 + 13x_6 \geq 60$$

$$35x_1 + 42x_2 + 18x_3 + 31x_4 + 56x_5 + 49x_6 \geq 150$$

$$37x_1 + 53x_2 + 28x_3 + 24x_4 + 29x_5 + 20x_6 \geq 125$$

2. *Technological limit:*

$$x_j \leq 1, \quad \text{for } j = 1, 2, \ldots, 6$$

3. *Nonnegativity:*

$$x_j \geq 0, \quad \text{for } j = 1, 2, \ldots, 6.$$

The operations research team used this model[1] to find the minimum-cost plan $(x_1, x_2, x_3, x_4, x_5, x_6) = (1, 0.623, 0.343, 1, 0.048, 1)$. Sensitivity analysis then was conducted, followed by detailed planning and managerial review. Soon after, this program for controlling air pollution was fully implemented by the company, and the citizens of Steeltown breathed deep sighs of relief.

---

[1] An equivalent formulation can express each decision variable in natural units for its abatement method; for example, x_1 and x_2 could represent the number of *feet* that the heights of the smokestacks are increased.

The four linear programming examples you have seen so far are but a small sampling of the uses of this technique. Many more illustrations are given in Chaps. 7 and 8; most involve business and industrial applications, but several others arise in different contexts. Chapter 7 focuses on certain special types of linear programming problems that provide many important applications. Chapter 8 considers some examples that are more difficult to formulate, and it also includes a case study involving the design of school attendance zones to achieve better racial balance. But before considering these topics, we next discuss how to solve linear programming problems.

3.5 Conclusions

Linear programming is a powerful technique for dealing with the problem of allocating limited resources among competing activities as well as other problems having a similar mathematical formulation. It has become a standard tool of great importance for numerous business and industrial organizations. Furthermore, almost any social organization is concerned with allocating resources in some context, and there is a growing recognition of the extremely wide applicability of this technique.

However, not all problems of allocating limited resources can be formulated to fit a linear programming model, even as a reasonable approximation. When one or more of the assumptions of linear programming is violated seriously, it may then be possible to apply another mathematical programming model instead, e.g., the models of integer programming (Chap. 13) or nonlinear programming (Chap. 14).

SELECTED REFERENCES

1. Anderson, David R., Dennis J. Sweeney, and Thomas A. Williams: *An Introduction to Management Science,* 5th ed., chaps. 2, 4, West, St. Paul, Minn., 1988.
2. Cook, Thomas M., and Robert A. Russell: *Introduction to Management Science,* 3d ed., chaps. 2–3, Prentice-Hall, Englewood Cliffs, N.J., 1985.
3. Gass, Saul I.: *Decision Making, Models and Algorithms, Part II,* Wiley, New York, 1980.
4. Salkin, Harvey M., and Jahar Saha (eds.): *Studies in Linear Programming,* North-Holland/American Elsevier, Amsterdam/New York, 1975.

PROBLEMS[1]

1.* Suppose you have just inherited $6,000 and you want to invest it. Upon hearing this news, two different friends have offered you an opportunity to become a partner in two different entrepreneurial ventures, one planned by each friend. In both cases, this investment would involve expending some of your time next summer as well as putting up cash. Becoming a *full* partner in the first friend's venture would require an investment of $5,000 and 400 hours, and your estimated profit (ignoring the value of your time) would be $4,500. The corresponding figures for the second friend's venture are $4,000 and 500 hours, with an estimated profit to you of $4,500. However, both friends are flexible and would allow you to come in at any

---

[1] Some additional *formulation* problems are given at the end of Chap. 8. Also note that at least partial answers to starred problems are given at the back of the book.

fraction of a full partnership you would like; your share of the profit would be proportional to this fraction.

Because you were looking for an interesting summer job anyway (maximum of 600 hours), you have decided to participate in one or both friends' ventures in whichever combination would maximize your total estimated profit. You now need to solve the problem of finding the best combination.

(a) Describe the analogy between this problem and the Wyndor Glass Co. problem discussed in Sec. 3.1. Then construct and fill in a table like Table 3.2 for this problem, identifying both the activities and the resources.

(b) Formulate the linear programming model for this problem.

(c) Solve this model graphically. What is your total estimated profit?

(d) Indicate why each of the four assumptions of linear programming (Sec. 3.3) appears to be reasonably satisfied for this problem. Is one assumption more doubtful than the others? If so, what should be done to take this into account?

2. A manufacturing firm has discontinued the production of a certain unprofitable product line. This act created considerable excess production capacity. Management is considering devoting this excess capacity to one or more of three products; call them products 1, 2, and 3. The available capacity on the machines that might limit output is summarized in the following table:

Machine Type	Available Time (in Machine Hours Per Week)
Milling machine	500
Lathe	350
Grinder	150

The number of machine hours required for each unit of the respective products is

Productivity Coefficient (in Machine Hours Per Unit)

Machine Type	Product 1	Product 2	Product 3
Milling machine	9	3	5
Lathe	5	4	0
Grinder	3	0	2

The Sales Department indicates that the sales potential for products 1 and 2 exceeds the maximum production rate and that the sales potential for product 3 is 20 units per week. The unit profit would be $50, $20, and $25, respectively, on products 1, 2, and 3. The objective is to determine how much of each product the firm should produce to maximize profit.

Formulate the linear programming model for this problem.

3.* Use the graphical procedure illustrated in Sec. 3.1 to solve the problem:

$$\text{Maximize} \quad Z = 2x_1 + x_2,$$

subject to

$$x_2 \leq 10$$

$$2x_1 + 5x_2 \leq 60$$

$$x_1 + x_2 \leq 18$$

$$3x_1 + x_2 \leq 44$$

and

$$x_1 \geq 0, \qquad x_2 \geq 0.$$

4. Use the graphical procedure illustrated in Sec. 3.1 to solve the problem:

$$\text{Maximize} \qquad Z = 10x_1 + 20x_2,$$

subject to

$$-x_1 + 2x_2 \leq 15$$

$$x_1 + x_2 \leq 12$$

$$5x_1 + 3x_2 \leq 45$$

and

$$x_1 \geq 0, \qquad x_2 \geq 0.$$

5. For each of the four assumptions of linear programming discussed in Sec. 3.3, write a one-paragraph analysis of how well you feel it applies to each of the following examples given in Sec. 3.4:

(a) Design of radiation therapy (Mary).
(b) Regional planning (Southern Confederation of Kibbutzim).
(c) Controlling air pollution (Nori & Leets Co.).

6. Consider a problem with two decision variables, x_1 and x_2, which represent the levels of activities 1 and 2, respectively. For each variable, the permissible values are 0, 1, and 2, where the feasible combinations of these values for the two variables are determined from a variety of constraints. The objective is to maximize a certain measure of performance denoted by Z. The values of Z for the possibly feasible values of (x_1, x_2) are estimated to be those given in the following table:

x_1 \ x_2	0	1	2
0	0	4	8
1	3	8	13
2	6	12	18

Based on this information, indicate whether this problem completely satisfies each of the four assumptions of linear programming. Justify your answers.

7. Consider the problem described at the beginning of Sec. 8.5, where the city of Middletown is using linear programming to redesign the school attendance zones for its high schools. The objective function used there is to minimize the total distance that students must travel, subject to constraints on the racial balance in the schools. Now suppose that the school board decides to change the objective function to minimizing the total *cost* of bussing the students. (However, the decision variables continue to be x_{ij} = number of students in tract i assigned to school j.) Each student assigned to a school more than a mile away will be given the opportunity to ride a bus. (However, some of these students may choose to get to and from school in some other way.) Each bus can carry 40 students. The daily cost of providing each bus is estimated to be $50 plus $1 for each student carried. A bus may transport students from more than one tract to try to fill the buses.

For each of the four assumptions of linear programming discussed in Sec. 3.3, write a one-paragraph analysis of how well it applies to the revised objective function.

8. Use the graphical procedure illustrated in Sec. 3.1 to solve the problem:

$$\text{Minimize} \quad Z = 15x_1 + 20x_2,$$

subject to

$$x_1 + 2x_2 \geq 10$$

$$2x_1 - 3x_2 \leq 6$$

$$x_1 + x_2 \geq 6$$

and

$$x_1 \geq 0, \quad x_2 \geq 0.$$

9. Use the graphical procedure illustrated in Sec. 3.1 and Fig. 3.5 to solve the problem:

$$\text{Minimize} \quad Z = 3x_1 + 2x_2,$$

subject to

$$x_1 + 2x_2 \leq 12$$

$$2x_1 + 3x_2 = 12$$

$$2x_1 + x_2 \geq 8$$

and

$$x_1 \geq 0, \quad x_2 \geq 0.$$

10. A farmer is raising pigs for market, and he wishes to determine the quantities of the available types of feed that should be given to each pig to meet certain nutritional requirements at a *minimum cost*. The number of units of each type of basic nutritional ingredient contained within a kilogram of each feed type is given in the following table, along with the daily nutritional requirements and feed costs:

Nutritional Ingredient	Kilogram of Corn	Kilogram of Tankage	Kilogram of Alfalfa	Minimum Daily Requirement
Carbohydrates	90	20	40	200
Protein	30	80	60	180
Vitamins	10	20	60	150
Cost (¢)	42	36	30	

Formulate the linear programming model for this problem.

11. A certain corporation has three branch plants with excess production capacity. All three plants have the capability for producing a certain new product, and management has decided to use some of the excess capacity in this way. This product can be made in three sizes—large, medium, and small—that yield a net unit profit of $420, $360, and $300, respectively. Plants 1, 2, and 3 have the excess capacity to produce 750, 900, and 450 units per day of this product, respectively, regardless of the size or combination of sizes involved.

The amount of available in-process storage space also imposes a limitation on the production rates of the new product. Plants 1, 2, and 3 have 13,000, 12,000, and 5,000 square feet of in-process storage space available for a day's production of this product. Each unit of the large, medium, and small sizes produced per day requires 20, 15, and 12 square feet, respectively.

Sales forecasts indicate that 900, 1,200, and 750 units of the large, medium, and small sizes, respectively, can be sold per day.

To maintain a uniform workload among the plants and to retain some flexibility, management has decided that the plants must use the same percentage of their excess capacity to produce the new product.

Management wishes to know how much of each of the sizes should be produced by each of the plants to maximize profit.

Formulate the linear programming model for this problem.

12. A farm family owns 125 acres of land and has $40,000 in funds available for investment. Its members can produce a total of 3,500 person-hours worth of labor during the winter months (mid-September to mid-May) and 4,000 person-hours during the summer. If any of these person-hours are not needed, younger members of the family will use them to work on a neighboring farm for $5/hour during the winter months and $6/hour during the summer.

Cash income may be obtained from three crops and two types of livestock: dairy cows and laying hens. No investment funds are needed for the crops. However, each cow will require an investment outlay of $1,200, and each hen will cost $9.

Each cow will require 1.5 acres of land, 100 person-hours of work during the winter months, and another 50 person-hours during the summer. Each cow will produce a net annual cash income of $1,000 for the family. The corresponding figures for each hen are: no acreage, 0.6 person-hour during the winter, 0.3 more person-hour during the summer, and an annual net cash income of $5. The chicken house can accommodate a maximum of 3,000 hens, and the size of the barn limits the herd to a maximum of 32 cows.

Estimated person-hours and income per acre planted in each of the three crops are

	Soybeans	Corn	Oats
Winter person-hours	20	35	10
Summer person-hours	50	75	40
Net annual cash income ($)	600	900	450

The family wishes to determine how much acreage should be planted in each of the crops and how many cows and hens should be kept to maximize its net cash income. Formulate the linear programming model for this problem.

13. A cargo plane has three compartments for storing cargo: front, center, and back. These compartments have capacity limits on both *weight* and *space,* as summarized below:

Compartment	Weight Capacity (Tons)	Space Capacity (Cubic Feet)
Front	12	7,000
Center	18	9,000
Back	10	5,000

Furthermore, the weight of the cargo in the respective compartments must be the same proportion of that compartment's weight capacity to maintain the balance of the airplane.

The following four cargoes have been offered for shipment on an upcoming flight as space is available:

Cargo	Weight (Tons)	Volume (Cubic Feet/Ton)	Profit ($/Ton)
1	20	500	320
2	16	700	400
3	25	600	360
4	13	400	290

Any portion of these cargoes can be accepted. The objective is to determine how much (if any) of each cargo should be accepted and how to distribute each among the compartments to maximize the total profit for the flight.

Formulate the linear programming model for this problem.

14. An investor has money-making activities A and B available at the beginning of each of the next five years (call them years 1 to 5). Each dollar invested in A at the beginning of a year returns $1.40 (a profit of $0.40) two years later (in time for immediate reinvestment). Each dollar invested in B at the beginning of a year returns $1.70 three years later.

In addition, money-making activities C and D will each be available at one time in the future. Each dollar invested in C at the beginning of year 2 returns $1.90 at the end of year 5. Each dollar invested in D at the beginning of year 5 returns $1.30 at the end of year 5.

The investor begins with $60,000 and wishes to know which investment plan maximizes the amount of money that can be accumulated by the beginning of year 6.

Formulate the linear programming model for this problem.

4

Solving Linear Programming Problems: The Simplex Method

We now are ready to begin studying the *simplex method,* the general procedure for solving linear programming problems. Developed by George Dantzig in 1947, it has proven to be a remarkably efficient method that is used routinely to solve huge problems on today's computers. Except for tiny problems, this method is always executed on a computer, and sophisticated software packages are widely available. Nevertheless, it is important to learn something about how the method works in order to understand how to perform *post-optimality analysis* (including sensitivity analysis) on the model. Therefore, this chapter describes and illustrates the main features of the simplex method.

The first section introduces the general nature of the simplex method, including its geometric interpretation. The following three sections then develop the procedure for solving any linear programming model that is in *our standard form* (as defined in Sec. 3.2) and has only *positive* right-hand sides (b_i) in the functional constraints. Certain details on resolving ties are deferred to Sec. 4.5, and Sec. 4.6 describes how to adapt this method to other model forms. We next discuss post-optimality analysis

(Sec. 4.7), and then conclude the chapter with a description of the computer implementation of the simplex method (Sec. 4.8).

4.1 The Essence of the Simplex Method

The simplex method actually is an **algorithm,** the first of many you will see in this book. Although you may not have heard this name used, you undoubtedly have encountered many algorithms before. For example, the familiar procedure for *long division* is an algorithm. So is the procedure for *calculating square roots*. In fact, any *iterative solution procedure* is an algorithm. Thus an algorithm is simply a process where a systematic procedure is repeated (iterated) over and over again until the desired result is obtained. Each time through the systematic procedure is called an *iteration*. (Can you see what the iteration is for the long division algorithm?) Consequently, an algorithm replaces one difficult problem by a series of easy ones.

In addition to iterations, algorithms also include a procedure for getting started and a criterion for determining when to stop, as summarized here:

Structure of Algorithms[1]

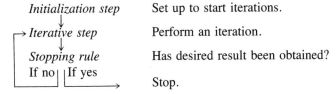

For most operations research algorithms, including the simplex method, the desired result mentioned in the stopping rule is that the current solution is optimal. In this case, the stopping rule actually is an *optimality test,* as shown here:

Structure of Most Operations Research Algorithms

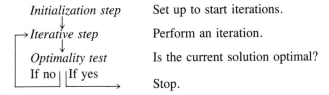

The simplex method is an *algebraic* procedure, where each iteration involves solving a system of equations to obtain a new trial solution for the optimality test. However, it also has a very useful *geometric* interpretation. To illustrate the general geometric concepts, we shall use the graphical solution to the Wyndor Glass Co. example presented in Sec. 3.1.

To refresh your memory, the graph for this example is repeated in Fig. 4.1. The five constraint lines and their points of intersection are highlighted in this figure because they are the keys to the analysis. In particular, these points of intersection are the **corner-point solutions** of the problem. The five that lie on the corners of the *feasible region*—(0, 0), (0, 6), (2, 6), (4, 3), (4, 0)—are the **corner-point feasible**

---

[1] Actually, the stopping rule usually is applied after the initialization step as well to see if *any* iterations are needed.

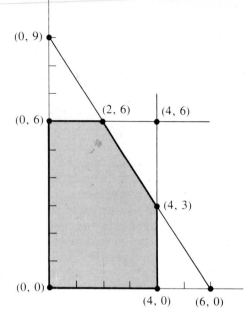

Figure 4.1
Constraint lines and
corner-point solutions
for the Wyndor Glass
Co. problem.

solutions. [The other three—(0, 9), (4, 6), (6, 0)—are called *corner-point infeasible solutions*.] Some of these corner-point feasible solutions are **adjacent** to each other in the sense that they are connected by a single edge (line segment) on the boundary of the feasible region; e.g., both (0, 6) and (4, 3) are adjacent to (2, 6).

Section 5.1 develops in detail the general properties of corner-point feasible solutions for linear programming problems of any size, as well as the relationships between these properties and the *algebra* of the simplex method presented in the next two sections. The three key properties[1] that form the foundation of the simplex method are summarized as follows.

Properties of Corner-Point Feasible Solutions

1a. If there is exactly one optimal solution, then it *must* be a corner-point feasible solution.

1b. If there are multiple optimal solutions, then at least two *must* be adjacent corner-point feasible solutions.

2. There are only a *finite* number of corner-point feasible solutions.

3. If a corner-point feasible solution has no *adjacent* corner-point feasible solutions that are *better* (as measured by Z), then there are no *better* corner-point feasible solutions anywhere; i.e., it is *optimal*.

Property 1 implies that the search for an optimal solution can be reduced to considering *only* the corner-point feasible solutions, so there are only a finite number of solutions to consider (Property 2). Property 3 provides a very convenient *optimality test*.

The simplex method exploits these three properties by examining only relatively few of the promising corner-point feasible solutions and stopping as soon as one of

---

[1] To ensure that the problem does, in fact, possess the optimal solution discussed in Properties 1 and 3, it is sufficient to assume that (1) the problem has feasible solutions and (2) the problem has a bounded feasible region. Assumption (2) also is needed to ensure that Property 1b holds.

them passes this optimality test. In particular, it repeatedly (iteratively) moves from the current corner-point feasible solution to a better adjacent corner-point feasible solution (which can be done very efficiently), until the current solution does not have any better adjacent corner-point feasible solutions. This procedure is summarized as follows.

Outline of the Simplex Method
1. *Initialization step:* Start at a corner-point feasible solution.
2. *Iterative step:* Move to a better adjacent corner-point feasible solution. (Repeat this step as often as needed.)
3. *Optimality test:* The current corner-point feasible solution is optimal when none of its adjacent corner-point feasible solutions are better.

This outline shows the essence of the simplex method, although the complete description in the next two sections does specify a convenient way of choosing the new solution in both the initialization and iterative steps. Using these choice rules, the simplex method proceeds as follows in the Wyndor Glass Co. example.

1. *Initialization step:* Start at $(0, 0)$.
2a. *Iteration 1:* Move from $(0, 0)$ to $(0, 6)$.
2b. *Iteration 2:* Move from $(0, 6)$ to $(2, 6)$.
3. *Optimality test:* Neither $(0, 6)$ nor $(4, 3)$ is better than $(2, 6)$, so stop. $(2, 6)$ is optimal.

4.2 Setting Up the Simplex Method

The preceding section stressed the geometric concepts that underlie the simplex method. However, this algorithm normally is run on a computer, which can follow only algebraic instructions. Therefore, it is necessary to translate the conceptually geometric procedure just described into a usable algebraic procedure. In this section, we introduce the *algebraic language* of the simplex method and relate it to the concepts of the preceding section.

In an algebraic procedure, it is much more convenient to deal with equations than with inequality relationships. Therefore, the first step in setting up the simplex method is to convert the functional *inequality constraints* into equivalent *equality constraints*. (The nonnegativity constraints can be left as inequalities because they are used only indirectly by the algorithm.) This conversion is done by introducing **slack variables.** To illustrate, consider the first functional constraint in the Wyndor Glass Co. example of Sec. 3.1,

$$x_1 \le 4.$$

The slack variable for this constraint is

$$x_3 = 4 - x_1,$$

which is just the slack between the two sides of the inequality. Thus

$$x_1 + x_3 = 4.$$

The original constraint $x_1 \le 4$ holds whenever $x_3 \ge 0$. Hence $x_1 \le 4$ is entirely *equivalent* to the set of constraints

$$x_1 + x_3 = 4$$

and $x_3 \geq 0$,

so these more convenient constraints are used instead.

By introducing slack variables in an identical fashion for the other functional constraints, the original linear programming model for the example can now be replaced by the *equivalent* model:

$$\text{Maximize} \quad Z = 3x_1 + 5x_2,$$

subject to

$$(1) \quad x_1 \qquad\quad + x_3 \qquad\qquad\quad = 4$$

$$(2) \qquad\qquad 2x_2 \qquad + x_4 \qquad\quad = 12$$

$$(3) \qquad 3x_1 + 2x_2 \qquad\qquad + x_5 = 18$$

and

$$x_j \geq 0, \qquad \text{for } j = 1, 2, \ldots, 5.$$

Although this problem is identical to the original, this form is much more convenient for algebraic manipulation and for identification of corner-point feasible solutions. We call this the **augmented form** of the problem, because the original form has been *augmented* by some additional variables needed to apply the simplex method.

The terminology used in the preceding section (corner-point solutions, etc.) applies to the original form of the problem. We now introduce the corresponding terminology for the augmented form.

> An **augmented solution** is a solution for the original variables that has been *augmented* by the corresponding values of the *slack variables*.

For example, augmenting the solution (3, 2) in the example yields the augmented solution (3, 2, 1, 8, 5) because the corresponding values of the slack variables are $x_3 = 1$, $x_4 = 8$, $x_5 = 5$.

> A **basic solution** is an *augmented* corner-point solution.[1]

To illustrate, consider the corner-point infeasible solution (4, 6) in the example. Augmenting it with the resulting values of the slack variables $x_3 = 0$, $x_4 = 0$, and $x_5 = -6$ yields the corresponding basic solution (4, 6, 0, 0, -6).

The fact that corner-point solutions (and so basic solutions) can be either feasible or infeasible implies the following definition:

> A **basic feasible solution** is an *augmented* corner-point feasible solution.

Thus the corner-point feasible solution (0, 6) in the example is equivalent to the basic feasible solution (0, 6, 4, 0, 6) for the problem in augmented form.

The only difference between basic solutions and corner-point solutions (or between basic feasible solutions and corner-point feasible solutions) is whether or not the values of the slack variables are included. For any basic solution, the corresponding corner-point solution is obtained simply by deleting the slack variables. Therefore, the geometric and algebraic relationships between these two solutions are very close, as described in Sec. 5.1.

Because the terms *basic solution* and *basic feasible solution* are very important

---

[1] When the original problem includes equality constraints, the basic solutions are just the augmented corner-point solutions that satisfy all these constraints.

parts of the standard vocabulary of linear programming, we now need to clarify their algebraic properties. For the augmented form of the example, notice that the system of functional constraints has two more variables (5) than equations (3). This fact gives us two *degrees of freedom* in solving the system, since any two variables can be chosen to be set equal to any arbitrary value in order to solve the three equations in terms of the remaining three variables (barring redundancies). The simplex method uses zero for this arbitrary value. The variables that are currently set to zero by the simplex method are called **nonbasic variables,** and the others are called **basic variables.** The resulting solution is called a *basic solution.* If all of the basic variables are nonnegative, the solution is called a *basic feasible solution.*

To illustrate these definitions, consider again the basic feasible solution (0, 6, 4, 0, 6). This solution was obtained before by augmenting the corner-point feasible solution (0, 6). However, another way to obtain this same solution is to choose x_1 and x_4 to be the two *nonbasic variables,* and so the two variables to be set equal to zero. The three equations then yield, respectively, $x_3 = 4$, $x_2 = 6$, and $x_5 = 6$ as the solution for the three *basic variables.* Because all three of these basic variables are nonnegative, this *basic solution* (0, 6, 4, 0, 6) is indeed a *basic feasible solution.*

Two basic feasible solutions are **adjacent** if *all but one* of their nonbasic variables are the same (so the same statement holds for their basic variables). Consequently, moving from the current basic feasible solution to an adjacent one involves switching one variable from nonbasic to basic and vice versa for one other variable.

To illustrate *adjacent basic feasible solutions,* consider one pair of adjacent corner-point feasible solutions in Fig. 4.1, (0, 0) and (0, 6). Their augmented solutions, (0, 0, 4, 12, 18) and (0, 6, 4, 0, 6), automatically are adjacent basic feasible solutions. However, you don't need to look at Fig. 4.1 to draw this conclusion. Another signpost is that their nonbasic variables, (x_1, x_2) and (x_1, x_4), are the same with just the one exception that x_2 has been replaced by x_4. Consequently, moving from (0, 0, 4, 12, 18) to (0, 6, 4, 0, 6) involves switching x_2 from nonbasic to basic and vice versa for x_4.

In general terms, the number of *nonbasic variables* in a basic solution always equals the number of *degrees of freedom* in the system of equations, and the number of *basic variables* always equals the number of *functional constraints.*

When dealing with the problem in augmented form, it is convenient to consider and manipulate the objective function equation at the same time as the new constraint equations. Therefore, before starting the simplex method, the problem needs to be rewritten once again in an equivalent way as

$$\text{Maximize} \quad Z,$$

subject to

$$(0) \qquad Z - 3x_1 - 5x_2 \qquad\qquad\qquad\quad = 0$$

$$(1) \qquad\qquad\qquad x_1 \qquad + x_3 \qquad\qquad = 4$$

$$(2) \qquad\qquad\qquad\qquad 2x_2 \qquad + x_4 \qquad = 12$$

$$(3) \qquad\qquad\qquad 3x_1 + 2x_2 \qquad\qquad + x_5 = 18$$

and

$$x_j \geq 0, \qquad \text{for } j = 1, 2, \ldots, 5.$$

It is just as if Eq. (0) actually were one of the original constraints, but because it already is in equality form, no slack variable is needed. With this interpretation, the basic solutions would be unchanged, except that Z would be viewed as a permanent additional basic variable.

Somewhat fortuitously, the model for the Wyndor Glass Co. problem fits *our standard form,* and all its functional constraints have positive right-hand sides (b_i). If this had not been the case, then additional adjustments would have been needed at this point before applying the simplex method. These details are deferred to Sec. 4.6, and we now turn to focusing on the simplex method itself.

4.3 The Algebra of the Simplex Method

The discussion in Sec. 4.1 of the essence of the simplex method did not get into the details of how the steps are performed. In particular, the following questions have not yet been answered completely (the parenthetical phrases restate the questions in the algebraic terminology of Sec. 4.2).

1. *Initialization step:* How is the initial *corner-point feasible solution* (basic feasible solution) selected?
2. *Iterative step:* When seeking to move to a better *adjacent corner-point feasible solution* (adjacent basic feasible solution),
   (a) How is the direction of movement selected? (Which nonbasic variable is selected to become basic?)
   (b) Where do we stop? (Which basic variable becomes nonbasic?)
   (c) How is the new solution identified?
3. *Optimality test:* How do we determine that the *current corner-point feasible solution* (basic feasible solution) has no *adjacent corner-point feasible solutions* (adjacent basic feasible solutions) that are better?

These are the questions addressed in this section. We continue to use the prototype example of Sec. 3.1, as rewritten at the end of Sec. 4.2, for illustrative purposes.

Initialization Step

The simplex method can start at any corner-point feasible solution (basic feasible solution), so it chooses a convenient one. Before considering slack variables, this choice is the *origin* (all original variables equal to zero), or $(x_1, x_2) = (0, 0)$ in the example.[1] Consequently, after the introduction of slack variables, the *original* variables are the *nonbasic* variables and the *slack* variables are the *basic* variables for the initial basic feasible solution. This choice is illustrated here, where the basic variables are shown in bold type.

$$(1) \qquad x_1 \qquad\ + x_3 \qquad\qquad\quad = 4$$

$$(2) \qquad\qquad\quad 2x_2 \qquad + x_4 \qquad = 12$$

$$(3) \qquad\qquad 3x_1 + 2x_2 \qquad\qquad + x_5 = 18$$

---

[1] Note that choosing the origin makes the left-hand side of all the original functional constraints equal to zero. Therefore, under the current assumptions about the form of the model, including $\leq$ functional constraints and positive right-hand sides, this corner-point solution automatically is *feasible.*

Because the nonbasic variables are set equal to zero, the remaining solution is read as if the nonbasic variables were not there, so $x_3 = 4$, $x_4 = 12$, and $x_5 = 18$, giving the **initial basic feasible solution** (0, 0, 4, 12, 18).

Notice that the reason this solution can be read immediately is that each equation has just one basic variable, which has a coefficient of 1, and this basic variable does not appear in any other equation. You will soon see that when the set of basic variables changes, the simplex method uses an algebraic procedure (Gaussian elimination) to convert the equations into this same convenient form for reading every subsequent basic feasible solution as well. This form is called **proper form from Gaussian elimination.**

Iterative Step

At each iteration, the simplex method moves from the current basic feasible solution to a better *adjacent* basic feasible solution. This movement involves converting one nonbasic variable into a basic variable (called the **entering basic variable**) and simultaneously converting a basic variable into a nonbasic variable (called the **leaving basic variable**), and then identifying the new basic feasible solution.

QUESTION 1: What is the criterion for selecting the *entering basic variable*?

The candidates for the entering basic variable are the n current nonbasic variables. The one chosen would be changed from a nonbasic to a basic variable, so its value would be increased from zero to some positive number and the others would be kept at zero. Since the new basic feasible solution is required to be an improvement (larger Z) over the current one, the rate of change in Z from increasing the entering basic variable must be a positive one. Using the current Eq. (0) to express Z just in terms of the nonbasic variables, the coefficient of each one is the rate at which Z would change as that variable is increased. The one that has the largest positive coefficient, and so would *increase Z at the fastest rate,* is chosen to be the entering basic variable.[1]

To illustrate, the two candidates for the entering basic variable in the example are the current nonbasic variables x_1 and x_2. Since the objective function already is written only in terms of these nonbasic variables, it can be considered just as is:

$$Z = 3x_1 + 5x_2.$$

Both variables have positive coefficients, so increasing either one would increase Z, but at the different rates of 3 and 5 per unit increase in the variable. Since $3 < 5$, the choice for the *entering basic variable* is x_2. Therefore, x_2 will be increased from zero while x_1 remains zero.

QUESTION 2: How is the *leaving basic variable* identified?

Ignoring slack variables, increasing x_2 from zero while keeping x_1 zero means that we are moving up the x_2 axis in Fig. 4.1. The *adjacent* corner-point feasible solution (0, 6) is reached by stopping at the first new *constraint line* ($2x_2 = 12$). We *must*

---

[1] Note that this criterion does not guarantee selecting the variable that would increase Z the most because the constraints may not allow increasing this variable as much as some of the others. However, the extra computations required to check this are not considered worthwhile.

stop there even though there is a corner-point solution at $(0, 9)$, because going further would give *infeasible* solutions that violate the $2x_2 \leq 12$ constraint.

For the problem in *augmented* form, feasible solutions must satisfy *both* the system of functional constraint equations *and* the nonnegativity constraints on *all* the variables (original variables and slack variables). Increasing x_2 from zero while keeping x_1 zero (nonbasic) means that some or all of the current basic variables (x_3, x_4, x_5) must change their values to keep the system of equations satisfied. Some of these variables will decrease as x_2 increases. The *adjacent* basic feasible solution is reached when the *first* of the basic variables (the *leaving basic variable*) reaches a value of zero. We *must* stop there to avoid going infeasible. Thus, when we have chosen the entering basic variable, the leaving basic variable is not a matter of choice. It must be the current basic variable whose nonnegativity constraint imposes the smallest upper bound on how much the entering basic variable can be increased, as illustrated next.

The possibilities for the leaving basic variable in the example are the current basic variables x_3, x_4, and x_5. The calculations for identifying which one must be the leaving basic variable are summarized in Table 4.1. Since x_1 is a nonbasic variable, $x_1 = 0$ in the second column of Table 4.1. The third column then indicates that (1) x_3 remains nonnegative ($= 4$) regardless of how much x_2 is increased; (2) $x_4 = 0$ when $x_2 = 6$ (whereas $x_4 > 0$ when $x_2 < 6$ and $x_4 < 0$ when $x_2 > 6$); and (3) $x_5 = 0$ when $x_2 = 9$ (whereas $x_5 > 0$ when $x_2 < 9$ and $x_5 < 0$ when $x_2 > 9$). Thus, the numbers calculated in the third column are the upper bound for x_2 before the corresponding basic variable in the first column would become negative. Since x_4 (the slack variable for the $2x_2 \leq 12$ constraint) imposes the smallest upper bound on x_2, the *leaving basic variable* is x_4, so $x_4 = 0$ (nonbasic) and $x_2 = 6$ (basic) in the new basic feasible solution.

QUESTION 3: How can the *new basic feasible solution* be identified most conveniently?

After identifying the entering and leaving basic variables (including the new value of the entering basic variable), all that needs to be done to identify the new basic feasible solution is to solve for the new values of the remaining basic variables. This solution could be obtained directly from Table 4.1. However, in order to get set up for the next iteration, the simplex method converts the system of equations into the same convenient *proper form from Gaussian elimination* as we had in the initialization step (namely, each equation has just one basic variable, which has a coefficient of 1, and this basic variable does not appear in any other equation). This conversion can be done by performing the following two kinds of algebraic operations:

Algebraic Operations for Solving a System of Linear Equations
1. Multiplying (or dividing) an equation by a nonzero constant.
2. Adding (or subtracting) a multiple of one equation to another equation.

Table 4.1 Calculations for Determining First Leaving Basic Variable for Wyndor Glass Co. Problem

Basic Variable	Equation	Upper Bound for x_2
x_3	$x_3 = 4 - x_1$	No limit
x_4	$x_4 = 12 - 2x_2$	$x_2 \leq \frac{12}{2} = 6 \leftarrow$ minimum
x_5	$x_5 = 18 - 3x_1 - 2x_2$	$x_2 \leq \frac{18}{2} = 9$

These operations are legitimate because they involve only (1) multiplying equals (both sides of an equation) by the same constant and (2) adding equals to equals. Therefore, a solution will satisfy the system of equations after such operations if and only if it did so before.

To illustrate, consider the original set of equations, where the *new* basic variables are shown in bold type (with Z playing the role of the basic variable in the objective function equation):

(0) $\qquad\qquad \mathbf{Z} - 3x_1 - 5\boldsymbol{x_2} \qquad\qquad\qquad = 0$

(1) $\qquad\qquad\qquad\quad x_1 \quad + \boldsymbol{x_3} \qquad\qquad = 4$

(2) $\qquad\qquad\qquad\qquad \mathbf{2x_2} \quad + x_4 \quad = 12$

(3) $\qquad\qquad\qquad\quad 3x_1 + \mathbf{2x_2} \qquad + \boldsymbol{x_5} = 18.$

Thus, x_2 has replaced x_4 as the basic variable in Eq. (2). We now need to solve this system of equations for Z, x_2, x_3, and x_5. Since x_2 has a coefficient of 2 in Eq. (2), this equation would be divided by 2 to give its new basic variable a coefficient of 1. (This is an example of algebraic operation 1.)

(2) $\qquad\qquad\qquad\qquad \boldsymbol{x_2} + \tfrac{1}{2}x_4 = 6.$

Next, x_2 must be eliminated from the other equations in which it appears. Using algebraic operation 2, this elimination is done as follows:

$$\text{New Eq. (0)} = \text{old Eq. (0)} + [5 \times \text{new Eq. (2)}],$$

so

$$
\begin{array}{r}
\mathbf{Z} - 3x_1 - 5\boldsymbol{x_2} \qquad\qquad = 0 \\
+ (\qquad\qquad 5\boldsymbol{x_2} + \tfrac{5}{2}x_4 = 30) \\
\hline
\mathbf{Z} - 3x_1 \qquad + \tfrac{5}{2}x_4 = 30.
\end{array}
$$

(0)

Now

$$\text{New Eq. (3)} = \text{old Eq. (3)} - [2 \times \text{new Eq. (2)}],$$

so

$$
\begin{array}{r}
3x_1 + \mathbf{2x_2} \qquad + \boldsymbol{x_5} = 18 \\
- (\qquad \mathbf{2x_2} + x_4 \qquad = 12) \\
\hline
3x_1 \qquad - x_4 + \boldsymbol{x_5} = 6.
\end{array}
$$

(3)

The resulting complete new set of equations is

(0) $\qquad\qquad \mathbf{Z} - 3x_1 \qquad\qquad + \tfrac{5}{2}x_4 \qquad = 30$

(1) $\qquad\qquad\qquad\quad x_1 \quad + \boldsymbol{x_3} \qquad\qquad = 4$

(2) $\qquad\qquad\qquad\qquad \boldsymbol{x_2} \quad + \tfrac{1}{2}x_4 \qquad = 6$

(3) $\qquad\qquad\qquad\quad 3x_1 \qquad\quad - x_4 + \boldsymbol{x_5} = 6.$

For purposes of illustration, exchange the locations of x_2 and x_4.

(0) $\qquad\qquad \mathbf{Z} - 3x_1 + \tfrac{5}{2}x_4 \qquad\qquad = 30$

(1) $\qquad\qquad\qquad\quad x_1 \quad + \boldsymbol{x_3} \qquad = 4$

(2)
$$\tfrac{1}{2}x_4 \quad + x_2 \quad = 6$$

(3)
$$3x_1 - x_4 \quad + x_5 = 6.$$

Now compare this last set of equations with the initial set obtained under the initialization step, and notice that it is indeed in the same convenient *proper form from Gaussian elimination* for immediately reading the current basic feasible solution after noting that the nonbasic variables (x_1 and x_4) equal zero. Thus we now have our new basic feasible solution, $(x_1, x_2, x_3, x_4, x_5) = (0, 6, 4, 0, 6)$, which yields $Z = 30$.

To place this algebraic procedure into broader perspective, we have just solved the original set of equations to obtain the general solution for Z, x_2, x_3, and x_5 in terms of x_1 and x_4. (This general solution can be expressed explicitly by moving x_1 and x_4 to the right-hand side of the new set of equations, but we won't bother to do this.) We then obtained a specific solution (the basic feasible solution) by setting x_1 and x_4 (the nonbasic variables) equal to zero. This procedure for obtaining the simultaneous solution of a system of linear equations is called the *Gauss-Jordan method of elimination,* or **Gaussian elimination** for short.[1] The key concept for this method is using the two kinds of algebraic operations to reduce the original system of equations to **proper form from Gaussian elimination,** where each basic variable has been eliminated from all but one equation (*its* equation) and has a coefficient of $+1$ in that equation. Once *proper form from Gaussian elimination* has been obtained, the solution for the basic variables can be read directly from the right-hand side of the equations.

If Gaussian elimination is not yet clear, you can find a more detailed description in Appendix 4.

Optimality Test

To determine whether the current basic feasible solution is optimal, the current Eq. (0) is used to rewrite the objective function just in terms of the current nonbasic variables,

$$Z = 30 + 3x_1 - \tfrac{5}{2}x_4.$$

Increasing either of these nonbasic variables from zero (while adjusting the values of the basic variables to continue satisfying the system of equations) would result in moving toward one of the two *adjacent* basic feasible solutions. Because x_1 has a *positive* coefficient, increasing x_1 would lead toward an adjacent basic feasible solution that is better than the current basic feasible solution, so the current solution is not optimal.

In general terms, the current basic feasible solution is optimal if and only if *all* of the nonbasic variables have *nonpositive* coefficients (≤ 0) in the current form of the objective function. This current form is obtained by bringing the x_j variables over to the right-hand side of the current Eq. (0) after all of the equations have been converted to *proper form from Gaussian elimination* [which eliminates basic variables from Eq. (0)]. Equivalently, the variables can be left on the left-hand side, in which case the optimality test is whether all of the nonbasic variables have *nonnegative* coefficients (≥ 0) in the current Eq. (0).

---

[1] Actually, there are some technical differences between the Gauss-Jordan method of elimination and Gaussian elimination, but we will not make this distinction.

The reason that the *current* form of the objective function is used for the optimality test instead of the *original* objective function is that the current form contains *all of the nonbasic variables* and *none of the basic variables*. All of the nonbasic variables are needed in order to be able to compare all of the adjacent basic feasible solutions with the current solution. The basic variables must not appear because their values may change when a nonbasic variable is increased from zero, in which case the coefficient of the nonbasic variable no longer indicates the rate of change of Z. Because of the constraint equations, the two forms of the objective function are *equivalent,* so the one that contains all the desired information is used.

The reason that Eq. (0) originally was appended to the system of constraint equations and then included in the process of Gaussian elimination was just so this new, more convenient form of the objective function could be obtained.

Before proceeding with the next iteration, it is now possible to give a meaningful summary of the simplex method. (Tie-breaking considerations are deferred to Sec. 4.5.)

Summary of the Simplex Method

1. **INITIALIZATION STEP:** Introduce slack variables. If the model is not in the form being assumed in this section, see Sec. 4.6 for the necessary adjustments. Otherwise, select the original variables to be the nonbasic variables (and thus equal to zero) and the slack variables to be the basic variables (and thus equal to the right-hand side) in the initial basic feasible solution. Go to the optimality test.

2. **ITERATIVE STEP**
Part 1: Determine the *entering basic variable:* Select the nonbasic variable that, when increased, would increase Z at the fastest rate. Do this by using the current Eq. (0) to express Z just in terms of the nonbasic variables and then selecting the nonbasic variable with the *largest* positive coefficient.[1]

Part 2: Determine the *leaving basic variable:* Select the basic variable that reaches zero first as the entering basic variable is increased. Each basic variable appears in just *its* equation, so this equation is used to determine when this basic variable reaches zero as the entering basic variable is increased. A formal algebraic procedure for doing this is to let e denote the subscript of the entering basic variable, let a'_{ie} denote its current coefficient in Eq. (i), and let b'_i denote the current right-hand side for this equation ($i = 1, 2, \ldots, m$). Then the upper bound for x_e in Eq. (i) is

$$x_e \leq \begin{cases} + \infty, & \text{if } a'_{ie} \leq 0 \\ \dfrac{b'_i}{a'_{ie}}, & \text{if } a'_{ie} > 0, \end{cases}$$

where this equation's basic variable reaches zero at this upper bound. Therefore, determine the equation with the *smallest* such upper bound, and select the basic variable in that equation as the leaving basic variable.

[1] Equivalently, the current Eq. (0) can be used directly, in which case the nonbasic variable with the largest *negative* coefficient would be selected. This is what is done in the tabular form of the simplex method presented in Sec. 4.4.

Part 3: Determine the *new basic feasible solution:* Starting from the current set of equations, solve for the basic variables and Z in terms of the nonbasic variables by Gaussian elimination (see Appendix 4). Set the nonbasic variables equal to zero; each basic variable (and Z) equals the new right-hand side of the one equation in which it appears (with a coefficient of 1).

3. OPTIMALITY TEST: Determine whether this solution is optimal: Check if Z can be increased by increasing any nonbasic variable. This determination can be made by rewriting the objective function just in terms of the nonbasic variables by bringing these variables to the right-hand side in the current Eq. (0) and then checking the sign of the coefficient of each nonbasic variable. If all these coefficients are nonpositive, then this solution is optimal, so stop.[1] Otherwise, go to the iterative step.

To illustrate, let us apply this summary to the next iteration for the example.

ITERATION 2 FOR EXAMPLE

Part 1: Because the current Eq. (0) yields $Z = 30 + 3x_1 - \frac{5}{2}x_4$, increasing only x_1 would increase Z; that is, x_1 has the largest (and only) positive coefficient. Therefore, x_1 is chosen as the new entering basic variable.

Part 2: The upper bounds on x_1 before the basic variable in the respective equations would become negative are shown in Table 4.2. Because x_5 gives the smallest upper bound for x_1, x_5 must be chosen as the leaving basic variable.

Part 3: After eliminating x_1 from all equations in the current set except Eq. (3), where x_1 replaces x_5 as the basic variable, the new set of equations in *proper form from Gaussian elimination* is

$$
\begin{array}{llr}
(0) & Z \quad\quad + \frac{3}{2}x_4 + \quad x_5 = 36 \\
(1) & x_3 + \frac{1}{3}x_4 - \frac{1}{3}x_5 = 2 \\
(2) & x_2 \quad + \frac{1}{2}x_4 \quad\quad = 6 \\
(3) & x_1 \quad\quad - \frac{1}{3}x_4 + \frac{1}{3}x_5 = 2.
\end{array}
$$

Therefore, the next basic feasible solution is $(x_1, x_2, x_3, x_4, x_5) = (2, 6, 2, 0, 0)$, yielding $Z = 36$.

Table 4.2 **Calculations for Determining Second Leaving Basic Variable for Wyndor Glass Co. Problem**

Basic Variable	Equation Number	Upper Bound for x_1
x_3	1	$x_1 \leq \frac{4}{1} = 4$
x_2	2	No limit
x_5	3	$x_1 \leq \frac{6}{3} = 2 \leftarrow$ minimum

[1]Equivalently, the current Eq. (0) can be used directly, in which case all these coefficients have to be nonnegative (≥ 0) for the solution to be optimal. This is what is done in the tabular form of the simplex method presented in Sec. 4.4.

OPTIMALITY TEST: Because the new form of the objective function is $Z = 36 - \frac{3}{2}x_4 - x_5$, so that the coefficient of neither nonbasic variable is positive, the current basic feasible solution just obtained must be optimal. Therefore, the desired solution to the original form of the problem is $x_1 = 2$, $x_2 = 6$, which yields $Z = 36$.

Continuing the Learning Process with Your OR COURSEWARE

This is the first of many points in the book where you may find it helpful to use your OR COURSEWARE (the diskettes packaged in the back of the book). This software includes a complete demonstration example of the simplex method in the algebraic form just presented. This vivid demonstration simultaneously displays both the algebra and the geometry of the simplex method as it dynamically evolves step by step. Like the many other demonstration examples accompanying other sections of the book (including the next section), this computer demonstration highlights concepts that are difficult to convey on the printed page.

Another feature of your OR COURSEWARE is a collection of routines for interactively executing the various algorithms presented throughout the book. One such routine is for the algebraic form of the simplex method. Like the others, this routine performs nearly all of the calculations while you make the decisions step by step, thereby enabling you to focus on concepts rather than getting bogged down in a lot of number crunching. Therefore, you probably will want to use this routine for your homework on this section. The software will help you get started by letting you know whenever you make a mistake on the first iteration of a problem. Follow the instructions, and then use the HELP command whenever you are unclear on which computer operation should be done next. You can return to the demonstration example whenever you need to review what to do on the next step of the simplex method, and then come back when ready to where you were in the problem. When you finish the problem, you can print out everything you have done for your homework by choosing the print command under the FILE menu.

Your OR COURSEWARE begins with a more complete introduction on how to use the software.

4.4 The Simplex Method in Tabular Form

The algebraic form of the simplex method presented in Sec. 4.3 may be the best one for learning the underlying logic of the algorithm. However, it is not the most convenient form for performing the required calculations. When you need to solve a problem by hand (or interactively with your OR COURSEWARE), we recommend the *tabular form* described in this section.[1]

The tabular form of the simplex method is *mathematically equivalent* to the algebraic form. However, instead of writing down each set of equations in full detail, we use a **simplex tableau** to record only the essential information, namely, (1) the coefficients of the variables, (2) the constants on the right-hand side of the equations, and (3) the basic variable appearing in each equation. This saves writing the symbols for the variables in each of the equations, but what is even more important is the fact

---

[1] A form more convenient for automatic execution on a computer is presented in Sec. 5.2.

that it permits highlighting the numbers involved in arithmetic calculations and recording the computations compactly.

To introduce the tabular form, we consider the *augmented form* of the Wyndor Glass Co. problem as presented at the end of Sec. 4.2. This system of equations [(0) to (3)] can be expressed as shown in Table 4.3. This table shows the layout for any *simplex tableau,* where the column on the left indicates which basic variable appears in each equation for the current basic feasible solution. [Although only the x_j variables are basic or nonbasic, Z plays the role of the basic variable for Eq. (0).] For example, the *Basic Variable* column of Table 4.3 indicates that the *initial* basic feasible solution has basic variables x_3, x_4, x_5, so the *nonbasic* variables are the ones not listed, x_1 and x_2. After setting $x_1 = 0$, $x_2 = 0$, the *Right Side* column gives the resulting solution for basic variables, so that the initial basic feasible solution is $(x_1, x_2, x_3, x_4, x_5) = (0, 0, 4, 12, 18)$ with $Z = 0$.

The reason why the *Right Side* column always gives the values of the basic variables in the current basic feasible solution is that the simplex method requires that the simplex tableau starting (or ending) each iteration be in *proper form from Gaussian elimination.* This form is where the column for each basic variable contains only *one nonzero* coefficient and this coefficient is a 1 in the row for this basic variable. (The columns for nonbasic variables can be anything.) Notice how the x_3, x_4, and x_5 columns (as well as the Z column) in Table 4.3 fit this special pattern. Consequently, each equation contains exactly one basic variable with a nonzero coefficient, where this coefficient is 1, so this basic variable equals the constant on the right-hand side of its equation. (Remember that the nonbasic variables equal zero.)

Under our current assumptions (stated at the beginning of the chapter) about the form of the original model, the initial simplex tableau is automatically in *proper form from Gaussian elimination.* When the simplex method moves from the current basic feasible solution to the next one, part 3 of the iterative step uses Gaussian elimination to restore this form for the new solution.

The simplex method develops a simplex tableau for each new basic feasible solution obtained until an optimal solution is reached. The procedure, which is just a tabular representation of the algebraic procedure presented in Sec. 4.3, is outlined next. (Tie-breaking considerations are deferred to Sec. 4.5.) We continue to use the Wyndor Glass Co. example for illustrative purposes.

INITIALIZATION STEP: Introduce slack variables. If the model is not in the form being assumed in this section, see Sec. 4.6 for the necessary adjustments. Otherwise, select the *original variables* to be the *initial nonbasic variables* (set equal to zero) and the *slack variables* to be the *initial basic variables.*

This selection yields the initial simplex tableau for the example already shown

Table 4.3 **Initial Simplex Tableau for Wyndor Glass Co. Problem**

Basic Variable	Eq. No.	Coefficient of						Right Side
		Z	x_1	x_2	x_3	x_4	x_5	
Z	0	1	-3	-5	0	0	0	0
x_3	1	0	1	0	1	0	0	4
x_4	2	0	0	2	0	1	0	12
x_5	3	0	3	2	0	0	1	18

in Table 4.3, so the initial basic feasible solution is (0, 0, 4, 12, 18). Go next to the optimality test to determine if this solution is optimal.

OPTIMALITY TEST: The current basic feasible solution is optimal if and only if *every* coefficient in Eq. (0) is nonnegative (≥ 0). If it is, stop; otherwise, go to the iterative step to obtain the next basic feasible solution, which involves changing one nonbasic variable to a basic variable (part 1) and vice versa (part 2) and then solving for the new solution (part 3).

The example has two negative coefficients in Eq. (0), -3 for x_1 and -5 for x_2, so go to the iterative step.

ITERATIVE STEP
Part 1: Determine the *entering basic variable* by selecting the variable (automatically a nonbasic variable) with the *negative coefficient* having the largest absolute value in Eq. (0). Put a box around the column below this coefficient, and call this the **pivot column.**

In the example, the largest (in absolute terms) negative coefficient is -5 for x_2 ($5 > 3$), so x_2 is to be changed to a basic variable. (This change is indicated in Table 4.4 by the box around the x_2 column below -5.)

Part 2: Determine the *leaving basic variable* by (a) picking out each coefficient in the pivot column that is strictly positive (>0), (b) dividing each of these coefficients into "right side" for the same row, (c) identifying the equation that has the *smallest* of these ratios, and (d) selecting the basic variable for this equation. (This basic variable is the one that reaches zero first as the entering basic variable is increased.) Put a box around this equation's row in the tableau to the right of the Z column, and call the boxed row the **pivot row.** (Hereafter, we continue to use the term **row** to refer just to a row of numbers to the right of the Z column, *including* the right-side number, and we label the rows by the numbers in the *Eq. No.* column.) Also call the one number that is in *both* boxes the **pivot number.**

The results of parts 1 and 2 for the example are shown in Table 4.4, where the **minimum ratio test** for determining the leaving basic variable is shown to the right of the tableau. The row 1 coefficient in the pivot column is 0, so the only two *strictly positive* coefficients are in rows 2 and 3. The ratios for these rows are 6 and 9, respectively, so the *minimum ratio* of 6 identifies row 2 as the *pivot row* (with 2 as the *pivot number*). Consequently, the *leaving basic variable* is x_4, the basic variable for row 2 shown in the first column.

Part 3: Determine the *new basic feasible solution* by constructing a new simplex tableau in *proper form from Gaussian elimination* below the current one. (The first

Table 4.4 Calculations to Determine First Leaving Basic Variable
for Wyndor Glass Co. Problem

Basic Variable	Eq. No.	Z	x_1	x_2	x_3	x_4	x_5	Right Side	Ratio
Z	0	1	-3	-5	0	0	0	0	
x_3	1	0	1	0	1	0	0	4	
x_4	2	0	0	2	0	1	0	12	$\rightarrow \frac{12}{2} = 6 \leftarrow$ minimum
x_5	3	0	3	2	0	0	1	18	$\rightarrow \frac{18}{2} = 9$

three columns are left unchanged except that the leaving basic variable in the *Basic Variable* column is replaced by the entering basic variable.) To change the coefficient of the new basic variable in the pivot row to 1, divide the entire row by the pivot number, so

$$\text{New pivot row} = \frac{\text{old pivot row}}{\text{pivot number}}.$$

For the example, because the *old pivot row* is the boxed row 2 in the first tableau of Table 4.5, and the *pivot number* is 2, applying this formula yields the *new pivot row* shown as row 2 in the second tableau in Table 4.5.

To complete the first iteration, we need to continue using Gaussian elimination to obtain a coefficient of 0 for the new basic variable x_2 in the other rows (including row 0) of this second tableau. Because row 1 already has a coefficient of 0 for x_2 in the first tableau, this row can be carried along to the second tableau without any change. However, rows 0 and 3 have *pivot column coefficients* of -5 and 2, respectively, so each of these rows needs to be changed by using the following formula:

New row = old row − (pivot column coefficient × new pivot row).

Alternatively, when the pivot column coefficient is *negative* (as for row 0), a more convenient form of this formula is

New row = old row + [(−pivot column coefficient) × new pivot row].

To illustrate, the missing rows for the second tableau of Table 4.5 are obtained as follows:

Row 0:
$$
\begin{array}{rrrrrrr}
& [-3 & -5 & 0 & 0 & 0, & 0] \\
+(5)\ [& 0 & 1 & 0 & \frac{1}{2} & 0, & 6] \\
\hline
\text{New row 0} = & [-3 & 0 & 0 & \frac{5}{2} & 0, & 30].
\end{array}
$$

Row 1: Unchanged because its pivot column coefficient is zero.

Row 3:
$$
\begin{array}{rrrrrrr}
& [& 3 & 2 & 0 & 0 & 1, & 18] \\
-(2)\ [& 0 & 1 & 0 & \frac{1}{2} & 0, & 6] \\
\hline
\text{New row 3} = & [& 3 & 0 & 0 & -1 & 1, & 6].
\end{array}
$$

These calculations yield the new tableau shown in Table 4.6 for iteration 1.

Table 4.5 **Simplex Tableaux for Wyndor Glass Co. Problem after Revising First Pivot Row**

Iteration	Basic Variable	Eq. No.	Z	x_1	x_2	x_3	x_4	x_5	Right Side
	Z	0	1	−3	−5	0	0	0	0
0	x_3	1	0	1	0	1	0	0	4
	x_4	2	0	0	2	0	1	0	12
	x_5	3	0	3	2	0	0	1	18
	Z	0	1						
1	x_3	1	0						
	x_2	2	0	0	1	0	$\frac{1}{2}$	0	6
	x_5	3	0						

Table 4.6 First Two Simplex Tableaux for Wyndor Glass Co. Problem

	Basic	Eq.				Coefficient of			Right
Iteration	Variable	No.	Z	x_1	x_2	x_3	x_4	x_5	Side
	Z	0	1	-3	-5	0	0	0	0
0	x_3	1	0	1	0	1	0	0	4
	x_4	2	0	0	2	0	1	0	12
	x_5	3	0	3	2	0	0	1	18
	Z	0	1	-3	0	0	$\frac{5}{2}$	0	30
1	x_3	1	0	1	0	1	0	0	4
	x_2	2	0	0	1	0	$\frac{1}{2}$	0	6
	x_5	3	0	3	0	0	-1	1	6

Because each basic variable always equals the right side of its equation, the new basic feasible solution is $(0, 6, 4, 0, 6)$, with $Z = 30$.

This work completes the iterative step, so we next return to the optimality test to check if the new basic feasible solution is optimal. Since the new row 0 still has a negative coefficient (-3 for x_1), the solution is not optimal, and so at least one more iteration is needed.

ITERATION 2 FOR EXAMPLE: The second iteration starts anew from the second tableau of Table 4.6 to find the next basic feasible solution. Following the instructions for parts 1 and 2, we find x_1 as the entering basic variable and x_5 as the leaving basic variable, as shown in Table 4.7.

Using the pivot number 3, the calculations to obtain the new tableau are

Row 3: Because this is the pivot row,

$$\text{New row 3} = \tfrac{1}{3}[\ \ 3 \quad 0 \quad 0 \quad -1 \quad 1, \quad 6]$$
$$= \quad [\ \ 1 \quad 0 \quad 0 \quad -\tfrac{1}{3} \quad \tfrac{1}{3}, \quad 2].$$

Row 0:

$$[-3 \quad 0 \quad 0 \quad \tfrac{5}{2} \quad 0, \quad 30]$$
$$+ 3[\ \ 1 \quad 0 \quad 0 \quad -\tfrac{1}{3} \quad \tfrac{1}{3}, \quad 2]$$
$$\text{New row 0} = \quad [\ \ 0 \quad 0 \quad 0 \quad \tfrac{3}{2} \quad 1, \quad 36].$$

Row 1:

$$[\ \ 1 \quad 0 \quad 1 \quad 0 \quad 0, \quad 4]$$
$$- \quad [\ \ 1 \quad 0 \quad 0 \quad -\tfrac{1}{3} \quad \tfrac{1}{3}, \quad 2]$$
$$\text{New row 1} = \quad [\ \ 0 \quad 0 \quad 1 \quad \tfrac{1}{3} \quad -\tfrac{1}{3}, \quad 2].$$

Row 2: Unchanged because its pivot column coefficient is zero.

Table 4.7 Calculations to Determine Second Leaving Basic Variable for Wyndor Glass Co. Problem

	Basic	Eq.				Coefficient of			Right	
Iteration	Variable	No.	Z	x_1	x_2	x_3	x_4	x_5	Side	Ratio
	Z	0	1	-3	0	0	$\frac{5}{2}$	0	30	
1	x_3	1	0	1	0	1	0	0	4	$\frac{4}{1} = 4$
	x_2	2	0	0	1	0	$\frac{1}{2}$	0	6	
	x_5	3	0	3	0	0	-1	1	6	$\frac{6}{3} = 2 \leftarrow$ minimum

Table 4.8 Complete Set of Simplex Tableaux for Wyndor Glass Co. Problem

Iteration	Basic Variable	Eq. No.	Z	x_1	x_2	x_3	x_4	x_5	Right Side
0	Z	0	1	-3	-5	0	0	0	0
	x_3	1	0	1	0	1	0	0	4
	x_4	2	0	0	2	0	1	0	12
	x_5	3	0	3	2	0	0	1	18
1	Z	0	1	-3	0	0	$\frac{5}{2}$	0	30
	x_3	1	0	1	0	1	0	0	4
	x_2	2	0	0	1	0	$\frac{1}{2}$	0	6
	x_5	3	0	3	0	0	-1	1	6
2	Z	0	1	0	0	0	$\frac{3}{2}$	1	36
	x_3	1	0	0	0	1	$\frac{1}{3}$	$-\frac{1}{3}$	2
	x_2	2	0	0	1	0	$\frac{1}{2}$	0	6
	x_1	3	0	1	0	0	$-\frac{1}{3}$	$\frac{1}{3}$	2

We now have the set of tableaux shown in Table 4.8. Therefore, the new basic feasible solution is (2, 6, 2, 0, 0), with $Z = 36$. Going to the optimality test, we find that this solution is *optimal* because none of the coefficients in row 0 is negative, so the algorithm is finished. Consequently, the optimal solution to the Wyndor Glass Co. problem (before introducing slack variables) is $x_1 = 2$, $x_2 = 6$.

Now compare Table 4.8 with the work done in Sec. 4.3 to verify that these two forms of the simplex method really are *equivalent*. Then note how the tabular form organizes the work being done in a considerably more convenient and compact form. We generally will use the tabular form hereafter.

4.5 Tie Breaking in the Simplex Method

You may have noticed in the preceding two sections that we never said what to do if the various choice rules of the simplex method do not lead to a clear-cut decision, either because of ties or other similar ambiguities. We discuss these details now.

Tie for the Entering Basic Variable

Part 1 of the iterative step chooses the nonbasic variable having the *negative* coefficient with the *largest absolute value* in the current Eq. (0) as the entering basic variable. Now suppose that two or more nonbasic variables are tied for having the largest negative coefficient (in absolute terms). For example, this would occur in the first iteration for the Wyndor Glass Co. problem (see Sec. 3.1) if its objective function were changed to $Z = 3x_1 + 3x_2$, so that the initial Eq. (0) becomes $Z - 3x_1 - 3x_2 = 0$. How should this tie be broken?

The answer is that the selection between these contenders may be made *arbitrarily*. The optimal solution will be reached eventually, regardless of the tied variable chosen, and there is no convenient method for predicting in advance which choice will lead there sooner. In this example, the simplex method happens to reach the optimal solution (2, 6) in three iterations with x_1 as the initial entering basic variable, versus two iterations if x_2 is chosen.

Now suppose that two or more basic variables tie for being the leaving basic variable in part 2 of the iterative step. Does it matter which one is chosen? Theoretically it does, and in a very critical way, because of the following sequence of events that could occur. First, all of the tied basic variables reach zero simultaneously as the entering basic variable is increased. Therefore, the one or ones *not* chosen to be the leaving basic variable also will have a value of zero in the new basic feasible solution. (Basic variables with a value of *zero* are called **degenerate,** and the same term is applied to the corresponding basic feasible solution.) Second, if one of these degenerate basic variables retains its value of zero until it is chosen at a subsequent iteration to be a leaving basic variable, the corresponding entering basic variable must also remain zero (since it cannot be increased without making the leaving basic variable negative), so the value of Z must remain unchanged. Third, if Z may remain the same rather than increase at each iteration, the simplex method may then go around in a loop, repeating the same sequence of solutions periodically rather than eventually increasing Z toward an optimal solution. In fact, examples have been artificially constructed so that they do become entrapped in just such a perpetual loop.

Fortunately, although a perpetual loop is theoretically possible, it has rarely been known to occur in practical problems. If a loop were to occur, one could always get out of it by changing the choice of the leaving basic variable. Furthermore, special rules[1] have been constructed for breaking ties so that such loops are always avoided. However, these rules have been virtually ignored in actual application, and they will not be repeated here. For your purposes, just break this kind of tie *arbitrarily* and proceed without worrying about the degenerate basic variables that result.

No Leaving Basic Variable—Unbounded Z

In part 2 of the iterative step there is one other possible outcome that we have not yet discussed, namely, that *no* variable qualifies to be the leaving basic variable.[2] This outcome would occur if the entering basic variable could be increased *indefinitely* without giving negative values to *any* of the current basic variables. In the tabular form, this means that *every* coefficient in the pivot column (excluding row 0) is either negative or zero. This situation is illustrated in Table 4.9 by deleting the last two functional constraints of the Wyndor Glass Co. problem (note the effect in Fig. 4.1).

Table 4.9 **Initial Simplex Tableau for Wyndor Glass Co.
Problem without Last Two Functional Constraints**

Basic Variable	Eq. No.	Coefficient of				Right Side	Ratio
		Z	x_1	x_2	x_3		
Z	0	1	-3	-5	0	0	
x_3	1	0	1	$\boxed{0}$	1	4	*No minimum*

[1] See, for example, A. Charnes: "Optimality and Degeneracy in Linear Programming," *Econometrica,* **20:**160–170, 1952.

[2] Note that the analogous case (no *entering* basic variable) cannot occur in part 1 of the iterative step, because the optimality test would stop the algorithm first by indicating that an optimal solution has been reached.

The interpretation of a tableau like the one shown in Table 4.9 is that the constraints do not prevent increasing the value of the objective function (Z) indefinitely, so the simplex method would stop with the message that Z is *unbounded*. Because even linear programming has not discovered a way of making infinite profits, the real message for practical problems is that a mistake has been made! The model probably has been misformulated, either by omitting relevant constraints or by stating them incorrectly. Alternatively, a computational mistake may have occurred.

Multiple Optimal Solutions

We mentioned in Sec. 3.2 (under the definition of **optimal solution**) that a problem can have more than one optimal solution. This fact was illustrated by changing the objective function in the Wyndor Glass Co. problem to $Z = 3x_1 + 2x_2$, so that every point on the line segment between (2, 6) and (4, 3) is optimal. We also noted in Sec. 4.1 that every such problem has at least two optimal corner-point feasible solutions (basic feasible solutions). By taking *weighted averages,* these solutions can be used to identify every other optimal solution (as described in Probs. 13 and 14).

The simplex method automatically stops after finding *one* optimal solution. However, for many applications of linear programming, there are intangible factors not incorporated into the model that can be used to make meaningful choices between solutions that are alternative optimal solutions according to the model. In such cases, these other optimal solutions should be identified as well. After the simplex method finds one optimal basic feasible solution, how do you recognize when there are others, and how do you find them? The answer is summarized as follows:

> Whenever a problem has more than one optimal basic feasible solution, at least one of the nonbasic variables has a coefficient of *zero* in the final Eq. (0), so increasing any such variable would not change the value of Z. Therefore, these other optimal basic

Table 4.10 **Complete Set of Simplex Tableaux to Obtain All Optimal Basic Feasible Solutions for Wyndor Glass Co. Problem with $c_2 = 2$**

Iteration	Basic Variable	Eq. No.	Coefficient of						Right Side	Solution Optimal?
			Z	x_1	x_2	x_3	x_4	x_5		
0	Z	0	1	-3	-2	0	0	0	0	No
	x_3	1	0	1	0	1	0	0	4	
	x_4	2	0	0	2	0	1	0	12	
	x_5	3	0	3	2	0	0	1	18	
1	Z	0	1	0	-2	3	0	0	12	No
	x_1	1	0	1	0	1	0	0	4	
	x_4	2	0	0	2	0	1	0	12	
	x_5	3	0	0	2	-3	0	1	6	
2	Z	0	1	0	0	**0**	0	1	18	Yes
	x_1	1	0	1	0	1	0	0	4	
	x_4	2	0	0	0	3	1	-1	6	
	x_2	3	0	0	1	$-\frac{3}{2}$	0	$\frac{1}{2}$	3	
Extra	Z	0	1	0	0	0	**0**	1	18	Yes
	x_1	1	0	1	0	0	$-\frac{1}{3}$	$\frac{1}{3}$	2	
	x_3	2	0	0	0	1	$\frac{1}{3}$	$-\frac{1}{3}$	2	
	x_2	3	0	0	1	0	$\frac{1}{2}$	0	6	

feasible solutions can be identified (if desired) by performing additional iterations of the simplex method, each time choosing a nonbasic variable with a zero coefficient as the entering basic variable.

To illustrate, consider again the case just mentioned, where the objective function in the Wyndor Glass Co. problem is changed to $Z = 3x_1 + 2x_2$. The simplex method obtains the first three tableaux shown in Table 4.10·and stops with an optimal basic feasible solution. However, because a nonbasic variable (x_3) then has a zero coefficient in row 0, we perform one more iteration in Table 4.10 to identify the other optimal basic feasible solution. Thus the two optimal basic feasible solutions are (4, 3, 0, 6, 0) and (2, 6, 2, 0, 0), each yielding $Z = 18$. Notice that the last tableau also has a *nonbasic* variable (x_4) with a zero coefficient in Eq. (0). This situation is inevitable because the extra iteration(s) does not change row 0, so each leaving basic variable necessarily retains its zero coefficient. Making x_4 an entering basic variable now would only lead back to the third tableau. (Check this.) Therefore, these two are the only basic feasible solutions that are optimal, and all *other* optimal solutions are a weighted average of these two. Specifically, let α and $(1 - \alpha)$ denote the *weights* on these two solutions, where α must be some number between 0 and 1. Then *every* optimal solution is given by the vector formula $\alpha(4, 3, 0, 6, 0) + (1 - \alpha)$ (2, 6, 2, 0, 0) for $0 \le \alpha \le 1$. [If the slack variables are ignored, this is just the formula for the line segment between (2, 6) and (4, 3) in Fig. 4.1.]

4.6 Adapting to Other Model Forms

Thus far we have presented the details of the simplex method under the assumption that the problem is in *our standard form* (*maximize Z subject to functional constraints in $\le$ form and nonnegativity constraints on all variables*), and that $b_i > 0$ for all $i = 1, 2, \ldots, m$. In this section we point out how to make the adjustments required for other legitimate forms of the linear programming model. You will see that all these adjustments can be made in the initialization step, so the rest of the simplex method can then be applied just as you have learned it already.

The only real problem that the other forms for functional constraints ($=$, $\ge$, or $b_i \le 0$) introduce is in identifying an *initial basic feasible solution*. Before, this initial solution was found very conveniently by letting the slack variables be the initial basic variables, so that each one just equals the *positive* right-hand side of its equation. Now, something else must be done. The standard approach that is used for all these cases is the **artificial variable technique.** This technique constructs a more convenient *revised problem* by introducing a dummy variable (called an *artificial variable*) into each constraint that needs one. This new variable is introduced just for the purpose of being the initial basic variable for that equation. The usual nonnegativity constraints are placed on these variables, and the objective function also is modified to impose an exorbitant penalty on their having values larger than zero. The iterations of the simplex method then automatically force the artificial variables to disappear (become zero), one at a time until they are all gone, after which the *real* problem is solved.

To illustrate the artificial variable technique, we first consider the case where the only nonstandard form in the problem is the presence of one or more equality constraints.

Equality Constraints

Any equality constraint,

$$a_{i1}x_1 + a_{i2}x_2 + \cdots + a_{in}x_n = b_i,$$

actually is equivalent to a pair of inequality constraints:

$$a_{i1}x_1 + a_{i2}x_2 + \cdots + a_{in}x_n \leq b_i$$

$$a_{i1}x_1 + a_{i2}x_2 + \cdots + a_{in}x_n \geq b_i.$$

However, rather than making this substitution and thereby increasing the number of constraints, it is more convenient to use the artificial variable technique described next.

Suppose that the Wyndor Glass Co. problem in Sec. 3.1 is modified to *require* that Plant 3 be used at full capacity. The only resulting change in the linear programming model is that the third constraint, $3x_1 + 2x_2 \leq 18$, instead becomes an *equality* constraint,

$$3x_1 + 2x_2 = 18.$$

Therefore, the feasible region for this problem (see Fig. 3.2) now consists of *just* the line segment connecting (2, 6) and (4, 3).

After introducing the slack variables still needed for the *inequality* constraints, the *augmented form* of the problem (see the end of Sec. 4.2) becomes

$$\begin{aligned}
(0) && Z - 3x_1 - 5x_2 && = 0 \\
(1) && x_1 \qquad + x_3 && = 4 \\
(2) && 2x_2 \qquad + x_4 && = 12 \\
(3) && 3x_1 + 2x_2 && = 18.
\end{aligned}$$

Unfortunately, these equations do not have an obvious initial basic feasible solution because there is no longer a slack variable to use as the initial basic variable for Eq. (3). The artificial variable technique circumvents this difficulty by introducing a nonnegative **artificial variable** (call it $\bar{x}_5$)[1] into this equation, just as if it were a slack variable! Thus the technique *revises* the problem by changing Eq. (3) to

$$(3) \qquad\qquad 3x_1 + 2x_2 + \bar{x}_5 = 18,$$

along with the nonnegativity constraint,

$$\bar{x}_5 \geq 0,$$

just as we had in the version of the Wyndor Glass Co. problem presented in Sec. 3.1. Proceeding as before, we now have an initial basic feasible solution (for the *revised problem*), $(x_1, x_2, x_3, x_4, \bar{x}_5) = (0, 0, 4, 12, 18)$.

The effect of introducing an artificial variable is to *enlarge* the feasible region. In this case, the feasible region expands from just the line segment connecting (2, 6) and (4, 3) to the entire shaded area shown in Fig. 3.2. A feasible solution for the *revised* problem (with $3x_1 + 2x_2 + \bar{x}_5 = 18$ and $\bar{x}_5 \geq 0$) is also feasible for the *original* problem (with $3x_1 + 2x_2 = 18$) if the artificial variable equals zero ($\bar{x}_5 = 0$).

Now suppose that the simplex method is permitted to proceed and obtain an *optimal* solution for the *revised* problem and that this solution happens to be *feasible*

---

[1] We shall always label the artificial variables by putting a bar over them.

for the *original* problem. It can then be concluded that this solution must also be *optimal* for the *original* problem, so we are finished. (The reason is that this solution is the best one in the *entire* feasible region for the revised problem, which includes the feasible region for the original problem.)

Unfortunately, there is no guarantee that the optimal solution to the revised problem also will be feasible for the original problem; that is, there is no guarantee until *another* revision is made. Using the **Big M method,** this new revision amounts to assigning such an *overwhelming penalty* to being outside the feasible region for the original problem that the optimal solution to the revised problem *must* lie within this region. Recall that the revised problem coincides with the original problem when $\bar{x}_5 = 0$. Therefore, if the original objective function, $Z = 3x_1 + 5x_2$, is changed to

$$Z = 3x_1 + 5x_2 - M\bar{x}_5,$$

where M denotes some *huge* positive number, then the maximum value of Z must occur when $\bar{x}_5 = 0$ ($\bar{x}_5$ cannot be negative). After a little more setting up (discussed next), applying the simplex method to *this* revised problem automatically leads to the desired solution.

Using this revised objective function, Eq. (0) becomes

(0) $Z - 3x_1 - 5x_2 + M\bar{x}_5 = 0,$

or in tabular form, the *preliminary* row 0 (call it R_0) becomes

$$R_0 = [-3 \quad -5 \quad 0 \quad 0 \quad M, \quad 0].$$

However, this R_0 *cannot* be used in the initial tableau for applying the simplex method because it is not in *proper form from Gaussian elimination*. This proper form requires, in part, that *every* basic variable (excluding Z, which only pretends to be a basic variable) has been eliminated from Eq. (0), whereas $\bar{x}_5$ now is a basic variable with a coefficient of M. Restoring this proper form is essential for both the *optimality test* and the procedure for determining the *entering basic variable*. This form normally is provided automatically by the Gaussian elimination part of the iterative step, and Gaussian elimination is used now to restore the form. Proceeding as if the column for the artificial variable ($\bar{x}_5$) were the pivot column and the equation containing this variable (row 3) were the pivot row, the calculations are shown below.

Row 0:	[	-3	-5	0	0	$M,$	0]
$-M$ [		3	2	0	0	1,	18]
New row 0 =		$[(-3M - 3),$	$(-2M - 5),$	0	0	0,	$-18M]$.

This completes the additional work required in the initialization step for problems of this type, and the rest of the simplex method proceeds just as before. The quantities involving M never appear anywhere except in row 0, so they need to be taken into account only in the optimality test and when determining an entering basic variable. One way of dealing with these quantities is to assign some particular (huge) numerical value to M and use the resulting numbers in row 0 in the usual way. However, this approach may result in significant round-off errors that invalidate the optimality test. Therefore, it is better to do what we have just shown, namely, express each coefficient in row 0 as a linear function $aM + b$ of the *symbolic* quantity M by separately recording and updating the current numerical value of (1) the *multiplicative* factor a and (2) the *additive* factor b. Because M is assumed to be so large that b always is negligible compared with aM when $a \neq 0$, the decisions in the optimality test and the choice of the entering basic variable are made by using just the *multipli-*

cative factors in the usual way. The one exception is when this use leads to a tie [where a tie for the optimality test means that the smallest multiplicative factor(s) equals zero], in which case the tie would be broken by using the corresponding *additive* factors.

Using this approach on the example yields the simplex tableaux shown in Table 4.11. The optimal solution ($x_1 = 2$, $x_2 = 6$) is the same as for the first version of the Wyndor Glass Co. problem (see Table 4.8 for its tableaux). However, a different sequence of basic feasible solutions was obtained because a comparison of the initial *multiplicative* factors ($3 > 2$) led to choosing x_1 rather than x_2 as the initial entering basic variable. If the equality constraint had been $3x_1 + 3x_2 = 18$ instead, both multiplicative factors would have been -3, so then a comparison of the *additive* factors ($5 > 3$) would have led to choosing x_2 as before.

This example involved only *one* equality constraint. If a linear programming model has more than one, each would be handled in just this same way. [If the right-hand side is negative, multiply through both sides by (-1) first.] Thus each equality constraint would be given an artificial variable to serve as its initial basic variable, each of these variables would be assigned a coefficient of $-M$ in the objective function [or $+M$ when the variable is brought to the left-hand side of Eq. (0)], and the resulting row 0 would have subtracted from it M times each equality constraint row.

The approach to other kinds of constraints requiring artificial variables is completely analogous. To illustrate the adjustments for a variety of different forms, we will use the model for designing Mary's radiation therapy, as presented in Sec. 3.4. For your convenience, this model is repeated below.

RADIATION THERAPY EXAMPLE

$$\text{Minimize} \quad Z = 0.4x_1 + 0.5x_2,$$

subject to

$$0.3x_1 + 0.1x_2 \leq 2.7$$

$$0.5x_1 + 0.5x_2 = 6$$

$$0.6x_1 + 0.4x_2 \geq 6$$

Table 4.11 **Complete Set of Simplex Tableaux for Wyndor Glass Co. Problem with an Equality Constraint**

Iteration	Basic Variable	Eq. No.	Z	x_1	x_2	x_3	x_4	$\bar{x}_5$	Right Side
0	Z	0	1	$(-3M-3)$	$(-2M-5)$	0	0	0	$-18M$
	x_3	1	0	1	0	1	0	0	4
	x_4	2	0	0	2	0	1	0	12
	$\bar{x}_5$	3	0	3	2	0	0	1	18
1	Z	0	1	0	$(-2M-5)$	$(3M+3)$	0	0	$-6M+12$
	x_1	1	0	1	0	1	0	0	4
	x_4	2	0	0	2	0	1	0	12
	$\bar{x}_5$	3	0	0	2	-3	0	1	6
2	Z	0	1	0	0	$-\frac{9}{2}$	0	$(M + \frac{5}{2})$	27
	x_1	1	0	1	0	1	0	0	4
	x_4	2	0	0	0	3	1	-1	6
	x_2	3	0	0	1	$-\frac{3}{2}$	0	$\frac{1}{2}$	3
3	Z	0	1	0	0	0	$\frac{3}{2}$	$(M + 1)$	36
	x_1	1	0	1	0	0	$-\frac{1}{3}$	$\frac{1}{3}$	2
	x_3	2	0	0	0	1	$\frac{1}{3}$	$-\frac{1}{3}$	2
	x_2	3	0	0	1	0	$\frac{1}{2}$	0	6

and
$$x_1 \geq 0, \qquad x_2 \geq 0.$$

The graphical solution for this example (originally presented in Fig. 3.7) is repeated here in a slightly different form in Fig. 4.2. The three lines in the figure, along with the two axes, constitute the five constraint lines of the problem. The dots lying at the intersection of a pair of constraint lines are the *corner-point solutions*. The only two corner-point *feasible* solutions are (6, 6) and (7.5, 4.5), and the feasible region is the line segment connecting these two points. The optimal solution is $(x_1, x_2) = (7.5, 4.5)$, with $Z = 5.25$.

We soon will show how the simplex method solves this problem in its entirety, including the sequence of corner-point solutions obtained. However, we first must

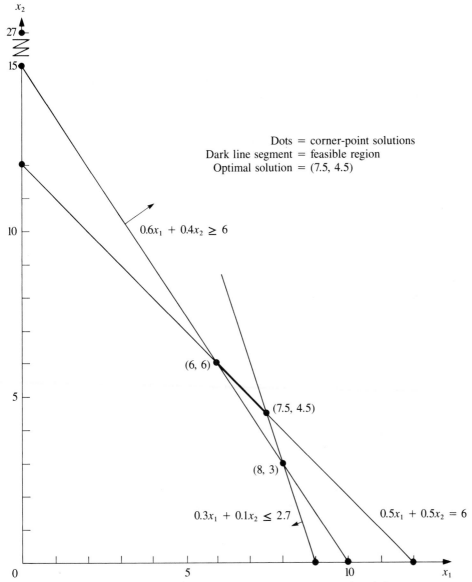

Dots = corner-point solutions
Dark line segment = feasible region
Optimal solution = (7.5, 4.5)

$0.6x_1 + 0.4x_2 \geq 6$

(6, 6)

(7.5, 4.5)

(8, 3)

$0.3x_1 + 0.1x_2 \leq 2.7$

$0.5x_1 + 0.5x_2 = 6$

Figure 4.2 Graphical display of the radiation therapy example and its corner-point solutions.

convert the model into an appropriate form for applying the simplex method. Two adjustments are to introduce a slack variable ($x_3 \geq 0$) into the first constraint,

$$0.3x_1 + 0.1x_2 \leq 2.7 \quad \rightarrow \quad 0.3x_1 + 0.1x_2 + x_3 = 2.7,$$

and an artificial variable ($\bar{x}_4 \geq 0$) into the equality constraint,

$$0.5x_1 + 0.5x_2 = 6 \quad \rightarrow \quad 0.5x_1 + 0.5x_2 + \bar{x}_4 = 6.$$

The temporary effect of introducing $\bar{x}_4$ into this constraint (before the Big M method forces $\bar{x}_4$ to be zero) is to allow solutions such that $0.5x_1 + 0.5x_2 \leq 6$. Considering the other constraints as well (see Fig. 4.2), the feasible region for the *revised* problem is thereby expanded to include the entire triangle whose vertices are (6, 6), (7.5, 4.5), and (8, 3).

The remaining adjustments for the other nonstandard forms in the model are described next.

$\geq$ Inequality Constraints

The direction of an inequality always is reversed when both sides are multiplied by (-1). As a result, any functional constraint of the $\geq$ form can be converted into an *equivalent* constraint of our standard $\leq$ form by changing the signs of all the numbers on both sides.

Using this approach for the third constraint of the radiation therapy example,

$$0.6x_1 + 0.4x_2 \quad \geq \quad 6$$

$$\rightarrow \quad -0.6x_1 - 0.4x_2 \quad \leq \quad -6$$

$$\rightarrow \quad -0.6x_1 - 0.4x_2 + x_5 = -6.$$

where x_5 is the slack variable for this constraint. However, one more change is still needed, as you will see next.

Negative Right-Hand Sides

You may recall that the simplex method was presented in the preceding sections under the assumption that $b_i > 0$ for all $i = 1, 2, \ldots, m$. This assumption enabled us to select the slack variables to be the *initial* basic variables (equal to the right-hand sides) and still obtain a *nondegenerate* basic *feasible* solution. We have since pointed out in Sec. 4.5 that degeneracy (basic variables equal to zero) does not need to be avoided. However, a negative right-hand side, such as in the third constraint,

$$-0.6x_1 - 0.4x_2 + x_5 = -6,$$

would give a negative value for the slack variable ($x_5 = -6$) in the initial solution (where $x_1 = 0$, $x_2 = 0$), which violates the nonnegativity constraint for this variable. Multiplying through the equation by (-1) makes the right-hand side positive:

$$0.6x_1 + 0.4x_2 - x_5 = 6,$$

but it also changes the coefficient of the slack variable to -1, so the variable still would be negative. However, in this form the constraint can be viewed as an *equality constraint* with a nonnegative right-hand side, so the *artificial variable technique* can be applied just as described earlier. If we let $\bar{x}_6$ be the nonnegative artificial variable for this constraint, its final form becomes

$$0.6x_1 + 0.4x_2 - x_5 + \bar{x}_6 = 6,$$

where $\bar{x}_6$ is used as the *initial basic variable* ($\bar{x}_6 = 6$) for this equation and x_5 begins as a nonbasic variable. The Big M method also would be applied just as before, as we shall demonstrate shortly.

As usual, introducing this artificial variable enlarges the feasible region. The original constraint allowed only solutions lying above or on the constraint boundary, $0.6x_1 + 0.4x_2 = 6$. Now it allows any solution lying below this constraint boundary as well, because both x_5 and $\bar{x}_6$ are constrained only to be nonnegative so their difference ($\bar{x}_6 - x_5$) can be *any* positive or negative number. *No* solutions are disallowed at all, so the temporary effect of introducing $\bar{x}_6$ has been to *eliminate* this constraint in the *revised* problem. (We keep the constraint in the system of equations only because it will become relevant again later, after the Big M method forces $\bar{x}_6$ to be zero.)

We now have twice revised the original problem by expanding its feasible region, first by introducing the artificial variable $\bar{x}_4$ into the equality constraint ($0.5x_1 + 0.5x_2 + \bar{x}_4 = 6$), and now by introducing $\bar{x}_6$. Consequently, the feasible region for the *revised* problem is the entire polyhedron in Fig. 4.2 whose vertices are $(0, 0)$, $(9, 0)$, $(7.5, 4.5)$, and $(0, 12)$.

You may have noticed that we took a somewhat circuitous route in converting the third constraint from its original form, $0.6x_1 + 0.4x_2 \geq 6$, to its final version, $0.6x_1 + 0.4x_2 - x_5 + \bar{x}_6 = 6$. In fact, we multiplied through the constraint by (-1) twice along the way! Now that you have seen the motivation leading to the final form, we should point out the following shortcut:

$$0.6x_1 + 0.4x_2 \qquad\qquad \geq 6$$

$$\rightarrow \quad 0.6x_1 + 0.4x_2 - x_5 \qquad = 0 \qquad (x_5 \geq 0)$$

$$\rightarrow \quad 0.6x_1 + 0.4x_2 - x_5 + \bar{x}_6 = 6 \qquad (x_5 \geq 0, \bar{x}_6 \geq 0).$$

In this form, x_5 is called a **surplus variable** because it subtracts the surplus of the left-hand side over the right-hand side to convert the inequality into an equivalent equation.

Minimization

One straightforward way of minimizing Z with the simplex method is to exchange the roles of the positive and negative coefficients in row 0 for both the optimality test and part 1 of the iterative step. However, rather than changing our instructions for the simplex method, we instead present the following simple way of converting any minimization problem into an equivalent maximization problem:

$$\textit{Minimizing} \qquad Z = \sum_{j=1}^{n} c_j x_j$$

is equivalent to

$$\textit{maximizing} \quad (-Z) = \sum_{j=1}^{n} (-c_j)x_j;$$

that is, the two formulations yield the same optimal solution(s).

The reason the two formulations are equivalent is that the smaller Z is, the larger $(-Z)$ is, so the solution that gives the *smallest* value of Z in the entire feasible region must also give the *largest* value of $(-Z)$ in this region.

Therefore, in the radiation therapy example, we make the following change in the formulation:

$$\text{Minimize} \qquad Z = 0.4x_1 + 0.5x_2$$
$$\rightarrow \quad \text{Maximize} \qquad (-Z) = -0.4x_1 - 0.5x_2.$$

After introducing artificial variables $(\bar{x}_4, \bar{x}_6)$ and then applying the Big M method, the corresponding conversion is

$$\text{Minimize} \qquad Z = 0.4x_1 + 0.5x_2 + M\bar{x}_4 + M\bar{x}_6$$
$$\rightarrow \quad \text{Maximize} \qquad (-Z) = -0.4x_1 - 0.5x_2 - M\bar{x}_4 - M\bar{x}_6.$$

Solving the Radiation Therapy Example

We now are nearly ready to apply the simplex method to the radiation therapy example. Using the maximization form just obtained, the entire system of equations is now

$$(0) \qquad -Z + 0.4x_1 + 0.5x_2 \qquad\quad + M\bar{x}_4 \qquad\quad + M\bar{x}_6 = 0$$

$$(1) \qquad\qquad 0.3x_1 + 0.1x_2 + x_3 \qquad\qquad\qquad\qquad = 2.7$$

$$(2) \qquad\qquad 0.5x_1 + 0.5x_2 \qquad\quad + \bar{x}_4 \qquad\qquad = 6$$

$$(3) \qquad\qquad 0.6x_1 + 0.4x_2 \qquad\qquad - x_5 + \bar{x}_6 = 6.$$

The basic variables $(x_3, \bar{x}_4, \bar{x}_6)$ for the *initial basic feasible solution* (for this *revised* problem) are shown in **boldface.**

Do you see any problem with this system of equations for starting the simplex method? Can the *optimality test* and the procedure for selecting an *entering basic variable* be applied to this Eq. (0)? No! The problem is that the system of equations is not yet in *proper form from Gaussian elimination* (where each basic variable has been eliminated from every equation except its own equation, for which it has a coefficient of $+1$). Equations (1) to (3) are fine, but the basic variables $\bar{x}_4$ and $\bar{x}_6$ still need to be eliminated from Eq. (0) by Gaussian elimination. Because $\bar{x}_4$ and $\bar{x}_6$ both have a coefficient of M, Eq. (0) needs to have subtracted from it *both M times Eq. (2) and M times Eq. (3). For example, the coefficient of x_1 in Eq. (0) becomes $0.4 - 0.5M - 0.6M = -1.1M + 0.4$, so that M is treated as a real (huge) number whose value is fixed but unspecified. The calculations for all of the coefficients (and the right-hand side) are summarized below, where the vectors are the relevant rows of the simplex tableau corresponding to the above system of equations.

$$
\begin{array}{lcccccccc}
\textit{Row 0:} & [& 0.4 & 0.5 & 0 & M & 0 & M, & 0 &] \\
& -M\,[& 0.5 & 0.5 & 0 & 1 & 0 & 0, & 6 &] \\
& -M\,[& 0.6 & 0.4 & 0 & 0 & -1 & 1, & 6 &] \\
\text{New row 0} = & & \multicolumn{7}{l}{[(-1.1M + 0.4), (-0.9M + 0.5), 0 \quad 0 \quad M \quad 0, \, -12M]}
\end{array}
$$

The resulting *initial simplex tableau,* ready to begin the simplex method, is shown at the top of Table 4.12. Applying the simplex method in just the usual way then yields the sequence of simplex tableaux shown in the rest of Table 4.12. For the

Table 4.12 The Big M Method for the Radiation Therapy Example

Iteration	Basic Variable	Eq. No.	Z	x_1	x_2	x_3	$\bar{x}_4$	x_5	$\bar{x}_6$	Right Side
0	Z	0	-1	$(-1.1M + 0.4)$	$(-0.9M + 0.5)$	0	0	M	0	$-12M$
	x_3	1	0	0.3	0.1	1	0	0	0	2.7
	$\bar{x}_4$	2	0	0.5	0.5	0	1	0	0	6
	$\bar{x}_6$	3	0	0.6	0.4	0	0	-1	1	6
1	Z	0	-1	0	$(-\frac{16}{30}M + \frac{11}{30})$	$(\frac{11}{3}M - \frac{4}{3})$	0	M	0	$-2.1M - 3.6$
	x_1	1	0	1	$\frac{1}{3}$	$\frac{10}{3}$	0	0	0	9
	$\bar{x}_4$	2	0	0	$\frac{1}{3}$	$-\frac{5}{3}$	1	0	0	1.5
	$\bar{x}_6$	3	0	0	0.2	-2	0	-1	1	0.6
2	Z	0	-1	0	0	$(-\frac{5}{3}M + \frac{7}{3})$	0	$(-\frac{5}{3}M + \frac{11}{6})$	$(\frac{5}{3}M - \frac{11}{6})$	$-0.5M - 4.7$
	x_1	1	0	1	0	$\frac{20}{3}$	0	$\frac{5}{3}$	$-\frac{5}{3}$	8
	$\bar{x}_4$	2	0	0	0	$\frac{5}{3}$	1	$\frac{5}{3}$	$-\frac{5}{3}$	0.5
	x_2	3	0	0	1	-10	0	-5	5	3
3	Z	0	-1	0	0	0.5	$(M - 1.1)$	0	M	-5.25
	x_1	1	0	1	0	5	-1	0	0	7.5
	x_5	2	0	0	0	1	0.6	1	-1	0.3
	x_2	3	0	0	1	-5	3	0	0	4.5

optimality test and the selection of the *entering basic variable* at each iteration, the quantities involving M are treated just as discussed in connection with Table 4.11. Specifically, whenever M is present, only its *multiplicative* factor is used, unless there is a tie, in which case the tie is broken by using the corresponding *additive* factors. Just such a tie occurs in the last selection of an entering basic variable (see the next-to-last tableau), where the coefficients of x_3 and x_5 in row 0 both have the same *multiplicative* factor, $-\frac{5}{3}$. Comparing the *additive* factors, $\frac{11}{6} < \frac{7}{3}$, leads to choosing x_5 as the entering basic variable.

Now see what the Big M method has done graphically in Fig. 4.2. Using just the original decision variables (x_1, x_2), the sequence of *corner-point solutions* obtained in Table 4.12 is

$$(0, 0) \rightarrow (9, 0) \rightarrow (8, 3) \rightarrow (7.5, 4.5).$$

For the first two corner-point solutions, both $\bar{x}_4$ and $\bar{x}_6$ are greater than zero, indicating that both $0.5x_1 + 0.5x_2 = 6$ and $0.6x_1 + 0.4x_2 \geq 6$ are violated. The Big M method then succeeds in driving $\bar{x}_6$ to zero at (8, 3), so that $0.6x_1 + 0.4x_2 \geq 6$ is satisfied. Next, $\bar{x}_4$ also is driven to zero at (7.5, 4.5), so that $0.5x_1 + 0.5x_2 = 6$ also is satisfied, and the first *feasible* solution for the *original* problem has been obtained. Fortuitously, this first feasible solution also is optimal, so no additional iterations are needed.

For other problems with artificial variables, it may be necessary to perform additional iterations to reach an optimal solution after obtaining the first feasible solution for the original problem. (This was the case for the example solved in Table 4.11.) Thus the Big M method can be thought of as having two phases. In the first phase, all of the artificial variables are driven to zero (because of the penalty of M per unit for being greater than zero) in order to reach an initial basic feasible solution for the original problem. In the second phase, all of the artificial variables are kept at zero (because of this same penalty) while the simplex method generates a sequence of basic feasible solutions leading to an optimal solution. The *two-phase method*

81

described next is a streamlined procedure for performing these two phases directly, without even introducing M explicitly.

The Two-Phase Method

For the radiation therapy example just solved in Table 4.12, the Big M method uses the following objective function (or its equivalent in maximization form) throughout the entire procedure.

Big M method: Minimize $\quad Z = 0.4x_1 + 0.5x_2 + M\bar{x}_4 + M\bar{x}_6.$

By contrast, the two-phase method is able to drop M by using two different objective functions.

Two-phase method:

Phase 1: Minimize $\quad Z = \bar{x}_4 + \bar{x}_6 \quad$ (until $\bar{x}_4 = 0, \bar{x}_6 = 0$).

Phase 2: Minimize $\quad Z = 0.4x_1 + 0.5x_2 \quad$ (with $\bar{x}_4 = 0, \bar{x}_6 = 0$).

Before solving the example in this way, let us summarize the general method.

SUMMARY OF THE TWO-PHASE METHOD

Initialization Step: Revise the constraints of the original problem by introducing artificial variables as needed to obtain an obvious initial basic feasible solution for the *revised* problem.

Phase 1: Use the simplex method to solve the linear programming problem:

Minimize $\quad Z = $ sum of artificial variables, subject to the revised constraints.

The optimal solution obtained for this problem (with $Z = 0$) will be a basic *feasible* solution for the *original* problem.

Phase 2: Drop the artificial variables (they are all zero now anyway).[1] Starting from the basic feasible solution obtained at the end of Phase 1, use the simplex method to solve the *original* problem.

Table 4.13 shows the result of applying *Phase 1* to the radiation therapy example. [Row 0 in the initial tableau is obtained by converting Minimize $Z = \bar{x}_4 + \bar{x}_6$ to Maximize $(-Z) = -\bar{x}_4 - \bar{x}_6$, and then using *Gaussian elimination* to eliminate $\bar{x}_4$ and $\bar{x}_6$ from $-Z + \bar{x}_4 + \bar{x}_6 = 0$.] In the next-to-last tableau, there is a tie for the *entering basic variable* between x_3 and x_5, which is broken arbitrarily in favor of x_3. The solution obtained at the end of Phase 1, then, is $(x_1, x_2, x_3, \bar{x}_4, x_5, \bar{x}_6) = (6, 6, 0.3, 0, 0, 0)$ or, after dropping $\bar{x}_4$ and $\bar{x}_6$, $(x_1, x_2, x_3, x_5) = (6, 6, 0.3, 0)$.

As claimed in the *Summary*, this solution from Phase 1 is indeed a basic feasible solution for the *original* problem because it is the solution (after setting $x_5 = 0$) to the original constraints in *augmented form*,

(1) $\qquad\qquad\qquad 0.3x_1 + 0.1x_2 + x_3 \qquad\quad = 2.7$

(2) $\qquad\qquad\qquad 0.5x_1 + 0.5x_2 \qquad\qquad\quad\ = 6$

(3) $\qquad\qquad\qquad 0.6x_1 + 0.4x_2 \qquad - x_5 = 6.$

---

[1] We are skipping over three other possibilities here: (1) artificial variables > 0 (discussed in the next subsection), (2) artificial variables that are degenerate basic variables, and (3) retaining the artificial variables as nonbasic variables in Phase 2 (and not allowing them to become basic) as an aid to subsequent sensitivity analysis. Your OR COURSEWARE allows you to explore these possibilities.

Table 4.13 **Phase 1 of the Two-Phase Method for the Radiation Therapy Example**

Iteration	Basic Variable	Eq. No.	Z	x_1	x_2	x_3	$\bar{x}_4$	x_5	$\bar{x}_6$	Right Side
	Z	0	-1	-1.1	-0.9	0	0	1	0	-12
0	x_3	1	0	0.3	0.1	1	0	0	0	2.7
	$\bar{x}_4$	2	0	0.5	0.5	0	1	0	0	6
	$\bar{x}_6$	3	0	0.6	0.4	0	0	-1	1	6
	Z	0	-1	0	$-\frac{16}{30}$	$\frac{11}{3}$	0	1	0	-2.1
1	x_1	1	0	1	$\frac{1}{3}$	$\frac{10}{3}$	0	0	0	9
	$\bar{x}_4$	2	0	0	$\frac{1}{3}$	$-\frac{5}{3}$	1	0	0	1.5
	$\bar{x}_6$	3	0	0	0.2	-2	0	-1	1	0.6
	Z	0	-1	0	0	$-\frac{5}{3}$	0	$-\frac{5}{3}$	$\frac{8}{3}$	-0.5
2	x_1	1	0	1	0	$\frac{20}{3}$	0	$\frac{5}{3}$	$-\frac{5}{3}$	8
	$\bar{x}_4$	2	0	0	0	$\frac{5}{3}$	1	$\frac{5}{3}$	$-\frac{5}{3}$	0.5
	x_2	3	0	0	1	-10	0	-5	5	3
	Z	0	-1	0	0	0	1	0	1	0
3	x_1	1	0	1	0	0	-4	-5	5	6
	x_3	2	0	0	0	1	$\frac{3}{5}$	1	-1	0.3
	x_2	3	0	0	1	0	6	5	-5	6

In fact, after deleting the $\bar{x}_4$ and $\bar{x}_6$ columns, Table 4.13 shows one way of using Gaussian elimination to solve this system of equations by reducing the system to the form displayed in the final tableau.

For *Phase 2*, $\bar{x}_4$ and $\bar{x}_6$ are dropped. To start the simplex method from the basic feasible solution, $(x_1, x_2, x_3, x_5) = (6, 6, 0.3, 0)$, rows 1–3 of the final tableau in Table 4.13 are already in *proper form from Gaussian elimination*. However, we now need to insert into row 0 the objective function for the original problem in this same proper form. The sequence of steps to obtain this new row 0 (including using rows 1 and 3 to eliminate x_1 and x_2 from row 0) is shown below:

$$\text{Minimize} \quad Z = \quad 0.4x_1 + 0.5x_2$$

$$\rightarrow \quad \text{Maximize} \quad (-Z) = -0.4x_1 - 0.5x_2$$

$$\rightarrow \quad (0) \quad -Z + 0.4x_1 + 0.5x_2 = 0.$$

$$
\begin{array}{ll}
\textit{Row 0:} & \quad [0.4 \quad 0.5 \quad 0 \quad \ 0, \quad \ 0 \ ] \\
& -0.4[1 \quad \ 0 \quad \ 0 \ -5, \quad \ 6 \ ] \\
& -0.5[0 \quad \ 1 \quad \ 0 \quad \ 5, \quad \ 6 \ ] \\
\textit{New row 0} = & \quad [0 \quad \ 0 \quad \ 0 \ -0.5, \ -5.4].
\end{array}
$$

The resulting initial simplex tableau for Phase 2 is shown at the top of Table 4.14. Applying the simplex method then leads in one iteration to the optimal solution shown in the second tableau, $(x_1, x_2, x_3, x_5) = (7.5, 4.5, 0, 0.3)$.

Now see what the two-phase method has done graphically in Fig. 4.2. Using just (x_1, x_2), the sequence of *corner-point solutions* obtained in Tables 4.13 and 4.14 is

Phase 1: $(0, 0) \rightarrow (9, 0) \rightarrow (8, 3) \rightarrow (6, 6)$
Phase 2: $(6, 6) \rightarrow (7.5, 4.5).$

Note that all of these solutions in Phase 1 are *infeasible* (except for the revised

Table 4.14 Phase 2 of the Two-Phase Method for the Radiation Therapy Example

| Iteration | Basic Variable | Eq. No. | Coefficient of | | | | | Right Side |
			Z	x_1	x_2	x_3	x_5	
0	Z	0	-1	0	0	0	-0.5	-5.4
	x_1	1	0	1	0	0	-5	6
	x_3	2	0	0	0	1	1	0.3
	x_2	3	0	0	1	0	5	6
1	Z	0	-1	0	0	0.5	0	-5.25
	x_1	1	0	1	0	5	0	7.5
	x_5	2	0	0	0	1	1	0.3
	x_2	3	0	0	1	-5	0	4.5

problem) until the last one. Phase 2 then deals only with corner-point *feasible* solutions.

If the tie for the entering basic variable in the next-to-last tableau of Table 4.13 had been broken in the other way, then Phase 1 would have gone directly from (8, 3) to (7.5, 4.5). After using (7.5, 4.5) to set up the initial simplex tableau for Phase 2, the *optimality test* would have revealed that this solution is optimal, so no iterations would be done.

It is interesting to compare the Big M and two-phase methods. Begin with their objective functions.

> *Big M method:* Minimize $Z = 0.4x_1 + 0.5x_2 + M\bar{x}_4 + M\bar{x}_6.$
> *Two-phase method:*
> > *Phase 1:* Minimize $Z = \bar{x}_4 + \bar{x}_6.$
> > *Phase 2:* Minimize $Z = 0.4x_1 + 0.5x_2.$

Because the $M\bar{x}_4$ and $M\bar{x}_6$ terms dominate the $0.4x_1$ and $0.5x_2$ terms in the objective function for the Big M method, this objective function is essentially equivalent to the Phase 1 objective function as long as $\bar{x}_4$ and/or $\bar{x}_6$ are greater than zero. Then, when both $\bar{x}_4 = 0$ and $\bar{x}_6 = 0$, the objective function for the Big M method becomes completely equivalent to the Phase 2 objective function.

Because of these virtual equivalencies in objective functions, the Big M and two-phase methods generally have the same sequence of basic feasible solutions. The one possible exception is when a tie for the entering basic variable occurs in Phase 1 of the two-phase method, as happened in the third tableau of Table 4.13. Notice that the first three tableaux of Tables 4.12 and 4.13 are almost identical, with the only difference being that the *multiplicative factors* of M in Table 4.12 become the sole quantities in the corresponding spots in Table 4.13. Consequently, the *additive factors* that broke the tie for the entering basic variable in the third tableau of Table 4.12 were not present to break this same tie in Table 4.13. The result for this example was an extra iteration for the two-phase method. Generally, however, the advantage of having the additive factors is minimal.

The two-phase method streamlines the Big M method by using only the multiplicative factors in Phase 1 and by dropping the artificial variables in Phase 2. (The Big M method could combine the multiplicative and additive factors by assigning an actual huge number to M, but this might create numerical instability problems.) For these reasons, the two-phase method is commonly used in computer codes.

So far in this section we have been concerned primarily with the fundamental problem of identifying an initial basic feasible solution when an obvious one is not available. You have seen how the artificial variable technique constructs an artificial problem and obtains an initial basic feasible solution for the revised problem instead. Either the Big M method or the two-phase method then enables the simplex method to begin its pilgrimage toward the basic feasible solutions, and ultimately toward the optimal solution, for the *original* problem.

However, you should be wary of a certain pitfall with this approach. There may be no obvious choice for the initial basic feasible solution for the very good reason that there are no feasible solutions at all! Nevertheless, by constructing an artificial feasible solution, there is nothing to prevent the simplex method from proceeding as usual and ultimately reporting a supposedly optimal solution.

Fortunately, the artificial variable technique provides the following signpost to indicate when this has happened:

> If the original problem has *no feasible solutions,* then either the Big M method or Phase 1 of the two-phase method yields a final solution that has at least one artificial variable *greater* than zero. Otherwise, they *all* equal zero.

To illustrate, let us change the first constraint in the radiation therapy example (see Fig. 4.2) as follows:

$$0.3x_1 + 0.1x_2 \leq 2.7 \quad \rightarrow \quad 0.3x_1 + 0.1x_2 \leq 1.8,$$

so that the problem no longer has any feasible solutions. Applying the Big M method just as before (see Table 4.12) yields the tableaux shown in Table 4.15. (Phase 1 of the two-phase method yields the same tableaux except that each expression involving M is replaced by just the multiplicative factor.) Hence the Big M method normally would be indicating that the optimal solution is (3, 9, 0, 0, 0, 0.6). However, since an artificial variable $\bar{x}_6 = 0.6 > 0$, the real message here is that the problem has no feasible solutions.

Table 4.15 The Big M Method for the Revision of the Radiation Therapy Example That Has No Feasible Solutions

Iteration	Basic Variable	Eq. No.	Z	x_1	x_2	x_3	$\bar{x}_4$	x_5	$\bar{x}_6$	Right Side
0	Z	0	-1	$(-1.1M + 0.4)$	$(-0.9M + 0.5)$	0	0	M	0	$-12M$
	x_3	1	0	0.3	0.1	1	0	0	0	1.8
	$\bar{x}_4$	2	0	0.5	0.5	0	1	0	0	6
	$\bar{x}_6$	3	0	0.6	0.4	0	0	-1	1	6
1	Z	0	-1	0	$(-\frac{16}{30}M + \frac{11}{30})$	$(\frac{11}{3}M - \frac{4}{3})$	0	M	0	$-5.4M - 2.4$
	x_1	1	0	1	$\frac{1}{3}$	$\frac{10}{3}$	0	0	0	6
	$\bar{x}_4$	2	0	0	$\frac{1}{3}$	$-\frac{5}{3}$	1	0	0	3
	$\bar{x}_6$	3	0	0	0.2	-2	0	-1	1	2.4
2	Z	0	-1	0	0	$(M + 0.5)$	$(1.6M - 1.1)$	M	0	$-0.6M - 5.7$
	x_1	1	0	1	0	5	-1	0	0	3
	x_2	2	0	0	1	-5	3	0	0	9
	$\bar{x}_6$	3	0	0	0	-1	-0.6	-1	1	0.6

In most practical problems, negative values for the decision variables would have no physical meaning, so it is necessary to include nonnegativity constraints in the formulations of their linear programming models. However, this is not always the case. To illustrate, suppose that the Wyndor Glass Co. problem is changed so that product 1 already is in production, and the first decision variable x_1 represents the *increase* in its production rate. Therefore, a negative value of x_1 would indicate that product 1 is to be cut back by that amount. Such reductions might be desirable to allow a larger production rate for the new, more profitable product 2, so negative values should be allowed for x_1 in the model.

Since the procedure for determining the *leaving basic variable* requires that all the variables have nonnegativity constraints, any problem containing variables allowed to be negative must be converted into an *equivalent* problem involving only nonnegative variables before applying the simplex method. Fortunately, this conversion can be done. The modification required for each variable depends upon whether or not it has a (negative) lower bound on the values allowed. Each of these two cases is now discussed.

VARIABLES WITH A BOUND ON THE NEGATIVE VALUES ALLOWED: Consider any decision variable x_j that is allowed to have negative values that satisfy a constraint of the form

$$x_j \geq L_j,$$

where L_j is some negative constant. This constraint can be converted into a nonnegativity constraint by making the change of variables,

$$x_j' = x_j - L_j, \qquad \text{so } x_j' \geq 0.$$

Thus $(x_j' + L_j)$ would be substituted for x_j throughout the model, so that the redefined decision variable x_j' cannot be negative.

To illustrate, suppose that the current production rate for product 1 in the Wyndor Glass Co. problem is 10. With the definition of x_1 just given, the complete model at this point is the same as that given in Sec. 3.1 except that the nonnegativity constraint, $x_1 \geq 0$, is replaced by

$$x_1 \geq -10.$$

To obtain the equivalent model needed for the simplex method, this decision variable would be redefined as the *total* production rate of product 1,

$$x_1' = x_1 + 10,$$

which yields the changes in the objective function and constraints as shown:

$Z = 3x_1 + 5x_2$	$Z = 3(x_1' - 10) + 5x_2$	$Z = -30 + 3x_1' + 5x_2$
$x_1 \leq 4$	$(x_1' - 10) \leq 4$	$x_1' \leq 14$
$2x_2 \leq 12$	$2x_2 \leq 12$	$2x_2 \leq 12$
$3x_1 + 2x_2 \leq 18$	$3(x_1' - 10) + 2x_2 \leq 18$	$3x_1' + 2x_2 \leq 48$
$x_1 \geq -10, x_2 \geq 0$	$(x_1' - 10) \geq -10, x_2 \geq 0$	$x_1' \geq 0, x_2 \geq 0$

With arrows ($\rightarrow$) between the three boxes.

VARIABLES WITH NO BOUND ON THE NEGATIVE VALUES ALLOWED: In the case where x_j does *not* have a lower bound constraint in the model formulated, another approach is required: x_j is replaced throughout the model by the *difference* of two new *nonnegative* variables,

$$x_j = x_j^+ - x_j^-, \qquad \text{where } x_j^+ \geq 0, \ x_j^- \geq 0.$$

Since x_j^+ and x_j^- can have any nonnegative values, this difference $(x_j^+ - x_j^-)$ can have *any* value (positive or negative), so it is a legitimate substitute for x_j in the model. But after such substitutions, the simplex method can proceed with just nonnegative variables.

The new variables, x_j^+ and x_j^-, have a simple interpretation. By the geometric definition of corner-point feasible solution (see Sec. 5.1), each basic feasible solution for the new form of the model necessarily has the property that *either* $x_j^+ = 0$ *or* $x_j^- = 0$ (or both). Therefore, at the optimal solution obtained by the simplex method,

$$x_j^+ = \begin{cases} x_j, & \text{if } x_j \geq 0 \\ 0, & \text{otherwise;} \end{cases}$$

$$x_j^- = \begin{cases} |x_j|, & \text{if } x_j \leq 0 \\ 0, & \text{otherwise;} \end{cases}$$

so that x_j^+ represents the *positive* part of the decision variable x_j and x_j^- its *negative* part (as suggested by the superscripts).

To illustrate this approach, let us use the same example as for the *bounded variable case*. However, now suppose that the $x_1 \geq -10$ constraint was not included in the original model because it clearly would not change the optimal solution. (In some problems, certain variables do not need explicit lower bound constraints because the functional constraints already prevent lower values.) Therefore, before applying the simplex method, x_1 would be replaced by the difference,

$$x_1 = x_1^+ - x_1^-, \qquad \text{where } x_1^+ \geq 0, \ x_1^- \geq 0,$$

as shown:

Maximize $Z = 3x_1 + 5x_2$
$x_1 \leq 4$
$2x_2 \leq 12$
$3x_1 + 2x_2 \leq 18$
$x_2 \geq 0$ (only)

$\rightarrow$

Maximize $Z = 3x_1^+ - 3x_1^- + 5x_2$
$x_1^+ - x_1^- \leq 4$
$2x_2 \leq 12$
$3x_1^+ - 3x_1^- + 2x_2 \leq 18$
$x_1^+ \geq 0, \ x_1^- \geq 0, \ x_2 \geq 0$

From a computational viewpoint, this approach has the disadvantage that the new equivalent model to be used has more variables than the original model. In fact, if *all* the original variables lack lower bound constraints, the new model will have *twice* as many variables. Fortunately, the approach can be modified slightly so that the number of variables is increased by only *one*, regardless of how many original variables need to be replaced. This modification is done by replacing each such variable x_j by

$$x_j = x_j' - x'', \qquad \text{where } x_j' \geq 0, \ x'' \geq 0,$$

instead, where x'' is the *same* variable for all relevant j. The interpretation of x'' in this case is that $-x''$ is the current value of the *largest* (in absolute terms) negative original variable, so that x_j' is the amount by which x_j exceeds this value. Thus the simplex method now can make some of the x_j' variables larger than zero even when $x'' > 0$.

We stressed in Secs. 2.3, 2.4, and 2.5 that *post-optimality analysis*—the analysis done *after* an optimal solution is obtained for the initial version of the model—constitutes a very major and very important part of most operations research studies. The fact that post-optimality analysis is very important is particularly true for typical linear programming applications. In this section, we focus on the role of the simplex method in performing this analysis.

Table 4.16 summarizes the typical steps in post-optimality analysis for linear programming studies. The last column of Table 4.16 identifies some algorithmic techniques that involve the simplex method. These techniques are introduced briefly here with the technical details deferred to later chapters.

Reoptimization

After having found an optimal solution for one version of a linear programming model, we frequently must solve again (often many times) for a slightly different version of the model. We nearly always have to solve again several times during the model debugging stage (described in Secs. 2.3 and 2.4), and we usually have to do so a large number of times during the later stages of post-optimality analysis as well.

One approach is simply to reapply the simplex method from scratch for each new version of the model, even though each run may require hundreds or even thousands of iterations for large problems. However, a *much more efficient* approach is to *reoptimize*. Reoptimization involves deducing how changes in the model get carried along to the *final* simplex tableau (as described in Secs. 5.3 and 6.6). This revised tableau and the optimal solution for the prior model are then used as the *initial tableau* and the *initial basic solution* for solving the new model. If this solution is feasible for the new model, then the simplex method is applied in the usual way, starting from this initial basic feasible solution. If the solution is not feasible, a related algorithm called the *dual simplex method* (described in Sec. 9.2) probably can be applied to find the new optimal solution,[1] starting from this initial basic solution.

Table 4.16 **Post-Optimality Analysis for Linear Programming**

Task	Purpose	Technique
Model debugging	Find errors and weaknesses in model	Reoptimization
Model validation	Demonstrate validity of final model	See Sec. 2.4
Final managerial decisions on resource allocations (the b_i)	Make appropriate division of organizational resources between activities under study and other important activities	Shadow prices
Evaluate estimates of model parameters	Determine crucial estimates that may affect optimal solution for further study	Sensitivity analysis
Evaluate trade-offs between model parameters	Determine best trade-off	Parametric linear programming

---

[1] The one requirement for using the dual simplex method here is that the *optimality test* still passes when applied to row 0 of the *revised* final tableau. If not, then still another algorithm called the *primal-dual method* can be used instead.

The big advantage of this **reoptimization technique** over resolving from scratch is that an optimal solution for the revised model probably is going to be *much* closer to the prior optimal solution than to an initial basic feasible solution constructed in the usual way for the simplex method. Therefore, assuming that the model revisions were modest, only a few iterations should be required to reoptimize instead of the hundreds or thousands that may be required when starting from scratch. In fact, the optimal solutions for the prior and revised models are frequently the same, in which case the reoptimization technique requires only one application of the optimality test and *no* iterations.

Shadow Prices

Recall (see Tables 3.2 and 3.3) that linear programming problems typically can be interpreted as allocating resources to activities, where the b_i represent the amounts of the respective resources being made available for the activities under consideration. In many cases, there may be some latitude in the amounts that will be made available. If so, the b_i used in the initial (validated) model actually may represent management's *tentative initial decision* on how much of the organization's resources will be provided to the activities considered in the model instead of to other important activities under the purview of management. From this broader perspective, some of the b_i can be increased in a revised model, but only if a sufficiently strong case can be made to management that this revision would be beneficial.

Consequently, information on the economic contribution of the resources to the measure of performance (Z) for the current study often would be extremely useful. The simplex method provides this information in the form of "shadow prices" for the respective resources.

> The **shadow price** for resource i (denoted by y_i^*) measures the *marginal value* of this resource, that is, the rate at which Z could be increased by (slightly) increasing the amount of this resource (b_i) being made available.[1] The simplex method identifies this shadow price by $y_i^* =$ coefficient of the ith slack variable in row 0 of the final simplex tableau.

To illustrate, for the Wyndor Glass Co. problem, the final tableau in Table 4.8 yields

$$y_1^* = 0 = \text{shadow price for resource 1,}$$

$$y_2^* = \tfrac{3}{2} = \text{shadow price for resource 2,}$$

$$y_3^* = 1 = \text{shadow price for resource 3,}$$

where these resources are the available production capacities of Plants 1, 2, and 3, respectively ($b_1 = 4$, $b_2 = 12$, and $b_3 = 18$). You can verify that these numbers are correct by checking in Figs. 3.2 and 3.3 that individually increasing each b_i by 1 indeed would increase the optimal value of Z by y_i^*. For example, Fig. 4.3 demonstrates this increase for resource 2 by reapplying the graphical procedure presented in Sec. 3.2. The optimal solution, (2, 6) with $Z = 36$, changes to $(\tfrac{5}{3}, \tfrac{13}{2})$ with $Z = 37\tfrac{1}{2}$

---

[1] The increase in b_i must be sufficiently small that the current set of basic variables remains optimal since this rate (marginal value) changes if the set of basic variables changes.

when b_2 is increased by 1 (from 12 to 13), so that

$$y_2^* = \Delta Z = 37\tfrac{1}{2} - 36 = \tfrac{3}{2}.$$

Figure 4.3 demonstrates that $y_2^* = \tfrac{3}{2}$ is the rate at which Z could be increased by increasing b_2 slightly. However, it also demonstrates the common phenomenon that this interpretation holds only for a small increase in b_2. Once b_2 is increased beyond 18, the optimal solution stays at $(0, 9)$ with no further increase in Z. (At that point, the set of basic variables in the optimal solution has changed, so a new final simplex tableau would be obtained with new shadow prices, including $y_2^* = 0$.)

Now note in Fig. 4.3 why $y_1^* = 0$. Because the constraint on resource 1, $x_1 \leq 4$, is *not binding* on the optimal solution, $(2, 6)$, there is a *surplus* of this resource. Therefore, increasing b_1 beyond 4 cannot yield a new optimal solution with a larger value of Z.

By contrast, the constraints on resources 2 and 3, $2x_2 \leq 12$ and $3x_1 + 2x_2 \leq 18$, are **binding constraints** (constraints that hold with equality at the optimal solution). Because the limited supply of these resources ($b_2 = 12$, $b_3 = 18$) *binds* Z from being increased further, they have *positive* shadow prices. Economists refer to such resources as *scarce goods*, whereas resources available in surplus (such as resource 1) are *free goods* (zero shadow price).

The kind of information provided by shadow prices clearly is valuable to management when it considers reallocations of resources within the organization. It also is very helpful when an increase in b_i can be achieved only by going outside the organization to purchase more of the resource in the marketplace. For example, suppose that Z represents *profit* and the unit profits of the activities (the c_j) include the costs (at regular prices) of all the resources consumed. Then a *positive* shadow price of y_i^* for resource i means that the total profit Z can be increased by y_i^* by purchasing one more unit of this resource at its regular price. Alternatively, if a *premium* price

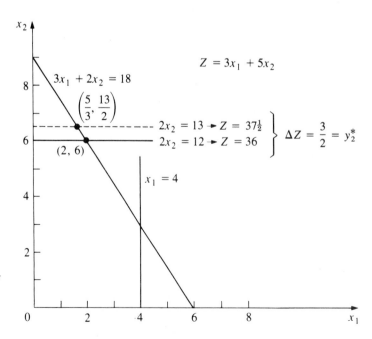

Figure 4.3
Illustration of shadow
price for resource 2
for Wyndor Glass
Co. problem.

must be paid for the resource in the marketplace, then y_i^* represents the *maximum* premium (excess over the regular price) that would be worth paying.

In the Wyndor Glass Co. problem, management has ruled out any expansion of the production capacity in the three plants at this time. Nevertheless, the available capacities allocated to the two new products can be increased by cutting back further on the current product line. The OR Department actually investigates this possibility as part of the sensitivity analysis study in Sec. 6.7.

The theoretical foundation for shadow prices is provided by the *duality theory* described in Chap. 6.

Sensitivity Analysis

When discussing the *certainty assumption* for linear programming at the end of Sec. 3.3, we pointed out that the values used for the model parameters (the a_{ij}, b_i, and c_j identified in Table 3.3) generally are just *estimates* of quantities whose true values will not become known until the linear programming study is implemented at some time in the future. The general purpose of sensitivity analysis is to identify the **sensitive parameters** (i.e., those that cannot be changed without changing the optimal solution), to try to estimate these parameters more closely, and then to select a solution that remains a good one over the range of likely values of the sensitive parameters.

How are the sensitive parameters identified? In the case of the b_i, you have just seen that this information is given by the shadow prices provided by the simplex method. In particular, if $y_i^* > 0$, then the optimal solution changes if b_i is changed, so b_i is a sensitive parameter. However, $y_i^* = 0$ implies that the optimal solution is not sensitive to at least small changes in b_i. Consequently, if the value used for b_i is an estimate of the amount of the resource that will be available (rather than a managerial decision), then the b_i that need to be estimated more closely are those with *positive* shadow prices—especially those with *large* shadow prices.

When there are just two variables, the sensitivity of the various parameters can be analyzed graphically. For example, in Fig. 4.3 (or Fig. 3.3), $c_1 = 3$ can be changed to any other value within the range from 0 to $7\frac{1}{2}$ without the optimal solution changing from (2, 6). (The reason is that any value of c_1 within this range keeps the slope of $Z = c_1x_1 + 5x_2$ between the slopes of the $2x_2 = 12$ and $3x_1 + 2x_2 = 18$ lines.) Similarly, if $c_2 = 5$ is the only parameter changed, it can have any value greater than 2 without affecting the optimal solution. Hence neither c_1 nor c_2 is a sensitive parameter.

The easiest way to analyze the sensitivity of each of the a_{ij} parameters graphically is to check if the corresponding constraint is *binding* on the optimal solution. Because $x_1 \leq 4$ is *not* a binding constraint, any sufficiently small change in its coefficients ($a_{11} = 1$, $a_{12} = 0$) is not going to change the optimal solution, so these are *not* sensitive parameters. On the other hand, both $2x_2 \leq 12$ and $3x_1 + 2x_2 \leq 18$ are *binding constraints,* so changing any one of their coefficients ($a_{21} = 0$, $a_{22} = 2$, $a_{31} = 3$, $a_{32} = 2$) is going to change the optimal solution, and therefore these are sensitive parameters.

Typically, greater attention is given to performing sensitivity analysis on the b_i and c_j parameters than on the a_{ij} parameters. On real problems with hundreds or thousands of constraints and variables, the effect of changing one a_{ij} is usually negligible, but changing one b_i or c_j can have real impact. Furthermore, in many cases, the a_{ij} values are determined by the *technology* being used (the a_{ij} are sometimes called *technological coefficients*), so there may be relatively little (or no) uncertainty about their final values. This is fortunate, because there are far more a_{ij} parameters than b_i and c_j parameters for large problems.

For problems with more than two variables, you cannot analyze the sensitivity of the parameters graphically as was just done for the Wyndor Glass Co. problem. However, you can extract the same kind of information from the simplex method. Getting this information requires using the *fundamental insight* described in Sec. 5.3 to deduce the changes that get carried along to the final simplex tableau as a result of changing the value of a parameter in the original model. The rest of the procedure is described in Sec. 6.6.

Parametric Linear Programming

Sensitivity analysis involves changing one parameter at a time in the original model to check its effect on the optimal solution. By contrast, **parametric linear programming** (or **parametric programming** for short) involves the systematic study of how the optimal solution changes as *many* of the parameters change *simultaneously* over some range. This study can provide a very useful extension of sensitivity analysis, e.g., to check the effect of "correlated" parameters that change together due to exogenous factors such as the state of the economy. However, a more important application is the investigation of *trade-offs* in parameter values. For example, if the c_j represent the unit profits of the respective activities, it may be possible to increase some of the c_j at the expense of decreasing others by an appropriate shifting of personnel and equipment among activities. Similarly, if the b_i represent the amounts of the respective resources being made available, it may be possible to increase some of the b_i by agreeing to accept decreases in some of the others.

In some applications, the main purpose of the study is to determine the most appropriate trade-off between two basic factors, such as *costs* and *benefits*. The usual approach is to express one of these factors in the objective function (e.g., minimize total cost) and incorporate the other into the constraints (e.g., benefits $\geq$ minimum acceptable level), as was done for the Nori & Leets Co. air pollution problem in Sec. 3.4. Parametric linear programming then enables systematic investigation of what happens when the initial tentative decision on the trade-off (e.g., the minimum acceptable level for the benefits) is changed by improving one factor at the expense of the other. This approach is illustrated by the case study in Sec. 8.5, where the two basic factors are the distance traveled by high school students and the degree of racial balance achieved in their schools.

The algorithmic technique for parametric linear programming is a natural extension of that for sensitivity analysis, so it too is based on the simplex method. The procedure is described in Sec. 9.3.

4.8 Computer Implementation

Computer codes for the simplex method now are widely available for essentially all modern computer systems. In fact, major computer manufacturers usually supply their customers with a rather sophisticated linear programming software package (Mathematical Programming System) that also includes many of the special procedures described in Chaps. 6, 7, and 9 (including the algorithmic techniques introduced in the preceding section). Other very good linear programming *systems* also have been developed by independent software development companies and service bureaus, and further progress continues to be made.

These production computer codes do not closely follow either the *algebraic form* or the *tabular form* of the simplex method presented in Secs. 4.3 and 4.4. These forms can be streamlined considerably for computer implementation. Therefore, the codes use instead a *matrix form* (usually called the *revised simplex method*) that is especially well suited for the computer. This form accomplishes exactly the same things as the algebraic or tabular form, but it does this while computing and storing *only* the numbers that are actually needed for the current iteration, and then it carries along the essential data in a more compact form. The *revised* simplex method is described in Sec. 5.2.

The available software packages are used routinely to solve surprisingly large linear programming problems. For example, a problem with 5,000 functional constraints and 10,000 variables usually can be solved in less than an hour on a mainframe computer of recent vintage.[1] Problems with several times this number of constraints and variables also have been successfully solved by the general simplex method. If the problem has some kind of special structure (as described in Chap. 7) that can be solved by a streamlined version of the simplex method, then even *much* larger sizes can sometimes be handled. For example, a problem with 100,000 functional constraints and 500,000 variables has been solved when all but about 1,000 of these constraints were of a special kind (*generalized upper bound constraints* discussed at the end of Sec. 7.5). Even this one is far from the largest when highly specialized types of linear programming problems (typically network flow problems) are considered.

Several factors affect how long it will take the general simplex method to solve a linear programming problem. The most important one is the *number of ordinary functional constraints*. In fact, computation time tends to be roughly proportional to the *cube* of this number, so that doubling this number may multiply the computation time by a factor of approximately 8. By contrast, the number of variables is a relatively minor factor.[2] Thus doubling the number of variables probably will not even double the computation time. A third factor of some importance is the *density* of the table of constraint coefficients (i.e., the *proportion* of the coefficients that are *not* zero), because this affects the computation time *per iteration*. One common rule of thumb for the *number of iterations* is that it tends to be roughly twice the number of functional constraints.

One difficulty in dealing with large linear programming problems is the tremendous amount of data involved. For example, a problem with just 1,000 functional constraints and variables would have 1 million constraint coefficients to be specified! Therefore, most experienced practitioners make extensive use of the computer for data-processing purposes both before and after applying the simplex method. Frequently, a **matrix generator program** will be written to convert the basic raw data into constraint coefficients in an appropriate format for the simplex method. The matrix

---

[1] On problems of this size, the computation time depends greatly upon the linear programming system being used because large savings can be achieved by using special techniques (e.g., *crashing techniques* for quickly finding an advanced initial basic feasible solution). When problems are resolved periodically after minor updating of the data, much time often is saved by using (or modifying) the last optimal solution to provide the initial basic solution for the new run.

[2] This statement assumes that the *revised* simplex method described in Sec. 5.2 is being used.

generator will do the arithmetic required in this conversion, repeat constraints of a recurring type, and fill in the zero coefficients (most of the coefficients usually are zeroes in large problems). It also should print out the key input data in an easily readable form so that they can be shown to various people for checking and correcting. Another useful function of a matrix generator is to scale the coefficients (by changing the units for the activities or resources) to approximately the same order of magnitude to avoid significant round-off error.

For many of the same reasons, it often is helpful to write an **output analyzer program** to convert the output of the simplex method into a useful form. An output analyzer (or **report writer**) has three major functions. Two of these are to compile and summarize relevant information for two of the *post-optimality analysis* tasks introduced in the preceding section, namely, *debugging the model* and *sensitivity analysis*. The third major purpose is to develop a well-organized report presenting the relevant information about the proposed solution in the vernacular of management.

4.9 New Developments

Two crucial events have been primarily responsible for the tremendous impact of linear programming in recent decades. One was the invention of the remarkably efficient simplex method in 1947 by George Dantzig.[1] The second crucial event was the *computer revolution* that makes it possible for the simplex method to solve such huge problems.

A Powerful New Algorithm

Now there has been a dramatic new development that promises to give further impetus to the impact of linear programming. In 1984, Narendra Karmarkar of AT&T Bell Laboratories published a landmark paper[2] announcing a new algorithm for solving huge linear programming problems. In contrast to the simplex method's approach of focusing on corner-point feasible solutions on the boundary of the feasible region, Karmarkar's algorithm is an **interior-point algorithm** that cuts through the interior of the feasible region to reach an optimal solution.

The initial claims were that this new algorithm can solve huge linear programming problems up to 50 times as fast as the simplex method. Not surprisingly, this dramatic announcement became front-page news in *The New York Times*, etc. However, because of proprietary considerations, no code was made available, and only sketchy details about the implementation of the algorithm were released, so it was not possible for others to check these claims. For the next four years, few additional details were forthcoming. Meanwhile, independent investigators were attempting to develop sophisticated computer implementations of the algorithm and reporting a variety of mixed results about comparisons with the simplex method. The entire operations research community was abuzz with excitement and controversy!

Then, in 1988, came another dramatic announcement. AT&T Bell Laboratories were releasing a powerful computer implementation of variants of Karmarkar's al-

---

[1] Since 1966, Professor Dantzig has been a colleague of the authors in the Department of Operations Research at Stanford University, and has continued to lead in the development of linear programming.

[2] Karmarkar, Narendra: "A New Polynomial-Time Algorithm for Linear Programming," *Combinatorica*, **4**: 373–395, 1984.

gorithm for commercial distribution. Called the AT&T KORBX Linear Programming System, it is a large-scale implementation of the algorithm on a parallel-vector mini-supercomputer (the Alliant FX8 supercomputer). Each installation of the entire system (including the dedicated computer) was available initially for approximately $8,900,000.

Coincident with this announcement, AT&T Bell Laboratories released substantial information on their computational experience with their system, including comparisons with a standard computer implementation of the simplex method (not necessarily the most efficient available for each problem run) called MINOS.[1] They reported successfully solving some very large problems with many thousand, or even many tens of thousands of functional constraints, including some too large for implementation by the simplex method. For some highly specialized problems, the sizes are even larger. For some of the larger problems, improvements in running time over the version of the simplex method used were great—factors of 10 to 50 were common.

So how does Karmarkar's interior-point algorithm compare in efficiency with the simplex method? This is a complicated question, because it requires a comparison for many different sizes and classes of linear programming problems. No definitive answer can be given at this time. We still need a disinterested party (or two competing parties) to conduct a comprehensive program of comparative testing with the most sophisticated computer implementations available for the two algorithms. Until this happens, the jury will remain out.

This much can be said now. Both algorithms are here to stay. We anticipate that they will play vital complementary roles in linear programming throughout your career. To define these roles, we need to point out a key advantage of each algorithm. The simplex method is ideally suited for *post-optimality analysis* as described in Sec. 4.7, whereas the interior-point algorithm cannot perform this analysis efficiently (except for obtaining shadow prices). On the other hand, the key advantage of the interior-point algorithm is that its rate of growth of computation time as the problem size grows frequently is somewhat less than for the simplex method. According to current evidence, the high setup time of the interior-point algorithm prevents it from being strongly competitive with the simplex method for relatively small problems (tens or perhaps hundreds of functional constraints). However, in at least some cases, it tends to gain ever-increasing superiority on larger and larger problems (thousands or tens of thousands of functional constraints). Experience may differ with different types of linear programming problems.

One other meaningful way of comparing the two algorithms is to examine their theoretical properties regarding *computational complexity*. Karmarkar has proven that the original version of his algorithm is a **polynomial time algorithm**; i.e., the time required to solve *any* linear programming problem can be bounded above by a polynomial function of the size of the problem. Pathological counterexamples have been constructed to demonstrate that the simplex method does not possess this property, so

---

[1] MINOS was developed largely in the Systems Optimization Laboratory of the Department of Operations Research of Stanford University. Perhaps better known as an optimizer for *nonlinear* programming, it is a state-of-the-art FORTRAN-based system that is widely used throughout the world. The performance of MINOS is controlled by a number of system parameters or "options" that enable fine-tuning for the characteristics of the particular problem being run. However, each option has a default value that should be appropriate for most problems, and these frequently were the values used by AT&T Bell Laboratories. We have been informed that some of the default values are different in a later version of MINOS to be more suitable for particularly challenging problems.

it is an **exponential time algorithm** (i.e., the required time can only be bounded above by an exponential function of problem size). This difference in *worst-case performance* is noteworthy. However, it tells us nothing about their comparison in *average performance* on real problems, which is the more crucial issue.

Based on all evidence now available, we currently anticipate the following roles for the two algorithms as we approach and enter the twenty-first century. The simplex method should continue to be the standard algorithm for the routine use of linear programming. However, Karmarkar's interior-point algorithm (or some of its variants and refinements) should gradually gain widespread use by heavy-duty users of linear programming dealing with relatively large problems. This algorithm converges toward an optimal solution without ever literally reaching it, so the procedure terminates with an arbitrarily close approximation of the desired solution. Therefore, when an interior-point algorithm is used to closely approximate an optimal solution, the solution obtained probably will be converted to the nearest basic feasible solution to serve as the *initial* solution for the simplex method to finish solving to optimality and then to conduct post-optimality analysis. The two algorithms thereby would become a *package* for dealing with some large problems, whereas the simplex method would be used by itself for routine cases.

Consequently, we believe that you probably will be involved with frequent use of the simplex method during your career, whereas you are less likely to use the interior-point algorithm as well. If you do use both algorithms, the computer system containing the interior-point algorithm will be a "black box" for quickly generating an optimal solution for the current model, so there is little need to know much about this algorithm. The "hands-on" work then comes with applying the simplex method for post-optimality analysis, so it is important that you be familiar with this algorithm. Consequently, our algorithmic focus in this part of the book is on the simplex method alone. We then give an elementary introduction to the nature of interior-point algorithms in Sec. 9.4, while foregoing mathematical details that are beyond the level of this book.

Improved Implementations of the Simplex Method

One of the by-products of the development of Karmarkar's interior-point algorithm has been a major renewal of efforts to improve the efficiency of computer implementations of the simplex method. This effort has been led by companies with a strong commercial interest in the simplex method. For example, IBM distributes MPSX/370, a widely used system based on the simplex method for use on IBM mainframes.

At the IBM Thomas J. Watson Research Center in Yorktown Heights, New York, a new experimental code, YKTLP, is being developed for the implementation of the simplex method on IBM mainframes, particularly the model 390 with Vector Facility. One feature of this code is the use of vector hardware to simultaneously compute the new coefficients of nonbasic variables in row 0 of the simplex tableau for the current iteration. When there are thousands of nonbasic variables, the benefit from such vector processing can be very substantial. Other opportunities to exploit vector processing to speed up each iteration also are being explored. This experimental code is already showing big improvements—factors of 10 to 50—over the standard MINOS implementation of the simplex method (the one used for the comparative testing by AT&T Bell Laboratories).

Linear Programming on Microcomputers

This is an exciting time to be introduced to linear programming for still another reason. We are now seeing an explosion in the capability of doing linear programming on *microcomputers*. You now can solve problems on a personal computer that only recently required the use of a mainframe computer, with all its inconveniences and expense.

There are now dozens of companies marketing linear programming software for microcomputers based on the simplex method. Many of these companies are located in the United States, but a considerable number are scattered around the world. The early emphasis was on educational software, but now many of the packages are suitable for commercial applications on problems of rather substantial size.

For example, the popular package LINDO can handle problems with under 2,000 functional constraints and 4,000 variables, provided the number of *nonzero* coefficients in the functional constraints does not exceed 32,000. Solving large problems usually requires additional memory, and the larger versions of some programs (such as LINDO) require a math coprocessor.

We pointed out in Sec. 4.8 that dealing with problems of such size normally requires making extensive use of the computer for data-processing purposes in constructing the model, etc. Mathematical modeling languages have now been developed for personal computers. For example, the GAMS/MINOS package is a combination of two well-known mainframe programs now available for IBM personal computers that offers a powerful algebraic modeling language for generating the constraints of the model automatically. (Yes, this is the same MINOS that is being used for comparative testing by the research laboratories of AT&T and IBM.) The English package XPRESS-LP offers a similar capability. MPL (Mathematical Programming Language) is a modeling system from Maximal Software in Iceland that is used as a front end for other linear programming packages.

The convenient data entry and editing features of spreadsheets also are very helpful in constructing linear programming models. Many of the current packages are spreadsheet-compatible, and several (e.g., VINO, What's Best?, and XA) actually perform the optimization within the spreadsheet program.

Some of the linear programming packages include extensions to other areas of mathematical programming. For example, LINDO includes integer programming (Chap. 13) and GAMS/MINOS includes nonlinear programming (Chap. 14).

Nearly all of the linear programming packages developed in the late 1980s were for IBM personal computers and IBM-compatibles. However, LINDO and Turbo-Simplex (from Maximal Software) also are available for the Macintosh computer. Macintosh-type computers have an architecture better suited for the graphics-based nature of many linear programming applications (including those involving networks), and we anticipate many more packages becoming available for the Macintosh in the future.

Your OR COURSEWARE (available for the first time in this edition) introduces you to the use of microcomputers for linear programming and other areas of operations research. However, this tutorial software is designed strictly for educational purposes while you deal with the tiny homework problems in this book. Later, when you are dealing with ''real'' problems, you will want to use a more powerful software package.

The above information about linear programming on microcomputers is very much up to date at this writing, but we need to point out its almost instant obsolescence. The explosion in the capability of doing linear programming on microcomputers is undoubtedly going to continue through the 1990s and beyond.

4.10 Conclusions

The simplex method is an efficient and reliable algorithm for solving linear programming problems. It also provides the basis for performing the various parts of post-optimality analysis very efficiently.

Although it has a useful geometric interpretation, the simplex method is an algebraic procedure. At each iteration, it moves from the current basic feasible solution to a better adjacent basic feasible solution by choosing both an entering basic variable and a leaving basic variable, and then using Gaussian elimination to solve a system of linear equations. When the current solution has no adjacent basic feasible solution that is better, the current solution is optimal and the algorithm stops.

We presented the full algebraic form of the simplex method to convey its logic, and then we streamlined the method to a more convenient tabular form. To set up for starting the simplex method, it is sometimes necessary to use artificial variables to obtain an initial basic feasible solution for a revised problem. If so, either the Big M method or the two-phase method is used to ensure that the simplex method obtains an optimal solution for the original problem.

Microcomputer software packages based on the simplex method now are widely available for dealing with problems of modest size. Mainframe programs are routinely used to solve and analyze problems with many hundreds or even thousands of functional constraints and variables.

Karmarkar's interior-point algorithm provides a powerful new tool for solving very large problems.

SELECTED REFERENCES

1. Bradley, Stephen P., Arnoldo C. Hax, and Thomas L. Magnanti: *Applied Mathematical Programming*, Addison-Wesley, Reading, Mass., 1977.
2. Chvátal, Vašek: *Linear Programming*, Freeman, San Francisco, 1983.
3. Dantzig, George B.: *Linear Programming and Extensions,* Princeton University Press, Princeton, N.J., 1963.
4. Gass, Saul I.: *Linear Programming: Methods and Applications*, 5th ed., McGraw-Hill, New York, 1985.
5. Orchard-Hays, William: *Advanced Linear Programming Computing Techniques*, McGraw-Hill, New York, 1968.
6. Sharma, Ramesh, ''The State of the Art of Linear Programming on Personal Computers,'' *Interfaces*, **18**(4): 49–58, July–August 1988.

PROBLEMS

(For all of the following problems, when you are asked to solve by the simplex method you can use the *interactive* routine in your OR COURSEWARE and then print out your work.)

1. Consider the linear programming model formulated for Prob. 1 of Chap. 3.

(*a*) Identify all the *corner-point feasible solutions* for this model.

(b) Solve by the simplex method in algebraic form.

(c) Solve by the simplex method in tabular form.

(d) Use the graphical solution to perform sensitivity analysis on this model; i.e., identify the *sensitive* parameters that cannot be changed without changing the optimal solution.

2. Consider the following problem.

$$\text{Maximize} \quad Z = x_1 + 2x_2,$$

subject to

$$x_1 + 3x_2 \leq 8 \quad \text{(resource 1)}$$

$$x_1 + x_2 \leq 4 \quad \text{(resource 2)}$$

and

$$x_1 \geq 0, \quad x_2 \geq 0.$$

(a) Solve this problem graphically. Identify all the *corner-point feasible solutions* for this model.

(b) Solve by the simplex method in algebraic form.

(c) Solve by the simplex method in tabular form.

(d) Identify the shadow prices for the resources from the final tableau for the simplex method. Demonstrate graphically that these shadow prices are the correct ones.

(e) Use the graphical solution to perform sensitivity analysis on this model; i.e., identify the *sensitive* parameters that cannot be changed without changing the optimal solution.

3. Follow the instructions of Prob. 2 for the following problem.

$$\text{Maximize} \quad Z = 3x_1 + 2x_2,$$

subject to

$$x_1 \qquad \leq 4 \quad \text{(resource 1)}$$

$$x_1 + 3x_2 \leq 15 \quad \text{(resource 2)}$$

$$2x_1 + x_2 \leq 10 \quad \text{(resource 3)}$$

and

$$x_1 \geq 0, \quad x_2 \geq 0.$$

4.* Consider the following problem.

$$\text{Maximize} \quad Z = 4x_1 + 3x_2 + 6x_3,$$

subject to

$$3x_1 + x_2 + 3x_3 \leq 30$$

$$2x_1 + 2x_2 + 3x_3 \leq 40$$

and

$$x_1 \geq 0, \quad x_2 \geq 0, \quad x_3 \geq 0.$$

(a) Solve by the simplex method in algebraic form.

(b) Solve by the simplex method in tabular form.

5. Use the simplex method to solve the following problem.

$$\text{Maximize} \quad Z = 2x_1 - x_2 + x_3,$$

subject to

$$3x_1 + x_2 + x_3 \leq 6$$

$$x_1 - x_2 + 2x_3 \leq 1$$

$$x_1 + x_2 - x_3 \leq 2$$

and

$$x_1 \geq 0, \quad x_2 \geq 0, \quad x_3 \geq 0.$$

6. Use the simplex method to solve the following problem.

$$\text{Maximize} \quad Z = -x_1 + x_2 + 2x_3,$$

subject to

$$x_1 + 2x_2 - x_3 \le 20$$

$$-2x_1 + 4x_2 + 2x_3 \le 60$$

$$2x_1 + 3x_2 + x_3 \le 50$$

and

$$x_1 \ge 0, \qquad x_2 \ge 0, \qquad x_3 \ge 0.$$

7. Consider the following problem.

$$\text{Maximize} \quad Z = 2x_1 - 2x_2 + 3x_3,$$

subject to

$$-x_1 + x_2 + x_3 \le 4 \quad \text{(resource 1)}$$

$$2x_1 - x_2 + x_3 \le 2 \quad \text{(resource 2)}$$

$$x_1 + x_2 + 3x_3 \le 12 \quad \text{(resource 3)}$$

and

$$x_1 \ge 0, \qquad x_2 \ge 0, \qquad x_3 \ge 0.$$

(a) Solve by the simplex method.
(b) Identify the shadow prices for the three resources and describe their significance.

8. Consider the following problem.

$$\text{Maximize} \quad Z = 2x_1 + 4x_2 - x_3,$$

subject to

$$3x_2 - x_3 \le 30 \quad \text{(resource 1)}$$

$$2x_1 - x_2 + x_3 \le 10 \quad \text{(resource 2)}$$

$$4x_1 + 2x_2 - 2x_3 \le 40 \quad \text{(resource 3)}$$

and

$$x_1 \ge 0, \qquad x_2 \ge 0, \qquad x_3 \ge 0.$$

(a) Solve by the simplex method.
(b) Identify the shadow prices for the three resources and describe their significance.

9. Consider the following problem.

$$\text{Maximize} \quad Z = 5x_1 + 4x_2 - x_3 + 3x_4,$$

subject to

$$3x_1 + 2x_2 - 3x_3 + x_4 \le 24 \quad \text{(resource 1)}$$

$$3x_1 + 3x_2 + x_3 + 3x_4 \le 36 \quad \text{(resource 2)}$$

and

$$x_1 \ge 0, \qquad x_2 \ge 0, \qquad x_3 \ge 0, \qquad x_4 \ge 0.$$

(a) Solve by the simplex method.
(b) Identify the shadow prices for the two resources and describe their significance.

10. Label each of the following statements as True or False, and then justify your answer by referring to specific statements (with page citations) in the chapter.

(a) The simplex method's rule for choosing the entering basic variable is used because it always leads to the *best* adjacent basic feasible solution (largest Z).

(b) The simplex method's rule for choosing the leaving basic variable is used because making another choice normally would yield a basic solution that is not feasible.

(c) When a linear programming model has an equality constraint, an artificial variable is introduced into this constraint in order to start the simplex method with an obvious initial basic solution that is feasible for the original model.

11. Consider the following problem.

$$\text{Maximize} \quad Z = 2x_1 + 4x_2 + 3x_3,$$

subject to

$$x_1 + 3x_2 + 2x_3 \leq 30$$

$$x_1 + x_2 + x_3 \leq 24$$

$$3x_1 + 5x_2 + 3x_3 \leq 60$$

and

$$x_1 \geq 0, \quad x_2 \geq 0, \quad x_3 \geq 0.$$

You are given the information that $x_1 > 0$, $x_2 = 0$, and $x_3 > 0$ in the optimal solution.

(a) Describe how one can use this information in order to *adapt* the simplex method to solve this problem in the minimum possible number of iterations (when starting from the usual initial feasible solution). Do *not* actually perform any iterations.

(b) Use the procedure developed in part (a) to solve this problem.

12. Consider the following problem.

$$\text{Maximize} \quad Z = 5x_1 + x_2 + 3x_3 + 4x_4,$$

subject to

$$x_1 - 2x_2 + 4x_3 + 3x_4 \leq 20$$

$$-4x_1 + 6x_2 + 5x_3 - 4x_4 \leq 40$$

$$2x_1 - 3x_2 + 3x_3 + 8x_4 \leq 50$$

and

$$x_1 \geq 0, \quad x_2 \geq 0, \quad x_3 \geq 0, \quad x_4 \geq 0.$$

Use the simplex method to demonstrate that Z is unbounded.

13. For this problem, we will use vector notation, $\mathbf{x} = (x_1, x_2, \ldots, x_n)$, to represent solutions more compactly. Consider N such solutions, $\mathbf{x}^{(1)}, \mathbf{x}^{(2)}, \ldots, \mathbf{x}^{(N)}$. A *weighted average* of these N solutions is defined to be a solution $\mathbf{x}$ such that

$$\mathbf{x} = \sum_{k=1}^{N} \alpha_k \mathbf{x}^{(k)},$$

where the weights $\alpha_1, \alpha_2, \ldots, \alpha_N$ are nonnegative and sum to 1. If the feasible region is bounded, then every feasible solution can be expressed as a weighted average of some of the corner-point feasible solutions (perhaps in more than one way). Similarly, after solutions are augmented with slack variables, every feasible solution can be expressed as a weighted average of some of the basic feasible solutions.

(a) Show that *any* weighted average of *any* set of *feasible* solutions must be a feasible solution (so that any weighted average of *corner-point* feasible solutions must be feasible).

(b) Use the result quoted in part (a) to show that *any* weighted average of basic feasible solutions must be a feasible solution.

14. Using the facts given in Prob. 13, show that the following statements must be true for any linear programming problem that has a bounded feasible region and multiple optimal solutions:

(a) Every weighted average of the optimal basic feasible solutions must be optimal.

(b) No *other* feasible solution can be optimal.

15. Consider the following problem.

$$\text{Maximize} \quad Z = x_1 + x_2 + x_3 + x_4,$$

subject to

$$x_1 + x_2 \leq 3$$

$$x_3 + x_4 \leq 2$$

and $\qquad x_j \geq 0, \qquad$ for $j = 1, 2, 3, 4.$

Use the simplex method to find *all* the optimal basic feasible solutions.

16.* Consider the following problem.

$$\text{Maximize} \quad Z = 2x_1 + 3x_2,$$

subject to

$$x_1 + 2x_2 \leq 4$$

$$x_1 + x_2 = 3$$

and

$$x_1 \geq 0, \qquad x_2 \geq 0.$$

(a) Solve this problem graphically.
(b) Using the Big *M* method, construct the complete first simplex tableau for the simplex method and identify the corresponding initial (artificial) basic feasible solution. Also identify the initial entering basic variable and the leaving basic variable.
(c) Solve by the simplex method.

17. Consider the following problem.

$$\text{Minimize} \quad Z = 3x_1 + 2x_2,$$

subject to

$$2x_1 + x_2 \geq 10$$

$$-3x_1 + 2x_2 \leq 6$$

$$x_1 + x_2 \geq 6$$

and

$$x_1 \geq 0, \qquad x_2 \geq 0.$$

(a) Solve this problem graphically.
(b) Using the Big *M* method, construct the complete first simplex tableau for the simplex method and identify the corresponding initial (artificial) basic feasible solution. Also identify the initial entering basic variable and the leaving basic variable.
(c) Solve by the simplex method.

18. Consider the following problem.

$$\text{Maximize} \quad Z = 2x_1 + 5x_2 + 3x_3,$$

subject to

$$x_1 - 2x_2 \qquad \geq 20$$

$$2x_1 + 4x_2 + x_3 = 50$$

and

$$x_1 \geq 0, \qquad x_2 \geq 0, \qquad x_3 \geq 0.$$

(a) Using the Big *M* method, construct the complete first simplex tableau for the simplex method and identify the corresponding initial (artificial) basic feasible solution. Also identify the initial entering basic variable and the leaving basic variable.
(b) Solve by the simplex method.
(c) Using the two-phase method, construct the complete first simplex tableau for Phase 1 and identify the corresponding initial (artificial) basic feasible solution. Also identify the initial entering basic variable and the leaving basic variable.
(d) Perform Phase 1.
(e) Construct the complete first simplex tableau for Phase 2.
(f) Perform Phase 2.
(g) Compare the sequence of basic feasible solutions obtained in part (b) with that in parts (d) and (f). Which of these solutions are *artificial* basic feasible solutions for just a revised problem and which are actually feasible for the *real* problem?

19. For the Big *M* method, explain why the simplex method never would choose an artificial variable to be an entering basic variable once all of the artificial variables are nonbasic.

20. Consider the following problem.

$$\text{Minimize} \quad Z = 2x_1 + x_2 + 3x_3,$$

subject to

$$5x_1 + 2x_2 + 7x_3 = 420$$

$$3x_1 + 2x_2 + 5x_3 \geq 280$$

and

$$x_1 \geq 0, \quad x_2 \geq 0, \quad x_3 \geq 0.$$

Using the two-phase method, solve by the simplex method.

21.* Consider the following problem.

$$\text{Maximize} \quad Z = -x_1 + 4x_2,$$

subject to

$$-3x_1 + x_2 \leq 6$$

$$x_1 + 2x_2 \leq 4$$

$$x_2 \geq -3$$

(no lower bound constraint for x_1).
(a) Solve this problem graphically.
(b) Reformulate this problem so that it has only two functional constraints and all variables have nonnegativity constraints.
(c) Solve by the simplex method.

22. Consider the following problem.

$$\text{Maximize} \quad Z = -x_1 + 2x_2 + x_3,$$

subject to

$$3x_2 + x_3 \leq 120$$

$$x_1 - x_2 - 4x_3 \leq 80$$

$$-3x_1 + x_2 + 2x_3 \leq 100$$

(no nonnegativity constraints).
(a) Reformulate this problem so all variables have nonnegativity constraints.
(b) Solve by the simplex method.

23. Consider the following problem.

$$\text{Minimize} \quad Z = 3x_1 + 2x_2 + 4x_3,$$

subject to

$$2x_1 + x_2 + 3x_3 = 60$$

$$3x_1 + 3x_2 + 5x_3 \geq 120$$

and

$$x_1 \geq 0, \quad x_2 \geq 0, \quad x_3 \geq 0.$$

(a) Using the Big M method, solve by the simplex method.
(b) Using the two-phase method, solve by the simplex method.
(c) Compare the sequence of basic feasible solutions obtained in parts (a) and (b). Which of these solutions are *artificial* basic feasible solutions for just a revised problem and which are actually feasible for the *real* problem?

24. This chapter has described the simplex method as applied to linear programming problems where the objective function is to be *maximized*. Section 4.6 then described how to convert a *minimization* problem into an equivalent maximization problem for applying the simplex method. Another option with minimization problems is to make a few modifications in the instructions for the simplex method given in the chapter in order to apply the algorithm directly.
(a) Describe what these modifications would need to be.

(b) Using the Big M method, apply the modified algorithm developed in part (a) to solve the following problem directly.

$$\text{Minimize} \quad Z = 3x_1 + 8x_2 + 5x_3,$$

subject to

$$3x_2 + 4x_3 \geq 70$$

$$3x_1 + 5x_2 + 2x_3 \geq 70$$

and

$$x_1 \geq 0, \quad x_2 \geq 0, \quad x_3 \geq 0.$$

25. Consider the following problem.

$$\text{Maximize} \quad Z = 4x_1 + 2x_2 + 3x_3 + 5x_4,$$

subject to

$$2x_1 + 3x_2 + 4x_3 + 2x_4 = 300$$

$$8x_1 + x_2 + x_3 + 5x_4 = 300$$

and

$$x_j \geq 0, \quad \text{for } j = 1, 2, 3, 4.$$

(a) Using the Big M method, construct the complete first simplex tableau for the simplex method and identify the corresponding initial (artificial) basic feasible solution. Also identify the initial entering basic variable and the leaving basic variable.

(b) Solve by the simplex method.

(c) Using the two-phase method, construct the complete first simplex tableau for Phase 1 and identify the corresponding initial (artificial) basic feasible solution. Also identify the initial entering basic variable and the leaving basic variable.

(d) Perform Phase 1.

(e) Construct the complete first simplex tableau for Phase 2.

(f) Perform Phase 2.

(g) Compare the sequence of the basic feasible solutions obtained in part (b) with that in parts (d) and (f). Which of these solutions are *artificial* basic feasible solutions for just a revised problem and which are actually feasible for the *real* problem?

26.* Consider the following problem.

$$\text{Minimize} \quad Z = 2x_1 + 3x_2 + x_3,$$

subject to

$$x_1 + 4x_2 + 2x_3 \geq 8$$

$$3x_1 + 2x_2 \qquad \geq 6$$

and

$$x_1 \geq 0, \quad x_2 \geq 0, \quad x_3 \geq 0.$$

(a) Reformulate this problem to fit *our standard form* for a linear programming model presented in Sec. 3.2.

(b) Using the Big M method, solve by the simplex method.

(c) Using the two-phase method, solve by the simplex method.

(d) Compare the sequence of basic feasible solutions obtained in parts (b) and (c). Which of these solutions are *artificial* basic feasible solutions for just a revised problem and which are actually feasible for the *real* problem?

27. Consider the following problem.

$$\text{Maximize} \quad Z = -2x_1 + x_2 - 4x_3 + 3x_4,$$

subject to

$$x_1 + x_2 + 3x_3 + 2x_4 \leq 4$$

$$x_1 \qquad - x_3 + x_4 \geq -1$$

$$2x_1 + x_2 \qquad \leq 2$$

$$x_1 + 2x_2 + x_3 + 2x_4 = 2$$

and $\qquad x_2 \geq 0, \qquad x_3 \geq 0, \qquad x_4 \geq 0$

(no nonnegativity constraint for x_1).

(a) Reformulate this problem (except for the equality constraint) to fit *our standard form* for a linear programming model presented in Sec. 3.2.

(b) Using the Big *M* method, construct the complete first simplex tableau for the simplex method and identify the corresponding initial (artificial) basic feasible solution. Also identify the initial entering basic variable and the leaving basic variable.

(c) Using the two-phase method, construct row 0 of the first simplex tableau for Phase 1.

(d) Use the *automatic* routine for the simplex method in your OR COURSEWARE to solve this problem.

28. Consider the following problem.

$$\text{Maximize} \qquad Z = 4x_1 + 5x_2 + 3x_3,$$

subject to

$$x_1 + x_2 + 2x_3 \geq 20$$

$$15x_1 + 6x_2 - 5x_3 \leq 50$$

$$x_1 + 3x_2 + 5x_3 \leq 30$$

and $\qquad x_1 \geq 0, \qquad x_2 \geq 0, \qquad x_3 \geq 0.$

Use the simplex method to demonstrate that this problem does not possess any feasible solutions.

29. Consider Prob. 20 of Chap. 8. The linear programming model for this problem has more than 5,000 functional constraints and more than 150,000 variables.

(a) There are more than 750,000,000 coefficients for these constraints, which creates a storage problem for a computer solution of the model. Considering that more than 99 percent of these coefficients are zeroes, recommend a way to alleviate this problem.

(b) Since the number of *nonzero* coefficients is well over 100,000, manually inputting these data into the computer would be excessively time-consuming. Considering that the number of items of basic raw data is much smaller, recommend a way to alleviate this problem.

5

The Theory of the Simplex Method

Chapter 4 introduced the basic mechanics of the simplex method. Now we shall delve a little deeper into this algorithm by examining some of its underlying theory. The first section develops the general geometric and algebraic properties that form the foundation of the simplex method. We then describe the *matrix form* of the simplex method (called the *revised simplex method*), which streamlines the procedure considerably for computer implementation. Next, we present a fundamental insight about a property of the simplex method that enables us to deduce how changes that are made in the original model get carried along to the final simplex tableau. This insight will provide the key to the important topics of Chap. 6 (duality theory and sensitivity analysis).

5.1 Foundations of the Simplex Method

Section 4.1 introduced *corner-point feasible solutions* and the key role they play in the simplex method. These geometric concepts were related to the algebra of the simplex method in Sec. 4.2. However, all of this was done in the context of the

106

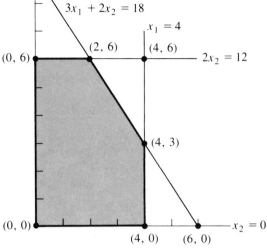

Figure 5.1 Constraint boundaries, constraint boundary equations, and corner-point solutions for the Wyndor Glass Co. problem.

Wyndor Glass Co. problem, which has only *two variables* and so has a straightforward geometric interpretation. How do these concepts generalize to higher dimensions when we deal with larger problems? We address this question in this section.

We begin by introducing some basic terminology for any linear programming problem with n variables (before introducing slack and artificial variables for initializing the simplex method). While we are doing this, you should find it helpful to refer to Fig. 5.1 to interpret these definitions in two dimensions ($n = 2$).

Terminology

It may seem intuitive that optimal solutions for any linear programming problem must lie on the boundary of the feasible region, and this is in fact a general property. Because boundary is a geometric concept, our initial definitions clarify how the boundary of the feasible region is identified algebraically.

> The **constraint boundary equation** for any constraint is obtained by replacing its ≤, =, or ≥ sign by an = sign.

Consequently, the form of a constraint boundary equation is $a_{i1}x_1 + a_{i2}x_2 + \cdots + a_{in}x_n = b_i$ for functional constraints and $x_j = 0$ for nonnegativity constraints. Each such equation defines a "flat" geometric shape (called a **hyperplane**) in n-dimensional space, analogous to the line in two-dimensional space and the plane in three-dimensional space. This hyperplane forms the **constraint boundary** for the corresponding constraint, because any points lying on one side of the constraint boundary violate the constraint whereas points on the constraint boundary satisfy the constraint. (Points on the *other* side also satisfy the constraint if it is an inequality constraint.)

For example, the Wyndor Glass Co. problem has five constraints (three functional constraints and two nonnegativity constraints), so it has the five *constraint boundary equations* shown in Fig. 5.1. Because $n = 2$, the *hyperplanes* defined by these constraint boundary equations are simply lines. Therefore, the *constraint boundaries* for the five constraints are the five lines shown in Fig. 5.1.

> The **boundary** of the feasible region contains just those feasible solutions that satisfy one or more of the constraint boundary equations.

Geometrically, any point on the boundary of the feasible region lies on one or more of the hyperplanes defined by the respective constraint boundary equations. Thus, in Fig. 5.1, the boundary consists of the five darker line segments.

Next, we give a general definition of *corner-point feasible solution* in n-dimensional space.

> A **corner-point feasible solution** is a feasible solution that does not lie on *any* line segment[1] connecting two *other* feasible solutions.

As this definition implies, a feasible solution that *does* lie on a line segment connecting two other feasible solutions is *not* a corner-point feasible solution. To illustrate when $n = 2$, consider Fig. 5.1. The point (2, 3) is *not* a corner-point feasible solution, because it lies on various such line segments, e.g., the line segment connecting (0, 3) and (4, 3). Similarly, (0, 3) is *not* a corner-point feasible solution, because it lies on the line segment connecting (0, 0) and (0, 6). However, (0, 0) *is* a corner-point feasible solution, because it is impossible to find two *other* feasible solutions that lie on completely opposite sides of (0, 0). (Try it.)

When the number of decision variables n is greater than 2 or 3, this definition for *corner-point feasible solution* is not a very convenient one for identifying such solutions. Therefore, it will prove most helpful to interpret these solutions algebraically. For the Wyndor Glass Co. example, each corner-point feasible solution in Fig. 5.1 lies at the intersection of two ($n = 2$) constraint lines; i.e., it is the *simultaneous solution* of a system of two constraint boundary equations. This situation is summarized in Table 5.1, where **defining equations** refer to the constraint boundary equations that yield (define) the indicated corner-point feasible solution. Similarly, for any linear programming problem, each corner-point feasible solution lies at the intersection of n constraint boundaries; i.e., it is the *simultaneous solution* of a system of n constraint boundary equations.

However, this is not to say that *every* set of n constraint boundary equations chosen from among the $(n + m)$ constraints (n nonnegativity and m functional constraints) yields a corner-point feasible solution. In particular, the simultaneous solution of such a system of equations might violate one or more of the other m constraints not chosen, in which case it is a corner-point *infeasible* solution. The example has three such solutions, as summarized in Table 5.2. (Check to see why they are infeasible.)

Furthermore, a system of n constraint boundary equations might have no solution at all. This occurs twice in the example, with the pairs of equations (1) $x_1 = 0$ and $x_1 = 4$ and (2) $x_2 = 0$ and $2x_2 = 12$. Such systems are of no interest to us.

---

[1] A formal definition of line segment is given in Appendix 1.

Table 5.1 Defining Equations for
Each Corner-Point Feasible Solution
for Wyndor Glass Co. Problem

Corner-Point Feasible Solution	Defining Equations
$(0, 0)$	$x_1 = 0$ $x_2 = 0$
$(0, 6)$	$x_1 = 0$ $2x_2 = 12$
$(2, 6)$	$2x_2 = 12$ $3x_1 + 2x_2 = 18$
$(4, 3)$	$3x_1 + 2x_2 = 18$ $x_1 = 4$
$(4, 0)$	$x_1 = 4$ $x_2 = 0$

The final possibility (which never occurs in the example) is that a system of n constraint boundary equations has multiple solutions because of redundant equations. You need not be concerned with this case either, because the simplex method circumvents its difficulties.

To summarize for the example, with five constraints and two variables, there are 10 pairs of constraint boundary equations. Five of these pairs became defining equations for corner-point feasible solutions (Table 5.1), three became defining equations for corner-point infeasible solutions (Table 5.2), and each of the final two pairs had no solution.

Adjacent Corner-Point Feasible Solutions

We now will focus on adjacent corner-point feasible solutions and their role in solving linear programming problems. Recall from Chap. 4 that, when ignoring slack and artificial variables, each iteration of the simplex method moves from the current corner-point feasible solution to an *adjacent* one. What is the *path* followed in this process? What really is meant by *adjacent* corner-point feasible solution? We first address these questions from a geometric viewpoint, and then turn to algebraic interpretations.

Table 5.2 Defining Equations for
Each Corner-Point Infeasible Solution
for Wyndor Glass Co. Problem

Corner-Point Infeasible Solution	Defining Equations
$(0, 9)$	$x_1 = 0$ $3x_1 + 2x_2 = 18$
$(4, 6)$	$2x_2 = 12$ $x_1 = 4$
$(6, 0)$	$3x_1 + 2x_2 = 18$ $x_2 = 0$

These questions are easy to answer when $n = 2$. In this case, the *boundary* of the feasible region consists of several connected *line segments* forming a *polygon*, as shown in Fig. 5.1 by the five darker line segments. These line segments are referred to as *edges* of the feasible region. Emanating out of each corner-point feasible solution are *two* such edges leading to an adjacent corner-point feasible solution at the other end. (Note in Fig. 5.1 how each corner-point feasible solution has two adjacent ones.) The path followed in an iteration is to move along one of these edges from one end to the other. In Fig. 5.1, the first iteration involves moving along the edge from $(0, 0)$ to $(0, 6)$, and then the next iteration moves along the edge from $(0, 6)$ to $(2, 6)$. As Table 5.1 illustrates, each of these moves to an adjacent corner-point feasible solution involves just one change in the set of defining equations (constraint boundaries on which the solution lies).

When $n = 3$, the answers are slightly more complicated. To help you visualize what is going on, Fig. 5.2 shows a three-dimensional drawing of a typical feasible region when $n = 3$, where the dots are the corner-point feasible solutions. This feasible region is a *polyhedron* rather than the polygon we had with $n = 2$ (Fig. 5.1), because the constraint boundaries now are *planes* rather than lines. The faces of the polyhedron form the *boundary* of the feasible region, where each face is the portion of a constraint boundary that satisfies the other constraints as well. Note that each corner-point feasible solution lies at the intersection of *three* constraint boundaries (perhaps including some of the $x_1 = 0$, $x_2 = 0$, and $x_3 = 0$ constraint boundaries for the nonnegativity constraints), and the solution also satisfies the other constraints. Such intersections that don't satisfy one or more of the other constraints yield corner-point *infeasible* solutions instead.

The darker line segment in Fig. 5.2 depicts the path of the simplex method on a typical iteration. The point $(2, 4, 3)$ is the *current* corner-point feasible solution to begin the iteration, and the point $(4, 2, 4)$ will be the new corner-point feasible solution at the end of the iteration. The point $(2, 4, 3)$ lies at the intersection of the $x_2 = 4$, $x_1 + x_2 = 6$, and $-x_1 + 2x_3 = 4$ constraint boundaries, so these three equations are the *defining equations* for this corner-point feasible solution. If the $x_2 = 4$ defining

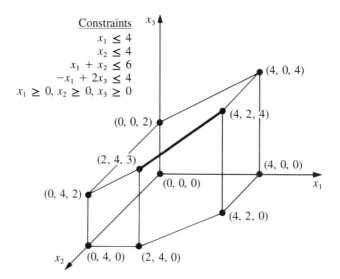

Constraints
$$x_1 \leq 4$$
$$x_2 \leq 4$$
$$x_1 + x_2 \leq 6$$
$$-x_1 + 2x_3 \leq 4$$
$$x_1 \geq 0, x_2 \geq 0, x_3 \geq 0$$

Figure 5.2 Feasible region and corner-point feasible solutions for a three-variable linear programming problem.

equation were removed, the intersection of the other two constraint boundaries (planes) would form a line. One segment of this line, shown as the dark line segment from $(2, 4, 3)$ to $(4, 2, 4)$ in Fig. 5.2, lies on the boundary of the feasible region, whereas the rest of the line is infeasible. This line segment is called an *edge* of the feasible region, and its endpoints, $(2, 4, 3)$ and $(4, 2, 4)$, are *adjacent* corner-point feasible solutions.

For $n = 3$, all of the *edges* of the feasible region are formed in this way as the feasible segment of the line lying at the intersection of two constraint boundaries, and the two endpoints of an edge are *adjacent* corner-point feasible solutions. In Fig. 5.2 there are 15 edges of the feasible region, and so there are 15 pairs of adjacent corner-point feasible solutions. For the current corner-point feasible solution $(2, 4, 3)$ there are *three* ways to remove one of its three defining equations to obtain an intersection of the other two constraint boundaries, so there are *three* edges emanating out of $(2, 4, 3)$. These edges lead to $(4, 2, 4)$, $(0, 4, 2)$, and $(2, 4, 0)$, so these are the corner-point feasible solutions that are adjacent to $(2, 4, 3)$.

For the next iteration, the simplex method chooses one of these three edges, say, the darker line segment in Fig. 5.2, and then moves along this edge away from $(2, 4, 3)$ until it reaches the first new constraint boundary, $x_1 = 4$, at its other endpoint. [We cannot continue further along this line to the next constraint boundary, $x_1 = 0$, because this leads to a corner-point *infeasible* solution, $(6, 0, 5)$.] The intersection of this first new constraint boundary with the two constraint boundaries forming the edge yields the *new* corner-point feasible solution $(4, 2, 4)$.

When $n > 3$, these same concepts generalize to higher dimensions, except the constraint boundaries now are *hyperplanes* instead of planes. Let us summarize.

> A corner-point feasible solution lies at the intersection of n constraint boundaries (and satisfies the other constraints as well). An **edge** of the feasible region is a feasible line segment that lies at the intersection of $(n - 1)$ constraint boundaries, where each endpoint lies on one additional constraint boundary (so that these endpoints are corner-point feasible solutions). Two corner-point feasible solutions are **adjacent** if the line segment connecting them is an edge of the feasible region. Emanating out of each corner-point feasible solution are n such edges, each one leading to one of the n adjacent corner-point feasible solutions. Each iteration of the simplex method moves from the current corner-point feasible solution to an adjacent one by moving along one of these n edges.

When you shift from a geometric viewpoint to an algebraic one, *intersection of constraint boundaries* changes to *simultaneous solution of constraint boundary equations*. The n constraint boundary equations yielding (defining) a corner-point feasible solution are its *defining equations*, where deleting one of these equations yields a line whose feasible segment is an edge of the feasible region.

We next analyze some key properties of corner-point feasible solutions, and then describe the implications of all of these concepts for interpreting the simplex method. However, while the above summary is fresh in your mind, let us give you a preview of its implications. When the simplex method chooses an entering basic variable, the geometric interpretation is that it is choosing one of the edges emanating out of the current corner-point feasible solution to move along. Increasing this variable from zero (and simultaneously changing the values of the other basic variables accordingly) corresponds to moving along this edge. Having the first basic variable reach zero (the leaving basic variable) corresponds to reaching the first new constraint boundary at the other end of this edge of the feasible region.

Properties of Corner-Point Feasible Solutions

In Sec. 4.1 we listed three key properties of corner-point feasible solutions that constitute the underlying principles of the simplex method. We now are in a position to explain why these properties (restated here) do indeed hold for *any* linear programming problem that has feasible solutions and a bounded feasible region.

> **Property 1:** (*a*) If there is exactly one optimal solution, then it *must* be a corner-point feasible solution. (*b*) If there are multiple optimal solutions, then at least two *must* be adjacent corner-point feasible solutions.

Property 1 is a rather intuitive one from a geometric viewpoint. First consider case (*a*), which is illustrated by the Wyndor Glass Co. problem (see Fig. 3.3 or 5.1) where the one optimal solution (2, 6) is indeed a corner-point feasible solution. Note that there is nothing special about this example that led to this result. For *any* problem having just one optimal solution, it always is possible to keep raising the objective function line (hyperplane) until it just touches one point (the optimal solution) at a corner of the feasible region.

The following algebraic viewpoint also clarifies why the property must hold in case (*a*). We will construct a *proof by contradiction* by assuming that the one optimal solution is *not* a corner-point feasible solution and then showing that this assumption leads to a contradiction and so cannot be true. The key step is to notice from the definition of *corner-point feasible solution* that this assumption implies that there must be two other feasible solutions such that the line segment connecting them contains the optimal solution. Let the vectors $\mathbf{x}^*$, $\mathbf{x}'$, $\mathbf{x}''$ denote the optimal solution and these two other feasible solutions, respectively, and let Z^*, Z_1, Z_2 denote their respective objective function values. Like each other point on the line segment connecting $\mathbf{x}'$ and $\mathbf{x}''$ (see Appendix 1),

$$\mathbf{x}^* = \alpha\mathbf{x}'' + (1 - \alpha)\mathbf{x}'$$

for some value of α such that $0 < \alpha < 1$. Thus

$$Z^* = \alpha Z_2 + (1 - \alpha)Z_1.$$

Since the weights, α and $(1 - \alpha)$, add up to 1, the only possibilities for how Z^*, Z_1, and Z_2 compare are (1) $Z^* = Z_1 = Z_2$, (2) $Z_1 < Z^* < Z_2$, and (3) $Z_1 > Z^* > Z_2$. The first possibility implies that $\mathbf{x}'$ and $\mathbf{x}''$ also are optimal, which contradicts the assumption that case (*a*) holds. Both the latter possibilities contradict the assumption that $\mathbf{x}^*$ is optimal. The resulting conclusion is that it is impossible to have a single optimal solution that is not a corner-point feasible solution.

Now consider case (*b*), which was demonstrated in Sec. 3.2 under the definition of *optimal solution* by changing the objective function in the example to $Z = 3x_1 + 2x_2$. What then happens in the graphical solution procedure is that the objective function line keeps getting raised until it contains the line segment connecting the two corner-point feasible solutions (2, 6) and (4, 3). The same thing would happen in higher dimensions except that now it would be an objective function *hyperplane* that keeps getting raised until it contains the line segment(s) connecting two (or more) adjacent corner-point feasible solutions. As a consequence, *all* optimal solutions can be obtained as weighted averages of optimal corner-point feasible solutions. (This situation is described further in Probs. 13 and 14 at the end of Chap. 4.)

The real significance of Property 1 is that it greatly simplifies the search for an optimal solution because now only corner-point feasible solutions need be considered. The magnitude of this simplification is emphasized in Property 2.

Property 2: There are only a *finite* number of corner-point feasible solutions.

This property certainly holds in Figs. 5.1 and 5.2, where there are just *five* and *ten* corner-point feasible solutions, respectively. To see why the number is finite in general, recall that each corner-point feasible solution is the simultaneous solution of a system of n out of the $(m + n)$ constraint boundary equations. The number of different combinations of $(m + n)$ equations taken n at a time is

$$\frac{(m + n)!}{m!n!},$$

which is a finite number. This number, in turn, is an *upper bound* on the number of corner-point feasible solutions. In Fig. 5.1, $m = 3$ and $n = 2$, so there are 10 different systems of two equations, but only half of them yield corner-point feasible solutions. In Fig. 5.2, $m = 4$ and $n = 3$, which gives 35 different systems of three equations, but only 10 yield corner-point feasible solutions.

Property 2 suggests that an optimal solution can be obtained just by exhaustive enumeration; i.e., find and compare all the finite number of corner-point feasible solutions. Unfortunately, there are finite numbers, and then there are finite numbers that (for all practical purposes) might as well be infinite. For example, a rather small linear programming problem with only $m = 50$ and $n = 50$ would have $(100!)/(50!)^2$ $\approx 10^{29}$ systems of equations to be solved! By contrast, the simplex method would need to examine only approximately 100 corner-point feasible solutions for a problem of this size. This tremendous savings can be obtained because of the optimality test provided by Property 3:

Property 3: If a corner-point feasible solution has no *adjacent* corner-point feasible solutions that are *better* (as measured by Z), then there are no *better* corner-point feasible solutions anywhere; i.e., it is *optimal*.

To illustrate Property 3, consider Fig. 5.1 (or Fig. 3.3) for the Wyndor Glass Co. example. For the corner-point feasible solution (2, 6), its adjacent corner-point feasible solutions are (0, 6) and (4, 3), and neither has a better value of Z than for (2, 6). This outcome implies that none of the *other* corner-point feasible solutions— (0, 0) and (4, 0)—can be better than (2, 6), so (2, 6) must be optimal.

By contrast, Fig. 5.3 shows a feasible region that can *never* occur for a linear programming problem but that does violate Property 3. The problem shown is identical to the Wyndor Glass Co. example (including the same objective function) *except* for the enlargement of the feasible region to the right of $(\frac{8}{3}, 5)$. Consequently, the adjacent corner-point feasible solutions for (2, 6) now are (0, 6) and $(\frac{8}{3}, 5)$, and again neither is better than (2, 6). However, *another* corner-point feasible solution (4, 5) now is better than (2, 6), thereby violating Property 3. The reason is that the boundary of the feasible region goes down from (2, 6) to $(\frac{8}{3}, 5)$, and then "bends outward" to (4, 5), beyond the objective function line passing through (2, 6).

The key point is that the kind of situation illustrated in Fig. 5.3 can never occur in linear programming. The feasible region in Fig. 5.3 implies that the $2x_2 \leq 12$ and

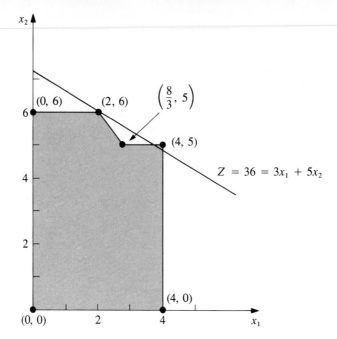

Figure 5.3
Modification of the
Wyndor Glass Co.
problem that
violates both linear
programming and
Property 3 for
corner-point feasible
solutions in linear
programming.

$3x_1 + 2x_2 \leq 18$ constraints apply for $0 \leq x_1 \leq \frac{8}{3}$. However, under the condition that $\frac{8}{3} \leq x_1 \leq 4$, the $3x_1 + 2x_2 \leq 18$ constraint is dropped and replaced by $x_2 \leq 5$. Such "conditional constraints" just are not allowed in linear programming.

The basic reason that Property 3 holds for any linear programming problem is that the feasible region always has the property of being a *convex set*, as defined in Appendix 1 and illustrated in several figures there. For two-variable linear programming problems, this convex property means that the *angle* inside the feasible region at *every* corner-point feasible solution is less than 180°. This property is illustrated in Fig. 5.1, where the angles at (0, 0), (0, 6), and (4, 0) are 90° and those at (2, 6) and (4, 3) are between 90° and 180°. By contrast, the feasible region in Fig. 5.3 is *not* a convex set, because the angle at $(\frac{8}{3}, 5)$ is more than 180°. This is the kind of "bending outward" at an angle greater than 180° that can never occur in linear programming. In higher dimensions, the same intuitive notion of "never bending outward" continues to apply.

To clarify the significance of a convex feasible region, consider the objective function hyperplane that passes through a corner-point feasible solution that has no adjacent corner-point feasible solutions that are better. [In the original Wyndor Glass Co. example, this hyperplane is the line passing through (2, 6) in Fig. 3.3.] All of these adjacent solutions [(0, 6) and (4, 3) in the example] must either lie on the hyperplane or lie on the unfavorable side (as measured by Z) of the hyperplane. The feasible region being convex means that its boundary cannot "bend outward" beyond an adjacent corner-point feasible solution to give another corner-point feasible solution that lies on the favorable side of the hyperplane, so Property 3 holds.

Extensions to the Augmented Form of the Problem

For any linear programming problem in *our standard form* (see Sec. 3.2), the appearance of the functional constraints after slack variables are introduced (see Sec. 4.2) is as follows:

$$\begin{array}{llr}
(1) & a_{11}x_1 + a_{12}x_2 + \cdots + a_{1n}x_n + x_{n+1} & = b_1 \\
(2) & a_{21}x_1 + a_{22}x_2 + \cdots + a_{2n}x_n \qquad\quad + x_{n+2} & = b_2 \\
& \qquad\qquad\qquad \vdots \\
(m) & a_{m1}x_1 + a_{m2}x_2 + \cdots + a_{mn}x_n \qquad\qquad\quad + x_{n+m} & = b_m,
\end{array}$$

where $x_{n+1}, x_{n+2}, \ldots, x_{n+m}$ are the slack variables. For other linear programming problems, Sec. 4.6 described how this same appearance (proper form from Gaussian elimination) can be obtained by introducing artificial variables, etc. Thus the original solutions $(x_1, x_2, \ldots, x_n)$ now are *augmented* by the corresponding values of the slack or artificial variables $(x_{n+1}, x_{n+2}, \ldots, x_{n+m})$. This augmentation led in Sec. 4.2 to defining **basic solutions** as *augmented corner-point solutions*, and **basic feasible solutions** as *augmented corner-point feasible solutions*. Consequently, the preceding three properties of corner-point feasible solutions also hold for basic feasible solutions.

Now let us clarify the algebraic relationships between basic solutions and corner-point solutions. Recall that each corner-point solution is the simultaneous solution of a system of n constraint boundary equations, which we called its *defining equations*. The key question is: "How do we tell whether a particular constraint boundary equation is one of the defining equations when the problem is in augmented form?" The answer, fortunately, is a simple one. Since there now are $(n + m)$ variables, one for each of the $(n + m)$ nonnegativity and functional constraints, each constraint has exactly one **indicating variable** that completely indicates (by whether its value is zero) whether that constraint's boundary equation is satisfied by the current solution. A summary appears in Table 5.3.

Thus whenever a constraint boundary equation is one of the defining equations for a corner-point solution, its indicating variable has a value of zero in the augmented form of the problem. Each such indicating variable is called a *nonbasic variable* for the corresponding basic solution. The resulting conclusions and terminology (already introduced in Sec. 4.2) are summarized next.

> Each **basic solution** has n **nonbasic variables** set equal to zero. The values of the remaining m variables (called **basic variables**) are the simultaneous solution of the system of m equations for the problem in augmented form (after setting the nonbasic variables to zero). This basic solution is the augmented corner-point solution whose n *defining equations* are those indicated by the nonbasic variables.

Now consider the basic *feasible* solutions. Note that the only requirements for a solution to be feasible in the augmented form of the problem are that it satisfy the system of equations and that *all* the variables be *nonnegative*.

Table 5.3 Indicating Variables for Constraint Boundary Equations*

Original Constraint (in Augmented Form)	Constraint Boundary Equation	Indicating Variable
$x_j \geq 0 \; (j = 1, 2, \ldots, n)$	$x_j = 0$	x_j
$\displaystyle\sum_{j=1}^{n} a_{ij}x_j + x_{n+i} = b_i$ $(i = 1, 2, \ldots, m)$	$\displaystyle\sum_{j=1}^{n} a_{ij}x_j = b_i$	x_{n+i}

* Indicating variable $= 0 \Rightarrow$ constraint boundary equation satisfied; indicating variable $\neq 0 \Rightarrow$ constraint boundary equation violated.

Table 5.4 **Indicating Variables for Constraint Boundary Equations of Wyndor Glass Co. Problem***

Original Constraint (in Augmented Form)	Constraint Boundary Equation	Indicating Variable
$x_1 \geq 0$	$x_1 = 0$	x_1
$x_2 \geq 0$	$x_2 = 0$	x_2
(1) $\quad x_1 + x_3 = 4$	$x_1 = 4$	x_3
(2) $\quad 2x_2 + x_4 = 12$	$2x_2 = 12$	x_4
(3) $\quad 3x_1 + 2x_2 + x_5 = 18$	$3x_1 + 2x_2 = 18$	x_5

* Indicating variable $= 0 \Rightarrow$ constraint boundary equation satisfied;
   indicating variable $\neq 0 \Rightarrow$ constraint boundary equation violated.

A **basic feasible solution** is a basic solution where all m basic variables are nonnegative (≥ 0). A basic feasible solution is said to be **degenerate** if any of these m variables equals zero.

Thus it is possible for a variable to be zero and still not be a nonbasic variable for the current basic feasible solution. (This case corresponds to a corner-point feasible solution that satisfies another constraint boundary equation in addition to its n defining equations.) Therefore, it is necessary to keep track of which is the current set of nonbasic variables (or the current set of basic variables) rather than relying upon their zero values.

We noted earlier that not every system of n constraint boundary equations yields a corner-point solution, either because the system has no solution or because it has multiple solutions. For analogous reasons, not every set of n nonbasic variables yields a basic solution. However, these cases are avoided by the simplex method.

To illustrate these definitions, consider the Wyndor Glass Co. example once more. Its constraint boundary equations and indicating variables are shown in Table 5.4.

Augmenting each of the corner-point feasible solutions (see Table 5.1) yields the *basic feasible solutions* listed in Table 5.5. This table places *adjacent* basic feasible solutions next to each other, except for the pair consisting of the first and last solutions listed. Notice that in each case the nonbasic variables necessarily are the indicating variables for the defining equations. Thus *adjacent* basic feasible solutions differ by

Table 5.5 **Basic Feasible Solutions for Wyndor Glass Co. Problem**

Corner-Point Feasible Solution	Defining Equations	Basic Feasible Solution	Nonbasic Variables
$(0, 0)$	$x_1 = 0$ $x_2 = 0$	$(0, 0, 4, 12, 18)$	x_1 x_2
$(0, 6)$	$x_1 = 0$ $2x_2 = 12$	$(0, 6, 4, 0, 6)$	x_1 x_4
$(2, 6)$	$2x_2 = 12$ $3x_1 + 2x_2 = 18$	$(2, 6, 2, 0, 0)$	x_4 x_5
$(4, 3)$	$3x_1 + 2x_2 = 18$ $x_1 = 4$	$(4, 3, 0, 6, 0)$	x_5 x_3
$(4, 0)$	$x_1 = 4$ $x_2 = 0$	$(4, 0, 0, 12, 6)$	x_3 x_2

Table 5.6 **Basic Infeasible Solutions for Wyndor Glass Co. Problem**

Corner-Point Infeasible Solution	Defining Equations	Basic Infeasible Solution	Nonbasic Variables
$(0, 9)$	$x_1 = 0$ $3x_1 + 2x_2 = 18$	$(0, 9, 4, -6, 0)$	x_1 x_5
$(4, 6)$	$2x_2 = 12$ $x_1 = 4$	$(4, 6, 0, 0, -6)$	x_4 x_3
$(6, 0)$	$3x_1 + 2x_2 = 18$ $x_2 = 0$	$(6, 0, -2, 12, 0)$	x_5 x_2

having just one different nonbasic variable. Also notice that each basic feasible solution necessarily is the resulting simultaneous solution of the system of equations for
the problem in augmented form (see Table 5.4) when the nonbasic variables are set
equal to zero.

Similarly, the other three corner-point solutions (see Table 5.2) yield the remaining basic solutions shown in Table 5.6.

The other two sets of nonbasic variables, (1) x_1 and x_3 and (2) x_2 and x_4, do not
yield a basic solution, because setting either pair of variables equal to zero leads to
having no solution for the system of equations (1)–(3) given in Table 5.4. This
conclusion parallels the observation we made early in this section that the corresponding sets of constraint boundary equations do not yield a solution.

The *simplex method* starts at a basic feasible solution and then iteratively moves
to a better adjacent basic solution until an optimal solution is reached. At each iteration, how is the adjacent basic feasible solution reached?

For the original form of the problem, recall that an adjacent corner-point solution
is reached from the current one by (1) deleting one constraint boundary (defining
equation) from the set of n constraint boundaries defining the current solution,
(2) moving away from the current solution in the *feasible* direction along the intersection of the remaining $(n - 1)$ constraint boundaries (an edge of the feasible region),
and (3) stopping when the *first* new constraint boundary (defining equation) is reached.

Equivalently, in our new terminology, the simplex method reaches an adjacent
basic feasible solution from the current one by (1) deleting one variable (the *entering
basic variable*) from the set of n nonbasic variables defining the current solution,
(2) moving away from the current solution by *increasing* this one variable from zero
(and adjusting the other basic variables to still satisfy the system of equations) while
keeping the remaining $(n - 1)$ nonbasic variables at zero, and (3) stopping when the
first of the basic variables (the *leaving basic variable*) reaches a value of zero (its
constraint boundary). With either interpretation, the choice among the n alternatives
in step (1) is made by selecting the one that would give the best rate of improvement
in Z (per unit increase in the entering basic variable) during step (2).

5.2 The Revised Simplex Method

The simplex method as described in Chap. 4 (hereafter called the *original simplex
method*) is a straightforward algebraic procedure. However, this way of executing the
algorithm (in either algebraic or tabular form) is not the most efficient computational

procedure for computers because it computes and stores many numbers that are not needed at the current iteration and that may not even become relevant for decision making at subsequent iterations. The only pieces of information relevant at each iteration are the coefficients of the nonbasic variables in Eq. (0), the coefficients of the entering basic variable in the other equations, and the right-hand side of the equations. It would be very useful to have a procedure that could obtain this information efficiently without computing and storing all the other coefficients.

As mentioned in Sec. 4.8, these considerations motivated the development of the *revised simplex method*. This method was designed to accomplish exactly the same things as the original simplex method, but in a way that is more efficient for execution on a computer. Thus it is a streamlined version of the original procedure. It computes and stores only the information that is currently needed, and it carries along the essential data in a more compact form.

The revised simplex method explicitly uses *matrix* manipulations, so it is necessary to describe the problem in matrix notation. (See Appendix 3 for a review of matrices.) To help you distinguish between matrices, vectors, and scalars, we consistently use **boldface CAPITAL** letters to represent matrices, **boldface lowercase** letters to represent vectors, and *italicized* letters in ordinary print to represent scalars. We also use a boldface zero (**0**) to denote a *null vector* (a vector whose elements all are zero) in either column or row form (which one should be clear from the context), whereas a zero in ordinary print (0) continues to represent the number zero.

Using matrices, our standard form for the general linear programming model given in Sec. 3.2 becomes

$$
\begin{aligned}
\text{Maximize} \quad & Z = \mathbf{cx}, \\
\text{subject to} \quad & \\
& \mathbf{Ax} \le \mathbf{b} \quad \text{and} \quad \mathbf{x} \ge \mathbf{0},
\end{aligned}
$$

where **c** is the row vector

$$
\mathbf{c} = [c_1, c_2, \ldots, c_n],
$$

x, b, and **0** are the column vectors such that

$$
\mathbf{x} = \begin{bmatrix} x_1 \\ x_2 \\ \vdots \\ x_n \end{bmatrix}, \quad
\mathbf{b} = \begin{bmatrix} b_1 \\ b_2 \\ \vdots \\ b_m \end{bmatrix}, \quad
\mathbf{0} = \begin{bmatrix} 0 \\ 0 \\ \vdots \\ 0 \end{bmatrix},
$$

and **A** is the matrix

$$
\mathbf{A} = \begin{bmatrix}
a_{11} & a_{12} & \cdots & a_{1n} \\
a_{21} & a_{22} & \cdots & a_{2n} \\
\vdots & \vdots & & \vdots \\
a_{m1} & a_{m2} & \cdots & a_{mn}
\end{bmatrix}.
$$

To obtain the *augmented form* of the problem, introduce the column vector of slack variables

$$\mathbf{x}_s = \begin{bmatrix} x_{n+1} \\ x_{n+2} \\ \vdots \\ x_{n+m} \end{bmatrix},$$

so that the constraints become

$$[\mathbf{A, I}] \begin{bmatrix} \mathbf{x} \\ \mathbf{x}_s \end{bmatrix} = \mathbf{b} \quad \text{and} \quad \begin{bmatrix} \mathbf{x} \\ \mathbf{x}_s \end{bmatrix} \geq \mathbf{0},$$

where $\mathbf{I}$ is the $m \times m$ identity matrix, and the null vector $\mathbf{0}$ now has $(n + m)$ elements.

Solving for a Basic Feasible Solution

Recall that the general approach of the simplex method is to obtain a sequence of *improving basic feasible solutions* until an optimal solution is reached. One of the key features of the revised simplex method involves the way in which it solves for each new basic feasible solution after identifying its basic and nonbasic variables. Given these variables, the resulting basic solution is the solution of the m equations

$$[\mathbf{A, I}] \begin{bmatrix} \mathbf{x} \\ \mathbf{x}_s \end{bmatrix} = \mathbf{b},$$

in which the n *nonbasic variables* from among the $(n + m)$ elements of

$$\begin{bmatrix} \mathbf{x} \\ \mathbf{x}_s \end{bmatrix}$$

are set equal to *zero*. Eliminating these n variables by equating them to zero leaves a set of m equations in m unknowns (the *basic variables*). This set of equations can be denoted by

$$\mathbf{Bx}_B = \mathbf{b},$$

where the **vector of basic variables,**

$$\mathbf{x}_B = \begin{bmatrix} x_{B1} \\ x_{B2} \\ \vdots \\ x_{Bm} \end{bmatrix},$$

is obtained by *eliminating* the *nonbasic variables* from

$$\begin{bmatrix} \mathbf{x} \\ \mathbf{x}_s \end{bmatrix},$$

and the **basis matrix,**

$$\mathbf{B} = \begin{bmatrix} B_{11} & B_{12} & \cdots & B_{1m} \\ B_{21} & B_{22} & \cdots & B_{2m} \\ \vdots & \vdots & & \vdots \\ B_{m1} & B_{m2} & \cdots & B_{mm} \end{bmatrix},$$

is obtained by *eliminating* the columns corresponding to *coefficients of nonbasic variables* from $[\mathbf{A}, \mathbf{I}]$. (In addition, the elements of $\mathbf{x}_B$ and, therefore, the columns of $\mathbf{B}$ may be placed in a different order when executing the simplex method.)

The simplex method introduces only basic variables such that $\mathbf{B}$ is *nonsingular,* so that $\mathbf{B}^{-1}$ always will exist. Therefore, to solve $\mathbf{Bx}_B = \mathbf{b}$, both sides would be premultiplied by $\mathbf{B}^{-1}$.

$$\mathbf{B}^{-1}\mathbf{Bx}_B = \mathbf{B}^{-1}\mathbf{b}.$$

Since $\mathbf{B}^{-1}\mathbf{B} = \mathbf{I}$, the desired solution for the basic variables is

$$\boxed{\mathbf{x}_B = \mathbf{B}^{-1}\mathbf{b}.}$$

Let $\mathbf{c}_B$ be the vector whose elements are the *objective function coefficients* (including zeroes for slack variables) for the corresponding elements of $\mathbf{x}_B$. The value of the objective function for this basic solution is then

$$\boxed{Z = \mathbf{c}_B\mathbf{x}_B = \mathbf{c}_B\mathbf{B}^{-1}\mathbf{b}.}$$

EXAMPLE: To illustrate this method of solving for a basic feasible solution, consider again the Wyndor Glass Co. problem presented in Sec. 3.1 and solved by the original simplex method in Table 4.8. In this case,

$$\mathbf{c} = [3, 5], \quad [\mathbf{A}, \mathbf{I}] = \begin{bmatrix} 1 & 0 & 1 & 0 & 0 \\ 0 & 2 & 0 & 1 & 0 \\ 3 & 2 & 0 & 0 & 1 \end{bmatrix}, \quad \mathbf{b} = \begin{bmatrix} 4 \\ 12 \\ 18 \end{bmatrix}, \quad \mathbf{x} = \begin{bmatrix} x_1 \\ x_2 \end{bmatrix}, \quad \mathbf{x}_s = \begin{bmatrix} x_3 \\ x_4 \\ x_5 \end{bmatrix}.$$

Referring to Table 4.8, the sequence of basic feasible solutions obtained by the simplex method (original or revised) is the following:

Iteration 0

$$\mathbf{x}_B = \begin{bmatrix} x_3 \\ x_4 \\ x_5 \end{bmatrix}, \quad \mathbf{B} = \begin{bmatrix} 1 & 0 & 0 \\ 0 & 1 & 0 \\ 0 & 0 & 1 \end{bmatrix} = \mathbf{B}^{-1}, \quad \text{so} \quad \begin{bmatrix} x_3 \\ x_4 \\ x_5 \end{bmatrix} = \begin{bmatrix} 1 & 0 & 0 \\ 0 & 1 & 0 \\ 0 & 0 & 1 \end{bmatrix} \begin{bmatrix} 4 \\ 12 \\ 18 \end{bmatrix} = \begin{bmatrix} 4 \\ 12 \\ 18 \end{bmatrix},$$

$$\mathbf{c}_B = [0, 0, 0], \quad \text{so } Z = [0, 0, 0] \begin{bmatrix} 4 \\ 12 \\ 18 \end{bmatrix} = 0.$$

Iteration 1

$$\mathbf{x}_B = \begin{bmatrix} x_3 \\ x_2 \\ x_5 \end{bmatrix}, \quad \mathbf{B} = \begin{bmatrix} 1 & 0 & 0 \\ 0 & 2 & 0 \\ 0 & 2 & 1 \end{bmatrix}, \quad \mathbf{B}^{-1} = \begin{bmatrix} 1 & 0 & 0 \\ 0 & \frac{1}{2} & 0 \\ 0 & -1 & 1 \end{bmatrix},$$

so

$$\begin{bmatrix} x_3 \\ x_2 \\ x_5 \end{bmatrix} = \begin{bmatrix} 1 & 0 & 0 \\ 0 & \frac{1}{2} & 0 \\ 0 & -1 & 1 \end{bmatrix} \begin{bmatrix} 4 \\ 12 \\ 18 \end{bmatrix} = \begin{bmatrix} 4 \\ 6 \\ 6 \end{bmatrix},$$

$$\mathbf{c}_B = [0, 5, 0], \quad \text{so } Z = [0, 5, 0] \begin{bmatrix} 4 \\ 6 \\ 6 \end{bmatrix} = 30.$$

$$\mathbf{x}_B = \begin{bmatrix} x_3 \\ x_2 \\ x_1 \end{bmatrix}, \qquad \mathbf{B} = \begin{bmatrix} 1 & 0 & 1 \\ 0 & 2 & 0 \\ 0 & 2 & 3 \end{bmatrix}, \qquad \mathbf{B}^{-1} = \begin{bmatrix} 1 & \frac{1}{3} & -\frac{1}{3} \\ 0 & \frac{1}{2} & 0 \\ 0 & -\frac{1}{3} & \frac{1}{3} \end{bmatrix},$$

so

$$\begin{bmatrix} x_3 \\ x_2 \\ x_1 \end{bmatrix} = \begin{bmatrix} 1 & \frac{1}{3} & -\frac{1}{3} \\ 0 & \frac{1}{2} & 0 \\ 0 & -\frac{1}{3} & \frac{1}{3} \end{bmatrix} \begin{bmatrix} 4 \\ 12 \\ 18 \end{bmatrix} = \begin{bmatrix} 2 \\ 6 \\ 2 \end{bmatrix},$$

$$\mathbf{c}_B = [0, 5, 3], \qquad \text{so } Z = [0, 5, 3] \begin{bmatrix} 2 \\ 6 \\ 2 \end{bmatrix} = 36.$$

Matrix Form of the Current Set of Equations

The last preliminary before summarizing the revised simplex method is to show the *matrix form* of the set of equations appearing in the simplex tableau for any iteration of the original simplex method.

For the *original* set of equations, the matrix form is

$$\begin{bmatrix} 1 & -\mathbf{c} & \mathbf{0} \\ 0 & \mathbf{A} & \mathbf{I} \end{bmatrix} \begin{bmatrix} Z \\ \mathbf{x} \\ \mathbf{x}_s \end{bmatrix} = \begin{bmatrix} 0 \\ \mathbf{b} \end{bmatrix}.$$

This set of equations also is exhibited in the first simplex tableau of Table 5.7.

The algebraic operations performed by the simplex method (multiply an equation by a constant and add a multiple of one equation to another equation) are expressed in matrix form by premultiplying both sides of the original set of equations by the appropriate matrix. This matrix would have the same elements as the identity matrix, *except* that each multiple for an algebraic operation would go into the spot needed to have the matrix multiplication perform this operation. Even after a series of algebraic operations over several iterations, we still can deduce what this matrix must be (symbolically) for the entire series by using what we already know about the right-hand side of the new set of equations. In particular, after any iteration, $\mathbf{x}_B = \mathbf{B}^{-1}\mathbf{b}$ and $Z = \mathbf{c}_B\mathbf{B}^{-1}\mathbf{b}$, so the right-hand side of the new set of equations has become

Table 5.7 **Initial and Later Simplex Tableaux in Matrix Form**

Iteration	Basic Variable	Eq. No.	Z	Coefficient of			Right Side
				Original Variables	Slack Variables		
0	Z	0	1	$-\mathbf{c}$	$\mathbf{0}$		0
	$\mathbf{x}_B$	$1 - m$	$\mathbf{0}$	$\mathbf{A}$	$\mathbf{I}$		$\mathbf{b}$
Any	Z	0	1	$\mathbf{c}_B\mathbf{B}^{-1}\mathbf{A} - \mathbf{c}$	$\mathbf{c}_B\mathbf{B}^{-1}$		$\mathbf{c}_B\mathbf{B}^{-1}\mathbf{b}$
	$\mathbf{x}_B$	$1 - m$	$\mathbf{0}$	$\mathbf{B}^{-1}\mathbf{A}$	$\mathbf{B}^{-1}$		$\mathbf{B}^{-1}\mathbf{b}$

$$\begin{bmatrix} Z \\ \mathbf{x}_B \end{bmatrix} = \begin{bmatrix} 1 & \mathbf{c}_B\mathbf{B}^{-1} \\ 0 & \mathbf{B}^{-1} \end{bmatrix}\begin{bmatrix} 0 \\ \mathbf{b} \end{bmatrix} = \begin{bmatrix} \mathbf{c}_B\mathbf{B}^{-1}\mathbf{b} \\ \mathbf{B}^{-1}\mathbf{b} \end{bmatrix}.$$

Because we perform the same series of algebraic operations on *both* sides of the original set of operations, we use this same matrix premultiplying the original right-hand side to premultiply the original left-hand side. Consequently, since

$$\begin{bmatrix} 1 & \mathbf{c}_B\mathbf{B}^{-1} \\ 0 & \mathbf{B}^{-1} \end{bmatrix}\begin{bmatrix} 1 & -\mathbf{c} & 0 \\ 0 & \mathbf{A} & \mathbf{I} \end{bmatrix} = \begin{bmatrix} 1 & \mathbf{c}_B\mathbf{B}^{-1}\mathbf{A} - \mathbf{c} & \mathbf{c}_B\mathbf{B}^{-1} \\ 0 & \mathbf{B}^{-1}\mathbf{A} & \mathbf{B}^{-1} \end{bmatrix},$$

the desired matrix form of the *set of equations after any iteration* is

$$\begin{bmatrix} 1 & \mathbf{c}_B\mathbf{B}^{-1}\mathbf{A} - \mathbf{c} & \mathbf{c}_B\mathbf{B}^{-1} \\ 0 & \mathbf{B}^{-1}\mathbf{A} & \mathbf{B}^{-1} \end{bmatrix}\begin{bmatrix} Z \\ \mathbf{x} \\ \mathbf{x}_s \end{bmatrix} = \begin{bmatrix} \mathbf{c}_B\mathbf{B}^{-1}\mathbf{b} \\ \mathbf{B}^{-1}\mathbf{b} \end{bmatrix}.$$

The second simplex tableau of Table 5.7 also exhibits this same set of equations.

EXAMPLE: To illustrate this matrix form for the current set of equations, consider the *final* set of equations resulting from iteration 2 for the Wyndor Glass Co. problem. Using the $\mathbf{B}^{-1}$ given for iteration 2,

$$\mathbf{B}^{-1}\mathbf{A} = \begin{bmatrix} 1 & \frac{1}{3} & -\frac{1}{3} \\ 0 & \frac{1}{2} & 0 \\ 0 & -\frac{1}{3} & \frac{1}{3} \end{bmatrix}\begin{bmatrix} 1 & 0 \\ 0 & 2 \\ 3 & 2 \end{bmatrix} = \begin{bmatrix} 0 & 0 \\ 0 & 1 \\ 1 & 0 \end{bmatrix},$$

$$\mathbf{c}_B\mathbf{B}^{-1} = [0, 5, 3]\begin{bmatrix} 1 & \frac{1}{3} & -\frac{1}{3} \\ 0 & \frac{1}{2} & 0 \\ 0 & -\frac{1}{3} & \frac{1}{3} \end{bmatrix} = [0, \tfrac{3}{2}, 1],$$

$$\mathbf{c}_B\mathbf{B}^{-1}\mathbf{A} - \mathbf{c} = [0, 5, 3]\begin{bmatrix} 0 & 0 \\ 0 & 1 \\ 1 & 0 \end{bmatrix} - [3, 5] = [0, 0].$$

Also, using the values of $\mathbf{x}_B = \mathbf{B}^{-1}\mathbf{b}$ and $Z = \mathbf{c}_B\mathbf{B}^{-1}\mathbf{b}$ calculated a few pages back, these results give the following set of equations:

$$\begin{bmatrix} 1 & 0 & 0 & 0 & \frac{3}{2} & 1 \\ 0 & 0 & 0 & 1 & \frac{1}{3} & -\frac{1}{3} \\ 0 & 0 & 1 & 0 & \frac{1}{2} & 0 \\ 0 & 1 & 0 & 0 & -\frac{1}{3} & \frac{1}{3} \end{bmatrix}\begin{bmatrix} Z \\ x_1 \\ x_2 \\ x_3 \\ x_4 \\ x_5 \end{bmatrix} = \begin{bmatrix} 36 \\ 2 \\ 6 \\ 2 \end{bmatrix},$$

as shown in the *final* simplex tableau in Table 4.8.

The Overall Procedure

There are two key implications from the matrix form of the current set of equations shown at the bottom of Table 5.7. The first is that *only* $\mathbf{B}^{-1}$ needs to be derived to be able to calculate all the numbers in the simplex tableau from the *original parameters*

(This implication is the essence of the **fundamental insight** described in the next section.) The second is that *any one* of these numbers (except $Z = \mathbf{c}_B\mathbf{B}^{-1}\mathbf{b}$) can be obtained by performing *only part* of a matrix multiplication. Therefore, the *required numbers* to perform an iteration of the simplex method can be obtained as needed *without* expending the computational effort to obtain *all* the numbers. These two key implications are incorporated into the following summary of the overall procedure.

Summary of Revised Simplex Method

1. *Initialization step:* Same as for original simplex method.
2. *Iterative step:*

   Part 1 Determine the entering basic variable: Same as for original simplex method.

   Part 2 Determine the leaving basic variable: Same as for original simplex method, except calculate *only* the numbers required to do this [the coefficients of the entering basic variable in every equation but Eq. (0), and then, for each strictly positive coefficient, the right-hand side of that equation].[1]

   Part 3 Determine the new basic feasible solution: Derive $\mathbf{B}^{-1}$ and set $\mathbf{x}_B = \mathbf{B}^{-1}\mathbf{b}$. (Calculating $\mathbf{x}_B$ is optional unless the optimality test finds it to be optimal.)
3. *Optimality test:* Same as for original simplex method, except calculate *only* the numbers required to do this test, i.e., the coefficients of the *nonbasic variables* in Eq. (0).

In part 3 of the iterative step, $\mathbf{B}^{-1}$ could be derived each time by using a standard computer routine for inverting a matrix. However, since $\mathbf{B}$ (and therefore $\mathbf{B}^{-1}$) changes so little from one iteration to the next, it is much more efficient to derive the new $\mathbf{B}^{-1}$ (denote it by $\mathbf{B}_{\text{new}}^{-1}$) from the $\mathbf{B}^{-1}$ at the preceding iteration (denote it by $\mathbf{B}_{\text{old}}^{-1}$). (For the *initial* basic feasible solution, $\mathbf{B} = \mathbf{I} = \mathbf{B}^{-1}$.) The method for doing this derivation is based directly upon the interpretation of the elements of $\mathbf{B}^{-1}$ (the coefficients of the slack variables in the current equations 1, 2, . . . , m) presented in the next section, as well as upon the procedure used by the original simplex method to obtain the new set of equations from the preceding set.

To describe this method formally, let

x_k = entering basic variable,

a'_{ik} = coefficient of x_k in current Eq. (i), for $i = 1, 2, \ldots, m$ (calculated in part 2 of the iterative step),

r = number of the equation containing the leaving basic variable.

Recall that the new set of equations [excluding Eq. (0)] can be obtained from the preceding set by subtracting a'_{ik}/a'_{rk} times Eq. (r) from Eq. (i), for all $i = 1, 2, \ldots, m$ *except $i = r$*, and then dividing Eq. (r) by a'_{rk}. Therefore, the element in row i and column j of $\mathbf{B}_{\text{new}}^{-1}$ is

---

[1] Because the value of $\mathbf{x}_B$ is the entire vector of right-hand sides except for Eq. (0), the relevant right-hand sides need not be calculated here if $\mathbf{x}_B$ was calculated in part 3 of the preceding iteration.

$$(\mathbf{B}_{\text{new}}^{-1})_{ij} = \begin{cases} (\mathbf{B}_{\text{old}}^{-1})_{ij} - \dfrac{a'_{ik}}{a'_{rk}} (\mathbf{B}_{\text{old}}^{-1})_{rj}, & \text{if } i \neq r \\[2ex] \dfrac{1}{a'_{rk}} (\mathbf{B}_{\text{old}}^{-1})_{rj}, & \text{if } i = r. \end{cases}$$

These formulas are expressed in matrix notation as

$$\mathbf{B}_{\text{new}}^{-1} = \mathbf{E}\mathbf{B}_{\text{old}}^{-1},$$

where the matrix $\mathbf{E}$ is an identity matrix except that its rth column is replaced by the vector

$$\boldsymbol{\eta} = \begin{bmatrix} \eta_1 \\ \eta_2 \\ \vdots \\ \eta_m \end{bmatrix}, \quad \text{where } \eta_i = \begin{cases} -\dfrac{a'_{ik}}{a'_{rk}}, & \text{if } i \neq r \\[2ex] \dfrac{1}{a'_{rk}}, & \text{if } i = r. \end{cases}$$

Thus $\mathbf{E} = [\mathbf{U}_1, \mathbf{U}_2, \ldots, \mathbf{U}_{r-1}, \boldsymbol{\eta}, \mathbf{U}_{r+1}, \ldots, \mathbf{U}_m]$, where the m elements of each of the $\mathbf{U}_i$ column vectors are 0 except for a 1 in the ith position.

EXAMPLE: We shall illustrate the revised simplex method by applying it to the Wyndor Glass Co. problem. The *initial* basic variables are the slack variables.

$$\mathbf{x}_B = \begin{bmatrix} x_3 \\ x_4 \\ x_5 \end{bmatrix}.$$

Iteration 1: Because the *initial* $\mathbf{B}^{-1} = \mathbf{I}$, no calculations are needed to obtain the numbers required to identify the *entering basic variable* x_2 ($-c_2 = -5 < -3 = -c_1$) and the *leaving basic variable* x_4 ($a_{12} = 0$, $b_2/a_{22} = \frac{12}{2} < \frac{18}{2} = b_3/a_{32}$, so $r = 2$). Thus the new set of basic variables is

$$\mathbf{x}_B = \begin{bmatrix} x_3 \\ x_2 \\ x_5 \end{bmatrix}.$$

To obtain the new $\mathbf{B}^{-1}$,

$$\boldsymbol{\eta} = \begin{bmatrix} -\dfrac{a_{12}}{a_{22}} \\[2ex] \dfrac{1}{a_{22}} \\[2ex] -\dfrac{a_{32}}{a_{22}} \end{bmatrix} = \begin{bmatrix} 0 \\ \frac{1}{2} \\ -1 \end{bmatrix},$$

so

$$\mathbf{B}^{-1} = \begin{bmatrix} 1 & 0 & 0 \\ 0 & \frac{1}{2} & 0 \\ 0 & -1 & 1 \end{bmatrix} \begin{bmatrix} 1 & 0 & 0 \\ 0 & 1 & 0 \\ 0 & 0 & 1 \end{bmatrix} = \begin{bmatrix} 1 & 0 & 0 \\ 0 & \frac{1}{2} & 0 \\ 0 & -1 & 1 \end{bmatrix},$$

so that
$$\begin{bmatrix} x_3 \\ x_2 \\ x_5 \end{bmatrix} = \begin{bmatrix} 1 & 0 & 0 \\ 0 & \frac{1}{2} & 0 \\ 0 & -1 & 1 \end{bmatrix} \begin{bmatrix} 4 \\ 12 \\ 18 \end{bmatrix} = \begin{bmatrix} 4 \\ 6 \\ 6 \end{bmatrix}.$$

To test whether this solution is optimal, we calculate the coefficients of the *nonbasic variables* (x_1 and x_4) in Eq. (0). Performing only the relevant parts of the matrix multiplications,

$$\mathbf{c}_B \mathbf{B}^{-1} \mathbf{A} - \mathbf{c} = [0, 5, 0] \begin{bmatrix} 1 & 0 & 0 \\ 0 & \frac{1}{2} & 0 \\ 0 & -1 & 1 \end{bmatrix} \begin{bmatrix} 1 & - \\ 0 & - \\ 3 & - \end{bmatrix} - [3, -] = [-3, -],$$

$$\mathbf{c}_B \mathbf{B}^{-1} = [0, 5, 0] \begin{bmatrix} - & 0 & - \\ - & \frac{1}{2} & - \\ - & -1 & - \end{bmatrix} = [-, \tfrac{5}{2}, -],$$

so the coefficients of x_1 and x_4 are -3 and $\frac{5}{2}$, respectively. Since x_1 has a negative coefficient, this solution is *not* optimal.

Iteration 2: Using these coefficients of the nonbasic variables, the next iteration begins by identifying x_1 as the *entering basic variable*. To determine the *leaving basic variable*, we must calculate the *other* coefficients of x_1:

$$\mathbf{B}^{-1}\mathbf{A} = \begin{bmatrix} 1 & 0 & 0 \\ 0 & \frac{1}{2} & 0 \\ 0 & -1 & 1 \end{bmatrix} \begin{bmatrix} 1 & - \\ 0 & - \\ 3 & - \end{bmatrix} = \begin{bmatrix} 1 & - \\ 0 & - \\ 3 & - \end{bmatrix}.$$

Using the right-side column for the current basic feasible solution (the value of $\mathbf{x}_B$) just given for iteration 1, the ratios $4/1 > 6/3$ indicate that x_5 is the leaving basic variable, so the new set of basic variables is

$$\mathbf{x}_B = \begin{bmatrix} x_3 \\ x_2 \\ x_1 \end{bmatrix}, \quad \text{with } \boldsymbol{\eta} = \begin{bmatrix} -\dfrac{a'_{11}}{a'_{31}} \\ -\dfrac{a'_{21}}{a'_{31}} \\ \dfrac{1}{a'_{31}} \end{bmatrix} = \begin{bmatrix} -\frac{1}{3} \\ 0 \\ \frac{1}{3} \end{bmatrix}.$$

Therefore, the new $\mathbf{B}^{-1}$ is

$$\mathbf{B}^{-1} = \begin{bmatrix} 1 & 0 & -\frac{1}{3} \\ 0 & 1 & 0 \\ 0 & 0 & \frac{1}{3} \end{bmatrix} \begin{bmatrix} 1 & 0 & 0 \\ 0 & \frac{1}{2} & 0 \\ 0 & -1 & 1 \end{bmatrix} = \begin{bmatrix} 1 & \frac{1}{3} & -\frac{1}{3} \\ 0 & \frac{1}{2} & 0 \\ 0 & -\frac{1}{3} & \frac{1}{3} \end{bmatrix},$$

so that
$$\begin{bmatrix} x_3 \\ x_2 \\ x_1 \end{bmatrix} = \begin{bmatrix} 1 & \frac{1}{3} & -\frac{1}{3} \\ 0 & \frac{1}{2} & 0 \\ 0 & -\frac{1}{3} & \frac{1}{3} \end{bmatrix} \begin{bmatrix} 4 \\ 12 \\ 18 \end{bmatrix} = \begin{bmatrix} 2 \\ 6 \\ 2 \end{bmatrix}.$$

Applying the optimality test, we find that the coefficients of the *nonbasic variables* (x_4 and x_5) in Eq. (0) are

$$\mathbf{c}_B \mathbf{B}^{-1} = [0, 5, 3] \begin{bmatrix} - & \frac{1}{3} & -\frac{1}{3} \\ - & \frac{1}{2} & 0 \\ - & -\frac{1}{3} & \frac{1}{3} \end{bmatrix} = [-, \tfrac{3}{2}, 1].$$

Because both coefficients ($\frac{3}{2}$ and 1) are nonnegative, the current solution ($x_1 = 2$, $x_2 = 6$, $x_3 = 2$, $x_4 = 0$, $x_5 = 0$) is optimal and the procedure terminates.

General Observations

Although the preceding pages describe the essence of the revised simplex method, we should point out that minor modifications may be made to improve the efficiency of its execution on computers. For example, $\mathbf{B}^{-1}$ may be obtained as the product of the previous $\mathbf{E}$ matrices. This modification requires storing only the $\boldsymbol{\eta}$ column of $\mathbf{E}$ and the number of the column, rather than the $\mathbf{B}^{-1}$ matrix, at each iteration. If magnetic tape must be used rather than core storage, this "product form" of the basis inverse may be the most efficient.

You should also note that the preceding discussion was limited to the case of linear programming problems fitting *our standard form* given in Sec. 3.2. However, the modifications for other forms are relatively straightforward. The *initialization step* would be conducted just as it would for the original simplex method (see Sec. 4.6). When this step involves introducing artificial variables to obtain an initial basic feasible solution (and thereby to obtain an *identity matrix* as the *initial basis matrix*), these variables would be included among the m elements of x_s.

Let us summarize the advantages of the revised simplex method over the original simplex method. One advantage is that the number of arithmetic computations may be reduced. This is especially true when the $\mathbf{A}$ matrix contains a large number of zero elements (which is usually the case for the large problems arising in practice). The amount of information that must be stored at each iteration is less, sometimes considerably so. The revised simplex method also permits the control of the round-off errors inevitably generated by computers. This control can be exercised by periodically obtaining the current $\mathbf{B}^{-1}$ by directly inverting $\mathbf{B}$. Furthermore, some of the post-optimality problems discussed in Sec. 4.7 can be handled more conveniently with the revised simplex method. For all these reasons, the revised simplex method is usually preferable to the original simplex method for computer execution.

5.3 A Fundamental Insight

We shall now focus on a property of the simplex method (in any form) that has been revealed by the *revised simplex method* in the preceding section. This fundamental insight provides the key to both duality theory and sensitivity analysis (Chap. 6), two very important parts of linear programming.

The insight involves the coefficients of the *slack* variables and the information they give. It is a direct result of the initialization step, where the ith slack variable (x_{n+i}) is given a coefficient of $+1$ in Eq. (i) and a coefficient of *zero* in *every other equation* [including Eq. (0)] for $i = 1, 2, \ldots, m$, as shown by the null vector $\mathbf{0}$ and the identity matrix $\mathbf{I}$ in the *Slack Variables* column for iteration 0 in Table 5.7.[1] The other key factor is that subsequent iterations change the initial equations *only* by:

 1. Multiplying (or dividing) an *entire* equation by a nonzero constant; or

---

[1] Throughout most of this section, we assume that the problem is in *our standard form*, with $b_i \geq 0$ for all $i = 1, 2, \ldots, m$, so that no additional adjustments are needed in the initialization step. We then adapt our conclusions to nonstandard forms late in the section.

2. Adding (or subtracting) a multiple of one *entire* equation to another *entire* equation.

As already described in the preceding section, a sequence of these kinds of algebraic operations is equivalent to premultiplying the initial simplex tableau by some matrix. (See Appendix 3 for a review of matrices.) The consequence can be summarized as follows.

> *Verbal Description of Fundamental Insight:* After any iteration, the coefficients of the *slack* variables in each equation immediately reveal how that equation has been obtained from the *initial* equations.

As one example of the importance of this insight, recall from Table 5.7 that the matrix formula for the *optimal* solution obtained by the simplex method is

$$\mathbf{x}_B = \mathbf{B}^{-1}\mathbf{b},$$

where $\mathbf{x}_B$ is the vector of basic variables, $\mathbf{B}^{-1}$ is the matrix of coefficients of *slack* variables for rows $1-m$ of the final tableau, and $\mathbf{b}$ is the vector of original right-hand sides (resource availabilities). (We soon will denote this particular $\mathbf{B}^{-1}$ by $\mathbf{S}^*$.) Post-optimality analysis normally includes an investigation of possible changes in $\mathbf{b}$. By using this formula, you can see exactly how the optimal basic feasible solution changes (or whether it becomes infeasible because of negative variables) as a function of $\mathbf{b}$. You do *not* have to reapply the simplex method over and over again for each new $\mathbf{b}$, because the coefficients of the slack variables tell all! In a similar fashion, this fundamental insight provides a tremendous computational saving for the rest of the sensitivity analysis as well.

To spell out the how and the why of this insight, let us look again at the Wyndor Glass Co. example. (Your OR COURSEWARE also includes another demonstration example.)

EXAMPLE: Table 5.8 shows the relevant portion of the simplex tableaux for demonstrating this fundamental insight. Darker lines have been drawn around the coefficients of the *slack* variables in the second and third tableaux because these are the

Table 5.8 Simplex Tableaux without Leftmost Columns for Wyndor Glass Co. Problem

Iteration	Coefficient of					Right Side
	x_1	x_2	x_3	x_4	x_5	
0	-3	-5	0	0	0	0
	1	0	1	0	0	4
	0	2	0	1	0	12
	3	2	0	0	1	18
1	$-\frac{3}{2}$	0	0	$\frac{5}{2}$	0	30
	1	0	1	0	0	4
	0	1	0	$\frac{1}{2}$	0	6
	3	0	0	-1	1	6
2	0	0	0	$\frac{3}{2}$	1	36
	0	0	1	$\frac{1}{3}$	$-\frac{1}{3}$	2
	0	1	0	$\frac{1}{2}$	0	6
	1	0	0	$-\frac{1}{3}$	$\frac{1}{3}$	2

crucial coefficients for applying the insight. To avoid clutter, we then identify the pivot row and pivot column by a single box around the pivot number only.

Iteration 1: To demonstrate the fundamental insight, our focus is on the algebraic operations performed by the simplex method while using Gaussian elimination to obtain the new basic feasible solution. If we divide the pivot row by the pivot number *last* rather than first, then the algebraic operations spelled out in Chap. 4 for iteration 1 are

$$
\begin{aligned}
\text{New row } 0 &= \text{old row } 0 + (\tfrac{5}{2}) \text{ old row } 2, \\
\text{New row } 1 &= \text{old row } 1 + (0) \text{ old row } 2, \\
\text{New row } 2 &= \qquad\quad (\tfrac{1}{2}) \text{ old row } 2, \\
\text{New row } 3 &= \text{old row } 3 + (-1) \text{ old row } 2.
\end{aligned}
$$

Ignoring row 0 for the moment, these algebraic operations amount to premultiplying rows 1–3 of the *initial* tableau by the first matrix shown below.

$$
\text{New rows 1–3} =
\begin{bmatrix} 1 & 0 & 0 \\ 0 & \tfrac{1}{2} & 0 \\ 0 & -1 & 1 \end{bmatrix}
\begin{bmatrix} 1 & 0 & 1 & 0 & 0 & 4 \\ 0 & 2 & 0 & 1 & 0 & 12 \\ 3 & 2 & 0 & 0 & 1 & 18 \end{bmatrix}
$$

$$
=
\begin{bmatrix} 1 & 0 & 1 & 0 & 0 & 4 \\ 0 & 1 & 0 & \tfrac{1}{2} & 0 & 6 \\ 3 & 0 & 0 & -1 & 1 & 6 \end{bmatrix}.
$$

Note how this first matrix is reproduced exactly as the coefficients of the *slack* variables in rows 1–3 of the new tableau, because the coefficients of the slack variables in rows 1–3 of the initial tableau form an *identity matrix*. Thus, just as stated in the verbal description of the fundamental insight, the coefficients of the *slack* variables in the new tableau do indeed provide a record of the algebraic operations performed.

This insight is not much to get excited about after just one iteration, since you can readily see from the initial tableau what the algebraic operations had to be, but it becomes invaluable after all of the iterations are completed.

For row 0, the algebraic operation performed amounts to the following matrix calculations, where now our focus is on the vector $[0, \tfrac{5}{2}, 0]$ that premultiplies rows 1–3 of the *initial* tableau.

$$
\text{New row } 0 = [-3, \quad -5 \mid 0, \quad 0, \quad 0 \mid 0] + [0, \quad \tfrac{5}{2}, \quad 0]
\begin{bmatrix} 1 & 0 & 1 & 0 & 0 & 4 \\ 0 & 2 & 0 & 1 & 0 & 12 \\ 3 & 2 & 0 & 0 & 1 & 18 \end{bmatrix}
$$

$$
= [-3, \quad 0, \quad \boxed{0, \quad \tfrac{5}{2}, \quad 0}, \quad 30].
$$

Note how this vector is reproduced exactly as the coefficients of the *slack* variables in row 0 of the new tableau, just as was claimed in the statement of the fundamental insight. (Once again, the reason is the *identity matrix* for the coefficients of the *slack* variables in rows 1–3 of the initial tableau, along with the zeroes for these coefficients in row 0 of the initial tableau.)

Iteration 2: The algebraic operations performed on the second tableau of Table 5.8 for iteration 2 are

$$
\begin{aligned}
\text{New row } 0 &= \text{old row } 0 + \quad (1) \text{ old row } 3, \\
\text{New row } 1 &= \text{old row } 1 + (-\tfrac{1}{3}) \text{ old row } 3,
\end{aligned}
$$

New row 2 = old row 2 + (0) old row 3,
New row 3 = ($\frac{1}{3}$) old row 3.

Ignoring row 0 for the moment, these operations amount to premultiplying rows 1–3 of this tableau by the matrix

$$\begin{bmatrix} 1 & 0 & -\frac{1}{3} \\ 0 & 1 & 0 \\ 0 & 0 & \frac{1}{3} \end{bmatrix}.$$

Writing this second tableau as the matrix product shown for iteration 1 (namely, the corresponding matrix times rows 1–3 of the *initial* tableau) then yields

$$\text{Final rows 1–3} = \begin{bmatrix} 1 & 0 & -\frac{1}{3} \\ 0 & 1 & 0 \\ 0 & 0 & \frac{1}{3} \end{bmatrix} \begin{bmatrix} 1 & 0 & 0 \\ 0 & \frac{1}{2} & 0 \\ 0 & -1 & 1 \end{bmatrix} \left[\begin{array}{ccc:ccc} 1 & 0 & 1 & 0 & 0 & 4 \\ 0 & 2 & 0 & 1 & 0 & 12 \\ 3 & 2 & 0 & 0 & 1 & 18 \end{array}\right]$$

$$= \begin{bmatrix} 1 & \frac{1}{3} & -\frac{1}{3} \\ 0 & \frac{1}{2} & 0 \\ 0 & -\frac{1}{3} & \frac{1}{3} \end{bmatrix} \left[\begin{array}{ccc:ccc} 1 & 0 & 1 & 0 & 0 & 4 \\ 0 & 2 & 0 & 1 & 0 & 12 \\ 3 & 2 & 0 & 0 & 1 & 18 \end{array}\right]$$

$$= \left[\begin{array}{cc|ccc|c} 0 & 0 & 1 & \frac{1}{3} & -\frac{1}{3} & 2 \\ 0 & 1 & 0 & \frac{1}{2} & 0 & 6 \\ 1 & 0 & 0 & -\frac{1}{3} & \frac{1}{3} & 2 \end{array}\right].$$

The first two matrices shown on the first line of these calculations summarize the algebraic operations of the second and first iterations, respectively. Their product, shown as the first matrix on the second line, then combines the algebraic operations of the two iterations. Note how this matrix is reproduced exactly as the coefficients of the *slack* variables in rows 1–3 of the new (final) tableau shown on the third line. What this portion of the tableau reveals is how the *entire* final tableau (except row 0) has been obtained from the *initial* tableau, namely,

Final row 1 = (1) initial row 1 + ($\frac{1}{3}$) initial row 2 + ($-\frac{1}{3}$) initial row 3,
Final row 2 = (0) initial row 1 + ($\frac{1}{2}$) initial row 2 + (0) initial row 3,
Final row 3 = (0) initial row 1 + ($-\frac{1}{3}$) initial row 2 + ($\frac{1}{3}$) initial row 3.

To see why these multipliers of the initial rows are correct, you would have to trace through all of the algebraic operations of both iterations. For example, why does final row 1 include ($\frac{1}{3}$) initial row 2, even though a multiple of row 2 has never been added directly to row 1? The reason is that initial row 2 was subtracted from initial row 3 in iteration 1, and then ($\frac{1}{3}$) old row 3 was subtracted from old row 1 in iteration 2.

However, there is no need for you to trace through. Even when the simplex method has gone through hundreds or thousands of iterations, the coefficients of the *slack* variables in the final tableau would reveal how this tableau has been obtained from the *initial* tableau. Furthermore, the same algebraic operations would give these same coefficients even if the values of some of the parameters in the original model (initial tableau) were changed, so these coefficients also reveal how the *rest* of the final tableau changes with changes in the initial tableau.

To complete this story for row 0, the fundamental insight reveals that the *entire* final row 0 can be calculated from the initial tableau by using just the coefficients of the *slack* variables in the final row 0, [0, $\frac{3}{2}$, 1]. This calculation is shown below, where the first vector is row 0 of the initial tableau and the matrix is rows 1–3 of the initial tableau.

$$\text{Final row } 0 = [-3, \quad -5 \mid 0, \quad 0, \quad 0 \mid 0] + [0, \tfrac{3}{2}, 1]\begin{bmatrix} 1 & 0 & 1 & 0 & 0 & 4 \\ 0 & 2 & 0 & 1 & 0 & 12 \\ 3 & 2 & 0 & 0 & 1 & 18 \end{bmatrix}$$

$$= [0, \quad 0, \boxed{0, \quad \tfrac{3}{2}, \quad 1}, \quad 36].$$

Note again how the vector premultiplying rows 1–3 of the initial tableau is reproduced exactly as the coefficients of the *slack* variables in the final row 0. These quantities must be identical because of the coefficients of the *slack* variables in the *initial* tableau (an identity matrix below a null vector). This conclusion is the row 0 part of the fundamental insight.

Mathematical Summary

Because its primary applications involve the *final* tableau, we shall now give a general mathematical expression for the fundamental insight just in terms of this tableau, using matrix notation. If you haven't read Sec. 5.2, you now need to know that the *parameters* of the model are given by the matrix $\mathbf{A} = \|a_{ij}\|$ and the vectors $\mathbf{b} = \|b_i\|$ and $\mathbf{c} = \|c_j\|$, as displayed at the beginning of that section.

The only other notation needed is summarized and illustrated in Table 5.9. Notice how the vector $\mathbf{t}$ (representing row 0) and the matrix $\mathbf{T}$ (representing the other rows) together correspond to the rows of the *initial* tableau in Table 5.8, whereas the vector $\mathbf{t}^*$ and matrix $\mathbf{T}^*$ together correspond to the rows of the *final* tableau in Table 5.8. This table also shows these vectors and matrices partitioned into three parts: The coefficients of the *original* variables, the coefficients of the *slack* variables (our focus), and the right-hand side. Once again, the notation distinguishes between parts of the *initial* tableau and the *final* tableau by using an asterisk only in the latter case.

Table 5.9 General Notation for Initial and Final Simplex Tableaux in Matrix Form, Illustrated by Wyndor Glass Co. Problem

Initial Tableau:

Row 0: $\mathbf{t} = [-3, -5 \mid 0, 0, 0 \mid 0] = [-\mathbf{c} \mid \mathbf{0} \mid 0]$.

Other rows: $\mathbf{T} = \begin{bmatrix} 1 & 0 & 1 & 0 & 0 & 4 \\ 0 & 2 & 0 & 1 & 0 & 12 \\ 3 & 2 & 0 & 0 & 1 & 18 \end{bmatrix} = [\mathbf{A} \mid \mathbf{I} \mid \mathbf{b}]$.

Combined: $\begin{bmatrix} \mathbf{t} \\ \mathbf{T} \end{bmatrix} = \begin{bmatrix} -\mathbf{c} & \mathbf{0} & 0 \\ \mathbf{A} & \mathbf{I} & \mathbf{b} \end{bmatrix}$.

Final Tableau:

Row 0: $\mathbf{t}^* = [0, 0 \mid 0, \tfrac{3}{2}, 1 \mid 36] = [\mathbf{z}^* - \mathbf{c} \mid \mathbf{y}^* \mid Z^*]$.

Other rows: $\mathbf{T}^* = \begin{bmatrix} 0 & 0 & 1 & \tfrac{1}{3} & -\tfrac{1}{3} & 2 \\ 0 & 1 & 0 & \tfrac{1}{2} & 0 & 6 \\ 1 & 0 & 0 & -\tfrac{1}{3} & \tfrac{1}{3} & 2 \end{bmatrix} = [\mathbf{A}^* \mid \mathbf{S}^* \mid \mathbf{b}^*]$.

Combined: $\begin{bmatrix} \mathbf{t}^* \\ \mathbf{T}^* \end{bmatrix} = \begin{bmatrix} \mathbf{z}^* - \mathbf{c} & \mathbf{y}^* & Z^* \\ \mathbf{A}^* & \mathbf{S}^* & \mathbf{b}^* \end{bmatrix}$.

For the coefficients of the *slack* variables (the middle part) in the *initial* tableau of Table 5.9, notice the null vector **0** in row 0 and the identity matrix **I** below, which provide the keys for the fundamental insight. The vector and matrix in the same location of the *final* tableau, **y*** and **S***, then play a prominent role in the equations for the fundamental insight. (This matrix was denoted by $\mathbf{B}^{-1}$ for *any* tableau in Sec. 5.2, but we now are letting **S*** denote this particular matrix for just the final tableau, where S stands for *Slack* variable coefficients.) **A** and **b** in the *initial* tableau turn into **A*** and **b*** in the *final* tableau. For row 0 of the *final* tableau, the coefficients of the *original* variables are **z*** − **c** (so the vector **z*** is what has been added to the vector of *initial* coefficients, −**c**), and the right-hand side Z^* denotes the *optimal* value of Z.

Now suppose that you are given the initial tableau, **t** and **T,** and just **y*** and **S*** from the final tableau. How can this information alone be used to calculate the rest of the final tableau? The answer is provided by Table 5.7. This table includes some information that is not directly relevant to our current discussion, namely, how **y*** and **S*** themselves can be calculated ($\mathbf{y}^* = \mathbf{c}_B\mathbf{B}^{-1}$ and $\mathbf{S}^* = \mathbf{B}^{-1}$) by knowing the current set of basic variables and so the current *basis matrix* **B**. However, the lower part of this table (which can represent either an intermediate or final simplex tableau) also shows how the rest of the tableau can be obtained from the coefficients of the slack variables, which is summarized as follows.

Fundamental Insight

1. $\mathbf{t}^* = \mathbf{t} + \mathbf{y}^*\mathbf{T} = [\mathbf{y}^*\mathbf{A} - \mathbf{c} \mid \mathbf{y}^* \mid \mathbf{y}^*\mathbf{b}].$
2. $\mathbf{T}^* = \mathbf{S}^*\mathbf{T} = [\mathbf{S}^*\mathbf{A} \mid \mathbf{S}^* \mid \mathbf{S}^*\mathbf{b}].$

We already used these two equations when dealing with iteration 2 for the Wyndor Glass Co. problem in the preceding subsection. In particular, the right-hand side of the expression for final row 0 for iteration 2 is just $\mathbf{t}^* + \mathbf{y}^*\mathbf{T}$, and the second line of the expression for final rows 1–3 is just **S*T**.

Now let us summarize the mathematical logic behind the two equations for the fundamental insight. To derive equation 2, recall that the entire sequence of algebraic operations performed by the simplex method (excluding those involving row 0) is equivalent to premultiplying **T** by some matrix, call it **M**. Therefore,

$$\mathbf{T}^* = \mathbf{MT},$$

but now we need to identify **M**. Writing out the component parts of **T** and **T***, this **T*** = **MT** equation becomes

$$[\mathbf{A}^* \mid \mathbf{S}^* \mid \mathbf{b}^*] = \mathbf{M}\,[\mathbf{A} \mid \mathbf{I} \mid \mathbf{b}]$$
$$= [\mathbf{MA} \mid \mathbf{M} \mid \mathbf{Mb}].$$

Because the middle (or any other) component of these equal matrices must be the same, it follows that **M** = **S***, so equation 2 is a valid equation.

Equation 1 is derived in a similar fashion by noting that the entire sequence of algebraic operations involving row 0 amounts to adding some linear combination of the rows in **T** to **t**, which is equivalent to adding to **t** some *vector* times **T**. Denoting this vector by **v**, we thereby have

$$\mathbf{t}^* = \mathbf{t} + \mathbf{vT},$$

but $\mathbf{v}$ still needs to be identified. Writing out the component parts of $\mathbf{t}$ and $\mathbf{t}^*$ yields

$$[\mathbf{z}^* - \mathbf{c} \mid \mathbf{y}^* \mid \mathbf{Z}^*] = [-\mathbf{c} \mid \mathbf{0} \mid \mathbf{0}] + \mathbf{v} [\mathbf{A} \mid \mathbf{I} \mid \mathbf{b}]$$

$$= [-\mathbf{c} + \mathbf{vA} \mid \mathbf{v} \mid \mathbf{vb}].$$

Equating the middle component of these equal vectors gives $\mathbf{v} = \mathbf{y}^*$, which validates equation 1.

Thus far, the fundamental insight has been described under the assumption that the original model is in *our standard form* described in Sec. 3.2. However, the above mathematical logic now reveals just what adjustments are needed for other forms of the original model. The key is the identity matrix $\mathbf{I}$ in the initial tableau, which turns into $\mathbf{S}^*$ in the final tableau. If some *artificial variables* must be introduced into the initial tableau to serve as initial basic variables, then it is the set of columns (appropriately ordered) for *all* of the initial basic variables (both slack and artificial) that form $\mathbf{I}$ in this tableau. The *same* columns in the final tableau provide $\mathbf{S}^*$ for the $\mathbf{T}^* = \mathbf{S}^*\mathbf{T}$ equation and $\mathbf{y}^*$ for the $\mathbf{t}^* = \mathbf{t} + \mathbf{y}^*\mathbf{T}$ equation. If Big M's were introduced into the preliminary row 0 as coefficients for artificial variables, then the $\mathbf{t}$ for the $\mathbf{t}^* = \mathbf{t} + \mathbf{y}^*\mathbf{T}$ equation is the row 0 for the initial tableau after algebraically eliminating these nonzero coefficients for basic variables. (Alternatively, the preliminary row 0 can be used for $\mathbf{t}$, but then these M's must be subtracted from the final row 0 to give $\mathbf{y}^*$.) (See Prob. 35.)

Applications

The *fundamental insight* has a variety of important applications in linear programming. One of these applications involves the *revised simplex method*. As described in the preceding section (see Table 5.7), this method used $\mathbf{S}^* = \mathbf{B}^{-1}$ and the initial tableau to calculate all the relevant numbers in the current tableau for *every* iteration. It goes even further than the fundamental insight by using $\mathbf{B}^{-1}$ to calculate $\mathbf{y}^*$ itself as $\mathbf{y}^* = \mathbf{c}_B\mathbf{B}^{-1}$.

Another application involves the interpretation of the *shadow prices* $(y_1^*, y_2^*, \ldots, y_m^*)$ described in Sec. 4.7. The fundamental insight reveals that Z^* (the value of Z for the optimal solution) is

$$Z^* = \mathbf{y}^*\mathbf{b} = \sum_{i=1}^{m} y_i^* b_i,$$

so, for example,

$$Z^* = 0b_1 + \tfrac{3}{2}b_2 + b_3$$

for the Wyndor Glass Co. problem. This equation immediately yields the interpretation for the y_i^* given in Sec. 4.7.

Another group of extremely important applications involves various *post-optimality tasks* (reoptimization technique, sensitivity analysis, parametric linear programming—described in Sec. 4.7) that involve investigating the effect of making one or more changes in the original model. In particular, suppose that the simplex method already has been applied to obtain an optimal solution (as well as $\mathbf{y}^*$ and $\mathbf{S}^*$) for the original model, and then these changes are made. If exactly the same sequence of algebraic operations were to be applied to the revised initial tableau, what would be the resulting changes in the *final* tableau? Because $\mathbf{y}^*$ and $\mathbf{S}^*$ don't change, the fundamental insight reveals the answer immediately.

For example, consider the change from $b_2 = 12$ to $b_2 = 13$ as illustrated in Fig. 4.3 for the Wyndor Glass Co. problem. It isn't necessary to *solve* for the new optimal solution $(x_1, x_2) = (\frac{5}{3}, \frac{13}{2})$ because the values of the basic variables in the final tableau ($\mathbf{b}^*$) are immediately revealed by the fundamental insight:

$$\begin{bmatrix} x_3 \\ x_2 \\ x_1 \end{bmatrix} = \mathbf{b}^* = \mathbf{S}^*\mathbf{b} = \begin{bmatrix} 1 & \frac{1}{3} & -\frac{1}{3} \\ 0 & \frac{1}{2} & 0 \\ 0 & -\frac{1}{3} & \frac{1}{3} \end{bmatrix} \begin{bmatrix} 4 \\ 13 \\ 18 \end{bmatrix} = \begin{bmatrix} \frac{7}{3} \\ \frac{13}{2} \\ \frac{5}{3} \end{bmatrix}.$$

There is an even easier way to make this calculation. Since the only change is in the *second* component of $\mathbf{b}$, which gets premultiplied by only the *second* column of $\mathbf{S}^*$, the *change* in $\mathbf{b}^*$ can be calculated as simply

$$\Delta\mathbf{b}^* = \begin{bmatrix} \frac{1}{3} \\ \frac{1}{2} \\ -\frac{1}{3} \end{bmatrix} \Delta b_2 = \begin{bmatrix} \frac{1}{3} \\ \frac{1}{2} \\ -\frac{1}{3} \end{bmatrix},$$

so the original values of the basic variables in the final tableau ($x_3 = 2$, $x_2 = 6$, $x_1 = 2$) now become

$$\begin{bmatrix} x_3 \\ x_2 \\ x_1 \end{bmatrix} = \begin{bmatrix} 2 \\ 6 \\ 2 \end{bmatrix} + \begin{bmatrix} \frac{1}{3} \\ \frac{1}{2} \\ -\frac{1}{3} \end{bmatrix} = \begin{bmatrix} \frac{7}{3} \\ \frac{13}{2} \\ \frac{5}{3} \end{bmatrix}.$$

(If any of these new values were *negative,* and thus infeasible, then the *reoptimization technique* described in Sec. 4.7 would be applied, starting from this revised final tableau.) Applying *incremental analysis* to the preceding equation for Z^* also immediately yields

$$\Delta Z = \Delta Z^* = \tfrac{3}{2} \Delta b_2 = \tfrac{3}{2}.$$

The fundamental insight can be applied to investigating other kinds of changes in the original model in a very similar fashion; it is the crux of the *sensitivity analysis* procedure described in the latter part of Chap. 6.

You also will see in the next chapter that the fundamental insight plays a key role in the very useful *duality theory* for linear programming.

5.4 Conclusions

Although the simplex method is an algebraic procedure, it is based on some fairly simple geometric concepts. These concepts enable the algorithm to examine only a relatively small number of basic feasible solutions before reaching and identifying an optimal solution.

The revised simplex method provides an effective way of adapting the simplex method for computer implementation.

The final simplex tableau includes complete information on how it can be algebraically reconstructed directly from the initial simplex tableau. This fundamental insight has some very important applications, especially for post-optimality analysis.

1. Dantzig, George B.: *Linear Programming and Extensions,* Princeton University Press, Princeton, N.J., 1963.
2. Gass, Saul I.: *Linear Programming,* 5th ed., McGraw-Hill, New York, 1985.
3. Luenberger, David G.: *Linear and Nonlinear Programming,* 2d ed., Addison-Wesley, Reading, Mass., 1984.
4. Murty, Katta G.: *Linear Programming,* 2d ed., Wiley, New York, 1983.
5. Schriver, Alexander: *Theory of Linear and Integer Programming,* Wiley, New York, 1986.
6. Solow, Daniel: *Linear Programming: An Introduction to Finite Improvement Algorithms,* Elsevier Science, New York, 1984.

PROBLEMS

1.* Consider the following problem.

$$\text{Maximize} \quad Z = 3x_1 + 2x_2,$$

subject to

$$2x_1 + x_2 \leq 6$$

$$x_1 + 2x_2 \leq 6$$

and

$$x_1 \geq 0, \qquad x_2 \geq 0.$$

(a) Solve this problem graphically. Identify the corner-point feasible solutions by circling them on the graph.

(b) Identify all the sets of two defining equations for this problem. For each one, solve (if a solution exists) for the corresponding corner-point solution, and classify it as a corner-point feasible solution or corner-point infeasible solution.

(c) Introduce slack variables in order to write the functional constraints in *augmented form*.

(d) For each set of defining equations from part (b), identify the *indicating variable* for each equation, display the equations from part (c) *after* deleting these indicating (nonbasic) variables, and give the resulting basic solution.

(e) Without executing the simplex method, use its geometric interpretation (and the objective function) to identify the path (sequence of corner-point feasible solutions) it would follow to reach the optimal solution. For each of these corner-point feasible solutions in turn, rewrite the corresponding information from part (d) and then identify the following decisions being made for the next iteration: (i) which *defining equation* is being deleted and which is being added; (ii) which *indicating variable* is being deleted (the entering basic variable) and which is being added (the leaving basic variable).

2. Follow the instructions of Prob. 1 for the linear programming model in Prob. 4 of Chap. 3.

3. Consider the following problem.

$$\text{Maximize} \quad Z = 2x_1 + 3x_2,$$

subject to

$$-3x_1 + x_2 \leq 1$$

$$4x_1 + 2x_2 \leq 20$$

$$4x_1 - x_2 \leq 10$$

$$-x_1 + 2x_2 \leq 5$$

and

$$x_1 \geq 0, \qquad x_2 \geq 0.$$

(a) Solve this problem graphically. Identify the corner-point feasible solutions by circling them on the graph.

(b) Develop a table giving each of the corner-point feasible solutions and the corresponding defining equations, basic feasible solution, and nonbasic variables. Calculate Z for each of these solutions and use just this information to identify the optimal solution.

(c) Develop the corresponding table for the corner-point infeasible solutions, and so on. Also identify the sets of defining equations and nonbasic variables that do not yield a solution.

4. Consider the following problem.

$$\text{Maximize} \quad Z = 2x_1 - x_2 + x_3,$$

subject to

$$3x_1 + x_2 + x_3 \leq 60$$

$$x_1 - x_2 + 2x_3 \leq 10$$

$$x_1 + x_2 - x_3 \leq 20$$

and

$$x_1 \geq 0, \quad x_2 \geq 0, \quad x_3 \geq 0.$$

After introducing slack variables and then performing one complete iteration of the simplex method, the following simplex tableau is obtained.

Iteration	Basic Variable	Eq. No.	Z	x_1	x_2	x_3	x_4	x_5	x_6	Right Side
	Z	0	1	0	-1	3	0	2	0	20
1	x_4	1	0	0	4	-5	1	-3	0	30
	x_1	2	0	1	-1	2	0	1	0	10
	x_6	3	0	0	2	-3	0	-1	1	10

(Header spanning columns x_1 through x_6: *Coefficient of*)

(a) Identify the *corner-point feasible solution* obtained at iteration 1.

(b) Identify the *constraint boundary equations* that define this corner-point feasible solution.

5. Consider the three-variable linear programming problem shown in Fig. 5.2.

(a) Construct a table like Table 5.1 giving the set of defining equations for each corner-point feasible solution.

(b) What are the defining equations for the corner-point *infeasible* solution (6, 0, 5)?

(c) Identify one of the systems of three constraint boundary equations that yield neither a corner-point feasible solution nor a corner-point infeasible solution. Explain why this occurs for this system.

6. Consider the linear programming problem given in Table 6.1 as the *dual problem* for the Wyndor Glass Co. example.

(a) Identify the 10 sets of defining equations for this problem. For each one, solve (if a solution exists) for the corresponding corner-point solution, and classify it as a corner-point feasible solution or corner-point infeasible solution.

(b) For each corner-point solution, give the corresponding basic solution and its set of nonbasic variables. (Compare with Table 6.9.)

7. Consider the following problem.

$$\text{Minimize} \quad Z = x_1 + 2x_2,$$

subject to

$$-x_1 + x_2 \leq 15$$

$$2x_1 + x_2 \leq 90$$

$$x_2 \geq 30$$

and

$$x_1 \geq 0, \quad x_2 \geq 0.$$

(a) Solve this problem graphically.

(b) Develop a table giving each of the corner-point feasible solutions and the corresponding defining equations, basic feasible solution, and nonbasic variables.

8. Reconsider Prob. 17 in Chap. 4.

(a) Identify the 10 sets of defining equations for this problem. For each one, solve (if a solution exists) for the corresponding corner-point solution, and classify it as a corner-point feasible solution or a corner-point infeasible solution.

(b) For each corner-point solution, give the corresponding basic solution and its set of nonbasic variables.

9. Reconsider Prob. 3 in Chap. 3.

(a) Identify the 15 sets of defining equations for this problem. For each one, solve (if a solution exists) for the corresponding corner-point solution, and classify it as a corner-point feasible solution or a corner-point infeasible solution.

(b) For each corner-point solution, give the corresponding basic solution and its set of nonbasic variables.

10.* Reconsider Prob. 23 in Chap. 4. Now you are given the information that the basic variables in the optimal solution are x_2 and x_3. Use this information to identify a system of three constraint boundary equations whose simultaneous solution must be this optimal solution. Then solve this system of equations to obtain this solution.

11. Reconsider Prob. 11 in Chap. 4. Using the given information and the theory of the simplex method, analyze the constraints of the problem in order to identify a system of three constraint boundary equations whose simultaneous solution must be the optimal solution (not augmented). Then solve this system of equations to obtain this solution.

12. Consider the following problem.

$$\text{Maximize} \quad Z = 2x_1 + 2x_2 + 3x_3,$$

subject to

$$2x_1 + x_2 + 2x_3 \leq 4$$

$$x_1 + x_2 + x_3 \leq 3$$

and

$$x_1 \geq 0, \quad x_2 \geq 0, \quad x_3 \geq 0.$$

Let x_4 and x_5 be the slack variables for the respective functional constraints. Starting with these two variables as the basic variables for the initial basic feasible solution, you now are given the information that the simplex method proceeds as follows to obtain the optimal solution in two iterations: (1) in iteration 1, the entering basic variable is x_3 and the leaving basic variable is x_4; (2) in iteration 2, the entering basic variable is x_2 and the leaving basic variable is x_5.

(a) Develop a three-dimensional drawing of the feasible region for this problem, and show the path followed by the simplex method.

(b) Give a geometric interpretation of why the simplex method followed this path.

(c) For each of the two edges of the feasible region traversed by the simplex method, give the equation of each of the two constraint boundaries on which it lies, and then give the equation of the additional constraint boundary at each endpoint.

(d) Identify the set of defining equations for each of the three corner-point feasible solutions (including the initial one) obtained by the simplex method. Use the defining equations to solve for these solutions.

(e) For each corner-point feasible solution obtained in part (d), give the corresponding basic feasible solution and its set of nonbasic variables. Explain how these nonbasic variables identify the defining equations obtained in part (d).

13. Consider the following problem.

$$\text{Maximize} \quad Z = 3x_1 + 4x_2 + 2x_3,$$

subject to

$$x_1 + x_2 + x_3 \leq 20$$

$$x_1 + 2x_2 + x_3 \leq 30$$

and

$$x_1 \geq 0, \quad x_2 \geq 0, \quad x_3 \geq 0.$$

Let x_4 and x_5 be the slack variables for the respective functional constraints. Starting with these two variables as the basic variables for the initial basic feasible solution, you now are given the information that the simplex method proceeds as follows to obtain the optimal solution in two iterations: (1) in iteration 1, the entering basic variable is x_2 and the leaving basic variable is x_5; (2) in iteration 2, the entering basic variable is x_1 and the leaving basic variable is x_4.

Follow the instructions of Prob. 12 for this situation.

14. By inspecting Fig. 5.2, explain why Property 1(b) for corner-point feasible solutions holds for this problem if it has the following objective function.

(a) Maximize $Z = x_3$.

(b) Maximize $Z = -x_1 + 2x_3$.

15. Consider the three-variable linear programming problem shown in Fig. 5.2.

(a) Explain in geometric terms why the set of solutions satisfying any individual constraint is a *convex set* as defined in Appendix 1.

(b) Use the conclusion in part (a) to explain why the entire feasible region (the set of solutions that simultaneously satisfies every constraint) is a *convex set*.

16. Suppose that the three-variable linear programming problem given in Fig. 5.2 has the objective function,

$$\text{Maximize} \quad Z = 3x_1 + 4x_2 + 3x_3.$$

Without using the algebra of the simplex method, apply just its geometric reasoning (including choosing the edge giving the maximum rate of increase of Z) to determine and explain the path it would follow in Fig. 5.2 from the origin to the optimal solution.

17. Consider the three-variable linear programming problem shown in Fig. 5.2.

(a) Construct a table like Table 5.4 giving the indicating variable for each constraint boundary equation and original constraint.

(b) For the corner-point feasible solution (2, 4, 3) and its three adjacent corner-point feasible solutions (4, 2, 4), (0, 4, 2), and (2, 4, 0), construct a table like Table 5.5 giving the corresponding defining equations, basic feasible solution, and nonbasic variables.

(c) Use the sets of defining equations from part (b) to demonstrate that (4, 2, 4), (0, 4, 2), and (2, 4, 0) indeed are adjacent to (2, 4, 3), but that none of these three corner-point feasible solutions are adjacent to each other. Then use the sets of nonbasic variables from part (b) to demonstrate the same thing.

18. The formula for the line passing through $(2, 4, 3)$ and $(4, 2, 4)$ in Fig. 5.2 can be written as

$$(2, 4, 3) + \alpha[(4, 2, 4) - (2, 4, 3)] = (2, 4, 3) + \alpha(2, -2, 1),$$

where $0 \le \alpha \le 1$ for just the line segment between these points. After augmenting with the slack variables x_4, x_5, x_6, x_7 for the respective functional constraints, this formula becomes

$$(2, 4, 3, 2, 0, 0, 0) + \alpha(2, -2, 1, -2, 2, 0, 0).$$

Use this formula directly to answer each of the following questions, and thereby relate the algebra and geometry of the simplex method as it goes through one iteration in moving from $(2, 4, 3)$ to $(4, 2, 4)$. (You are given the information that it is moving along this line segment.)

(a) What is the entering basic variable?
(b) What is the leaving basic variable?
(c) What is the new basic feasible solution?

19. Consider a two-variable mathematical programming problem that has the feasible region shown on the graph, where the six dots correspond to corner-point feasible solutions. The problem has a linear objective function, and the two dashed lines are objective function lines passing through the optimal solution $(4, 5)$ and the second-best corner-point feasible solution $(2, 5)$. Note that the nonoptimal solution $(2, 5)$ is better than both of its adjacent corner-point feasible solutions, which violates Property 3 in Sec. 5.1 for corner-point feasible solutions in linear programming.

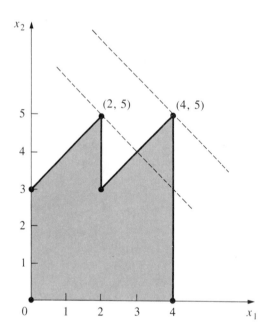

Demonstrate that this problem *cannot* be a linear programming problem by constructing the feasible region that would result if the six line segments on the boundary were constraint boundaries for *linear programming constraints*.

20. Consider the following problem.

$$\text{Maximize} \quad Z = 8x_1 + 4x_2 + 6x_3 + 3x_4 + 9x_5,$$

subject to

$$x_1 + 2x_2 + 3x_3 + 3x_4 \qquad \le 180 \quad \text{(resource 1)}$$

$$4x_1 + 3x_2 + 2x_3 + \ x_4 + \ x_5 \le 270 \quad \text{(resource 2)}$$

$$x_1 + 3x_2 \qquad + \ x_4 + 3x_5 \le 180 \quad \text{(resource 3)}$$

and

$$x_j \ge 0 \qquad (j = 1, \dots, 5).$$

You are given the facts that the basic variables in the optimal solution are x_3, x_1, and x_5, and that

$$\begin{bmatrix} 3 & 1 & 0 \\ 2 & 4 & 1 \\ 0 & 1 & 3 \end{bmatrix}^{-1} = \tfrac{1}{27} \begin{bmatrix} 11 & -3 & 1 \\ -6 & 9 & -3 \\ 2 & -3 & 10 \end{bmatrix}.$$

(a) Use the given information to identify the optimal solution.

(b) Use the given information to identify the *shadow prices* for the three resources.

21.* Use the revised simplex method to solve the following problem.

$$\text{Maximize} \qquad Z = 5x_1 + 8x_2 + 7x_3 + 4x_4 + 6x_5,$$

subject to

$$2x_1 + 3x_2 + 3x_3 + 2x_4 + 2x_5 \le 20$$

$$3x_1 + 5x_2 + 4x_3 + 2x_4 + 4x_5 \le 30$$

and

$$x_j \ge 0, \qquad \text{for } j = 1, 2, 3, 4, 5.$$

22. Use the revised simplex method to solve the linear programming model given in Prob. 4, Chap. 4.

23. Reconsider Prob. 1. For the sequence of corner-point solutions identified in part (e), construct the *basis matrix* **B** for each of the corresponding basic feasible solutions. For each one, invert **B** manually and use this $\mathbf{B}^{-1}$ to calculate the current solution and then perform the next iteration (or demonstrate that the current solution is optimal).

24. Use the revised simplex method to solve the linear programming model given in Prob. 2, Chap. 4.

25. Use the revised simplex method to solve the linear programming model given in Prob. 7, Chap. 4.

26. Use the revised simplex method to solve each of the following linear programming models:

(a) Model given in Prob. 4, Chap. 3.

(b) Model given in Prob. 9, Chap. 4.

27.* Consider the following problem.

$$\text{Maximize} \qquad Z = x_1 - x_2 + 2x_3,$$

subject to

$$2x_1 - 2x_2 + 3x_3 \le 5$$

$$x_1 + x_2 - x_3 \le 3$$

$$x_1 - x_2 + x_3 \le 2$$

and

$$x_1 \ge 0, \qquad x_2 \ge 0, \qquad x_3 \ge 0.$$

Let x_4, x_5, and x_6 denote the slack variables for the respective constraints. After applying the simplex method, a portion of the final simplex tableau is as follows:

Basic Variable	Eq. No.	Z	x_1	x_2	x_3	x_4	x_5	x_6	Right Side
Z	0	1				1	1	0	
x_2	1	0				1	3	0	
x_6	2	0				0	1	1	
x_3	3	0				1	2	0	

(a) Use the *fundamental insight* presented in Sec. 5.3 to identify the missing numbers in the final simplex tableau. Show your calculations.

(b) Identify the defining equations of the corner-point feasible solution corresponding to the optimal basic feasible solution in the final simplex tableau.

28. Consider the following problem.

$$\text{Maximize} \quad Z = 4x_1 + 3x_2 + x_3 + 2x_4,$$

subject to

$$4x_1 + 2x_2 + x_3 + x_4 \leq 5$$

$$3x_1 + x_2 + 2x_3 + x_4 \leq 4$$

and

$$x_1 \geq 0, \quad x_2 \geq 0, \quad x_3 \geq 0, \quad x_4 \geq 0.$$

Let x_5 and x_6 denote the slack variables for the respective constraints. After applying the simplex method, a portion of the final simplex tableau is as follows:

Basic Variable	Eq. No.	Z	x_1	x_2	x_3	x_4	x_5	x_6	Right Side
Z	0	1					1	1	
x_2	1	0					1	-1	
x_4	2	0					-1	2	

(a) Use the *fundamental insight* presented in Sec. 5.3 to identify the missing numbers in the final simplex tableau. Show your calculations.

(b) Identify the defining equations of the corner-point feasible solution corresponding to the optimal basic feasible solution in the final simplex tableau.

29. Consider the following problem.

$$\text{Maximize} \quad Z = 6x_1 + x_2 + 2x_3,$$

subject to

$$2x_1 + 2x_2 + \tfrac{1}{2}x_3 \leq 2$$

$$-4x_1 - 2x_2 - \tfrac{3}{2}x_3 \leq 3$$

$$x_1 + 2x_2 + \tfrac{1}{2}x_3 \leq 1$$

and

$$x_1 \geq 0, \quad x_2 \geq 0, \quad x_3 \geq 0.$$

Let x_4, x_5, and x_6 denote the slack variables for the respective constraints. After applying the simplex method, a portion of the final simplex tableau is as follows:

Basic Variable	Eq. No.	Coefficient of							Right Side
		Z	x_1	x_2	x_3	x_4	x_5	x_6	
Z	0	1				2	0	2	
x_5	1	0				1	1	2	
x_3	2	0				-2	0	4	
x_1	3	0				1	0	-1	

Use the *fundamental insight* presented in Sec. 5.3 to identify the missing numbers in the final simplex tableau. Show your calculations.

30. For iteration 2 of the example in Sec. 5.3, the following expression was shown.

$$\text{Final row } 0 = [-3, \quad -5 \mid 0, \quad 0, \quad 0 \mid 0] + [0, \quad \tfrac{3}{2}, \quad 1]\begin{bmatrix} 1 & 0 & 1 & 0 & 0 & 4 \\ 0 & 2 & 0 & 1 & 0 & 12 \\ 3 & 2 & 0 & 0 & 1 & 18 \end{bmatrix}.$$

Derive this expression by combining the algebraic operations (in matrix form) for iterations 1 and 2 that affect row 0.

31. Consider the following problem.

$$\text{Maximize} \quad Z = x_1 - x_2 + 2x_3,$$

subject to

$$x_1 + x_2 + 3x_3 \le 15$$

$$2x_1 - x_2 + x_3 \le 2$$

$$-x_1 + x_2 + x_3 \le 4$$

and

$$x_1 \ge 0, \quad x_2 \ge 0, \quad x_3 \ge 0.$$

Let x_4, x_5, and x_6 denote the slack variables for the respective constraints. After applying the simplex method, a portion of the final simplex tableau is as follows:

Basic Variable	Eq. No.	Coefficient of							Right Side
		Z	x_1	x_2	x_3	x_4	x_5	x_6	
Z	0	1				0	$\tfrac{3}{2}$	$\tfrac{1}{2}$	
x_4	1	0				1	-1	-2	
x_3	2	0				0	$\tfrac{1}{2}$	$\tfrac{1}{2}$	
x_2	3	0				0	$-\tfrac{1}{2}$	$\tfrac{1}{2}$	

(a) Use the *fundamental insight* presented in Sec. 5.3 to identify the missing numbers in the final simplex tableau. Show your calculations.

(b) Identify the defining equations of the corner-point feasible solution corresponding to the optimal basic feasible solution in the final simplex tableau.

32. Consider the following problem.

$$\text{Maximize} \quad Z = 20x_1 + 6x_2 + 8x_3,$$

subject to

$$8x_1 + 2x_2 + 3x_3 \leq 200$$

$$4x_1 + 3x_2 \qquad \leq 100$$

$$2x_1 \qquad + x_3 \leq 50$$

$$x_3 \leq 20$$

and

$$x_1 \geq 0, \qquad x_2 \geq 0, \qquad x_3 \geq 0.$$

Let x_4, x_5, x_6, and x_7 denote the slack variables for the first through fourth constraints, respectively. Suppose that after some number of iterations of the simplex method a portion of the current simplex tableau is as follows:

Basic	Eq.				Coefficient of					Right
Variable	No.	Z	x_1	x_2	x_3	x_4	x_5	x_6	x_7	Side
Z	0	1				$\frac{9}{4}$	$\frac{1}{2}$	0	0	
x_1	1	0				$\frac{3}{16}$	$-\frac{1}{8}$	0	0	
x_2	2	0				$-\frac{1}{4}$	$\frac{1}{2}$	0	0	
x_6	3	0				$-\frac{3}{8}$	$\frac{1}{4}$	1	0	
x_7	4	0				0	0	0	1	

(a) Use the *fundamental insight* presented in Sec. 5.3 to identify the missing numbers in the current simplex tableau. Show your calculations.

(b) Indicate which of these missing numbers would be generated by the *revised simplex method* in order to perform the next iteration.

(c) Identify the defining equations of the corner-point feasible solution corresponding to the basic feasible solution in the current simplex tableau.

33. Most of the description of the *fundamental insight* presented in Sec. 5.3 assumes that the problem is in *our standard form*. Now consider each of the following other forms, where the additional adjustments in the initialization step are those presented in Sec. 4.6, including using artificial variables and the Big M method where appropriate. Describe the resulting adjustments in the *fundamental insight*.

(a) Equality constraints.

(b) Negative right-hand sides.

(c) Variables allowed to be negative (with no lower bound).

34. Reconsider Prob. 18 in Chap. 4. For this model that is not in *our standard form*, construct the complete first simplex tableau for the simplex method, and then identify the columns that will contain **S*** for applying the *fundamental insight* in the *final* tableau. Explain why these are the appropriate columns.

35. Consider the following problem.

$$\text{Minimize} \quad Z = 2x_1 + 3x_2 + 2x_3,$$

subject to

$$x_1 + 4x_2 + 2x_3 \geq 8$$

$$3x_1 + 2x_2 \qquad \geq 6$$

and

$$x_1 \geq 0, \qquad x_2 \geq 0, \qquad x_3 \geq 0.$$

Let x_4 and x_6 be the surplus variables for the first and second constraints, respectively. Let $\bar{x}_5$ and $\bar{x}_7$ be the corresponding artificial variables. After making the adjustments described in Sec. 4.6 for this model form when using the Big M method, the initial simplex tableau ready to apply the simplex method is as follows:

Basic Variable	Eq. No.	Z	x_1	x_2	x_3	x_4	$\bar{x}_5$	x_6	$\bar{x}_7$	Right Side
					Coefficient of					
Z	0	-1	$(-4M + 2)$	$(-6M + 3)$	$(-2M + 2)$	M	0	M	0	$-14M$
$\bar{x}_5$	1	0	1	4	2	-1	1	0	0	8
$\bar{x}_7$	2	0	3	2	0	0	0	-1	1	6

After applying the simplex method, a portion of the final simplex tableau is as follows:

Basic Variable	Eq. No.	Z	x_1	x_2	x_3	x_4	$\bar{x}_5$	x_6	$\bar{x}_7$	Right Side
					Coefficient of					
Z	0	-1					$(M - 0.5)$		$(M - 0.5)$	
x_2	1	0					0.3		-0.1	
x_1	2	0					-0.2		0.4	

(a) Based on the above tableaux, use the *fundamental insight* presented in Sec. 5.3 to identify the missing numbers in the final simplex tableau. Show your calculations.

(b) Examine the mathematical logic presented in Sec. 5.3 to validate the *fundamental insight* (see the $\mathbf{T^*} = \mathbf{MT}$ and $\mathbf{t^*} = \mathbf{t} + \mathbf{vT}$ equations and the subsequent derivations of $\mathbf{M}$ and $\mathbf{v}$). This logic assumes that the original model *fits our standard form*, whereas the current problem does not fit this form. Show how, with minor adjustments, this same logic applies to the current problem when $\mathbf{t}$ is row 0 and $\mathbf{T}$ is rows 1–2 in the initial simplex tableau given above. Derive $\mathbf{M}$ and $\mathbf{v}$ for this problem.

(c) When applying the $\mathbf{t^*} = \mathbf{t} + \mathbf{vT}$ equation, another option is to use $\mathbf{t} = [2\ 3\ 2\ 0\ M\ 0\ M,\ 0]$, which is the *preliminary* row 0 before algebraically eliminating the nonzero coefficients of the initial basic variables, $\bar{x}_5$ and $\bar{x}_7$. Repeat part (b) for this equation with this new $\mathbf{t}$. After deriving the new $\mathbf{v}$, show that this equation yields the same final row 0 for this problem as the equation derived in part (b).

(d) Identify the defining equations of the corner-point feasible solution corresponding to the optimal basic feasible solution in the final simplex tableau.

36. Consider the following problem.

$$\text{Maximize} \quad Z = 2x_1 + 4x_2 + 3x_3,$$

subject to

$$x_1 + 3x_2 + 2x_3 = 20$$

$$x_1 + 5x_2 \geq 10$$

and

$$x_1 \geq 0, \quad x_2 \geq 0, \quad x_3 \geq 0.$$

Let $\bar{x}_4$ be the artificial variable for the first constraint. Let x_5 and $\bar{x}_6$ be the surplus variable and artificial variable, respectively, for the second constraint.

You are now given the information that a portion of the *final* simplex tableau is as follows:

Basic Variable	Eq. No.	Z	x_1	x_2	x_3	$\bar{x}_4$	x_5	$\bar{x}_6$	Right Side
Z	0	1				$M + 2$	0	M	
x_1	1	0				1	0	0	
x_5	2	0				1	1	-1	

(a) Extend the *fundamental insight* presented in Sec. 5.3 to identify the missing numbers in the final simplex tableau. Show your calculations.

(b) Identify the defining equations of the corner-point feasible solution corresponding to the optimal solution in the final simplex tableau.

37. Consider the following problem.

$$\text{Maximize} \quad Z = 3x_1 + 7x_2 + 2x_3,$$

subject to

$$-2x_1 + 2x_2 + x_3 \leq 10$$

$$3x_1 + x_2 - x_3 \leq 20$$

and

$$x_1 \geq 0, \quad x_2 \geq 0, \quad x_3 \geq 0.$$

You are given the fact that the basic variables in the optimal solution are x_1 and x_3.

(a) Introduce slack variables, and then use the given information to find the optimal solution directly by Gaussian elimination (see Appendix 4).

(b) Extend the work in part (a) to find the *shadow prices*.

(c) Use the given information to identify the defining equations of the optimal corner-point feasible solution, and then solve these equations to obtain the optimal solution.

(d) Construct the *basis matrix* **B** for the optimal basic feasible solution, invert **B** manually, and then use this $\mathbf{B}^{-1}$ to solve for the optimal solution and the *shadow prices* (**y***). Then apply the optimality test for the *revised simplex method* to verify that this solution is optimal.

(e) Given $\mathbf{B}^{-1}$ and **y*** from part (d), use the *fundamental insight* presented in Sec. 5.3 to construct the complete final simplex tableau.

6

Duality Theory and Sensitivity Analysis

One of the most important discoveries in the early development of linear programming was the concept of duality and its many important ramifications. This discovery revealed that every linear programming problem has associated with it another linear programming problem called the **dual**. The relationships between the dual problem and the original problem (called the **primal**) prove to be extremely useful in a variety of ways. For example, you soon will see that the shadow prices described in Sec. 4.7 actually are provided by the optimal solution for the dual problem. We shall describe many other valuable applications of duality theory in this chapter as well.

One of the key roles of duality theory is that of the interpretation and implementation of *sensitivity analysis*. As we already mentioned in Secs. 2.3, 3.3, and 4.7, sensitivity analysis is a very important part of almost every linear programming study. Because some or all of the parameter values used in the original model are just *estimates* of future conditions, the effect on the optimal solution if other conditions prevail instead needs to be investigated. Furthermore, certain parameter values (such as resource amounts) may represent *managerial decisions,* in which case the choice

145

of the parameter values may be the main issue to be studied, which would be done through sensitivity analysis.

For greater clarity, the first three sections discuss duality theory under the assumption that the *primal* linear programming problem is in *our standard form* (but with no restriction that the b_i need to be positive). Other forms are then discussed in Sec. 6.4. We begin the chapter by introducing the essence of duality theory and its applications. We then describe the economic interpretation of the dual problem (Sec. 6.2) and delve deeper into the relationships between the primal and dual problems (Sec. 6.3). Section 6.5 focuses on the role of duality theory in sensitivity analysis. The basic procedure for sensitivity analysis (which is based on the fundamental insight of Sec. 5.3) is summarized in Sec. 6.6 and illustrated in Sec. 6.7.

6.1 The Essence of Duality Theory

Using *our standard form* for the *primal problem* at the left (perhaps after conversion from another form), its *dual problem* has the form shown to the right.

<table>
<tr><td align="center">*Primal problem*</td><td align="center">*Dual problem*</td></tr>
<tr><td>

Maximize $\quad Z = \sum_{j=1}^{n} c_j x_j,$

subject to

$\sum_{j=1}^{n} a_{ij} x_j \leq b_i, \quad \text{for } i = 1, 2, \ldots, m$

and

$\quad x_j \geq 0, \quad \text{for } j = 1, 2, \ldots, n.$

</td><td>

Minimize $\quad y_0 = \sum_{i=1}^{m} b_i y_i,$

subject to

$\sum_{i=1}^{m} a_{ij} y_i \geq c_j, \quad \text{for } j = 1, 2, \ldots, n$

and

$\quad y_i \geq 0, \quad \text{for } i = 1, 2, \ldots, m.$

</td></tr>
</table>

Thus the dual problem uses exactly the same *parameters* as the primal problem, but in different locations. To highlight the comparison, now see these same two problems in *matrix notation* (as introduced at the beginning of Sec. 5.2), where **c** and $\mathbf{y} = [y_1, y_2, \ldots, y_m]$ are row vectors but **b** and **x** are column vectors.

<table>
<tr><td align="center">*Primal problem*</td><td align="center">*Dual problem*</td></tr>
<tr><td>

Maximize $\quad Z = \mathbf{cx},$

subject to

$\qquad \mathbf{Ax} \leq \mathbf{b}$

and

$\qquad \mathbf{x} \geq \mathbf{0}.$

</td><td>

Minimize $\quad y_0 = \mathbf{yb},$

subject to

$\qquad \mathbf{yA} \geq \mathbf{c}$

and

$\qquad \mathbf{y} \geq \mathbf{0}.$

</td></tr>
</table>

To illustrate, the primal and dual problems for the Wyndor Glass Co. example of Sec. 3.1 are shown in Table 6.1 in matrix form.

The **primal-dual table** for linear programming (Table 6.2) also helps to highlight the correspondence between the two problems. It shows all the linear programming parameters (the a_{ij}, b_i, and c_j) and how they are used to construct the two problems. All the headings for the primal problem are horizontal, whereas the headings for the dual problem are read by turning the book sideways. We suggest that you begin by looking at each problem *individually* by covering up the headings for the

Table 6.1 **Primal and Dual Problems for Wyndor Glass Co. Example**

Primal problem

$$\text{Maximize} \quad Z = [3, 5]\begin{bmatrix} x_1 \\ x_2 \end{bmatrix},$$

subject to

$$\begin{bmatrix} 1 & 0 \\ 0 & 2 \\ 3 & 2 \end{bmatrix}\begin{bmatrix} x_1 \\ x_2 \end{bmatrix} \leq \begin{bmatrix} 4 \\ 12 \\ 18 \end{bmatrix}$$

and •

$$\begin{bmatrix} x_1 \\ x_2 \end{bmatrix} \geq \begin{bmatrix} 0 \\ 0 \end{bmatrix}.$$

Dual problem

$$\text{Minimize} \quad y_0 = [y_1, y_2, y_3]\begin{bmatrix} 4 \\ 12 \\ 18 \end{bmatrix},$$

subject to

$$[y_1, y_2, y_3]\begin{bmatrix} 1 & 0 \\ 0 & 2 \\ 3 & 2 \end{bmatrix} \geq [3, 5]$$

and

$$[y_1, y_2, y_3] \geq [0, 0, 0].$$

other problem with your hands. Then, after you see what the table is saying for the individual problems, compare them.

Particularly notice in Table 6.2 how (1) the parameters for a *constraint* in either problem are the coefficients of a *variable* in the other problem, and (2) the coefficients for the *objective function* of either problem are the *right sides* for the other problem. Thus there is a direct correspondence between these entities in the two problems, as summarized in Table 6.3. These correspondences are a key to some of the applications of duality theory, including sensitivity analysis.

Table 6.2 **Primal-Dual Table for Linear Programming, Illustrated by Wyndor Glass Co. Example**

(a) General case

			Primal Problem					
			Coefficient of			Right Side		Coefficients for Objective Function (Minimize)
			x_1	x_2 $\cdots$	x_n			
Dual Problem	Coefficient of	y_1	a_{11}	a_{12} $\cdots$	a_{1n}	$\leq b_1$		
		y_2	a_{21}	a_{22} $\cdots$	a_{2n}	$\leq b_2$		
		y_m	a_{m1}	a_{m2} $\cdots$	a_{mn}	$\leq b_m$		
	Right Side		VI c_1	VI c_2 $\cdots$	VI c_n			

Coefficients for Objective Function (Maximize)

(b) Wyndor Glass Co. example

	x_1	x_2	
y_1	1	0	$\leq$ 4
y_2	0	2	$\leq$ 12
y_3	3	2	$\leq$ 18
	VI	VI	
	3	5	

Table 6.3 Correspondence between
Entities in Primal and Dual Problems

One Problem	Other Problem
Constraint i ⟷	variable i
Objective function ⟷	right sides

Origin of the Dual Problem

Duality theory is based directly on the *fundamental insight* (particularly with regard to row 0) presented in Sec. 5.3. To see why, we continue to use the notation introduced in Table 5.9 for row 0 of the *final* tableau, except for replacing Z^* by y_0 and dropping the asterisks from $\mathbf{z}^*$ and $\mathbf{y}^*$ when referring to *any* tableau. Thus, at *any* given iteration of the simplex method for the primal problem, the current numbers in row 0 are denoted as shown in the (partial) tableau given in Table 6.4. Also recall [see Eq. (1) in the *Mathematical Summary* subsection of Sec. 5.3] that the fundamental insight led to the following relationships between these quantities and the parameters of the original model:

$$y_0 = \mathbf{yb} = \sum_{i=1}^{m} b_i y_i,$$

$$\mathbf{z} = \mathbf{yA}, \qquad \text{so } z_j = \sum_{i=1}^{m} a_{ij} y_i, \qquad \text{for } j = 1, 2, \ldots, n.$$

The remaining key is to express what the simplex method tries to accomplish (according to the optimality test) in terms of these symbols. Specifically, it seeks a set of basic variables, and the corresponding basic feasible solution, such that *all* coefficients in row 0 are *nonnegative*. It then stops with this optimal solution. This goal is expressed symbolically as follows:

Condition for Optimality: $\qquad z_j - c_j \geq 0, \qquad \text{for } j = 1, 2, \ldots, n$

$$y_i \geq 0, \qquad \text{for } i = 1, 2, \ldots, m.$$

After substituting the preceding expression for z_j, the condition for optimality says that the simplex method can be interpreted as seeking values for $y_1, y_2, \ldots, y_m$ such that

$$y_0 = \sum_{i=1}^{m} b_i y_i,$$

subject to

$$\sum_{i=1}^{m} a_{ij} y_i \geq c_j, \quad \text{for } j = 1, 2, \ldots, n$$

and

$$y_i \geq 0, \quad \text{for } i = 1, 2, \ldots, m.$$

Table 6.4 Notation for Entries in Row 0 of Simplex Tableau

| Iteration | Basic Variable | Eq. No. | Z | Coefficient of | | | | | | | Right Side |
				x_1	x_2	$\cdots$	x_n	x_{n+1}	x_{n+2} $\cdots$ x_{n+m}	
Any	Z	0	1	$(z_1 - c_1)$	$(z_2 - c_2)$	$\cdots$	$(z_n - c_n)$	y_1	y_2 $\cdots$ y_m	y_0

But, except for lacking an objective for y_0, this problem is precisely the *dual problem*! To complete the formulation, let us now explore what the missing objective should be.

Since y_0 is just the current value of Z, and since the objective for the primal problem is to maximize Z, a natural first reaction is that y_0 should be maximized also. However, this is not correct for the following rather subtle reason: The only *feasible* solutions for this new problem are those that satisfy the condition for *optimality* for the primal problem. Therefore, it is *only* the optimal solution for the primal problem that corresponds to a feasible solution for this new problem. As a consequence, the optimal value of Z in the primal problem is the *minimum* feasible value of y_0 in the new problem, so y_0 should be minimized. (The full justification for this conclusion is provided by the relationships we develop in Sec. 6.3.) Adding this objective of minimizing y_0 gives the *complete* dual problem.

Consequently, the dual problem may be viewed as a restatement in linear programming terms of the *goal* of the simplex method, namely, to reach a solution for the primal problem that *satisfies the optimality test*. *Before* this goal has been reached, the corresponding **y** in row 0 (coefficients of slack variables) of the current tableau must be *infeasible* for the *dual problem*. However, *after* the goal is reached, the corresponding **y** must be an *optimal solution* (labeled **y***) for the *dual problem,* because it is a feasible solution that attains the minimum feasible value of y_0. This optimal solution $(y_1^*, y_2^*, \ldots, y_m^*)$ provides for the primal problem the *shadow prices* that were described in Sec. 4.7. Furthermore, this optimal y_0 is just the optimal value of Z, so the *optimal objective function values are equal* for the two problems. This fact also implies that $\mathbf{cx} \leq \mathbf{yb}$ for any **x** and **y** that are *feasible* for the primal and dual problems, respectively.

To illustrate, Table 6.5 shows row 0 for the respective iterations when the simplex method is applied to the Wyndor Glass Co. example. In each case, row 0 is partitioned into three parts: the coefficients of the *original* variables (x_1, x_2), the coefficients of the *slack* variables (x_3, x_4, x_5), and the right-hand side (value of Z). Each row 0 identifies a solution for the dual problem, as shown to its right in Table 6.5. Included are the values of

$$z_1 - c_1 = y_1 + 3y_3 - 3,$$

$$z_2 - c_2 = 2y_2 + 2y_3 - 5,$$

Table 6.5 Respective Row 0's and Corresponding Dual Solutions for Wyndor Glass Co. Example

| Iteration | Primal Problem | | | | | | Dual Problem | | | | | |
	Row 0						y_1	y_2	y_3	$z_1 - c_1$	$z_2 - c_2$	y_0
0	$[-3,$	-5	$0,$	$0,$	0	$0]$	0	0	0	-3	-5	0
1	$[-3,$	0	$0,$	$\frac{5}{2},$	0	$30]$	0	$\frac{5}{2}$	0	-3	0	30
2	$[\ 0,$	0	$0,$	$\frac{3}{2},$	1	$36]$	0	$\frac{3}{2}$	1	0	0	36

the surplus variables for the functional constraints of the dual problem, $y_1 + 3y_3 \geq$ 3 and $2y_2 + 2y_3 \geq 5$. Thus a negative value for either surplus variable indicates that the corresponding constraint is violated. Also included is the value of the dual objective function, $y_0 = 4y_1 + 12y_2 + 18y_3$. As displayed in Table 6.4, *all* of these quantities are identified by row 0.

For the *initial* row 0, Table 6.5 shows that the corresponding dual solution, $(y_1, y_2, y_3) = (0, 0, 0)$, is infeasible because both surplus variables are negative. The first iteration succeeds in eliminating one of these negative values, but not the other. After two iterations, the *optimality test* is satisfied for the primal problem because all of the dual variables and surplus variables are nonnegative. This dual solution, $(y_1^*, y_2^*, y_3^*) = (0, \frac{3}{2}, 1)$, is optimal (as could be verified by applying the simplex method directly to the dual problem), so the optimal value of Z and y_0 is $Z^* = 36 = y_0^*$.

Summary of Primal-Dual Relationships

Now let us summarize the newly discovered key relationships between the primal and dual problems.

> **Weak duality property:** If $\mathbf{x}$ is a *feasible* solution for the *primal problem* and $\mathbf{y}$ is a *feasible* solution for the *dual problem,* then
>
> $$\mathbf{cx} \leq \mathbf{yb}.$$

For example, for the Wyndor Glass Co. problem, one feasible solution is (using the superscript T to denote the *transpose operation* described in Appendix 3) $\mathbf{x} = [3, 3]^T$, which yields $Z = \mathbf{cx} = 24$, and one feasible solution for the dual problem is $\mathbf{y} = [1, 1, 2]$, which yields a larger objective function value, $y_0 = \mathbf{yb} = 52$. For *any* such pair of feasible solutions, this inequality must hold because the *maximum* feasible value of $Z = \mathbf{cx}$ (36) *equals* the *minimum* feasible value of the dual objective function $y_0 = \mathbf{yb}$, which is our next property.

> **Strong duality property:** If $\mathbf{x}^*$ is an *optimal* solution for the *primal problem* and $\mathbf{y}^*$ is an *optimal* solution for the *dual problem,* then
>
> $$\mathbf{cx}^* = \mathbf{y}^*\mathbf{b}.$$

> **Complementary solutions property:** At each iteration, the simplex method simultaneously identifies a corner-point feasible solution $\mathbf{x}$ for the *primal problem* and a **complementary solution** $\mathbf{y}$ for the *dual problem* (found in row 0, coefficients of the slack variables), where
>
> $$\mathbf{cx} = \mathbf{yb}.$$

> If $\mathbf{x}$ is *not optimal* for the primal problem, then $\mathbf{y}$ is *not feasible* for the dual problem.

To illustrate the *complementary solutions property,* after one iteration for the Wyndor Glass Co. problem, $\mathbf{x} = [0, 6]^T$ and $\mathbf{y} = [0, \frac{5}{2}, 0]$, with $\mathbf{cx} = 30 = \mathbf{yb}$.

Complementary optimal solutions property: At the final iteration, the simplex method simultaneously identifies an optimal solution $\mathbf{x}^*$ for the *primal problem* and a **complementary optimal solution** $\mathbf{y}^*$ for the *dual problem* (found in row 0, coefficients of the slack variables), where

$$\mathbf{cx}^* = \mathbf{y}^*\mathbf{b}.$$

The y_i^* are the *shadow prices* for the primal problem.

For the example, the final iteration yields $\mathbf{x}^* = [2, 6]^T$ and $\mathbf{y}^* = [0, \frac{3}{2}, 1]$, with $\mathbf{cx}^* = 36 = \mathbf{y}^*\mathbf{b}$.

We shall take a closer look at some of these properties in Sec. 6.3. There you will see that the *complementary solutions property* can be extended considerably further. In particular, after slack and surplus variables are introduced to augment the respective problems, every *basic* solution in the primal problem has a complementary *basic* solution in the dual problem. We already have noted that the simplex method identifies the values of the surplus variables for the dual problem as the $(z_j - c_j)$ in Table 6.4. This result then leads to an additional *complementary slackness property* that relates the basic variables in one problem to the nonbasic variables in the other (Tables 6.7 and 6.8), but more about that later.

In Sec. 6.4, after describing how to construct the dual problem when the primal problem is *not in our standard form*, we discuss another very useful property, which is summarized as follows:

> **Symmetry property:** For *any* primal problem and its dual problem, all relationships between them must be *symmetric* because the dual of this dual problem is this primal problem.

Therefore, all of the preceding properties hold regardless of which of the two problems is labeled as the primal problem. (The direction of the inequality for the *weak duality property* does require that the *primal* problem be expressed or reexpressed in *maximization* form and the *dual* problem in *minimization* form.) Consequently, the simplex method can be applied to either problem, and it simultaneously will identify complementary solutions (ultimately a complementary optimal solution) for the other problem.

Applications

As we have just implied, one important application of duality theory is that the *dual* problem can be solved directly by the simplex method in order to identify an optimal solution for the primal problem. We discussed in Sec. 4.8 that the number of functional constraints affects the computational effort of the simplex method far more than the number of variables. If $m > n$, so that the dual problem has fewer functional constraints (n) than the primal problem (m), then applying the simplex method directly to the dual problem instead of the primal problem probably will achieve a substantial reduction in computational effort.

The *weak* and *strong duality properties* describe key relationships between the primal and dual problems. One useful application is for evaluating a proposed solution for the primal problem. For example, suppose that $\mathbf{x}$ is a feasible solution that has been proposed for implementation, and that a feasible solution $\mathbf{y}$ has been found by inspection for the dual problem such that $\mathbf{cx} = \mathbf{yb}$. In this case, $\mathbf{x}$ must be *optimal* without even applying the simplex method! Even if $\mathbf{cx} < \mathbf{yb}$, then $\mathbf{yb}$ still provides an *upper bound* on the optimal value of Z, so if $(\mathbf{yb} - \mathbf{cx})$ is small, intangible factors favoring $\mathbf{x}$ may lead to its selection without further ado.

One of the key applications of the *complementary solutions property* is its use in the *dual simplex method* presented in Sec. 9.2. This algorithm operates on the *primal* problem exactly as if the *simplex method* were being applied simultaneously to the *dual* problem, which can be done because of this property. Because the roles of *row 0* and the *right side* in the simplex tableau have been reversed, the dual simplex method requires that row 0 *begins and remains nonnegative* while the right side *begins* with some *negative* values (subsequent iterations strive to reach a nonnegative right side). Consequently, this algorithm occasionally is used because it is more convenient to set up the initial tableau in this form than in the form required by the simplex method. Furthermore, it frequently is used for *reoptimization* (discussed in Sec. 4.7), because changes in the original model lead to the revised final tableau fitting this form. This situation is common for certain types of *sensitivity analysis*, as you will see later in the chapter.

In general terms, duality theory plays a central role in sensitivity analysis. This role is the topic of Sec. 6.5.

Another important application is its use in the economic interpretation of the dual problem and the resulting insights for analyzing the primal problem. You already have seen one example when we discussed *shadow prices* in Sec. 4.7. The next section describes how this interpretation extends to the entire dual problem and then to the simplex method.

6.2 Economic Interpretation of Duality

The economic interpretation of duality is based directly upon the typical interpretation for the primal problem (linear programming problem in *our standard form*) presented in Sec. 3.2. To refresh your memory, we have summarized this interpretation of the primal problem in Table 6.6.

Interpretation of the Dual Problem

To see how this interpretation of the primal problem leads to an economic interpretation for the dual problem,[1] note in Table 6.4 that y_0 is the value of Z (total profit) at the current iteration. Because

$$y_0 = b_1 y_1 + b_2 y_2 + \cdots + b_m y_m,$$

each $b_i y_i$ can thereby be interpreted as the current *contribution to profit* by having b_i units of resource i available for the primal problem. Thus

> y_i is interpreted as the *contribution to profit* per unit of resource i ($i = 1, 2, \ldots, m$), when the current set of basic variables is used to obtain the primal solution.

---

[1] Actually, several slightly different interpretations have been proposed. The one presented here seems to us to be the most useful because it also directly interprets what the simplex method does in the primal problem.

Table 6.6 Economic Interpretation of Primal Problem

Quantity	Interpretation
x_j	Level of activity j $(j = 1, 2, \ldots, n)$
c_j	Unit profit from activity j
Z	Total profit from all activities
b_i	Amount of resource i available $(i = 1, 2, \ldots, m)$
a_{ij}	Amount of resource i consumed by each unit of activity j

In other words, the y_i (or y_i^* in the optimal solution) are just the *shadow prices* discussed in Sec. 4.7.

This interpretation of the dual variables leads to our interpretation of the overall dual problem. Specifically, since each unit of activity j in the primal problem consumes a_{ij} units of resource i,

$\sum_{i=1}^{m} a_{ij}y_i$ is interpreted as the current *contribution to profit* of the mix of resources that would be consumed if one unit of activity j were used ($j = 1, 2, \ldots, n$).

This same mix of resources (and more) probably can be used in other ways as well, but no alternative use should be considered if it is less profitable than one unit of activity j. Since c_j is interpreted as the unit profit from activity j, each functional constraint in the dual problem is interpreted as follows:

$\sum_{i=1}^{m} a_{ij}y_i \geq c_j$ says that the actual *contribution to profit* of the above mix of resources must be at least as much as if they were used by one unit of activity j; otherwise, we would not be making the best possible use of these resources.

Similarly, the interpretation of the nonnegativity constraints is the following:

$y_i \geq 0$ says that the *contribution to profit* of resource i ($i = 1, 2, \ldots, m$) must be nonnegative; otherwise, it would be better not to use this resource at all.

The objective,

$$\text{Minimize} \quad y_0 = \sum_{i=1}^{m} b_i y_i,$$

can be viewed as minimizing the total implicit value of the resources consumed by the activities.

This interpretation can be sharpened somewhat by differentiating between *basic* and *nonbasic* variables in the primal problem for any given basic feasible solution $(x_1, x_2, \ldots, x_{n+m})$. Recall that the *basic* variables (the only variables whose values can be nonzero) *always* have a coefficient of *zero* in row 0. Therefore, referring again to Table 6.4 and the accompanying equation for z_j,

$$\sum_{i=1}^{m} a_{ij}y_i = c_j, \quad \text{if } x_j > 0 \quad (j = 1, 2, \ldots, n),$$

$$y_i = 0, \quad \text{if } x_{n+i} > 0 \quad (i = 1, 2, \ldots, m).$$

(This is one version of the *complementary slackness property* discussed in the next section.) The economic interpretation of the first statement is that whenever an activity j operates at a strictly positive level ($x_j > 0$), the marginal value of the resources it consumes *must equal* (as opposed to exceeding) the unit profit from this activity. The second statement implies that the marginal value of resource i is *zero* ($y_i = 0$) whenever the supply of this resource is not exhausted by the activities ($x_{n+i} > 0$). In economic terminology, such a resource is a "free good"; the price of goods that are oversupplied must drop to zero by the law of supply and demand. This fact is what justifies interpreting the objective for the dual problem as minimizing the total implicit value of the resources *consumed*, rather than the resources *allocated*.

Interpretation of the Simplex Method

The interpretation of the dual problem also provides an economic interpretation of what the simplex method does in the primal problem. The *goal* of the simplex method is to find how to use the available resources in the most profitable feasible way. To attain this goal we must reach a basic feasible solution that satisfies all the *requirements* on profitable use of the resources (the constraints of the dual problem). These requirements comprise the *condition for optimality* for the algorithm. For any given basic feasible solution, the requirements (dual constraints) associated with the *basic* variables are automatically satisfied (with equality). However, those associated with *nonbasic* variables may or may not be satisfied.

In particular, if an *original variable* x_j is nonbasic so that activity j is not used, then the *current contribution to profits* of the resources that would be required to undertake each unit of activity j,

$$\sum_{i=1}^{m} a_{ij} y_i,$$

may be either smaller ($<$) or larger ($\geq$) than the unit profit c_j obtainable from the activity. If it is smaller, so $(z_j - c_j) < 0$ in row 0 of the simplex tableau, then these resources can be used more profitably by initiating this activity. If it is larger, then these resources already are being assigned elsewhere in a more profitable way, so they should not be diverted to activity j.

Similarly, if a *slack variable* x_{n+i} is nonbasic so that the total allocation b_i of resource i is being used, then y_i is the *current contribution to profit* of this resource on a marginal basis. Hence, if $y_i < 0$, profit can be increased by cutting back on the use of this resource (i.e., increasing x_{n+i}). If $y_i \geq 0$, it is worthwhile to continue fully using this resource.

Therefore, what the simplex method does is to examine all the nonbasic variables in the current basic feasible solution to see which ones can provide a *more profitable use of the resources* by being increased. If *none* can, so that no feasible shifts or reductions in the current proposed use of the resources can increase profit, the current solution must be optimal. If one or more can, the simplex method selects the variable that, if increased by 1, would *improve the profitability* of the use of the resources the most. It then actually increases this variable (the *entering basic variable*) as much as it can until the marginal values of the resources change. This increase results in a new basic feasible solution with a new row 0 (dual solution), and the whole process is repeated.

To solidify your understanding of this interpretation of the simplex method, we suggest that you apply it to the Wyndor Glass Co. problem, using both Fig. 3.2 and Table 4.8. (See Prob. 6.)

The economic interpretation of the dual problem considerably expands our ability to analyze the primal problem. However, you already have seen in Sec. 6.1 that this interpretation is just one ramification of the relationships between the two problems. In the next section, we delve into these relationships more deeply.

6.3 Primal-Dual Relationships

Because the dual problem is a linear programming problem, it also has corner-point solutions. Furthermore, by using the augmented form of the problem, we can express these corner-point solutions as basic solutions. Because the functional constraints have the $\geq$ form, this augmented form is obtained by *subtracting* the surplus (rather than adding the slack) from the left-hand side of each constraint j ($j = 1, 2, \ldots, n$).[1] This surplus is

$$z_j - c_j = \sum_{i=1}^{m} a_{ij} y_i - c_j, \qquad \text{for } j = 1, 2, \ldots, n.$$

Thus $(z_j - c_j)$ plays the role of the *surplus variable* for constraint j (or its slack variable if the constraint is multiplied through by -1). Therefore, augmenting each corner-point solution $(y_1, y_2, \ldots, y_m)$ yields a basic solution $(y_1, y_2, \ldots, y_m, z_1 - c_1, z_2 - c_2, \ldots, z_n - c_n)$ by using this expression for $(z_j - c_j)$. Since the augmented form has n functional constraints and $(n + m)$ variables, each basic solution has n basic variables and m nonbasic variables. (Note how m and n reverse their previous roles here because, as Table 6.3 indicates, dual constraints correspond to primal variables and dual variables correspond to primal constraints.)

Complementary Basic Solutions

One of the important relationships between the primal and dual problems is a direct correspondence between their basic solutions. The key to this correspondence is row 0 of the simplex tableau for the primal basic solution, such as shown in Table 6.4 or 6.5. Such a row 0 can be obtained for *any* primal basic solution, feasible or not, by using the formulas given in the bottom part of Table 5.7.

Note again in Tables 6.4 and 6.5 how a complete solution for the dual problem (including the surplus variables) can be read directly from row 0. Thus, because of its coefficient in row 0, each variable in the primal problem has an associated variable in the dual problem, as summarized in Table 6.7.

A key insight here is that the dual solution read from row 0 must also be a basic solution! The reason is that the m basic variables for the primal problem are required

Table 6.7 Association between Variables in Primal and Dual Problems

Primal Variable	Associated Dual Variable
(Original variable) x_j	$(z_j - c_j)$ (surplus variable), $j = 1, 2, \ldots, n$
(Slack variable) x_{n+i}	y_i (original variable), $i = 1, 2, \ldots, m$

[1] You might wonder why we do not also introduce *artificial variables* into these constraints as discussed in Sec. 4.6. The reason is that these variables have no purpose other than to change the feasible region temporarily as a convenience in starting the simplex method. We are not interested now in applying the simplex method to the dual problem, and we do not want to change its feasible region.

to have a coefficient of *zero* in row 0, which thereby requires the m associated dual variables to be zero, i.e., *nonbasic* variables for the dual problem. The values of the remaining n (basic) variables then will be the simultaneous solution to the system of equations given at the beginning of the section. In matrix form, this system of equations is $z - c = yA - c$, and the fundamental insight of Sec. 5.3 actually identifies its solution for $z - c$ and y as being the corresponding entries in row 0.

Because of the *symmetry property* quoted in Sec. 6.1 (and the direct association between variables shown in Table 6.7), the correspondence between basic solutions in the primal and dual problems is a symmetric one. Furthermore, a pair of complementary basic solutions has the same objective function value, shown as y_0 in Table 6.4.

Let us now summarize our conclusions about the correspondence between primal and dual basic solutions, where the first property extends the *complementary solutions property* of Sec. 6.1 to the *augmented forms* of the two problems and then to *any* basic solution (feasible or not) in the primal problem.

> **Complementary basic solutions property:** Each *basic* solution in the *primal problem* has a **complementary basic solution** in the *dual problem,* where their respective objective function values (Z and y_0) are equal. Given row 0 of the simplex tableau for the primal basic solution, the complementary dual basic solution (y, $z - c$) is found as shown in Table 6.4.

The next property shows how to identify the *basic* and *nonbasic* variables in this complementary basic solution.

> **Complementary slackness property:** Using the association between variables given in Table 6.7, the variables in the primal basic solution and the complementary dual basic solution satisfy the **complementary slackness** relationship shown in Table 6.8. Furthermore, this relationship is a symmetric one, so that these two basic solutions are complementary to each other.

The reason for using the name *complementary slackness* for this latter property is that it says (in part) that for each pair of associated variables, if one of them has *slack* in its nonnegativity constraint (a basic variable > 0), then the other one must have *no slack* (a nonbasic variable $= 0$). We mentioned in Sec. 6.2 that this property has a useful economic interpretation for linear programming problems.

EXAMPLE: To illustrate these two properties, again consider the Wyndor Glass Co. problem of Sec. 3.1. All eight of its basic solutions (five feasible and three infeasible) are shown in Tables 5.5 and 5.6 along with the corresponding corner-point solutions.

Table 6.8 **Complementary Slackness Relationship for Complementary Basic Solutions**

Primal Variable	Associated Dual Variable	
Basic	Nonbasic	(m variables)
Nonbasic	Basic	(n variables)

Thus its dual problem (see Table 6.1) also must have eight basic solutions, each complementary to one of these primal solutions, as shown in Table 6.9.

The three basic feasible solutions obtained by the simplex method for the primal problem are the first, fifth, and sixth primal solutions shown in Table 6.9. You already saw in Table 6.5 how the complementary basic solutions for the dual problem can be read directly from row 0, starting with the coefficients of the slack variables and then the original variables. The other dual basic solutions also could be identified in this way by constructing row 0 for each of the other primal basic solutions, using the formulas given in the bottom part of Table 5.7.

Alternatively, for each primal basic solution, the *complementary slackness property* can be used to identify the basic and nonbasic variables for the complementary dual basic solution, so that the system of equations given at the beginning of the section can be solved directly to obtain this complementary solution. For example, consider the next-to-last primal basic solution in Table 6.9, where x_1, x_2, and x_5 are basic variables. Using Tables 6.7 and 6.8, we see that the complementary slackness property implies that $(z_1 - c_1)$, $(z_2 - c_2)$, and y_3 are nonbasic variables for the complementary dual basic solution. Setting these variables equal to zero in the dual problem equations, $y_1 + 3y_3 - (z_1 - c_1) = 3$ and $2y_2 + 2y_3 - (z_2 - c_2) = 5$, immediately yields $y_1 = 3$, $y_2 = \frac{5}{2}$.

Finally, notice that Table 6.9 demonstrates that $(0, \frac{3}{2}, 1, 0, 0)$ is the optimal solution for the dual problem, because it is the basic *feasible* solution with minimal y_0 (36).

Relationships between Complementary Basic Solutions

We now turn our attention to the relationships between complementary basic solutions, beginning with their *feasibility* relationships. The middle columns in Table 6.9 provide some valuable clues. For the pairs of complementary solutions, notice how the yes or no answers on feasibility also satisfy a complementary relationship in most cases. In particular, with one exception, whenever one solution is feasible, the other is not. (It also is possible for *neither* solution to be feasible, as happened with the third pair.) The one exception is the sixth pair, where the primal solution is known to be optimal. The explanation is suggested by the $Z = y_0$ column. Because the sixth dual solution also is optimal (by the *complementary optimal solutions property*), with $y_0 = 36$, then the first five dual solutions *cannot be feasible* because $y_0 < 36$ (remember that the dual problem objective is to *minimize* y_0). By the same token, the last two primal solutions *cannot be feasible* because $Z > 36$.

Table 6.9 Complementary Basic Solutions for Wyndor Glass Co. Example

	Primal Problem				*Dual Problem*	
No.	Basic Solution	Feasible?	$Z = y_0$	Feasible?	Basic Solution	
1	$(0, 0, 4, 12, 18)$	Yes	0	No	$(0, 0, 0, -3, -5)$	
2	$(4, 0, 0, 12, 6)$	Yes	12	No	$(3, 0, 0, 0, -5)$	
3	$(6, 0, -2, 12, 0)$	No	18	No	$(0, 0, 1, 0, -3)$	
4	$(4, 3, 0, 6, 0)$	Yes	27	No	$(-\frac{3}{2}, 0, \frac{5}{2}, 0, 0)$	
5	$(0, 6, 4, 0, 6)$	Yes	30	No	$(0, \frac{5}{2}, 0, -3, 0)$	
6	$(2, 6, 2, 0, 0)$	Yes	36	Yes	$(0, \frac{3}{2}, 1, 0, 0)$	
7	$(4, 6, 0, 0, -6)$	No	42	Yes	$(3, \frac{5}{2}, 0, 0, 0)$	
8	$(0, 9, 4, -6, 0)$	No	45	Yes	$(0, 0, \frac{5}{2}, \frac{9}{2}, 0)$	

Table 6.10 **Classification of Basic Solutions**

		Satisfies Condition for Optimality?	
		Yes	No
Feasible?	Yes	Optimal	Suboptimal
	No	Superoptimal	Neither feasible nor superoptimal

This explanation is further supported by the *strong duality property* that optimal primal and dual solutions have $Z = y_0$.

Next, let us state the *extension* of the *complementary optimal solutions property* of Sec. 6.1 for the *augmented forms* of the two problems.

Complementary optimal basic solutions property: Each *optimal* basic solution in the *primal problem* has a **complementary optimal basic solution** in the dual problem, where their respective objective function values (Z and y_0) are equal.[1] Given row 0 of the simplex tableau for the optimal primal solution, the complementary optimal dual solution ($\mathbf{y^*}$, $\mathbf{z^*} - \mathbf{c}$) is found as shown in Table 6.4.

To review the reasoning behind this property, note that the dual solution ($\mathbf{y^*}$, $\mathbf{z^*} - \mathbf{c}$) must be feasible for the dual problem because the *condition for optimality* for the primal problem requires that *all* these dual variables (including surplus variables) be *nonnegative*. Since this solution is *feasible*, it must be *optimal* for the dual problem by the *weak duality property*.

Basic solutions can be classified according to whether or not they satisfy each of two conditions. One is the *condition for feasibility,* namely, whether *all* the variables (including slack variables) in the augmented solution are *nonnegative*. The other is the *condition for optimality,* namely, whether *all* the coefficients in row 0 (i.e., all the variables in the complementary basic solution) are *nonnegative*. Our names for the different types of basic solutions are summarized in Table 6.10. For example, in Table 6.9, primal basic solutions 1, 2, 4, and 5 are suboptimal, 6 is optimal, 7 and 8 are superoptimal, and 3 is neither feasible nor superoptimal.

Using these definitions, the general relationships between complementary basic solutions are summarized in Table 6.11. The resulting range of possible (common) values for the objective functions ($Z = y_0$) for the first three pairs given in Table 6.11 (the last pair can have any value) is shown in Fig. 6.1. Thus, while the simplex method is dealing directly with suboptimal basic solutions and working toward optimality in the primal problem, it is simultaneously dealing indirectly with complementary superoptimal solutions and working toward feasibility in the dual problem. Conversely, it sometimes is more convenient (or necessary) to work directly with

---

[1] Because of the *symmetry property,* it thereby follows that if either problem possesses at least one optimal solution, the other must also. The only ways in which both problems can have no optimal solutions are when (1) both problems have no feasible solutions, or (2) one problem has no feasible solutions and the other problem has an unbounded feasible region that permits improving the objective function value indefinitely in the favorable direction.

Table 6.11 **Relationships between Complementary Basic Solutions**

Primal Basic Solution	Complementary Dual Basic Solution
Suboptimal	Superoptimal
Optimal	Optimal
Superoptimal	Suboptimal
Neither feasible nor superoptimal	Neither feasible nor superoptimal

superoptimal basic solutions and to move toward feasibility in the primal problem, which is the purpose of the *dual simplex method* described in Sec. 9.2.

These relationships prove very useful, particularly in sensitivity analysis, as you will see later in the chapter.

6.4 Adapting to Other Primal Forms

Thus far it has been assumed that the model for the primal problem is in *our standard form*. However, we indicated at the beginning of the chapter that *any* linear programming problem, whether in our standard form or not, possesses a dual problem. Therefore, this section focuses on how the dual problem changes for other primal forms.

Each *nonstandard* form was discussed in Sec. 4.6, and we pointed out how it is possible to convert each one into an *equivalent* standard form if so desired. These

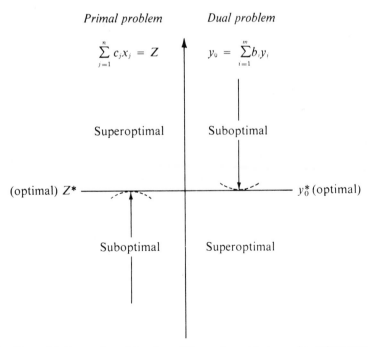

Figure 6.1 Range of possible values of $Z = y_0$ for certain types of complementary basic solutions.

Table 6.12 **Conversions to Standard Form for Linear Programming Models**

Nonstandard Form	Equivalent Standard Form
Minimize Z	Maximize $(-Z)$
$\displaystyle\sum_{j=1}^{n} a_{ij}x_j \geq b_i$	$\displaystyle -\sum_{j=1}^{n} a_{ij}x_j \leq -b_i$
$\displaystyle\sum_{j=1}^{n} a_{ij}x_j = b_i$	$\displaystyle\sum_{j=1}^{n} a_{ij}x_j \leq b_i \quad and \quad -\sum_{j=1}^{n} a_{ij}x_j \leq -b_i$
x_j unconstrained in sign	$(x_j^+ - x_j^-), \quad x_j^+ \geq 0, \quad x_j^- \geq 0$

conversions are summarized in Table 6.12. Hence you always have the option of converting any model into *our standard form* and *then* constructing its dual problem in the usual way. To illustrate, we do this for our standard *dual problem* (it must have a dual also) in Table 6.13. Note that what we end up with is just our standard *primal problem!* Since *any* pair of primal and dual problems can be converted into these forms, this fact demonstrates the following key property of primal-dual relationships:

Symmetry property: For any primal problem and its dual problem, all relationships between them must be *symmetric,* because the dual of this dual problem is this primal problem.

As a result, all the statements made earlier in the chapter about the relationships of the dual problem to the primal problem also hold in reverse.

Another consequence of the symmetry property is that it is immaterial which problem is called the primal and which is called the dual. In practice, you might see a linear programming problem fitting our standard form being referred to as the *dual*

Table 6.13 **Constructing the Dual of the Dual Problem**

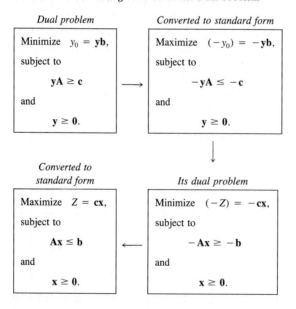

problem. The convention is that the model formulated to fit the actual problem is called the *primal problem,* regardless of its form.

Our illustration of how to construct the dual problem for a nonstandard primal problem did not involve either equality constraints or variables unconstrained in sign. Actually, for these two forms, a shortcut is available. It is possible to show [see Probs. 20 and 17(c)] that an *equality constraint* in the primal problem should be treated just like an ≤ constraint in constructing the dual problem *except* that the nonnegativity constraint for the corresponding dual variable should be *deleted* (i.e., this variable is unconstrained in sign). By the symmetry property, *deleting a nonnegativity constraint* in the primal problem affects the dual problem only by changing the corresponding constraint into an *equality constraint.*

Because of these shortcuts, it is necessary only to convert the *primal problem* into the form shown in *either* column of Table 6.14. You then construct its *dual problem* in the usual way, using the form shown in the *other* column. The double-sided arrows in Table 6.14 show the specific correspondences between the two forms that must be followed. Specifically, an inequality functional constraint in one problem corresponds to including a nonnegativity constraint in the other, whereas an equality constraint in one problem corresponds to deleting a nonnegativity constraint in the other. Beware of mixing the forms in the two columns (e.g., maximize Z with ≥ constraints) for defining the primal problem. Such mixing is not allowed for the purpose of constructing the dual problem.

To illustrate this procedure, consider the *radiation therapy example* presented in Sec. 3.4. (Its model is shown on p. 39.) Let this model be our primal problem. Before finding its dual problem, we need to convert the primal problem into one of the two allowable forms shown in Table 6.14. Let us illustrate it both ways.

The radiation therapy model already almost fits the form in the *second* column of Table 6.14. The only discrepancy is that the first functional constraint, $0.3x_1 + 0.1x_2 \leq 2.7$, is in ≤ form rather than ≥ or =. However, multiplying through the constraint by (-1) converts it into ≥ form, thereby yielding the allowable form of the model shown on the left side of Table 6.15. Its dual problem then is constructed in just the usual way (as summarized in Table 6.2a) *except* for following the form in the *first* column of Table 6.14. The result is shown on the right side of Table 6.15. Note that the nonnegativity constraint has been deleted for the *second* variable (y_2) because the *second* functional constraint in the primal problem is an *equality* constraint.

Table 6.14 Corresponding Primal-Dual Forms

Primal Problem (or Dual Problem)	Dual Problem (or Primal Problem)
Maximize Z (or y_0)	Minimize y_0 (or Z)
Constraint i	Variable y_i (or x_i)
≤ form ⟷	→ $y_i \geq 0$
= form ⟷	→ $y_i \geq 0$ deleted
Variable x_j (or y_j)	Constraint j
$x_j \geq 0$ ⟷	→ ≥ form
$x_j \geq 0$ deleted ⟷	→ = form

Table 6.15 One Primal-Dual Form for the Radiation Therapy Example

Primal problem	Dual problem
Minimize $Z = 0.4x_1 + 0.5x_2$, subject to $\quad -0.3x_1 - 0.1x_2 \geq -2.7$ $\quad\ \ 0.5x_1 + 0.5x_2 = 6$ $\quad\ \ 0.6x_1 + 0.4x_2 \geq 6$ and $\qquad x_1 \geq 0, \qquad x_2 \geq 0.$	Maximize $y_0 = -2.7y_1 + 6y_2 + 6y_3$, subject to $\quad -0.3y_1 + 0.5y_2 + 0.6y_3 \leq 0.4$ $\quad -0.1y_1 + 0.5y_2 + 0.4y_3 \leq 0.5$ and $\qquad y_1 \geq 0, \qquad y_3 \geq 0$ $\qquad$ (y_2 unconstrained in sign).

Equivalently, the form in the *first* column of Table 6.14 can be used instead to set up the primal problem. (This form is needed anyway to apply the simplex method as presented in Chap. 4.) Using the first two conversions in Table 6.12, this approach leads to the form of the primal problem shown on the left side of Table 6.16. The corresponding dual form from the *second* column of Table 6.14 is then used to construct the dual problem shown on the right side of Table 6.16.

Note that the two versions of the primal and dual problems in Tables 6.15 and 6.16 are *completely equivalent*, where $y_2' = -y_2$. This equivalency is inevitable because the differences involve substituting only equivalent forms.

When the simplex method is applied to the nonstandard primal forms, and when the artificial variable technique (perhaps supplemented by the Big M method) is used to adapt to them, the duality interpretation of row 0 of the simplex tableau must be adjusted somewhat. The reason is that the artificial variables and M's *revise* the primal problem, which thereby changes its dual problem, so the complementary basic solutions shown in row 0 are for this *revised dual problem*. However, *after* the artificial variables have been eliminated (made nonbasic) so that the current solution is a legitimate basic feasible solution for the *original* primal problem, row 0 can still be used to identify the complementary basic solution for the original dual problem. We describe how to do this next.

Suppose that we use the form in the first column of Table 6.14. For each *equality* constraint i, its artificial variable plays the role of a slack variable, *except* that M has been added initially to the coefficient of this variable in row 0. Therefore, the current

Table 6.16 The Other Primal-Dual Form for the Radiation Therapy Example

Primal problem	Dual problem
Maximize $(-Z) = -0.4x_1 - 0.5x_2$, subject to $\quad\ \ 0.3x_1 + 0.1x_2 \leq 2.7$ $\quad\ \ 0.5x_1 + 0.5x_2 = 6$ $\quad -0.6x_1 - 0.4x_2 \leq -6$ and $\qquad x_1 \geq 0, \qquad x_2 \geq 0.$	Minimize $y_0 = 2.7y_1 + 6y_2' - 6y_3$, subject to $\quad 0.3y_1 + 0.5y_2' - 0.6y_3 \geq -0.4$ $\quad 0.1y_1 + 0.5y_2' - 0.4y_3 \geq -0.5$ and $\qquad y_1 \geq 0, \qquad y_3 \geq 0$ $\qquad$ (y_2' unconstrained in sign).

value of the corresponding dual variable y_i is the current coefficient of this artificial variable *minus* M. If a $\leq$ constraint has a negative right-hand side initially (perhaps because it was converted from a $\geq$ constraint) so that it has been given an artificial variable, the dual variable corresponding to this constraint still equals the coefficient of its *slack* variable. The coefficient of the artificial variable would be ignored in this case. Finally, if a variable x_j is unconstrained in sign so that it has been replaced by the difference of two nonnegative variables $(x_j^+ - x_j^-)$, then the coefficient of x_j^+, denoted by $(z_j^+ - c_j)$, would be used just as for x_j. In other words,

$$z_j^+ = \sum_{i=1}^{m} a_{ij} y_i.$$

The coefficient of x_j^-, namely, $(z_j^- + c_j) = -(z_j^+ - c_j)$, would be ignored. Except for these cases, the coefficients in row 0 would be used just as before (see Sec. 6.3) to give the values of the corresponding dual variables.

To illustrate this procedure, we ask you to refer to the set of simplex tableaux given in Table 4.12 for the radiation therapy example. The first three tableaux still have artificial variables as basic variables. However, this is not the case for the *final* tableau, so we can use its row 0 to identify the *optimal* solution for the dual problem shown in Table 6.16. The first primal constraint is a standard one, so y_1 is just the coefficient of the first slack variable (x_3), or $y_1 = 0.5$. The second constraint is an equality constraint, so we refer to the coefficient of its artificial variable $(\bar{x}_4)$ to obtain

$$y_2' = (M - 1.1) - M = -1.1.$$

The third constraint has a negative right-hand side, so we use the coefficient of its slack variable (x_5) to yield $y_3 = 0$. As before, the surplus variables for the dual problem equal the coefficients of the original variables, x_1 and x_2, so $(z_1 - c_1) = 0$, $(z_2 - c_2) = 0$. This completes the optimal dual basic solution,

$$(y_1, y_2', y_3, z_1 - c_1, z_2 - c_2) = (0.5, -1.1, 0, 0, 0).$$

6.5 The Role of Duality Theory in Sensitivity Analysis

As described further in the next two sections, *sensitivity analysis* basically involves investigating the effect on the optimal solution of making changes in the values of the model parameters (the a_{ij}, b_i, and c_j). However, changing parameter values in the *primal* problem also changes the corresponding values in the *dual* problem. Therefore, you have your choice of which problem to use to investigate each change. Because of the primal-dual relationships presented in Secs. 6.1 and 6.3 (especially the *complementary basic solutions property*), it is easy to move back and forth between the two problems as desired. In some cases, it is more convenient to analyze the dual problem *directly* in order to determine the complementary effect on the primal problem. We begin by considering two such cases.

Changes in the Coefficients of a Nonbasic Variable

Suppose that the changes made in the original model occur in the coefficients of a variable that was *nonbasic* in the original optimal solution. What is the effect of these changes on this solution? Is it still feasible? Is it still optimal?

Because the variable involved is nonbasic (value of zero), changing its coefficients cannot affect the feasibility of the solution. Therefore, the open question in this case is whether it is still *optimal*. As Tables 6.10 and 6.11 indicate, an *equivalent* question is whether the complementary basic solution for the *dual* problem is still *feasible* after making these changes. Since these changes affect the dual problem by changing only one constraint, this question can be answered simply by checking whether this complementary basic solution still satisfies this revised constraint.

We shall illustrate this case in the corresponding subsection of Sec. 6.7 after developing a relevant example.

Introduction of a New Variable

As indicated in Table 6.6, the decision variables in the model typically represent the level of the various activities under consideration. In some situations, these activities were selected from a larger group of *possible* activities, where the remaining activities were not included in the original model because they seemed less attractive. Or perhaps these other activities did not come to light until after the original model was formulated and solved. Either way, the key question is whether any of these previously unconsidered activities are sufficiently worthwhile to warrant initiation. In other words, would adding any of these activities to the model change the original optimal solution?

Adding another activity amounts to introducing a new variable, with the appropriate coefficients in the functional constraints and objective function, into the model. The only resulting change in the *dual* problem is to add a *new constraint* (see Table 6.3).

After these changes are made, would the original optimal solution, along with the new variable equal to *zero* (nonbasic), still be *optimal* for the primal problem? As for the preceding case, an *equivalent* question is whether the complementary basic solution for the *dual* problem is still *feasible*. And, as before, this question can be answered simply by checking whether this complementary basic solution satisfies *one* constraint, which in this case is the *new* constraint for the dual problem.

To illustrate, suppose for the Wyndor Glass Co. problem of Sec. 3.1 that a possible *third* new product now is being considered for inclusion in the product line. Letting x_{new} represent the production rate for this product, the resulting *revised* model is shown as follows:

$$\text{Maximize} \quad Z = 3x_1 + 5x_2 + 4x_{new},$$

$$\text{subject to} \quad x_1 \qquad\quad + 2x_{new} \leq 4$$

$$2x_2 + 3x_{new} \leq 12$$

$$3x_1 + 2x_2 + \quad x_{new} \leq 18$$

$$\text{and} \quad x_1 \geq 0, \quad x_2 \geq 0, \quad x_{new} \geq 0.$$

After introducing slack variables, the *original* optimal solution for this problem without x_{new} (see Table 4.8) was $(x_1, x_2, x_3, x_4, x_5) = (2, 6, 2, 0, 0)$. Is this solution, along with $x_{new} = 0$, still optimal?

To answer this question, check the complementary basic solution for the dual problem, which Table 6.9 (and Table 4.8) identifies as

$$(y_1, y_2, y_3, z_1 - c_1, z_2 - c_2) = (0, \tfrac{3}{2}, 1, 0, 0).$$

Since this solution was optimal for the original dual problem, it certainly satisfies the original dual constraints shown in Table 6.1. But does it satisfy the one new dual constraint,

$$2y_1 + 3y_2 + y_3 \geq 4?$$

Plugging in this solution,

$$2(0) + 3(\tfrac{3}{2}) + (1) \geq 4$$

is satisfied, so this dual solution is still *feasible* (and thus still optimal). Consequently, the original primal solution (2, 6, 2, 0, 0), along with $x_{new} = 0$, is still *optimal,* so this third possible new product should *not* be added to the product line.

This approach also makes it very easy to conduct sensitivity analysis on the coefficients of the new variable added to the primal problem. By simply checking the new dual constraint, you can immediately see how far any of these parameter values can be changed before they affect the *feasibility* of the dual solution and so the *optimality* of the primal solution.

Other Applications

Already we have discussed two other key applications of duality theory to sensitivity analysis, namely, *shadow prices* and the *dual simplex method*. As described in Secs. 4.7 and 6.2, the optimal dual solution ($y_1^*, y_2^*, \ldots, y_m^*$) provides the shadow prices for the respective resources that indicate how Z would change if (small) changes were made in the b_i (the resource amounts). The resulting analysis will be illustrated in some detail in Sec. 6.7.

In more general terms, the economic interpretation of the dual problem and of the simplex method presented in Sec. 6.2 provides some useful insights for sensitivity analysis.

When we investigate the effect of changing the b_i or the a_{ij} (for basic variables), the *original* optimal solution may become a *superoptimal* basic solution instead (see Table 6.10). If you then want to *reoptimize* to identify the new optimal solution, the *dual simplex method* (discussed at the end of Secs. 6.1 and 6.3) should be applied, starting from this basic solution.

We mentioned in Sec. 6.1 that it sometimes is more efficient to solve the *dual* problem directly by the simplex method in order to identify an optimal solution for the *primal* problem. When the solution has been found in this way, sensitivity analysis for the primal problem then is conducted by applying the procedure described in the next two sections directly to the *dual* problem and then inferring the complementary effects on the *primal* problem (e.g., see Table 6.11). This approach to sensitivity analysis is relatively straightforward because of the close primal-dual relationships described in Secs. 6.1 and 6.3. (See Prob. 47.)

6.6 The Essence of Sensitivity Analysis

The work of the operations research team usually is not even nearly done when the simplex method has been successfully applied to identify an optimal solution for the model. As we pointed out at the end of Sec. 3.3, one assumption of linear programming is that all the parameters of the model (the a_{ij}, b_i, and c_j) are *known constants*.

Actually, the parameter values used in the model normally are just *estimates* based on a *prediction of future conditions*. The data obtained to develop these estimates often are rather crude or nonexistent, so that the parameters in the original formulation may represent little more than quick rules of thumb provided by harassed line personnel. They may even represent deliberate overestimates or underestimates to protect the interests of the estimators.

Thus the successful manager and operations research staff will maintain a healthy skepticism about the original numbers coming out of the computer and will view them in many cases as only a starting point for further analysis of the problem. An "optimal" solution is optimal only with respect to the specific model being used to represent the real problem, and such a solution becomes a reliable guide for action only after it has been verified as performing well for other reasonable representations of the problem as well. Furthermore, the model parameters (particularly the b_i) sometimes are set as a result of *managerial policy decisions* (e.g., the amount of certain resources to be made available to the activities), and these decisions should be reviewed after seeing their potential consequences.

For these reasons it is important to perform **sensitivity analysis** to investigate the effect on the optimal solution provided by the simplex method if the parameters take on other possible values. Usually there will be some parameters that can be assigned any reasonable value without affecting the optimality of this solution. However, there may also be parameters with likely alternative values that would yield a new optimal solution. This situation is particularly serious if the original solution would then have a substantially inferior value of the objective function, or perhaps even be infeasible! Therefore, the basic objective of sensitivity analysis is to identify these particularly *sensitive* parameters, so that special care can then be taken to estimate them more closely and to select a solution that performs well for most of their likely values.

Sensitivity analysis would require an exorbitant computational effort if it were necessary to reapply the simplex method from the beginning to investigate each new change in a parameter value. Fortunately, the *fundamental insight* discussed in Sec. 5.3 virtually eliminates computational effort. The basic idea is that the fundamental insight *immediately* reveals just how any changes in the original model would change the numbers in the final simplex tableau (assuming that the *same* sequence of algebraic operations originally performed by the simplex method were to be *duplicated*). Therefore, after making a few simple calculations to revise this tableau, we can check easily whether the original optimal basic feasible solution is now nonoptimal (or infeasible). If so, this solution would be used as the initial basic solution to restart the simplex method (or dual simplex method) to find the new optimal solution, if desired. If the changes in the model are not major, only a very few iterations should be required to reach the new optimal solution from this "advanced" initial basic solution.

To describe this procedure more specifically, consider the following situation. The simplex method already has been used to obtain an optimal solution to a linear programming model with specified values for the b_i, c_j, and a_{ij} parameters. To initiate sensitivity analysis, one or more of the parameters now is changed. After making the changes, let $\bar{b}_i$, $\bar{c}_j$, and $\bar{a}_{ij}$ denote the values of the various parameters. Thus, in matrix notation,

$$\mathbf{b} \to \bar{\mathbf{b}}, \qquad \mathbf{c} \to \bar{\mathbf{c}}, \qquad \mathbf{A} \to \bar{\mathbf{A}},$$

for the *revised* model.

The first step is to revise the *final* simplex tableau to reflect these changes. Continuing to use the notation presented in Table 5.9, as well as the accompanying formulas for the *fundamental insight* [(1) $\mathbf{t^*} = \mathbf{t} + \mathbf{y^*T}$ and (2) $\mathbf{T^*} = \mathbf{S^*T}$], the revised final tableau is calculated from y^* and $\mathbf{S^*}$ (which have not changed) and the new initial tableau, as shown in Table 6.17.

To illustrate, suppose that the original model for the Wyndor Glass Co. problem of Sec. 3.1 is revised as shown at the right.

Original model	Revised model
Maximize $Z = [3, 5]\begin{bmatrix} x_1 \\ x_2 \end{bmatrix}$,	Maximize $Z = [4, 5]\begin{bmatrix} x_1 \\ x_2 \end{bmatrix}$,
subject to	subject to
$\begin{bmatrix} 1 & 0 \\ 0 & 2 \\ 3 & 2 \end{bmatrix}\begin{bmatrix} x_1 \\ x_2 \end{bmatrix} \leq \begin{bmatrix} 4 \\ 12 \\ 18 \end{bmatrix}$	$\begin{bmatrix} 1 & 0 \\ 0 & 2 \\ 2 & 2 \end{bmatrix}\begin{bmatrix} x_1 \\ x_2 \end{bmatrix} \leq \begin{bmatrix} 4 \\ 24 \\ 18 \end{bmatrix}$
and	and
$\mathbf{x} \geq \mathbf{0}.$	$\mathbf{x} \geq \mathbf{0}.$

Thus the changes from the original model are $c_1 = 3 \rightarrow 4$, $a_{31} = 3 \rightarrow 4$, and $b_2 = 12 \rightarrow 24$. Figure 6.2 shows the graphical effect of these changes. For the original model, the simplex method already has identified the optimal corner-point feasible solution as (2, 6), lying at the intersection of the two constraint boundaries shown as dashed lines, $2x_2 = 12$ and $3x_1 + 2x_2 = 18$. Now the revision of the model has shifted both of these constraint boundaries as shown by the dark lines, $2x_2 = 24$ and $2x_1 + 2x_2 = 18$. Consequently, the previous corner-point feasible solution (2, 6) now shifts to the new intersection $(-3, 12)$, which is a corner-point *infeasible* solution for the revised model. The procedure described in the preceding paragraphs finds this shift *algebraically* (in augmented form). Furthermore, it does so in a manner that is very efficient even for huge problems where graphical analysis is impossible.

To carry out this procedure, we begin by displaying the parameters of the revised model in matrix form:

$$\mathbf{\bar{c}} = [4, 5],$$

$$\mathbf{\bar{A}} = \begin{bmatrix} 1 & 0 \\ 0 & 2 \\ 2 & 2 \end{bmatrix}, \qquad \mathbf{\bar{b}} = \begin{bmatrix} 4 \\ 24 \\ 18 \end{bmatrix}.$$

Table 6.17 **Revised Final Simplex Tableau Resulting from Changes in Original Model**

	Eq. No.	Z	Coefficient of			Right Side
			Original Variables	Slack Variables		
New initial tableau	0	1	$-\mathbf{\bar{c}}$	**0**		0
	$1 - m$	**0**	$\mathbf{\bar{A}}$	**I**		$\mathbf{\bar{b}}$
Revised final tableau	0	1	$\mathbf{z^*} - \mathbf{\bar{c}} = \mathbf{y^*\bar{A}} - \mathbf{\bar{c}}$	$\mathbf{y^*}$		$Z^* = \mathbf{y^*\bar{b}}$
	$1 - m$	**0**	$\mathbf{A^*} = \mathbf{S^*\bar{A}}$	$\mathbf{S^*}$		$\mathbf{b^*} = \mathbf{S^*\bar{b}}$

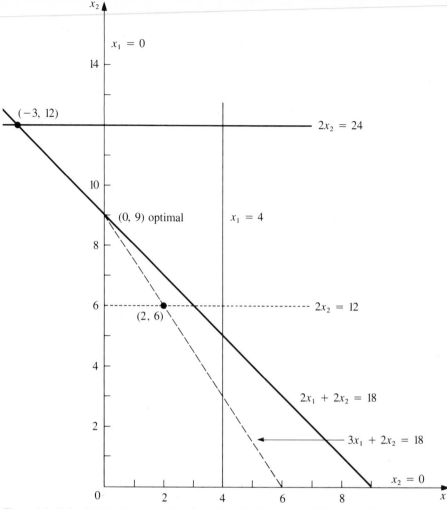

Figure 6.2 Shift of the final corner-point solution from $(2, 6)$ to $(-3, 12)$ for the revision of the Wyndor Glass Co. problem where $c_1 = 3 \to 4$, $a_{31} = 3 \to 2$, and $b_2 = 12 \to 24$.

The resulting new *initial* simplex tableau is shown at the top of Table 6.18. Below this tableau is the *original* final tableau (as first given in Table 4.8). We have drawn dark boxes around the portions of this final tableau that the changes in the model definitely *do not change,* namely, the coefficients of the *slack* variables in both row 0 ($\mathbf{y}^*$) and the rest of the rows ($\mathbf{S}^*$). Thus,

$$\mathbf{y}^* = [0, \tfrac{3}{2}, 1], \qquad \mathbf{S}^* = \begin{bmatrix} 1 & \tfrac{1}{3} & -\tfrac{1}{3} \\ 0 & \tfrac{1}{2} & 0 \\ 0 & -\tfrac{1}{3} & \tfrac{1}{3} \end{bmatrix}.$$

These coefficients of the *slack* variables necessarily are unchanged with the same algebraic operations originally performed by the simplex method because the coefficients of these same variables in the *initial* tableau are unchanged.

However, because other portions of the *initial* tableau have changed, there will be changes in the rest of the *final* tableau as well. Using the formulas in Table 6.17, the revised numbers in the rest of the *final* tableau are calculated as follows:

$$\mathbf{z^*} - \bar{\mathbf{c}} = [0, \tfrac{3}{2}, 1]\begin{bmatrix} 1 & 0 \\ 0 & 2 \\ 2 & 2 \end{bmatrix} - [4, 5] = [-2, 0], \quad Z^* = [0, \tfrac{3}{2}, 1]\begin{bmatrix} 4 \\ 24 \\ 18 \end{bmatrix} = 54,$$

$$\mathbf{A^*} = \begin{bmatrix} 1 & \tfrac{1}{3} & -\tfrac{1}{3} \\ 0 & \tfrac{1}{2} & 0 \\ 0 & -\tfrac{1}{3} & \tfrac{1}{3} \end{bmatrix}\begin{bmatrix} 1 & 0 \\ 0 & 2 \\ 2 & 2 \end{bmatrix} = \begin{bmatrix} \tfrac{1}{3} & 0 \\ 0 & 1 \\ \tfrac{2}{3} & 0 \end{bmatrix},$$

$$\mathbf{b^*} = \begin{bmatrix} 1 & \tfrac{1}{3} & -\tfrac{1}{3} \\ 0 & \tfrac{1}{2} & 0 \\ 0 & -\tfrac{1}{3} & \tfrac{1}{3} \end{bmatrix}\begin{bmatrix} 4 \\ 24 \\ 18 \end{bmatrix} = \begin{bmatrix} 6 \\ 12 \\ -2 \end{bmatrix}.$$

The resulting *revised* final tableau is shown at the bottom of Table 6.18.

Actually, we can substantially streamline these calculations for obtaining the revised final tableau. Because none of the coefficients of x_2 changed in the *original* model (tableau), none of them can change in the *final* tableau, so we can delete their calculation. Several other original parameters (a_{11}, a_{21}, b_1, b_3) also were not changed, so another shortcut is to calculate only the *incremental changes* in the final tableau in terms of the incremental changes in the initial tableau, ignoring those terms in the vector or matrix multiplication that involve *zero* change in the initial tableau. In particular, the only incremental changes in the *initial* tableau are $\Delta c_1 = 1$, $\Delta a_{31} = -1$, and $\Delta b_2 = 12$, so these are the only terms that need be considered. This streamlined approach is shown below, where a zero or dash appears in each spot where no calculation is needed.

$$\Delta(\mathbf{z^*} - \mathbf{c}) = \mathbf{y^*}\,\Delta\mathbf{A} - \Delta\mathbf{c} = [0, \tfrac{3}{2}, 1]\begin{bmatrix} 0 & - \\ 0 & - \\ -1 & - \end{bmatrix} - [1, -] = [-2, -].$$

$$\Delta Z^* = \mathbf{y^*}\,\Delta\mathbf{b} = [0, \tfrac{3}{2}, 1]\begin{bmatrix} 0 \\ 12 \\ 0 \end{bmatrix} = 18.$$

$$\Delta\mathbf{A^*} = \mathbf{S^*}\,\Delta\mathbf{A} = \begin{bmatrix} 1 & \tfrac{1}{3} & -\tfrac{1}{3} \\ 0 & \tfrac{1}{2} & 0 \\ 0 & -\tfrac{1}{3} & \tfrac{1}{3} \end{bmatrix}\begin{bmatrix} 0 & - \\ 0 & - \\ -1 & - \end{bmatrix} = \begin{bmatrix} \tfrac{1}{3} & - \\ 0 & - \\ -\tfrac{1}{3} & - \end{bmatrix}.$$

$$\Delta\mathbf{b^*} = \mathbf{S^*}\,\Delta\mathbf{b} = \begin{bmatrix} 1 & \tfrac{1}{3} & -\tfrac{1}{3} \\ 0 & \tfrac{1}{2} & 0 \\ 0 & -\tfrac{1}{3} & \tfrac{1}{3} \end{bmatrix}\begin{bmatrix} 0 \\ 12 \\ 0 \end{bmatrix} = \begin{bmatrix} 4 \\ 6 \\ -4 \end{bmatrix}.$$

Adding these increments to the original quantities in the final tableau (middle of Table 6.18) then yields the revised final tableau (bottom of Table 6.18).

This *incremental analysis* also provides a useful general insight, namely, that changes in the final tableau must be *proportional* to each change in the initial tableau. We illustrate in the next section how this property enables us to use linear interpolation or extrapolation to determine the range of values for a given parameter over which the final basic solution remains both feasible and optimal.

After obtaining the *revised* final simplex tableau, we next convert the tableau to *proper form from Gaussian elimination* (as needed). In particular, the basic variable for row i must have a coefficient of 1 in that row and a coefficient of *zero* in *every*

Table 6.18 Obtaining the Revised Final Simplex Tableau for the Revised Wyndor Glass Co. Problem

	Basic Variable	Eq. No.	Z	x_1	x_2	x_3	x_4	x_5	Right Side
						Coefficient of			
New initial tableau	Z	0	1	-4	-5	0	0	0	0
	x_3	1	0	1	0	1	0	0	4
	x_4	2	0	0	2	0	1	0	24
	x_5	3	0	2	2	0	0	1	18
Final tableau for original model	Z	0	1	0	0	0	$\frac{3}{2}$	1	36
	x_3	1	0	0	0	1	$\frac{1}{3}$	$-\frac{1}{3}$	2
	x_2	2	0	0	1	0	$\frac{1}{2}$	0	6
	x_1	3	0	1	0	0	$-\frac{1}{3}$	$\frac{1}{3}$	2
Revised final tableau	Z	0	1	-2	0	0	$\frac{3}{2}$	1	54
	x_3	1	0	$\frac{1}{3}$	0	1	$\frac{1}{3}$	$-\frac{1}{3}$	6
	x_2	2	0	0	1	0	$\frac{1}{2}$	0	12
	x_1	3	0	$\frac{2}{3}$	0	0	$-\frac{1}{3}$	$\frac{1}{3}$	-2

other row (including row 0) for the tableau to be in the proper form for identifying and evaluating the current basic solution. Therefore, if the changes have violated this requirement (which can occur only if the original constraint coefficients of a basic variable have been changed), further changes must be made to restore this form. This restoration is done by using Gaussian elimination, i.e., by successively applying part 3 of the iterative step for the simplex method (see Chap. 4) as if each violating basic variable were an entering basic variable. Note that these algebraic operations may also cause further changes in the right-side column, so that the current basic solution can be read from this column only when proper form from Gaussian elimination has been fully restored.

For the example, the revised final simplex tableau shown in the top half of Table 6.19 is not in proper form from Gaussian elimination because of the column for the basic variable x_1. Specifically, the coefficient of x_1 in *its* row (row 3) is $\frac{2}{3}$ instead of 1, and it has *nonzero* coefficients (-2 and $\frac{1}{3}$) in rows 0 and 1. To restore proper form, row 3 is multiplied by $\frac{3}{2}$; then, 2 times this new row 3 is added to row 0 and $\frac{1}{3}$ of the new row 3 is subtracted from row 1. This yields the *proper form from Gaussian elimination* shown in the bottom half of Table 6.19, which now can be used to identify the new values for the current (previously optimal) basic solution,

$$(x_1, x_2, x_3, x_4, x_5) = (-3, 12, 7, 0, 0).$$

Because x_1 is negative, this basic solution no longer is feasible. However, it is *superoptimal* (see Table 6.10) because *all* the coefficients in row 0 still are *non-negative*. Therefore, the *dual simplex method* can be used to *reoptimize* (if desired), starting from this basic solution. Referring to Fig. 6.2 (and ignoring slack variables), the dual simplex method uses just one iteration to move from the corner-point solution $(-3, 12)$ to the optimal corner-point feasible solution $(0, 9)$. (It is often useful in sensitivity analysis to identify the solutions that are optimal for some set of likely values of the model parameters and then to determine which of these solutions most *consistently* performs well for the various likely parameter values.)

Table 6.19 Converting the Revised Final Simplex Tableau to Proper Form from Gaussian Elimination for the Revised Wyndor Glass Co. Problem

	Basic Variable	Eq. No.	Z	x_1	x_2	x_3	x_4	x_5	Right Side
Revised final tableau	Z	0	1	-2	0	0	$\frac{3}{2}$	1	54
	x_3	1	0	$\frac{1}{3}$	0	1	$\frac{1}{3}$	$-\frac{1}{3}$	6
	x_2	2	0	0	1	0	$\frac{1}{2}$	0	12
	x_1	3	0	$\frac{2}{3}$	0	0	$-\frac{1}{3}$	$\frac{1}{3}$	-2
Converted to proper form	Z	0	1	0	0	0	$\frac{1}{2}$	2	48
	x_3	1	0	0	0	1	$\frac{1}{2}$	$-\frac{1}{2}$	7
	x_2	2	0	0	1	0	$\frac{1}{2}$	0	12
	x_1	3	0	1	0	0	$-\frac{1}{2}$	$\frac{1}{2}$	-3

If the basic solution $(-3, 12, 7, 0, 0)$ had been *neither* feasible nor superoptimal (i.e., if the tableau had negative entries in *both* the right-side column and row 0), artificial variables could have been introduced to convert the tableau to the proper form for an *initial* simplex tableau.[1]

When testing to see how *sensitive* the original optimal solution is to the various parameters of the model, the common approach is to check each parameter individually, changing its value from the initial estimate to other possibilities in the *range of likely values* (including the endpoints of this range). After the *sensitive* parameters have been identified, then some combinations of simultaneous changes of these parameters may be investigated. *Each* time one (or more) of the parameters is changed, the procedure described and illustrated here would be applied. Let us now summarize this procedure.

Summary of Procedure for Sensitivity Analysis

1. *Revision of model:* Make the desired change or changes in the model to be investigated next.
2. *Revision of final tableau:* Use the fundamental insight to determine the resulting changes in the final simplex tableau.
3. *Conversion to proper form from Gaussian elimination:* Convert this tableau to the proper form for identifying and evaluating the current basic solution by applying (as necessary) Gaussian elimination.
4. *Feasibility test:* Test this solution for feasibility by checking whether all its basic variable values in the right-side column of the tableau still are non-negative.
5. *Optimality test:* Test this solution for optimality (if feasible) by checking whether all its nonbasic variable coefficients in row 0 of the tableau still are nonnegative.
6. *Reoptimization:* If this solution fails either test, the new optimal solution can be obtained (if desired) by using the current tableau as the initial simplex tableau (making any necessary conversions) for the simplex method or dual simplex method.

---

[1] There also exists a primal-dual algorithm that can be directly applied to such a simplex tableau without any conversion.

In the next section, we shall discuss and illustrate the application of this procedure to each of the major categories of revisions in the original model. This discussion will involve, in part, expanding upon the example introduced in this section
for investigating changes in the Wyndor Glass Co. model. In fact, we shall begin by
individually checking each of the preceding changes. At the same time, we shall
integrate some of the applications of duality theory to sensitivity analysis discussed
in Sec. 6.5.

6.7 Applying Sensitivity Analysis

Sensitivity analysis often begins with the investigation of the effect of changes in the
b_i, the amount of resource i $(i = 1, 2, \ldots, m)$ being made available for the activities
under consideration. The reason is that there generally is more flexibility in setting
and adjusting these values than there is for the other parameters of the model. As
already discussed in Secs. 4.7 and 6.2, the economic interpretation of the dual variables (the y_i) as *shadow prices* is extremely useful for deciding which changes should
be considered.

Case 1—Changes in the b_i

Suppose that the only changes in the current model are that one or more of the b_i
parameters $(i = 1, 2, \ldots, m)$ has been changed. In this case, the *only* resulting
changes in the final simplex tableau are in the right-side column. Therefore, both the
conversion to proper form from Gaussian elimination and the *optimality test* steps of
the general procedure can be skipped.

EXAMPLE: Sensitivity analysis is begun for the *original* Wyndor Glass Co. problem
of Sec. 3.1 by examining the optimal values of the y_i dual variables ($y_1^* = 0$, $y_2^* =
\frac{3}{2}$, $y_3^* = 1$). These *shadow prices* give the *marginal value* of each resource i for the
activities (two new products) under consideration. As discussed in Sec. 4.7 (see Fig.
4.3), the total profit from these activities can be increased \$1.50/minute for each
additional unit of resource 2 (production capacity in Plant 2) that is made available.
This increase in profit holds for relatively small changes that do not affect the feasibility of the current basic solution (and so do not affect the values of the y_i^*).
Consequently, the OR Department has investigated the marginal profitability
from the other current uses of this resource to determine if any are less than
\$1.50/minute. This investigation reveals that one old product is far less profitable.
The production rate for this product already has been reduced to the minimum amount
that would justify its marketing expenses. However, it can be discontinued altogether,
which would provide an additional 12 units of resource 2 for the new products. Thus
the next step is to determine what profit could be obtained from the new products if
this shift were to be made. This shift changes b_2 from 12 to 24 in the linear programming model. Figure 6.3 shows the graphical effect of this change, including the shift
in the final corner-point solution from $(2, 6)$ to $(-2, 12)$. (Note that this figure differs
from Fig. 6.2 because the constraint, $3x_1 + 2x_2 \leq 18$, has not been changed here.)
When the fundamental insight (Table 6.17) is applied, the effect of this change
on the original final simplex tableau (middle of Table 6.18) is found to be that the

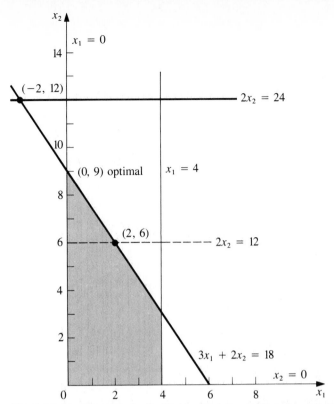

Figure 6.3 Feasible region for the Wyndor Glass Co. problem after changing just b_2 to 24.

entries in the right-side column change to the following values:

$$Z^* = \mathbf{y}^*\overline{\mathbf{b}} = [0, \tfrac{3}{2}, 1] \begin{bmatrix} 4 \\ 24 \\ 18 \end{bmatrix} = 54,$$

$$\mathbf{b}^* = \mathbf{S}^*\overline{\mathbf{b}} = \begin{bmatrix} 1 & \tfrac{1}{3} & -\tfrac{1}{3} \\ 0 & \tfrac{1}{2} & 0 \\ 0 & -\tfrac{1}{3} & \tfrac{1}{3} \end{bmatrix} \begin{bmatrix} 4 \\ 24 \\ 18 \end{bmatrix} = \begin{bmatrix} 6 \\ 12 \\ -2 \end{bmatrix}, \quad \text{so} \quad \begin{bmatrix} x_3 \\ x_2 \\ x_1 \end{bmatrix} = \begin{bmatrix} 6 \\ 12 \\ -2 \end{bmatrix}.$$

Equivalently, because the only change in the original model is $\Delta b_2 = 24 - 12 = 12$, *incremental analysis* can be used to calculate these same values more quickly as follows:

$$\Delta Z^* = \tfrac{3}{2}(12) = 18, \quad \text{so } Z^* = 36 + 18 = 54,$$

$$\Delta b_1^* = \tfrac{1}{3}(12) = 4, \quad \text{so } b_1^* = 2 + 4 = 6,$$

$$\Delta b_2^* = \tfrac{1}{2}(12) = 6, \quad \text{so } b_2^* = 6 + 6 = 12,$$

$$\Delta b_3^* = -\tfrac{1}{3}(12) = -4, \quad \text{so } b_3^* = 2 - 4 = -2,$$

where the *original* values of these quantities are obtained from the right-side column in the original final tableau (middle of Table 6.18). The resulting *revised final tableau* corresponds completely to this original final tableau except for replacing the right-side column with these new values.

Therefore, the current (previously optimal) basic solution has become

$$(x_1, x_2, x_3, x_4, x_5) = (-2, 12, 6, 0, 0),$$

which fails the *feasibility test* because of the negative value. The dual simplex method now can be applied, starting with this revised simplex tableau, to find the new optimal solution. This method leads in just one iteration to the new final simplex tableau shown in Table 6.20. (Alternatively, the simplex method could be applied from the beginning, which also would lead to this final tableau in just one iteration in this case.) This tableau indicates that the new optimal solution is

$$(x_1, x_2, x_3, x_4, x_5) = (0, 9, 4, 6, 0),$$

with $Z = 45$, thereby providing an increase in profit from the new products of \$9/minute over the previous $Z = 36$. The fact that $x_4 = 6$ indicates that 6 of the 12 additional units of resource 2 are unused by this solution.

Although $\Delta b_2 = 12$ proved to be too large an increase in b_2 to retain feasibility (and so optimality) with the basic solution where x_1, x_2, and x_3 are the basic variables (middle of Table 6.18), the above *incremental analysis* shows immediately just how large an increase is feasible. In particular, note that

$$b_1^* = 2 + \tfrac{1}{3} \Delta b_2,$$

$$b_2^* = 6 + \tfrac{1}{2} \Delta b_2,$$

$$b_3^* = 2 - \tfrac{1}{3} \Delta b_2,$$

where these three quantities are the values of x_3, x_2, and x_1, respectively, for this basic solution. The solution remains feasible, and so optimal, as long as all three quantities remain *nonnegative*. To determine the range of values of b_2 over which all three quantities remain nonnegative, set each quantity to zero, solve for Δb_2, and choose the positive and negative values of Δb_2 closest to zero, namely,

$$-6 \leq \Delta b_2 \leq 6, \quad \text{or} \quad 6 \leq b_2 \leq 18.$$

Thus, $6 \leq b_2 \leq 18$ is the **allowable range** for b_2 over which the original final basic solution (with new values for the basic variables) remains feasible, and so optimal,

Table 6.20 Revised Data for Wyndor Glass Co. Problem after Changing Just b_2

| | | | Final Simplex Tableau after Reoptimization | | | | | | |
| | | | | | Coefficient of | | | | |
Model parameters			Basic Variable	Eq. No.	Z	x_1	x_2	x_3	x_4	x_5	Right Side
$c_1 = 3, c_2 = 5 \ (n = 2)$			Z	0	1	$\tfrac{9}{2}$	0	0	0	$\tfrac{5}{2}$	45
$a_{11} = 1, a_{12} = 0, b_1 = 4$			x_3	1	0	1	0	1	0	0	4
$a_{21} = 0, a_{22} = 2, b_2 = 24$			x_2	2	0	$\tfrac{3}{2}$	1	0	0	$\tfrac{1}{2}$	9
$a_{31} = 3, a_{32} = 2, b_3 = 18$			x_4	3	0	-3	0	0	1	-1	6

as long as this is the only change in the original model. (Many linear programming software packages use this technique for automatically generating the *allowable range* for each b_i, and a similar technique for each c_j.)

Based on the results with $b_2 = 24$, the relatively unprofitable old product will be discontinued and the unused 6 units of resource 2 will be saved for some future use. Since y_3^* still is positive, a similar study is made of the possibility of changing the allocation of resource 3, but the resulting decision is to retain the current allocation. Therefore, the current linear programming model at this point has the parameter values and optimal solution shown in Table 6.20.

Case 2a—Changes in the Coefficients of a Nonbasic Variable

Consider a particular variable x_j (fixed j) that is a *nonbasic* variable in the optimal solution shown by the final simplex tableau (so that x_j is *not* included in the list of basic variables in the first column of this tableau). Case 2a is where the only changes in the current model are that one or more of the coefficients of this variable—c_j, a_{1j}, a_{2j}, ... , a_{mj}—have been changed.

As described at the beginning of Sec. 6.5, duality theory provides a very convenient way of checking these changes. In particular, if the *complementary* basic solution $\mathbf{y}^*$ in the dual problem still satisfies the single dual constraint that has changed, then the original optimal solution in the primal problem *remains optimal* as is. Conversely, if $\mathbf{y}^*$ violates this dual constraint, then this primal solution is *no longer optimal*.

If the optimal solution has changed and you wish to find the new one, you can do so rather easily. Simply apply the fundamental insight to revise the x_j column (the only one that has changed) in the final simplex tableau. With the current basic solution no longer optimal, the new value of $(z_j^* - c_j)$ now will be the one negative coefficient in row 0, so restart the simplex method with x_j as the *initial entering basic variable*.

Note that this procedure is a streamlined version of the general procedure summarized at the end of Sec. 6.6. Steps 3 and 4 (*conversion to proper form from Gaussian elimination* and *feasibility test*) have been deleted as irrelevant, because the only column being changed in the revision of the final tableau (before reoptimization) is for the *nonbasic* variable x_j. Step 5 (*optimality test*) has been replaced by a quicker test of optimality to be performed right after step 1 (*revision of model*). It is only if this test reveals that the optimal solution has changed, and you wish to find the new one, that steps 2 and 6 (*revision of final tableau* and *reoptimization*) are needed.

EXAMPLE: Since x_1 is nonbasic in the current optimal solution (see Table 6.20) for the Wyndor Glass Co. problem, the next step in its sensitivity analysis is to check whether any reasonable changes in the estimates of the coefficients of x_1 could still make it advisable to introduce product 1. The set of changes that goes as far as realistically possible to make product 1 more attractive would be to reset $c_1 = 4$ and $a_{31} = 2$ (as was done in Sec. 6.6).

This change in a_{31} revises the feasible region from that shown in Fig. 6.3 to the corresponding region in Fig. 6.2 when $3x_1 + 2x_2 = 18$ is replaced by $2x_1 + 2x_2 = 18$. (Ignore the $2x_2 = 12$ line, because the $2x_2 \leq 12$ constraint already has been replaced by $2x_2 \leq 24$.) The change in c_1 revises the objective function from $Z = 3x_1 + 5x_2$ to $Z = 4x_1 + 5x_2$. By using Fig. 6.2 to draw the objective function line, $Z = 45 = 4x_1 + 5x_2$, through the current optimal solution $(0, 9)$, you can verify that $(0, 9)$ *remains optimal* after these changes in a_{31} and c_1.

To use duality theory to draw this same conclusion, observe that the changes in c_1 and a_{31} lead to a single revised constraint for the dual problem (see Table 6.1). Both this revised constraint and the current $\mathbf{y^*}$ (coefficients of the slack variables in row 0 of Table 6.20) are shown below.

$$y_1 + 3y_3 \geq 3 \quad \rightarrow \quad y_1 + 2y_3 \geq 4,$$

$$y_1^* = 0, \qquad y_2^* = 0, \qquad y_3^* = \tfrac{5}{2}.$$

Note that $\mathbf{y^*}$ *still* satisfies the revised constraint, so the current primal solution (Table 6.20) is still optimal.

This approach of examining the revised dual constraint makes it easy to see just how much the parameters involved can be changed before the current optimal solution would become nonoptimal. For example, with $a_{31} = 2$, so that $y_1^* + 2y_3^* = 5$, the **allowable range** for c_1 without changing the optimal solution is $c_1 \leq 5$. Similarly, with the original value of a_{31} ($a_{31} = 3$), so that $y_1^* + 3y_3^* = \tfrac{15}{2}$, the *allowable range* is $c_1 \leq \tfrac{15}{2}$, so c_1 can be increased by as much as $\tfrac{9}{2}$ ($\Delta c_1 \leq \tfrac{9}{2}$) above its original value of $c_1 = 3$. This latter allowable range also can be obtained directly from Table 6.20, which gives $\tfrac{9}{2}$ as the coefficient of x_1 in row 0 of the final tableau. When the only parameter change from Table 6.20 is an increase in c_1, Table 6.17 indicates that the *only* resulting change in the final tableau is that this coefficient becomes $\tfrac{9}{2} - \Delta c_1$, so $\Delta c_1 \leq \tfrac{9}{2}$ is the allowable change permitted by the optimality test, and $c_1 \leq 3 + \tfrac{9}{2} = \tfrac{15}{2}$ is the allowable range for c_1. (This is the method used by many linear programming software packages to obtain the *allowable range* for the c_j corresponding to nonbasic variables.)

Because any larger changes in the original estimates of the coefficients of x_1 would be unrealistic, the OR Department concludes that these coefficients are *insensitive* parameters in the current model. Therefore, they will be kept fixed at their best estimates shown in Table 6.20, $c_1 = 3$ and $a_{31} = 3$, for the remainder of the sensitivity analysis.

Case 2b—Introduction of a New Variable

After solving for the optimal solution, we may discover that the linear programming formulation did not consider all the attractive alternative activities. Considering a new activity requires introducing a new variable with the appropriate coefficients into the objective function and constraints of the current model—which is Case 2b.

The convenient way to deal with this case is to treat it just as if it were Case 2a! This is done by pretending that the new variable x_j actually was in the original model with *all* its coefficients equal to *zero* (so that they still are zero in the *final* simplex tableau) and that x_j is a *nonbasic* variable in the current basic feasible solution. Therefore, if we change these zero coefficients to their actual values for the new variable, the procedure (including any reoptimization) does indeed become identical to that for Case 2a.

In particular, all you have to do to check whether the current solution still is optimal is check whether the *complementary* basic solution $\mathbf{y^*}$ satisfies the one new dual constraint that corresponds to the new variable in the primal problem. We already have described this approach and then illustrated it for the Wyndor Glass Co. problem in Sec. 6.5.

Case 3—Changes in the Coefficients of a Basic Variable

Now suppose that the variable x_j (fixed j) under consideration is a *basic* variable in the optimal solution shown by the final simplex tableau (so x_j appears in the first column of this tableau). Case 3 assumes that the only changes in the current model are in the coefficients of this variable.

Case 3 differs from Case 2a because of the requirement that a simplex tableau be in *proper form from Gaussian elimination*. This requirement allows the column for a *nonbasic* variable to be anything, so it does not affect Case 2a. However, for Case 3, it requires that the basic variable x_j must have a coefficient of 1 in its row of the simplex tableau and a coefficient of *zero* in *every* other row (including row 0). Therefore, after the changes in the x_j column of the *final* simplex tableau have been calculated,[1] it probably will be necessary to apply *Gaussian elimination* to restore this form, as was illustrated in Table 6.19. This step in turn probably will change the value of the current basic solution and may make it either infeasible or nonoptimal. Consequently, all the steps of the overall procedure summarized at the end of Sec. 6.6 are required for Case 3.

EXAMPLE: Because x_2 is a basic variable in Table 6.20 for the Wyndor Glass Co. problem, sensitivity analysis of its coefficients fits Case 3. Given the current optimal solution $(x_1 = 0, x_2 = 9)$, product 2 is the *only* new product that should be introduced, and its production rate should be relatively large. Therefore, the key question now is whether the initial estimates that led to the coefficients of x_2 in the current model could have *overestimated* the attractiveness of product 2 so much that they invalidate this conclusion. This question can be tested by checking the *most pessimistic* set of reasonable estimates for these coefficients, which turns out to be $\bar{c}_2 = 3$, $\bar{a}_{22} = 3$, and $\bar{a}_{32} = 4$.

The graphical effect of these changes is that the feasible region changes from the one shown in Fig. 6.3 to the one in Fig. 6.4. The optimal solution in Fig. 6.3 is $(x_1, x_2) = (0, 9)$, which is the corner-point solution lying at the intersection of the $x_1 = 0$ and $3x_1 + 2x_2 = 18$ constraint boundaries. With the revision of the constraints, the corresponding corner-point solution in Fig. 6.4 is $(0, \frac{9}{2})$. However, this solution no longer is optimal, because the revised objective function of $Z = 3x_1 + 3x_2$ now yields a new optimal solution of $(x_1, x_2) = (4, \frac{3}{2})$.

Now let us see how we draw these same conclusions *algebraically*. Because the only changes in the model are in the coefficients of x_2, the *only* resulting changes in the *final* simplex tableau (Table 6.20) are in the x_2 column. Therefore, the formulas in Table 6.17 are used to recompute just this column.

$$\mathbf{z} - \bar{\mathbf{c}} = \mathbf{y}^*\mathbf{A}^* - \bar{\mathbf{c}} = [0, 0, \tfrac{5}{2}]\begin{bmatrix} - & 0 \\ - & 3 \\ - & 4 \end{bmatrix} - [-, 3] = [-, 7].$$

$$\mathbf{A}^* = \mathbf{S}^*\bar{\mathbf{A}} = \begin{bmatrix} 1 & 0 & 0 \\ 0 & 0 & \tfrac{1}{2} \\ 0 & 1 & -1 \end{bmatrix}\begin{bmatrix} - & 0 \\ - & 3 \\ - & 4 \end{bmatrix} = \begin{bmatrix} - & 0 \\ - & 2 \\ - & 1 \end{bmatrix}.$$

---

[1] For the relatively sophisticated reader, we should point out a possible pitfall for Case 3 that would be discovered at this point. Specifically, the changes in the *initial* tableau can destroy the linear independence of the columns of coefficients of basic variables. This event occurs only if the unit coefficient of the basic variable x_j in the *final* tableau has been changed to *zero* at this point, in which case more extensive simplex method calculations must be used for Case 3.

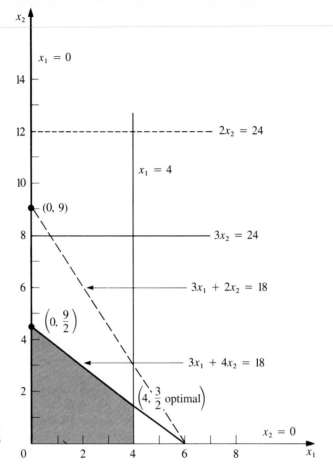

Figure 6.4 Feasible
region for the Case 3
example.

(Equivalently, *incremental analysis* with $\Delta c_2 = -2$, $\Delta a_{22} = 1$, and $\Delta a_{32} = 2$ can
be used in the same way to obtain this column.)

The resulting *revised* final tableau is shown at the top of Table 6.21. Note that
the new coefficients of the basic variable x_2 do not have the required values, so the
conversion-to-proper-form-from-Gaussian-elimination step must be applied next. This
step involves dividing row 2 by 2, subtracting 7 times the new row 2 from row 0,
and adding the new row 2 to row 3.

The resulting second tableau in Table 6.21 gives the new value of the current
basic solution, namely, $x_3 = 4$, $x_2 = \frac{9}{2}$, $x_4 = \frac{21}{2}$ ($x_1 = 0$, $x_5 = 0$). Since all these
variables are *nonnegative,* the solution is still *feasible.* However, because of the
negative coefficient of x_1 in row 0, we know that it is *no longer optimal.* Therefore,
the simplex method would be applied to this tableau, with this solution as the initial
basic feasible solution, to find the new optimal solution. The initial entering basic
variable is x_1, with x_3 as the leaving basic variable. Just one iteration is needed in
this case to reach the new optimal solution: $x_1 = 4$, $x_2 = \frac{3}{2}$, $x_4 = \frac{39}{2}$ ($x_3 = 0$,
$x_5 = 0$), as shown in the last tableau of Table 6.21.

Now look again in Fig. 6.4 at this new optimal solution $(4, \frac{3}{2})$. This solution is
optimal under the current *pessimistic* estimate that $c_2 = 3$, so $Z = 3x_1 + 3x_2$.
However, under the *original* estimate that $c_2 = 5$, the solution $(0, \frac{9}{2})$ would be

Table 6.21 Sensitivity Analysis Procedure Applied to Case 3 Example

	Basic Variable	Eq. No.	Z	x_1	x_2	x_3	x_4	x_5	Right Side
					Coefficient of				
Revised final tableau	Z	0	1	$\frac{9}{2}$	7	0	0	$\frac{5}{2}$	45
	x_3	1	0	1	0	1	0	0	4
	x_2	2	0	$\frac{3}{2}$	2	0	0	$\frac{1}{2}$	9
	x_4	3	0	-3	-1	0	1	-1	6
Converted to proper form	Z	0	1	$-\frac{3}{4}$	0	0	0	$\frac{3}{4}$	$\frac{27}{2}$
	x_3	1	0	1	0	1	0	0	4
	x_2	2	0	$\frac{3}{4}$	1	0	0	$\frac{1}{4}$	$\frac{9}{2}$
	x_4	3	0	$-\frac{9}{4}$	0	0	1	$-\frac{3}{4}$	$\frac{21}{2}$
New final tableau after reoptimization (only one iteration of the simplex method needed in this case)	Z	0	1	0	0	$\frac{3}{4}$	0	$\frac{3}{4}$	$\frac{33}{2}$
	x_1	1	0	1	0	1	0	0	4
	x_2	2	0	0	1	$-\frac{3}{4}$	0	$\frac{1}{4}$	$\frac{3}{2}$
	x_4	3	0	0	0	$\frac{9}{4}$	1	$-\frac{3}{4}$	$\frac{39}{2}$

optimal. Because of the constraint line, $3x_1 + 4x_2 = 18$, the crossover point from one optimal solution to the other is at $c_2 = 4$. If c_2 were less than 3, then $(4, \frac{3}{2})$ remains optimal as long as $c_2 \geq 0$. Therefore, the *allowable range* for c_2 without changing the optimal solution $(4, \frac{3}{2})$ is $0 \leq c_2 \leq 4$.

Table 6.22 shows how this *allowable range* for c_2 is calculated algebraically, where only the relevant portions of the simplex tableaux (row 0 and the row for x_2) are displayed. The starting point (first tableau displayed) is the final tableau from the bottom of Table 6.21. The key then is to apply steps 1, 2, 3, and 5 of the sensitivity analysis procedure (Sec. 6.6) when $c_2 = 3$ is increased or decreased by a small amount, where $\Delta c_2 = \pm 1$ is the convenient amount. The second and third tableaux in Table 6.22 show the effect of $\Delta c_2 = 1$, whereas the fourth and fifth tableaux repeat the procedure for $\Delta c_2 = -1$. (Note that the second and fourth tableaux differ from the first only in that the coefficient of x_2 in row 0 changes from 0 to $-\Delta c_2$.) With

Table 6.22 Obtaining Allowable Range for c_2 without Changing New Optimal Solution for Case 3 Example

	Basic Variable	Eq. No.	Z	x_1	x_2	x_3	x_4	x_5	Right Side
					Coefficient of				
Final tableau	Z	0	1	0	0	$\frac{3}{4}$	0	$\frac{3}{4}$	$\frac{33}{2}$
	x_2	2	0	0	1	$-\frac{3}{4}$	0	$\frac{1}{4}$	$\frac{3}{2}$
Revised final tableau when $\Delta c_2 = 1$	Z	0	1	0	-1	$\frac{3}{4}$	0	$\frac{3}{4}$	$\frac{33}{2}$
	x_2	2	0	0	1	$-\frac{3}{4}$	0	$\frac{1}{4}$	$\frac{3}{2}$
Converted to proper form	Z	0	1	0	0	0	0	1	18
	x_2	2	0	0	1	$-\frac{3}{4}$	0	$\frac{1}{4}$	$\frac{3}{2}$
Revised final tableau when $\Delta c_2 = -1$	Z	0	1	0	1	$\frac{3}{4}$	0	$\frac{3}{4}$	$\frac{33}{2}$
	x_2	2	0	0	1	$-\frac{3}{4}$	0	$\frac{1}{4}$	$\frac{3}{2}$
Converted to proper form	Z	0	1	0	0	$\frac{3}{2}$	0	$\frac{1}{2}$	15
	x_2	2	0	0	1	$-\frac{3}{4}$	0	$\frac{1}{4}$	$\frac{3}{2}$

$\Delta c_2 = 1$, the coefficient of x_3 in row 0 decreases from $\frac{3}{4}$ (first tableau) to 0 (third tableau), which indicates that $\Delta c_2 = 1$ is the crossover point above which the current basic solution no longer would be optimal. With $\Delta c_2 = -1$, the only number in row 0 that decreases is the coefficient of x_5. The decrease is from $\frac{3}{4}$ (first tableau) to $\frac{1}{2}$ (fifth tableau), which is only one-third of the way to 0. By linear extrapolation, $\Delta c_2 = -3$ thereby is the crossover point below which the current basic solution no longer is optimal. Therefore, $-3 \le \Delta c_2 \le 1$ (or $0 \le c_2 \le 4$) is the *allowable range* without changing the optimal solution.

All of this analysis suggests that c_2, a_{22}, and a_{32} are relatively sensitive parameters. However, additional data for estimating them more closely can be obtained only by conducting a pilot run. Therefore, the OR Department recommends that production of product 2 be initiated immediately on a small scale ($x_2 = \frac{3}{2}$) and that this experience be used to guide the decision on whether the remaining production capacity should be allocated to product 2 or product 1.

Case 4—Introduction of a New Constraint

The last case is one in which a new constraint must be introduced into the model after it has already been solved. This case may occur because the constraint was overlooked initially or because new considerations have arisen since the model was formulated originally. Another possibility is that the constraint was deleted purposely to decrease computational effort because it appeared to be less restrictive than other constraints already in the model, but now this impression needs to be checked with the optimal solution actually obtained.

To see if the current optimal solution would be affected by a new constraint, all you have to do is check directly whether the optimal solution satisfies the constraint. If it does, then it would still be the *best feasible solution* (i.e., the optimal solution), even if the constraint were added to the model. The reason is that a new constraint can only eliminate some previously feasible solutions without adding any new ones.

If the new constraint does eliminate the current optimal solution, and if you want to find the new solution, then introduce this constraint into the *final* simplex tableau (as an additional row) *just* as if this were the *initial* tableau, where the usual additional variable (slack variable or artificial variable) is designated to be the basic variable for this new row. Because the new row probably will have *nonzero* coefficients for some of the other basic variables, the *conversion-to-proper-form-from-Gaussian-elimination* step is applied next, and then the *reoptimization* step is applied in the usual way.

Just as for some of the preceding cases, this procedure for Case 4 is a streamlined version of the general procedure summarized at the end of Sec. 6.6. The only question to be addressed for this case is whether the previously optimal solution still is *feasible*, so step 5 (*optimality test*) has been deleted. Step 4 (*feasibility test*) has been replaced by a much quicker test of feasibility (does the previously optimal solution satisfy the new constraint?) to be performed right after step 1 (*revision of model*). It is only if this test provides a negative answer, and you wish to reoptimize, that steps 2, 3, and 6 are used (*revision of final tableau, conversion to proper form from Gaussian elimination,* and *reoptimization*).

EXAMPLE: To illustrate this case, suppose that the new constraint,

$$2x_1 + 3x_2 \le 24,$$

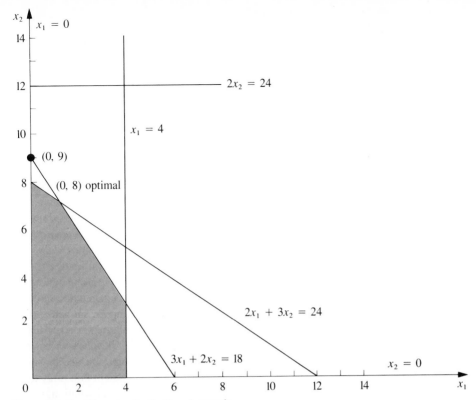

Figure 6.5 Feasible region for the Case 4 example.

is introduced into the model given in Table 6.20. The graphical effect is shown in Fig. 6.5. The previous optimal solution $(0, 9)$ violates the new constraint, so the optimal solution changes to $(0, 8)$.

To analyze this example algebraically, note that $(0, 9)$ yields $2x_1 + 3x_2 = 27 > 24$, so this previous optimal solution is no longer feasible. To find the new optimal solution, add the new constraint to the current final simplex tableau as just described, with the slack variable x_6 as its initial basic variable. This step yields the first tableau shown in Table 6.23. The *conversion-to-proper-form-from-Gaussian-elimination* step then requires subtracting three times row 2 from the new row, which identifies the current basic solution: $x_3 = 4$, $x_2 = 9$, $x_4 = 6$, $x_6 = -3$ $(x_1 = 0, x_5 = 0)$, as shown in the second tableau. Applying the dual simplex method to this tableau then leads in just one iteration (more are sometimes needed) to the new optimal solution in the last tableau of Table 6.23.

Systematic Sensitivity Analysis—Parametric Programming

So far we have described how to test specific changes in the model parameters. Another common approach to sensitivity analysis is to vary one or more parameters continuously over some interval(s) to see when the optimal solution changes.

For example, with the Wyndor Glass Co. problem, rather than beginning by testing the specific change from $b_2 = 12$ to $\bar{b}_2 = 24$, we might instead set

$$\bar{b}_2 = 12 + \theta,$$

Table 6.23 Sensitivity Analysis Procedure Applied to Case 4 Example

	Basic Variable	Eq. No.	Coefficient of							Right Side
			Z	x_1	x_2	x_3	x_4	x_5	x_6	
Revised final tableau	Z	0	1	$\frac{9}{2}$	0	0	0	$\frac{5}{2}$	0	45
	x_3	1	0	1	0	1	0	0	0	4
	x_2	2	0	$\frac{3}{2}$	1	0	0	$\frac{1}{2}$	0	9
	x_4	3	0	-3	0	0	1	-1	0	6
	x_6	New	0	2	3	0	0	0	1	24
Converted to proper form	Z	0	1	$\frac{9}{2}$	0	0	0	$\frac{5}{2}$	0	45
	x_3	1	0	1	0	1	0	0	0	4
	x_2	2	0	$\frac{3}{2}$	1	0	0	$\frac{1}{2}$	0	9
	x_4	3	0	-3	0	0	1	-1	0	6
	x_6	New	0	$-\frac{5}{2}$	0	0	0	$-\frac{3}{2}$	1	-3
New final tableau after reoptimization (only one iteration of dual simplex method needed in this case)	Z	0	1	$\frac{1}{3}$	0	0	0	0	$\frac{5}{3}$	40
	x_3	1	0	1	0	1	0	0	0	4
	x_2	2	0	$\frac{2}{3}$	1	0	0	0	$\frac{1}{3}$	8
	x_4	3	0	$-\frac{1}{3}$	0	0	1	0	$-\frac{2}{3}$	8
	x_5	New	0	$\frac{5}{3}$	0	0	0	1	$-\frac{2}{3}$	2

and then vary θ continuously from 0 to 12 (the maximum value of interest). The geometric interpretation in Fig. 6.3 is that the $2x_2 = 12$ constraint line is being shifted upward to $2x_2 = 12 + \theta$, with θ being increased from 0 to 12. The result is that the original optimal corner-point feasible solution (2, 6) shifts up the $3x_1 + 2x_2 = 18$ constraint line toward $(-2, 12)$. This corner-point solution remains optimal as long as it is still feasible ($x_1 \geq 0$), after which (0, 9) becomes the optimal solution.

The algebraic calculations of the effect of having $\Delta b_2 = \theta$ are directly analogous to those for the Case 1 example where $\Delta b_2 = 12$. In particular, by using the expressions for Z^* and $\mathbf{b}^*$ given in Table 6.17, the middle tableau in Table 6.18 indicates that the corresponding optimal solution is

$$Z = 36 + \tfrac{3}{2}\theta$$

$$x_3 = 2 + \tfrac{1}{3}\theta$$

$$x_2 = 6 + \tfrac{1}{2}\theta \qquad (x_4 = 0,\ x_5 = 0)$$

$$x_1 = 2 - \tfrac{1}{3}\theta$$

for θ small enough that this solution still is feasible, i.e., for $\theta \leq 6$. For $\theta > 6$, the dual simplex method yields the tableau shown in Table 6.20, with $Z = 45$, $x_3 = 4$, $x_2 = 9$, but $x_4 = -6 + \theta$ ($x_1 = 0$, $x_5 = 0$). This information can then be used (along with other data not incorporated into the model on the effect of increasing b_2) to decide whether to retain the original optimal solution and, if not, how much to increase b_2.

In a similar way, we can investigate the effect on the optimal solution of varying several parameters simultaneously. When we vary just b_i parameters, we express the *new* value $\bar{b}_i$ in terms of the *original* value b_i as follows:

$$\bar{b}_i = b_i + \alpha_i\theta, \qquad \text{for } i = 1, 2, \ldots, m,$$

where the α_i are input constants specifying the desired rate of increase (positive or negative) of the corresponding right-hand side as θ is increased.

For example, suppose that it is possible to shift some of the production of a current Wyndor Glass Co. product from Plant 2 to Plant 3, thereby increasing b_2 by decreasing b_3. Also suppose that b_3 decreases twice as fast as b_2 increases. Then

$$\bar{b}_2 = 12 + \theta$$

$$\bar{b}_3 = 18 - 2\theta,$$

where the (nonnegative) value of θ measures the amount of production shifted. (Thus $\alpha_1 = 0$, $\alpha_2 = 1$, and $\alpha_3 = -2$ in this case.) Referring to Fig. 6.3, the geometric interpretation is that, as θ is increased from 0, the $2x_2 = 12$ constraint line is being pushed up to $2x_2 = 12 + \theta$ (ignore the $2x_2 = 24$ line) and *simultaneously* the $3x_1 + 2x_2 = 18$ constraint line is being pushed down to $3x_1 + 2x_2 = 18 - 2\theta$. The original optimal corner-point feasible solution (2, 6) lies at the intersection of the $2x_2 = 12$ and $3x_1 + 2x_2 = 18$ lines, so shifting these lines causes this corner-point solution to shift. However, with the objective function of $Z = 3x_1 + 5x_2$, this corner-point solution will remain optimal as long as it is still feasible ($x_1 \geq 0$).

An algebraic investigation of simultaneously changing b_2 and b_3 in this way again involves using the formulas in Table 6.17 to calculate the resulting changes in the *final* tableau (middle of Table 6.18). The only quantities that can change are in the right-side column, namely,

$$Z^* = \mathbf{y}^*\bar{\mathbf{b}} = [0, \tfrac{3}{2}, 1] \begin{bmatrix} 4 \\ 12 + \theta \\ 18 - 2\theta \end{bmatrix} = 36 - \tfrac{1}{2}\theta,$$

$$\mathbf{b}^* = \mathbf{S}^*\mathbf{b}^* = \begin{bmatrix} 1 & \tfrac{1}{3} & -\tfrac{1}{3} \\ 0 & \tfrac{1}{2} & 0 \\ 0 & -\tfrac{1}{3} & \tfrac{1}{3} \end{bmatrix} \begin{bmatrix} 4 \\ 12 + \theta \\ 18 - 2\theta \end{bmatrix} = \begin{bmatrix} 2 + \theta \\ 6 + \tfrac{1}{2}\theta \\ 2 - \theta \end{bmatrix}.$$

Therefore, the optimal solution becomes

$$Z = 36 - \tfrac{1}{2}\theta$$

$$x_3 = 2 + \theta$$

$$x_2 = 6 + \tfrac{1}{2}\theta$$

$$x_1 = 2 - \theta$$

$$(x_4 = 0, \quad x_5 = 0)$$

for θ small enough that this solution still is feasible, i.e., for $\theta \leq 2$. (Check this conclusion in Fig. 6.3.) However, the fact that Z decreases as θ increases from 0 indicates that the best choice for θ is $\theta = 0$, so none of the possible shifting of production should be done.

The approach to varying several c_j parameters simultaneously is similar. In this case, we express the new value $\bar{c}_j$ in terms of the *original* value of c_j as follows:

$$\bar{c}_j = c_j + \alpha_j\theta, \quad \text{for } j = 1, 2, \ldots, n,$$

where the α_j are input constants specifying the desired rate of increase (positive or negative) of c_j as θ is increased.

To illustrate this case, reconsider the sensitivity analysis of c_1 and c_2 for the Wyndor Glass Co. problem that was performed earlier in this section. Starting with the version of the model presented in Table 6.20 and Fig. 6.3, we *separately* considered the effect of changing c_1 from 3 to 4 (its most optimistic estimate) and c_2 from 5 to 3 (its most pessimistic estimate). Now we can *simultaneously* consider both changes, as well as various intermediate cases with smaller changes, by setting

$$\bar{c}_1 = 3 + \theta, \qquad c_2 = 5 - 2\theta,$$

where the value of θ measures the *fraction* of the maximum possible change that is made. The result is to replace the original objective function, $Z = 3x_1 + 5x_2$, by a *function* of θ,

$$Z(\theta) = (3 + \theta)x_1 + (5 - 2\theta)x_2,$$

so the optimization now can be performed for any desired (fixed) value of θ between 0 and 1. By checking the effect as θ increases from 0 to 1, we can determine just when and how the optimal solution changes as the error in the original estimates of these parameters increases.

Considering these changes simultaneously is especially appropriate if there are factors that cause the parameters to change together. Are the two products competitive in some sense, so that a larger-than-expected unit profit for one implies a smaller-than-expected unit profit for the other? Are they both affected by some exogenous factor, such as the advertising emphasis of a competitor? Is it possible to simultaneously change both unit profits through appropriate shifting of personnel and equipment?

Referring to the feasible region shown in Fig. 6.3, the geometric interpretation of changing the objective function from $Z = 3x_1 + 5x_2$ to $Z(\theta) = (3 + \theta)x_1 + (5 - 2\theta)x_2$ is that we are changing the *slope* of the original objective function line, $Z = 45 = 3x_1 + 5x_2$, that can be drawn through the optimal solution (0, 9). If θ is increased enough, this slope will change sufficiently that the optimal solution will switch from (0, 9) to another corner-point feasible solution (4, 3). (Check graphically whether this occurs for $\theta \leq 1$.)

The algebraic procedure for dealing simultaneously with these two changes, $\Delta c_1 = \theta$ and $\Delta c_2 = -2\theta$, is similar to that shown in Table 6.22 (which dealt with $\Delta c_2 = \pm 1$ for another version of the model). Although the changes now are expressed in terms of θ rather than being specific numerical amounts, θ is treated just like an unknown number. Displaying just the relevant rows of the tableaux involved (row 0 and the row for the basic variable x_2), the results of this procedure are shown in Table 6.24. The first tableau shown is just the *final* tableau for the current version of the model (before changing c_1 and c_2) as given in Table 6.20. Referring to the formulas in Table 6.17, the only changes in the *revised* final tableau shown next are that Δc_1

Table 6.24 Dealing with $\Delta c_1 = \theta$ and $\Delta c_2 = -2\theta$ for the Model of Table 6.20

	Basic Variable	Eq. No.	Z	x_1	x_2	x_3	x_4	x_5	Right Side
					Coefficient of				
Final tableau	Z	0	1	$\frac{9}{2}$	0	0	0	$\frac{5}{2}$	45
	x_2	2	0	$\frac{3}{2}$	1	0	0	$\frac{1}{2}$	9
Revised final tableau when $\Delta c_1 = \theta$ and $\Delta c_2 = -2\theta$	$Z(\theta)$	0	1	$\frac{9}{2} - \theta$	2θ	0	0	$\frac{5}{2}$	45
	x_2	2	0	$\frac{3}{2}$	1	0	0	$\frac{1}{2}$	9
Converted to proper form	$Z(\theta)$	0	1	$\frac{9}{2} - 4\theta$	0	0	0	$\frac{5}{2} - \theta$	$45 - 18\theta$
	x_2	2	0	$\frac{3}{2}$	1	0	0	$\frac{1}{2}$	9

and Δc_2 are subtracted from the row 0 coefficients of x_1 and x_2, respectively. To convert this tableau to *proper form from Gaussian elimination,* we subtract 2θ times row 2 from row 0, which yields the last tableau shown. The expressions in terms of θ for the coefficients of nonbasic variables (x_1 and x_5) in row 0 of this tableau show that the current basic feasible solution remains optimal for $\theta \leq \frac{9}{8}$. Because $\theta = 1$ is the maximum realistic value of θ, this indicates that c_1 and c_2 together are *insensitive* parameters with respect to the model of Table 6.20. There is no need to try to estimate these parameters more closely unless other parameters change (as occurred for the Case 3 example).

As we discussed in Sec. 4.7, this way of continuously varying several parameters simultaneously is referred to as *parametric linear programming.* Section 9.3 presents the complete parametric linear programming procedure (including identifying new optimal solutions for larger values of θ) when just c_j parameters are being varied and then when just b_i parameters are being varied. Some linear programming software packages also include routines for varying just the coefficients of a single variable or varying just the parameters of a single constraint. In addition to the other applications discussed in Sec. 4.7, these procedures provide a convenient way of conducting sensitivity analysis systematically.

6.8 Conclusions

Every linear programming problem has associated with it a *dual* linear programming problem. There are a number of very useful relationships between the original (primal) problem and its dual problem that enhance our ability to analyze the primal problem. For example, the economic interpretation of the dual problem gives *shadow prices* that measure the marginal value of the resources in the primal problem, as well as providing an interpretation of the simplex method. Because the simplex method can be applied directly to either problem in order to solve both of them simultaneously, considerable computational effort sometimes can be saved by dealing directly with the dual problem. Duality theory, including the *dual simplex method* for working with superoptimal basic solutions, also plays a major role in *sensitivity analysis*.

The values used for the parameters of a linear programming model generally are just *estimates*. Therefore, sensitivity analysis needs to be performed to investigate what happens if these estimates are wrong. The *fundamental insight* of Sec. 5.3 provides the key for performing this investigation efficiently. The general objectives are to identify the *sensitive* parameters that affect the optimal solution, to try to estimate these sensitive parameters more closely, and then to select a solution that remains good over the range of likely values of the sensitive parameters. This analysis is a very important part of most linear programming studies.

SELECTED REFERENCES

1. Bradley, Stephen P., Arnoldo C. Hax, and Thomas L. Magnanti: *Applied Mathematical Programming,* Addison-Wesley, Reading, Mass., 1977.

2. Dantzig, George B.: *Linear Programming and Extensions,* Princeton University Press, Princeton, N.J., 1963.

3. Eppen, Gary D., F. J. Gould, and Charles Schmidt: *Quantitative Concepts for Management,* 3d ed., Prentice-Hall, Englewood Cliffs, N.J., 1988.

4. Gale, David: *The Theory of Linear Economic Models,* McGraw-Hill, New York, 1960.

5. Luenberger, David G.: *Introduction to Linear and Nonlinear Programming,* 2d ed., Addison-Wesley, Reading, Mass., 1984.

6. Murty, Katta: *Linear Programming,* 2d ed., Wiley, New York, 1983.

PROBLEMS

1. Construct the primal-dual table and the dual problem for each of the following linear programming models fitting our standard form:
(*a*) Model given in Prob. 3 of Chap. 4.
(*b*) Model given in Prob. 9 of Chap. 4.

2.* Construct the dual problem for each of the following linear programming models fitting our standard form:
(*a*) Model given in Prob. 4 of Chap. 3.
(*b*) Model given in Prob. 7 of Chap. 4.

3. Consider the linear programming model given in Prob. 12 of Chap. 4.
(*a*) Construct the primal-dual table and the dual problem for this model.
(*b*) What does the fact that Z is unbounded for this model imply about its dual problem?

4. For each of the following linear programming models, give your recommendation on the more efficient way (probably) for obtaining an optimal solution: (1) applying the simplex method directly to this primal problem or (2) applying the simplex method directly to the dual problem instead. Explain.

(*a*) Maximize $Z = 10x_1 - 4x_2 + 7x_3,$

subject to

$$3x_1 - x_2 + 2x_3 \leq 25$$
$$x_1 - 2x_2 + 3x_3 \leq 25$$
$$5x_1 + x_2 + 2x_3 \leq 40$$
$$x_1 + x_2 + x_3 \leq 90$$
$$2x_1 - x_2 + x_3 \leq 20$$

and $x_1 \geq 0,$ $x_2 \geq 0,$ $x_3 \geq 0.$

(*b*) Maximize $Z = 2x_1 + 5x_2 + 3x_3 + 4x_4 + x_5,$

subject to

$$x_1 + 3x_2 + 2x_3 + 3x_4 + x_5 \leq 6$$
$$4x_1 + 6x_2 + 5x_3 + 7x_4 + x_5 \leq 15$$

and $x_j \geq 0,$ for $j = 1, 2, 3, 4, 5.$

5. Consider the following problem.

Maximize $Z = -x_1 - 2x_2 - x_3,$

subject to

$$x_1 + x_2 + 2x_3 \leq 1$$
$$2x_1 - x_3 \leq 1$$

and $x_1 \geq 0,$ $x_2 \geq 0,$ $x_3 \geq 0.$

(*a*) Construct the dual problem.
(*b*) Use duality theory to show that the optimal solution to the primal problem has $Z \leq 0.$

6. Consider the simplex tableaux for the Wyndor Glass Co. problem given in Table 4.8. For each tableau, give the economic interpretation of the following items:

(a) Each of the coefficients of the slack variables (x_3, x_4, x_5) in row 0.

(b) Each of the coefficients of the original variables (x_1, x_2) in row 0.

(c) The resulting choice for the entering basic variable (or the decision to stop after the final tableau).

7.* Consider the following problem.

$$\text{Maximize} \quad Z = 6x_1 + 8x_2,$$

subject to

$$5x_1 + 2x_2 \leq 20$$

$$x_1 + 2x_2 \leq 10$$

and

$$x_1 \geq 0, \qquad x_2 \geq 0.$$

(a) Construct the dual problem for this primal problem.

(b) Solve both the primal problem and the dual problem graphically. Identify the corner-point feasible solutions and corner-point infeasible solutions for both problems. Calculate the objective function values for all these solutions.

(c) Use the information obtained in part (b) to construct a table listing the complementary basic solutions and so forth for these problems. (Use the same column headings as for Table 6.9.)

(d) Solve the primal problem by the simplex method. After each iteration (including iteration 0), identify the basic feasible solution for this problem and the complementary basic solution for the dual problem. Also identify the corresponding corner-point solutions.

8. Consider the model with two functional constraints and two variables given in Prob. 2 of Chap. 4. Follow the instructions of Prob. 7 above for this model.

9. Consider the primal and dual problems for the Wyndor Glass Co. example given in Table 6.1. Using Tables 5.5, 5.6, 6.8, and 6.9, construct a new table giving the eight sets of nonbasic variables for the primal problem in column 1, the corresponding sets of associated variables for the dual problem in column 2, and the set of nonbasic variables for each complementary basic solution of the dual problem in column 3. Explain why this table demonstrates the *complementary slackness property* for this example.

10. Suppose that a primal problem has a *degenerate* basic feasible solution (one or more basic variables equal to zero) as its optimal solution. What does this degeneracy imply about the dual problem? Why? Is the converse also true?

11. Consider the following problem.

$$\text{Maximize} \quad Z = 2x_1 - 4x_2,$$

subject to

$$x_1 - x_2 \leq 1$$

and

$$x_1 \geq 0, \qquad x_2 \geq 0.$$

(a) Construct the dual problem, and then find its optimal solution by inspection.

(b) Use the *complementary slackness property* and the optimal solution for the dual problem to find the optimal solution for the primal problem.

(c) Suppose that c_1, the coefficient of x_1 in the primal objective function, actually can have any value in the model. For what values of c_1 does the dual problem have *no* feasible solutions? For these values, what does duality theory then imply about the primal problem?

12. Consider the following problem.

$$\text{Maximize} \quad Z = 2x_1 + 7x_2 + 4x_3,$$

subject to

$$x_1 + 2x_2 + x_3 \leq 10$$

$$3x_1 + 3x_2 + 2x_3 \leq 10$$

and

$$x_1 \geq 0, \quad x_2 \geq 0, \quad x_3 \geq 0.$$

(a) Construct the dual problem for this primal problem.

(b) Use the dual problem to demonstrate that the optimal value of Z for the primal problem cannot exceed 25.

(c) It has been conjectured that x_2 and x_3 should be the basic variables for the optimal solution of the primal problem. Show that this conjecture is not true by directly deriving this basic solution (and Z) using *Gaussian elimination* (see Appendix 4). Simultaneously derive and identify the complementary basic solution for the dual problem.

(d) Solve the dual problem graphically. Use this solution to identify the basic variables and the nonbasic variables for the optimal solution of the primal problem. Directly derive this solution, using *Gaussian elimination*.

13. Reconsider the model of part (b) of Prob. 4.

(a) Construct its dual problem.

(b) Solve this dual problem graphically.

(c) Use the result from part (b) to identify the nonbasic variables and basic variables for the optimal basic solution for the *primal* problem.

(d) Use the results from part (c) to obtain the optimal basic solution for the primal problem *directly* by using *Gaussian elimination* to solve for its basic variables, starting from the initial system of equations [excluding Eq. (0)] constructed for the simplex method.

(e) Use the results from part (c) to identify the defining equations (see Sec. 5.1) for the optimal corner-point solution for the primal problem, and then use these equations to find this solution.

14. Consider the model given in Prob. 37 of Chap. 5.

(a) Construct the dual problem.

(b) Use the given information about the basic variables in the optimal primal solution to identify the nonbasic variables and basic variables for the optimal dual solution.

(c) Use the results from part (b) to identify the defining equations (see Sec. 5.1) for the optimal corner-point solution for the dual problem, and then use these equations to find this solution.

(d) Solve the dual problem graphically to verify your results from part (c).

15. Consider the model given in Prob. 3 of Chap. 3.

(a) Construct the dual problem for this model.

(b) Use the fact that $(x_1, x_2) = (13, 5)$ is optimal for the primal problem to identify the nonbasic variables and basic variables for the optimal basic solution for the dual problem.

(c) Identify the optimal basic solution for the dual problem by *directly* deriving Eq. (0) corresponding to the optimal primal solution identified in part (b). Derive this equation by using *Gaussian elimination* (see Appendix 4).

(d) Use the results from part (b) to identify the defining equations (see Sec. 5.1) for the optimal corner-point solution for the dual problem. Verify your optimal dual solution from part (c) by checking to see that it satisfies this system of equations.

16. Suppose that you also want information about the *dual* problem when you apply the *revised simplex method* (see Sec. 5.2) to the primal problem in *our standard form*.

(a) How would you identify the *optimal solution* for the dual problem?

(b) After obtaining the basic feasible solution at each iteration, how would you identify the *complementary basic solution* in the dual problem?

17. Consider the primal and dual problems in *our standard form* presented in matrix notation at the beginning of Sec. 6.1. Use *only* this definition of the dual problem for a primal problem in this form to prove each of the following results.

(a) The *weak duality property* presented in Sec. 6.1.

(b) If the primal problem has an *unbounded* feasible region that permits increasing Z indefinitely, then the dual problem has *no* feasible solutions.

(c) If the functional constraints for the primal problem, $\mathbf{Ax} \leq \mathbf{b}$, are changed to $\mathbf{Ax} = \mathbf{b}$, the only resulting change in the dual problem is to *delete* the nonnegativity constraints, $\mathbf{y} \geq \mathbf{0}$.

18.* Construct the dual problem for the linear programming problem given in Prob. 26 of Chap. 4.

19. For each of the following linear programming models, convert this primal problem into one of the two forms given in Table 6.14 and then construct its dual problem:

(a) Model given in Prob. 17 of Chap. 4.

(b) Model given in Prob. 18 of Chap. 4.

(c) Model given in Prob. 27 of Chap. 4.

20. Consider the model with equality constraints given in Prob. 25 of Chap. 4.

(a) Construct its dual problem by using the corresponding primal-dual form given in Table 6.14.

(b) Demonstrate that the answer in part (a) is correct (i.e., equality constraints yield dual variables without nonnegativity constraints) by first converting the primal problem to *our standard form* (see Table 6.12), then constructing its dual problem, and then converting this dual problem to the form obtained in part (a).

21.* Consider the model without nonnegativity constraints given in Prob. 22 of Chap. 4.

(a) Construct its dual problem by using the corresponding primal-dual form given in Table 6.14.

(b) Demonstrate that the answer in part (a) is correct (i.e., variables without nonnegativity constraints yield equality constraints in the dual problem) by first converting the primal problem to *our standard form* (see Table 6.12), then constructing its dual problem, and then converting this dual problem to the form obtained in part (a).

22. Consider the dual problem for the Wyndor Glass Co. example given in Table 6.1. Demonstrate that *its* dual problem is the primal problem given in Table 6.1 by going through the conversion steps given in Table 6.13.

23. Consider the primal and dual problems in *our standard form* presented in matrix notation at the beginning of Sec. 6.1. Let $\mathbf{y}^*$ denote the optimal solution for this dual problem. Suppose that $\mathbf{b}$ is then replaced by $\bar{\mathbf{b}}$. Let $\bar{\mathbf{x}}$ denote the optimal solution for the new primal problem.

Prove that
$$\mathbf{c}\bar{\mathbf{x}} \leq \mathbf{y}^*\bar{\mathbf{b}}.$$

24.* Consider the following problem.

$$\text{Maximize} \quad Z = 3x_1 + x_2 + 4x_3,$$

subject to

$$6x_1 + 3x_2 + 5x_3 \leq 25$$

$$3x_1 + 4x_2 + 5x_3 \leq 20$$

and

$$x_1 \geq 0, \quad x_2 \geq 0, \quad x_3 \geq 0.$$

The corresponding *final* set of equations yielding the optimal solution is

(0) $\qquad Z + 2x_2 \qquad + \frac{1}{5}x_4 + \frac{3}{5}x_5 = 17.$

(1) $\qquad x_1 - \frac{1}{3}x_2 \qquad + \frac{1}{3}x_4 - \frac{1}{3}x_5 = \frac{5}{3}.$

(2) $\qquad x_2 + x_3 - \frac{1}{5}x_4 + \frac{2}{5}x_5 = 3.$

(a) Identify the optimal solution from this set of equations.

(b) Construct the dual problem.

(c) Identify the optimal solution for the dual problem from the final set of equations. Verify this solution by solving the dual problem graphically.

(d) Suppose that the original problem is changed to

$$\text{Maximize} \quad Z = 3x_1 + 3x_2 + 4x_3,$$

$$\text{subject to} \qquad\qquad 6x_1 + 2x_2 + 5x_3 \le 25$$

$$3x_1 + 3x_2 + 5x_3 \le 20$$

$$\text{and} \qquad\qquad x_1 \ge 0, \qquad x_2 \ge 0, \qquad x_3 \ge 0.$$

Use duality theory to determine whether the previous optimal solution is still optimal.

(e) Use the *fundamental insight* presented in Sec. 5.3 to identify the new coefficients of x_2 in the final set of equations after it has been adjusted for the changes in the original problem given in part (d).

(f) Now suppose that the only change in the original problem is that a new variable x_{new} has been introduced into the model as follows:

$$\text{Maximize} \quad Z = 3x_1 + x_2 + 4x_3 + 2x_{new},$$

$$\text{subject to} \qquad\qquad 6x_1 + 3x_2 + 5x_3 + 3x_{new} \le 25$$

$$3x_1 + 4x_2 + 5x_3 + 2x_{new} \le 20$$

$$\text{and} \qquad\qquad x_1 \ge 0, \qquad x_2 \ge 0, \qquad x_3 \ge 0, \qquad x_{new} \ge 0.$$

Use duality theory to determine whether the previous optimal solution, along with $x_{new} = 0$, is still optimal.

(g) Use the *fundamental insight* presented in Sec. 5.3 to identify the coefficients of x_{new} as a nonbasic variable in the final set of equations resulting from the introduction of x_{new} into the original model as shown in part (f).

25. Consider the model of Prob. 35. Use duality theory directly to determine whether the current basic solution remains optimal after each of the following independent changes.

(a) The change in part (e) of Prob. 35.

(b) The change in part (g) of Prob. 35.

26. Consider the model of Prob. 38. Use duality theory directly to determine whether the current basic solution remains optimal after each of the following independent changes.

(a) The change in part (c) of Prob. 38.

(b) The change in part (f) of Prob. 38.

27. Consider the model of Prob. 39. Use duality theory directly to determine whether the current basic solution remains optimal after each of the following independent changes.

(a) The change in part (b) of Prob. 39.

(b) The change in part (d) of Prob. 39.

28. Reconsider the model of Prob. 24. You are now to conduct sensitivity analysis by *independently* investigating each of the following six changes in the original model. For each change, use the sensitivity analysis procedure to revise the given *final* set of equations (in tableau form) and convert it to the proper form for identifying and evaluating the current basic solution. Then test this solution for feasibility and for optimality. (Do not reoptimize.)

(a) Change the right-hand side of constraint 1 to $b_1 = 15$.
(b) Change the right-hand side of constraint 2 to $b_2 = 5$.
(c) Change the coefficient of x_2 in the objective function to $c_2 = 4$.
(d) Change the coefficient of x_3 in the objective function to $c_3 = 3$.
(e) Change the coefficient of x_2 in constraint 2 to $a_{22} = 1$.
(f) Change the coefficient of x_1 in constraint 1 to $a_{11} = 10$.

29. Consider the following problem.

$$\text{Maximize} \quad Z = 3x_1 + x_2 + 2x_3,$$

subject to

$$x_1 - x_2 + 2x_3 \leq 20$$

$$2x_1 + x_2 - x_3 \leq 10$$

and

$$x_1 \geq 0, \qquad x_2 \geq 0, \qquad x_3 \geq 0.$$

Let x_4 and x_5 denote the slack variables for the respective functional constraints. After applying the simplex method, the final simplex tableau is

Basic Variable	Eq. No.	Coefficient of						Right Side
		Z	x_1	x_2	x_3	x_4	x_5	
Z	0	1	8	0	0	3	4	100
x_3	1	0	3	0	1	1	1	30
x_2	2	0	5	1	0	1	2	40

(a) Perform sensitivity analysis to determine which of the 11 parameters of the model are *sensitive* parameters in the sense that any change in just that parameter's value will change the optimal solution.
(b) Find the *allowable range* for each c_j without changing the optimal solution.
(c) Find the *allowable range* for each b_i without changing the optimal set of basic variables.

30. For the problem given in Table 6.20, find the *allowable range* for c_2 without changing the optimal solution. Show your work algebraically, using the tableau given in Table 6.20. Then justify your answer from a geometric viewpoint, referring to Fig. 6.3.

31. For the *original* Wyndor Glass Co. problem, use the last tableau in Table 4.8 to do the following.
(a) Find the *allowable range* for each b_i without changing the optimal set of basic variables.
(b) Find the *allowable range* for c_1 and c_2 without changing the optimal solution.

32. For the Case 4 example presented in Sec. 6.7, use the last tableau in Table 6.23 to do the following.
(a) Find the *allowable range* for each b_i without changing the optimal set of basic variables.
(b) Find the *allowable range* for c_1 and c_2 without changing the optimal solution.

33. Consider the following problem.

$$\text{Maximize} \quad Z = 2x_1 + 5x_2,$$

subject to

$$x_1 + 2x_2 \leq 10$$

$$x_1 + 3x_2 \leq 12$$

and $\qquad$ $x_1 \geq 0, \qquad x_2 \geq 0.$

Let x_3 and x_4 denote the slack variables for the respective functional constraints. After we apply the simplex method, the final simplex tableau is

| Basic | Eq. | Coefficient of | | | | | | Right |
Variable	No.	Z	x_1	x_2	x_3	x_4		Side
Z	0	1	0	0	1	1		22
x_1	1	0	1	0	3	-2		6
x_2	2	0	0	1	-1	1		2

While doing *post-optimality analysis,* you learn that all four b_i and c_j values used in the original model just given are accurate only to within ± 50 percent. In other words, their ranges of *likely values* are $5 \leq b_1 \leq 15$, $6 \leq b_2 \leq 18$, $1 \leq c_1 \leq 3$, and $2.5 \leq c_2 \leq 7.5$. Your job now is to perform sensitivity analysis to determine for each parameter whether this uncertainty is likely to affect the optimal solution. Specifically, for each parameter, determine the *allowable range* of values for which the current basic solution (perhaps with new values for the basic variables) will remain optimal. Then divide up the range of *likely values* between these *allowable values* and *other values* for which the current basic solution will no longer be optimal.

(a) Perform this sensitivity analysis *graphically* on the original model.
(b) Now perform this sensitivity analysis as described and illustrated in Sec. 6.7 for b_1 and c_1.
(c) Repeat part (b) for b_2.
(d) Repeat part (b) for c_2.

34. Consider the following problem.

$$\text{Maximize} \qquad Z = 3x_1 + 4x_2 + 8x_3,$$

subject to

$$2x_1 + 3x_2 + 5x_3 \leq 9$$

$$x_1 + 2x_2 + 3x_3 \leq 5$$

and $\qquad$ $x_1 \geq 0, \qquad x_2 \geq 0, \qquad x_3 \geq 0.$

Let x_4 and x_5 denote the slack variables for the respective functional constraints. After we apply the simplex method, the final simplex tableau is

| Basic | Eq. | Coefficient of | | | | | | Right |
Variable	No.	Z	x_1	x_2	x_3	x_4	x_5	Side
Z	0	1	0	1	0	1	1	14
x_1	1	0	1	-1	0	3	-5	2
x_3	2	0	0	1	1	-1	2	1

While doing *post-optimality analysis,* you learn that some of the parameter values used in the original model just given are just rough estimates, where the range of *likely values* in each case is within ± 50 percent of the value used here. For each of these following parameters, perform sensitivity analysis to determine whether this uncertainty is likely to affect the optimal solution. Specifically, for each parameter, determine the *allowable range* of values for which the current basic solution (perhaps with new values for the basic variables) will remain optimal. Then divide up the range of *likely values* between these *allowable values* and *other values* for which the current basic solution will no longer be optimal.

    (a) The parameter b_2.
    (b) The parameter c_2.
    (c) The parameter a_{22}.
    (d) The parameter c_3.
    (e) The parameter a_{12}.
    (f) The parameter b_1.

35.* Consider the following problem.

$$\text{Maximize} \quad Z = -5x_1 + 5x_2 + 13x_3,$$

subject to

$$-x_1 + x_2 + 3x_3 \leq 20$$

$$12x_1 + 4x_2 + 10x_3 \leq 90$$

and

$$x_j \geq 0 \quad (j = 1, 2, 3).$$

If we let x_4 and x_5 be the slack variables for the respective constraints, the simplex method yields the following *final* set of equations:

$$(0) \qquad\qquad Z \qquad\quad + 2x_3 + 5x_4 \qquad = 100.$$

$$(1) \qquad\qquad -x_1 + x_2 + 3x_3 + \quad x_4 \qquad = \quad 20.$$

$$(2) \qquad\qquad 16x_1 \qquad\quad - 2x_3 - 4x_4 + x_5 = \quad 10.$$

Now you are to conduct sensitivity analysis by *independently* investigating each of the following nine changes in the original model. For each change, use the sensitivity analysis procedure to revise this set of equations (in tableau form) and convert it to proper form from Gaussian elimination for identifying and evaluating the current basic solution. Then test this solution for feasibility and for optimality. (Do not reoptimize.)

    (a) Change the right-hand side of constraint 1 to

$$b_1 = 30.$$

    (b) Change the right-hand side of constraint 2 to

$$b_2 = 70.$$

    (c) Change the right-hand sides to

$$\begin{bmatrix} b_1 \\ b_2 \end{bmatrix} = \begin{bmatrix} 10 \\ 100 \end{bmatrix}.$$

    (d) Change the coefficient of x_3 in the objective function to

$$c_3 = 8.$$

    (e) Change the coefficients of x_1 to

$$\begin{bmatrix} c_1 \\ a_{11} \\ a_{21} \end{bmatrix} = \begin{bmatrix} -2 \\ 0 \\ 5 \end{bmatrix}.$$

    (f) Change the coefficients of x_2 to

$$\begin{bmatrix} c_2 \\ a_{12} \\ a_{22} \end{bmatrix} = \begin{bmatrix} 6 \\ 2 \\ 5 \end{bmatrix}.$$

(g) Introduce a new variable x_6 with coefficients

$$\begin{bmatrix} c_6 \\ a_{16} \\ a_{26} \end{bmatrix} = \begin{bmatrix} 10 \\ 3 \\ 5 \end{bmatrix}.$$

(h) Introduce a new constraint $2x_1 + 3x_2 + 5x_3 \le 50$. (Denote its slack variable by x_6.)

(i) Change constraint 2 to

$$10x_1 + 5x_2 + 10x_3 \le 100.$$

36.* Reconsider the model of Prob. 35. Suppose that we now want to apply parametric linear programming analysis to this problem. Specifically, the right-hand sides of the functional constraints are changed to

$$20 + 2\theta \quad \text{(for constraint 1)}$$

$$90 - \theta \quad \text{(for constraint 2)},$$

where θ can be assigned any positive or negative values.

Express the basic solution (and Z) corresponding to the original optimal solution as a function of θ. Determine the lower and upper bounds on θ before this solution would become infeasible.

37. Consider the example for Case 3 of sensitivity analysis in Sec. 6.7, where $\bar{c}_2 = 3$, $\bar{a}_{22} = 3, \bar{a}_{32} = 4$, and where the other parameters are given in Table 6.20. Starting from the resulting final tableau given at the bottom of Table 6.21, construct a table like Table 6.24 to perform *parametric linear programming analysis,* where

$$c_1 = 3 + \theta, \qquad c_2 = 3 + 2\,\theta.$$

How far can θ be increased above 0 before the current basic solution is no longer optimal?

38. Consider the following problem.

$$\text{Maximize} \quad Z = 2x_1 - x_2 + x_3,$$

subject to

$$3x_1 + x_2 + x_3 \le 60$$

$$x_1 - x_2 + 2x_3 \le 10$$

$$x_1 + x_2 - x_3 \le 20$$

and

$$x_1 \ge 0, \qquad x_2 \ge 0, \qquad x_3 \ge 0.$$

Let x_4, x_5, and x_6 denote the slack variables for the respective constraints. After we apply the simplex method, the final simplex tableau is

Basic Variable	Eq. No.	Z	x_1	x_2	x_3	x_4	x_5	x_6	Right Side
Z	0	1	0	0	$\frac{3}{2}$	0	$\frac{3}{2}$	$\frac{1}{2}$	25
x_4	1	0	0	0	1	1	-1	-2	10
x_1	2	0	1	0	$\frac{1}{2}$	0	$\frac{1}{2}$	$\frac{1}{2}$	15
x_2	3	0	0	1	$-\frac{3}{2}$	0	$-\frac{1}{2}$	$\frac{1}{2}$	5

Now you are to conduct sensitivity analysis by *independently* investigating each of the following six changes in the original model. For each change, use the sensitivity analysis procedure to revise this set of equations (in tableau form) and convert it to proper form from Gaussian elimination for identifying and evaluating the current basic solution. Then test this solution for feasibility and for optimality. (Do not reoptimize.)

(a) Change the right-hand sides

$$\text{from} \quad \begin{bmatrix} b_1 \\ b_2 \\ b_3 \end{bmatrix} = \begin{bmatrix} 60 \\ 10 \\ 20 \end{bmatrix} \quad \text{to} \quad \begin{bmatrix} b_1 \\ b_2 \\ b_3 \end{bmatrix} = \begin{bmatrix} 70 \\ 20 \\ 10 \end{bmatrix}.$$

(b) Change the coefficients of x_1

$$\text{from} \quad \begin{bmatrix} c_1 \\ a_{11} \\ a_{21} \\ a_{31} \end{bmatrix} = \begin{bmatrix} 2 \\ 3 \\ 1 \\ 1 \end{bmatrix} \quad \text{to} \quad \begin{bmatrix} c_1 \\ a_{11} \\ a_{21} \\ a_{31} \end{bmatrix} = \begin{bmatrix} 1 \\ 2 \\ 2 \\ 0 \end{bmatrix}.$$

(c) Change the coefficients of x_3

$$\text{from} \quad \begin{bmatrix} c_3 \\ a_{13} \\ a_{23} \\ a_{33} \end{bmatrix} = \begin{bmatrix} 1 \\ 1 \\ 2 \\ -1 \end{bmatrix} \quad \text{to} \quad \begin{bmatrix} c_3 \\ a_{13} \\ a_{23} \\ a_{33} \end{bmatrix} = \begin{bmatrix} 2 \\ 3 \\ 1 \\ -2 \end{bmatrix}.$$

(d) Change the objective function to $Z = 3x_1 - 2x_2 + 3x_3$.
(e) Introduce a new constraint $3x_1 - 2x_2 + x_3 \le 30$. (Denote its slack variable by x_7.)
(f) Introduce a new variable x_8 with coefficients

$$\begin{bmatrix} c_8 \\ a_{18} \\ a_{28} \\ a_{38} \end{bmatrix} = \begin{bmatrix} -1 \\ -2 \\ 1 \\ 2 \end{bmatrix}.$$

39. Consider the following problem.

$$\text{Maximize} \quad Z = 2x_1 + 7x_2 - 3x_3,$$

subject to

$$x_1 + 3x_2 + 4x_3 \le 30$$

$$x_1 + 4x_2 - x_3 \le 10$$

and

$$x_1 \ge 0, \quad x_2 \ge 0, \quad x_3 \ge 0.$$

Letting x_4 and x_5 be the slack variables for the respective constraints, the simplex method yields the following *final* set of equations:

$$(0) \qquad Z + x_2 + x_3 \qquad + 2x_5 = 20.$$

$$(1) \qquad - x_2 + 5x_3 + x_4 - x_5 = 20.$$

$$(2) \qquad x_1 + 4x_2 - x_3 \qquad + x_5 = 10.$$

Now you are to conduct sensitivity analysis by *independently* investigating each of the following seven changes in the original model. For each change, use the sensitivity analysis procedure to revise this set of equations (in tableau form) and convert it to proper form from Gaussian elimination for identifying and evaluating the current basic solution. Then test this solution for feasibility and for optimality. (Do not reoptimize.)

(a) Change the right-hand sides to

$$\begin{bmatrix} b_1 \\ b_2 \end{bmatrix} = \begin{bmatrix} 20 \\ 30 \end{bmatrix}.$$

(b) Change the coefficients of x_3 to

$$\begin{bmatrix} c_3 \\ a_{13} \\ a_{23} \end{bmatrix} = \begin{bmatrix} -2 \\ 3 \\ -2 \end{bmatrix}.$$

(c) Change the coefficients of x_1 to

$$\begin{bmatrix} c_1 \\ a_{11} \\ a_{21} \end{bmatrix} = \begin{bmatrix} 4 \\ 3 \\ 2 \end{bmatrix}.$$

(d) Introduce a new variable x_6 with coefficients

$$\begin{bmatrix} c_6 \\ a_{16} \\ a_{26} \end{bmatrix} = \begin{bmatrix} 3 \\ 1 \\ 2 \end{bmatrix}.$$

(e) Change the objective function to $Z = x_1 + 5x_2 - 2x_3$.
(f) Introduce a new constraint $3x_1 + 2x_2 + 3x_3 \le 25$.
(g) Change constraint 2 to $x_1 + 2x_2 + 2x_3 \le 35$.

40. Reconsider the model of Prob. 39. Suppose that we now want to apply parametric linear programming analysis to this problem. Specifically, the right-hand sides of the functional constraints are changed to

$$30 + 3\theta \quad \text{(for constraint 1)}$$

$$10 - \theta \quad \text{(for constraint 2)},$$

where θ can be assigned any positive or negative values.

Express the basic solution (and Z) corresponding to the original optimal solution as a function of θ. Determine the lower and upper bounds on θ before this solution would become infeasible.

41. Consider the following problem.

$$\text{Maximize} \quad Z = 2x_1 - x_2 + x_3,$$

subject to

$$3x_1 - 2x_2 + 2x_3 \le 15$$

$$-x_1 + x_2 + x_3 \le 3$$

$$x_1 - x_2 + x_3 \le 4$$

and

$$x_1 \ge 0, \quad x_2 \ge 0, \quad x_3 \ge 0.$$

If we let x_4, x_5, and x_6 be the slack variables for the respective constraints, the simplex method yields the following *final* set of equations:

(0) $\quad Z \quad + 2x_3 + x_4 + x_5 \quad\quad = 18.$

(1) $\quad\quad x_2 + 5x_3 + x_4 + 3x_5 \quad\quad = 24.$

(2) $\quad\quad\quad 2x_3 \quad\quad + x_5 + x_6 = 7.$

(3) $\quad x_1 \quad + 4x_3 + x_4 + 2x_5 \quad\quad = 21.$

Now you are to conduct sensitivity analysis by *independently* investigating each of the following eight changes in the original model. For each change, use the sensitivity analysis procedure to revise this set of equations (in tableau form) and convert it to proper form from Gaussian elimination for identifying and evaluating the current basic solution. Then test this solution for feasibility and for optimality. (Do not reoptimize.)

(a) Change the right-hand sides to

$$
\begin{bmatrix} b_1 \\ b_2 \\ b_3 \end{bmatrix} = \begin{bmatrix} 10 \\ 4 \\ 2 \end{bmatrix}.
$$

(b) Change the coefficient of x_3 in the objective function to $c_3 = 2$.
(c) Change the coefficient of x_1 in the objective function to $c_1 = 3$.
(d) Change the coefficients of x_3 to

$$
\begin{bmatrix} c_3 \\ a_{13} \\ a_{23} \\ a_{33} \end{bmatrix} = \begin{bmatrix} 4 \\ 3 \\ 2 \\ 1 \end{bmatrix}.
$$

(e) Change the coefficients of x_1 and x_2 to

$$
\begin{bmatrix} c_1 \\ a_{11} \\ a_{21} \\ a_{31} \end{bmatrix} = \begin{bmatrix} 1 \\ 1 \\ -2 \\ 3 \end{bmatrix} \quad \text{and} \quad \begin{bmatrix} c_2 \\ a_{12} \\ a_{22} \\ a_{32} \end{bmatrix} = \begin{bmatrix} -2 \\ -1 \\ 3 \\ 2 \end{bmatrix},
$$

respectively.

(f) Change the objective function to $Z = 5x_1 + x_2 + 3x_3$.
(g) Change constraint 1 to $2x_1 - x_2 + 4x_3 \leq 12$.
(h) Introduce a new constraint $2x_1 + x_2 + 3x_3 \leq 60$.

42. Reconsider part (d) of Prob. 41. Use duality theory directly to determine whether the original optimal solution is still optimal.

43. Reconsider the model of Prob. 41. Suppose that you now have the option of making trade-offs in the profitability of the first two activities, whereby the objective function coefficient of x_1 can be increased by any amount by simultaneously decreasing the objective function coefficient of x_2 by the same amount. Thus the alternative choices of the objective function are

$$
Z(\theta) = (2 + \theta)x_1 - (1 + \theta)x_2 + x_3,
$$

where any nonnegative value of θ can be chosen.

Construct a table like Table 6.24 to perform parametric linear programming analysis on this problem. Determine the upper bound on θ before the original optimal solution would become nonoptimal. Then determine the best choice of θ over this range.

44. Consider the following problem.

$$\text{Maximize} \quad Z = 10x_1 + 4x_2,$$

subject to

$$3x_1 + x_2 \leq 30$$

$$2x_1 + x_2 \leq 25$$

and

$$x_1 \geq 0, \qquad x_2 \geq 0.$$

Let x_3 and x_4 denote the slack variables for the respective functional constraints. After applying the simplex method, the final simplex tableau is

Basic Variable	Eq. No.	Z	x_1	x_2	x_3	x_4	Right Side
Z	0	1	0	0	2	2	110
x_2	1	0	0	1	−2	3	15
x_1	2	0	1	0	1	−1	5

Now suppose that *both* of the following changes are made simultaneously in the original model:

(1) The first constraint is changed to $4x_1 + x_2 \leq 40$.

(2) Parametric programming is introduced to change the objective function to the alternative choices of

$$Z(\theta) = (10 - 2\theta)x_1 + (4 + \theta)x_2,$$

where any nonnegative value of θ can be chosen.

(a) Construct the resulting revised final tableau (as a function of θ), and then convert this tableau to proper form from Gaussian elimination. Use this tableau to identify the new optimal solution that applies for either $\theta = 0$ or sufficiently small values of θ.

(b) What is the upper bound on θ before this optimal solution would become nonoptimal?

(c) Over the range of θ from zero to this upper bound, which choice of θ gives the largest value of the objective function?

45. Consider the following problem.

$$\text{Maximize} \quad Z = 9x_1 + 8x_2 + 5x_3,$$

subject to

$$2x_1 + 3x_2 + x_3 \leq 4$$

$$5x_1 + 4x_2 + 3x_3 \leq 11$$

and

$$x_1 \geq 0, \qquad x_2 \geq 0, \qquad x_3 \geq 0.$$

Let x_4 and x_5 denote the slack variables for the respective functional constraints. After we apply the simplex method, the final simplex tableau is

Basic	Eq.				Coefficient of			Right
Variable	No.	Z	x_1	x_2	x_3	x_4	x_5	Side
Z	0	1	0	2	0	2	1	19
x_1	1	0	1	5	0	3	-1	1
x_3	2	0	0	-7	1	-5	2	2

(a) Suppose that a new technology has become available for conducting the first activity considered in this problem. If the new technology were to be adopted to replace the existing one, the coefficients of x_1 in the model would change

$$\text{from} \quad \begin{bmatrix} c_1 \\ a_{11} \\ a_{21} \end{bmatrix} = \begin{bmatrix} 9 \\ 2 \\ 5 \end{bmatrix} \quad \text{to} \quad \begin{bmatrix} c_1 \\ a_{11} \\ a_{21} \end{bmatrix} = \begin{bmatrix} 18 \\ 3 \\ 6 \end{bmatrix}.$$

Use the sensitivity analysis procedure to investigate the potential effect and desirability of adopting the new technology. Specifically, assuming it were adopted, construct the resulting revised final tableau, convert this tableau to proper form from Gaussian elimination, and then reoptimize (if necessary) to find the new optimal solution.

(b) Now suppose that you have the option of mixing the old and new technologies for conducting the first activity. Let θ denote the fraction of the technology used that is from the new technology, so $0 \leq \theta \leq 1$. Given θ, the coefficients of x_1 in the model become

$$\begin{bmatrix} c_1 \\ a_{11} \\ a_{21} \end{bmatrix} = \begin{bmatrix} 9 + 9\theta \\ 2 + \theta \\ 5 + \theta \end{bmatrix}.$$

Construct the resulting revised final tableau (as a function of θ), and convert this tableau to proper form from Gaussian elimination. Use this tableau to identify the current basic solution as a function of θ. Over the allowable values of $0 \le \theta \le 1$, give the range of values of θ for which this solution is both feasible and optimal. What is the best choice of θ within this range?

46. Consider the following problem.

$$\text{Maximize} \quad Z = 3x_1 + 5x_2 + 2x_3,$$

subject to

$$-2x_1 + 2x_2 + x_3 \le 5$$

$$3x_1 + x_2 - x_3 \le 10$$

and

$$x_1 \ge 0, \quad x_2 \ge 0, \quad x_3 \ge 0.$$

Let x_4 and x_5 be the slack variables for the respective functional constraints. After applying the simplex method, the final simplex tableau is

Basic Variable	Eq. No.	Z	Coefficient of x_1	x_2	x_3	x_4	x_5	Right Side
Z	0	1	0	20	0	9	7	115
x_1	1	0	1	3	0	1	1	15
x_3	2	0	0	8	1	3	2	35

Parametric linear programming analysis now is to be applied simultaneously to the objective function and right-hand sides, where the model in terms of the new parameter is the following:

$$\text{Maximize} \quad Z(\theta) = (3 + 2\theta)x_1 + (5 + \theta)x_2 + (2 - \theta)x_3,$$

subject to

$$-2x_1 + 2x_2 + x_3 \le 5 + 6\theta$$

$$3x_1 + x_2 - x_3 \le 10 - 8\theta$$

and

$$x_1 \ge 0, \quad x_2 \ge 0, \quad x_3 \ge 0.$$

Construct the resulting revised final tableau (as a function of θ), and convert this tableau to proper form from Gaussian elimination. Use this tableau to identify the current basic solution as a function of θ. For $\theta \ge 0$, give the range of values of θ for which this solution is both feasible and optimal. What is the best choice of θ within this range?

47. Consider the following problem.

$$\text{Minimize} \quad y_0 = 5y_1 + 4y_2,$$

subject to

$$4y_1 + 3y_2 \ge 4$$

$$2y_1 + y_2 \ge 3$$

$$y_1 + 2y_2 \ge 1$$

$$y_1 + y_2 \ge 2$$

and

$$y_1 \ge 0, \quad y_2 \ge 0.$$

Because this *primal* problem has more functional constraints than variables, suppose that the simplex method has been applied directly to its *dual* problem. If we let x_5 and x_6 denote the slack variables for this dual problem, the resulting *final* simplex tableau is

Basic Variable	Eq. No.	Z	x_1	x_2	x_3	x_4	x_5	x_6	Right Side
Z	0	1	3	0	2	0	1	1	9
x_2	1	0	1	1	-1	0	1	-1	1
x_4	2	0	2	0	3	1	-1	2	3

For each of the following independent changes in the original primal model, you now are to conduct sensitivity analysis by *directly* investigating the effect on the *dual* problem and then inferring the complementary effect on the primal problem. For each change, apply the procedure for sensitivity analysis summarized at the end of Sec. 6.6 to the *dual* problem (do *not* reoptimize), and then give your conclusions as to whether the current basic solution for the *primal* problem still is feasible and whether it still is optimal. Then check your conclusions by a direct *graphical* analysis of the *primal* problem.

(a) Change the objective function to $y_0 = 3y_1 + 5y_2$.

(b) Change the right-hand sides of the functional constraints to 3, 5, 2, and 3, respectively.

(c) Change the first constraint to $2y_1 + 4y_2 \geq 7$.

(d) Change the second constraint to $5y_1 + 2y_2 \geq 10$.

48. Consider the Wyndor Glass Co. problem described in Sec. 3.1. Suppose that, in addition to considering the introduction of two new products, management now is also considering changing the production rate of a certain old product that is still profitable. Refer to Table 3.1. The capacity used per unit production rate of this old product is 1, 4, and 3 for Plants 1, 2, and 3, respectively. Therefore, if we let θ denote the *change* (positive or negative) in the production rate of this old product, the right-hand side of the three functional constraints in Sec. 3.1 becomes $(4 - \theta)$, $(12 - 4\theta)$, and $(18 - 3\theta)$, respectively. Thus choosing a negative value of θ would free additional capacity for producing more of the two new products, whereas a positive value would have the opposite effect.

(a) Use a *parametric linear programming* formulation to determine the effect of different choices of θ on the optimal solution for the product mix of the two new products given in the final tableau of Table 4.8. In particular, use the *fundamental insight* of Sec. 5.3 to obtain expressions for Z and the basic variables (x_3, x_2, and x_1) in terms of θ, assuming that θ is sufficiently close to zero that this "final" basic solution still is feasible and thus optimal for the given value of θ.

(b) Now consider the broader question of the choice of θ along with the product mix for the two new products. What is the break-even unit profit for the old product (in comparison with the two new products) below which its production rate should be decreased ($\theta < 0$) in favor of the new products and above which its production rate should be increased ($\theta > 0$) instead?

(c) If the unit profit is above this break-even point, how much can the old product's production rate be increased before the "final" basic feasible solution would become infeasible?

(d) If the unit profit is below this break-even point, how much can the old product's production rate be decreased (assuming its previous rate was larger than this decrease) before the "final" basic feasible solution would become infeasible?

49. Consider the following problem.

$$\text{Maximize} \quad Z = 2x_1 - x_2 + 3x_3,$$

subject to

$$x_1 + x_2 + x_3 = 3$$

$$x_1 - 2x_2 + x_3 \geq 1$$

$$2x_2 + x_3 \leq 2$$

and $\qquad x_1 \geq 0, \qquad x_2 \geq 0, \qquad x_3 \geq 0.$

Suppose that the Big M method (see Sec. 4.6) is used to obtain the initial (artificial) basic feasible solution. Let $\bar{x}_4$ be the artificial slack variable for the first constraint, x_5 the slack (surplus) variable for the second constraint, $\bar{x}_6$ the artificial variable for the second constraint, and x_7 the slack variable for the third constraint. The corresponding *final* set of equations yielding the optimal solution is

$$(0) \qquad Z + 5x_2 \qquad + (M + 2)\bar{x}_4 \qquad + M\bar{x}_6 + x_7 = 8$$

$$(1) \qquad x_1 - x_2 \qquad + \qquad \bar{x}_4 \qquad - x_7 = 1$$

$$(2) \qquad 2x_2 + x_3 \qquad + x_7 = 2$$

$$(3) \qquad 3x_2 \qquad + \qquad \bar{x}_4 + x_5 - \bar{x}_6 \qquad = 2.$$

Suppose that the original objective function is changed to $Z = 2x_1 + 3x_2 + 4x_3$, and that the original third constraint is changed to $2x_2 + x_3 \leq 1$. Use the sensitivity analysis procedure to revise the final set of equations (in tableau form) and convert it to proper form from Gaussian elimination for identifying and evaluating the current basic solution. Then test this solution for feasibility and for optimality. Reoptimize (if needed), starting from this final tableau, to find the new optimal solution.

7

Special Types of Linear Programming Problems

Chapter 3 emphasized the wide applicability of linear programming. We continue to broaden our horizons in this chapter by discussing some particularly important types of linear programming problems. These special types share several key characteristics. The first is that they all arise frequently in a variety of contexts. They also tend to require a very large number of constraints and variables, so a straightforward computer application of the simplex method may require an exorbitant computational effort. Fortunately, another characteristic is that most of the a_{ij} coefficients in the constraints are zeroes, and the relatively few nonzero coefficients appear in a distinctive pattern. As a result, it has been possible to develop special *streamlined* versions of the simplex method that achieve dramatic computational savings by exploiting this *special structure* of the problem. Therefore, it is important to become sufficiently familiar with these special types of problems so that you can recognize them when they arise and apply the proper computational procedure.

To describe special structures, we shall introduce the table (matrix) of constraint coefficients shown in Table 7.1, where a_{ij} is the coefficient of the jth variable in the

Table 7.1 Table of Constraint Coefficients
for Linear Programming

$$A = \begin{bmatrix} a_{11} & a_{12} & \cdots & a_{1n} \\ a_{21} & a_{22} & \cdots & a_{2n} \\ \vdots & \vdots & & \vdots \\ a_{m1} & a_{m2} & \cdots & a_{mn} \end{bmatrix}$$

*i*th functional constraint. Later, portions of the table containing only coefficients equal to zero will be indicated by leaving them blank, whereas blocks containing nonzero coefficients will be shaded darker.

Probably the most important special type of linear programming problem is the so-called *transportation problem,* and we shall describe it first. Its special solution procedure also will be presented, partially to illustrate the kind of streamlining of the simplex method that can be obtained by exploiting the special structure in the problem. Next we shall present two special types of linear programming problems (the *transshipment problem* and the *assignment problem*) that are closely related to the transportation problem. (All three of these problems can be represented by *networks*, so they will be considered again from that viewpoint in Sec. 10.6.)

Sections 7.5 to 7.7 discuss some special types of linear programming problems that can be characterized by where the *blocks of nonzero coefficients* appear in the table of constraint coefficients. One type frequently arises in multidivisional organizations. A second arises in multitime period problems. A third combines the first two types.

One of the practical problems involved in the application of linear programming is the uncertainty about what the values of the model parameters will turn out to be when the adopted solution actually is implemented. Occasionally, the degree of uncertainty is so great that some or all of the model parameters need to be treated explicitly as *random variables*. Sections 7.8 and 7.9 present two special formulations, *stochastic programming* and *chance-constrained programming,* for this problem of *linear programming under uncertainty.*

7.1 The Transportation Problem

Prototype Example

One of the main products of the P & T COMPANY is canned peas. The peas are prepared at three canneries (near Bellingham, Washington; Eugene, Oregon; and Albert Lea, Minnesota) and then shipped by truck to four distributing warehouses in the western United States (Sacramento, California; Salt Lake City, Utah; Rapid City, South Dakota; and Albuquerque, New Mexico), as shown in Fig. 7.1. Because the shipping costs are a major expense, management is initiating a study to reduce them as much as possible. For the upcoming season, an estimate has been made of what the output will be from each cannery, and each warehouse has been allocated a certain amount from the total supply of peas. This information (in units of truckloads), along with the shipping cost per truckload for each cannery-warehouse combination, is given in Table 7.2. Thus there are a total of 300 truckloads to be shipped. The problem now is to determine which plan for assigning these shipments to the various cannery-warehouse combinations would *minimize the total shipping cost.*

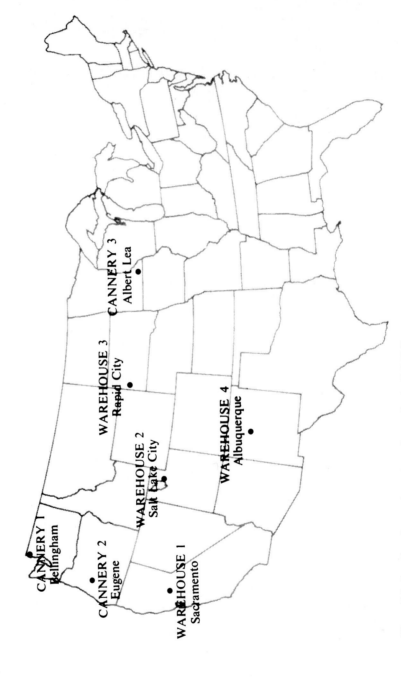

Figure 7.1 Location of canneries and warehouses for the P & T Co.

CANNERY 1
Bellingham

CANNERY 2
Eugene

CANNERY 3
Albert Lea

WAREHOUSE 3
Rapid City

WAREHOUSE 2
Salt Lake City

WAREHOUSE 4
Albuquerque

WAREHOUSE 1
Sacramento

Table 7.2 **Shipping Data for P & T Co.**

	Shipping Cost ($) Per Truckload				
	Warehouse				
	1	2	3	4	Output
1	464	513	654	867	75
Cannery 2	352	416	690	791	125
3	995	682	388	685	100
Allocation	80	65	70	85	

This is actually a linear programming problem of the *transportation problem type*. To formulate the model, let Z denote total shipping cost, and let x_{ij} ($i = 1, 2, 3$; $j = 1, 2, 3, 4$) be the number of truckloads to be shipped from cannery i to warehouse j. Thus the objective is to choose the values of these 12 decision variables (the x_{ij}) so as to

Minimize $\quad Z = 464x_{11} + 513x_{12} + 654x_{13} + 867x_{14} + 352x_{21} + 416x_{22}$
$$+ 690x_{23} + 791x_{24} + 995x_{31} + 682x_{32} + 388x_{33} + 685x_{34},$$

subject to the constraints

$$x_{11} + x_{12} + x_{13} + x_{14} \qquad\qquad\qquad\qquad\qquad\qquad = 75$$
$$x_{21} + x_{22} + x_{23} + x_{24} \qquad\qquad\qquad = 125$$
$$x_{31} + x_{32} + x_{33} + x_{34} = 100$$
$$x_{11} \qquad\qquad + x_{21} \qquad\qquad + x_{31} \qquad\qquad = 80$$
$$x_{12} \qquad\qquad + x_{22} \qquad\qquad + x_{32} \qquad\qquad = 65$$
$$x_{13} \qquad\qquad + x_{23} \qquad\qquad + x_{33} \qquad = 70$$
$$x_{14} \qquad\qquad + x_{24} \qquad\qquad + x_{34} = 85$$

and $\qquad\qquad x_{ij} \geq 0 \qquad (i = 1, 2, 3; j = 1, 2, 3, 4).$

Table 7.3 shows the constraint coefficients. As you will see next, it is the special

Table 7.3 **Table of Constraint Coefficients for the P & T Co.**

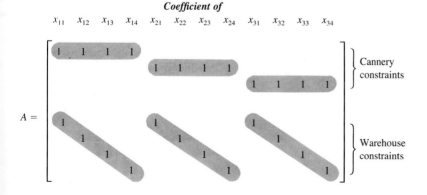

structure in the pattern of these coefficients that distinguishes this problem as a transportation problem, not its context.

By the way, the optimal solution for this problem is $x_{11} = 0$, $x_{12} = 20$, $x_{13} = 0$, $x_{14} = 55$, $x_{21} = 80$, $x_{22} = 45$, $x_{23} = 0$, $x_{24} = 0$, $x_{31} = 0$, $x_{32} = 0$, $x_{33} = 70$, $x_{34} = 30$. When you learn the optimality test that appears in Sec. 7.2, you will be able to verify this yourself (see Prob. 9).

The Transportation Problem Model

To describe the general model for the transportation problem, we need to use terms that are considerably less specific than those for the components of the prototype example. In particular, the general transportation problem is concerned (literally or figuratively) with distributing *any* commodity from *any* group of supply centers, called **sources,** to *any* group of receiving centers, called **destinations,** in such a way as to minimize the total distribution cost. The correspondence in terminology between the prototype example and the general problem is summarized in Table 7.4.

Thus, in general, source i ($i = 1, 2, \ldots, m$) has a supply of s_i units to distribute to the destinations, and destination j ($j = 1, 2, \ldots, n$) has a demand for d_j units to be received from the sources. A basic assumption is that the cost of distributing units from source i to destination j is directly proportional to the number distributed, where c_{ij} denotes the cost per unit distributed. As for the prototype example, these input data can be summarized conveniently in the **cost and requirements table** shown in Table 7.5.

Letting Z be total distribution cost and x_{ij} ($i = 1, 2, \ldots, m; j = 1, 2, \ldots, n$) be the number of units to be distributed from source i to destination j, the linear programming formulation of this problem becomes

$$\text{Minimize} \quad Z = \sum_{i=1}^{m} \sum_{j=1}^{n} c_{ij}x_{ij},$$

subject to

$$\sum_{j=1}^{n} x_{ij} = s_i, \quad \text{for } i = 1, 2, \ldots, m$$

$$\sum_{i=1}^{m} x_{ij} = d_j, \quad \text{for } j = 1, 2, \ldots, n$$

and

$$x_{ij} \geq 0, \quad \text{for all } i \text{ and } j.$$

Table 7.4 **Terminology for the Transportation Problem**

Prototype Example	General Problem
Truckloads of canned peas	Units of a commodity
Three canneries	m sources
Four warehouses	n destinations
Output from cannery i	s_i supply from source i
Allocation to warehouse j	d_j demand at destination j
Shipping cost per truckload from cannery i to warehouse j	c_{ij} cost per unit distributed from source i to destination j

**Table 7.5 Cost and Requirements Table for
the Transportation Problem**

		Cost Per Unit Distributed				
		Destination				
		1	2	$\cdots$	n	Supply
	1	c_{11}	c_{12}	$\cdots$	c_{1n}	s_1
	2	c_{21}	c_{22}	$\cdots$	c_{2n}	s_2
Source	.	.	.	$\cdots$	.	.
	.	.	.		.	.
	m	c_{m1}	c_{m2}	$\cdots$	c_{mn}	s_m
Demand		d_1	d_2	$\cdots$	d_n	

Note that the resulting table of constraint coefficients has the special structure shown in Table 7.6. *Any* linear programming problem that fits this special formulation is of the transportation problem type, regardless of its physical context. In fact, there have been numerous applications unrelated to transportation that have been fitted to this special structure, as we shall illustrate in the next example. (The *assignment problem* described in Sec. 7.4 is an additional example.) This is one of the reasons why the transportation problem is generally considered the most important special type of linear programming problem.

For many applications, the supply and demand quantities in the model (the s_i and d_j) have integer values, and implementation will require that the distribution quantities (the x_{ij}) also have integer values. Fortunately, because of the special structure shown in Table 7.6, such problems have the following property.

> **Integer solutions property:** For transportation problems where every s_i and d_j has an integer value, all the basic variables (allocations) in *every* basic feasible solution (including an optimal one) also have *integer* values.

The solution procedure described in Sec. 7.2 deals only with basic feasible solutions, so it automatically will obtain an *integer* optimal solution for this case.

Table 7.6 Table of Constraint Coefficients for the Transportation Problem

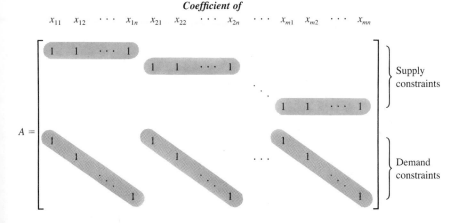

Therefore, it is unnecessary to add a constraint to the model that the x_{ij} must have integer values.

However, in order to have an optimal solution of any kind, a transportation problem must possess *feasible* solutions. The following property indicates when this will occur.

> **Feasible solutions property:** A necessary and sufficient condition for a transportation problem to have any feasible solutions is that
>
> $$\sum_{i=1}^{m} s_i = \sum_{j=1}^{n} d_j.$$

This property may be verified by observing that the constraints require that both

$$\sum_{i=1}^{m} s_i \quad \text{and} \quad \sum_{j=1}^{n} d_j \quad \text{equal} \quad \sum_{i=1}^{m} \sum_{j=1}^{n} x_{ij}.$$

This condition that the total supply must equal the total demand merely requires that the system be in balance. If the problem has physical significance and this condition is not met, it usually means that either s_i or d_j actually represents a *bound* rather than an exact requirement. If this is the case, a fictitious "source" or "destination" (called the **dummy source** or the **dummy destination**) can be introduced to take up the slack in order to convert the inequalities into equalities and satisfy the feasibility condition. The next two examples illustrate how to do this conversion, as well as how to fit some other common variations into the transportation problem formulation.

EXAMPLE—PRODUCTION SCHEDULING: The NORTHERN AIRPLANE COM-PANY builds commercial airplanes for various airline companies around the world. The last stage in the production process is to produce the jet engines and then to install them (a very fast operation) in the completed airplane frame. The company has been working under some contracts to deliver a considerable number of airplanes in the near future, and the production of the jet engines for these planes must now be scheduled for the next 4 months.

To meet the contracted dates for delivery, the company must supply engines for installation in the quantities indicated in the second column of Table 7.7. Thus the cumulative number of engines produced by the end of months 1, 2, 3, and 4 must be at least 10, 25, 50, and 70, respectively.

The facilities that will be available for producing the engines vary according to other production, maintenance, and renovation work scheduled during this period.

Table 7.7 **Production Scheduling Data for Northern Airplane Co.**

Month	Scheduled Installations	Maximum Production	Unit Cost* of Production	Unit Cost* of Storage
1	10	25	1.08	0.015
2	15	35	1.11	0.015
3	25	30	1.10	0.015
4	20	10	1.13	

* Cost is expressed in millions of dollars.

The resulting monthly differences in the maximum number that can be produced and the cost (in millions of dollars) of producing each one are given in the third and fourth columns of Table 7.7.

Because of the variations in production costs, it may well be worthwhile to produce some of the engines a month or more before they are scheduled for installation, and this possibility is being considered. The drawback is that such engines must be stored until the scheduled installation (the airplane frames will not be ready early) at a storage cost of $15,000/month (including interest on expended capital) for each engine,[1] as shown in the last column of Table 7.7.

The production manager wants a schedule developed for the number of engines to be produced in each of the 4 months so that the total of the production and storage costs will be minimized.

One way to formulate a mathematical model for this problem is to let x_j be the number of jet engines to be produced in month j, for $j = 1, 2, 3, 4$. By using only these four decision variables, the problem can be formulated as a linear programming problem that does *not* fit the transportation problem type. (See Prob. 20.)

On the other hand, by adopting a different viewpoint, we can instead formulate the problem as a transportation problem that requires *much* less effort to solve. This viewpoint will describe the problem in terms of sources and destinations and then identify the corresponding x_{ij}, c_{ij}, s_i, and d_j. (See if you can do this before reading further.)

Because the units being distributed are jet engines, each of which is to be scheduled for production in a particular month and then installed in a particular (perhaps different) month,

$\text{Source } i = \text{production of jet engines in month } i \quad (i = 1, 2, 3, 4)$

$\text{Destination } j = \text{installation of jet engines in month } j \quad (j = 1, 2, 3, 4)$

$x_{ij} = \text{number of engines produced in month } i \text{ for installation in month } j$

$c_{ij} = \text{cost associated with each unit of } x_{ij}$

$\quad = \begin{cases} \text{cost per unit for production and any storage, if } i \leq j \\ ?, \text{ if } i > j \end{cases}$

$s_i = ?$

$d_j = \text{number of scheduled installations in month } j.$

The corresponding (incomplete) cost and requirements table is given in Table 7.8. Thus it remains to identify the missing costs and the supplies.

Since it is impossible to produce engines in one month for installation in an earlier month, x_{ij} must be zero if $i > j$. Therefore, there is no real cost that can be associated with such x_{ij}. Nevertheless, in order to have a well-defined transportation problem to which a standard software package (solution procedure of Sec. 7.2) can be applied, it is necessary to assign some value for the unidentified costs. Fortunately, we can use the *Big M method* introduced in Sec. 4.6 to assign this value. Thus we assign a *very* large number (denoted by M for convenience) to the unidentified cost

---

[1] For modeling purposes, assume that this storage cost is incurred at the *end of the month* for just those engines that are being held over into the next month. Thus engines that are produced in a given month for installation in the same month are assumed to incur no storage cost.

Table 7.8 Incomplete Cost and Requirements Table
for the Northern Airplane Co.

		Cost Per Unit Distributed				
		Destination				
		1	2	3	4	Supply
Source	1	1.080	1.095	1.110	1.125	?
	2	?	1.110	1.125	1.140	?
	3	?	?	1.100	1.115	?
	4	?	?	?	1.130	?
Demand		10	15	25	20	

entries in Table 7.8 to force the corresponding values of x_{ij} to be zero in the final solution.

The numbers that need to be inserted into the supply column of Table 7.8 are not obvious because the "supplies," the amount produced in the respective months, are not fixed quantities. In fact, the objective is to solve for the most desirable values of these production quantities. Nevertheless, it is necessary to assign some fixed number to every entry in the table, including those in the supply column, to have a transportation problem. A clue is provided by the fact that although the supply constraints are not present in the usual form, these constraints do exist in the form of upper bounds on the amount that can be supplied, namely,

$$x_{11} + x_{12} + x_{13} + x_{14} \leq 25,$$

$$x_{21} + x_{22} + x_{23} + x_{24} \leq 35,$$

$$x_{31} + x_{32} + x_{33} + x_{34} \leq 30,$$

$$x_{41} + x_{42} + x_{43} + x_{44} \leq 10.$$

The only change from the standard model for the transportation problem is that these constraints are in the form of inequalities instead of equations.

To convert these inequalities to equations in order to fit the transportation problem model, we use the familiar device of *slack variables* as introduced in Sec. 4.2. In this context, the slack variables are allocations to a single *dummy destination* that represent the *unused production capacity* in the respective months. This change permits the supply in the transportation problem formulation to be the total production capacity. in the given month. Furthermore, because the demand for the dummy destination is the total unused capacity, this demand is

$$(25 + 35 + 30 + 10) - (10 + 15 + 25 + 20) = 30.$$

With this demand included, the sum of the supplies now equals the sum of the demands, which is the condition given by the *feasible solutions property* for having feasible solutions.

The cost entries associated with the dummy destination should be *zero* because there is no cost incurred by a fictional allocation. (Cost entries of M would be *inappropriate* for this column because we do not want to force the corresponding values of x_{ij} to be zero. In fact, these values need to sum to 30.)

Table 7.9 **Complete Cost and Requirements Table for the
Northern Airplane Co.**

		\multicolumn{5}{c}{Cost Per Unit Distributed}	
		\multicolumn{5}{c}{Destination}	
		1	2	3	4	5(D)	Supply
Source	1	1.080	1.095	1.110	1.125	0	25
	2	M	1.110	1.125	1.140	0	35
	3	M	M	1.100	1.115	0	30
	4	M	M	M	1.130	0	10
Demand		10	15	25	20	30	

The resulting final cost and requirements table is given in Table 7.9, with the dummy destination labeled as destination 5(D). Using this formulation, it is quite easy to find the optimal production schedule by the solution procedure described in Sec. 7.2. (See Prob. 19 and its answer in the back of the book.)

EXAMPLE—DISTRIBUTION OF WATER RESOURCES: The METRO WATER DIS-TRICT is an agency that administers the distribution of water in a certain large geographic region. The region is fairly arid, so the District must purchase and bring in water from outside the region. The sources of this imported water are the Colombo, Sacron, and Calorie Rivers. The District then resells the water to users in its region. Its main customers are the water departments of the cities of Berdoo, Los Devils, San Go, and Hollyglass.

It is possible to supply any of these cities with water brought in from any of the three rivers, with the exception that no provision has been made to supply Holly-glass with Calorie River water. However, because of the geographic layouts of the viaducts and the cities in the region, the cost to the District of supplying water depends upon both the source of the water and the city being supplied. The variable cost per acre foot of water (in dollars) for each combination of river and city is given in Table 7.10. Despite these variations, the price per acre foot charged by the District is independent of the source of the water and is the same for all cities.

The management of the District is now faced with the problem of how to allocate the available water during the upcoming summer season. Using units of 1 million acre feet, the amounts available from the three rivers are given in the right-hand column of Table 7.10. The District is committed to providing a certain minimum

Table 7.10 **Water Resources Data for Metro Water District**

\multicolumn{6}{c}{Cost ($) Per Acre Foot}	
City River	Berdoo	Los Devils	San Go	Hollyglass	Supply
Colombo River	16	13	22	17	50
Sacron River	14	13	19	15	60
Calorie River	19	20	23	—	50
Min. needed	30	70	0	10	(in units of
Requested	50	70	30	∞	million acre feet)

amount to meet the essential needs of each city (with the exception of San Go, which
has an independent source of water), as shown in the *Min. needed* row of the table.
The *Requested* row indicates that Los Devils desires no more than the minimum
amount, but that Berdoo would like to buy as much as 20 more, San Go would buy
up to 30 more, and Hollyglass will take as much as it can get.

Management wishes to allocate *all* the available water from the three rivers to
the four cities in such a way as to at least meet the essential needs of each city while
minimizing the total cost to the District.

Formulation: Table 7.10 already is close to the proper form for a cost and requirements table, with the rivers being the sources and the cities being the destinations.
However, the one basic difficulty is that it is not clear what the demands at the
destinations should be. The amount to be received at each destination (except Los
Devils) actually is a decision variable, with both a lower and an upper bound. This
upper bound is the amount requested unless the request exceeds the total supply
remaining after meeting the minimum needs of the other cities, in which case this
remaining supply becomes the upper bound. Thus insatiably thirsty Hollyglass has an
upper bound of

$$(50 + 60 + 50) - (30 + 70 + 0) = 60.$$

Unfortunately, just like the other numbers in the cost and requirements table of
a transportation problem, the demand quantities must be *constants,* not bounded decision variables. To begin resolving this difficulty, temporarily suppose that it is not
necessary to satisfy the minimum needs, so that the upper bounds are the only constraints on amounts to be allocated to the cities. In this circumstance, can the requested
allocations be viewed as the demand quantities for a transportation problem formulation? After one adjustment, yes! (Do you see already what the needed adjustment
is?)

The situation is analogous to Northern Airplane Co.'s production scheduling
problem, where there was *excess supply capacity*. Now there is *excess demand capacity*. Consequently, rather than introducing a *dummy destination* to "receive" the
unused supply capacity, the adjustment needed here is to introduce a *dummy source*
to "send" the *unused demand capacity*. The imaginary supply quantity for this dummy
source would be the amount by which the sum of the demands exceeds the sum of
the real supplies:

$$(50 + 70 + 30 + 60) - (50 + 60 + 50) = 50.$$

This formulation yields the cost and requirements table shown in Table 7.11,
which uses units of million acre feet and million dollars. The cost entries in the *Dummy*
row are *zero* because there is no cost incurred by the fictional allocations from this
dummy source. On the other hand, a huge unit cost of M is assigned to the Calorie
River–Hollyglass spot. The reason is that Calorie River water cannot be used to supply
Hollyglass and assigning a cost of M will prevent any such allocation.

Now let us see how we can take each city's minimum needs into account in
this kind of formulation. Because San Go has no minimum need, it is already all set.
Similarly, the formulation for Hollyglass does not require any adjustments because
its demand (60) exceeds the dummy source's supply (50) by 10, so the amount
supplied to Hollyglass from the *real* sources will be *at least 10* in any feasible solution.
Consequently, its minimum need of 10 from the rivers is guaranteed. (If this coinci-

Table 7.11 Cost and Requirements Table without Minimum Needs for Metro Water District

		Cost Per Unit Distributed				
		Destination				
		Berdoo	Los Devils	San Go	Hollyglass	Supply
Source	Colombo R.	16	13	22	17	50
	Sacron R.	14	13	19	15	60
	Calorie R.	19	20	23	M	50
	Dummy	0	0	0	0	50
Demand		50	70	30	60	

dence had not occurred, Hollyglass would need the same adjustments that we shall have to make for Berdoo.)

Los Devils' minimum need equals its requested allocation, so its *entire* demand of 70 must be filled from the real sources rather than the dummy source. This requirement calls for the Big M method! Assigning a huge unit cost of M to the allocation from the dummy source to Los Devils ensures that this allocation will be zero in an optimal solution.

Finally, consider Berdoo. In contrast to the case of Hollyglass, the dummy source has an adequate (fictional) supply to "provide" at least some of Berdoo's minimum need in addition to its extra requested amount. Therefore, since Berdoo's minimum need is 30, adjustments must be made to prevent the dummy source from contributing more than 20 to Berdoo's total demand of 50. This adjustment is accomplished by splitting Berdoo into two destinations, one having a demand of 30 with a unit cost of M for any allocation from the dummy source and the other having a demand of 20 with a unit cost of zero for the dummy source allocation. This formulation gives the final cost and requirements table shown in Table 7.12.

This problem will be solved in the next section to illustrate the solution procedure presented there.

7.2 A Streamlined Simplex Method for the Transportation Problem

Because the transportation problem is just a special type of linear programming problem, it can be solved by applying the simplex method as described in Chap. 4.

Table 7.12 Cost and Requirements Table for Metro Water District

			Cost Per Unit Distributed					
			Destination					
			B.(min.)	B.(extra)	L.D.	S.G.	H.	
			1	2	3	4	5	Supply
Source	Col. R.	1	16	16	13	22	17	50
	Sac. R.	2	14	14	13	19	15	60
	Cal. R.	3	19	19	20	23	M	50
	Dummy	4(D)	M	0	M	0	0	50
Demand			30	20	70	30	60	

Table 7.13 Original Simplex Tableau before Applying Simplex Method to Transportation Problem

Basic Variable	Eq. No.	Z	Coefficient of $\cdots x_{ij} \cdots z_i \cdots z_{m+j} \cdots$			Right Side
Z	0	-1	c_{ij}	M	M	0
	1					
	$\vdots$					
z_i	i	0	1	1		s_i
	$\vdots$					
z_{m+j}	$m+j$	0	1		1	d_j
	$\vdots$					
	$m+n$					

However, you will see in this section that some tremendous computational shortcuts can be obtained in this method by exploiting the special structure shown in Table 7.6. We shall refer to this streamlined procedure as the **transportation simplex method.**

As you read on, note particularly how the special structure is exploited to achieve great computational savings. Then bear in mind that comparable savings sometimes can be achieved by exploiting other types of special structures as well, including those described later in the chapter.

Setting Up the Transportation Simplex Method

To highlight the streamlining achieved by the transportation simplex method, let us first review how the general (unstreamlined) simplex method would set up the transportation problem in tabular form. After constructing the table of constraint coefficients (see Table 7.6), converting the objective function to maximization form, and using the Big M method to introduce artificial variables $z_1, z_2, \ldots, z_{m+n}$ into the $(m + n)$ respective equality constraints (see Sec. 4.6), typical columns of the simplex tableau would have the form shown in Table 7.13, where all entries *not shown* in these columns are *zeroes*. [The one remaining adjustment before the first iteration of the simplex method is algebraically to eliminate the nonzero coefficients of the initial (artificial) basic variables in row 0.]

After any subsequent iteration, row 0 then would have the form shown in Table 7.14. Because of the pattern of zeroes and ones for the coefficients in Table 7.13, the *fundamental insight* presented in Sec. 5.3 implies that the u_i and v_j would have the following interpretation:

Table 7.14 Row 0 of Simplex Tableau When Applying Simplex Method to Transportation Problem

Basic Variable	Eq. No.	Z	Coefficient of $\cdots x_{ij} \cdots z_i \cdots z_{m+j} \cdots$			Right Side
Z	0	-1	$c_{ij} - u_i - v_j$	$M - u_i$	$M - v_j$	$-\sum_{i=1}^{m} s_i u_i - \sum_{j=1}^{n} d_j v_j$

u_i = multiple of *original* row i that has been subtracted (directly or indirectly) from *original* row 0 by simplex method during all iterations leading to current simplex tableau.

v_j = multiple of *original* row $(m + j)$ that has been subtracted (directly or indirectly) from *original* row 0 by simplex method during all iterations leading to current simplex tableau.

You might recognize the u_i and v_j from Chap. 6 as being the *dual variables*.[1] If x_{ij} is a nonbasic variable, $(c_{ij} - u_i - v_j)$ is interpreted as the rate at which Z would change as x_{ij} is increased.

To lay the groundwork for simplifying this setup, recall what information is needed by the simplex method. In the initialization step, an initial basic feasible solution must be obtained, which is done artificially by introducing artificial variables as the initial basic variables and setting them equal to the s_i and d_j. The optimality test and part 1 of the iterative step (selecting an entering basic variable) require knowing the current row 0, which is obtained by subtracting a certain multiple of another row from the preceding row 0. Part 2 (determining the leaving basic variable) must identify the basic variable that reaches zero first as the entering basic variable is increased, which is done by comparing the current coefficients of the entering basic variable and the corresponding right side. Part 3 must determine the new basic feasible solution, which is found by subtracting certain multiples of one row from the other rows in the current simplex tableau.

Now, how does the *transportation simplex method* obtain the same information in much simpler ways? This story will unfold fully in the coming pages, but here are some preliminary answers.

First, *no artificial variables* are needed, because a simple and convenient procedure (with several variations) is available for constructing an initial basic feasible solution.

Second, the current row 0 can be obtained *without using any other row* simply by calculating the current values of the u_i and v_j directly. Since each basic variable must have a coefficient of zero in row 0, the current u_i and v_j are obtained by solving the set of equations

$$c_{ij} - u_i - v_j = 0 \qquad \text{for each } i \text{ and } j \text{ such that } x_{ij} \text{ is a basic variable,}$$

which can be done in a straightforward way. (Note how the special structure in Table 7.13 makes this convenient way of obtaining row 0 possible by yielding $c_{ij} - u_i - v_j$ as the coefficient of x_{ij} in Table 7.14.)

Third, the leaving basic variable can be identified in a simple way without (explicitly) using the coefficients of the entering basic variable. The reason is that the special structure of the problem makes it easy to see how the solution must change as the entering basic variable is increased. As a result, the new basic feasible solution also can be identified immediately *without any algebraic manipulations* on the rows of the simplex tableau.

The grand conclusion is that *almost the entire simplex tableau* (and the work of maintaining it) *can be eliminated*! Besides the input data (the c_{ij}, s_i, and d_j values), the only information needed by the transportation simplex method is the current basic

---

[1] It would be easier to recognize these variables as dual variables by relabeling all these variables as y_i and then changing all the signs in row 0 of Table 7.14 by converting the objective function back to its original minimization form.

Table 7.15 **Format of Transportation Simplex Tableau**

		Destination				Supply	u_i
		1	2	$\cdots$	n		
	1	c_{11}	c_{12}	$\cdots$	c_{1n}	s_1	
	2	c_{21}	c_{22}	$\cdots$	c_{2n}	s_2	
Source	$\vdots$	$\vdots$	$\vdots$	$\vdots$	$\vdots$	$\vdots$	
	m	c_{m1}	c_{m2}	$\cdots$	c_{mn}	s_m	
Demand		d_1	d_2	$\cdots$	d_n	Z =	
v_j							

Additional information to be added in each cell:

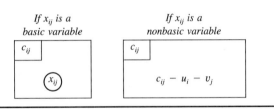

feasible solution,[1] the current values of the u_i and v_j, and the resulting values of $(c_{ij} - u_i - v_j)$ for nonbasic variables x_{ij}. When you solve a problem by hand, it is convenient to record this information for each iteration in a **transportation simplex tableau,** such as shown in Table 7.15. [Note carefully that the values of x_{ij} and $(c_{ij} - u_i - v_j)$ are distinguished in these tableaux by circling the former but not the latter.]

You can gain a fuller appreciation for the great difference in efficiency and convenience between the simplex and the transportation simplex methods by applying them both to the same small problem (see Prob. 22). However, the difference becomes even more pronounced for large problems that must be solved on a computer. This pronounced difference is suggested somewhat by comparing the sizes of the simplex and the transportation simplex tableaux. Thus, for a transportation problem having *m* sources and *n* destinations, the simplex tableau would have $(m + n + 1)$ rows and $(m + 1)(n + 1)$ columns (excluding those to the left of the x_{ij} columns), and the transportation simplex tableau would have *m* rows and *n* columns (excluding the two extra informational rows and columns). Now try plugging in various values for *m* and *n* (for example, $m = 10$ and $n = 100$ would be a rather typical middle-sized transportation problem), and note how the ratio of the number of cells in the simplex tableau to the number in the transportation simplex tableau increases as *m* and *n* increase.

---

[1] Since nonbasic variables are automatically zero, the current basic feasible solution is fully identified by recording just the values of the basic variables. We shall use this convention from now on.

Recall that the objective of the initialization step is to obtain an initial basic feasible solution. Because all the functional constraints in the transportation problem are *equality* constraints, the simplex method would obtain this solution by introducing artificial variables and using them as the initial basic variables, as described in Sec. 4.6. The resulting basic solution actually is feasible only for a revised version of the problem, so a number of iterations are needed to drive these artificial variables to zero in order to reach the real basic feasible solutions. The transportation simplex method bypasses all this by instead using a simpler procedure to directly construct a real basic feasible solution on a transportation simplex tableau.

Before outlining this procedure, we need to point out that the number of basic variables in any basic solution of a transportation problem is one fewer than you might expect. Ordinarily, there is one basic variable for each functional constraint in a linear programming problem. For transportation problems with m sources and n destinations, the number of functional constraints is $m + n$. However,

$$\text{number of basic variables} = m + n - 1.$$

The reason is that the functional constraints are equality constraints, and this set of $(m + n)$ equations has one *extra* (or *redundant*) equation that can be deleted without changing the feasible region; i.e., any one of the constraints is automatically satisfied whenever the other $m + n - 1$ constraints are satisfied. (This fact can be verified by showing that any supply constraint exactly equals the sum of the demand constraints minus the sum of the *other* supply constraints, and that any demand equation also can be reproduced by summing the supply equations and subtracting the other demand equations. See Prob. 23.) Therefore, any *basic feasible solution* appears on a transportation simplex tableau with exactly $(m + n - 1)$ circled *nonnegative* allocations, where the sum of the allocations for each row or column equals its supply or demand.[1]

The procedure for constructing an initial basic feasible solution selects the $(m + n - 1)$ basic variables one at a time. After each selection, a value that will satisfy one additional constraint (thereby eliminating that constraint's row or column from further consideration for providing allocations) is assigned to that variable. Thus, after $(m + n - 1)$ selections, an entire basic solution has been constructed in such a way as to satisfy all the constraints. A number of different criteria have been proposed for selecting the basic variables. We present and illustrate three of these criteria here after outlining the general procedure.

General Procedure[2] for Constructing an Initial Basic Feasible Solution

To Begin: All source rows and destination columns of the transportation simplex tableau are initially under consideration for providing a basic variable (allocation).

---

[1] However, note that any feasible solution with $(m + n - 1)$ nonzero variables is *not necessarily* a basic solution because it might be the weighted average of two or more degenerate basic feasible solutions (i.e., basic feasible solutions having some basic variables equal to zero). We need not be concerned about mislabeling such solutions as being basic, however, because the transportation simplex method constructs only legitimate basic feasible solutions.

[2] In Sec. 4.1 we pointed out that the simplex method is an example of the algorithms (iterative solution procedures) so prevalent in operations research work. Note that this procedure also is an algorithm, where each successive execution of the (four) steps constitutes an iteration.

Step 1: From among the rows and columns still under consideration, select the next basic variable (allocation) according to some criterion.

Step 2: Make that allocation large enough to exactly use up the remaining supply in its row or the remaining demand in its column (whichever is smaller).

Step 3: Eliminate that row or column (whichever had the smaller remaining supply or demand) from further consideration. (If the row and column have the *same* remaining supply and demand, then arbitrarily select the *row* as the one to be eliminated. The column will be used later to provide a *degenerate* basic variable, i.e., a circled allocation of zero.)

Step 4: If only one row or only one column remains under consideration, then the procedure is completed by selecting every *remaining* variable (i.e., those variables that were neither previously selected to be basic nor eliminated from consideration by eliminating their row or column) associated with that row or column to be basic with the only feasible allocation. Otherwise, return to step 1.

Alternative Criteria for Step 1

1. *Northwest corner rule:* Begin by selecting x_{11} (i.e., start in the *northwest corner* of the transportation simplex tableau). Thereafter, if x_{ij} was the last basic variable selected, then next select $x_{i,j+1}$ (i.e., move one column to the *right*) if source i has any supply remaining. Otherwise, next select $x_{i+1,j}$ (i.e., move one row *down*).

EXAMPLE: To make this description more concrete, we now illustrate the general procedure on the Metro Water District problem (see Table 7.12) with the northwest corner rule being used in step 1. Because $m = 4$ and $n = 5$ in this case, the procedure would find an initial basic feasible solution having $m + n - 1 = 8$ basic variables.

As shown in Table 7.16, the first allocation is $x_{11} = 30$, which exactly uses up the demand in column 1 (and eliminates this column from further consideration). This first iteration leaves a supply of 20 remaining in row 1, so next select $x_{1,1+1} = x_{12}$ to be a basic variable. Because this supply is no larger than the demand of 20 in column 2, all of it is allocated, $x_{12} = 20$, and this row is eliminated from further consideration. (Row 1 is chosen for elimination rather than column 2 because of the parenthetical instruction in step 3.) Therefore, select $x_{1+1,2} = x_{22}$ next. Because the remaining demand of 0 in column 2 is less than the supply of 60 in row 2, allocate $x_{22} = 0$ and eliminate column 2.

Continuing in this manner, we eventually obtain the entire *initial basic feasible solution* shown in Table 7.16, where the circled numbers are the values of the basic variables ($x_{11} = 30, \ldots, x_{45} = 50$) and all the other variables (x_{13}, etc.) are nonbasic variables equal to zero. Arrows have been added to show the order in which the basic variables (allocations) were selected. The value of Z for this solution is

$$Z = 16(30) + 16(20) + \cdots + 0(50) = 2470 + 10M.$$

2. *Vogel's approximation method:* For each row and column remaining under consideration, calculate its **difference,** which is defined as *the arithmetic difference between the smallest and next-to-the-smallest unit cost* (c_{ij}) *still remaining in that row or column.* In that row or column having the *largest difference,* select the variable having the *smallest remaining unit cost.* (Ties for the largest difference may be broken arbitrarily.)

Table 7.16 Initial Basic Feasible Solution from Northwest Corner Rule

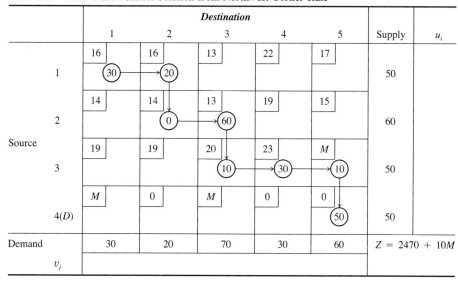

		Destination 1	Destination 2	Destination 3	Destination 4	Destination 5	Supply	u_i
Source	1	16 (30)	16 (20)	13	22	17	50	
	2	14	14 (0)	13 (60)	19	15	60	
	3	19	19	20 (10)	23 (30)	M (10)	50	
	4(D)	M	0	M	0	0 (50)	50	
Demand		30	20	70	30	60	$Z = 2470 + 10M$	
v_j								

EXAMPLE: Now let us apply the general procedure to the Metro Water District problem by using the criterion for Vogel's approximation method to select the next basic variable in step 1. With this criterion, it is more convenient to work with cost and requirements tables (rather than with complete transportation simplex tableaux), beginning with the one shown in Table 7.12. At each iteration, after calculating and displaying the *difference* for every row and column remaining under consideration, the largest difference is circled and the smallest unit cost in its row or column is enclosed in a box. The resulting selection (and value) of the variable having this unit cost as the next basic variable is indicated in the lower right-hand corner of the current table, along with the row or column thereby being eliminated from further consideration (see steps 2 and 3 of the general procedure). The table for the next iteration is exactly the same except for deleting this row or column and subtracting the last allocation from its supply or demand (whichever remains).

Applying this procedure to the Metro Water District problem yields the sequence of cost and requirements tables shown in Table 7.17, where the resulting initial basic feasible solution consists of the eight basic variables (allocations) given in the lower right-hand corner of the respective cost and requirements tables.

This example illustrates two relatively subtle features of the general procedure that warrant special attention. First, note that the final iteration selects *three* variables (x_{31}, x_{32}, and x_{33}) to become basic instead of the single selection made at the other iterations. The reason is that only *one* row (row 3) remains under consideration at this point. Therefore, step 4 of the general procedure says to select *every* remaining variable associated with row 3 to be basic.

Second, note that the allocation of $x_{23} = 20$ at the next-to-last iteration exhausts *both* the remaining supply in its row *and* the remaining demand in its column. However, rather than eliminate both the row and column from further consideration, step 3 says to eliminate *only the row,* saving the column to provide a *degenerate* basic variable later. Column 3 is, in fact, used for just this purpose at the final iteration when $x_{33} = 0$ is selected as one of the basic variables. For another illustration of this

Table 7.17 **Initial Basic Feasible Solution from Vogel's Approximation Method**

		Destination						Row
		1	2	3	4	5	Supply	Difference
Source	1	16	16	13	22	17	50	3
	2	14	14	13	19	15	60	1
	3	19	19	20	23	M	50	0
	4(D)	M	0	M	$\boxed{0}$	0	50	0
Demand		30	20	70	30	60	\multicolumn{2}{l}{Select $x_{44} = 30$}
Column difference		2	14	0	⑲	15	\multicolumn{2}{l}{Eliminate column 4}

		Destination					Row
		1	2	3	5	Supply	Difference
Source	1	16	16	13	17	50	3
	2	14	14	13	15	60	1
	3	19	19	20	M	50	0
	4(D)	M	0	M	$\boxed{0}$	20	0
Demand		30	20	70	60	\multicolumn{2}{l}{Select $x_{45} = 20$}
Column difference		2	14	0	⑮	\multicolumn{2}{l}{Eliminate row 4(D)}

		Destination					Row
		1	2	3	5	Supply	Difference
Source	1	16	16	$\boxed{13}$	17	50	③
	2	14	14	13	15	60	1
	3	19	19	20	M	50	0
Demand		30	20	70	40	\multicolumn{2}{l}{Select $x_{13} = 50$}
Column difference		2	2	0	2	\multicolumn{2}{l}{Eliminate row 1}

		Destination					Row
		1	2	3	5	Supply	Difference
Source	2	14	14	13	$\boxed{15}$	60	1
	3	19	19	20	M	50	0
Demand		30	20	20	40	\multicolumn{2}{l}{Select $x_{25} = 40$}
Column difference		5	5	7	$(M - 15)$	\multicolumn{2}{l}{Eliminate column 5}

		Destination				Row
		1	2	3	Supply	Difference
Source	2	14	14	$\boxed{13}$	20	1
	3	19	19	20	50	0
Demand		30	20	20	\multicolumn{2}{l}{Select $x_{23} = 20$}
Column difference		5	5	⑦	\multicolumn{2}{l}{Eliminate row 2}

		Destination			Supply
		1	2	3	
Source	3	19	19	20	50
Demand		30	20	0	

Select $x_{31} = 30$

$x_{32} = 20$ $Z = 2,460$

$x_{33} = 0$

same phenomenon, see Table 7.16 where the allocation of $x_{12} = 20$ results in eliminating only row 1, so that column 2 is saved to provide a degenerate basic variable, $x_{22} = 0$, at the next iteration.

Although a zero allocation might seem irrelevant, it actually plays an important role. You will see soon that the transportation simplex method must know *all* ($m + n - 1$) basic variables, including those with value zero, in the current basic feasible solution.

3. *Russell's approximation method:* For each source row i remaining under consideration, determine its $\bar{u}_i$, which is the *largest* unit cost (c_{ij}) still remaining in that row. For each destination column j remaining under consideration, determine its $\bar{v}_j$, which is the *largest* unit cost (c_{ij}) still remaining in that column. For each variable x_{ij} not previously selected in these rows and columns, calculate $\Delta_{ij} = c_{ij} - \bar{u}_i - \bar{v}_j$. Select the variable having the *largest* (in absolute terms) *negative* value of Δ_{ij}. (Ties may be broken arbitrarily.)

EXAMPLE: Using the criterion for Russell's approximation method in step 1, we again apply the general procedure to the Metro Water District problem (see Table 7.12). The results, including the sequence of basic variables (allocations), are shown in Table 7.18.

At iteration 1, the largest unit cost in row 1 is $\bar{u}_1 = 22$, the largest in column 1 is $\bar{v}_1 = M$, and so forth. Thus

$$\Delta_{11} = c_{11} - \bar{u}_1 - \bar{v}_1 = 16 - 22 - M = -6 - M.$$

Calculating all the Δ_{ij} for $i = 1, 2, 3, 4$ and $j = 1, 2, 3, 4, 5$ shows that $\Delta_{45} = 0 - 2M$ has the *largest negative* value, so $x_{45} = 50$ is selected as the first basic variable (allocation). This allocation exactly uses up the supply in row 4, so this row is eliminated from further consideration.

Note that eliminating this row changes $\bar{v}_1$ and $\bar{v}_3$ for the next iteration. Therefore, the second iteration requires recalculating the Δ_{ij} with $j = 1, 3$, as well as eliminating $i = 4$. The largest negative value now is

$$\Delta_{15} = 17 - 22 - M = -5 - M,$$

so $x_{15} = 10$ becomes the second basic variable (allocation), eliminating column 5 from further consideration.

Table 7.18 Initial Basic Feasible Solution from Russell's Approximation Method

Iteration	$\bar{u}_1$	$\bar{u}_2$	$\bar{u}_3$	$\bar{u}_4$	$\bar{v}_1$	$\bar{v}_2$	$\bar{v}_3$	$\bar{v}_4$	$\bar{v}_5$	Largest Negative Δ_{ij}	Allocation
1	22	19	M	M	M	19	M	23	M	$\Delta_{45} = -2M$	$x_{45} = 50$
2	22	19	M		19	19	20	23	M	$\Delta_{15} = -5 - M$	$x_{15} = 10$
3	22	19	23		19	19	20	23		$\Delta_{13} = -29$	$x_{13} = 40$
4		19	23		19	19	20	23		$\Delta_{23} = -26$	$x_{23} = 30$
5		19	23		19	19		23		$\Delta_{21} = -24*$	$x_{21} = 30$
6										Irrelevant	$x_{31} = 0$
											$x_{32} = 20$
											$x_{34} = 30$
											$Z = 2,570$

* Tie with $\Delta_{22} = -24$ broken arbitrarily.

The subsequent iterations proceed similarly, but you may want to test your understanding by verifying the remaining allocations given in Table 7.18. As with the other procedures in this (and other) section, you should find your OR COURSE-WARE useful for doing the calculations involved and illuminating the approach.

Comparison of Alternative Criteria for Step 1

Now let us compare these three criteria for selecting the next basic variable. The main virtue of the northwest corner rule is that it is quick and easy. However, because it pays no attention to unit costs (c_{ij}), usually the solution obtained will be far from optimal. (Note in Table 7.16 that $x_{35} = 10$ even though $c_{35} = M$.) Expending a little more effort to find a good initial basic feasible solution might greatly reduce the number of iterations then required by the transportation simplex method to reach an optimal solution (see Probs. 7 and 12). Finding such a solution is the objective of the other two criteria.

Vogel's approximation method has been a popular criterion for many years,[1] partially because it is relatively easy to implement by hand. Because *difference* represents the minimum extra unit cost incurred by failing to make an allocation to the cell having the smallest unit cost in that row or column, this criterion does take costs into account in an effective way.

Russell's approximation method provides another excellent criterion[2] that is still quick to implement on a computer (but not manually). Although more experimentation is required to determine which is more effective *on the average,* this criterion *frequently* does obtain a better solution than Vogel's. (For the example, Vogel's approximation method happened to find the optimal solution with $Z = 2,460$, whereas Russell's misses slightly with $Z = 2,570$.) For a large problem, it may be worthwhile to apply *both* criteria and then use the better solution to start the iterations of the transportation simplex method.

One distinct advantage of Russell's approximation method is that it is patterned directly after part 1 of the iterative step for the transportation simplex method (as you will see soon), which somewhat simplifies the overall computer code. In particular, the $\bar{u}_i$ and $\bar{v}_j$ have been defined in such a way that the relative values of the $(c_{ij} - \bar{u}_i - \bar{v}_j)$ *estimate* the relative values of the $(c_{ij} - u_i - v_j)$ that will be obtained when the transportation simplex method reaches an optimal solution.

We now shall use the initial basic feasible solution obtained in Table 7.18 by Russell's approximation method to illustrate the remainder of the transportation simplex method. Thus, our *initial transportation simplex tableau* (before solving for the u_i and v_j) is the one shown in Table 7.19 (see p. 224).

The next step is to check whether this initial solution is optimal by applying the *optimality test.*

Optimality Test

Using the notation of Table 7.14, we can reduce the standard optimality test for the simplex method (see Sec. 4.3) to the following for the transportation problem:

---

[1] Reinfeld, N. V., and W. R. Vogel: *Mathematical Programming,* Prentice-Hall, Englewood Cliffs, N.J., 1958.

[2] Russell, Edward J.: "Extension of Dantzig's Algorithm to Finding an Initial Near-Optimal Basis for the Transportation Problem," *Operations Research,* **17**:187–191, 1969.

Optimality test: A basic feasible solution is optimal if and only if $(c_{ij} - u_i - v_j) \geq 0$ for every (i, j) such that x_{ij} is nonbasic.[1]

Thus the only work required by the optimality test is the derivation of the values of the u_i and v_j for the current basic feasible solution and then the calculation of these $(c_{ij} - u_i - v_j)$.

Since $(c_{ij} - u_i - v_j)$ is required to be zero if x_{ij} is a basic variable, the u_i and v_j satisfy the set of equations

$$c_{ij} = u_i + v_j \qquad \text{for each } (i, j) \text{ such that } x_{ij} \text{ is basic.}$$

There are $(m + n - 1)$ basic variables, and so there are $(m + n - 1)$ of these equations. Since the number of unknowns (the u_i and v_j) is $(m + n)$, one of these variables can be assigned a value arbitrarily without violating the equations. (The rule we shall adopt is to select the u_i that has the largest number of allocations in its row and assign it the value of zero.) Because of the simple structure of these equations, it is then very simple to solve for the remaining variables algebraically.

To demonstrate, we give each equation that corresponds to a basic variable in our initial basic feasible solution.

x_{31}: $19 = u_3 + v_1$.   Set $u_3 = 0$, so $v_1 = 19$,

x_{32}: $19 = u_3 + v_2$. $v_2 = 19$,

x_{34}: $23 = u_3 + v_4$. $v_4 = 23$,

x_{21}: $14 = u_2 + v_1$.   Know $v_1 = 19$, so $u_2 = -5$.

x_{23}: $13 = u_2 + v_3$.   Know $u_2 = -5$, so $v_3 = 18$.

x_{13}: $13 = u_1 + v_3$.   Know $v_3 = 18$, so $u_1 = -5$.

x_{15}: $17 = u_1 + v_5$.   Know $u_1 = -5$, so $v_5 = 22$.

x_{45}: $0 = u_4 + v_5$.   Know $v_5 = 22$, so $u_4 = -22$.

Setting $u_3 = 0$ (since row 3 of Table 7.19 has the largest number of allocations, 3) and moving down the equations one at a time immediately gives the derivation of values for the unknowns shown to the right of the equations.

Once you get the hang of it, you probably will find it even more convenient to solve these equations without writing them down by working directly on the transportation simplex tableau. Thus, in Table 7.19 you would begin by writing in the value $u_3 = 0$ and then picking out the circled allocations (x_{31}, x_{32}, x_{34}) in that row. For each one you would set $v_j = c_{3j}$ and then look for circled allocations (except in row 3) in these columns (x_{21}). Mentally calculate $u_2 = c_{21} - v_1$, pick out x_{23}, set $v_3 = c_{23} - u_2$, and so on until you have filled in all the values for the u_i and v_j. (Try it.) Then calculate and fill in the value of $(c_{ij} - u_i - v_j)$ for each nonbasic variable x_{ij} (i.e., for each cell without a circled allocation), and you will have the completed initial transportation simplex tableau shown in Table 7.20.

We are now in a position to apply the optimality test by checking the value of the $(c_{ij} - u_i - v_j)$ given in Table 7.20. Because two of these values, $(c_{25} - u_2 - v_5) = -2$ and $(c_{44} - u_4 - v_4) = -1$, are negative, we conclude that the current

---

[1] The one exception is that two or more equivalent degenerate basic feasible solutions (i.e., identical solutions having different degenerate basic variables equal to zero) can be optimal with only some of these basic solutions satisfying the optimality test. This exception is illustrated later in the example (see the identical solutions in the last two tableaux of Table 7.23, where only the latter solution satisfies the criterion for optimality).

Table 7.19 **Initial Transportation Simplex Tableau (before Obtaining the $c_{ij} - u_i - v_j$) from Russell's Approximation Method**

Iteration 0		Destination 1	Destination 2	Destination 3	Destination 4	Destination 5	Supply	u_i
Source	1	16	16	13 (40)	22	17 (10)	50	
	2	14 (30)	14	13 (30)	19	15	60	
	3	19 (0)	19 (20)	20	23 (30)	*M*	50	
	4(*D*)	*M*	0	*M*	0	0 (50)	50	
Demand		30	20	70	30	60	$Z = 2{,}570$	
v_j								

basic feasible solution is not optimal. Therefore, the transportation simplex method must next go to the iterative step to find a better basic feasible solution.

Iterative Step

As with the full-fledged simplex method, the iterative step for this streamlined version must determine an *entering basic variable* (part 1), a *leaving basic variable* (part 2), and then identify the resulting *new basic feasible solution* (part 3).

Table 7.20 **Completed Initial Transportation Simplex Tableau**

Iteration 0		Destination 1	Destination 2	Destination 3	Destination 4	Destination 5	Supply	u_i
Source	1	16; +2	16; +2	13 (40)	22; +4	17 (10)	50	−5
	2	14 (30)	14; 0	13 (30)	19; +1	15; −2	60	−5
	3	19 (0)	19 (20)	20; +2	23 (30)	*M*; $M - 22$	50	0
	4(*D*)	*M*; $M + 3$	0; +3	*M*; $M + 4$	0; −1	0 (50)	50	−22
Demand		30	20	70	30	60	$Z = 2{,}570$	
v_j		19	19	18	23	22		

PART 1: Since $(c_{ij} - u_i - v_j)$ represents the rate at which the objective function would change as the nonbasic variable x_{ij} is increased, the entering basic variable must have a *negative* $(c_{ij} - u_i - v_j)$ to decrease the total cost Z. Thus the candidates in Table 7.20 are x_{25} and x_{44}. To choose between the candidates, select the one having the *largest* (in absolute terms) *negative* value of $(c_{ij} - u_i - v_j)$ to be the *entering basic variable*, which is x_{25} in this case.

PART 2: Increasing the entering basic variable from zero sets off a *chain reaction* of compensating changes in other basic variables (allocations) in order to continue satisfying the supply and demand constraints. The first basic variable to be decreased to zero then becomes the *leaving basic variable*.

With x_{25} as the entering basic variable, the chain reaction in Table 7.20 is the relatively simple one summarized in Table 7.21. (We shall always indicate the entering basic variable by placing a boxed + sign in its cell.) Thus increasing x_{25} requires decreasing x_{15} by the same amount to restore the demand of 60 in column 5, which in turn requires increasing x_{13} by this amount to restore the supply of 50 in row 1, which in turn requires decreasing x_{23} by this amount to restore the demand of 70 in column 3. This decrease in x_{23} successfully completes the chain reaction because it also restores the supply of 60 in row 2. (Equivalently, we could have started the chain reaction by restoring this supply in row 2 with the decrease in x_{23}, and then increase x_{13} and decrease x_{15}.)

The net result is that cells (2, 5) and (1, 3) become **recipient cells,** each receiving its additional allocation from one of the **donor cells,** (1, 5) and (2, 3). (These cells are indicated in Table 7.21 by the + and − signs.) Note that cell (1, 5) had to be the donor cell for column 5 rather than cell (4, 5), because cell (4, 5) would have no recipient cell in row 4 to continue the chain reaction. [Similarly, if the chain reaction had been started in row 2 instead, cell (2, 1) could not be the donor cell for this row because the chain reaction could not then be completed successfully after necessarily choosing cell (3, 1) as the next recipient cell and either cell (3, 2) or (3, 4) as its donor cell.]

Each donor cell decreases its allocation by exactly the same amount that the entering basic variable (and other recipient cells) is increased. Therefore, the donor

Table 7.21 Part of Initial Transportation Simplex Tableau Showing the Chain Reaction Caused by Increasing the Entering Basic Variables x_{25}

		Destination			
		3	4	5	Supply
	1	13 (40)+	22 +4	17 (10)−	50
Source	2	13 (30)−	19 +1	15 + −2	60
		⋮	⋮	⋮	
Demand		70	30	60	

cell that starts with the smallest allocation—cell $(1, 5)$ in this case (since $10 < 30$ in Table 7.21)—must reach a zero allocation first as the entering basic variable x_{25} is increased. Thus x_{15} becomes the *leaving basic variable*.

In general, there always is just *one* chain reaction (in either direction) that can be completed successfully to maintain feasibility when the entering basic variable is increased from zero. This chain reaction can be identified by selecting among the cells having a basic variable: first the donor cell in the *column* having the entering basic variable, then the recipient cell in the row having this donor cell, then the donor cell in the column having this recipient cell, and so on until the chain reaction yields a donor cell in the *row* having the entering basic variable. When a column or row has more than one additional basic variable cell, it may be necessary to trace them all further to see which one must be selected to be the donor or recipient cell. (All but this one eventually will reach a dead end in a row or column having no additional basic variable cell.) After identifying the chain reaction, the donor cell having the *smallest* allocation automatically provides the leaving basic variable. (In the case of a tie for the donor cell having the smallest allocation, any one can be chosen arbitrarily to provide the leaving basic variable.)

PART 3: The *new basic feasible solution* is identified simply by adding the value of the leaving basic variable (before any change) to the allocation for each recipient cell and subtracting *this same amount* from the allocation for each donor cell. In Table 7.21 the value of the leaving basic variable x_{15} is 10, so this portion of the transportation simplex tableau changes as shown in Table 7.22 for the new solution. (Since x_{15} is nonbasic in the new solution, its new allocation of zero is no longer shown in this new tableau.)

We can now highlight a useful interpretation of the $(c_{ij} - u_i - v_j)$ quantities derived during the optimality test. Because of the shift of 10 allocation units from the donor cells to the recipient cells (shown in Tables 7.21 and 7.22), the total cost changes by

$$\Delta Z = 10(15 - 17 + 13 - 13) = 10(-2) = 10(c_{25} - u_2 - v_5).$$

Thus the effect of increasing the entering basic variable x_{25} from zero has been a cost change at the rate of -2 per unit increase in x_{25}. This is precisely what the value of $(c_{25} - u_2 - v_5) = -2$ in Table 7.20 indicates would happen. In fact, another (but

Table 7.22 **Part of Second Transportation Simplex Tableau Showing the Changes in the Basic Feasible Solution**

		Destination			
		3	4	5	Supply
Source	1 · · ·	13 (50)	22	17	50
	2 · · ·	13 (20)	19	15 (10)	60
	· · ·	⋮	⋮	⋮	
Demand		70	30	60	

less efficient) way of deriving $(c_{ij} - u_i - v_j)$ for each nonbasic variable x_{ij} is to identify the chain reaction caused by increasing this variable from 0 to 1 and then to calculate the resulting cost change. This intuitive interpretation sometimes is useful for checking calculations during the optimality test.

Before completing the solution of the Metro Water District problem, let us now summarize the rules for the transportation simplex method.

Summary of Transportation Simplex Method

1. *Initialization step:* Construct an initial basic feasible solution by the procedure outlined earlier in this section. Go to the optimality test.
2. *Iterative step:*

    Part 1. Determine the entering basic variable: Select the nonbasic variable x_{ij} having the *largest* (in absolute terms) *negative* value of $(c_{ij} - u_i - v_j)$.

    Part 2. Determine the leaving basic variable: Identify the chain reaction required to retain feasibility when the entering basic variable is increased. From among the donor cells, select the basic variable having the *smallest* value.

    Part 3. Determine the new basic feasible solution: Add the value of the leaving basic variable to the allocation for each recipient cell. Subtract this value from the allocation for each donor cell.
3. *Optimality test:* Derive the u_i and v_j by selecting the row having the largest number of allocations and setting its $u_i = 0$ and then solving the set of equations $c_{ij} = u_i + v_j$ for each (i, j) such that x_{ij} is basic. If $(c_{ij} - u_i - v_j) \geq 0$ for every (i, j) such that x_{ij} is *nonbasic,* then the current solution is optimal, so stop. Otherwise, go to the iterative step.

Continuing to apply this procedure to the Metro Water District problem yields the complete set of transportation simplex tableaux shown in Table 7.23. Since all the $(c_{ij} - u_i - v_j)$ are nonnegative in the fourth tableau, the optimality test identifies the set of allocations in this tableau as being optimal, which concludes the algorithm.

It would be good practice for you to derive the values of the u_i and v_j given in the second, third, and fourth tableaux. Try doing this by working directly on the tableaux. Also check out the chain reactions in the second and third tableaux, which are somewhat more complicated than the one you already have seen in Table 7.21.

You should note three special points that are illustrated by this example. First, the initial basic feasible solution is *degenerate* because the basic variable $x_{31} = 0$. However, this *degenerate* basic variable causes no complication, because cell (3, 1) becomes a *recipient cell* in the second tableau, which increases x_{31} to a value greater than zero.

Second, another *degenerate* basic variable (x_{34}) arises in the third tableau because the basic variables for *two* donor cells in the second tableau, cells (2, 1) and (3, 4), *tie* for having the smallest value (30). (This tie is broken arbitrarily by selecting x_{21} as the leaving basic variable; if x_{34} had been selected instead, then x_{21} would have become the degenerate basic variable.) This degenerate basic variable does appear to create a complication subsequently, because cell (3, 4) becomes a *donor cell* in the third tableau but has nothing to donate! Fortunately, such an event actually gives no cause for concern. Since zero is the amount to be added to or subtracted from the allocations for the recipient and donor cells, these allocations do not change. However,

Table 7.23 Complete Set of Transportation Simplex Tableaux for the Metro Water District Problem

Iteration 0		Destination 1	2	3	4	5	Supply	u_i
Source	1	16 / +2	16 / +2	13 / (40) +	22 / +4	17 / (10) −	50	−5
	2	14 / (30)	14 / 0	13 / (30) −	19 / +1	15 / + −2	60	−5
	3	19 / (0)	19 / (20)	20 / +2	23 / (30)	M / M − 22	50	0
	4(D)	M / M + 3	0 / +3	M / M + 4	0 / −1	0 / (50)	50	−22
Demand		30	20	70	30	60	Z = 2,570	
v_j		19	19	18	23	22		

Iteration 1		Destination 1	2	3	4	5	Supply	u_i
Source	1	16 / +2	16 / +2	13 / (50)	22 / +4	17 / +2	50	−5
	2	14 / (30) −	14 / 0	13 / (20)	19 / +1	15 / (10) +	60	−5
	3	19 / (0) +	19 / (20)	20 / +2	23 / (30) −	M / M − 20	50	0
	4(D)	M / M + 1	0 / +1	M / M + 2	0 / + −3	0 / (50) −	50	−20
Demand		30	20	70	30	60	Z = 2,550	
v_j		19	19	18	23	20		

the degenerate basic variable does become the leaving basic variable, so it is replaced by the entering basic variable as the circled allocation of zero in the fourth tableau. This change in the set of basic variables changes the values of the u_i and v_j. Therefore, if any of the $(c_{ij} - u_i - v_j)$ had been negative in the fourth tableau, the algorithm would have gone on to make *real* changes in the allocations (whenever all donor cells have nondegenerate basic variables).

Iteration 2

Source	Destination 1	2	3	4	5	Supply	u_i
1	16 +5	16 +5	13 (50)	22 +7	17 +2	50	−8
2	14 +3	14 +3	13 (20) −	19 +4	15 (40) +	60	−8
3	19 (30)	19 (20)	20 [+] −1	23 (0) −	M M − 23	50	0
4(D)	M M + 4	0 +4	M M + 2	0 (30) +	0 (20) −	50	−23
Demand	30	20	70	30	60	Z = 2,460	
v_j	19	19	21	23	23		

Iteration 3

Source	Destination 1	2	3	4	5	Supply	u_i
1	16 +4	16 +4	13 (50)	22 +7	17 +2	50	−7
2	14 +2	14 +2	13 (20)	19 +4	15 (40)	60	−7
3	19 (30)	19 (20)	20 (0)	23 +1	M M − 22	50	0
4(D)	M M + 3	0 +3	M M + 2	0 (30)	0 (20)	50	−22
Demand	30	20	70	30	60	Z = 2,460	
v_j	19	19	20	22	22		

Third, because none of the $(c_{ij} - u_i - v_j)$ turned out to be negative in the fourth tableau, the *equivalent* set of allocations in the third tableau is optimal also. Thus the algorithm executed one more iteration than necessary. This extra iteration is a flaw that occasionally arises in both the transportation simplex method and the simplex method because of *degeneracy,* but it is not sufficiently serious to warrant any adjustments in these algorithms.

7.3 The Transshipment Problem

One requirement of the transportation problem is advance knowledge of the method of distribution of units from each source i to each destination j, so that the corresponding cost per unit (c_{ij}) can be determined. Sometimes, however, the best method of distribution is not clear because of the possibility of **transshipments,** whereby shipments would go through intermediate transfer points (which might be other sources or destinations). For example, rather than shipping a special cargo directly from port 1 to port 3, it may be cheaper to include it with regular cargoes from port 1 to port 2 and then from port 2 to port 3.

Such possibilities for transshipments could be investigated in advance to determine the cheapest route from each source to each destination. However, this might be a very complicated and time-consuming task if there are many possible intermediate transfer points. Therefore, it may be much more convenient to let a computer algorithm solve *simultaneously* for the amount to ship from each source to each destination *and* the route to follow for each shipment so as to minimize the total shipping cost.

This extension of the transportation problem to include the routing decisions is referred to as the **transshipment problem.**

Fortunately, there is a simple way to reformulate the transshipment problem to fit it back into the format of the transportation problem. Thus the *transportation simplex method* also can be used to solve the transshipment problem.

To clarify the structure of the transshipment problem and the nature of this reformulation, we shall now extend the prototype example for the transportation problem to include transshipments.

Prototype Example

After further investigation, the P & T COMPANY (see Sec. 7.1) has found that it can cut costs by discontinuing its own trucking operation and using common carriers instead to truck its canned peas. Since no single trucking company serves the entire area containing all the canneries and warehouses, many of the shipments will need to be transferred to another truck at least once along the way. These transfers can be made at intermediate canneries or warehouses, or at five other locations (Butte, Montana; Boise, Idaho; Cheyenne, Wyoming; Denver, Colorado; and Omaha, Nebraska) referred to as *junctions,* as shown in Fig. 7.2. The shipping cost per truckload between each of these points is given in Table 7.24, where a dash indicates that a direct shipment is not possible.

For example, a truckload of peas can still be sent from cannery 1 to warehouse 4 by direct shipment at a cost of $871. However, another possibility, shown below, is to ship the truckload from cannery 1 to junction 2, transfer it to a truck going to warehouse 2, and then transfer it again to go to warehouse 4, at a cost of only ($286 + $207 + $341) = $834.

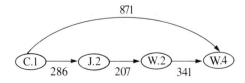

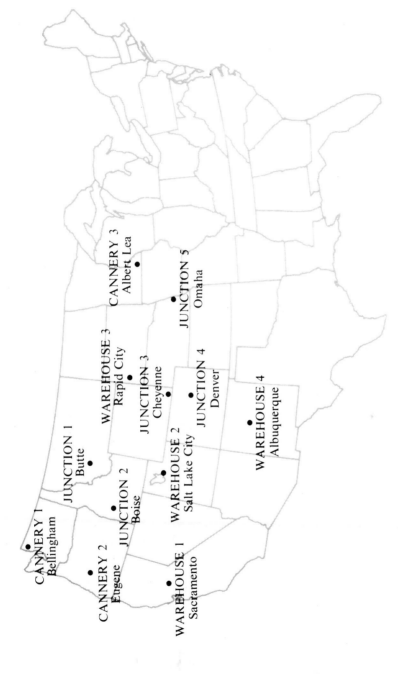

Figure 7.2 Location of canneries, warehouses, and junctions for the P & T Co.

231

Table 7.24 Independent Trucking Data for P & T Co.

		Cannery			Junction					Warehouse				Output
From \ **To**		1	2	3	1	2	3	4	5	1	2	3	4	
Cannery	1		$146	—	$324	$286	—	—	—	$452	$505	—	$871	75
	2	$146		—	$373	$212	$570	$609	—	$335	$407	$688	$784	125
	3	—	—		$658	—	$405	$419	$158	—	$685	$359	$673	100
Junction	1	$322	$371	$656		$262	$398	$430	—	$503	$234	$329	—	
	2	$284	$210	—	$262		$406	$421	$644	$305	$207	$464	$558	
	3	—	$569	$403	$398	$406		$ 81	$272	$597	$253	$171	$282	
	4	—	$608	$418	$431	$422	$ 81		$287	$613	$280	$236	$229	
	5	—	—	$158	—	$647	$274	$288		$831	$501	$293	$482	
Warehouse	1	$453	$336	—	$505	$307	$599	$615	$831		$359	$706	$587	
	2	$505	$407	$683	$235	$208	$254	$281	$500	$357		$362	$341	
	3	—	$687	$357	$329	$464	$171	$236	$290	$705	$362		$457	
	4	$868	$781	$670	—	$558	$282	$229	$480	$587	$340	$457		
Allocation										80	65	70	85	

Shipping Cost Per Truckload

This possibility is only one of many indirect ways of shipping a truckload from cannery 1 to warehouse 4 that needs to be considered, if indeed this cannery should send anything to this warehouse. The overall problem is to determine how the output from all the canneries should be shipped to meet the warehouse allocations and minimize the total shipping cost.

Now let us see how this *transshipment problem* can be reformulated as a transportation problem. The basic idea is to interpret the individual truck trips (as opposed to complete journeys for truckloads) as being the shipment from a source to a destination, and so label *all* 12 locations (canneries, junctions, and warehouses) as being *both* potential *destinations* and potential *sources* for these shipments. To illustrate this interpretation, consider the above example where a truckload of peas is shipped from cannery 1 to warehouse 4 by being *transshipped* through junction 2 and then warehouse 2. The first truck trip for this shipment has cannery 1 as its source and junction 2 as its destination, but then junction 2 becomes the source for the second truck trip with warehouse 2 as its destination. Finally, warehouse 2 becomes the source for the third trip with this same shipment, where warehouse 4 then is the destination. In a similar fashion, any of the 12 locations can become a source, a destination, or both, for truck trips.

Thus, for the reformulation as a transportation problem, we have 12 sources and 12 destinations. The c_{ij} unit costs for the resulting *cost and requirements table* shown in Table 7.25 are just the shipping costs per truckload already given in Table 7.24. The impossible shipments indicated by dashes in Table 7.24 are assigned a huge unit cost of M. Because each location is both a source and a destination, the diagonal elements in the cost and requirements table represent the unit cost of a shipment from a given location *to itself*. The costs of these fictional shipments going nowhere are zero.

To complete the reformulation of this transshipment problem as a transportation problem, we now need to explain how to obtain the demand and supply quantities in Table 7.25. The number of truckloads transshipped through a location should be included in both the demand for that location as a destination and the supply for that

Table 7.25 Cost and Requirements Table for the P & T Co. Transshipment Problem Formulated as a Transportation Problem

			(Canneries)			(Junctions)					(Warehouses)				
			1	2	3	4	5	6	7	8	9	10	11	12	Supply
		1	0	146	M	324	286	M	M	M	452	505	M	871	375
	(Canneries)	2	146	0	M	373	212	570	609	M	335	407	688	784	425
		3	M	M	0	658	M	405	419	158	M	685	359	673	400
		4	322	371	656	0	262	398	430	M	503	234	329	M	300
		5	284	210	M	262	0	406	421	644	305	207	464	558	300
Source	(Junctions)	6	M	569	403	398	406	0	81	272	597	253	171	282	300
		7	M	608	418	431	422	81	0	287	613	280	236	229	300
		8	M	M	158	M	647	274	288	0	831	501	293	482	300
		9	453	336	M	505	307	599	615	831	0	359	706	587	300
	(Warehouses)	10	505	407	683	235	208	254	281	500	357	0	362	341	300
		11	M	687	357	329	464	171	236	290	705	362	0	457	300
		12	868	781	670	M	558	282	229	480	587	340	457	0	300
Demand			300	300	300	300	300	300	300	300	380	365	370	385	

location as a source. Since we do not know this number in advance, we instead add a safe upper bound on this number to *both* the original demand and supply for that location (shown as allocation and output in Table 7.24) and then introduce the *same* slack variable into its demand and supply constraints. (This single slack variable thereby serves the role of both a dummy source and a dummy destination.) Since it never would pay to return a truckload to be transshipped through the same location more than once, a safe upper bound on this number for *any* location is the *total number of truckloads* (300), so we shall use 300 as the upper bound. The slack variable for both constraints for location i would be x_{ii}, the (fictional) number of truckloads shipped from this location to itself. Thus, $(300 - x_{ii})$ is the real number of truckloads transshipped through location i.

Adding 300 to each of the allocation and demand quantities in Table 7.24 (where blanks are zeroes) now gives us the complete cost and requirements table shown in Table 7.25 for the transportation problem formulation of our transshipment problem. Therefore, using the transportation simplex method to obtain an optimal solution for this transportation problem provides an optimal shipping plan (ignoring the x_{ii}) for the P & T Company.

General Features

Our prototype example illustrates all the general features of the transshipment problem and its relationship to the transportation problem. Thus the transshipment problem can be described in general terms as being concerned with how to allocate and route units (truckloads of canned peas in the example) from *supply centers* (canneries) to *receiving centers* (warehouses) via intermediate *transshipment points* (junctions, other supply centers, and other receiving centers). In addition to transshipping units, each supply center generates a given net surplus of units to be distributed, and each receiving center absorbs a given net deficit, whereas the junctions neither generate nor absorb

233

any units. (The problem has feasible solutions only if the total net surplus generated at the supply centers *equals* the total net deficit to be absorbed at the receiving centers.)

A direct shipment may be impossible ($c_{ij} = M$) for certain pairs of locations. In addition, certain supply centers and receiving centers may not be able to serve as transshipment points at all. In the reformulation of the transshipment problem as a transportation problem, the easiest way to deal with any such center is to delete its *column* (for a supply center) or its *row* (for a receiving center) in the cost and requirements table, and then add *nothing* to its *original* supply or demand quantity.

A positive cost c_{ij} is incurred for each unit sent *directly* from location i (a supply center, junction, or receiving center) to another location j. The objective is to determine the plan for allocating and routing the units that minimizes the total cost.

The resulting mathematical model for the transshipment problem (see Prob. 27) has a special structure slightly different from that for the transportation problem. As in the latter case, it has been found that some applications that have nothing to do with transportation can be fitted to this special structure. However, regardless of the physical context of the application, this model always can be reformulated as an equivalent transportation problem in the manner illustrated by the prototype example.

7.4 The Assignment Problem

The **assignment problem** is the special type of linear programming problem where the resources are being allocated to the activities on a *one-to-one basis*. Thus each resource or *assignee* (e.g., an employee, machine, or time slot) is to be assigned uniquely to a particular activity or *assignment* (e.g., a task, site, or event). There is a cost c_{ij} associated with assignee i ($i = 1, 2, \ldots, n$) performing assignment j ($j = 1, 2, \ldots, n$), so that the objective is to determine how all the assignments should be made in order to minimize the total cost.

Prototype Example

The JOB SHOP COMPANY has purchased three new machines of different types. There are four available locations in the shop where a machine could be installed. Some of these locations are more desirable than others for particular machines because of their proximity to work centers that would have a heavy work flow to and from these machines. Therefore, the objective is to assign the new machines to the available locations to minimize the total cost of materials handling. The estimated cost per unit time of materials handling involving each of the machines is given in Table 7.26 for

Table 7.26 **Materials-Handling Cost Data for the Job Shop Co.**

		Location			
		1	2	3	4
Machine	1	13	16	12	11
	2	15	—	13	20
	3	5	7	10	6

the respective locations. Location 2 is not considered suitable for machine 2. There would be no work flow between the new machines.

To formulate this problem as an assignment problem, we must introduce a *dummy machine* for the extra location. Also, an extremely large cost M should be attached to the assignment of machine 2 to location 2 to prevent this assignment in the optimal solution. The resulting assignment problem *cost table* is shown in Table 7.27. This cost table contains all of the necessary data for solving the problem. The optimal solution is to assign machine 1 to location 4, machine 2 to location 3, and machine 3 to location 1, for a total cost of 29. The *dummy* machine is assigned to location 2, so this location is available for some future *real* machine.

We shall discuss how this solution is obtained after formulating the mathematical model for the general assignment problem.

The Assignment Problem Model and Solution Procedures

The mathematical model for the assignment problem uses the following decision variables:

$$x_{ij} = \begin{cases} 1, & \text{if assignee } i \text{ performs assignment } j \\ 0, & \text{if not} \end{cases}$$

for $i = 1, 2, \ldots, n$ and $j = 1, 2, \ldots, n$. Thus each x_{ij} is a *binary variable* (0 or 1). As discussed at length in the chapter on integer programming (Chap. 13), binary variables are important in operations research for representing *yes-or-no decisions*. In this case, the yes-or-no decisions are: Should assignee i perform assignment j?

Letting Z denote total cost, the assignment problem model is

$$\text{Minimize} \quad Z = \sum_{i=1}^{n} \sum_{j=1}^{n} c_{ij} x_{ij},$$

subject to

$$\sum_{j=1}^{n} x_{ij} = 1, \quad \text{for } i = 1, 2, \ldots, n$$

$$\sum_{i=1}^{n} x_{ij} = 1, \quad \text{for } j = 1, 2, \ldots, n$$

and

$$x_{ij} \geq 0, \quad \text{for all } i \text{ and } j$$

$$(x_{ij} \text{ binary, for all } i \text{ and } j).$$

Table 7.27 **Cost Table for the Job Shop Co. Assignment Problem**

		Assignment (Location)			
		1	2	3	4
	1	13	16	12	11
Assignee	2	15	M	13	20
(Machine)	3	5	7	10	6
	4(D)	0	0	0	0

The first set of functional constraints specifies that each assignee performs exactly one assignment, whereas the second set requires each assignment to be performed by exactly one assignee. *If we delete the parenthetical restriction that the x_{ij} be binary,* the model clearly is a special type of linear programming problem and so can be readily solved. Fortunately, for reasons about to unfold, we *can* delete this restriction. (This deletion is the reason the assignment problem appears in this chapter rather than the integer programming chapter.)

Now compare this model (without the binary restriction) with the *transportation problem model* presented in the second subsection of Sec. 7.1 (including Table 7.6). Note how similar their structures are. In fact, the assignment problem is just a special type of transportation problem where the *sources* now are *assignees* and the *destinations* now are *assignments,* and where

number of sources (m) = number of destinations (n),
every supply $s_i = 1$,
every demand $d_j = 1$.

Now focus on the *integer solutions property* in the subsection on the transportation problem model. Because every s_i and d_j are integers $(= 1)$ now, this property implies that *every basic feasible solution* (including an optimal one) is *integer* for an assignment problem. The functional constraints of the assignment problem model prevent any variable from being greater than one, and the nonnegativity constraints prevent values less than zero. Therefore, by deleting the binary restriction to enable solving an assignment problem as a linear programming problem, the resulting basic feasible solutions obtained (including the final optimal solution) *automatically* will satisfy the binary restriction anyway.

For any particular assignment problem, practitioners normally do not bother writing out the full mathematical model. It is simpler to formulate the problem by filling out a *cost table* (e.g., Table 7.27), including identifying the assignees and assignments, since this table contains all the essential data in a far more compact form.

Because the assignment problem is a special type of transportation problem, one convenient way to solve any particular assignment problem is to apply the *transportation simplex method* described in Sec. 7.2. This approach requires converting the cost table to a *cost and requirements table* for the equivalent transportation problem, as shown in Table 7.28a.

For example, Table 7.28b shows the *cost and requirements table* for the Job Shop Co. problem that is obtained from the *cost table* of Table 7.27. When the transportation simplex method is applied to this transportation problem formulation, the resulting optimal solution has basic variables $x_{13} = 0$, $x_{14} = 1$, $x_{23} = 1$, $x_{31} = 1$, $x_{41} = 0$, $x_{42} = 1$, $x_{43} = 0$. (You are asked to verify this solution in Prob. 32.) The *degenerate* basic variables $(x_{ij} = 0)$ and the assignment for the *dummy machine* $(x_{42} = 1)$ don't mean anything for the original problem, so the *real* assignments are machine 1 to location 4, machine 2 to location 3, and machine 3 to location 1.

It is no coincidence that this optimal solution provided by the transportation simplex method has so many *degenerate* basic variables. For *any* assignment problem, the transportation problem formulation shown in Table 7.28a has $m = n$. Transportation problems in general have $(m + n - 1)$ basic variables (allocations), so every basic feasible solution for this particular kind of transportation problem has $(2n - 1)$

Table 7.28 Cost and Requirements Table for the Assignment Problem Formulated as a
Transportation Problem, Illustrated by the Job Shop Co. Example

(a) *General Case*

		Cost Per Unit Distributed				
			Destination			
		1	2	$\cdots$	n	Supply
	1	c_{11}	c_{12}	$\cdots$	c_{1n}	1
Source	2	c_{21}	c_{22}	$\cdots$	c_{2n}	1
	$\vdots$	$\vdots$	$\vdots$		$\vdots$	$\vdots$
$m = n$		c_{n1}	c_{n2}	$\cdots$	c_{nn}	1
Demand		1	1	$\cdots$	1	

(b) *Job Shop Co. Example*

		Cost Per Unit Distributed				
		Destination (location)				
		1	2	3	4	Supply
	1	13	16	12	11	1
Source	2	15	M	13	20	1
(machine)	3	5	7	10	6	1
	4(D)	0	0	0	0	1
Demand		1	1	1	1	

basic variables, but exactly n of these $x_{ij} = 1$. Therefore, there always are $(n - 1)$ *degenerate* basic variables ($x_{ij} = 0$). As discussed at the end of Sec. 7.2, degenerate basic variables do not cause any major complication in the execution of the algorithm. However, they do frequently cause *wasted iterations,* where nothing changes (same allocations) except for the labeling of which allocations of zero correspond to degenerate basic variables rather than nonbasic variables. These wasted iterations are a major drawback for applying the transportation simplex method in this kind of situation, where *always* are so many degenerate basic variables.

Another drawback of the transportation simplex method here is that it is purely a *general-purpose* algorithm for solving all transportation problems. Therefore, it does nothing to exploit the additional *special structure* in this special type of transportation problem ($m = n$, every $s_i = 1$, and every $d_j = 1$). Specialized algorithms have been developed to fully streamline the procedure for solving just assignment problems. These algorithms operate directly on the *cost table,* and do not bother with degenerate basic variables. When a computer code is available for one of these algorithms, it generally should be used in preference to the transportation simplex method. We outline the key ideas for a specialized algorithm at the end of the section, after developing another example that will be used to illustrate the algorithm.

Example—Assigning Products to Plants

The BETTER PRODUCTS COMPANY has decided to initiate the production of four new products, using three plants that currently have excess production capacity. The products require a comparable production effort per unit, so the available production capacity of the plants is measured by the number of units of any product that can be produced per day, as given in the last column of Table 7.29. The bottom row gives the required production rate per day to meet projected sales. Each plant can produce any of these products, *except* that Plant 2 *cannot* produce product 3. However, the variable costs per unit of each product differ from plant to plant, as shown in the main body of Table 7.29.

Management now needs to make a decision on how to split up the production of the products among plants. Two kinds of options are available.

Table 7.29 Data for the Better Products Co. Problem

Plant \ Product	Unit Cost 1	2	3	4	Capacity Available
1	41	27	28	24	75
2	40	29	—	23	75
3	37	30	27	21	45
Production rate	20	30	30	40	

Option 1: Permit *product splitting,* where the same product is produced in more than one plant.

Option 2: Prohibit *product splitting.*

This second option imposes a constraint that can only increase the cost of an optimal solution based on Table 7.29. On the other hand, the key advantage of Option 2 is that it eliminates some *hidden costs* associated with product splitting that are not reflected in Table 7.29, including extra setup, distribution, and administration. Therefore, management wants both options analyzed before making a final decision. For Option 2, management further specifies that every plant should be assigned at least one of the products.

We will formulate and solve the model for each option in turn, where Option 1 leads to a *transportation problem* and Option 2 leads to an *assignment problem.*

FORMULATION OF OPTION 1: With product splitting permitted, Table 7.29 can be converted directly to a cost and requirements table for a transportation problem. The plants become the *sources* and the products become the *destinations* (or vice versa), so the *supplies* are the available production capacities and the *demands* are the required production rates. Only two changes need to be made in Table 7.29. First, because Plant 2 cannot produce product 3, such an allocation is prevented by assigning it a huge unit cost of M. Second, the total capacity ($75 + 75 + 45 = 195$) exceeds the total required production ($20 + 30 + 30 + 40 = 120$), so a *dummy destination* with a demand of 75 is needed to balance these two quantities. The resulting cost and requirements table is shown in Table 7.30.

Table 7.30 Cost and Requirements Table for the Transportation Problem Formulation of Option 1 for the Better Products Co. Problem

		Cost Per Unit Distributed Destination (Product) 1	2	3	4	5(D)	Supply
Source (plant)	1	41	27	28	24	0	75
	2	40	29	M	23	0	75
	3	37	30	27	21	0	45
Demand		20	30	30	40	75	

The optimal solution for this transportation problem has basic variables (allocations) $x_{12} = 30$, $x_{13} = 30$, $x_{15} = 15$, $x_{24} = 15$, $x_{25} = 60$, $x_{31} = 20$, and $x_{34} = 25$, so

Plant 1 produces all of products 2 and 3,
Plant 2 produces half of product 4,
Plant 3 produces half of product 4 and all of product 1.

The total cost is $Z = 3,270$.

FORMULATION OF OPTION 2: Without product splitting, each product must be assigned to just one plant. Therefore, the products can be interpreted as the *assignments* for an *assignment problem,* where the plants are the *assignees.*

Management has specified that every plant should be assigned at least one of the products. There are more products (four) than plants (three), so one of the plants will need to be assigned two products. Plant 3 has only enough excess capacity to produce one product (see Table 7.29), so *either* Plant 1 or Plant 2 will take the extra product.

To make this assignment of an extra product possible within an assignment problem formulation, Plants 1 and 2 each are split into *two* assignees, as shown in Table 7.31.

The number of assignees (now five) must equal the number of assignments (now four), so a *dummy assignment* (product) is introduced into Table 7.31 as 5(*D*). The role of this dummy assignment is to provide the *fictional* second product to either Plant 1 or Plant 2, whichever one receives only one *real* product. There is no cost for producing a *fictional* product so, as usual, the cost entries for the dummy assignment are zero. The one exception is the entry of M in the last row of Table 7.31. The reason for M here is that Plant 3 must receive a *real* product, so the Big M method is needed to prevent the assignment of the fictional product to Plant 3.

The remaining cost entries in Table 7.31 are *not* the *unit* costs shown in Table 7.29 or 7.30. For an assignment problem, the cost c_{ij} is the *total* cost associated with assignee i performing assignment j. For Table 7.31, the *total* cost (per day) for Plant i to produce product j is the unit cost of production *times* the number of units produced (per day), where these two quantities for the multiplication are given separately in Table 7.29. (As in Table 7.30, M again is used to prevent the infeasible assignment of product 3 to Plant 2.)

**Table 7.31 Cost Table for the Assignment
Problem Formulation of Option 2 for the Better
Products Co. Problem**

		Assignment (Product)				
		1	2	3	4	5(*D*)
	1*a*	820	810	840	960	0
	1*b*	820	810	840	960	0
Assignee	2*a*	800	870	*M*	920	0
(Plant)	2*b*	800	870	*M*	920	0
	3	740	900	810	840	*M*

The optimal solution for this assignment problem is

Plant 1 produces products 2 and 3,
Plant 2 produces product 1,
Plant 3 produces product 4,

where the dummy assignment is given to Plant 2. The total cost is $Z = 3,290$.

As usual, one way to obtain this optimal solution is to convert the cost table of Table 7.31 to a cost and requirements table for the equivalent transportation problem (see Table 7.28) and then apply the transportation simplex method. Because of the identical rows in Table 7.31, this approach can be streamlined by combining the *five* assignees into *three* sources with supplies 2, 2, and 1, respectively. (See Prob. 31.) This streamlining also decreases by two the number of *degenerate* basic variables in every basic feasible solution.

Now look back and compare this solution to the one obtained for Option 1, which included the splitting of product 4 between Plants 2 and 3. The allocations are somewhat different for the two solutions, but the total costs are virtually the same ($Z = 3,270$ for Option 1 versus $Z = 3,290$ for Option 2). Therefore, considering the hidden costs associated with product splitting, management decided to adopt the Option 2 solution.

An Algorithm for the Assignment Problem

Earlier in this section, we pointed out that the transportation simplex method can be used to solve assignment problems but that a *specialized* algorithm designed for such problems should be more efficient. We now will describe a basic algorithm (sometimes called the **Hungarian algorithm**) of this type. We will focus just on the key ideas without filling in all the details needed for a complete computer implementation. (Chap. 6 of Selected Reference 9 provides additional information about solving assignment problems.)

The algorithm operates directly on the *cost table* for the problem. More precisely, it converts the original cost table into a series of *equivalent* cost tables until it reaches one where an optimal solution is obvious. This final equivalent cost table is one consisting of only *positive* or *zero* elements where all the assignments can be made to the zero element positions. Since the total cost cannot be negative, this set of assignments with a zero total cost is clearly optimal. The question remaining is how to convert the original cost table into this form.

The key to this conversion is the fact that one can add or subtract any constant from every element of a row or column of the cost table without really changing the problem. That is, an optimal solution for the new cost table must also be optimal for the old one, and conversely.

To illustrate these ideas, consider the cost table for the *Job Shop Co. problem* given in Table 7.27. To convert this cost table into an *equivalent* cost table, suppose that we subtract 11 from every element in row 1, which yields:

	1	2	3	4
1	2	5	1	0
2	15	M	13	20
3	5	7	10	6
4(D)	0	0	0	0

Since any feasible solution must have exactly one assignment in row 1, the total cost for the new table must always be exactly 11 less than for the old table. Hence, the solution which minimizes total cost for one table must also minimize total cost for the other.

Notice that, whereas the original cost table had only strictly positive elements in the first three rows, the new table has a zero element in row 1. Since the objective is to obtain enough strategically located zero elements to yield a complete set of assignments, this process should be continued on the other rows and columns. Negative elements are to be avoided, so the constant to be subtracted should be the minimum element in the row or column. Doing this for rows 2 and 3 yields the following equivalent cost table:

	1	2	3	4
1	2	5	1	[0]
2	2	M	[0]	7
3	[0]	2	5	1
4(D)	0	[0]	0	0

This cost table has all the zero elements required for a complete set of assignments, as shown by the four boxes, so these four assignments constitute an *optimal solution* (as claimed earlier for this problem). The total cost for this optimal solution is seen in Table 7.27 to be $Z = 29$, which is just the sum of amounts that have been subtracted from rows 1, 2, and 3.

Unfortunately, an optimal solution is not always obtained quite so easily, as we now illustrate with the assignment problem formulation of Option 2 for the Better Products Co. problem shown in Table 7.31.

Because this problem's cost table already has zero elements in every row but the last one, let us begin the process of converting to *equivalent* cost tables by subtracting the minimum element in each *column* from every entry in that column. The result is shown below.

	1	2	3	4	5(D)
1a	80	0	30	120	0
1b	80	0	30	120	0
2a	60	60	M	80	0
2b	60	60	M	80	0
3	0	90	0	0	M

Now *every* row and column has at least one zero element, but a complete set of assignments with zero elements is *not* possible this time. In fact, the maximum number of assignments that can be made in zero element positions is only 3. (Try it.) Therefore, one more idea must be implemented to finish solving this problem that was not needed for the first example.

This idea involves a new way of creating *additional* positions with zero elements without creating any negative elements. Rather than subtracting a constant from a *single* row or column, we now add or subtract a constant from a *combination* of rows and columns.

This procedure begins by drawing a set of lines through some of the rows and columns in such a way as to *cover all the zeros*. This is preferably done with a *minimum* number of lines, as shown below.

	1	2	3	4	5(D)
1a	80	0	30	120	0
1b	80	0	30	120	0
2a	60	60	M	80	0
2b	60	60	M	80	0
3	0	90	0	0	M

Notice that the minimum element not crossed out is 30 in the two top positions in column 3. Therefore, subtracting 30 from every element in the entire table, i.e., from every row or from every column, will create a new zero element in these two positions. Then, in order to restore the previous zero elements and eliminate negative elements, add 30 to each row or column with a line covering it—row 3 and columns 2 and 5(D). The result is given below.

	1	2	3	4	5(D)
1a	50	0	0	90	0
1b	50	0	0	90	0
2a	30	60	M	50	0
2b	30	60	M	50	0
3	0	120	0	0	M

A shortcut for obtaining this cost table from the preceding one is to subtract 30 from just the elements without a line through them and then add 30 to every element that lies at the intersection of two lines.

With this new cost table, it now is possible to make four assignments to zero element positions, but still not five. (Try it.) Therefore, we repeat the above procedure, where four lines (the same number as the maximum number of assignments) now are the minimum needed to cover all zeroes. One way of doing this is shown below.

	1	2	3	4	5(D)
1a	50	0	0	90	0
1b	50	0	0	90	0
2a	30	60	M	50	0
2b	30	60	M	50	0
3	0	120	0	0	M

The minimum element not covered by a line is again 30, but now in the first position in rows 2a and 2b. Therefore, we subtract 30 from every *uncovered* element and add

30 to every *doubly* *covered* element (except for ignoring elements of M), which gives the following equivalent cost table.

	1	2	3	4	5(D)
1a	50	$\boxed{0}$	0	90	30
1b	50	0	$\boxed{0}$	90	30
2a	$\boxed{0}$	30	M	20	0
2b	0	30	M	20	$\boxed{0}$
3	0	120	0	$\boxed{0}$	M

This table actually has several ways of making a complete set of assignments to zero element positions (several optimal solutions), including the one shown by the five boxes. The resulting total cost is seen in Table 7.31 to be

$$Z = 810 + 840 + 800 + 0 + 840 = 3{,}290.$$

7.5 Multidivisional Problems

Another important class of linear programming problems having an exploitable special structure consists of **multidivisional problems.** Their special feature is that they involve coordinating the decisions of the separate divisions of a large organization. Because the divisions operate with considerable autonomy, the problem is *almost* decomposable into separate problems, where each division is concerned only with optimizing its own operation. However, some overall coordination is required in order to best divide certain organizational resources among the divisions.

    As a result of this special feature, the *table of constraint coefficients* for multidivisional problems has the **block angular structure** shown in Table 7.32. (Recall that shaded blocks represent the only portions of the table that have *any* nonzero a_{ij} coefficients.) Thus each smaller block contains the coefficients of the constraints for one *subproblem,* namely, the problem of optimizing the operation of a division considered by itself. The long block at the top gives the coefficients of the **linking**

Table 7.32 **Table of Constraint Coefficients for Multidivisional Problems**

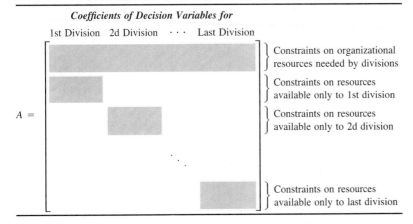

constraints for the *master problem,* namely, the problem of coordinating the activities of the divisions by dividing organizational resources among them so as to obtain an overall optimal solution for the entire organization.

Because of their nature, multidivisional problems frequently are very large, containing many hundreds or even thousands of constraints and variables. Therefore, it may be necessary to exploit the special structure in order to be able to solve such a problem with a reasonable expenditure of computer time, or even to solve it at all! The **decomposition principle** (described in Sec. 9.5) provides an effective way of exploiting the special structure.

Conceptually, this streamlined version of the simplex method can be thought of as having each division solve its subproblem and sending this solution as its proposal

Table 7.33 Data for the Good Foods Corp. Multidivisional Problem

Divisional Data				*Subproblem*

Processed Foods Division

Product / Resource	Resource Usage/Unit 1	2	3	Amount Available
1	2	4	3	10
2	7	3	6	15
3	5	0	3	12
ΔZ/unit	8	5	6	
Level	x_1	x_2	x_3	

Maximize $Z_1 = 8x_1 + 5x_2 + 6x_3,$

subject to

$$2x_1 + 4x_2 + 3x_3 \leq 10$$
$$7x_1 + 3x_2 + 6x_3 \leq 15$$
$$5x_1 \qquad\quad + 3x_3 \leq 12$$

and

$$x_1 \geq 0, \qquad x_2 \geq 0, \qquad x_3 \geq 0.$$

Canned Foods Division

Product / Resource	Resource Usage/Unit 4	5	6	Amount Available
4	3	1	2	7
5	2	4	3	9
ΔZ/unit	9	7	9	
Level	x_4	x_5	x_6	

Maximize $Z_2 = 9x_4 + 7x_5 + 9x_6,$

subject to

$$3x_4 + x_5 + 2x_6 \leq 7$$
$$2x_4 + 4x_5 + 3x_6 \leq 9$$

and

$$x_4 \geq 0, \qquad x_5 \geq 0, \qquad x_6 \geq 0.$$

Frozen Foods Division

Product / Resource	Resource Usage/Unit 7	8	Amount Available
6	8	5	25
7	7	9	30
8	6	4	20
ΔZ/unit	6	5	
Level	x_7	x_8	

Maximize $Z_3 = 6x_7 + 5x_8,$

subject to

$$8x_7 + 5x_8 \leq 25$$
$$7x_7 + 9x_8 \leq 30$$
$$6x_7 + 4x_8 \leq 20$$

and

$$x_7 \geq 0, \qquad x_8 \geq 0.$$

Data for Organizational Resources

Product / Resource	Resource Usage/Unit 1	2	3	4	5	6	7	8	Amount Available
Corn	5	3	0	2	0	3	4	6	30
Potatoes	2	0	4	3	7	0	1	0	20

to "headquarters" (the master problem), where negotiators then coordinate the proposals from all the divisions to find an optimal solution for the overall organization. If the subproblems are of manageable size and the master problem is not too large (not more than 50 to 100 constraints), this approach is successful in solving some *extremely* large multidivisional problems. It is particularly worthwhile when the total number of constraints is quite large (at least several hundred) and there are more than a few subproblems.

Prototype Example

The GOOD FOODS CORPORATION is a very large producer and distributor of food products. It has three main divisions: the Processed Foods Division, the Canned Foods Division, and the Frozen Foods Division. Because costs and market prices change frequently in the food industry, Good Foods periodically uses a corporate linear programming model to revise the production rates for its various products in order to use its available production capacities in the most profitable way. This model is similar to that for the Wyndor Glass Co. problem (see Sec. 3.1), but on a much larger scale, having hundreds of constraints and variables. (Since our space is limited, we shall describe a simplified version of this model that combines the products or resources by types.)

   The corporation grows its own high-quality corn and potatoes, and these basic food materials are the only ones currently in short supply that are used by all the divisions. Except for these organizational resources, each division uses only its own resources and thus could determine its optimal production rates autonomously. The data for each division and the corresponding *subproblem* involving just its products and resources are given in Table 7.33 (where Z represents profit in millions of dollars per month), along with the data for the organizational resources.

   The resulting linear programming problem for the corporation is

$$\text{Maximize} \quad Z = 8x_1 + 5x_2 + 6x_3 + 9x_4 + 7x_5 + 9x_6 + 6x_7 + 5x_8,$$

subject to

$$5x_1 + 3x_2 \qquad + 2x_4 \qquad + 3x_6 + 4x_7 + 6x_8 \le 30$$
$$2x_1 \qquad + 4x_3 + 3x_4 + 7x_5 \qquad + x_7 \qquad \le 20$$
$$2x_1 + 4x_2 + 3x_3 \qquad\qquad \le 10$$
$$7x_1 + 3x_2 + 6x_3 \qquad\qquad \le 15$$
$$5x_1 \qquad + 3x_3 \qquad\qquad \le 12$$
$$3x_4 + x_5 + 2x_6 \qquad \le 7$$
$$2x_4 + 4x_5 + 3x_6 \qquad \le 9$$
$$8x_7 + 5x_8 \le 25$$
$$7x_7 + 9x_8 \le 30$$
$$6x_7 + 4x_8 \le 20$$

and
$$x_j \ge 0, \quad \text{for } j = 1, 2, \dots, 8.$$

   Note how the corresponding table of constraint coefficients shown in Table 7.34 fits the special structure for multidivisional problems given in Table 7.32. Therefore,

Table 7.34 **Table of Constraint
Coefficients for the Good Foods
Corp. Multidivisional Problem**

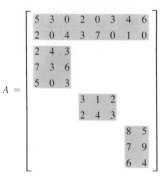

the Good Foods Corp. can indeed solve this problem (or a more detailed version of it) by the streamlined version of the simplex method provided by the decomposition principle.

Important Special Cases

Some even simpler forms of the special structure exhibited in Table 7.32 arise quite frequently. Two particularly common forms are shown in Table 7.35.

The first form occurs when some or all of the variables can be divided into groups such that the *sum* of the variables in *each* group must not exceed a specified upper bound for that group (or perhaps must equal a specified constant). Constraints of this form,

$$x_{j1} + x_{j2} + \cdots + x_{jk} \le b_i$$

$$\text{(or} \quad x_{j1} + x_{j2} + \cdots + x_{jk} = b_i),$$

usually are called either *generalized upper bound constraints* (**GUB constraints** for short) or *group constraints*. Although Table 7.35 shows each GUB constraint as involving *consecutive* variables, this is not necessary. For example,

$$x_1 + x_5 + x_9 \le 1$$

is a GUB constraint, as is

$$x_8 + x_3 + x_6 = 20.$$

Table 7.35 **Table of Constraint Coefficients for Important Special Cases of the Structure for Multidivisional Problems Given in Table 7.32**

Generalized Upper Bounds   Upper Bounds

The second form shown in Table 7.35 occurs when some or all of the *individual* variables must not exceed a specified upper bound for that variable. These constraints,

$$x_j \le b_i,$$

normally are referred to as **upper bound constraints.** For example, both

$$x_1 \le 1 \quad \text{and} \quad x_2 \le 5$$

are upper bound constraints.

Either *GUB* or *upper bound constraints* may occur because of the multidivisional nature of the problem. However, we should emphasize that they often arise in many other contexts as well. In fact, you already have seen several examples containing such constraints as summarized below.

Note in Table 7.6 that all supply constraints in the *transportation problem* actually are *GUB constraints*. (Table 7.6 fits the form in Table 7.35 by placing the supply constraints below the demand constraints.) In addition, the demand constraints also are *GUB constraints*, but ones not involving *consecutive* variables.

In the Southern Confederation of Kibbutzim *regional planning problem* (see Sec. 3.4), the constraints involving usable land for each kibbutz and total acreage for each crop all are GUB constraints.

The technological limit constraints in the Nori & Leets Co. *air pollution problem* (see Sec. 3.4) are upper bound constraints, as are two of the three functional constraints in the Wyndor Glass Co. *product mix problem* (see Sec. 3.1).

Because of the prevalence of *GUB* and *upper bound constraints*, special techniques have been developed for streamlining the way in which the simplex method deals with them. (The technique for upper bound constraints is described in Sec. 9.1, and the one for GUB constraints[1] is quite similar.) If there are many such constraints, these techniques can drastically reduce the computation time for a problem.

7.6 Multitime Period Problems

Any successful organization must plan ahead and take into account probable changes in its operating environment. For example, predicted future changes in sales because of *seasonal* variations or *long run* trends in demand might affect how the firm should operate currently. Such situations frequently lead to the formulation of *multitime period* linear programming problems for planning several time periods (e.g., days, months, or years) into the future. Just as for multidivisional problems, multitime period problems are *almost* decomposable into separate subproblems, where each subproblem in this case is concerned with optimizing the operation of the organization during one of the time periods. However, some overall planning is required to coordinate the activities in the different time periods.

The resulting special structure for multitime period problems is shown in Table 7.36. Each approximately square block gives the coefficients of the constraints for one *subproblem* concerned with optimizing the operation of the organization during a particular time period considered by itself. Each oblong block then contains the coefficients of the **linking variables** for those activities that affect two or more time

---

[1] Dantzig, George B., and Richard M. Van Slyke: "Generalized Upper Bounded Techniques for Linear Programming," *Journal of Computer and Systems Sciences,* **1**: 213–226, 1967.

periods. For example, the linking variables may describe inventories that are retained at the end of one time period for use in some later time period, as we shall illustrate in the prototype example.

As with multidivisional problems, the multiplicity of subproblems often causes multitime period problems to have a very large number of constraints and variables, so again a method for exploiting the *almost decomposable* special structure of these problems is needed. Fortunately, the *same* method can be used for both types of problems! The idea is to reorder the variables in the multitime period problem to first list all the *linking variables*, as shown in Table 7.37, and then to construct its dual problem. This dual problem exactly fits the *block angular structure* shown in Table 7.32. (For this reason the special structure in Table 7.37 is referred to as the **dual angular structure**.) Therefore, the *decomposition principle* for multidivisional problems can be used to solve this dual problem. Since directly applying even this streamlined version of the simplex method to the dual problem automatically identifies an optimal solution for the primal problem as a by-product, this provides an efficient way of solving many large multitime period problems.

Prototype Example

The WOODSTOCK COMPANY operates a large warehouse that buys and sells lumber. Since the price of lumber changes during the different seasons of the year, the company sometimes builds up a large stock when prices are low and then stores the lumber for sale later at a higher price. The manager feels that there is considerable room for increasing profits by improving the scheduling of purchases and sales, so he has hired a team of operations research consultants to develop the most profitable schedule.

Table 7.36 **Table of Constraint Coefficients for Multitime Period Problems**

Table 7.37 Table of Constraint Coefficients for Multitime Period Problems after Reordering the Variables

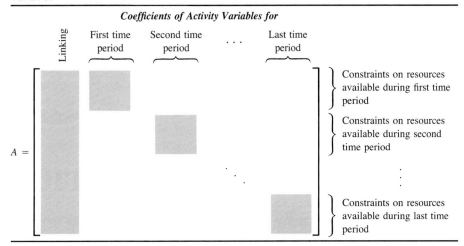

Since the company buys lumber in large quantities, its purchase price is slightly less than its selling price in each season. These prices are shown in Table 7.38, along with the maximum amount that can be sold during each season. The lumber would be purchased at the beginning of a season and sold throughout the season. If the lumber purchased is to be stored for sale in a later season, a handling cost of $7 per 1,000 board feet is incurred, as well as a storage cost (including interest on capital tied up) of $10 per 1,000 board feet for each season stored. A maximum of 2 million board feet can be stored in the warehouse at any one time. (This includes lumber purchased for sale in the same period.) Since lumber should not age too long before sale, the manager wants it all sold by the end of autumn (before the low winter prices go into effect).

The team of operations research consultants concluded that this problem should be formulated as a linear programming problem of the multitime period type. Numbering the seasons (1 = winter, 2 = spring, 3 = summer, 4 = autumn) and letting x_i be the number of 1,000 board feet purchased in season i, y_i be the number sold in season i, and z_{ij} be the number stored in season i for sale in season j, this formulation is

$$\text{Maximize} \quad Z = -410x_1 + 425y_1 - 17z_{12} - 27z_{13} - 37z_{14} - 430x_2 + 440y_2$$
$$- 17z_{23} - 27z_{24} - 460x_3 + 465y_3 - 17z_{34} - 450x_4 + 455y_4,$$

Table 7.38 Price Data for the Woodstock Company

Season	Purchase Price†	Selling Price†	Maximum Sales‡
Winter	410	425	1,000
Spring	430	440	1,400
Summer	460	465	2,000
Autumn	450	455	1,600

†Prices are in dollars per thousand board feet.
‡Sales are in thousand board feet.

$$
\begin{aligned}
x_1 - y_1 - z_{12} - z_{13} - z_{14} &= 0 \\
x_1 &\le 2000 \\
y_1 &\le 1000 \\
z_{12} \qquad\qquad + x_2 - y_2 - z_{23} - z_{24} &= 0 \\
z_{12} \qquad\qquad\qquad - y_2 &\le 0 \\
z_{12} + z_{13} + z_{14} + x_2 &\le 2000 \\
y_2 &\le 1400 \\
z_{13} \qquad + z_{23} \qquad + x_3 - y_3 - z_{34} &= 0 \\
z_{13} \qquad + z_{23} \qquad\quad - y_3 &\le 0 \\
z_{13} + z_{14} \qquad + z_{23} + z_{24} + x_3 &\le 2000 \\
y_3 &\le 2000 \\
z_{14} \qquad\qquad + z_{24} \qquad + z_{34} + x_4 - y_4 &= 0 \\
y_4 &\le 1600
\end{aligned}
$$

and $\quad x_i \ge 0, \quad y_i \ge 0, \quad z_{ij} \ge 0, \quad$ for $i = 1, 2, 3, 4$, and $j = 2, 3, 4$.

Thus this formulation contains four *subproblems*, where the subproblem for season i is obtained by deleting all variables except x_i and y_i from the overall problem. The storage variables (the z_{ij}) then provide the *linking variables* that interrelate these four time periods. Therefore, after reordering the variables to first list these linking variables, the corresponding table of constraint coefficients has the form shown in Table 7.39, where *all* blanks are *zeroes*. Since this form fits the *dual angular structure* given in Table 7.37, the streamlined solution procedure for this kind of special structure can be used to solve the problem (or much larger versions of it).

7.7 Multidivisional Multitime Period Problems

You saw in the preceding two sections how decentralized decision making can lead to *multidivisional problems* and how a changing operating environment can lead to

Table 7.39 Table of Constraint Coefficients for the Woodstock Company Multitime Period Problem after Reordering the Variables

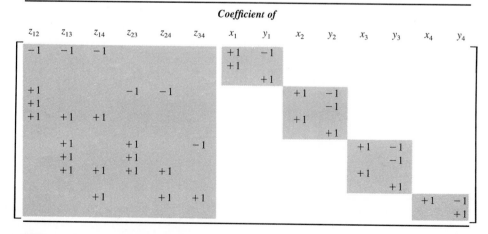

						Coefficient of							
z_{12}	z_{13}	z_{14}	z_{23}	z_{24}	z_{34}	x_1	y_1	x_2	y_2	x_3	y_3	x_4	y_4
-1	-1	-1				+1	-1						
						+1							
							+1						
+1			-1	-1				+1	-1				
+1									-1				
+1	+1	+1						+1					
									+1				
	+1		+1		-1					+1	-1		
	+1		+1								-1		
	+1	+1	+1	+1						+1			
											+1		
		+1		+1	+1							+1	-1
													+1

Table 7.40 **Table of Constraint Coefficients for Multidivisional Multitime Period Problems**

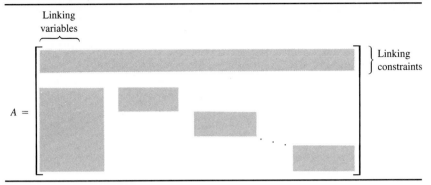

multitime period problems. We discussed these two situations separately to focus on their individual special structures. However, we should now emphasize that it is fairly common for problems to possess *both* characteristics simultaneously. For example, because costs and market prices change frequently in the food industry, the Good Foods Corp. might want to expand their multidivisional problem to consider the effect of such predicted changes several time periods into the future. This would allow the model to indicate how to most profitably stock up on materials when costs are low and store portions of the food products until prices are more favorable. Similarly, if the Woodstock Co. also owns several other warehouses, it might be advisable to expand their model to include and coordinate the activities of these divisions of their organization. (Also see Prob. 46 for another way in which the Woodstock Co. problem might expand to include the multidivisional structure.)

The combined special structure for such *multidivisional multitime period problems* is shown in Table 7.40. It contains many *subproblems* (the approximately square blocks), each of which is concerned with optimizing the operation of one division during one of the time periods considered in isolation. However, it also includes *both* linking constraints and linking variables (the oblong blocks). The *linking constraints* coordinate the divisions by making them share the organizational resources available during one or more time periods. The *linking variables* coordinate the time periods by representing activities that affect the operation of a particular division (or possibly different divisions) during two or more time periods.

One way of exploiting the combined special structure of these problems is to apply an extended version of the *decomposition principle* for multidivisional problems. This involves treating everything but the linking constraints as one large subproblem and then using this decomposition principle to coordinate the solution for this subproblem with the *master problem* defined by the linking constraints. Since this large subproblem has the *dual angular structure* shown in Table 7.37, it would be solved by the special solution procedure for multitime period problems, which again involves using this decomposition principle.

Other procedures for exploiting this combined special structure also have been developed.[1] More experimentation is still needed to test the relative efficiency of the available procedures.

---

[1]For further information, see Chap. 5 of Selected Reference 7 at the end of this chapter.

7.8 Stochastic Programming

One of the common problems in the practical application of linear programming is the difficulty of determining the proper values of the model parameters (the c_j, a_{ij}, and b_i). The true values of these parameters may not become known until after a solution has been chosen and implemented. This can sometimes be attributed solely to the inadequacy of the investigation. However, the values these parameters take on often are influenced by random events that are impossible to predict. In short, some or all of the model parameters may be *random variables*.

When these random variable parameters have relatively small variances, the standard approach is to perform sensitivity analysis as described in Chap. 6. However, if some of the parameters have relatively large variances, this approach is not very adequate. What is needed is a way of formulating the problem so that the optimization will directly take the *uncertainty* into account.

Some such approaches for *linear programming under uncertainty* have been developed. These formulations can be classified into two types, stochastic programming and chance-constrained programming, which are described in this and the next section, respectively. The main distinction between these types is that *stochastic programming* requires all constraints to hold with probability 1, whereas *chance-constrained programming* permits a *small* probability of violating any functional constraint. The former type was given its name because it is particularly applicable when the values of the decision variables are chosen at two or more different points in time (i.e., *stochastically*), although the latter type also can be adapted to this kind of *multistage problem*. The general approach for dealing with both types is to *reformulate* them as new *equivalent* linear programming problems where the certainty assumption *is* satisfied, and then solving by the simplex method. This clever reformulation for each type is the key to its practicality.

Focusing now on stochastic programming, we will introduce its main ideas only, largely through simple illustrative examples, rather than developing a complete formal description.

If some or all of the c_j are random variables, then

$$Z = \sum_{j=1}^{n} c_j x_j$$

also is a random variable for any given solution. Since it is meaningless to maximize a random variable, Z must be replaced by some deterministic function. There are many possible choices for this function, each of which may be very reasonable under certain circumstances. Perhaps the most natural choice, and certainly the most widely used, is the expected value of Z,

$$E(Z) = \sum_{j=1}^{n} E(c_j) x_j.$$

Similarly, the functional constraints

$$\sum_{j=1}^{n} a_{ij} x_j \leq b_i, \qquad \text{for } i = 1, 2, \ldots, m$$

must be reinterpreted if any of the a_{ij} and b_i are random variables. One interpretation is that a solution is considered feasible only if it satisfies all the constraints for *all*

252

possible combinations of the parameter values. This is the interpretation assumed in this section, although it is soon modified to allow certain random variable parameters to become known before values are assigned to certain x_j.

One danger with this strict interpretation of feasibility is that there may well not exist *any* solution that satisfies all the constraints for *every* possible combination of the parameter values. If so, a more liberal interpretation can be used, such as the one given in the next section.

The remainder of the section is devoted to elaborating on how stochastic programming implements its interpretation of feasibility for two categories of problems.

One-Stage Problems

A *one-stage problem* is one where the values for all the x_j must be chosen simultaneously (i.e., at one stage) before learning which value has been taken on by any of the random variable parameters. This is in contrast to the *multistage problems* considered later where the decision making is done over two or more stages while observing the values taken on by some of the random variable parameters.

The formulation for one-stage problems is relatively straightforward. Consider first the case where the a_{ij} and b_i that are random variables are *mutually independent*. Then each of these a_{ij} and b_i with multiple possible values would be replaced by its *most restrictive value* for its constraint; i.e., functional constraint i becomes

$$\sum_{j=1}^{n} (\max a_{ij})x_j \leq \min b_i,$$

where max a_{ij} is the *largest* value that the random variable a_{ij} can take on and min b_i is the *smallest* value that the random variable b_i can take on. By replacing the random variables with these *constants*, the new constraint ensures that the original constraint will be satisfied for *every* possible combination of values for the random variable parameters. Furthermore, the new constraint satisfies the *certainty assumption* of linear programming discussed in Sec. 3.3, so the reformulated problem can be solved by the simplex method.

For example, consider the constraint,

$$a_{11}x_1 + a_{12}x_2 \leq b_1,$$

where a_{11}, a_{12}, and b_1 all are independent random variables having the following ranges of possible values:

$$1 \leq a_{11} \leq 2, \qquad 2 \leq a_{12} \leq 3, \qquad 4 \leq b_1 \leq 5.$$

To reformulate to satisfy the certainty assumption of linear programming, this constraint should be replaced by

$$2x_1 + 3x_2 \leq 4.$$

Reformulating a constraint in this manner is more restrictive than necessary if the random variable parameters are *jointly dependent* in a way that prevents the parameters from *simultaneously* achieving their most restrictive values. A case of special interest is where, at least as an approximation, the problem can be described as having a relatively small number of possible *scenarios* for how the problem will unfold over time, where each scenario provides certain *fixed* values for all the parameters. Which scenario will occur may depend on some exogenous factor, such as the state of the economy, or the market's reception to new products, or the extent of progress on new technological advances.

For this kind of situation, the original constraint with random variables would be replaced by a set of new constraints, where each new constraint would have the parameter values that correspond to one of the scenarios. For example, consider again the constraint,

$$a_{11}x_1 + a_{12}x_2 \le b_1,$$

but suppose now that a_{11}, a_{12}, and b_1 each are random variables that have just the two possible values shown below:

$$a_{11} = 1 \text{ or } 2, \qquad a_{12} = 2 \text{ or } 3, \qquad b_1 = 4 \text{ or } 5.$$

Further suppose that there are just two scenarios, where each one dictates which of the two values each random variable will take on, as follows:

$$\text{Scenario 1: } a_{11} = 1, a_{12} = 3, b_1 = 4.$$

$$\text{Scenario 2: } a_{11} = 2, a_{12} = 2, b_1 = 5.$$

In this case, the original constraint with random variables would be replaced by the two new constraints,

$$x_1 + 3x_2 \le 4$$

$$2x_1 + 2x_2 \le 5.$$

This approach does have the drawback of increasing the number of functional constraints, which substantially increases the computation time for the simplex method. This drawback can become quite serious if a large number of scenarios need to be considered.

Multistage Problems

We now consider problems where the decisions on the values of the x_j are made at two or more points in time (stages). That is, some of the x_j are *first-stage variables*, others are *second-stage variables*, and so on. For example, this occurs when scheduling the production of some products over several time periods, where each x_j gives the production level for one of the products in one of the time periods.

Although the decisions are made in stages, they still need to be considered jointly in one model because the activities involved are consuming the same limited resources. However, the overall optimization makes the decisions for later stages *conditional* upon what happens at preceding stages, namely, the values taken on by some of the random variable parameters (typically the constraint coefficients for the variables associated with the preceding stages). Therefore, the stochastic programming approach enables adjusting the decisions for later stages based on unfolding circumstances.

The key idea for the stochastic programming formulation here is to replace each original decision variable beyond the first stage by a set of new decision variables, where each new decision variable represents the original decision under one of the possible circumstances that could prevail at that point.

To illustrate this approach, consider the problem,

$$\text{Maximize} \quad Z = 3x_1 + 7x_2 + 11x_3,$$

subject to
$$a_{11}x_1 + a_{12}x_2 + a_{13}x_3 \leq 100$$

and
$$x_1 \geq 0, \quad x_2 \geq 0, \quad x_3 \geq 0,$$

where a_{11}, a_{12}, and a_{13} are independent random variables such that

$$a_{11} = \begin{cases} 1, & \text{with probability } \frac{1}{2} \\ 2, & \text{with probability } \frac{1}{2} \end{cases}$$

$$a_{12} = \begin{cases} 3, & \text{with probability } \frac{1}{2} \\ 4, & \text{with probability } \frac{1}{2} \end{cases}$$

$$a_{13} = \begin{cases} 5, & \text{with probability } \frac{1}{2} \\ 6, & \text{with probability } \frac{1}{2}, \end{cases}$$

and where x_1, x_2, and x_3 are the decision variables for stages 1, 2, and 3, respectively. The value taken on by a_{11} will be known before the value of x_2 must be chosen, and the value taken on by a_{12} will be known before the value of x_3 must be chosen.

The stochastic programming formulation for this example replaces x_2 by the set of new decision variables,

$$x_{21} = \text{value chosen for } x_2 \text{ if } a_{11} = 1$$

$$x_{22} = \text{value chosen for } x_2 \text{ if } a_{11} = 2,$$

and then replaces x_3 by the set of new decision variables,

$$x_{31} = \text{value chosen for } x_3 \text{ if } a_{11} = 1, a_{12} = 3$$

$$x_{32} = \text{value chosen for } x_3 \text{ if } a_{11} = 1, a_{12} = 4$$

$$x_{33} = \text{value chosen for } x_3 \text{ if } a_{11} = 2, a_{12} = 3$$

$$x_{34} = \text{value chosen for } x_3 \text{ if } a_{11} = 2, a_{12} = 4.$$

The resulting reformulated problem is

$$\text{Maximize} \quad E(Z) = 3x_1 + 7(\tfrac{1}{2})(x_{21} + x_{22}) + 11(\tfrac{1}{4})(x_{31} + x_{32} + x_{33} + x_{34}),$$

subject to
$$x_1 + 3x_{21} + 6x_{31} \leq 100$$

$$x_1 + 4x_{21} + 6x_{32} \leq 100$$

$$2x_1 + 3x_{22} + 6x_{33} \leq 100$$

$$2x_1 + 4x_{22} + 6x_{34} \leq 100$$

and
$$x_1 \geq 0 \quad \text{and all } x_{ij} \geq 0,$$

which is an ordinary linear programming problem that can be solved by the simplex method. Note that each of the four functional constraints represents one of the four possible combinations of values for a_{11} and a_{12}. The reason that all four constraints have $a_{13} = 6$ and there are not four additional constraints with $a_{13} = 5$ is that 6 is the *most restrictive value* of a_{13} for this *last-stage* parameter. In the objective function, the multipliers of $\frac{1}{2}$ and $\frac{1}{4}$ arise because these are the probabilities of the combinations of parameter values that result in using the respective variables (x_{21}, x_{22}, and then x_{31}, x_{32}, x_{33}, x_{34}) for determining the value of x_2 or x_3.

This example also illustrates how the stochastic programming approach greatly increases the size of the model to be solved, especially if the number of stages and the number of possible combinations of values for the random variable parameters are large. This problem is avoided by the approach described in the next section.

7.9 Chance-Constrained Programming

The preceding section presented the stochastic programming approach to *linear programming under uncertainty*. Chance-constrained programming provides another way of dealing with this problem. This alternative approach may be used when it is highly desirable, but not absolutely essential, that the functional constraints hold.

When some or all of the parameters of the model are random variables, the stochastic programming formulation requires that all the functional constraints must hold for *all* possible combinations of values for these random variable parameters. By contrast, the chance-constrained programming formulation requires only that each constraint must hold for most of these combinations. More precisely, this formulation replaces the original linear programming constraints,

$$\sum_{j=1}^{n} a_{ij}x_j \le b_i, \qquad \text{for } i = 1, 2, \ldots, m,$$

by

$$P\left\{\sum_{j=1}^{n} a_{ij}x_j \le b_i\right\} \ge \alpha_i, \qquad \text{for } i = 1, 2, \ldots, m,$$

where the α_i are specified constants between zero and one (although they are normally chosen to be reasonably close to one). Therefore, a nonnegative solution, (x_1, x_2, $\cdots$, x_n), is considered to be feasible if and only if

$$P\left\{\sum_{j=1}^{n} a_{ij}x_j \le b_i\right\} \ge \alpha_i, \qquad \text{for } i = 1, 2, \ldots, m.$$

Each complementary probability, $1 - \alpha_i$, represents the allowable risk that the random variables will take on values such that

$$\sum_{j=1}^{n} a_{ij}x_j > b_i.$$

Thus, the objective is to select the "best" nonnegative solution that "probably" will turn out to satisfy each of the original constraints when the random variables (the a_{ij}, b_i, and c_j) take on their values.

There are many possible expressions for the objective function when some of the c_j are random variables, and several of these have been explored elsewhere[1] in the context of chance-constrained programming. However, only the one assumed in the preceding section, namely, the expected value function, is considered here.

[1] A. Charnes and W. W. Cooper, "Deterministic Equivalents for Optimizing and Satisficing under Chance Constraints," *Operations Research*, **11**:18–39 (1963).

No procedure is now available for solving the general chance-constrained (linear) programming problem. However, certain important special cases are solvable. The one discussed here is where: (1) all the a_{ij} parameters are constants, so that only some or all of the c_j and b_i are random variables, (2) the probability distribution of the b_i is a known multivariate normal distribution, and (3) c_j is statistically independent of b_i ($j = 1, 2, \ldots, n; i = 1, 2, \ldots, m$).

As in the preceding section, it is initially assumed that all of the x_j must be determined before learning the value taken on by any of the random variables. Then, after the approach for this case is developed, the more general case where this assumption is dropped will be discussed.

One-Stage Problems

The chance-constrained programming problem considered here fits the linear programming model format except for the constraints,

$$P\left\{\sum_{j=1}^{n} a_{ij}x_j \leq b_i\right\} \geq \alpha_i, \quad \text{for } i = 1, 2, \ldots, m.$$

Therefore, the goal is to convert these constraints into legitimate linear programming constraints, so that the simplex method can be used to solve the problem. This can be done under the stated assumptions, as shown below.

To begin, notice that

$$P\left\{\sum_{j=1}^{n} a_{ij}x_j \leq b_i\right\} = P\left\{\frac{\sum_{j=1}^{n} a_{ij}x_j - E(b_i)}{\sigma_{b_i}} \leq \frac{b_i - E(b_i)}{\sigma_{b_i}}\right\},$$

where $E(b_i)$ and σ_{b_i} are the mean and standard deviation of b_i, respectively. Since b_i is assumed to have a normal distribution, $[b_i - E(b_i)]/\sigma_{b_i}$ must also be normal with mean zero and standard deviation one. In Table A5.1 in Appendix 5, K_α is taken to be the constant such that

$$P\{Y \geq K_\alpha\} = \alpha,$$

where α is any given number between zero and one, and where Y is the random variable whose probability distribution is normal with mean zero and standard deviation one. Table A5.1 gives K_α for various values of α. For example,

$$K_{0.90} = -1.28, K_{0.95} = -1.645, \text{ and } K_{0.99} = -2.33.$$

Therefore, it now follows that

$$P\left\{K_{\alpha_i} \leq \frac{b_i - E(b_i)}{\sigma_{b_i}}\right\} = \alpha_i.$$

Note that this probability would be increased if K_{α_i} were replaced by a number $< K_{\alpha_i}$. Hence,

$$P\left\{\frac{\sum_{j=1}^{n} a_{ij}x_j - E(b_i)}{\sigma_{b_i}} \leq \frac{b_i - E(b_i)}{\sigma_{b_i}}\right\} \geq \alpha_i$$

for a given solution if and only if

$$\frac{\sum_{j=1}^{n} a_{ij}x_j - E(b_i)}{\sigma_{b_i}} \leq K_{\alpha_i}.$$

Rewriting both expressions in an equivalent form, the conclusion is that

$$P\left\{\sum_{j=1}^{n} a_{ij}x_j \leq b_i\right\} \geq \alpha_i$$

$$\sum_{j=1}^{n} a_{ij}x_j \leq E(b_i) + K_{\alpha_i}\sigma_{b_i},$$

so that this probability constraint can be replaced by this linear programming constraint. The fact that these constraints are equivalent is illustrated by Fig. 7.3.

To summarize, the chance-constrained programming problem considered above can be reduced to the following equivalent linear programming problem.

$$\text{Maximize} \quad E(Z) = \sum_{j=1}^{n} E(c_j)x_j,$$

subject to $\quad \sum_{j=1}^{n} a_{ij}x_j \leq E(b_i) + K_{\alpha_i}\sigma_{b_i}, \quad \text{for } i = 1, 2, \ldots, m,$

and $\quad\quad\quad\quad\quad x_j \geq 0, \quad \text{for } j = 1, 2, \ldots, n.$

Multistage Problems

We now will consider multistage problems such as discussed in the preceding section, where decisions beyond the first stage take into account the value taken on by certain random variable parameters at preceding stages. In our current context, we assume that some of the b_i become known before some of the x_j values must be chosen.

We need to formulate and solve problems of this type in such a way that the final decision on the x_j is partially based on the new information that has become available. The chance-constrained programming approach to this situation is to solve for each x_j as an explicit function of the b_i whose values become known before a

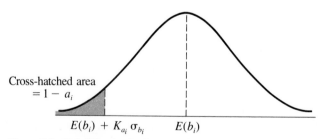

Cross-hatched area
$= 1 - a_i$

$E(b_i) + K_{a_i}\,\sigma_{b_i} \quad\quad\quad E(b_i)$

Figure 7.3 Probability density function of b_i.

value must be assigned to x_j. From a computational standpoint, it is convenient to deal with linear functions of the b_i, thereby leading to what are called *linear decision rules* for the x_j. In particular, let

$$x_j = \sum_{k=1}^{m} d_{jk}b_k + y_j, \qquad \text{for } j = 1, 2, \ldots, n,$$

where the d_{jk} are specified constants (where $d_{jk} = 0$ whenever the value taken on by b_k is not known before a value must be assigned to x_j), and where the y_j are decision variables.[1] (These equations are often written in matrix form as $\mathbf{x} = \mathbf{Db} + \mathbf{y}$.) The proper choice of the d_{jk} depends very much on the nature of the individual problem (if indeed it can be formulated reasonably in this way). An example is given later which illustrates how the d_{jk} are chosen.

Given the d_{jk}, it is only necessary to solve for the y_j. Then, when the time comes to assign a value to x_j, this value is obtained from the above equation. The details on how to solve for the y_j are given below.

The first step is to substitute

$$\left(\sum_{k=1}^{m} d_{jk}b_k + y_j \right) \qquad \text{for } x_j \qquad (\text{for } j = 1, 2, \ldots, n)$$

throughout the original chance-constrained programming model. The objective function becomes

$$E(Z) = E\left[\sum_{j=1}^{n} c_j \left(\sum_{k=1}^{n} d_{jk}b_k + y_j \right) \right]$$

$$= \sum_{j=1}^{n} \sum_{k=1}^{m} d_{jk}E(c_j)E(b_k) + \sum_{j=1}^{n} E(c_j)y_j.$$

Since

$$\sum_{j=1}^{n} \sum_{k=1}^{m} d_{jk}E(c_j)E(b_k)$$

is a constant, it can be dropped from the objective function, so that the new objective becomes

$$\text{Maximize} \qquad \sum_{j=1}^{n} E(c_j)y_j.$$

Since

$$\sum_{j=1}^{n} a_{ij}x_j = \sum_{j=1}^{n} a_{ij} \left(\sum_{k=1}^{m} d_{jk}b_k + y_j \right)$$

$$= \sum_{j=1}^{n} \sum_{k=1}^{m} a_{ij}d_{jk}b_k + \sum_{j=1}^{n} a_{ij}y_j,$$

[1] Another common type of linear decision rule in chance-constrained programming is to let

$$x_j = \sum_{k=1}^{m} b_k d_{jk}, \qquad \text{for } j = 1, 2, \ldots, n,$$

where d_{jk} is a *decision variable* if b_k becomes known before a value must be assigned to x_j and is zero otherwise. This case is considered in Problem 49.

the constraints,

$$P\left\{\sum_{j=1}^{n} a_{ij}x_j \leq b_i\right\} \geq \alpha_i, \qquad \text{for } j = 1, 2, \ldots, m,$$

become $\quad P\left\{\sum_{j=1}^{n} a_{ij}y_j \leq b_i - \sum_{j=1}^{n}\sum_{k=1}^{m} a_{ij}d_{jk}b_k\right\} \geq \alpha_i, \qquad \text{for } j = 1, 2, \ldots, m.$

The next step is to reduce these constraints to linear programming constraints. This is done just as before since the fundamental nature of the constraints has not been changed. Because

$$\left(b_i - \sum_{j=1}^{n}\sum_{k=1}^{m} a_{ij}d_{jk}b_k\right)$$

is a linear function of normal random variables, it must also be a normally distributed random variable. Let μ_i and σ_i denote the mean and standard deviation, respectively, of

$$\left(b_i - \sum_{j=1}^{n}\sum_{k=1}^{m} a_{ij}d_{jk}b_k\right).$$

Thus, $\qquad \mu_i = E(b_i) - \sum_{j=1}^{n}\sum_{k=1}^{m} a_{ij}d_{jk}E(b_k),$

and, if the b_k are mutually independent,

$$\sigma_i^2 = \sum_{\substack{k=1 \\ k \neq i}}^{m} \left[\sum_{j=1}^{n} a_{ij}d_{jk}\right]^2 \sigma_{b_k}^2 + \left[1 - \sum_{j=1}^{n} a_{ij}d_{ji}\right]^2 \sigma_{b_i}^2.$$

(Lacking independence, covariance terms would be included.) It then follows as before that these constraints are equivalent to the linear programming constraints,

$$\sum_{j=1}^{n} a_{ij}y_j \leq \mu_i + K_{\alpha_i}\sigma_i, \qquad \text{for } j = 1, 2, \ldots, m.$$

It usually makes sense for the individual problem to add the restriction that

$$y_j \geq 0, \qquad \text{for } j = 1, 2, \ldots, n.$$

The model consisting of the new objective function and these constraints can then be solved by the simplex method.

To illustrate the way in which linear decision rules may arise, consider the problem of scheduling the production output for a given product over the next n time periods. Let x_j ($j = 1, 2, \ldots, n$) be the total number of units produced in time periods 1 through j, so that $(x_j - x_{j-1})$ is the output in period j. Thus, the x_j are the decision variables. Let S_j ($j = 1, 2, \ldots, n$) be the total number of units sold in time periods 1 through j. Assuming sales cannot be predicted exactly in advance, the S_j are random variables such that the value taken on by S_j becomes known at the end of period j. Assume that the S_j are normally distributed.

Suppose that the firm's management places a high priority on not alienating customers by a late delivery of their purchases. Hence, assuming no initial inventory, the x_j should be chosen such that it is almost certain that $x_j \geq S_j$. Therefore, one set of constraints that should be included in the mathematical model is

$$P\{x_j \geq S_j\} \geq \alpha_j, \qquad \text{for } j = 1, 2, \ldots, n,$$

where the α_j are selected numbers close to one.

However, rather than solving for the x_j directly at the outset, the problem should be solved in such a way that the information on cumulative sales can be used as it becomes available. Suppose that the final decision on x_j need not be made until the beginning of period j. It would be highly desirable to take into account the value taken on by S_{j-1} before assigning a value to x_j. Therefore, let

$$x_j = S_{j-1} + y_j, \qquad \text{for } j = 1, 2, \ldots, n \text{ (where } S_0 = 0),$$

and then solve only for the y_j at the outset.

To express this example in the notation used earlier, the constraints should be written as

$$P\{-x_i \leq -S_i\} \geq \alpha_i, \qquad \text{for } i = 1, 2, \ldots, m(m = n),$$

so that $b_i = -S_i$. Hence,

$$x_j = \sum_{k=1}^{m} d_{jk} b_k + y_j = -b_{j-1} + y_j,$$

so that $d_{j(j-1)} = -1$ and $d_{jk} = 0$ for $k \neq j - 1$. Since y_j is just the number of units of the product that is available for immediate delivery in period j, it is natural to impose the additional restriction that $y_j \geq 0$ for $j = 1, 2, \ldots, n$. Therefore, assuming that the remainder of the model also fits the linear programming format, this particular problem can be formulated and solved by the general procedure described in this section.

7.10 Conclusions

The linear programming model encompasses a wide variety of specific types of problems. The general simplex method is a powerful algorithm that can solve surprisingly large versions of any of these problems. However, some of these problem types have such simple formulations that they can be solved much more efficiently by *streamlined* versions of the simplex method that exploit their *special structure*. These streamlined versions can cut down tremendously on the computer time required for large problems, and they sometimes make it computationally feasible to solve huge problems. This is particularly true for *transportation* and *transshipment problems, assignment problems,* and problems with many *upper bound* or *GUB constraints*. For general *multidivisional problems, multitime period problems,* or *combinations* of the two, the setup times are sufficiently large for their streamlined procedures that they should be used selectively only on large problems.

We shall reexamine the special structure of the transportation, transshipment, and assignment problems once again in Sec. 10.6. There we shall see that these problems are special cases of an important class of linear programming problems known as the *minimum cost flow problem*. This problem has the interpretation of minimizing the cost for the flow of goods through a network. This network interpretation will add further insight into the structure of these three problems introduced in this chapter.

Stochastic programming and *chance-constrained programming* provide useful ways of dealing with linear programming problems where the uncertainty assumption is so badly violated that some or all of the model parameters must be treated explicitly as random variables.

Much research continues to be devoted to developing streamlined solution procedures for special types of linear programming problems, including some not discussed here. At the same time there is widespread interest in applying linear programming to optimize the operation of complicated large-scale systems, including social systems. The resulting formulations usually have special structures that can be exploited. Recognizing and exploiting special structures has become a very important factor in the successful application of linear programming.

We shall turn our attention in the next chapter to some other important considerations in applying linear programming.

SELECTED REFERENCES

1. Bazaraa, Mokhtar S., and John J. Jarvis: *Linear Programming and Network Flows,* Wiley, New York, 1977.

2. Bradley, Stephen P., Arnoldo C. Hax, and Thomas L. Magnanti: *Applied Mathematical Programming,* chap. 12, Addison-Wesley, Reading, Mass., 1977.

3. Dantzig, George B.: *Linear Programming and Extensions,* chaps. 14–23, Princeton University Press, Princeton, N.J., 1963.

4. Driebeek, Norman J.: *Applied Linear Programming,* Addison-Wesley, Reading, Mass., 1969.

5. Gass, Saul I.: *Linear Programming: Methods and Applications,* 5th ed., McGraw-Hill, New York, 1985.

6. Geoffrion, Arthur M.: "Elements of Large-Scale Mathematical Programming," *Management Science,* **16**:652–691, 1970.

7. Lasdon, Leon S.: *Optimization Theory for Large Systems,* Macmillan, New York, 1970.

8. Murty, Katta: *Linear Programming,* 2d ed., Wiley, New York, 1983.

9. Spivey, W. Allen, and Robert M. Thrall: *Linear Optimization,* chaps. 6, 7, 10, Holt, Rinehart & Winston, New York, 1970.

PROBLEMS

1.* A company has three plants producing a certain product that is to be shipped to four distribution centers. Plants 1, 2, and 3 produce 12, 17, and 11 shipments per month, respectively. Each distribution center needs to receive 10 shipments per month. The distance from each plant to the respective distributing centers is given in miles as follows:

		Distribution Center			
		1	2	3	4
Plant	1	800	1,300	400	700
	2	1,100	1,400	600	1,000
	3	600	1,200	800	900

The freight cost for each shipment is $100 plus 50 cents/mile.

How much should be shipped from each plant to each of the distribution centers to minimize the total shipping cost?

(a) Formulate this problem as a *transportation problem* by constructing the appropriate cost and requirements table.

(b) Use the *northwest corner rule* to obtain an initial basic feasible solution.

(c) Starting with the initial basic feasible solution from part (b), use the *transportation simplex method* to obtain an optimal solution.

2. Tom would like 3 pints of home brew today and an additional 4 pints of home brew tomorrow. Dick is willing to sell a maximum of 5 pints total at a price of $3.00/pint today and $2.70/pint tomorrow. Harry is willing to sell a maximum of 4 pints total at a price of $2.90/pint today and $2.80/pint tomorrow.

Tom wishes to know what his purchases should be to minimize his cost while satisfying his thirst requirements.

(a) Formulate the *linear programming* model for this problem, and construct the initial simplex tableau (see Chaps. 3 and 4).

(b) Formulate this problem as a *transportation problem* by constructing the appropriate cost and requirements table.

(c) Starting with the *northwest corner rule,* use the *transportation simplex method* to solve the problem as formulated in part (b).

3. A contractor has to haul gravel to three building sites. He can purchase as much as 18 tons at a gravel pit in the north of the city and 14 tons at one in the south. He needs 10, 5, and 10 tons at sites 1, 2, and 3, respectively. The purchase price per ton at each gravel pit and the hauling cost per ton are given in the table below.

Site Pit	Hauling Cost Per Ton			Price Per Ton
	1	2	3	
North	3	6	5	10
South	6	3	4	12

The contractor wishes to determine how much to haul from each pit to each site in order to minimize the total cost for purchasing and hauling gravel.

(a) Formulate a *linear programming* model for this problem. Using the Big M method, construct the initial simplex tableau ready to apply the simplex method (but do not actually solve).

(b) Now formulate this problem as a *transportation problem* by constructing the appropriate cost and requirements table. Compare the size of this table (and the corresponding transportation simplex tableaux) used by the transportation simplex method with the size of the simplex tableaux from part (a) that would be needed by the simplex method.

(c) The contractor notices that he can supply sites 1 and 2 completely from the north pit and site 3 completely from the south pit. Use the *optimality test* (but no iterations) of the transportation simplex method to check whether the corresponding basic feasible solution is optimal.

(d) Starting with the *northwest corner rule,* use the *transportation simplex method* to solve the problem as formulated in part (b).

(e) As usual, let c_{ij} denote the unit cost associated with source i and destination j as given in the cost and requirements table constructed in part (b). For the optimal solution obtained in part (d), suppose that the value of c_{ij} for each basic variable x_{ij} is fixed at the value given in the cost and requirements table, but that the value of c_{ij} for each nonbasic variable x_{ij} possibly can be altered through bargaining because the site manager wants to pick up the business. Use sensitivity analysis to determine the *allowable range* of values for each of the latter c_{ij} independently such that this optimal solution will remain optimal.

4. A corporation has decided to produce three new products. Five branch plants now have excess product capacity. The unit manufacturing cost of the first product would be $31, $29, $32, $28, and $29, in Plants 1, 2, 3, 4, and 5, respectively. The unit manufacturing cost of the second product would be $45, $41, $46, $42, and $43 in Plants 1, 2, 3, 4, and 5, respectively. The unit manufacturing cost of the third product would be $38, $35, and $40 in Plants 1, 2, and 3, respectively, whereas Plants 4 and 5 do not have the capability for producing this product. Sales forecasts indicate that 300, 500, and 400 units of products 1, 2, and 3, respectively, should be produced per day. Plants 1, 2, 3, 4, and 5 have the capacity to produce 200, 300, 200, 300, and 500 units daily, respectively, regardless of the product or combinations of products involved. Assume that any plant having the capability and capacity to produce them can produce any combination of the products in any quantity.

Management wishes to know how to allocate the new products to the plants to minimize total manufacturing cost.

(a) Formulate this problem as a *transportation problem* by constructing the appropriate cost and requirements table.

(b) Starting with *Vogel's approximation method,* use the *transportation simplex method* to solve the problem as formulated in part (a).

5.* Suppose that England, France, and Spain produce all the wheat, barley, and oats in the world. The world demand for wheat requires 125 million acres of land devoted to wheat production. Similarly, 60 million acres of land are required for barley and 75 million acres of land for oats. The total amount of land available for these purposes in England, France, and Spain is 70 million acres, 110 million acres, and 80 million acres, respectively. The number of hours of labor needed in England, France, and Spain to produce an acre of wheat is 18 hours, 13 hours, and 16 hours, respectively. The number of hours of labor needed in England, France, and Spain to produce an acre of barley is 15 hours, 12 hours, and 12 hours, respectively. The number of hours of labor needed in England, France, and Spain to produce an acre of oats is 12 hours, 10 hours, and 16 hours, respectively. The labor cost per hour in producing wheat is $3.00, $2.40, and $3.30 in England, France, and Spain, respectively. The labor cost per hour in producing barley is $2.70, $3.00, and $2.80 in England, France, and Spain, respectively. The labor cost per hour in producing oats is $2.30, $2.50, and $2.10 in England, France, and Spain, respectively. The problem is to allocate land use in each country so as to meet the world food requirement and minimize the total labor cost.

(a) Formulate this problem as a *transportation problem* by constructing the appropriate cost and requirements table.

(b) Starting with the *northwest corner rule,* use the *transportation simplex method* to solve this problem.

6. A firm producing a single product has three plants and four customers. The three plants will produce 6, 8, and 4 units, respectively, during the next time period. The firm has made a commitment to sell 4 units to customer 1, 6 units to customer 2, and at least 2 units to customer 3. Both customers 3 and 4 also want to buy as many of the remaining units as possible. The net profit associated with shipping a unit from plant i for sale to customer j is given by the following table:

		Customer			
		1	2	3	4
Plant	1	8	7	5	2
	2	5	2	1	3
	3	6	4	3	5

Management wishes to know how many units to sell to customers 3 and 4 and how many units to ship from each of the plants to each of the customers to maximize profit.

(a) Formulate this problem as a *transportation problem* by constructing the appropriate cost and requirements table.

(b) Starting with *Vogel's approximation method*, use the *transportation simplex method* to solve the problem as formulated in part (a).

7. Plans need to be made for the energy systems for a new building. The three possible sources of energy are electricity, natural gas, and a solar heating unit.

Energy needs in the building are for electricity, water heating, and space heating, where the daily requirements (all measured in the same units) are

Electricity	20 units
Water heating	10 units
Space heating	30 units.

The size of the roof limits the solar heater to 30 units, but there is no limit to the electricity and natural gas available. Electricity needs can be met only by purchasing electricity (at a cost of $50 per unit). Both other energy needs can be met by any source or combination of sources. The unit costs are

	Electricity	Natural Gas	Solar Heater
Water heating	$90	$60	$30
Space heating	$80	$50	$40

The objective is to minimize the total cost of meeting the energy needs.

(a) Formulate this problem as a *transportation problem* by constructing the appropriate cost and requirements table.

(b) Use the *northwest corner rule* to obtain an initial basic feasible solution for the problem as formulated in part (a).

(c) Starting with the initial basic feasible solution from part (b), use the *transportation simplex method* to obtain an optimal solution.

(d) Use *Vogel's approximation method* to obtain an initial basic feasible solution for the problem as formulated in part (a).

(e) Starting with the initial basic feasible solution from part (d), use the *transportation simplex method* to obtain an optimal solution. Compare the number of iterations required by the transportation simplex method here and in part (c).

8. A company has two plants producing a certain product that is to be shipped to three distribution centers. The unit production costs are the same at the two plants, and the shipping cost (in hundreds of dollars) per unit of the product is shown for each combination of plant and distribution center as follows:

Distribution Center

		1	2	3
Plant	A	8	7	4
	B	6	8	5

A total of 60 units is to be produced and shipped per week. Each plant can produce and ship any amount up to a maximum of 50 units per week, so there is considerable flexibility on how to divide the total production between the two plants so as to reduce shipping costs.

Management's objective is to determine how much should be produced at each plant, and then what the overall shipping pattern should be in order to minimize total shipping cost.

(a) Assume that each distribution center must receive exactly 20 units per week. Formulate this problem as a *transportation problem* by constructing the appropriate cost and requirements table.

(b) Starting with the *northwest corner rule*, use the *transportation simplex method* to solve the problem as formulated in part (a).

(c) Now assume that any distribution center may receive any quantity between 10 and 30 units per week in order to further reduce total shipping cost, provided only that the total shipped to all three distribution centers must still equal 60 units per week. Formulate this problem as a *transportation problem* by constructing the appropriate cost and requirements table.

(d) Starting with *Vogel's approximation method,* use the *transportation simplex method* to solve the problem as formulated in part (c).

(e) Now assume that distribution centers 1, 2, and 3 must receive exactly 10, 20, and 30 units per week, respectively. For administrative convenience, management has decided that each distribution center will be supplied totally by a single plant, so that one plant will supply one distribution center and the other plant will supply the other two distribution centers. The choice of these assignments of plants to distribution centers is to be made solely on the basis of minimizing total shipping cost. Formulate this problem as an *assignment problem*.

(f) Starting with *Russell's approximation method,* use the *transportation simplex method* to solve the problem as formulated in part (e).

9. Consider the prototype example for the transportation problem (the P & T Co. problem) presented at the beginning of Sec. 7.1. Verify that the solution given there actually is optimal by applying just the *optimality test* portion of the transportation simplex method to this solution.

10. Consider the transportation problem formulation of Option 1 for the Better Products Co. problem presented in Table 7.30. Verify that the optimal solution given in Sec. 7.4 actually is optimal by applying just the *optimality test* portion of the transportation simplex method to this solution.

11. Consider the transportation problem having the following cost and requirements table:

		Destination					
		1	2	3	4	5	Supply
	1	8	6	3	7	5	20
Source	2	5	M	8	4	7	30
	3	6	3	9	6	8	30
	4(D)	0	0	0	0	0	20
Demand		25	25	20	10	20	

After several iterations of the transportation simplex method, the following transportation simplex tableau is obtained:

	Destination					Supply	u_i
	1	2	3	4	5		
1	8	6	3 (20)	7	5	20	
2	5 (25)	M	8	4 (5)	7	30	
Source 3	6	3 (25)	9	6 (5)	8	30	
4(D)	0	0 (0)	0 (0)	0	0 (20)	20	
Demand	25	25	20	10	20		
v_j							

Continue the transportation simplex method for *two more* iterations. After two iterations, state whether the solution is optimal and, if so, why.

12.* Consider the transportation problem having the following cost and requirements table:

		Destination				Supply
		1	2	3	4	
	1	3	7	6	4	5
Source	2	2	4	3	2	2
	3	4	3	8	5	3
Demand		3	3	2	2	

Use each of the following criteria to obtain an initial basic feasible solution. In each case apply the *transportation simplex method,* starting with this initial solution, to obtain an optimal solution. Compare the resulting number of iterations for the transportation simplex method.

(*a*) Northwest corner rule.
(*b*) Vogel's approximation method.
(*c*) Russell's approximation method.

13. Consider the transportation problem having the following cost and requirements table:

		Destination				Supply
		1	2	3	4	
	1	7	4	1	4	1
	2	4	6	7	2	1
Source	3	8	5	4	6	1
	4	6	7	6	3	1
Demand		1	1	1	1	

(a) Notice that this problem has three special characteristics: (1) number of sources = number of destinations, (2) each supply = 1, and (3) each demand = 1. Transportation problems with these characteristics are of a special type called the *assignment problem* (as described in Sec. 7.4). Use the *integer solutions property* to explain why this type of transportation problem can be interpreted as assigning sources to destinations on a one-to-one basis.

(b) How many basic variables are there in every basic feasible solution? How many of these are *degenerate* basic variables ($=0$)?

(c) Use the *northwest corner rule* to obtain an initial basic feasible solution.

(d) Construct an initial basic feasible solution by applying the general procedure for the initialization step of the transportation simplex method. However, rather than using one of the three criteria for step 1 presented in Sec. 7.2, use the following criterion for selecting the next basic variable.

Minimum cost criterion: From among the rows and columns still under consideration, select the variable x_{ij} having the smallest unit cost c_{ij} to be the next basic variable. (Ties may be broken arbitrarily.)

(e) Starting with the initial basic feasible solution from part (c), use the *transportation simplex method* to obtain an optimal solution.

14. Reconsider the transportation problem given in Prob. 13. Starting from a certain initial basic feasible solution, the transportation simplex method yields the following *final* transportation simplex tableau:

		Destination 1	Destination 2	Destination 3	Destination 4	Supply	u_i
Source	1	7	4 (0) 1	1 (1)	4 (0)	1	0
	2	4 (1)	6 4	7 8	2 (0)	1	-2
	3	8 1	5 (1)	4 2	6 1	1	1
	4	6 1	7 4	6 6	3 (1)	1	-1
Demand		1	1	1	1	Z = 13	
v_j		6	4	1	4		

Adapt the sensitivity analysis procedure for general linear programming presented in Secs. 6.6 and 6.7 to *independently* investigate each of the two changes in the *original* model indicated below by deducing the resulting change or changes in the above *final* transportation simplex tableau. Use this approach to determine if the basic feasible solution in this tableau is still optimal.

(a) Change $c_{11} = 7$ to $c_{11} = 5$.

(b) Change $c_{13} = 1$ to $c_{13} = 3$.

15. Consider the transportation problem having the following cost and requirements table:

	Destination				
Source	1	2	3	4	Supply
1	5	6	4	2	10
2	2	M	1	3	20
3	3	4	2	1	20
4	2	1	3	2	10
Demand	20	10	10	20	

(a) Use the *northwest corner rule* to construct an initial basic feasible solution.
(b) Starting with the initial basic solution from part (a), use the *transportation simplex method* to obtain an optimal solution.

16. Consider the transportation problem having the following cost and requirements table:

	Destination			
Source	1	2	3	Supply
1	6	3	5	4
2	4	M	7	3
3	3	4	3	2
Demand	4	2	3	

(a) Use *Vogel's approximation method* to select the *first* basic variable for an initial basic feasible solution.
(b) Use *Russell's approximation method* to select the *first* basic variable for an initial basic feasible solution.
(c) Use the *northwest corner rule* to construct a complete initial basic feasible solution.
(d) Starting with the initial basic feasible solution from part (c), use the *transportation simplex method* to obtain an optimal solution.

17. Consider the transportation problem having the following cost and requirements table:

	Destination					
Source	1	2	3	4	5	Supply
1	2	4	6	5	7	4
2	7	6	3	M	4	6
3	8	7	5	2	5	6
4	0	0	0	0	0	4
Demand	4	4	2	5	5	

Use each of the following criteria to obtain an initial basic feasible solution. Compare the values of the objective function for these solutions.
(a) Northwest corner rule.
(b) Vogel's approximation method.
(c) Russell's approximation method.

(*d*) Use the best of these solutions to initialize the *transportation simplex method* to obtain an optimal solution.

18. Consider the transportation problem having the following cost and requirements table:

		\multicolumn{6}{c}{Destination}	
		1	2	3	4	5	6	Supply
	1	13	10	22	29	18	0	5
	2	14	13	16	21	M	0	6
Source	3	3	0	M	11	6	0	7
	4	18	9	19	23	11	0	4
	5	30	24	34	36	28	0	3
Demand		3	5	4	5	6	2	

Use each of the following criteria to obtain an initial basic feasible solution. Compare the values of the objective function for these solutions.

(*a*) Northwest corner rule.

(*b*) Vogel's approximation method.

(*c*) Russell's approximation method.

(*d*) Use the best of these solutions to initialize the *transportation simplex method* and then obtain the optimal solution.

19.* Use the *transportation simplex method* to solve the *Northern Airplane Co.* production scheduling problem as it is formulated in Table 7.9.

20. Consider the Northern Airplane Co. production scheduling problem presented in Sec. 7.1 (see Table 7.7). Formulate this problem as a general *linear programming* problem by letting the decision variables be x_j = number of jet engines to be produced in month j ($j = 1, 2, 3, 4$). Construct the *initial simplex tableau* for this formulation, and then contrast the size (number of rows and columns) of this tableau and the *transportation simplex tableaux* for the transportation problem formulation of the problem (see Table 7.9).

21. The BUILD-EM-FAST COMPANY has agreed to supply its best customer with three widgits during *each* of the next 3 weeks, even though producing them will require some overtime work. The relevant production data are as follows:

Week	Maximum Production, Regular Time	Maximum Production, Overtime	Production Cost Per Unit, Regular Time
1	2	2	$300
2	2	1	$500
3	1	2	$400

The cost per unit produced with overtime for each week is $100 more than for regular time. The cost of storage is $50 per unit for each week it is stored. There is already an inventory of two widgits on hand currently, but the company does not want to retain any widgits in inventory after the 3 weeks.

Management wants to know how many units should be produced in each week in order to maximize profit.

(*a*) Formulate this problem as a *transportation problem* by constructing the appropriate cost and requirements table.

(*b*) Use the *transportation simplex method* to solve this problem.

22. Consider the transportation problem having the following cost and requirements table:

		Destination		Supply
		1	2	
Source	1	8	5	4
	2	6	4	2
Demand		3	3	

(a) Using your choice of a criterion from Sec. 7.2 for obtaining the initial basic feasible solution, solve this problem manually by the *transportation simplex method*. (Keep track of your time.)

(b) Reformulate this problem as a general *linear programming* problem, and then solve it manually by the *simplex method*. [Keep track of how long part (b) takes you, and contrast it with the computation time for part (a).]

23. Consider the general linear programming formulation of the transportation problem (see Table 7.6). Verify the claim in Sec. 7.2 that the set of $(m + n)$ functional constraint equations (m supply constraints and n demand constraints) has one *redundant* equation; i.e., any one equation can be reproduced from a linear combination of the other $(m + n - 1)$ equations.

24.* Suppose that the air freight charge per ton between seven particular locations is given by the following table (except where no direct air freight service is available):

Location	1	2	3	4	5	6	7
1	—	21	50	62	93	77	—
2	21	—	17	54	67	—	48
3	50	17	—	60	98	67	25
4	62	54	60	—	27	—	38
5	93	67	98	27	—	47	42
6	77	—	67	—	47	—	35
7	—	48	25	38	42	35	—

A certain corporation must ship a certain perishable commodity from locations 1–3 to locations 4–7. A total of 70, 80, and 50 tons of this commodity are to be sent from locations 1, 2, and 3, respectively. A total of 30, 60, 50, and 60 tons are to be sent to locations 4, 5, 6, and 7, respectively. Shipments can be sent through intermediate locations at a cost equal to the sum of the costs for each of the legs of the journey. The problem is to determine the shipping plan that minimizes the total freight cost.

(a) Describe how this problem fits into the format of the general *transshipment problem*.

(b) Reformulate this problem as an equivalent *transportation problem* by constructing the appropriate cost and requirements table.

(c) Use *Vogel's approximation method* to obtain an initial basic feasible solution for the problem formulated in part (b). Describe the corresponding shipping pattern.

(d) Use the *transportation simplex method* to obtain an optimal solution for the problem formulated in part (b). Describe the corresponding optimal shipping pattern.

25. Consider the airline company problem described in Prob. 2 at the end of Chap. 10.

(a) Describe how this problem can be fitted into the format of the *transshipment problem*.

(b) Reformulate this problem as an equivalent *transportation problem* by constructing the appropriate cost and requirements table.

(c) Use *Vogel's approximation method* to obtain an initial basic feasible solution for the problem formulated in part (b).

(d) Use the *transportation simplex method* to obtain an optimal solution for the problem formulated in part (b).

26. A student about to enter college away from home has decided that she will need an automobile during the next 4 years. But since funds are going to be very limited, she wants to do this in the cheapest possible way. However, considering both the initial purchase price and the operating and maintenance costs, it is not clear whether she should purchase a very old car or just a moderately old car. Furthermore, it is not clear whether she should plan to trade in her car at least once during the 4 years, before the costs become too high.

The relevant data *each* time she purchases a car are

	Purchase Price	Operating and Maintenance Costs for Ownership Year				Trade-in Value at End of Ownership Year			
		1	2	3	4	1	2	3	4
Very old car	$1,200	$1,900	$2,200	$2,500	$2,800	$700	$500	$400	$300
Moderately old car	$4,500	$1,000	$1,300	$1,700	$2,300	$2,500	$1,800	$1,300	$1,000

If the student trades in a car during the next 4 years, she would do it at the end of a year (during the summer) on another car of one of these two kinds. She definitely plans to trade in her car at the end of the 4 years on a much newer model. However, she needs to determine which plan for purchasing and (perhaps) trading in cars during the 4 years would minimize the *total* net cost for the 4 years.

(a) Describe how this problem can be fitted into the format of the *transshipment problem.*

(b) Reformulate this problem as an equivalent *transportation problem* by constructing the appropriate cost and requirements table.

(c) Use *Russell's approximation method* to obtain an initial basic feasible solution for the problem as formulated in part (b).

(d) Use the *transportation simplex method* to obtain an optimal solution for the problem formulated in part (b).

27. Without using x_{ii} variables to introduce fictional shipments from a location to itself, formulate the *linear programming* model for the general *transshipment problem* described at the end of Sec. 7.3. Identify the special structure of this model by constructing its *table of constraint coefficients* (similar to Table 7.6) that shows the location and values of the nonzero coefficients.

28. Four cargo ships will be used for shipping goods from one port to four other ports (labeled 1, 2, 3, 4). Any ship can be used for making any one of these four trips. However, because of differences in the ships and cargoes, the total cost of loading, transporting, and unloading the goods for the different ship–port combinations varies considerably, as shown in the following table:

		Port			
		1	2	3	4
Ship	1	5	4	6	7
	2	6	6	7	5
	3	7	5	7	6
	4	5	4	6	6

The objective is to assign the ships to ports on a one-to-one basis in such a way as to minimize the total cost for all four shipments.

   (a) Describe how this problem fits into the general format for the *assignment problem*.

   (b) Reformulate this problem as an equivalent *transportation problem* by constructing the appropriate cost and requirements table.

   (c) Use the *northwest corner rule* to obtain an initial basic feasible solution for the problem as formulated in part (b).

   (d) Starting with the initial basic feasible solution from part (c), use the *transportation simplex method* to obtain an optimal set of assignments for the original problem.

   (e) Are there other optimal solutions in addition to the one obtained in part (d)? If so, use the *transportation simplex method* to identify them.

   (f) Use the *assignment problem algorithm* presented in Sec. 7.4 to solve this problem.

29. Reconsider Prob. 4. Suppose that the sales forecasts have been revised downward to 240, 400, and 320 units per day of products 1, 2, and 3, respectively. Thus each plant now has the capacity to produce all that is required of any one product. Therefore, management has decided that each new product should be assigned to only one plant and that no plant should be assigned more than one product (so that three plants are each to be assigned one product, and two plants are to be assigned none). The objective is to make these assignments so as to minimize the *total* cost of producing these amounts of the three products.

   (a) Formulate this problem as an *assignment problem* by constructing the appropriate cost table.

   (b) Reformulate this assignment problem as an equivalent *transportation problem* by constructing the appropriate cost and requirements table.

   (c) Starting with *Vogel's approximation method*, use the *transportation simplex method* to solve the problem as formulated in part (b).

   (d) Use the *assignment problem algorithm* presented in Sec. 7.4 to solve this problem.

30.* The coach of a certain swim team needs to assign swimmers to a 200-yard medley relay team to send to the Junior Olympics. Since most of his best swimmers are very fast in more than one stroke, it is not clear which swimmer should be assigned to each of the four strokes. The five fastest swimmers and the best times (in seconds) they have achieved in each of the strokes (for 50 yards) are

Stroke	Carl	Chris	David	Tony	Ken
Backstroke	37.7	32.9	33.8	37.0	35.4
Breaststroke	43.4	33.1	42.2	34.7	41.8
Butterfly	33.3	28.5	38.9	30.4	33.6
Freestyle	29.2	26.4	29.6	28.5	31.1

The coach wishes to determine how to assign four swimmers to the four different strokes to minimize the sum of the corresponding best times.

   (a) Formulate this problem as an *assignment problem*.

   (b) Reformulate this assignment problem as an equivalent *transportation problem* by constructing the appropriate cost and requirements table.

   (c) Starting with *Vogel's approximation method*, use the *transportation simplex method* to solve the problem as formulated in part (b).

   (d) Use the *assignment problem algorithm* presented in Sec. 7.4 to solve this problem.

31. Consider the assignment problem formulation of Option 2 for the Better Products Co. problem presented in Table 7.31.

   (a) Reformulate this problem as an equivalent *transportation problem* with *three* sources and five destinations by constructing the appropriate cost and requirements table.

   (b) Convert the optimal solution given in Sec. 7.4 for this assignment problem into a complete basic feasible solution (including degenerate basic variables) for the transportation problem formulated in part (a). Specifically, apply the "General Procedure for Constructing an Initial Basic Feasible Solution" given in Sec. 7.2. For each iteration of the procedure, rather than using any of the three alternative criteria presented for step 1, select the next basic variable to correspond to the next assignment of a plant to a product given in the optimal solution. When only one row or only one column remains under consideration, use step 4 to select the remaining basic variables.

(c) Verify that the optimal solution given in Sec. 7.4 for this assignment problem actually is optimal by applying just the *optimality test* portion of the transportation simplex method to the complete basic feasible solution obtained in part (b).

(d) Use the *northwest corner rule* to obtain an initial basic feasible solution for the problem as formulated in part (a).

(e) Starting with the initial basic feasible solution from part (d), use the *transportation simplex method* to obtain an optimal solution for the problem as formulated in part (a). Compare this optimal basic feasible solution with the one obtained in part (b).

(f) Now reformulate this assignment problem as an equivalent *transportation problem* with *five* sources and five destinations by constructing the appropriate cost and requirements table. Compare this transportation problem with the one formulated in part (a).

(g) Repeat part (b) for the problem as formulated in part (f). Compare the basic feasible solution obtained with the one from part (c).

32. Starting with *Vogel's approximation method,* use the *transportation simplex method* to solve the Job Shop Co. assignment problem as formulated in Table 7.28b. (As stated in Sec. 7.4, the resulting optimal solution has basic variables $x_{13} = 0$, $x_{14} = 1$, $x_{23} = 1$, $x_{31} = 1$, $x_{41} = 0$, $x_{42} = 1$, $x_{43} = 0$.)

33. Reconsider Prob. 3. Now suppose that trucks (and their drivers) need to be hired to do the hauling, where each truck can only be used to haul gravel from a single pit to a single site. Each truck can haul 5 tons (and costs five times the hauling cost per ton given earlier). Only full trucks would be used to supply each site.

(a) Formulate this problem as an *assignment problem* by constructing the appropriate cost table, including identifying the assignees and assignments.

(b) Reformulate this assignment problem as an equivalent *transportation problem* with *two* sources and *three* destinations by constructing the appropriate cost and requirements table.

(c) Starting with the *northwest corner rule,* use the *transportation simplex method* to solve the problem as formulated in part (b).

(d) Use the *assignment problem algorithm* presented in Sec. 7.4 to solve this problem.

34. Consider the transportation problem formulation and solution of the Metro Water District problem presented in Secs. 7.1 and 7.2 (see Tables 7.12 and 7.23). Adapt the sensitivity analysis procedure presented in Sec. 6.6 to conduct sensitivity analysis on this problem by *independently* investigating each of the following four changes in the original model. For each change, revise the final transportation simplex tableau as needed for identifying and evaluating the current basic solution. Then test this solution for feasibility and for optimality. (Do not reoptimize.)

(a) Change c_{34} from 23 to $c_{34} = 20$.

(b) Change c_{23} from 13 to $c_{23} = 16$.

(c) Decrease the supply from source 2 to 50 and decrease the demand at destination 5 to 50.

(d) Increase the supply at source 2 to 80 and increase the demand at destination 2 to 40.

35. Consider the assignment problem having the following cost table:

		Job		
		1	2	3
Person	1	M	8	7
	2	7	6	4
	3(D)	0	0	0

(a) Reformulate this problem as an equivalent *transportation problem* by constructing the appropriate cost and requirements table.

(b) Use *Vogel's approximation method* to obtain an initial basic feasible solution for the problem as formulated in part (a).

(c) Starting with the initial basic feasible solution from part (b), use the *transportation simplex method* to obtain an optimal solution for the problem as formulated in part (a).

(d) Use the *assignment problem algorithm* presented in Sec. 7.4 to solve this problem.

36. Consider the assignment problem having the following cost table:

		Assignment			
		1	2	3	4
Assignee	A	4	1	0	1
	B	1	3	4	0
	C	3	2	1	3
	D	2	2	3	0

(a) Reformulate this problem as an equivalent *transportation problem* by constructing the appropriate cost and requirements table.

(b) Use the *northwest corner rule* to obtain an initial basic feasible solution for the problem as formulated in part (a).

(c) Starting with the initial basic feasible solution from part (b), use the *transportation simplex method* to obtain an optimal solution for the problem as formulated in part (a).

(d) Use the *assignment problem algorithm* presented in Sec. 7.4 to solve this problem.

37. Consider the assignment problem having the following cost table:

		Assignment			
		1	2	3	4
Assignee	A	4	6	5	5
	B	7	4	5	6
	C	4	7	6	4
	D	5	3	4	7

(a) Reformulate this problem as an equivalent *transportation problem* by constructing the appropriate cost and requirements table.

(b) Use the *northwest corner rule* to obtain an initial basic feasible solution for the problem as formulated in part (a).

(c) Starting with the initial basic feasible solution from part (b), use the *transportation simplex method* to obtain an optimal solution for the problem formulated in part (a).

(d) Use the *assignment problem algorithm* presented in Sec. 7.4 to solve this problem.

38. Consider the *linear programming* model for the general *assignment problem* given in Sec. 7.4. Construct the *table of constraint coefficients* for this model. Compare this table with the one for the general *transportation problem* (Table 7.6). In what ways does the general assignment problem have more *special structure* than the general transportation problem?

39.* Describe how the Wyndor Glass Co. problem formulated in Sec. 3.1 can be interpreted as a *multidivisional* linear programming problem. Identify the variables and constraints for the *master problem* and each *subproblem*.

40. Consider the following linear programming problem.

$$\text{Maximize} \quad Z = 2x_1 + 4x_2 + 3x_3 + 2x_4 - 5x_5 + 3x_6,$$

subject to

$$3x_1 + 2x_2 + 3x_3 \leq 30$$

$$2x_5 - x_6 \leq 20$$

$$5x_1 - 2x_2 + 3x_3 + 4x_4 + 2x_5 + x_6 \leq 20$$

$$3 \leq x_4 \leq 15$$

$$2x_5 + 3x_6 \leq 40$$

$$5x_1 - x_3 \leq 30$$

$$2x_1 + 4x_2 + 2x_4 + 3x_6 \leq 60$$

$$-x_1 + 2x_2 + x_3 \geq 20$$

and

$$x_j \geq 0, \quad \text{for} \quad j = 1, 2, \ldots, 6.$$

(a) Rewrite this problem in a form that demonstrates that it possesses the special structure for *multidivisional problems*. Identify the variables and constraints for the *master problem* and each *subproblem*.
(b) Construct the corresponding *table of constraint coefficients* having the *block angular structure* shown in Table 7.32. (Include only nonzero coefficients, and draw a box around each block of these coefficients to emphasize this structure.)

41. Consider the following *table of constraint coefficients* for a linear programming problem:

Coefficient of

Constraint	x_1	x_2	x_3	x_4	x_5	x_6	x_7
1		1		1		1	
2			1				
3	4	3	-2	2	4		1
4			2			4	
5	1			1			
6		5	3		1	-2	4
7						1	
8		2			1		3
9	2			4			

(a) Show how this table can be converted into the *block angular structure for multidivisional* linear programming as shown in Table 7.32 (with three subproblems in this case) by reordering the variables and constraints appropriately.
(b) Identify the *upper bound constraints* and *GUB constraints* for this problem.

42. A corporation has two divisions (the Eastern Division and the Western Division) that operate semiautonomously, with each developing and marketing its own products. However, to coordinate their product lines and to promote efficiency, the divisions compete at the corporate level for investment funds for new *product development projects*. In particular, each division submits its proposals to corporate headquarters in September for new major projects to be undertaken the following year, and available funds are then allocated in such a way as to maximize the estimated total net discounted profits that will eventually result from the projects.

For the upcoming year, each division is proposing three new major projects. Each project can be undertaken at any level, where the estimated net discounted profit would be *proportional* to the level. The relevant data on the projects are summarized as follows:

	Eastern Division Project			Western Division Project		
	1	2	3	1	2	3
Level	x_1	x_2	x_3	x_4	x_5	x_6
Required investment (in millions of dollars)	$16x_1$	$7x_2$	$13x_3$	$8x_4$	$20x_5$	$10x_6$
Net profitability	$7x_1$	$3x_2$	$5x_3$	$4x_4$	$7x_5$	$5x_6$
Facility restriction	$10x_1 + 3x_2 + 7x_3 \leq 50$			$6x_4 + 13x_5 + 9x_6 \leq 45$		
Labor restriction	$4x_1 + 2x_2 + 5x_3 \leq 30$			$3x_4 + 8x_5 + 2x_6 \leq 25$		

A total of $150,000,000 is budgeted for investment in these projects.

(a) Formulate this problem as a *multidivisional* linear programming problem.

(b) Construct the corresponding *table of constraint coefficients* having the *block angular structure* shown in Table 7.32.

43. Consider the following *table of constraint coefficients* for a linear programming problem:

Constraint	x_1	x_2	x_3	x_4	x_5	x_6	x_7	x_8	x_9	x_{10}
1	3	1								
2	1	2	−1							
3				1	5					
4				1	2	−1	−1	−1		
5					1					
6			1			1	1	1	3	2
7								2	−1	1

Show how this table can be converted into the *dual angular* structure for *multitime period* linear programming shown in Table 7.37 (with three time periods in this case) by reordering the variables and constraints appropriately.

44. Consider the *Wyndor Glass Co.* problem described in Sec. 3.1 (see Table 3.1). Suppose that decisions have been made to discontinue additional products in the future and to initiate other new products. Therefore, the capacity available in each of the three plants will be different than shown in Table 3.1 after the first year. Furthermore, the unit profit (exclusive of storage costs) that can be realized from the sale of the two products being analyzed will vary from year to year as market conditions change. Therefore, it may be worthwhile to *store* some of the units produced in 1 year for sale in a later year. The storage costs involved would be approximately $2 per unit per year for either product.

The relevant data for the next 3 years are summarized below.

		Capacity Available in Year		
		1	2	3
	1	4	6	3
Plant	2	12	12	10
	3	18	24	15
Unit profit, product 1		$3	$4	$5
Unit profit, product 2		$5	$4	$8

The capacities used by each product remains the same for each year as shown in Table 3.1. The objective is to determine how much of each product to produce in each year and what portion to store for sale in each subsequent year to maximize the total profit over the 3 years.

(a) Formulate this problem as a *multitime period* linear programming problem.

(b) Construct the corresponding *table of constraint coefficients* having the dual angular structure shown in Table 7.37.

45. Consider the following *table of constraint coefficients* for a linear programming problem:

Constraint	x_1	x_2	x_3	x_4	x_5	x_6	x_7	x_8	x_9	x_{10}
1	2			3				1		
2		1	1				2	2		
3	5	−1	2	−1	−1		−3			4
4						1		−1		
5		−1			2			−2	5	3
6	1			1						
7	2	1		3		2		1	−1	
8		−1	2				1	−1		
9						1			2	1
10		−1		4					1	5

Show how this table can be converted into the form for *multidivisional multitime period* problems shown in Table 7.40 (with two linking constraints, two linking variables, and four subproblems in this case) by reordering the variables and constraints appropriately.

46. Consider the *Woodstock Company* multitime period problem described in Sec. 7.6 (see Table 7.38). Suppose that the company has decided to expand its operations to also buy, store, and sell *plywood* in this warehouse. For the upcoming year, the relevant data for *raw lumber* are still as given in Sec. 7.6. The corresponding price data for plywood are

Season	Purchase Price†	Selling Price†	Maximum Sales‡
Winter	680	705	800
Spring	715	730	1,200
Summer	760	770	1,500
Autumn	740	750	100

†Prices are in dollars per 1,000 board feet.
‡Sales are in 1,000 board feet.

For plywood stored for sale in a later season, the handling cost is $6 per 1,000 board feet, and the storage cost is $18 per 1,000 board feet. The storage capacity of 2 million board feet now applies to the *total* for raw lumber and plywood. Everything should still be sold by the end of autumn.

The objective now is to determine the most profitable schedule for buying and selling raw lumber *and* plywood.

(*a*) Formulate this problem as a *multidivisional multitime period* linear programming problem.

(*b*) Construct the corresponding *table of constraint coefficients* having the form shown in Table 7.40.

47. Consider the following problem.

$$\text{Maximize} \quad Z = 20x_1 + 30x_2 + 25x_3,$$

subject to

$$3x_1 + 2x_2 + x_3 \leq b_1$$

$$2x_1 + 4x_2 + 2x_3 \leq b_2$$

$$x_1 + 3x_2 + 5x_3 \leq b_3$$

and

$$x_j \geq 0, \quad \text{for } j = 1, 2, 3,$$

where b_1, b_2, and b_3 are random variables. Assume that the probability distribution of each of these random variables is such that it can take on any one of three possible values. These values are (29, 30, 31) for b_1, (48, 50, 52) for b_2, and (57, 60, 63) for b_3. In each case, the probability of the middle value is $1/2$, whereas each of the other two values has a probability of $1/4$. The random variables are statistically independent. Suppose that the constraints are required to hold with probability 1.

(*a*) Reformulate this problem as an equivalent ordinary linear programming problem.

(*b*) Suppose that the value taken on by b_1 will be known when a value must be assigned to x_2, and both b_1 and b_2 will be known when x_3 must be specified. Use the stochastic programming approach to formulate an equivalent ordinary linear programming problem that maximizes $E(Z)$ while taking this information into account.

48. Reconsider Problem 47. Suppose, after further analysis, it is decided that b_1, b_2, and b_3 each actually has a normal distribution, with a mean and standard deviation of (30, 1), (50, 2), and (60, 3) respectively. Therefore, a chance-constrained programming approach is to be used instead, where the first, second, and third constraints are required to hold with probability 0.975, 0.95, and 0.90, respectively.

(*a*) Consider the solution, $(x_1, x_2, x_3) = (2\frac{1}{3}, 7\frac{1}{3}, 6\frac{1}{3})$. What are the probabilities that the respective original constraints will be satisfied by this solution? Is this solution feasible? What is the probability that *all* the original constraints will be satisfied by this solution?

(*b*) Reformulate this chance-constrained programming problem as an equivalent ordinary linear programming problem.

(*c*) Suppose that [as in part (b) of Problem 47] the value taken on by b_1 will be known when a value must be assigned to x_2, and both b_1 and b_2 will be known when x_3 must be specified. Use the linear decision rules,

$$x_2 = \tfrac{1}{4}b_1 - y_2,$$

$$x_3 = \tfrac{1}{2}b_1 + \tfrac{1}{2}b_2 - y_3,$$

in order to formulate an equivalent ordinary linear programming problem that maximizes $E(Z)$ while taking this information into account.

49. Consider the chance-constrained programming constraint,

$$P\left\{ \sum_{j=1}^{n} a_{ij}x_j \leq b_i \right\} \geq \alpha_i.$$

(a) Suppose that, in addition to b_i, the a_{ij} also are (independent) random variables whose probability distributions are normal with known mean $E(a_{ij})$ and variance $\mathrm{Var}(a_{ij})$. Convert this constraint into an equivalent deterministic nonlinear constraint.

(b) Suppose that the x_j are expressed as linear decision rules of the form,

$$x_j = \sum_{k=1}^{m} b_k \, d_{jk}, \qquad \text{for } j = 1, 2, \ldots, n,$$

where each d_{jk} is a *decision variable* if the value taken on by b_k will be known when a value must be assigned to x_j, and is zero otherwise. Assume that the b_k are independent random variables with known normal distributions, and that the a_{ij} are constants. Convert this constraint into an equivalent constraint of the form obtained in part (a).

8

Formulating Linear Programming Models, Including Goal Programming

Chapter 3 introduced the general nature of linear programming problems, and Chaps. 4, 5, and 6 described how to solve and analyze them. Then Chap. 7 discussed some particularly important special types of linear programming problems. However, these chapters have presented only a portion of the story. The most successful users of linear programming report that one of the most crucial areas of their work is *building the model*. Many of the most noteworthy applications of linear programming involve problems whose natural formulation does not even resemble a linear programming model. It is only through some relatively sophisticated *formulation techniques* that the problems can be *reformulated* to fit linear programming and its exceptionally powerful solution procedures. To provide you with a more complete perspective about the application of linear programming, this chapter focuses on describing and illustrating some of the most useful formulation techniques.

The first section describes how to deal with variables and linear functions that can be either positive or negative but with different unit costs. This description leads into the key topic of *goal programming* (Sec. 8.2), where the *single objective* that is characteristic of linear programming is *replaced by several goals* toward which we must strive simultaneously. The formulation technique of Sec. 8.1, however, enables us to convert such a problem back into the linear programming format. Section 8.3 deals with a fairly similar problem, where there are *several objective functions* and the one with the *smallest value* is to be maximized. Another formulation technique is introduced to show us how to restore the linear programming format in this case.

All three of these sections also illustrate an additional, widely used formulation technique, namely, the introduction of **auxiliary variables.** In contrast to decision variables, auxiliary variables do not represent the decisions to be made. Instead, auxiliary variables simply are extra variables that are helpful for formulating the model. This technique arises again in Sec. 8.4, which presents some examples of relatively difficult formulations. Section 8.5 then concludes with a case study (school rezoning to achieve racial balance) that pulls together some of the key ideas from this chapter and the preceding ones.

8.1 Variables or Linear Functions with Positive and Negative Components

Variables with Positive and Negative Components

As we discussed at the end of Sec. 4.6, it sometimes is necessary to deal with variables that are allowed to be either positive or negative. When there is no bound on the negative values allowed, each such variable (say, x_j) can be replaced throughout the model by the *difference* of two new *nonnegative* variables (say, x_j^+ and x_j^-), so that

$$x_j = x_j^+ - x_j^-, \qquad \text{where } x_j^+ \geq 0, x_j^- \geq 0.$$

We interpreted x_j^+ as representing the *positive component* of x_j, and x_j^- as its *negative component*. In particular,

$$x_j^+ = \begin{cases} +x_j & \text{if } x_j \geq 0 \\ 0 & \text{if } x_j \leq 0, \end{cases}$$

$$x_j^- = \begin{cases} 0 & \text{if } x_j \geq 0 \\ -x_j & \text{if } x_j \leq 0, \end{cases}$$

for all basic feasible solutions, because such solutions necessarily have the property that *either* $x_j^+ = 0$ or $x_j^- = 0$ (or both). Therefore, when the simplex method is applied to the model after substituting $(x_j^+ - x_j^-)$ for x_j, it never will have both x_j^+ and x_j^- as basic variables at the same time. (We shall continue to use this notation with plus and minus superscripts throughout the chapter to represent the *positive* and *negative components* of *any* quantity, regardless of whether the quantity is the value of a *variable* or a *function*.)

The effect of the choice of value for x_j may be quite different for positive and negative values. For example, suppose that x_j represents the *inventory level* of a particular product. If $x_j > 0$ (so $x_j^+ > 0$ and $x_j^- = 0$), the costs incurred include storage expenses and interest charges on the capital tied up in this inventory. On the other hand, $x_j < 0$ (so $x_j^- > 0$ and $x_j^+ = 0$) means that a shortage of x_j^- has occurred. The costs in this case result from lost sales, both now (if customers won't wait) and in the future (disgruntled customers won't return). Because of this difference between

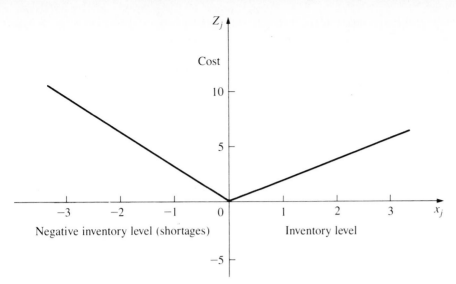

Figure 8.1 Illustration of inventory cost violating the proportionality assumption of linear programming.

the positive and negative cases, the cost of x_j is not simply proportional to x_j, so the proportionality assumption of linear programming (discussed in Sec. 3.3) is violated for this example. The violation of the proportionality assumption is illustrated in Fig. 8.1, where, instead of a single straight line passing through the origin (the proportionality assumption), the unit cost of holding inventory (positive x_j) per unit time is \$2, whereas the unit cost of shortages (negative x_j) per unit time is \$3 (instead of $-\$2$).

Fortunately, as long as the proportionality assumption holds for the positive and negative cases *considered separately*, the objective function can be reformulated in a linear programming format by using x_j^+ and x_j^-. Let

$$Z_j = \text{contribution of } x_j \text{ to the objective function } Z.$$

For appropriate constants, c_j^+ and c_j^-,

$$\text{if} \quad Z_j = \begin{cases} c_j^+ x_j & \text{for } x_j \geq 0, \\ c_j^-(-x_j) & \text{for } x_j \leq 0, \end{cases} \quad \text{then} \quad Z_j = c_j^+ x_j^+ + c_j^- x_j^-.$$

For example, in Fig. 8.1, $c_j^+ = 2$ and $c_j^- = 3$, which yields

$$Z_j = \begin{cases} 2x_j & \text{for } x_j \geq 0, \\ 3(-x_j) & \text{for } x_j \leq 0, \end{cases} \quad \text{so that} \quad Z_j = 2x_j^+ + 3x_j^-.$$

The one restriction on the use of this technique is that c_j^+ and c_j^- must satisfy the following relationship:

$$c_j^+ + c_j^- \geq 0 \qquad \text{when minimizing } Z,$$

$$c_j^+ + c_j^- \leq 0 \qquad \text{when maximizing } Z.$$

[When this relationship does not hold, the $Z_j = c_j^+ x_j^+ + c_j^- x_j^-$ contribution to Z would create an *unbounded* Z in the favorable direction simply by adding the *same* large positive number (unbounded in size) to both x_j^+ and x_j^-. Adding the same number to x_j^+ and x_j^- does not change the value of $x_j = x_j^+ - x_j^-$.]

Because it arises with some frequency in practical applications, an important special case of this technique is where $c_j^+ = c_j^-$ (call this common value c_j), so that Z_j is simply proportional to the absolute value of x_j, $|x_j|$. To satisfy the preceding restriction on c_j^+ and c_j^-, assume that $c_j \geq 0$ when minimizing Z or $c_j \leq 0$ when maximizing Z. Note that

$$|x_j| = x_j^+ + x_j^-.$$

Therefore, if $Z_j = c_j|x_j|$, then $Z_j = c_j(x_j^+ + x_j^-)$.

To contrast this case with the one considered at the end of Sec. 4.6 where the proportionality assumption is satisfied,

if $Z_j = c_j x_j$, then $Z_j = c_j(x_j^+ - x_j^-)$.

Linear Functions with Positive and Negative Components

The measure of performance also can behave as illustrated in Fig. 8.1 when the abscissa value is given by a *linear function* instead of a single variable. In fact, the same example of *inventory level* frequently arises naturally in the model as a linear function of the decision variables. You will see this occur in the second example of Sec. 8.4, where the decision variables for each time period j (or t in Sec. 8.4) are the production level P_j and work force level W_j. However, it is necessary to incorporate the inventory level into the model in order to include the inventory costs in the objective function. To lay the groundwork for doing this incorporation, we introduce an **auxiliary variable** x_j (or I_j in Sec. 8.4) for each time period j to represent the inventory level at the *end* of the period; then we express this variable as a linear function of the appropriate decision variables. In this case,

$$x_j = x_{j-1} + P_j - S_j,$$

where S_j is the forecasted *sales level* (a given constant) for time period j.

How is this linear function, $x_{j-1} + P_j - S_j$, incorporated into the model? If we use the notation with $+$ and $-$ superscripts introduced at the beginning of the section, $(x_{j-1} + P_j - S_j)^+$ and $(x_{j-1} + P_j - S_j)^-$ represent the *positive* and *negative components*, respectively, of this function. In other words,

$$(x_{j-1} + P_j - S_j)^+ = \begin{cases} +(x_{j-1} + P_j - S_j) & \text{if } (x_{j-1} + P_j - S_j) \geq 0 \\ 0 & \text{if } (x_{j-1} + P_j - S_j) \leq 0, \end{cases}$$

$$(x_{j-1} + P_j - S_j)^- = \begin{cases} 0 & \text{if } (x_{j-1} + P_j - S_j) \geq 0 \\ -(x_{j-1} + P_j - S_j) & \text{if } (x_{j-1} + P_j - S_j) \leq 0. \end{cases}$$

Now we introduce the additional auxiliary variables, x_j^+ and x_j^-, defined as

$$x_j^+ = (x_{j-1} + P_j - S_j)^+,$$

$$x_j^- = (x_{j-1} + P_j - S_j)^-.$$

If we define c_j^+ and c_j^- as in the preceding subsection (and with the same restriction on their values), the contribution of the inventory cost in period j to the objective function again is

$$Z_j = c_j^+ x_j^+ + c_j^- x_j^-.$$

However, the crucial difference from the preceding subsection is that since the $x_j = x_j^+ - x_j^-$ variables are not the decision variables included in the original model, the *definitions* of x_j^+ and x_j^- must be incorporated *directly* into the linear programming model. It is not enough to simply record the definitions, as we just did, because the simplex method considers only the objective function and constraints that constitute the model. Since

$$x_j = x_{j-1} + P_j - S_j \quad \text{and} \quad x_j = x_j^+ - x_j^- \quad \text{for each } j,$$

x_j^+ and x_j^- can be incorporated directly by adding the *equality constraints,*

$$x_j^+ - x_j^- = x_{j-1}^+ - x_{j-1}^- + P_j - S_j \quad \text{for each } j,$$

to the model. (The variables on the right-hand side of these constraints should be moved to the left-hand side for proper form.) These additional constraints ensure that x_j^+ and x_j^- will take on appropriate values, given the values assigned to the decision variables by the simplex method.

This technique of introducing *auxiliary variables* and then using *equality constraints* to define them in the model is a very common one in a variety of applications.

You will see the above inventory example of this formulation technique worked out in the context of a complete model in Sec. 8.4.

Perhaps the most important application of this technique is to *goal programming,* which is described next.

8.2 Goal Programming

We have assumed throughout the preceding chapters that the objectives of the organization conducting the linear programming study can be encompassed within a single overriding objective, such as maximizing total profit or minimizing total cost. However, this assumption is not always realistic. In fact, as we discussed in Sec. 2.1, studies have found that the management of American corporations frequently focuses on a variety of other objectives—e.g., to maintain stable profits, increase (or maintain) one's share of the market, diversify products, maintain stable prices, improve worker morale, maintain family control of the business, and increase company prestige. *Goal programming* provides a way of striving toward several such objectives *simultaneously.*

The basic approach of **goal programming** is to establish a specific numeric *goal* for each of the objectives, formulate an objective function for each objective, and then seek a solution that minimizes the (weighted) sum of deviations of these objective functions from their respective goals.

There are two cases to be considered. One, called **nonpreemptive goal programming,** is where all of the goals are of *roughly comparable importance.* The other, called **preemptive goal programming,** is where there is a *hierarchy of priority levels* for the goals, so that the goals of *primary* importance receive first-priority attention, those of *secondary* importance receive second-priority attention, and so forth (if there are more than two priority levels).

We begin with an example that illustrates the basic features of *nonpreemptive* goal programming and then discuss the *preemptive* case.

Prototype Example for Nonpreemptive Goal Programming

The DEWRIGHT COMPANY is considering three new products to replace current models that are being discontinued, so their OR Department has been assigned the task of determining which mix of these products should be produced. Management wants primary consideration given to three factors: long-run profit, stability in the work force, and the level of capital investment that would be required now for new equipment. In particular, they have established the goals of (1) achieving a long-run profit (net present value) of at least $125,000,000 from these products, (2) maintaining the current employment level of 4,000 employees, and (3) holding the capital invest-ment to less than $55,000,000. However, management realizes that it probably won't be possible to attain all of these goals simultaneously, so they have discussed their priorities with the OR Department. This discussion has led to setting *penalty weights* of 5 for missing the profit goal (per million dollars under), 2 for going over the employment goal (per hundred employees), 4 for going under this same goal, and 3 for exceeding the capital investment goal (per million dollars over). Each new prod-uct's contribution to profit, employment level, and capital investment level is *pro-portional* to the rate of production. These contributions per unit rate of production are shown in Table 8.1, along with the goals and penalty weights.

FORMULATION: The Dewright Company problem includes all three possible types of goals: a *lower, one-sided goal* (long-run profit); a *two-sided goal* (employment level); and an *upper, one-sided goal* (capital investment). Letting the decision vari-ables x_1, x_2, x_3 be the production rates of products 1, 2, and 3, respectively, these goals can be stated as

$$12x_1 + 9x_2 + 15x_3 \geq 125 \quad \text{(Profit goal)}$$

$$5x_1 + 3x_2 + 4x_3 = 40 \quad \text{(Employment goal)}$$

$$5x_1 + 7x_2 + 8x_3 \leq 55 \quad \text{(Investment goal).}$$

Note that these three relationships are *not* constraints. It is not even expected that all of them can be satisfied simultaneously. The right-hand sides are not fixed constants with no flexibility. Instead, they are *managerial goals* to be approached as closely as possible. More precisely, given the penalty weights in the last column of Table 8.1, the overall objective becomes

$$\text{Minimize} \quad Z = 5(12x_1 + 9x_2 + 15x_3 - 125)^-$$

$$+ 2(5x_1 + 3x_2 + 4x_3 - 40)^+$$

Table 8.1 **Data for Dewright Co. Nonpreemptive Goal Programming Problem**

| | Unit Contribution | | | | | Penalty |
| | **Product** | | | | | |
Factor	1	2	3	Goal	(Units)	Weight
Long-run profit	12	9	15	≥ 125	(millions of dollars)	5
Employment level	5	3	4	$= 40$	(hundreds of employees)	2(+), 4(−)
Capital investment	5	7	8	≤ 55	(millions of dollars)	3

$$+4(\ 5x_1\ +\ 3x_2\ +\ 4x_3\ -\ 40)^-$$

$$+3(\ 5x_1\ +\ 7x_2\ +\ 8x_3\ -\ 55)^+.$$

This expression uses the notation of $+$ and $-$ as superscripts introduced in Sec. 8.1 to represent the *positive* and *negative components* of the function inside the parentheses. For example, consider the following two cases involving the employment level, $5x_1\ +\ 3x_2\ +\ 4x_3$, relative to its goal of 40.

(1)   If $(5x_1\ +\ 3x_2\ +\ 4x_3\ -\ 40) = +10$,

then $(5x_1\ +\ 3x_2\ +\ 4x_3\ -\ 40)^+ = +10$,

$$(5x_1\ +\ 3x_2\ +\ 4x_3\ -\ 40)^- = 0.$$

(2)   If $(5x_1\ +\ 3x_2\ +\ 4x_3\ -\ 40) = -5$,

then $(5x_1\ +\ 3x_2\ +\ 4x_3\ -\ 40)^+ = 0$,

$$(5x_1\ +\ 3x_2\ +\ 4x_3\ -\ 40)^- = +5.$$

Thus, in the first case, the contribution of the second and third terms to Z is $2(10) + 4(0) = 20$, and in the second case it is $2(0) + 4(5) = 20$ again. Because management considers *overshooting* the employment level goal per unit to be *half* as serious as *undershooting* this goal per unit (penalty weights of 2 versus 4), overshooting by 10 provides the same total penalty as undershooting by 5.

Unfortunately, Z is *not* a linear function, because each of the four terms has the *nonlinear* form illustrated in Fig. 8.1 (with a zero slope on one side of the origin), where the value of the abscissa is given by the linear function inside the parentheses. Therefore, the simplex method *cannot* be applied to solve the model in this form. However, it can be applied after the model is *reformulated* to fit the linear programming format. Reformulating requires using the formulation technique presented in the preceding section.

In particular, the first step is to introduce the new *auxiliary variables*,

$$y_1 = 12x_1 + 9x_2 + 15x_3 - 125,$$

$$y_2 = 5x_1 + 3x_2 + 4x_3 - 40,$$

$$y_3 = 5x_1 + 7x_2 + 8x_3 - 55,$$

as well as their *positive* and *negative components*,

$$y_1 = y_1^+ - y_1^-, \quad \text{where} \quad y_1^+ \geq 0, y_1^- \geq 0,$$

$$y_2 = y_2^+ - y_2^-, \quad \text{where} \quad y_2^+ \geq 0, y_2^- \geq 0,$$

$$y_3 = y_3^+ - y_3^-, \quad \text{where} \quad y_3^+ \geq 0, y_3^- \geq 0.$$

Because there is no penalty for *exceeding* the profit goal of 125 or being *under* the investment goal of 55, neither y_1^+ nor y_3^- should appear in the objective function representing the total penalty for deviations from the goals. However, it is possible (and even desirable) to have $y_1^+ > 0$ and $y_3^- > 0$, so both of these variables should appear (along with $y_1^-, y_2^+, y_2^-,$ and y_3^+) in the *equality constraints* that define the relationship between these six auxiliary variables and the three original decision variables (x_1, x_2, x_3). Using the penalty weights shown in Table 8.1 then leads to the following *linear programming* formulation of this goal programming problem:

$$\text{Minimize} \quad Z = 5y_1^- + 2y_2^+ + 4y_2^- + 3y_3^+,$$

subject to

$$12x_1 + 9x_2 + 15x_3 - (y_1^+ - y_1^-) = 125$$

$$5x_1 + 3x_2 + 4x_3 - (y_2^+ - y_2^-) = 40$$

$$5x_1 + 7x_2 + 8x_3 - (y_3^+ - y_3^-) = 55$$

and

$$x_j \geq 0, \quad y_k^+ \geq 0, \quad y_k^- \geq 0 \quad (j = 1, 2, 3; \, k = 1, 2, 3).$$

(If the original problem had any actual linear programming *constraints,* such as constraints on fixed amounts of certain resources being available, these would be included in the model.)

Applying the simplex method to this formulation yields an optimal solution, $x_1 = \frac{25}{3}$, $x_2 = 0$, $x_3 = \frac{5}{3}$, with $y_1^+ = 0$, $y_1^- = 0$, $y_2^+ = \frac{25}{3}$, $y_2^- = 0$, $y_3^+ = 0$, $y_3^- = 0$. Therefore, $y_1 = 0$, $y_2 = \frac{25}{3}$, $y_3 = 0$, so the first and third goals are fully satisfied, but the employment level goal of 40 is exceeded by $8\frac{1}{3}$ (833 employees). The resulting penalty for deviating from the goals is $Z = 16\frac{2}{3}$.

Preemptive Goal Programming

The preceding example assumes that all of the goals are of roughly comparable importance. Now consider the case of *preemptive* goal programming, where there is a hierarchy of priority levels for the goals. Such a case arises when one or more of the goals clearly is far more important than the others. Thus the initial focus should be on achieving as closely as possible these *first-priority* goals. The other goals also might naturally divide further into *second-priority* goals, *third-priority* goals, and so on. After we find an optimal solution with respect to the first-priority goals, we can break any ties for the optimal solution by considering the second-priority goals. Any ties that remain after this reoptimization can be broken by considering the third-priority goals, and so on.

When we deal with goals on the *same* priority level, our approach is just like the one described for *nonpreemptive* goal programming. Any of the same three types of goals (lower one-sided, two-sided, upper one-sided) can arise. Different *penalty weights* for deviations from different goals still can be included, if desired. The formulation technique of Sec. 8.1 again is used to reformulate this portion of the problem to fit the linear programming format.

One way of solving the *overall* problem is to solve a *sequence* of linear programming problems. We shall call this procedure the **sequential procedure.**

At the first stage of the sequential procedure, the only goals included in the linear programming model are the first-priority goals, and the simplex method is applied in the usual way. If the resulting optimal solution is *unique,* we adopt it immediately without considering any additional goals.

However, if there are *multiple* optimal solutions with the same optimal value of Z (call it Z*), we move to the second stage by adding the second-priority goals to the model. If Z* = 0, the *auxiliary variables* representing the *deviations from first-priority goals* now can be completely deleted from the model, where the equality constraints that contain these variables are replaced by the mathematical expressions (inequalities or equations) for these goals to ensure that they continue to be fully achieved. On the other hand, if $Z^* > 0$, the second-stage model simply adds the second-priority goals to the first-stage model (as if these additional goals actually were first-priority goals), but then it also adds the constraint that the *first-stage objective function* must equal Z^* (which enables us again to delete the terms involving first-priority goals from the second-stage objective function). After we apply the simplex method again, we repeat the same process for any lower-priority goals.

It also is possible to duplicate the work of the *sequential procedure* with just *one run* of the simplex method if a slight modification is first made in the algorithm. We shall call this procedure the **streamlined procedure.**

If there are just *two* priority levels, the modification for the streamlined procedure is one you already have seen, namely, the form of the *Big M method* illustrated throughout Sec. 4.6. In this form, instead of replacing M throughout the model by some huge positive number before running the simplex method, we retain the *symbolic* quantity M in the sequence of simplex tableaux. Each coefficient in row 0 (for each iteration) is some linear function, $aM + b$, where a is the current *multiplicative factor* and b is the current *additive factor*. The usual decisions based on these coefficients (entering basic variable and optimality test) now are based solely on the *multiplicative* factors, *except* that any ties would be broken by using the *additive* factors.

The linear programming formulation for the streamlined procedure with two priority levels would include *all* of the goals in the model in the usual manner, but with *basic penalty weights* of M and 1 assigned to deviations from *first-priority* and *second-priority* goals, respectively. If different penalty weights are desired within the same priority level, these basic penalty weights then are multiplied by the individual penalty weights assigned within the level.

When there are more than two priority levels (say, p of them), the *streamlined procedure* generalizes in a straightforward way. The *basic penalty weights* for the respective levels now are $M_1, M_2, \ldots, M_{p-1}, 1$, where M_1 represents a number that is vastly larger than M_2, M_2 is vastly larger than $M_3, \ldots$, and M_{p-1} is vastly larger than 1. Each coefficient in row 0 of each simplex tableau is now a linear function of all of these quantities, where the *multiplicative* factor of M_1 is used to make the necessary decisions, with *tiebreakers* beginning with the *multiplicative* factor of M_2 and ending with the *additive* factor.

We shall now illustrate both the *sequential procedure* and the *streamlined procedure* by modifying the Dewright Company problem.

Example for Preemptive Goal Programming

Faced with the unpleasant recommendation to increase the company's work force by more than 20 percent, the management of the DEWRIGHT COMPANY has reconsidered the original formulation of the problem that was summarized in Table 8.1. This increase in the work force probably would be a rather temporary one, so the very high cost of training 833 new employees would be largely wasted, and the large (undoubtedly well-publicized) layoffs would make it more difficult for the company to attract high-quality employees in the future. Consequently, management has concluded that a very high priority should be placed on avoiding an increase in the work force. Furthermore, management has learned that raising *more than* $55,000,000 for capital investment for the new products would be extremely difficult, so a very high priority also should be placed on avoiding capital investment above this level.

Table 8.2 **Revised Formulation for Dewright Co. Preemptive Goal Programming Problem**

Priority Level	Factor	Goal	Penalty Weight
First priority	Employment level	≤ 40	$2M$
	Capital investment	≤ 55	$3M$
Second priority	Long-run profit	≥ 125	5
	Employment level	≥ 40	4

Based on these considerations, management has concluded that a *preemptive goal programming* approach now should be used, where the two goals just discussed should be the *first-priority* goals, and the other two original goals (exceeding $125,000,000 in long-run profit and avoiding a *decrease* in the employment level) should be the *second-priority* goals. Within the two priority levels, the relative penalty weights still should be the same as given in the last column of Table 8.1. This reformulation is summarized in Table 8.2. (The portions of Table 8.1 that are not included in Table 8.2 are *unchanged*.)

SEQUENTIAL PROCEDURE: At the first stage of the *sequential procedure*, only the two *first-priority* goals are included in the linear programming model. Therefore, we can drop the common factor M for their penalty weights shown in Table 8.2. Proceeding just as for the nonpreemptive model if these were the only goals, the resulting linear programming model is

$$\text{Minimize} \quad Z = 2y_2^+ + 3y_3^+,$$

subject to

$$5x_1 + 3x_2 + 4x_3 - (y_2^+ - y_2^-) = 40$$
$$5x_1 + 7x_2 + 8x_3 - (y_3^+ - y_3^-) = 55$$

and

$$x_j \geq 0, \quad y_k^+ \geq 0, \quad y_k^- \geq 0 \quad (j = 1, 2, 3; k = 2, 3).$$

(For ease of comparison with the *nonpreemptive* model with all four goals, we have kept the same subscripts on the auxiliary variables.)

Using the simplex method (or inspection), an optimal solution for this linear programming model has $y_2^+ = 0$ and $y_3^+ = 0$, with $Z = 0$ (so $Z^* = 0$), because there are innumerable solutions for (x_1, x_2, x_3) that satisfy the relationships,

$$5x_1 + 3x_2 + 4x_3 \leq 40$$

$$5x_1 + 7x_2 + 8x_3 \leq 55,$$

as well as the nonnegativity constraints. Therefore, these two first-priority goals should be used as *constraints* hereafter. Using them as constraints will force y_2^+ and y_3^+ to remain *zero* and thereby disappear from the model automatically.

If we drop y_2^+ and y_3^+ but add the *second-priority* goals, the *second-stage* linear programming model becomes

$$\text{Minimize} \quad Z = 5y_1^- + 4y_2^-,$$

subject to

$$12x_1 + 9x_2 + 15x_3 - (y_1^+ - y_1^-) \qquad\qquad = 125$$
$$5x_1 + 3x_2 + 4x_3 \qquad\qquad + y_2^- \qquad = 40$$
$$5x_1 + 7x_2 + 8x_3 \qquad\qquad\qquad + y_3^- = 55$$

and

$$x_j \geq 0, \quad y_k^+ \geq 0, \quad y_k^- \geq 0 \quad (j = 1, 2, 3; k = 1, 2, 3).$$

Applying the simplex method to this model yields the unique optimal solution, $x_1 = 5$, $x_2 = 0$, $x_3 = 3\frac{3}{4}$, $y_1^+ = 0$, $y_1^- = 8\frac{3}{4}$, $y_2^- = 0$, $y_3^- = 0$, with $Z = 43\frac{3}{4}$.

Because this solution is unique (*or* because there are no more priority levels), the procedure can now stop, with $(x_1, x_2, x_3) = (5, 0, 3\frac{3}{4})$ as the optimal solution for the *overall* problem. This solution fully achieves both *first-priority* goals, as well as one of the *second-priority* goals (no decrease in employment level), and it falls short by just $8\frac{3}{4}$ of the other *second-priority* goal (long-run profit ≥ 125).

STREAMLINED PROCEDURE: Using the *streamlined procedure* instead of the *sequential procedure*, we work with just *one* linear programming model that includes *all* of the goals, as follows:

$$\text{Minimize} \quad Z = 5y_1^- + 2My_2^+ + 4y_2^- + 3My_3^+,$$

subject to

$$12x_1 + 9x_2 + 15x_3 - (y_1^+ - y_1^-) = 125$$

$$5x_1 + 3x_2 + 4x_3 - (y_2^+ - y_2^-) = 40$$

$$5x_1 + 7x_2 + 8x_3 - (y_3^+ - y_3^-) = 55$$

and

$$x_j \geq 0, \quad y_k^+ \geq 0, \quad y_k^- \geq 0 \quad (j = 1, 2, 3; k = 1, 2, 3).$$

Because this model uses M to symbolize a huge positive number, the simplex method should be applied as described and illustrated throughout Sec. 4.6. Doing this naturally yields the same unique optimal solution just obtained by the *sequential procedure*.

8.3 Maximizing the Minimum Progress toward All Objectives

Goal programming is one very useful tool for dealing with problems where several objectives must be considered simultaneously. However, it does require establishing goals for all of the objectives, and it is not always possible to do this in a meaningful way. In particular, some objectives are *open-ended* and one wants to make as much progress toward them as possible. To put it another way, for open-ended objectives there is no minimum standard (goal) such that you would be relatively indifferent about the amount of progress made beyond this standard. (For example, many managers consider the objective of maximizing profit to be of this type.) With open-ended objectives, you may also want to make progress on *all* of the objectives *simultaneously*. In this case, it may be appropriate to *maximize the minimum progress toward all objectives*.

To formulate this approach, suppose that there are K objectives,

$$Z_1 = \sum_{j=1}^{n} c_{j1} x_j \quad \text{(Objective 1)}$$

$$Z_2 = \sum_{j=1}^{n} c_{j2} x_j \quad \text{(Objective 2)}$$

$$\vdots$$

$$Z_K = \sum_{j=1}^{n} c_{jK} x_j \quad \text{(Objective } K\text{).}$$

We wish to increase together the values of all of these individual objective functions. Therefore, the *overall objective function* for the model becomes

$$\text{Maximize} \quad Z = \text{minimum} \{Z_1, Z_2, \ldots, Z_K\},$$

so an optimal solution for $(x_1, x_2, \ldots, x_n)$ is one that makes the *smallest* Z_k ($k = 1, 2, \ldots, K$) as large as possible.

This overall objective function certainly does not fit into a linear programming format. Now let us see how the problem can be *reformulated* into this format.

We begin by introducing an *auxiliary variable* z to represent the minimum value among the K objectives,

$$z = \text{minimum} \{Z_1, Z_2, \ldots, Z_K\}.$$

Introducing this auxiliary variable enables us to write the overall objective function as

$$\text{Maximize} \quad Z = z,$$

which is a legitimate linear programming objective function (one variable with a coefficient of $+1$ and all other coefficients zero).

The remaining question is how to incorporate the *definition* of z directly into a linear programming model. The definition implies that

$$z \le \sum_{j=1}^{n} c_{j1} x_j$$

$$z \le \sum_{j=1}^{n} c_{j2} x_j$$

$$\vdots$$

$$z \le \sum_{j=1}^{n} c_{jK} x_j,$$

where these inequalities are legitimate linear programming *constraints* (after bringing all variables to the left-hand side for proper form). Furthermore, the definition also implies that one or more of these constraints (the one with the *smallest* right-hand side) will hold with *equality*. Therefore, z is simply the *largest* quantity that satisfies all K of these constraints, which condition is already ensured by maximizing $Z = z$.

Consequently, the *equivalent* linear programming model is

$$\text{Maximize} \quad Z = z,$$

$$\text{subject to} \quad \sum_{j=1}^{n} c_{jk} x_j - z \ge 0, \quad \text{for } k = 1, 2, \ldots, K$$

$$x_j \ge 0, \quad \text{for } j = 1, 2, \ldots, n,$$

and   any other linear programming constraints in the original model.

If it is clear that z will turn out to be nonnegative, a nonnegativity constraint can be included in the model for this variable as well.

If the Z_k are not measured in common units, they should be multiplied by the appropriate constants to convert them to a common unit of measurement.

When the objectives are to be minimized rather than maximized, the *overall objective function* for the original model would change to

$$\text{Minimize} \quad Z = \text{maximum } \{Z_1, Z_2, \ldots, Z_K\}.$$

The *equivalent* linear programming model then is

$$\text{Minimize} \quad Z = z,$$

subject to
$$\sum_{j=1}^{n} c_{jk}x_j - z \leq 0, \quad \text{for } k = 1, 2, \ldots, K$$

$$x_j \geq 0, \quad \text{for } j = 1, 2, \ldots, n,$$

and   any other linear programming constraints in the original model.

Prototype Example

An international relief agency, the FOOD AND AGRICULTURE ORGANIZATION, is sending agricultural experts to two underdeveloped countries whose greatest need is to increase their food production by improving their agricultural techniques. Therefore, the experts will be used to develop pilot projects and training programs to demonstrate and teach these techniques. However, the number of such projects that can be undertaken is restricted by the limited availability of three required resources: equipment, experts, and money. The question is how many projects should be undertaken in each of the countries in order to make the best possible use of the resources.

It has been estimated that *each full project* undertaken in country 1 eventually would increase the food production in this country sufficiently to feed 2,000 additional people. The corresponding estimate for country 2 is for an increase that would feed an additional 3,000 people. The two countries differ in the mix of resources needed for projects. These data are summarized in Table 8.3. It is feasible to consider projects at fractional levels as well as whole projects. We assume that fractions of projects will affect the data of Table 8.3 proportionally.

Because both countries are in desperate need, the Food and Agriculture Organization is determined to increase the food production in *both* countries as much as possible. Therefore, it has chosen the overall objective of maximizing the *minimum increase in food production* in the two countries.

Table 8.3 **Data for Food and Agriculture Organization Problem**

Resource	Amount Used Per Project		Amount Available
	Country 1	Country 2	
Equipment	0	5	20
Experts	1	2	10
Money	60	20	300 (thousands of dollars)
People fed	2,000	3,000	

FORMULATION: The decision variables, x_1 and x_2, are the number of projects to be undertaken in countries 1 and 2, respectively. There are *two* objectives in this case— to increase the food production in country 1 and to increase the food production in country 2. Their objective functions are

$$Z_1 = 2{,}000x_1 \qquad \text{(Objective 1)}$$

$$Z_2 = 3{,}000x_2 \qquad \text{(Objective 2).}$$

Therefore, using Table 8.3 to construct the constraints, the overall model is

$$\text{Maximize} \quad Z = \text{minimum } \{Z_1, Z_2\}$$

$$= \text{minimum } \{2{,}000x_1, 3{,}000x_2\},$$

subject to

$$5x_2 \le 20$$

$$x_1 + 2x_2 \le 10$$

$$60x_1 + 20x_2 \le 300$$

and

$$x_1 \ge 0, \qquad x_2 \ge 0.$$

To *reformulate* this model to fit the linear programming format, we introduce an *auxiliary variable z,* defined as

$$z = \text{minimum } \{Z_1, Z_2\} = \text{minimum } \{2{,}000x_1, 3{,}000x_2\},$$

so that z represents the *minimum increase in food production* in the two countries. For example, if $(x_1, x_2) = (4, 2)$, so that $Z_1 = 8{,}000$ and $Z_2 = 6{,}000$, then $z = 6{,}000$ as the minimum of the two quantities. Following the approach described earlier in this section to incorporate the *definition* of z directly into the model, the resulting *equivalent* linear programming model is

$$\text{Maximize} \quad Z = z,$$

subject to

$$2{,}000x_1 \qquad\qquad - z \ge \quad 0$$

$$3{,}000x_2 - z \ge \quad 0$$

$$5x_2 \qquad \le \ \ 20$$

$$x_1 + \quad 2x_2 \qquad \le \ \ 10$$

$$60x_1 + \quad 20x_2 \qquad \le 300$$

and

$$x_1 \ge 0, \qquad x_2 \ge 0, \qquad z \ge 0.$$

Applying the simplex method (which does not differentiate between *decision* variables and *auxiliary* variables) yields the optimal solution,

$$x_1 = \tfrac{45}{11}, \qquad \text{so} \qquad Z_1 = 8{,}182$$

$$x_2 = \tfrac{30}{11}, \qquad \text{so} \qquad Z_2 = 8{,}182$$

$$z = 8{,}182.$$

Consequently, an additional 8,182 people will be fed in *each* of the two countries.

8.4 Some Formulation Examples

We now present two examples that illustrate the kinds of challenging formulation problems that frequently are encountered in real applications of linear programming.

Reclaiming Solid Wastes

The SAVE-IT COMPANY operates a reclamation center that collects four types of solid waste materials and then treats them so that they can be amalgamated into a salable product. Three different grades of this product can be made, depending upon the mix of the materials used. Although there is some flexibility in the mix for each grade, quality standards do specify a minimum or maximum percentage (by weight) of certain materials allowed in that product grade. These specifications are given in Table 8.4 along with the cost of amalgamation and the selling price for each grade.

The reclamation center collects its solid waste materials from some regular sources and so is normally able to maintain a steady production rate for treating these materials. Table 8.5 gives the quantities available for collection and treatment each week, as well as the cost of treatment, for each type of material.

The problem facing the company is to determine just how much of each product grade to produce *and* the exact mix of materials to be used for each grade so as to maximize the total weekly profit (total sales income minus the total costs of *both* amalgamation and treatment).

FORMULATION: Before attempting to construct a linear programming model, we must give careful consideration to the proper definition of the decision variables. Although this definition is often obvious, it sometimes becomes the crux of the entire formulation. After clearly identifying what information is really desired and the most convenient form for conveying this information by means of decision variables, we

Table 8.4 **Product Data for Save-It Co.**

Grade	Specification	Amalgamation Cost ($) Per Pound	Selling Price ($) Per Pound
A	Not more than 30% of material 1 Not less than 40% of material 2 Not more than 50% of material 3	3.00	8.50
B	Not more than 50% of material 1 Not less than 10% of material 2	2.50	7.00
C	Not more than 70% of material 1	2.00	5.50

Table 8.5 **Solid Waste Materials Data for Save-It Co.**

Material	Pounds/Week Available	Treatment Cost ($) Per Pound
1	3,000	3
2	2,000	6
3	4,000	4
4	1,000	5

can develop the objective function and the constraints on the values of these decision variables.

In this particular problem, the decisions to be made are well defined, but the appropriate means of conveying this information may require some thought. (Try it and see if you first obtain the following *inappropriate* choice of decision variables.)

Because one set of decisions concerns the *amount* of each product grade to be produced, it would seem natural to define one set of decision variables accordingly. Proceeding tentatively along this line, define

y_i = number of pounds of product grade i produced per week   ($i = A, B, C$).

The mixture of each grade is identified by the proportion of each material in the product. This identification would suggest defining the other set of decision variables as

z_{ij} = proportion of material j in product grade i ($i = A, B, C; j = 1, 2, 3, 4$).

However, Table 8.5 gives both the treatment cost and the availability of the materials by *quantity* (pounds) rather than *proportion*, so it is this *quantity* information that needs to be recorded in the objective function and in some of the constraints. For material j ($j = 1, 2, 3, 4$),

$$\text{Quantity of material } j \text{ used} = z_{Aj}y_A + z_{Bj}y_B + z_{Cj}y_C.$$

Unfortunately, this expression is *not* a linear function because it involves products of variables. Therefore, a linear programming model cannot be constructed with these decision variables.

Fortunately, there is another way of defining the decision variables that will fit the linear programming format. (Do you see how to do it?) It is accomplished by merely replacing each *product* of the old decision variables by a single variable! In other words, define

$x_{ij} = z_{ij}y_i$ (for $i = A, B, C; j = 1, 2, 3, 4$)

  = number of pounds of material j allocated to product grade i per week,

and then let the x_{ij} be the decision variables. The total amount of product grade i produced per week is then $x_{i1} + x_{i2} + x_{i3} + x_{i4}$. The proportion of material j in product grade i is $x_{ij}/(x_{i1} + x_{i2} + x_{i3} + x_{i4})$. Therefore, this choice of decision variables conveys all the necessary information and proves to be well suited to the construction of the following linear programming model. (Note particularly how the mixture constraints involving the *nonlinear* proportion function are written in a linear form.)

The total profit Z is given by

$$Z = 5.5(x_{A1} + x_{A2} + x_{A3} + x_{A4}) + 4.5(x_{B1} + x_{B2} + x_{B3} + x_{B4})$$

$$+ 3.5(x_{C1} + x_{C2} + x_{C3} + x_{C4}) - 3(x_{A1} + x_{B1} + x_{C1})$$

$$- 6(x_{A2} + x_{B2} + x_{C2}) - 4(x_{A3} + x_{B3} + x_{C3}) - 5(x_{A4} + x_{B4} + x_{C4}).$$

Thus, after combining common terms, the model becomes

Maximize $Z = 2.5x_{A1} - 0.5x_{A2} + 1.5x_{A3} + 0.5x_{A4} + 1.5x_{B1} - 1.5x_{B2} + 0.5x_{B3}$

$$- 0.5x_{B4} + 0.5x_{C1} - 2.5x_{C2} - 0.5x_{C3} - 1.5x_{C4},$$

subject to the following constraints:

1. *Availability of materials:*

$$x_{A1} + x_{B1} + x_{C1} \leq 3{,}000$$

$$x_{A2} + x_{B2} + x_{C2} \leq 2{,}000$$

$$x_{A3} + x_{B3} + x_{C3} \leq 4{,}000$$

$$x_{A4} + x_{B4} + x_{C4} \leq 1{,}000.$$

2. *Mixture specifications:*

$$x_{A1} \leq 0.3(x_{A1} + x_{A2} + x_{A3} + x_{A4})$$

$$x_{A2} \geq 0.4(x_{A1} + x_{A2} + x_{A3} + x_{A4})$$

$$x_{A3} \leq 0.5(x_{A1} + x_{A2} + x_{A3} + x_{A4}).$$

$$x_{B1} \leq 0.5(x_{B1} + x_{B2} + x_{B3} + x_{B4})$$

$$x_{B2} \geq 0.1(x_{B1} + x_{B2} + x_{B3} + x_{B4}).$$

$$x_{C1} \leq 0.7(x_{C1} + x_{C2} + x_{C3} + x_{C4}).$$

and

3. *Nonnegativity:*

$$x_{ij} \geq 0, \qquad \text{for } i = A, B, C; j = 1, 2, 3, 4.$$

This formulation completes the model, except that the constraints for the mixture specifications need to be rewritten in the proper form for a linear programming model by bringing all variables to the left-hand side and combining terms, as follows:

2. *Mixture specifications:*

$$0.7x_{A1} - 0.3x_{A2} - 0.3x_{A3} - 0.3x_{A4} \leq 0$$

$$-0.4x_{A1} + 0.6x_{A2} - 0.4x_{A3} - 0.4x_{A4} \geq 0$$

$$-0.5x_{A1} - 0.5x_{A2} + 0.5x_{A3} - 0.5x_{A4} \leq 0.$$

$$0.5x_{B1} - 0.5x_{B2} - 0.5x_{B3} - 0.5x_{B4} \leq 0$$

$$-0.1x_{B1} + 0.9x_{B2} - 0.1x_{B3} - 0.1x_{B4} \geq 0.$$

$$0.3x_{C1} - 0.7x_{C2} - 0.7x_{C3} - 0.7x_{C4} \leq 0.$$

Production and Employment Scheduling[1]

The BOOMBUST COMPANY faces an unstable sales market and so must frequently make adjustments of some kind to compensate for predicted changes in the level of sales. When sales are increasing, these adjustments take the form of increasing the

---

[1] This example is based on a model that was first developed by Fred Hanssmann and Sidney W. Hess in ''A Linear Programming Approach to Production and Employment Scheduling,'' *Management Technology*, **1**:46–52, 1960.

work force (hiring), having the existing work force work overtime, or using up existing (or future) inventories. Similarly, when sales are dropping, the company decreases its work force (lays people off), underutilizes its current work force, or builds up inventories. All these alternatives are costly in some way, especially when they are used to extremes. Consequently, the company often uses some combination of these possible adjustments. However, it is very difficult to determine just which combination is least expensive, particularly when a series of adjustments is being planned to meet a series of predicted changes in sales.

Therefore, management has asked the OR Department to study this problem and develop a systematic procedure for production and employment scheduling that will minimize the total cost of meeting the projected sales. The procedure should provide a month-by-month schedule over the upcoming 12 months, for planning purposes, but then the procedure should be *reapplied each month* to update the schedule based on the latest sales forecasts.

The Marketing Division provides updated forecasts each month on the total volume of projected sales for the company in each of the next 12 months. The decisions to be made are concerned with the total *work-force level* (number of employees) and the *production rate* to be scheduled for each of these 12 months. These decisions, in turn, determine the net *inventory level* (amount stored minus back-orders) for these months. These quantities are denoted as follows for month t ($t = 1, 2, \ldots, 12$):

$$S_t = \text{sales forecast,}$$

$$W_t = \text{work-force level,}$$

$$P_t = \text{production rate,}$$

$$I_t = \text{inventory level at the end of the month.}$$

Because the company produces more than one product, S_t, P_t, and I_t each represents the *total* quantity aggregated over all the products, expressed in the common unit of dollar value.

To relate W_t and P_t, it is estimated that 10 employees are required on the average to produce one unit of production per month without working overtime, so that

$$W_t = 10P_t \quad \text{if the work force is fully utilized on regular time only,}$$
$$W_t < 10P_t \quad \text{if overtime is used,}$$
$$W_t > 10P_t \quad \text{if the work force is underutilized on regular time.}$$

Various kinds of costs need to be taken into account in the model. Using the notation introduced in Sec. 8.1 ($+$ and $-$ as superscripts), we summarize these costs (in units of thousands of dollars) in Table 8.6.

Table 8.6 **Cost Data for Boombust Co. Problem**

Type of Cost	Amount	Origin of Costs
Hiring cost	$4(W_t - W_{t-1})^+$	Training, reorganization
Layoff cost	$(W_t - W_{t-1})^-$	Severance pay, reorganization, low morale
Regular payroll	$5W_t$	Wages, fringe benefits
Overtime cost	$7(10P_t - W_t)^+$	Premium wages
Inventory cost	$2I_t^+$	Storage expenses, interest on capital tied up
Shortage cost	$3I_t^-$	Customer dissatisfaction, lost future sales

Note that each of the cost functions in Table 8.6 (except the regular payroll) is a *nonlinear* function of the quantity involved because the function has the form illustrated in Fig. 8.1 (but with a *zero* slope on one side of the origin). Therefore, letting Z be the total cost over all 12 months, the "natural" formulation of the model is the *nonlinear* programming problem.

$$\text{Minimize} \quad Z = \sum_{t=1}^{12} \{4(W_t - W_{t-1})^+ + (W_t - W_{t-1})^- + 5W_t + 7(10P_t - W_t)^+$$

$$+ 2I_t^+ + 3I_t^-\},$$

subject to
$$I_t = I_{t-1} + P_t - S_t$$

and
$$W_t \geq 0, \quad P_t \geq 0 \qquad \text{for } t = 1, 2, \ldots, 12,$$

where the initial inventory level I_0 and work-force level W_0 are given.

Now let us see how the OR Department *reformulated* this problem to fit the linear programming format.

FORMULATION: As in the preceding example, the key to achieving a linear programming formulation of the problem is the appropriate definition of the decision variables. In this case, finding the appropriate definition involves combining two formulation techniques that were initially presented in Sec. 8.1. First, we introduce *auxiliary variables* (x_t and y_t) to represent the quantities, $(W_t - W_{t-1})$ and $(10P_t - W_t)$, so that

$$x_t = W_t - W_{t-1}, \qquad y_t = 10P_t - W_t,$$

for $t = 1, 2, \ldots, 12$. Next, because each of these variables and the I_t are *variables with positive and negative components*, we replace *each* of them by the difference of two new *nonnegative* auxiliary variables as per the following summary:

$$x_t = x_t^+ - x_t^-, \qquad \text{so} \qquad x_t^+ = (W_t - W_{t-1})^+$$

$$x_t^- = (W_t - W_{t-1})^-,$$

$$y_t = y_t^+ - y_t^-, \qquad \text{so} \qquad y_t^+ = (10P_t - W_t)^+,$$

$$I_t = I_t^+ - I_t^-,$$

where $\quad x_t^+ \geq 0, \qquad x_t^- \geq 0, \qquad y_t^+ \geq 0, \qquad y_t^- \geq 0, \qquad I_t^+ \geq 0, \qquad I_t^- \geq 0.$

The objective function then becomes a *linear* function,

$$Z = \sum_{t=1}^{12} \{4x_t^+ + x_t^- + 5W_t + 7y_t^+ + 2I_t^+ + 3I_t^-\}.$$

The set of constraints for the linear programming formulation can be constructed simply by using the constraints for the preceding nonlinear programming problem (after substituting $I_t^+ - I_t^-$ for I_t for $t = 1, 2, \ldots, 12$), and then incorporating the *definition* of the other nonnegative auxiliary variables into the model by introducing *additional* equality constraints. Consequently, the complete linear programming model[1] is

$$\text{Minimize} \quad Z = \sum_{t=1}^{12} \{4x_t^+ + x_t^- + 5W_t + 7y_t^+ + 2I_t^+ + 3I_t^-\},$$

subject to

$$\left. \begin{array}{l} I_t^+ - I_t^- = I_{t-1}^+ - I_{t-1}^- + P_t - S_t \\ x_t^+ - x_t^- = W_t - W_{t-1} \\ y_t^+ - y_t^- = 10P_t - W_t, \end{array} \right\} \quad \text{for } t = 1, 2, \ldots, 12,$$

and

$$W_t \geq 0, \quad P_t \geq 0, \quad x_t^+ \geq 0, \quad x_t^- \geq 0, \quad y_t^+ \geq 0, \quad y_t^- \geq 0, \quad I_t^+ \geq 0, \quad I_t^- \geq 0$$

$$(t = 1, 2, \ldots, 12),$$

except that the variables appearing in the right-hand sides of the functional constraints still need to be transferred to the left-hand side for proper form. (Also, in the first constraint, when $t = 1$, $I_0^+ - I_0^-$ should be replaced by the known constant I_0.)

8.5 A Case Study—School Rezoning to Achieve Racial Balance[2]

The CITY OF MIDDLETOWN has three high schools, two of them attended primarily by white students and the other attended primarily by black students. Therefore, the Middletown school board has decided to redesign the school attendance zones to reduce the racial isolation in these schools. The new zones will apply only to students entering high school in the future, so the goal is to achieve reasonable racial balance in 3 years without substantially increasing the distances that the students must travel to school.

The school district superintendent has read some articles about how operations research has been used to greatly aid the comprehensive planning of efficient zoning designs. On her recommendation, the school board has hired a team of operations research consultants to conduct the study and make recommendations.

The consultants begin by defining and gathering the relevant data. For this purpose they divide the city geographically into 10 tracts. Since the current junior high population represents the anticipated high school population in 3 years, they then determine the number of white students and black students now in junior high from each tract. The distance the students must travel to school is a fundamental consideration, so they also determine the distance (in miles) from the center of each tract to each school. All this information appears in Table 8.7, along with the maximum number of students that can be assigned to each school.

The consultants next begin formulating a *mathematical model* for the problem. In this case (as for many practical problems), the objective is not too well defined.

---

[1] It is possible to reformulate this model further to reduce the number of variables, but the number of functional constraints remains the same, so the resulting reduction in computational effort turns out to be minor.

[2] Although this case study is a hypothetical one, it is similar to several actual studies. The theory is based primarily on a paper by L. B. Hickman and H. M. Taylor, "School Rezoning to Achieve Racial Balance: A Linear Programming Approach," *J. Socio-Econ. Planning Sci.*, **3**:127–134, 1969–1970.

Table 8.7 Data for Middletown Study

Tract	No. of Whites	No. of Blacks	Distance School 1	Distance School 2	Distance School 3
1	300	150	1.2	1.5	3.3
2	400	0	2.6	4.0	5.5
3	200	300	0.7	1.1	2.8
4	0	500	1.8	1.3	2.0
5	200	200	1.5	0.4	2.3
6	100	350	2.0	0.6	1.7
7	250	200	1.2	1.4	3.1
8	300	200	3.5	2.3	1.2
9	150	250	3.2	1.2	0.7
10	350	100	3.8	1.8	1.0
	School capacity:		1,500	2,000	1,300

Instead, only *two basic considerations* have been articulated (racial balance and distance traveled to school), and the goal is to achieve a reasonable *trade-off* between them. A common approach in this kind of situation is to express one consideration in the objective function and the other in the constraints. Thus there is a choice between optimizing the racial balance subject to constraints on distance traveled or optimizing the distance traveled subject to constraints on racial balance. Because it is easier to express distance traveled in the objective function, and because it seems more reasonable to (eventually) set minimal standards on racial balance for the constraints, the consultants choose the latter alternative.

However, the objective of "optimizing the distance traveled" needs to be stated more precisely. One possibility is to *minimize the maximum distance* that any student must travel, using the corresponding formulation technique presented in Sec. 8.3, but this objective might lead to many students having to travel the maximum distance. Another more convenient objective that may yield a better overall result is to *minimize the sum of the distances traveled* by all students. If this leads to a few unacceptable inequities, they can be eliminated during the *sensitivity analysis* phase by introducing constraints on distance traveled by groups of students having excessive distances in the original optimal solution. Therefore, the structure chosen for the model is to *minimize total distance traveled* subject to constraints on racial balance and any other required constraints.

Ultimately, decisions must be made about which individual students to assign to the respective schools. However, these detailed decisions on how to draw the boundaries of the school attendance zones can be worked out after the broader decisions on how many students from each tract to assign to each school are made. Therefore, the *decision variables* chosen for the model are

$$x_{ij} = \text{number of students in tract } i \text{ assigned}$$
$$\text{to school } j \quad (i = 1, 2, \ldots, 10; j = 1, 2, 3).$$

Rather than breaking these variables down further into the number of white students and the number of black students to be assigned, the consultants made a *simplifying assumption* that the racial mixture in each tract will be maintained in the assignments to the respective schools.

The resulting formulation of the model using Table 8.7 is as follows:

$$\text{Minimize} \quad Z = 1.2x_{11} + 1.5x_{12} + \cdots + 1.0x_{10,3},$$

subject to the following constraints:

1. *Tract assignment:*

$$x_{11} + x_{12} + x_{13} = 450$$

$$x_{21} + x_{22} + x_{23} = 400$$

$$\vdots$$

$$x_{10,1} + x_{10,2} + x_{10,3} = 450.$$

2. *School capacity:*

$$x_{11} + x_{21} + \cdots + x_{10,1} \leq 1{,}500$$

$$x_{12} + x_{22} + \cdots + x_{10,2} \leq 2{,}000$$

$$x_{13} + x_{23} + \cdots + x_{10,3} \leq 1{,}300.$$

3. *Nonnegativity:*

$$x_{ij} \geq 0, \quad \text{for } i = 1, 2, \ldots, 10; j = 1, 2, 3.$$

and

4. *Racial balance:*

Still to be developed.

The racial balance constraints need to specify that the fraction of students of a given race in a given school must fall within certain limits. After discussing the issue with the school board and noting that the entire student population is equally divided between whites and blacks, it is decided that the same limits should apply to all the schools and that these limits should be *symmetric* with respect to the races. Thus, for each school and either race, the fraction of students should fall within the limits

$$\tfrac{1}{2} - \theta \leq \text{fraction} \leq \tfrac{1}{2} + \theta,$$

so that θ represents the maximum allowable deviation from an equal distribution of races in a school.

However, the school board members do not wish to specify a value for θ until they can see the consequences of their decision in terms of the distances the students must travel. (Remember that they want to achieve a reasonable trade-off between these two considerations.) Therefore, the consultants conclude that they should use *parametric programming* (see Secs. 4.7, 6.7, and 9.3) to determine how the optimal solution changes over the entire range of possible values of θ ($0 \leq \theta \leq \tfrac{1}{2}$).

To express the racial balance constraints mathematically, we must first express the fraction of students of each race in each school in terms of the decision variables. For example,

$$\text{Fraction of white students in school 1} = \frac{(\tfrac{300}{450})x_{11} + (\tfrac{400}{400})x_{21} + \cdots + (\tfrac{350}{450})x_{10,1}}{x_{11} + x_{21} + \cdots + x_{10,1}},$$

where each coefficient in the numerator is simply the number of white students in that tract divided by the total number of students in that tract (see Table 8.7). Thus the *lower limit* constraint on this fraction is

$$L \le \frac{(\frac{2}{3})x_{11} + x_{21} + \cdots + (\frac{7}{9})x_{10,1}}{x_{11} + x_{21} + \cdots + x_{10,1}},$$

where

$$L = \tfrac{1}{2} - \theta.$$

Because constraints in this form require the use of less efficient *nonlinear* programming algorithms, the consultants next convert these constraints into an equivalent form that fits the *linear* programming format. This conversion is done by multiplying both sides by the denominator of the right-hand side to obtain

$$L(x_{11} + x_{21} + \cdots + x_{10,1}) \le \tfrac{2}{3}x_{11} + x_{21} + \cdots + \tfrac{7}{9}x_{10,1},$$

and then subtracting this right-hand side from both sides to obtain

$$(L - \tfrac{2}{3})x_{11} + (L - 1)x_{21} + \cdots + (L - \tfrac{7}{9})x_{10,1} \le 0.$$

This same approach is used to develop the lower limit constraints for all six fractions (one for each combination of race and school), which is summarized as follows:

4. *Racial balance:*

$$(L - \tfrac{2}{3})x_{11} + (L - 1)x_{21} + \cdots + (L - \tfrac{7}{9})x_{10,1} \le 0$$

$$(L - \tfrac{1}{3})x_{11} + (L - 0)x_{21} + \cdots + (L - \tfrac{2}{9})x_{10,1} \le 0$$

$$(L - \tfrac{2}{3})x_{12} + (L - 1)x_{22} + \cdots + (L - \tfrac{7}{9})x_{10,2} \le 0$$

$$(L - \tfrac{1}{3})x_{12} + (L - 0)x_{22} + \cdots + (L - \tfrac{2}{9})x_{10,2} \le 0$$

$$(L - \tfrac{2}{3})x_{13} + (L - 1)x_{23} + \cdots + (L - \tfrac{7}{9})x_{10,3} \le 0$$

$$(L - \tfrac{1}{3})x_{13} + (L - 0)x_{23} + \cdots + (L - \tfrac{2}{9})x_{10,3} \le 0.$$

This approach also could be used to develop the corresponding *upper limit* constraints representing the requirement that each fraction $\le \frac{1}{2} + \theta$. However, because

Fraction of white students $= 1 -$ fraction of black students,

the preceding lower limit constraints on both types of fractions *guarantee* that the upper limit requirements are satisfied also. Therefore, no additional constraints are needed for the model.

One flaw in this formulation is that the x_{ij} (as well as the corresponding numbers of white students and black students from tract i assigned to school j) are allowed to take on *noninteger* values (the *divisibility* assumption of linear programming discussed in Sec. 3.3). However, considering the large numbers of students involved, the consultants feel that there will be no difficulty in adjusting a noninteger optimal solution to integer values during the subsequent analysis. They know from experience that a linear programming formulation has major computational advantages over an *integer programming* formulation, so this approximation seems well worthwhile.

The stage now is set to begin the computational phase of the study. When L is sufficiently small (that is, θ is sufficiently close to $\frac{1}{2}$), the racial balance constraints

Table 8.8 **Cost and Requirements Table for Transportation Problem Formulation of Middletown Problem Without Racial Balance Constraints**

Distance Per Student

		Destination			
		School 1	School 2	School 3	Supply
	Tract 1	1.2	1.5	3.3	450
	Tract 2	2.6	4.0	5.5	400
	Tract 3	0.7	1.1	2.8	500
	Tract 4	1.8	1.3	2.0	500
	Tract 5	1.5	0.4	2.3	400
Source	Tract 6	2.0	0.6	1.7	450
	Tract 7	1.2	1.4	3.1	450
	Tract 8	3.5	2.3	1.2	500
	Tract 9	3.2	1.2	0.7	400
	Tract 10	3.8	1.8	1.0	450
	Dummy 11(*D*)	0	0	0	300
Demand		1,500	2,000	1,300	

have no effect and can be deleted. The consultants also note that the problem without these constraints can be formulated as a *transportation problem* (the special type of linear programming problem described in Sec. 7.1), as shown in Table 8.8. Therefore, rather than using the simplex method, they begin by applying the much more efficient *transportation simplex method* (see Sec. 7.2) to this formulation. The resulting optimal solution has basic variables $x_{11} = 450$, $x_{21} = 400$, $x_{31} = 500$, $x_{42} = 500$, $x_{52} = 400$, $x_{62} = 450$, $x_{71} = 150$, $x_{72} = 300$, $x_{83} = 500$, $x_{92} = 50$, $x_{93} = 350$, $x_{10,3} = 450$. $x_{11,2} = 300$, with $Z = 4,965$. (Notice that this solution already is an integer solution, which always occurs with transportation problems that have integer supplies and demands.)

The next step is to determine when this solution also is optimal for the original model with the racial balance constraints included. This determination is made by checking how large L can be made before the solution violates any of the racial balance constraints, which turns out to be $L \leq 0.285$. Because the solution is feasible for this range of values of L, it must also be optimal for these values.

Given this information, the consultants next use *parametric programming* to determine how the optimal solution changes as L is increased continuously to $\frac{1}{2}$, beginning with the preceding solution at $L = 0.285$. The result is a continually changing optimal solution where each variable is expressed as a function of L. (This approach can be thought of as applying the *sensitivity analysis procedure* described in Sec. 6.6 on a continuing basis to determine the effect of introducing the racial balance constraints as needed and of changing the coefficients of the variables in these constraints.)

However, the consultants feel that the results in this form would be too complex to be considered effectively by the school board. Therefore, after a careful examination of the results, they select a relatively small number of interesting alternatives—$L = 0.285, 0.30, 0.35, 0.40$—which represent a cross section of trade-offs between racial balance and distance traveled. These alternatives are analyzed in detail, and appropriate refinements are made in the "optimal solution" obtained from the model. The

Table 8.9 Summary of Results for Middletown Problem

θ	Optimal Z	Average Distance Traveled (Miles)	Percentage			
			<1 Mile	1.0–1.4 Miles	1.5–1.9 Miles	≥ 2 Miles
0.215	4,965	1.103	37.8	53.3	0	8.9
0.20	4,983	1.107	38.9	51.1	1.1	8.9
0.15	5,063	1.125	38.9	41.1	11.1	8.9
0.10	5,182	1.152	38.9	36.1	16.1	8.9

consultants then present their basic data and conclusions for the four alternatives to the school board, as summarized in Table 8.9 (with $\theta = 0.5 - L$).

After considerable deliberation, the school board members choose the $\theta = 0.15$ alternative. However, they modify this alternative slightly to avoid reassigning a very small proportion of one tract to a new school. The resulting master plan allocates tracts 2, 3, and 7 to school 1, tracts 1, 4, 5, and 6 to school 2, and tracts 8 and 9 to school 3, with tract 10 split as follows: $x_{10,2} = 50$, $x_{10,3} = 400$. Because this plan yields $\theta = 0.155$, the school board officially announces the new policy: that either race should form *at least one-third* the student body of any high school. They then instruct the superintendent to have her staff implement this policy, using the master plan as a basis for detailed planning.

8.6 Conclusions

This chapter has described and illustrated some particularly useful formulation techniques for building linear programming models. This material provides a good background for you, but the best teacher in this area is experience! Our goal has been to provide you with a solid foundation for dealing with real problems and for *continuing* to learn the art of linear programming.

SELECTED REFERENCES

1. Beale, E. M. L.: *Introduction to Optimization,* chaps. 8–9, Wiley, New York, 1988.

2. Bradley, Stephen P., Arnoldo C. Hax, and Thomas L. Magnanti: *Applied Mathematical Programming,* chaps. 5–7, Addison-Wesley, Reading, Mass., 1977.

3. Driebeek, Norman J.: *Applied Linear Programming,* Addison-Wesley, Reading, Mass., 1969.

4. Ignizio, James P.: *Goal Programming and Extensions,* Heath, Lexington, Mass., 1976.

5. Lawrence, K., and S. Zanakis: *Production Planning and Scheduling: Mathematical Programming Applications,* Industrial Engineering and Management Press, Atlanta, Ga., 1984.

6. Lee, Sang M.: *Goal Programming for Decision Analysis,* Auerbach, Philadelphia, 1972.

7. Ozan, Turgut: *Applied Mathematical Programming for Engineering and Production Management,* Prentice-Hall, Englewood Cliffs, N.J., 1986.

8. Salkin, Harvey M., and Jahar Saha (eds.): *Studies in Linear Programming,* North-Holland/American Elsevier, Amsterdam/New York, 1975.

9. Shapiro, Roy D.: *Optimization Models for Planning and Allocation: Text and Cases in Mathematical Programming,* Wiley, New York, 1984.

10. Steuer, Ralph E.: *Multiple Criteria Optimization: Theory, Computation, and Application,* Wiley, New York, 1985.

11. Tilanus, C. B., O. B. DeGans, and J. K. Lenstra (eds.): *Quantitative Methods in Management: Case Studies of Failures and Successes,* Wiley, New York, 1986.

12. Williams, H. P.: *Model Building in Mathematical Programming,* 2d ed., Wiley, Chichester, England, and New York, 1985.

PROBLEMS

1. Consider the following problem.

$$\text{Minimize} \quad Z = |x_1| + 2|x_2|,$$

subject to

$$x_1 + x_2 \geq 4$$

$$-x_1 + x_2 \geq 6$$

$$-x_1 - 3x_2 \geq -22$$

(no nonnegativity constraints).

(a) Use the technique presented in Sec. 8.1 to formulate the linear programming model for this problem.

(b) Use the *simplex method* to solve the model as formulated in part (a).

(c) Solve the original problem *graphically* by considering *each* of the four quadrants separately.

2. Consider the following problem.

$$\text{Minimize} \quad Z = f_1(x_1) + f_2(x_2),$$

subject to

$$2x_1 + 3x_2 \geq 9$$

$$x_1 + 2x_2 \leq 9$$

$$-x_2 \leq 1$$

(no nonnegativity constraints), where

$$f_1(x_1) = \begin{cases} 3x_1 & \text{if} \quad x_1 \geq 0 \\ x_1 & \text{if} \quad x_1 \leq 0, \end{cases}$$

$$f_2(x_2) = \begin{cases} 4x_2 & \text{if} \quad x_2 \geq 0 \\ 3x_2 & \text{if} \quad x_2 \leq 0. \end{cases}$$

(a) Use the technique presented in Sec. 8.1 to formulate the linear programming model for this problem.

(b) Use the simplex method to solve the problem as formulated in part (a).

(c) Solve the original problem graphically by considering *each* of the four quadrants separately.

3. Consider the following problem.

$$\text{Minimize} \quad Z = |x_1 - 2x_2 + x_3|,$$

subject to

$$2x_1 + 3x_2 + 4x_3 \geq 40$$

$$7x_1 + 5x_2 + 3x_3 \geq 70$$

and

$$x_1 \geq 0, \qquad x_2 \geq 0, \qquad x_3 \geq 0.$$

Use the technique presented in Sec. 8.1 to formulate the linear programming model for this problem.

4. Consider the following problem.

$$\text{Minimize} \quad Z = f_1(3x_1 - 2x_2) + f_2(3x_2 - 4x_3),$$

subject to

$$6x_1 + 7x_2 + 5x_3 \geq 30$$

$$8x_1 + 5x_2 + 9x_3 \geq 40$$

and

$$x_1 \geq 0, \quad x_2 \geq 0, \quad x_3 \geq 0,$$

where

$$f_1(3x_1 - 2x_2) = \begin{cases} 3(3x_1 - 2x_2) & \text{if} \quad 3x_1 - 2x_2 \geq 0 \\ -5(3x_1 - 2x_2) & \text{if} \quad 3x_1 - 2x_2 \leq 0, \end{cases}$$

$$f_2(3x_2 - 4x_3) = \begin{cases} 4(3x_2 - 4x_3) & \text{if} \quad 3x_2 - 4x_3 \geq 0 \\ -2(3x_2 - 4x_3) & \text{if} \quad 3x_2 - 4x_3 \leq 0. \end{cases}$$

Use the technique presented in Sec. 8.1 to formulate the linear programming model for this problem.

5. The Research and Development Division of a certain company has developed three new products. The problem is to decide which mix of these products should be produced. Management wants primary consideration given to three factors: long-run profit, stability in the work force, and achieving an increase in the company's earnings next year. In particular, using the units given in the following table, they want to

$$\text{Maximize} \quad Z = 2P - 5C - 3D,$$

where P = total (discounted) profit over the life of the new products,

C = change (in either direction) in the current level of employment,

D = decrease (if any) in next year's earnings from the current year's level.

The amount of any increase in earnings does not enter into Z, because management is concerned primarily with just achieving some increase to keep the stockholders happy. (It has mixed feelings about a large increase that then would be difficult to surpass in subsequent years.)

The impact of each of the new products (per unit rate of production) on each of these factors is shown in the following table:

	Unit Contribution				
	Product				
Factor	1	2	3	Goal	(Units)
Long-run profit	20	15	25	None	(millions of dollars)
Employment level	6	4	5	= 50	(hundreds of employees)
Earnings next year	8	7	5	≥ 75	(millions of dollars)

Except for certain additional constraints not described here, use the *goal programming technique* to formulate the linear programming model for this problem.

6. Reconsider the Middletown case study presented in Sec. 8.5. Suppose that the objective is *changed* to *minimize racial imbalance* subject to a constraint on distance traveled and other necessary constraints. Racial imbalance is defined as the *sum* (over the three high schools) of *the absolute difference between the number of white students and the number of black students at each high school*. The constraint on distance traveled is that the average distance traveled by students to school *must not* exceed 1.15 miles.

Describe how this problem fits into the framework of nonpreemptive goal programming by identifying the goals involved, and then use the *goal programming technique* to formulate the new linear programming model.

7.* Consider a *preemptive goal programming* problem with three priority levels, just one goal for each priority level, and just two activities to contribute toward these goals, as summarized in the following table:

Unit Contribution

Priority Level	Activity 1	Activity 2	Goal
First priority	1	2	≤ 20
Second priority	1	1	$= 15$
Third priority	2	1	≥ 40

(a) Use the *goal programming technique* to formulate one complete linear programming model for this problem.

(b) Construct the inital simplex tableau for applying the *streamlined procedure*. Identify the *initial basic feasible solution* and the *initial entering basic variable*, but do not proceed further.

(c) Starting from (b), use the *streamlined procedure* to solve the problem.

(d) Use the logic of *preemptive goal programming* to solve the problem *graphically* by focusing on just the two decision variables. Explain the logic used.

(e) Use the *sequential procedure* to solve this problem. After using the *goal programming technique* to formulate the linear programming model (including auxiliary variables) at each stage, solve the model *graphically* by focusing on just the two decision variables. Identify *all* optimal solutions obtained for each stage.

8. Redo Prob. 7 with the following revised table:

Unit Contribution

Priority Level	Activity 1	Activity 2	Goal
First priority	1	1	≤ 20
Second priority	1	1	≥ 30
Third priority	1	2	≥ 50

9. A certain developing country has 15,000,000 acres of publicly controlled agricultural land in active use. Its government currently is planning a way to divide this land among three basic crops (labeled 1, 2, and 3) next year. A certain percentage of each of these crops is exported in order to obtain badly needed foreign capital (dollars), and the rest of each of these crops is used to feed the populace. Raising these crops also provides employment for a signifi-cant proportion of the population. Therefore, the main factors to be considered in allocating the land to these crops are (1) the amount of foreign capital generated, (2) the number of citizens fed, and (3) the number of citizens employed in raising these crops. The following table shows how much each 1,000 acres of each crop contributes toward these factors, and the last column gives the goal established by the government for each of these factors.

Contribution Per 1,000 Acres

Factor	Crop 1	Crop 2	Crop 3	Goal
Foreign capital	$3,000	$5,000	$4,000	$\geq \$70,000,000$
Citizens fed	150	75	100	$\geq 1,750,000$
Citizens employed	10	15	12	$= 200,000$

(a) In evaluating the relative seriousness of *not* achieving these goals, the government has concluded that the following deviations from the goals should be considered *equally undesirable:* (1) each $100 under the foreign capital goal, (2) each person under the citizens fed goal, and (3) each deviation of one (in either direction) from the citizens employed goal. Use the *goal programming technique* to formulate the linear programming model for this problem.

(b) Now suppose that the government concludes that the importance of the various goals differs greatly so that a *preemptive goal programming* approach should be used. In particular, the first-priority goal is *citizens fed* ≥ 1,750,000, the second-priority goal is *foreign capital* ≥ $70,000,000, and the third-priority goal is *citizens employed* = 200,000. Use the *goal programming technique* to formulate one complete linear programming model for this problem.

(c) Use the *streamlined procedure* to solve the problem as formulated in part (b).

(d) Use the *sequential procedure* to solve the problem as presented in part (b).

10. Reconsider Prob. 9. Suppose now that the third-priority goal actually is a *lower one-sided goal* (desire ≥ 200,000 citizens employed). The government feels that it is critical to come at least close to satisfying *all* of these goals. Therefore, the decision has been made to *reformulate* the problem to adopt the overriding objective of *maximizing the minimum progress* (on a percentage basis) toward all these goals.

Use the technique presented in Sec. 8.3 to formulate the linear programming model for this new problem.

11. One of the most important problems in the field of *statistics* is the *linear regression problem.* Roughly speaking, this problem involves fitting a straight line to statistical data represented by points—$(x_1, y_1), (x_2, y_2), \ldots, (x_n, y_n)$—on a graph. If we denote the line by $y = a + bx$, the objective is to choose the constants a and b to provide the "best" fit according to some criterion. The criterion usually used is the *method of least squares,* but there are other interesting criteria where linear programming can be used to solve for the optimal values of a and b.

For each of the following criteria, formulate the linear programming model for this problem.

(a) Minimize the sum of the absolute deviations of the data from the line; that is,

$$\text{Minimize} \quad \sum_{i=1}^{n} |y_i - (a + bx_i)|.$$

(*Hint:* Note that this problem can be viewed as a nonpreemptive goal programming problem where each data point represents a "goal" for the regression line.)

(b) Minimize the maximum absolute deviation of the data from the line; that is,

$$\text{Minimize} \quad \max_{i=1, 2, \ldots, n} |y_i - (a + bx_i)|.$$

12. Reconsider the Middletown case study described in Sec. 8.5. Suppose that the objective is changed from minimizing the sum of the distances traveled by all students to minimizing the *maximum* over the tracts of the total distance traveled by the students in each respective tract, subject to the same constraints (including racial balance constraints) as before. Formulate the new linear programming model for this problem.

13. Reconsider the Middletown case study described in Sec. 8.5. Suppose that bussing must be provided for all students traveling more than 1.5 miles (but to no others), and that the school board has adopted the objective of minimizing the total cost of this bussing, subject to the same constraints (including racial balance constraints) as before. Assuming that this cost is proportional to the sum of the distance traveled by all *bussed* students, formulate the new objective function for this problem.

14. The school board of a certain city has made the decision to close one of its middle schools (sixth, seventh, and eighth grades) at the end of this school year and reassign all of next year's middle school students to the three remaining middle schools. The school district provides bussing for all middle school students who must travel more than approximately a mile, so the school board wants a plan for reassigning the students that will minimize the total bussing cost. The cost per student of bussing from each of the six residential areas of the city to each of the schools is shown in the following table (along with other basic data for next year), where 0 indicates that bussing is not needed and a dash indicates an infeasible assignment.

Area	No. of Students	% 6th Grade	% 7th Grade	% 8th Grade	Bussing Cost Per Student		
					School 1	School 2	School 3
1	450	32	38	30	3	0	7
2	600	37	28	35	0	4	5
3	550	31	32	37	6	3	2
4	350	28	39	33	2	5	—
5	500	39	34	27	0	—	4
6	450	33	29	38	5	3	0
			School capacity:		900	1,100	1,000

The school board also has imposed the restriction that each grade must constitute between 30 and 35 percent of each school's population. The above table shows the percentage of each area's middle school population for next year that falls into each of the three grades. The school attendance zone boundaries can be drawn so as to split any given area among more than one school, but assume that the percentages shown in the table will continue to hold for any partial assignment of an area to a school.

Formulate a linear programming model for determining how many students should be assigned from each area to each school.

15. A company desires to blend a new alloy of 40 percent tin, 35 percent zinc, and 25 percent lead from several available alloys having the following properties:

Property	Alloy				
	1	2	3	4	5
Percentage tin	60	25	45	20	50
Percentage zinc	10	15	45	50	40
Percentage lead	30	60	10	30	10
Cost ($/lb)	22	20	25	24	27

The objective is to determine the proportions of these alloys that should be blended to produce the new alloy at a minimum cost. Formulate the linear programming model for this problem.

16. At the beginning of the fall semester, the director of the computer facility of a certain university is confronted with the problem of assigning different working hours to her operators. Because all the operators are currently enrolled in the university, her main concern is to make certain that the operators' working times are not so excessive that they would interfere with study times.

There are six operators (four men and two women). They all have different wage rates because of differences in their experience with computers and in their programming ability. The following table shows their wage rates, along with the maximum number of hours that each can work each day.

Operators	Wage Rate	Maximum Hours of Availability				
		Mon.	Tue.	Wed.	Thurs.	Fri.
K. C.	$6.00/hour	6	0	6	0	6
D. H.	$6.10/hour	0	6	0	6	0
H. B.	$5.90/hour	4	8	4	0	4
S. C.	$5.80/hour	5	5	5	0	5
K. S.	$6.80/hour	3	0	3	8	0
N. K.	$7.30/hour	0	0	0	6	2

Because of a tight budget, the director has to minimize cost. Her decision is that the operators with the highest wage rates should work the least possible number of hours, except that this number should not be so low as to impair his or her knowledge of the operation. This level is set arbitrarily at 8 hours per week for the male operators and 7 hours per week for the female operators (K. S., N. K.).

The computer facility is to be open for operation from 8 A.M. to 10 P.M. Monday through Friday with exactly one operator on duty during these hours. On Saturdays and Sundays, the computer is to be operated by other staff.

Formulate a linear programming model so that the director can determine the number of hours she should assign to each operator on each day.

17. A lumber company has three sources of wood and five markets to be supplied. The annual availability of wood at sources 1, 2, and 3 is 10, 20, and 15 million board feet, respectively. The amount that can be sold annually at markets 1, 2, 3, 4, and 5 is 7, 12, 9, 10, and 8 million board feet, respectively.

In the past the company has shipped the wood by train. However, because shipping costs have been increasing, the alternative of using ships to make some of the deliveries is being investigated. This alternative would require the company to invest in some ships. Except for these investment costs, the shipping costs in thousands of dollars per million board feet by rail and by water (when feasible) would be the following for each route:

	Unit Cost by Rail					Unit Cost by Ship				
	Market					Market				
Source	1	2	3	4	5	1	2	3	4	5
1	61	72	45	55	66	31	38	24	—	35
2	69	78	60	49	56	36	43	28	24	31
3	59	66	63	61	47	—	33	36	32	26

The capital investment (in thousands of dollars) in ships required for each million board feet to be transported annually by ship along each route is given as follows:

	Investment for Ships				
	Market				
Source	1	2	3	4	5
1	275	303	238	—	285
2	293	318	270	250	265
3	—	283	275	268	240

Considering the expected useful life of the ships and the time value of money, the equivalent uniform annual cost of these investments is one-tenth the amount given in the table. The company is able to raise only $6,750,000 to invest in ships. The objective is to determine the overall shipping plan that minimizes the total equivalent uniform annual cost while meeting this investment budget and the sales demand at the markets. Formulate the linear programming model for this problem.

18. A company needs to lease warehouse storage space over the next 5 months. Just how much space will be required in each of these months is known. However, since these space requirements are quite different, it may be most economical to lease only the amount needed each month on a month-by-month basis. On the other hand, the additional cost for leasing space for additional months is much less than for the first month, so it may be less expensive to lease the maximum amount needed for the entire 5 months. Another option is the intermediate approach of changing the total amount of space leased (by adding a new lease and/or having an old lease expire) at least once but not every month.

The space requirement (in thousands of square feet) and the leasing costs (in hundreds of dollars) for the various leasing periods are as follows:

Month	Required Space	Leasing Period (Months)	Cost Per 1,000 Sq. Ft. Leased
1	30	1	650
2	20	2	1,000
3	40	3	1,350
4	10	4	1,600
5	50	5	1,900

The objective is to minimize the total leasing cost for meeting the space requirements. Formulate the linear programming model for this problem.

19. A spaceship to take astronauts to Mars and back is being designed. This spaceship will have three compartments, each with its own independent life support system. The key element in each of these life support systems is a small *oxidizer* unit that triggers a chemical process for producing oxygen. However, these units cannot be tested in advance, and only some of them succeed in triggering this chemical process. Therefore, it is important to have several backup units for each system. Because the requirements are different for the three compartments, the units needed for each one have somewhat different characteristics. A decision must now be made on the number of units to be provided for each compartment, taking into account design limitations on the *total* amount of *space, weight,* and *cost* that can be allocated to these units for the entire spaceship. The following table summarizes these limitations, as well as the characteristics of the individual units for each compartment:

Compartment	Space (Cu. In.)	Weight (Lb.)	Cost ($)	Probability of Failure
1	40	15	40,000	0.30
2	50	20	45,000	0.40
3	30	10	35,000	0.20
Limitation	500	200	500,000	

If all the units fail in just one or two of the compartments, the astronauts can occupy the remaining compartment(s) and continue their space voyage but with some loss in the amount of scientific information they can obtain. However, if all units fail in all three compartments, the astronauts can still return the spaceship safely, but the whole voyage must be completely aborted at great expense. Therefore, the objective is to *minimize the probability* of all units failing, subject to the preceding limitations and the further restriction that each compartment has a probability of no more than 0.05 that all its units fail. Formulate the linear programming model for this problem. (*Hint:* Use logarithms.)

20. A large paper manufacturing company has 10 paper mills and a large number (say, 1,000) of customers to be supplied. It uses three alternative types of machines and four types of raw materials to make five different types of paper. Therefore, the company needs to develop a detailed production-distribution plan on a monthly basis, with an objective of minimizing the total cost of producing and distributing the paper during the month. Specifically, it is necessary to determine jointly the amount of each type of paper to be made at each paper mill on each type of machine *and* the amount of each type of paper to be shipped from each paper mill to each customer.

The relevant data can be expressed symbolically as

D_{jk} = number of units of paper type k demanded by customer j,

r_{klm} = number of units of raw material m needed to produce one unit of paper type k on machine type l,

R_{im} = number of units of raw material m available at paper mill i,

c_{kl} = number of capacity units of machine type l that will produce one unit of paper type k,

C_{il} = number of capacity units of machine type l available at paper mill i,

P_{ikl} = production cost for each unit of paper type k produced on machine type l at paper mill i,

T_{ijk} = transportation cost for each unit of paper type k shipped from paper mill i to customer j.

(*a*) Using these symbols, formulate the linear programming model for this problem.
(*b*) Considering the special structure of this model, give your recommendation on how it should be solved.

21. One measure of the quality of the water in a river is its dissolved oxygen (D.O.) concentration. This measure is of interest partly because certain minimum concentration levels of D.O. are necessary to permit fish and other aquatic animals to survive. A large portion of the waste released into streams is organic material. This material is a source of nutrients for many organisms found in streams. In the process of utilizing the organic material, the organisms withdraw the D.O. contained in the stream. Thus the larger the amount of these wastes, the larger the biochemical oxygen demand (B.O.D.).

Consider the following river system consisting of two tributaries leading into the main stream. The daily flow rate of water at cities 1 to 4 is known to be f_1, f_2, $f_1 + f_2$, $f_1 + f_2$, respectively. Water at city 1 or city 2 requires 1 day to reach city 3, and water at city 3 requires 1 day to reach city 4. Let D_i be the known D.O. concentration of the water just above city i ($i = 1, 2$). Similarly, let B_i be the known waste concentration of the water, measured by its B.O.D. concentration, just above city i ($i = 1, 2$). Wastes are discharged from cities 1 to 3 into the stream in known amounts w_1, w_2, w_3 per day. (These are negligible in comparison with f_1, f_2.)

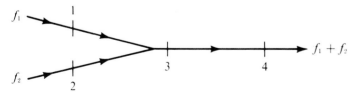

If the waste discharged from city i ($i = 1, 2, 3$) is untreated, its B.O.D. concentration would be U_i. However, if appropriate treatment processes are used, the B.O.D. concentration can be lowered to any level between L_i and U_i. The cost of reducing the B.O.D. concentration from U_i is c_i per unit reduction. However, some treatment is necessary at some or all of these cities to achieve at least a minimum standard S for the D.O. concentration at cities 3 and 4. The problem is to choose the B.O.D. concentration of wastes discharged from cities 1 to 3 that minimizes the cost of meeting this standard.

The following biochemical model has been developed to help solve problems of this type. Suppose that river water has a daily flow rate of f and a B.O.D. concentration of b and then has waste discharged at a rate of w per day with a B.O.D. concentration of x. The effect immediately downstream is to raise the B.O.D. concentration of the water to

$$v_0 = \frac{bf + xw}{f + w} \approx b + \left(\frac{w}{f}\right) x.$$

There is no immediate effect on the D.O. concentration. However, both the D.O. concentration and the B.O.D. concentration would change gradually downstream. In particular, if u_0 and v_0 are, respectively, the current D.O. and B.O.D. concentrations of the water at a particular location on a river, and if no waste is added to this water, then the respective concentrations u_1, v_1 of D.O. and B.O.D. in this same water 1 day downstream become

$$u_1 = \alpha + \beta u_0 - \gamma v_0$$

$$v_1 = \delta + \varepsilon v_0,$$

where α, β, γ, δ, ε are positive constants reflecting the various underlying physical and biochemical processes. Formulate the linear programming model for this problem.

9

Other Algorithms for Linear Programming

The key to the extremely widespread use of linear programming is the availability of an exceptionally efficient algorithm—the simplex method—that will routinely solve the large-sized problems that typically arise in practice. However, the simplex method is only part of the arsenal of algorithms regularly used by linear programming practitioners. Chapter 7 described special types of linear programming problems for which *streamlined* versions of the simplex method are available (as illustrated by the *transportation simplex method* in Sec. 7.2). Section 4.8 mentioned that production computer codes *adapt* the simplex method to a more convenient matrix form presented in Sec. 5.2. Sections 4.7 and 6.6 pointed out how certain *modifications* or *extensions* of the simplex method are particularly useful for sensitivity analysis. Thus all these algorithms are *variants* of the simplex method as it was presented in Chap. 4. Consequently, they also are exceptionally efficient.

This chapter includes four particularly important algorithms based on the simplex method. In particular, the next three sections present the *upper bound technique* (a streamlined version of the simplex method for dealing with variables having upper bounds), the *dual simplex method* (a modification particularly useful for sensitivity analysis), and *parametric programming* (an extension for systematic sensitivity anal-

315

ysis). Section 9.5 describes the *decomposition principle* for streamlining the simplex method to exploit the block angular structure that characterize multidivisional problems.

Section 4.9 introduced an exciting new advancement in linear programming—the development of a powerful new type of algorithm that moves through the interior of the feasible region. We describe this *interior-point approach* further in Sec. 9.4.

9.1 The Upper Bound Technique

At the end of Sec. 7.5 we discussed the fact that it is common for some or all of the *individual* x_j variables to have *upper bound constraints*

$$x_j \leq u_j,$$

where u_j is a positive constant representing the maximum *feasible* value of x_j. (The right-hand side of Table 7.35 shows the resulting *special structure* of the functional constraints.) However, we also pointed out in Sec. 4.8 that the most important determinant of computation time for the simplex method is the *number of functional constraints,* whereas the number of *nonnegativity* constraints is relatively unimportant. Therefore, having a large number of upper bound constraints among the functional constraints greatly increases computational effort.

The *upper bound technique* avoids this increased effort by removing the upper bound constraints from the functional constraints and treating them separately, essentially like nonnegativity constraints. Removing the upper bound constraints in this way causes no problems as long as none of the variables get increased over its upper bound. The only time the simplex method increases some of the variables is when the entering basic variable is increased to obtain a new basic feasible solution. Therefore, the upper bound technique simply applies the simplex method in the usual way to the *remainder* of the problem (i.e., without the upper bound constraints) but with the one additional restriction that each new basic feasible solution is required to satisfy the upper bound constraints in addition to the usual lower bound (nonnegativity) constraints.

To implement this idea, note that a decision variable x_j with an upper bound constraint ($x_j \leq u_j$) can always be replaced by

$$x_j = u_j - y_j,$$

where y_j would then be the decision variable. In other words, you have a choice between letting the decision variable be the *amount above zero* (x_j) or the *amount below* u_j ($y_j = u_j - x_j$). (We shall refer to x_j and y_j as *complementary* decision variables.) Because

$$0 \leq x_j \leq u_j,$$

it also follows that

$$0 \leq y_j \leq u_j.$$

Thus at any point during the simplex method you can either

1. Use x_j, where $0 \leq x_j \leq u_j$;

or

2. Replace x_j by $(u_j - y_j)$, where $0 \le y_j \le u_j$.

The upper bound technique uses the following rule to make this choice:

Rule: Begin with choice 1.
Whenever $x_j = 0$, use choice 1, so x_j is *nonbasic*.
Whenever $x_j = u_j$, use choice 2, so $y_j = 0$ is *nonbasic*.
Switch choices only when the other extreme value of x_j is reached.

Therefore, whenever a *basic* variable reaches its upper bound, you should switch choices and use its *complementary* decision variable as the new nonbasic variable (the leaving basic variable) for identifying the new basic feasible solution. Thus the one substantive modification being made in the simplex method is in the rule for selecting the *leaving basic variable*.

Recall that the simplex method selects as the leaving basic variable the one that would be the first to become infeasible by going negative as the entering basic variable is increased. The modification now made is to select instead the variable that would be the first to become infeasible *in any way,* either by going negative or by going over the upper bound, as the entering basic variable is increased. (Notice that one possibility is that the *entering* basic variable may become infeasible first by going over its upper bound, so that its *complementary* decision variable becomes the leaving basic variable.) If the leaving basic variable reaches zero, then proceed as usual with the simplex method. However, if it reaches its upper bound instead, then switch choices and make its complementary decision variable the leaving basic variable.

To illustrate, consider the problem:

$$\text{Maximize} \quad Z = 2x_1 + x_2 + 2x_3,$$

subject to

$$4x_1 + x_2 \quad\;\; = 12$$

$$-2x_1 \quad\;\; + x_3 = \;\; 4$$

and

$$0 \le x_1 \le \;\; 4$$

$$0 \le x_2 \le 15$$

$$0 \le x_3 \le \;\; 6.$$

Thus all three variables have upper bound constraints ($u_1 = 4$, $u_2 = 15$, $u_3 = 6$).

The two equality constraints are already in *proper form from Gaussian elimination* for identifying the initial basic feasible solution ($x_1 = 0$, $x_2 = 12$, $x_3 = 4$), and none of the variables in this solution exceed its upper bound, so x_2 and x_3 can be used as the initial basic variables without introducing artificial variables. However, these variables then need to be eliminated algebraically from the objective function to obtain the initial Eq. (0), as follows:

$$
\begin{array}{rl}
Z - 2x_1 - x_2 - 2x_3 =& 0 \\
+ \quad (4x_1 + x_2) \qquad\quad =& 12) \\
+ 2(-2x_1 \qquad + x_3) =& 4) \\
\hline
Z - 2x_1 \qquad\qquad\quad\;\; =& 20.
\end{array}
$$

(0)

Table 9.1 Equations and Calculations for Initial Leaving Basic Variable in Example for Upper Bound Technique

Initial Set of Equations	Maximum Feasible Value of x_1
(0) $\quad Z - 2x_1 \qquad\qquad = 20$	$x_1 \le 4 \qquad$ (since $u_1 = 4$)
(1) $\qquad 4x_1 + x_2 \qquad = 12$	$x_1 \le \dfrac{12}{4} = 3$
(2) $\qquad -2x_1 \qquad + x_3 = 4$	$x_1 \le \dfrac{6-4}{2} = 1 \leftarrow \min$
	(because $u_3 = 6$)

To start the first iteration, this initial Eq. (0) indicates that the initial *entering basic variable* is x_1. Since the upper bound constraints are not to be included, the *entire* initial set of equations and the corresponding calculations for selecting the leaving basic variables are those shown in Table 9.1. The second column shows how much the entering basic variable x_1 can be *increased* from zero before some basic variable (including x_1) becomes infeasible. The maximum value given next to Eq. (0) is just the upper bound constraint for x_1. For Eq. (1), since the coefficient of x_1 is *positive, increasing* x_1 to 3 decreases the basic variable in this equation (x_2) from 12 to its *lower* bound of *zero*. For Eq. (2), since the coefficient of x_1 is *negative, increasing* x_1 to 1 *increases* the basic variable in this equation (x_3) from 4 to its *upper* bound of 6.

Because this last maximum value of x_1 is the smallest, x_3 provides the *leaving* basic variable. However, because x_3 reached its *upper* bound, replace x_3 by $(6 - y_3)$ so that $y_3 = 0$ becomes the new *nonbasic* variable for the next basic feasible solution and x_1 becomes the new basic variable in Eq. (2). This replacement leads to the following changes in this equation:

$$(2) \quad -2x_1 + \quad x_3 \quad = \quad 4$$
$$\rightarrow \quad -2x_1 + (6 - y_3) = \quad 4$$
$$\rightarrow \quad -2x_1 - \quad y_3 \quad = -2$$
$$\rightarrow \quad x_1 + \quad \tfrac{1}{2}y_3 \quad = \quad 1.$$

Therefore, after eliminating x_1 algebraically from the other equations, the *second* complete set of equations becomes

$$(0) \qquad Z \quad + \quad y_3 = 22$$
$$(1) \qquad x_2 - 2y_3 = 8$$
$$(2) \qquad x_1 + \tfrac{1}{2}y_3 = 1.$$

The resulting basic feasible solution is $x_1 = 1$, $x_2 = 8$, $y_3 = 0$. By the *optimality test*, it also is an optimal solution, so $x_1 = 1$, $x_2 = 8$, $x_3 = 6 - y_3 = 6$ is the desired solution to the original problem.

9.2 The Dual Simplex Method

The *dual simplex method* can be thought of as the *mirror image* of the simplex method. This interpretation is best explained by referring to Tables 6.10 and 6.11 and Fig. 6.1. The simplex method deals directly with *suboptimal* basic solutions and moves toward an optimal solution by striving to satisfy the *optimality test*. By contrast, the dual simplex method deals directly with *superoptimal* basic solutions and moves toward an optimal solution by striving to achieve *feasibility*. Furthermore, the dual simplex method deals with a problem as if the simplex method were being applied simultaneously to its dual problem. If we make their *initial* basic solutions *complementary,* the two methods move in complete sequence, obtaining *complementary* basic solutions with each iteration.

The dual simplex method is very useful in certain special types of situations. Ordinarily it is easier to find an initial basic feasible solution than an initial superoptimal basic solution. However, it is occasionally necessary to introduce many *artificial* variables to construct an initial basic feasible solution artificially. In such cases it may be easier to begin with a superoptimal basic solution and use the dual simplex method. Furthermore, fewer iterations may be required when it is not necessary to drive many artificial variables to zero.

As we mentioned several times in Chap. 6 as well as in Sec. 4.7, another important primary application of the dual simplex method is its use in conjunction with sensitivity analysis. Suppose that an optimal solution has been obtained by the simplex method but that it becomes necessary (or of interest for sensitivity analysis) to make minor changes in the model. If the formerly optimal basic solution is *no longer feasible* (but still satisfies the optimality test), you can immediately apply the dual simplex method by starting with this *superoptimal* basic solution. Applying the dual simplex method in this way usually leads to the new optimal solution much more quickly than solving the new problem from the beginning with the simplex method.

The rules for the dual simplex method are very similar to those for the simplex method. In fact, once they are started, the only difference between them is in the criteria used for selecting the entering and the leaving basic variables and for stopping the algorithm.

To start the dual simplex method, we must have all the coefficients in Eq. (0) *nonnegative* (so that the basic solution is superoptimal). The basic solutions will be infeasible (except for the last one) only because some of the variables are negative. The method continues to decrease the value of the objective function, always retaining *nonnegative coefficients* in Eq. (0), until all the *variables* are nonnegative. Such a basic solution is feasible (it satisfies all the equations) and is, therefore, optimal by the simplex method criterion of nonnegative coefficients in Eq. (0).

The details of the dual simplex method are summarized below.

Summary of Dual Simplex Method

1. *Initialization step:* Introduce slack variables as needed to construct a set of equations describing the problem. Find a basic solution such that the coefficients in Eq. (0) are zero for basic variables and nonnegative for nonbasic variables. Go to the feasibility test.

2. *Iterative step:*

   Part 1. Determine the leaving basic variable: Select the basic variable with the *largest negative value.*

Part 2. Determine the entering basic variable: Select the nonbasic variable whose coefficient in Eq. (0) reaches zero first as an increasing multiple of the equation containing the leaving basic variable is added to Eq. (0). This selection is made by checking the nonbasic variables with *negative coefficients* in that equation (the one containing the leaving basic variable) and selecting the one with the smallest ratio of the Eq. (0) coefficient to the absolute value of the coefficient in that equation.

Part 3. Determine the new basic solution: Starting from the current set of equations, solve for the basic variables in terms of the nonbasic variables by Gaussian elimination (see Appendix 4). When we set the nonbasic variables equal to zero, each basic variable (and Z) equals the new right-hand side of the one equation in which it appears (with a coefficient of $+1$).

3. *Feasibility test:* Determine whether this solution is feasible (and therefore optimal): Check to see whether all the basic variables are *nonnegative*. If they are, then this solution is feasible, and therefore optimal, so stop. Otherwise, go to the iterative step.

To fully understand the dual simplex method, you must realize that the method proceeds just as if the *simplex method* were being applied to the complementary basic solutions in the *dual problem*. (In fact, this interpretation was the motivation for constructing the method as it is.) Part 1, determining the leaving basic variable, is equivalent to determining the entering basic variable in the dual problem. The variable with the largest negative value corresponds to the largest negative coefficient in Eq. (0) of the dual problem (see Table 6.3). Part 2, determining the entering basic variable, is equivalent to determining the leaving basic variable in the dual problem. The coefficient in Eq. (0) that reaches zero first corresponds to the variable in the dual problem that reaches zero first. The two criteria for stopping the algorithm are also complementary.

We shall now illustrate the dual simplex method by applying it to the *dual problem* for the Wyndor Glass Co. (see Table 6.1). Normally this method is applied directly to the problem of concern (a primal problem). However, we have chosen this problem because you have already seen the simplex method applied to *its* dual problem (namely, the primal problem[1]) in Table 4.8 so you can compare the two. To facilitate the comparison, we shall continue to denote the decision variables in the problem being solved by y_i rather than x_j.

In *maximization* form, the problem to be solved is

$$\text{Maximize} \quad Z = -4y_1 - 12y_2 - 18y_3,$$

subject to

$$y_1 \qquad\quad + 3y_3 \geq 3$$

$$2y_2 + 2y_3 \geq 5$$

and

$$y_1 \geq 0, \qquad y_2 \geq 0, \qquad y_3 \geq 0.$$

After the functional constraints are converted to $\leq$ form and the slack variables are introduced, the initial set of equations is that shown for iteration 0 in Table 9.2. Notice that all the coefficients in Eq. (0) are nonnegative, so the solution is optimal if it is feasible.

---

[1] Recall that the *symmetry property* in Secs. 6.1 and 6.4 points out that the dual of a dual problem is the original primal problem.

Table 9.2 **Dual Simplex Method Applied to Wyndor Glass Co. Dual Problem**

Iteration	Basic Variable	Eq. No.	Z	y_1	y_2	y_3	y_4	y_5	Right Side
0	Z	0	1	4	12	18	0	0	0
	y_4	1	0	-1	0	-3	1	0	-3
	y_5	2	0	0	-2	-2	0	1	-5
1	Z	0	1	4	0	6	0	6	-30
	y_4	1	0	-1	0	-3	1	0	-3
	y_2	2	0	0	1	1	0	$-\frac{1}{2}$	$\frac{5}{2}$
2	Z	0	1	2	0	0	2	6	-36
	y_3	1	0	$\frac{1}{3}$	0	1	$-\frac{1}{3}$	0	1
	y_2	2	0	$-\frac{1}{3}$	1	0	$\frac{1}{3}$	$-\frac{1}{2}$	$\frac{3}{2}$

The initial basic solution is $y_1 = 0$, $y_2 = 0$, $y_3 = 0$, $y_4 = -3$, $y_5 = -5$, with $Z = 0$, which is not feasible because of the negative values. The leaving basic variable is y_5 ($5 > 3$), and the entering basic variable is y_2 ($\frac{12}{2} < \frac{18}{2}$), which leads to the second set of equations, labeled as iteration 1 in Table 9.2. The corresponding basic solution is $y_1 = 0$, $y_2 = \frac{5}{2}$, $y_3 = 0$, $y_4 = -3$, $y_5 = 0$, with $Z = -30$, which is not feasible.

The next leaving basic variable is y_4, and the entering basic variable is y_3 ($\frac{6}{3} < \frac{4}{1}$), which leads to the final set of equations in Table 9.2. The corresponding basic solution is $y_1 = 0$, $y_2 = \frac{3}{2}$, $y_3 = 1$, $y_4 = 0$, $y_5 = 0$, with $Z = -36$, which is feasible and therefore optimal.

Notice that the optimal solution for the *dual* of this problem[1] is $x_1^* = 2$, $x_2^* = 6$, $x_3^* = 2$, $x_4^* = 0$, $x_5^* = 0$, as was obtained in Table 4.8 by the simplex method. We suggest that you now trace through Tables 9.2 and 4.8 simultaneously and compare the complementary steps for the two mirror-image methods.

9.3 Parametric Linear Programming

At the end of Sec. 6.7 we described *parametric linear programming* and its use for conducting sensitivity analysis systematically by gradually changing various model parameters simultaneously. We shall now present the algorithmic procedure, first for the case where the c_j parameters are being changed and then where the b_i parameters are varied.

Systematic Changes in the c_j Parameters

For the case where the c_j parameters are being changed, the *objective function* of the ordinary linear programming model,

$$Z = \sum_{j=1}^{n} c_j x_j,$$

---

[1] The *complementary optimal basic solutions property* presented in Sec. 6.3 indicates how to read the optimal solution for the dual problem from row 0 of the final simplex tableau for the primal problem. This same conclusion holds regardless of whether the simplex method or the dual simplex method is used to obtain the final tableau.

is replaced by

$$Z(\theta) = \sum_{j=1}^{n} (c_j + \alpha_j\theta)x_j,$$

where the α_j are given input constants representing the *relative* rates at which the coefficients are to be changed. Therefore, gradually increasing θ from zero changes the coefficients at these relative rates.

The values assigned to the α_j may represent interesting simultaneous changes of the c_j for systematic sensitivity analysis of the effect of increasing the magnitude of these changes. They may also be based on how the coefficients (e.g., unit profits) would change together with respect to some factor measured by θ. This factor might be uncontrollable, e.g., the *state of the economy*. However, it may also be under the control of the decision maker, e.g., the amount of personnel and equipment to shift from some of the activities to others.

For any given value of θ, the optimal solution of the corresponding linear programming problem can be obtained by the simplex method. This solution may have been obtained already for the original problem where $\theta = 0$. However, the objective is to *find the optimal solution* of the modified linear programming problem [maximize $Z(\theta)$ subject to the original constraints] *as a function of θ*. Therefore, the solution procedure needs to be able to determine when and how the optimal solution changes (if it does) as θ increases from zero to any specified positive number.

Figure 9.1 illustrates how $Z^*(\theta)$, the objective function value for the optimal solution (given θ), changes as θ increases. $Z^*(\theta)$ always has this *piecewise linear* and *convex*[1] form (see Prob. 24). The corresponding optimal solution changes (as θ increases) *just* at the values of θ where the slope of the $Z^*(\theta)$ function changes. Thus Fig. 9.1 depicts a problem where three different solutions are optimal for different values of θ, one for $0 \le \theta \le \theta_1$, the second for $\theta_1 \le \theta \le \theta_2$, and the third for $\theta \ge \theta_2$. Because the value of each x_j remains the same within each of these intervals for θ, the value of $Z^*(\theta)$ varies with θ only because the *coefficients* of the x_j are changing as a linear function of θ.

The solution procedure is based directly upon the sensitivity analysis procedure for investigating changes in the c_j parameters (cases 2a and 3, Sec. 6.7). As described in the last subsection of Sec. 6.7, the only basic difference with parametric linear

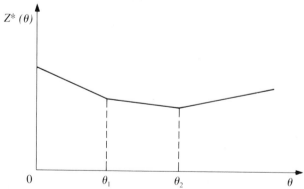

Figure 9.1 Objective function value for an optimal solution as a function of θ for parametric linear programming with systematic changes in the c_j parameters.

[1] See Appendix 1 for a definition and discussion of *convex* functions.

programming is that the changes now are expressed in terms of θ rather than as specific numbers.

To illustrate, suppose that $\alpha_1 = 2$ and $\alpha_2 = -1$ for the original Wyndor Glass Co. problem (see Sec. 3.1 and Table 4.8), so that

$$Z(\theta) = (3 + 2\theta)x_1 + (5 - \theta)x_2.$$

Beginning with the *final* simplex tableau for $\theta = 0$ (Table 4.8), its Eq. (0),

(0) $$Z + \tfrac{3}{2}x_4 + x_5 = 36,$$

would first have these changes from the *original* ($\theta = 0$) coefficients added into it on the left-hand side:

(0) $$Z - 2\theta x_1 + \theta x_2 + \tfrac{3}{2}x_4 + x_5 = 36.$$

Because both x_1 and x_2 are basic variables [appearing in Eqs. (3) and (2), respectively], they both need to be eliminated algebraically from Eq. (0):

$$Z - 2\theta x_1 + \theta x_2 + \tfrac{3}{2}x_4 + x_5 = 36$$

$$+2\theta \text{ times Eq. (3)}$$

$$- \theta \text{ times Eq. (2)}$$

(0) $$\overline{Z + (\tfrac{3}{2} - \tfrac{7}{6}\theta)x_4 + (1 + \tfrac{2}{3}\theta)x_5 = 36 - 2\theta.}$$

The *optimality test* says that the current basic feasible solution will *remain* optimal as long as these coefficients of the nonbasic variables remain *nonnegative:*

$$\tfrac{3}{2} - \tfrac{7}{6}\theta \geq 0, \qquad \text{for } 0 \leq \theta \leq \tfrac{9}{7},$$

$$1 + \tfrac{2}{3}\theta \geq 0, \qquad \text{for all } \theta \geq 0.$$

Therefore, after increasing θ past $\theta = \tfrac{9}{7}$, x_4 would need to be the *entering basic variable* for another *iteration* of the simplex method to find the new optimal solution. Then θ would be increased further until another coefficient goes negative, and so on until θ has been increased as far as desired.

This entire procedure is now summarized, and the example is completed in Table 9.3.

Summary of Parametric Programming Procedure for
Systematic Changes in the c_j Parameters

Step 1: Solve the problem with $\theta = 0$ by the simplex method.

Step 2: Use the sensitivity analysis procedure (cases 2a and 3, Sec. 6.7) to introduce the $\Delta c_j = \alpha_j\theta$ changes into Eq. (0).

Step 3: Increase θ until one of the nonbasic variables has its coefficient in Eq. (0) go negative (or until θ has been increased as far as desired).

Step 4: Use this variable as the entering basic variable for an iteration of the simplex method to find the new optimal solution. Return to step 3.

Systematic Changes in the b_i Parameters

For the case where the b_i parameters change systematically, the one modification made in the original linear programming model is that b_i is replaced by $(b_i + \alpha_i\theta)$, for

Table 9.3 The c_j Parametric Programming Procedure Applied to Wyndor Glass Co. Example

Range of θ	Basic Variable	Eq. No.	Z	x_1	x_2	x_3	x_4	x_5	Right Side	Optimal Solution
	$Z(\theta)$	0	1	0	0	0	$\dfrac{9-7\theta}{6}$	$\dfrac{3+2\theta}{3}$	$36-2\theta$	$x_4 = 0$
										$x_5 = 0$
$0 \le \theta \le \dfrac{9}{7}$	x_3	1	0	0	0	1	$\dfrac{1}{3}$	$-\dfrac{1}{3}$	2	$x_3 = 2$
	x_2	2	0	0	1	0	$\dfrac{1}{2}$	0	6	$x_2 = 6$
	x_1	3	0	1	0	0	$-\dfrac{1}{3}$	$\dfrac{1}{3}$	2	$x_1 = 2$
	$Z(\theta)$	0	1	0	0	$\dfrac{-9+7\theta}{2}$	0	$\dfrac{5-\theta}{2}$	$27+5\theta$	$x_3 = 0$
										$x_5 = 0$
$\dfrac{9}{7} \le \theta \le 5$	x_4	1	0	0	0	3	1	-1	6	$x_4 = 6$
	x_2	2	0	0	1	$-\dfrac{3}{2}$	0	$\dfrac{1}{2}$	3	$x_2 = 3$
	x_1	3	0	1	0	1	0	0	4	$x_1 = 4$
	$Z(\theta)$	0	1	0	$-5+\theta$	$3+2\theta$	0	0	$12+8\theta$	$x_2 = 0$
										$x_3 = 0$
$\theta \ge 5$	x_4	1	0	0	2	0	1	0	12	$x_4 = 12$
	x_5	2	0	0	2	-3	0	1	6	$x_5 = 6$
	x_1	3	0	1	0	1	0	0	4	$x_1 = 4$

$i = 1, 2, \ldots , m$, where the α_i are given input constants. Thus the problem becomes

$$\text{Maximize} \quad Z(\theta) = \sum_{j=1}^{n} c_j x_j,$$

$$\text{subject to} \quad \sum_{j=1}^{n} a_{ij} x_j \le b_i + \alpha_i \theta, \quad \text{for } i = 1, 2, \ldots , m,$$

$$\text{and} \quad x_j \ge 0, \quad \text{for } j = 1, 2, \ldots , n.$$

The goal is to identify the optimal solution as a function of θ.

With this formulation, the corresponding objective function value $Z^*(\theta)$ always has the *piecewise linear* and *concave*[1] form shown in Fig. 9.2. (See Prob. 25.) The set of basic variables in the optimal solution still changes (as θ increases) *only* where the slope of $Z^*(\theta)$ changes. However, in contrast to the preceding case, the values of these variables now change as a (linear) function of θ between the slope changes. The reason is that increasing θ changes the right-hand sides in the *initial* set of equations, which then causes changes in the right-hand sides in the *final* set of equations, i.e., in the values of the final set of basic variables. Figure 9.2 depicts a problem with three sets of basic variables that are optimal for different values of θ, one for $0 \le \theta \le \theta_1$, the second for $\theta_1 \le \theta \le \theta_2$, and the third for $\theta \ge \theta_2$. Within each of these

---

[1] See Appendix 1 for a definition and discussion of *concave* functions.

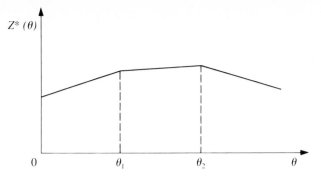

Figure 9.2 Objective function value for an optimal solution as a function of θ for parametric linear programming with systematic changes in the b_i parameters.

intervals for θ, the value of $Z^*(\theta)$ varies with θ despite the fixed coefficients c_j because the values of the x_j are changing.

The following solution procedure summary is very similar to that just presented for systematic changes in the c_j parameters. The reason is that changing the b_i is equivalent to changing the coefficients in the objective function of the *dual* model. Therefore, the procedure for the primal problem is exactly *complementary* to applying simultaneously the procedure for systematic changes in the c_j parameters to the *dual* problem. Consequently, the *dual simplex method* (see Sec. 9.2) now would be used to obtain each new optimal solution, and the applicable sensitivity analysis case (see Sec. 6.7) now is case 1, but these differences are the only major differences.

Summary of Parametric Programming Procedure for
Systematic Changes in the b_i Parameters

Step 1: Solve the problem with $\theta = 0$ by the simplex method.

Step 2: Use the sensitivity analysis procedure (case 1, Sec. 6.7) to introduce the $\Delta b_i = \alpha_i \theta$ changes into the right-side column.

Step 3: Increase θ until one of the basic variables has its value in the right-side column go negative (or until θ has been increased as far as desired).

Step 4: Use this variable as the leaving basic variable for an iteration of the dual simplex method to find the new optimal solution. Return to step 3.

To illustrate this procedure in a way that demonstrates its *duality* relationship with the procedure for systematic changes in the c_j parameters, we shall now apply it to the *dual* problem for the Wyndor Glass Co. (see Table 6.1). In particular, suppose that $\alpha_1 = 2$ and $\alpha_2 = -1$ so that the functional constraints become

$$y_1 \quad + 3y_3 \geq 3 + 2\theta, \quad \text{or} \quad -y_1 \quad - 3y_3 \leq -3 - 2\theta$$

$$2y_2 + 2y_3 \geq 5 - \theta, \quad \text{or} \quad -2y_2 - 2y_3 \leq -5 + \theta.$$

Thus the *dual* of *this* problem is just the example considered in Table 9.3.

This problem with $\theta = 0$ has already been solved in Table 9.2, so we begin with the *final* simplex tableau given there. Using the sensitivity analysis procedure

Table 9.4 The b_i Parametric Programming Procedure Applied to Dual of Wyndor Glass Co. Example

Range of θ	Basic Variable	Eq. No.	Z	y_1	y_2	y_3	y_4	y_5	Right Side	Optimal Solution
	$Z(\theta)$	0	1	2	0	0	2	6	$-36 + 2\theta$	$y_1 = y_4 = y_5 = 0$
$0 \le \theta \le \dfrac{9}{7}$	y_3	1	0	$\dfrac{1}{3}$	0	1	$-\dfrac{1}{3}$	0	$\dfrac{3 + 2\theta}{3}$	$y_3 = \dfrac{3 + 2\theta}{3}$
	y_2	2	0	$-\dfrac{1}{3}$	1	0	$\dfrac{1}{3}$	$-\dfrac{1}{2}$	$\dfrac{9 - 7\theta}{6}$	$y_2 = \dfrac{9 - 7\theta}{6}$
	$Z(\theta)$	0	1	0	6	0	4	3	$-27 - 5\theta$	$y_2 = y_4 = y_5 = 0$
$\dfrac{9}{7} \le \theta \le 5$	y_3	1	0	0	1	1	0	$-\dfrac{1}{2}$	$\dfrac{5 - \theta}{2}$	$y_3 = \dfrac{5 - \theta}{2}$
	y_1	2	0	1	-3	0	-1	$\dfrac{3}{2}$	$\dfrac{-9 + 7\theta}{2}$	$y_1 = \dfrac{-9 + 7\theta}{2}$
	$Z(\theta)$	0	1	0	12	6	4	0	$-12 - 8\theta$	$y_2 = y_3 = y_4 = 0$
$\theta \ge 5$	y_5	1	0	0	-2	-2	0	1	$-5 + \theta$	$y_5 = -5 + \theta$
	y_1	2	0	1	0	3	-1	0	$3 + 2\theta$	$y_1 = 3 + 2\theta$

for case 1, Sec. 6.7, we find that the entries in the right-side column of this tableau change to the values given below.

$$y_0^* = \mathbf{y^*\bar{b}} = [2,\ 6]\begin{bmatrix} -3 - 2\theta \\ -5 + \theta \end{bmatrix} = -36 + 2\theta,$$

$$\mathbf{b^*} = \mathbf{S^*\bar{b}} = \begin{bmatrix} -\dfrac{1}{3} & 0 \\ \dfrac{1}{3} & -\dfrac{1}{2} \end{bmatrix}\begin{bmatrix} -3 - 2\theta \\ -5 + \theta \end{bmatrix} = \begin{bmatrix} 1 + \dfrac{2\theta}{3} \\ 3 - \dfrac{7\theta}{6} \end{bmatrix}.$$

Therefore, the two basic variables in this tableau,

$$y_3 = \frac{3 + 2\theta}{3} \quad \text{and} \quad y_2 = \frac{9 - 7\theta}{6},$$

remain nonnegative for $0 \le \theta \le \frac{9}{7}$. Increasing θ past $\theta = \frac{9}{7}$ requires making y_2 a *leaving basic variable* for another *iteration* of the dual simplex method, and so on, as summarized in Table 9.4.

We suggest that you now trace through Tables 9.3 and 9.4 simultaneously to note the *duality* relationship between the two procedures.

9.4 An Interior-Point Algorithm

In Sec. 4.9 we discussed a dramatic new development in linear programming, the invention by Narendra Karmarkar of AT&T Bell Laboratories of a powerful new algorithm for solving huge linear programming problems. We now introduce the nature of Karmarkar's approach by describing a relatively elementary variant (the "affine" or "affine-scaling" variant) of his algorithm.[1]

Throughout this section we shall focus on Karmarkar's main ideas on an intuitive level while avoiding mathematical details. In particular, we shall bypass certain details that are needed for the full implementation of the algorithm (e.g., how to find an initial feasible trial solution) but are not central to a basic conceptual understanding. The ideas to be described can be summarized as follows:

Concept 1: Shoot through the *interior* of the feasible region toward an optimal solution.

Concept 2: Move in a direction that improves the objective function value at the fastest possible rate.

Concept 3: Transform the feasible region to place the current trial solution near its center, thereby enabling a large improvement when implementing Concept 2.

To illustrate these ideas throughout the section, we shall use the following example:

$$\text{Maximize} \quad Z = x_1 + 2x_2,$$

subject to

$$x_1 + x_2 \leq 8$$

and

$$x_1 \geq 0, \quad x_2 \geq 0.$$

This problem is depicted graphically in Fig. 9.3, where the optimal solution is seen to be $(x_1, x_2) = (0, 8)$ with $Z = 16$.

The Relevance of the Gradient for Concepts 1 and 2

The algorithm begins with an initial trial solution that (like all subsequent trial solutions) lies in the *interior* of the feasible region. Thus, for the example, the solution must not lie on any of the three lines ($x_1 = 0$, $x_2 = 0$, $x_1 + x_2 = 8$) that form the boundary of this region in Fig. 9.3. We have arbitrarily chosen $(x_1, x_2) = (2, 2)$ to be this initial trial solution.

To begin implementing Concepts 1 and 2, note in Fig. 9.3 that the direction of movement from (2, 2) that increases Z at the fastest possible rate is *perpendicular* to (and toward) the objective function line, $Z = 16 = x_1 + 2x_2$. We have shown this direction by the arrow from (2, 2) to (3, 4). Using vector addition,

$$(3, 4) = (2, 2) + (1, 2),$$

where the vector (1, 2) is the **gradient** of the objective function. (We will discuss *gradients* further in Sec. 14.5 in the broader context of *nonlinear programming,* where algorithms similar to Karmarkar's have long been used.) The components of (1, 2) are just the coefficients in the objective function. Thus, with one subsequent modifi-

---

[1] The basic approach for this variant actually was proposed in 1967 by a Russian mathematician, I. I. Dikin, and then rediscovered soon after the appearance of Karmarkar's work by a number of researchers, including E. R. Barnes, T. M. Cavalier, and A. L. Soyster. Also see Vanderbei, Robert J., Marc S. Meketon, and Barry A. Freedman: "A Modification of Karmarkar's Linear Programming Algorithm," *Algorithmica,* **1**(4) (Special Issue on New Approaches to Linear Programming): 395–407, 1986.

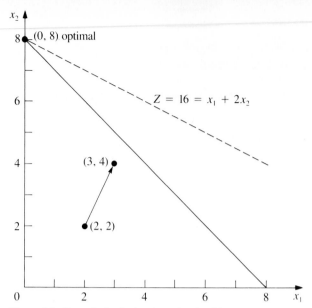

Figure 9.3 Example for the interior-point algorithm.

cation, the gradient $(1, 2)$ defines the ideal direction in which to move, where the question of the *distance to move* will be considered later.

The algorithm actually operates on linear programming problems after they have been rewritten in *augmented form*. Letting x_3 be the slack variable for the functional constraint of the example, this form is

$$\text{Maximize} \quad Z = x_1 + 2x_2,$$

subject to
$$x_1 + x_2 + x_3 = 8$$
and
$$x_1 \geq 0, \qquad x_2 \geq 0, \qquad x_3 \geq 0.$$

Using matrix notation (slightly different from Chap. 5), the augmented form can be written in general as

$$\text{Maximize} \quad Z = \mathbf{c}^\mathrm{T}\mathbf{x},$$

subject to
$$\mathbf{A}\mathbf{x} = \mathbf{b}$$
and
$$\mathbf{x} \geq \mathbf{0},$$

where
$$\mathbf{c} = \begin{bmatrix} 1 \\ 2 \\ 0 \end{bmatrix}, \quad \mathbf{x} = \begin{bmatrix} x_1 \\ x_2 \\ x_3 \end{bmatrix}, \quad \mathbf{A} = [1, 1, 1], \quad \mathbf{b} = [8], \quad \mathbf{0} = \begin{bmatrix} 0 \\ 0 \\ 0 \end{bmatrix}$$

for the example. Note that $\mathbf{c}^\mathrm{T} = [1, 2, 0]$ now is the *gradient* of the objective function.

The augmented form of the example is depicted graphically in Fig. 9.4. The feasible region now consists of the triangle with vertices $(8, 0, 0)$, $(0, 8, 0)$, $(0, 0, 8)$. Points in the *interior* of this feasible region are those where $x_1 > 0$, $x_2 > 0$, and $x_3 > 0$. Each of these three $x_j > 0$ conditions has the effect of forcing (x_1, x_2) away from one of the three lines forming the boundary of the feasible region in Fig. 9.3.

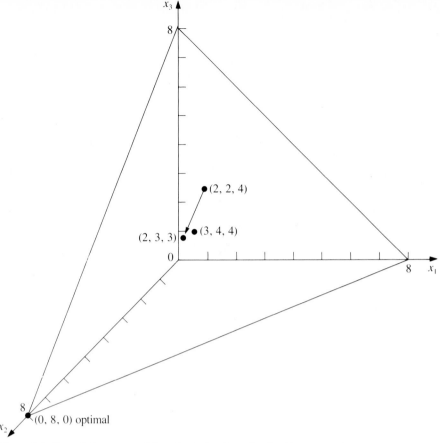

Figure 9.4 Example in augmented form for the interior-point algorithm.

Using the Projected Gradient to Implement Concepts 1 and 2

In augmented form, the initial trial solution for the example is $(x_1, x_2, x_3) = (2, 2, 4)$. Adding the gradient $(1, 2, 0)$ leads to

$$(3, 4, 4) = (2, 2, 4) + (1, 2, 0).$$

However, now there is a complication. The algorithm cannot move from $(2, 2, 4)$ toward $(3, 4, 4)$, because $(3, 4, 4)$ is infeasible! When $x_1 = 3$ and $x_2 = 4$, then $x_3 = 8 - x_1 - x_2 = 1$ instead of 4. The point $(3, 4, 4)$ lies on the near side as you look down on the feasible triangle in Fig. 9.4. Therefore, to remain feasible, the algorithm (indirectly) *projects* the point $(3, 4, 4)$ down onto the feasible triangle by dropping a line that is *perpendicular* to this triangle. This perpendicular line intersects the triangle at $(2, 3, 3)$. Because

$$(2, 3, 3) = (2, 2, 4) + (0, 1, -1),$$

the **projected gradient** of the objective function (the gradient projected onto the feasible region) is $(0, 1, -1)$. It is this projected gradient that defines the direction of movement for the algorithm, as shown by the arrow in Fig. 9.4.

A formula is available for computing the projected gradient directly. Defining the *projection matrix* **P** as

$$\mathbf{P} = \mathbf{I} - \mathbf{A}^T(\mathbf{A}\,\mathbf{A}^T)^{-1}\,\mathbf{A},$$

the *projected gradient* (in column form) is

$$\mathbf{c}_p = \mathbf{P}\,\mathbf{c}.$$

Thus, for the example,

$$\mathbf{P} = \begin{bmatrix} 1 & 0 & 0 \\ 0 & 1 & 0 \\ 0 & 0 & 1 \end{bmatrix} - \begin{bmatrix} 1 \\ 1 \\ 1 \end{bmatrix}\left([1\ \ 1\ \ 1]\begin{bmatrix} 1 \\ 1 \\ 1 \end{bmatrix}\right)^{-1}[1\ \ 1\ \ 1]$$

$$= \begin{bmatrix} 1 & 0 & 0 \\ 0 & 1 & 0 \\ 0 & 0 & 1 \end{bmatrix} - \tfrac{1}{3}\begin{bmatrix} 1 \\ 1 \\ 1 \end{bmatrix}[1\ \ 1\ \ 1]$$

$$= \begin{bmatrix} 1 & 0 & 0 \\ 0 & 1 & 0 \\ 0 & 0 & 1 \end{bmatrix} - \tfrac{1}{3}\begin{bmatrix} 1 & 1 & 1 \\ 1 & 1 & 1 \\ 1 & 1 & 1 \end{bmatrix} = \begin{bmatrix} \tfrac{2}{3} & -\tfrac{1}{3} & -\tfrac{1}{3} \\ -\tfrac{1}{3} & \tfrac{2}{3} & -\tfrac{1}{3} \\ -\tfrac{1}{3} & -\tfrac{1}{3} & \tfrac{2}{3} \end{bmatrix},$$

so

$$\mathbf{c}_p = \begin{bmatrix} \tfrac{2}{3} & -\tfrac{1}{3} & -\tfrac{1}{3} \\ -\tfrac{1}{3} & \tfrac{2}{3} & -\tfrac{1}{3} \\ -\tfrac{1}{3} & -\tfrac{1}{3} & \tfrac{2}{3} \end{bmatrix}\begin{bmatrix} 1 \\ 2 \\ 0 \end{bmatrix} = \begin{bmatrix} 0 \\ 1 \\ -1 \end{bmatrix}.$$

Moving from (2, 2, 4) in the direction of the projected gradient (0, 1, −1) involves increasing α from zero in the formula,

$$\mathbf{x} = \begin{bmatrix} 2 \\ 2 \\ 4 \end{bmatrix} + 4\alpha\mathbf{c}_p = \begin{bmatrix} 2 \\ 2 \\ 4 \end{bmatrix} + 4\alpha\begin{bmatrix} 0 \\ 1 \\ -1 \end{bmatrix},$$

where the coefficient 4 is used simply to give an upper bound of 1 for α to maintain feasibility (all $x_j \geq 0$). Thus α measures the fraction used of the distance that could be moved before leaving the feasible region.

How large should α be made for moving to the next trial solution? Because the increase in Z is proportional to α, a value close to the upper bound of 1 is good for giving a relatively large step toward optimality on the current iteration. However, the problem with a value too close to 1 is that the next trial solution then is jammed against a constraint boundary, thereby making it difficult to take large improving steps during subsequent iterations. It is very helpful for trial solutions to be near the center of the feasible region (or at least the portion of the feasible region in the vicinity of an optimal solution), and not too close to any constraint boundary. With this in mind, Karmarkar has stated for his algorithm that a value as large as $\alpha = 0.25$ should be "safe." In practice, much larger values (for example, $\alpha = 0.9$) sometimes are used. For purposes of this example (and the problems at the end of the chapter), we have chosen $\alpha = 0.5$.

A Centering Scheme for Implementing Concept 3

We now have just one more step to complete the description of the algorithm, namely, a special scheme for transforming the feasible region to place the current trial solution near its center. We have just described the benefit of having the trial solution near the center, but another important benefit of this centering scheme is that it keeps turning the direction of the projected gradient to point more nearly toward an optimal solution as the algorithm converges toward this solution.

The basic idea of the centering scheme is straightforward—simply change the scale (units) for each of the variables so that the trial solution becomes equidistant from the constraint boundaries in the new coordinate system. (Karmarkar's original algorithm uses a more sophisticated centering scheme.)

For the example, there are three constraint boundaries in Fig. 9.3, each one corresponding to a zero value for one of the three variables of the problem in augmented form, namely, $x_1 = 0$, $x_2 = 0$, and $x_3 = 0$. In Fig. 9.4, see how these three constraint boundaries intersect the $\mathbf{A}x = \mathbf{b}$ ($x_1 + x_2 + x_3 = 8$) plane to form the boundary of the feasible region. The initial trial solution is $(x_1, x_2, x_3) = (2, 2, 4)$, so this solution is two units away from the $x_1 = 0$ and $x_2 = 0$ constraint boundaries and four units away from the $x_3 = 0$ constraint boundary, when using the units of the respective variables. However, whatever these units are in each case, they are quite arbitrary and can be changed as desired without changing the problem. Therefore, let us rescale the variables as follows:

$$\tilde{x}_1 = \frac{x_1}{2}, \qquad \tilde{x}_2 = \frac{x_2}{2}, \qquad \tilde{x}_3 = \frac{x_3}{4}$$

in order to make the current trial solution of $(x_1, x_2, x_3) = (2, 2, 4)$ become

$$(\tilde{x}_1, \tilde{x}_2, \tilde{x}_3) = (1, 1, 1).$$

In these new coordinates, the problem becomes

$$\text{Maximize} \qquad Z = 2\tilde{x}_1 + 4\tilde{x}_2,$$

subject to
$$2\tilde{x}_1 + 2\tilde{x}_2 + 4\tilde{x}_3 = 8$$

and
$$\tilde{x}_1 \geq 0, \qquad \tilde{x}_2 \geq 0, \qquad \tilde{x}_3 \geq 0,$$

as depicted graphically in Fig. 9.5.

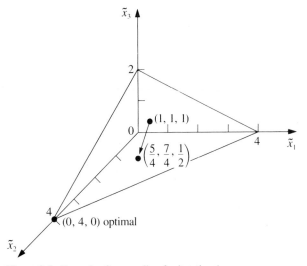

Figure 9.5 Example after rescaling for iteration 1.

Note that the trial solution $(1, 1, 1)$ in Fig. 9.5 is equidistant from the three constraint boundaries, $\tilde{x}_1 = 0$, $\tilde{x}_2 = 0$, and $\tilde{x}_3 = 0$. For each subsequent iteration as well, the problem is rescaled again to achieve this same property, so that the current trial solution always is $(1, 1, 1)$ in the current coordinates.

Summary and Illustration of the Algorithm

Now let us summarize and illustrate the algorithm by going through the first iteration for the example, then giving a summary of the general procedure, and then applying this summary to a second iteration.

ITERATION 1: Given the initial trial solution, $(x_1, x_2, x_3) = (2, 2, 4)$, let $\mathbf{D}$ be the corresponding *diagonal matrix,* so that

$$\mathbf{D} = \begin{bmatrix} 2 & 0 & 0 \\ 0 & 2 & 0 \\ 0 & 0 & 4 \end{bmatrix}.$$

The rescaled variables then are the components of

$$\tilde{\mathbf{x}} = \mathbf{D}^{-1}\mathbf{x} = \begin{bmatrix} \dfrac{1}{2} & 0 & 0 \\ 0 & \dfrac{1}{2} & 0 \\ 0 & 0 & \dfrac{1}{4} \end{bmatrix} \begin{bmatrix} x_1 \\ x_2 \\ x_3 \end{bmatrix} = \begin{bmatrix} \dfrac{x_1}{2} \\ \dfrac{x_2}{2} \\ \dfrac{x_3}{4} \end{bmatrix}.$$

In these new coordinates, $\mathbf{A}$ and $\mathbf{c}$ have become

$$\tilde{\mathbf{A}} = \mathbf{AD} = \begin{bmatrix} 1 & 1 & 1 \end{bmatrix} \begin{bmatrix} 2 & 0 & 0 \\ 0 & 2 & 0 \\ 0 & 0 & 4 \end{bmatrix} = \begin{bmatrix} 2 & 2 & 4 \end{bmatrix},$$

$$\tilde{\mathbf{c}} = \mathbf{Dc} = \begin{bmatrix} 2 & 0 & 0 \\ 0 & 2 & 0 \\ 0 & 0 & 4 \end{bmatrix} \begin{bmatrix} 1 \\ 2 \\ 0 \end{bmatrix} = \begin{bmatrix} 2 \\ 4 \\ 0 \end{bmatrix}.$$

Therefore, the *projection matrix* is

$$\mathbf{P} = \mathbf{I} - \tilde{\mathbf{A}}^T(\tilde{\mathbf{A}}\tilde{\mathbf{A}}^T)^{-1}\tilde{\mathbf{A}}$$

$$= \begin{bmatrix} 1 & 0 & 0 \\ 0 & 1 & 0 \\ 0 & 0 & 1 \end{bmatrix} - \begin{bmatrix} 2 \\ 2 \\ 4 \end{bmatrix} \left(\begin{bmatrix} 2 & 2 & 4 \end{bmatrix} \begin{bmatrix} 2 \\ 2 \\ 4 \end{bmatrix} \right)^{-1} \begin{bmatrix} 2 & 2 & 4 \end{bmatrix}$$

$$= \begin{bmatrix} 1 & 0 & 0 \\ 0 & 1 & 0 \\ 0 & 0 & 1 \end{bmatrix} - \frac{1}{24}\begin{bmatrix} 4 & 4 & 8 \\ 4 & 4 & 8 \\ 8 & 8 & 16 \end{bmatrix} = \begin{bmatrix} \frac{5}{6} & -\frac{1}{6} & -\frac{1}{3} \\ -\frac{1}{6} & \frac{5}{6} & -\frac{1}{3} \\ -\frac{1}{3} & -\frac{1}{3} & \frac{1}{3} \end{bmatrix},$$

so that the *projected gradient* is

$$\mathbf{c}_p = \mathbf{P}\tilde{\mathbf{c}} = \begin{bmatrix} \frac{5}{6} & -\frac{1}{6} & -\frac{1}{3} \\ -\frac{1}{6} & \frac{5}{6} & -\frac{1}{3} \\ -\frac{1}{3} & -\frac{1}{3} & \frac{1}{3} \end{bmatrix} \begin{bmatrix} 2 \\ 4 \\ 0 \end{bmatrix} = \begin{bmatrix} 1 \\ 3 \\ -2 \end{bmatrix}.$$

Define v as the *absolute value* of the *negative* component of $\mathbf{c}_p$ having the *largest* absolute value, so that $v = |-2| = 2$ in this case. Consequently, in the *current* coordinates, the algorithm now moves from the current trial solution, $(\tilde{x}_1, \tilde{x}_2, \tilde{x}_3) = (1, 1, 1)$, to the next trial solution,

$$\tilde{\mathbf{x}} = \begin{bmatrix} 1 \\ 1 \\ 1 \end{bmatrix} + \frac{\alpha}{v} \mathbf{c}_p = \begin{bmatrix} 1 \\ 1 \\ 1 \end{bmatrix} + \frac{0.5}{2} \begin{bmatrix} 1 \\ 3 \\ -2 \end{bmatrix} = \begin{bmatrix} \frac{5}{4} \\ \frac{7}{4} \\ \frac{1}{2} \end{bmatrix},$$

as shown in Fig. 9.5. (The definition of v has been chosen to make the smallest component of $\tilde{\mathbf{x}}$ equal to zero when $\alpha = 1$ in this equation for the next trial solution.) In the *original* coordinates, this solution is

$$\begin{bmatrix} x_1 \\ x_2 \\ x_3 \end{bmatrix} = \mathbf{D}\tilde{\mathbf{x}} = \begin{bmatrix} 2 & 0 & 0 \\ 0 & 2 & 0 \\ 0 & 0 & 4 \end{bmatrix} \begin{bmatrix} \frac{5}{4} \\ \frac{7}{4} \\ \frac{1}{2} \end{bmatrix} = \begin{bmatrix} \frac{5}{2} \\ \frac{7}{2} \\ 2 \end{bmatrix}.$$

This completes the iteration, and this new solution will be used to start the next iteration.

These steps can be summarized as follows for any iteration.

Summary of the Interior-Point Algorithm

Step 1: Given the current trial solution $(x_1, x_2, \ldots, x_n)$, set

$$\mathbf{D} = \begin{bmatrix} x_1 & 0 & 0 & \cdots & 0 \\ 0 & x_2 & 0 & \cdots & 0 \\ 0 & 0 & x_3 & \cdots & 0 \\ \vdots & \vdots & \vdots & & \vdots \\ 0 & 0 & 0 & \cdots & x_n \end{bmatrix}.$$

Step 2: Calculate $\tilde{\mathbf{A}} = \mathbf{AD}$ and $\tilde{\mathbf{c}} = \mathbf{Dc}$.

Step 3: Calculate $\mathbf{P} = \mathbf{I} - \tilde{\mathbf{A}}^{\mathsf{T}}(\tilde{\mathbf{A}}\tilde{\mathbf{A}}^{\mathsf{T}})^{-1} \tilde{\mathbf{A}}$ and $\mathbf{c}_p = \mathbf{P}\tilde{\mathbf{c}}$.

Step 4: Identify the negative component of $\mathbf{c}_p$ having the largest absolute value, and set v equal to this absolute value. Then calculate

$$\tilde{\mathbf{x}} = \begin{bmatrix} 1 \\ 1 \\ \vdots \\ 1 \end{bmatrix} + \frac{\alpha}{v} \mathbf{c}_p,$$

where α is a selected constant between 0 and 1 (e.g., $\alpha = 0.5$).

Step 5: Calculate $\mathbf{x} = \mathbf{D}\tilde{\mathbf{x}}$ as the trial solution for the next iteration (step 1). (If this trial solution is virtually unchanged from the preceding one, then the algorithm has virtually converged to an optimal solution, so stop.)

Now let us apply this summary to iteration 2 for the example.

ITERATION 2:

Step 1: Given the current trial solution, $(x_1, x_2, x_3) = (\frac{5}{2}, \frac{7}{2}, 2)$, set

$$\mathbf{D} = \begin{bmatrix} \frac{5}{2} & 0 & 0 \\ 0 & \frac{7}{2} & 0 \\ 0 & 0 & 2 \end{bmatrix}.$$

(Note that the rescaled variables are

$$\begin{bmatrix} \tilde{x}_1 \\ \tilde{x}_2 \\ \tilde{x}_3 \end{bmatrix} = \mathbf{D}^{-1} \mathbf{x} = \begin{bmatrix} \frac{2}{5} & 0 & 0 \\ 0 & \frac{2}{7} & 0 \\ 0 & 0 & \frac{1}{2} \end{bmatrix} \begin{bmatrix} x_1 \\ x_2 \\ x_3 \end{bmatrix} = \begin{bmatrix} \frac{2}{5}x_1 \\ \frac{2}{7}x_2 \\ \frac{1}{2}x_3 \end{bmatrix},$$

so that the basic feasible solutions in these new coordinates are

$$\tilde{\mathbf{x}} = \mathbf{D}^{-1} \begin{bmatrix} 8 \\ 0 \\ 0 \end{bmatrix} = \begin{bmatrix} \frac{16}{5} \\ 0 \\ 0 \end{bmatrix}, \qquad \tilde{\mathbf{x}} = \mathbf{D}^{-1} \begin{bmatrix} 0 \\ 8 \\ 0 \end{bmatrix} = \begin{bmatrix} 0 \\ \frac{16}{7} \\ 0 \end{bmatrix},$$

and $\qquad \tilde{\mathbf{x}} = \mathbf{D}^{-1} \begin{bmatrix} 0 \\ 0 \\ 8 \end{bmatrix} = \begin{bmatrix} 0 \\ 0 \\ 4 \end{bmatrix},$

as depicted in Fig. 9.6.)

Step 2:

$$\tilde{\mathbf{A}} = \mathbf{AD} = [\frac{5}{2}, \frac{7}{2}, 2] \qquad \text{and} \qquad \tilde{\mathbf{c}} = \mathbf{Dc} = \begin{bmatrix} \frac{5}{2} \\ 7 \\ 0 \end{bmatrix}.$$

Step 3:

$$\mathbf{P} = \begin{bmatrix} \frac{13}{18} & -\frac{7}{18} & -\frac{2}{9} \\ -\frac{7}{18} & \frac{41}{90} & -\frac{14}{45} \\ -\frac{2}{9} & -\frac{14}{45} & \frac{37}{45} \end{bmatrix} \qquad \text{and} \qquad \mathbf{c}_p = \begin{bmatrix} -\frac{11}{12} \\ \frac{133}{60} \\ -\frac{41}{15} \end{bmatrix}.$$

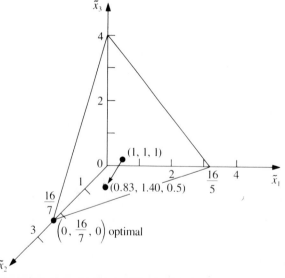

Figure 9.6 Example after rescaling for iteration 2.

Step 4: $\left|-\frac{41}{15}\right| > \left|-\frac{11}{12}\right|$, so $v = \frac{41}{15}$, so

$$\tilde{\mathbf{x}} = \begin{bmatrix} 1 \\ 1 \\ 1 \end{bmatrix} + \frac{0.5}{\frac{41}{15}} \begin{bmatrix} -\frac{11}{12} \\ \frac{133}{60} \\ -\frac{41}{15} \end{bmatrix} = \begin{bmatrix} \frac{273}{328} \\ \frac{461}{328} \\ \frac{1}{2} \end{bmatrix} \approx \begin{bmatrix} 0.83 \\ 1.40 \\ 0.50 \end{bmatrix}.$$

Step 5:

$$\mathbf{x} = \mathbf{D}\tilde{\mathbf{x}} = \begin{bmatrix} \frac{1365}{656} \\ \frac{3227}{656} \\ 1 \end{bmatrix} \approx \begin{bmatrix} 2.08 \\ 4.92 \\ 1.00 \end{bmatrix}$$

is the trial solution for iteration 3.

Since there is little to be learned by repeating these calculations for additional iterations, we shall stop here. However, we do show in Fig. 9.7 the reconfigured feasible region after rescaling based on the trial solution just obtained for iteration 3. As always, the rescaling has placed the trial solution at $(\tilde{x}_1, \tilde{x}_2, \tilde{x}_3) = (1, 1, 1)$, equidistant from the $\tilde{x}_1 = 0$, $\tilde{x}_2 = 0$, and $\tilde{x}_3 = 0$ constraint boundaries. Note in Figs. 9.5, 9.6, and 9.7 how the sequence of iterations and rescaling have the effect of "sliding" the optimal solution toward $(1, 1, 1)$ while the other basic feasible solutions tend to slide away. Eventually, after enough iterations, the optimal solution will lie very near $(\tilde{x}_1, \tilde{x}_2, \tilde{x}_3) = (0, 1, 0)$ after rescaling, while the other two basic feasible

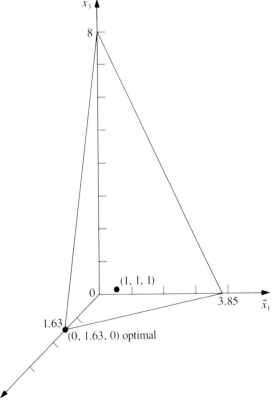

Figure 9.7 Example after rescaling for iteration 3.

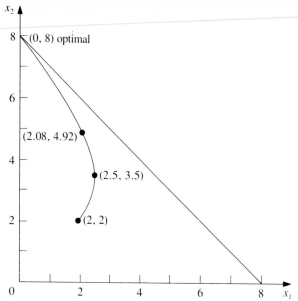

Figure 9.8 Trajectory of the interior-point algorithm for the example in the original x_1-x_2 coordinate system.

solutions will be *very* far from the origin on the $\tilde{x}_1$ and $\tilde{x}_3$ axes. Step 5 of that iteration then will yield a solution in the original coordinates very near the optimal solution of $(x_1, x_2, x_3) = (0, 8, 0)$.

Figure 9.8 shows the progress of the algorithm in the original x_1-x_2 coordinate system before augmenting the problem. The three points—$(x_1, x_2) = (2, 2)$, $(2.5, 3.5)$, and $(2.08, 4.92)$—are the trial solutions for initiating iterations 1, 2, and 3, respectively. We then have drawn a smooth curve through and beyond these points to show the trajectory of the algorithm in subsequent iterations as it approaches (x_1, x_2) $= (0, 8)$.

The functional constraint for this particular example happened to be an inequality constraint. However, equality constraints cause no difficulty for the algorithm, since it deals with the constraints only after any necessary augmenting has been done to convert them to equality form, $\mathbf{Ax} = \mathbf{b}$, anyway. To illustrate, suppose that the only change in the example is that the constraint, $x_1 + x_2 \leq 8$, is changed to $x_1 + x_2 = 8$. Thus the feasible region in Fig. 9.3 changes to just the line segment between $(8, 0)$ and $(0, 8)$. Given an initial feasible trial solution in the *interior* ($x_1 > 0$ and $x_2 > 0$) of this line segment—say, $(x_1, x_2) = (4, 4)$—the algorithm can proceed just as presented in the five-step summary with just the two variables and $\mathbf{A} = [1 \quad 1]$. For each iteration, the *projected gradient* points along this line segment in the direction of $(0, 8)$. With $\alpha = \frac{1}{2}$, iteration 1 leads from $(4, 4)$ to $(2, 6)$, iteration 2 leads from $(2, 6)$ to $(1, 7)$, etc. (Problem 29 asks you to verify these results.)

Although either version of the example has only one functional constraint, having more than one leads to just one change in the procedure as already illustrated (other than more extensive calculations). Having a single functional constraint in the example meant that $\mathbf{A}$ had only a single row, so the $(\tilde{\mathbf{A}}\tilde{\mathbf{A}}^\mathsf{T})^{-1}$ term in step 3 only involved taking the reciprocal of the number obtained from the vector product $(\tilde{\mathbf{A}}\tilde{\mathbf{A}}^\mathsf{T})$. Multiple functional constraints means that $\mathbf{A}$ has multiple rows, so then the $(\tilde{\mathbf{A}}\tilde{\mathbf{A}}^\mathsf{T})^{-1}$

term involves finding the *inverse* of the matrix obtained from the matrix product $(\tilde{\mathbf{A}}\tilde{\mathbf{A}}^T)$.

To conclude, we need to add a comment to place the algorithm into better perspective. For our extremely small example, the algorithm requires relatively extensive calculations, and then, after many iterations, obtains only an approximation of the optimal solution. By contrast, the graphical procedure of Sec. 3.1 finds the optimal solution in Fig. 9.3 immediately, and the *simplex method* requires only one quick iteration. However, do not let this contrast fool you into downgrading the efficiency of the interior-point algorithm. This algorithm is designed for dealing with *big* problems having many hundreds or thousands of functional constraints. The simplex method typically requires thousands of iterations on such problems. By "shooting" through the interior of the feasible region, the interior-point algorithm tends to require a substantially smaller number of iterations (although with considerably more work per iteration). Therefore, as discussed in Sec. 4.9, interior-point algorithms similar to the one presented here should play an important role in the future of linear programming.

9.5 The Decomposition Principle for Multidivisional Problems

In Sec. 7.5, we discussed the special class of linear programming problems called *multidivisional problems* and their special *block angular structure* (see Table 7.32). We also mentioned that the streamlined version of the simplex method called the *decomposition principle* provides an effective way of exploiting this special structure to solve very large problems. (This approach also is applicable to the *dual* of the class of *multitime period problems* presented in Sec. 7.6.) We shall describe and illustrate this procedure after reformulating (decomposing) the problem in a way that enables the algorithm to exploit its special structure.

A Useful Reformulation (Decomposition) of the Problem

The basic approach is to reformulate the problem in a way that greatly reduces the number of functional constraints and then to apply the *revised simplex method* (see Sec. 5.2). Therefore, we need to begin by giving the *matrix form* of multidivisional problems:

$$\text{Maximize} \quad Z = \mathbf{c}\mathbf{x},$$

subject to
$$\mathbf{A}\mathbf{x} \leq \mathbf{b}\dagger \quad \text{and} \quad \mathbf{x} \geq \mathbf{0},$$

where the **A** matrix has the block angular structure

$$\mathbf{A} = \begin{bmatrix} \mathbf{A}_1 & \mathbf{A}_2 & \cdots & \mathbf{A}_N \\ \mathbf{A}_{N+1} & \mathbf{0} & \cdots & \mathbf{0} \\ \mathbf{0} & \mathbf{A}_{N+2} & \cdots & \mathbf{0} \\ \vdots & \vdots & & \vdots \\ \mathbf{0} & \mathbf{0} & \cdots & \mathbf{A}_{2N} \end{bmatrix}$$

†The following discussion would not be changed substantially if $\mathbf{A}\mathbf{x} = \mathbf{b}$.

where the $\mathbf{A}_i$ $(i = 1, 2, \ldots, 2N)$ are matrices, and the $\mathbf{0}$ are null matrices. Expanding, this can be rewritten as

$$\text{Maximize} \quad Z = \sum_{j=1}^{N} \mathbf{c}_j \mathbf{x}_j,$$

subject to
$$[\mathbf{A}_1, \mathbf{A}_2, \ldots, \mathbf{A}_N, \mathbf{I}]\begin{bmatrix} \mathbf{x} \\ \mathbf{x}_s \end{bmatrix} = \mathbf{b}_0, \quad \begin{bmatrix} \mathbf{x} \\ \mathbf{x}_s \end{bmatrix} \geq \mathbf{0},$$

$$\mathbf{A}_{N+j}\mathbf{x}_j \leq \mathbf{b}_j \quad \text{and} \quad \mathbf{x}_j \geq \mathbf{0}, \quad \text{for } j = 1, 2, \ldots, N,$$

where $\mathbf{c}_j$, $\mathbf{x}_j$, $\mathbf{b}_0$, and $\mathbf{b}_j$ are vectors such that $\mathbf{c} = [\mathbf{c}_1, \mathbf{c}_2, \ldots, \mathbf{c}_N]$,

$$\mathbf{x} = \begin{bmatrix} \mathbf{x}_1 \\ \mathbf{x}_2 \\ \cdot \\ \cdot \\ \cdot \\ \mathbf{x}_N \end{bmatrix}, \quad \mathbf{b} = \begin{bmatrix} \mathbf{b}_0 \\ \mathbf{b}_1 \\ \cdot \\ \cdot \\ \cdot \\ \mathbf{b}_N \end{bmatrix},$$

and where $\mathbf{x}_s$ is the vector of slack variables for the first set of constraints.

This structure suggests that it may be possible to solve the overall problem by doing little more than solving the N *subproblems* of the form

$$\text{Maximize} \quad Z_j = \mathbf{c}_j \mathbf{x}_j,$$

subject to
$$\mathbf{A}_{N+j}\mathbf{x}_j \leq \mathbf{b}_j \quad \text{and} \quad \mathbf{x}_j \geq \mathbf{0},$$

thereby greatly reducing computational effort. After some reformulation, this approach can indeed be used.

Assume that the set of feasible solutions for each subproblem is a bounded set (i.e., none of the variables can approach infinity). Although a more complicated version of the approach can still be used otherwise, this assumption will simplify the discussion.

The set of points $\mathbf{x}_j$ such that $\mathbf{x}_j \geq \mathbf{0}$ and $\mathbf{A}_{N+j}\mathbf{x}_j \leq \mathbf{b}_j$ constitutes a *convex set* with a finite number of *extreme points* (the *corner-point feasible solutions* for the subproblem having these constraints).[1] Therefore, under the assumption that the set is bounded, any point in the set can be represented as a weighted average of the extreme points. To express this mathematically, let n_j be the number of extreme points, and denote these points by $\mathbf{x}_{jk}^*$ for $k = 1, 2, \ldots, n_j$. Then any solution $\mathbf{x}_j$ to subproblem j which satisfies the constraints $\mathbf{A}_{N+j}\mathbf{x}_j \leq \mathbf{b}_j$ and $\mathbf{x}_j \geq \mathbf{0}$ also satisfies the equation

$$\mathbf{x}_j = \sum_{k=1}^{n_j} \rho_{jk}\mathbf{x}_{jk}^*$$

for some combination of the ρ_{jk} such that

$$\sum_{k=1}^{n_j} \rho_{jk} = 1$$

and $\rho_{jk} \geq 0$ $(k = 1, 2, \ldots, n_j)$. Furthermore, this is not true for any $\mathbf{x}_j$ that is not a feasible solution for subproblem j. (You may have shown these facts for Prob. 13, Chap. 4.)

---

[1] See Appendix 1 for a definition and discussion of *convex sets* and *extreme points*.

Therefore, this equation for $\mathbf{x}_j$ and the constraints on the ρ_{jk} provide a method for representing the feasible solutions to subproblem j without using any of the original constraints. Hence, the overall problem can now be reformulated with far fewer constraints as

$$\text{Maximize} \quad Z = \sum_{j=1}^{N} \sum_{k=1}^{n_j} (\mathbf{c}_j \mathbf{x}_{jk}^*)\rho_{jk},$$

subject to

$$\sum_{j=1}^{N} \sum_{k=1}^{n_j} (\mathbf{A}_j \mathbf{x}_{jk}^*)\rho_{jk} + \mathbf{x}_s = \mathbf{b}_0, \ \mathbf{x}_s \geq \mathbf{0}, \ \sum_{k=1}^{n_j} \rho_{jk} = 1, \quad \text{for } j = 1, 2, \ldots, N,$$

and $\quad \rho_{jk} \geq 0, \quad \text{for } j = 1, 2, \ldots, N \quad \text{and} \quad k = 1, 2, \ldots, n_j.$

This formulation is completely equivalent to the one given earlier. However, since it has far fewer constraints, it should be solvable with much less computational effort. The fact that the number of variables (which are now the ρ_{jk} and the elements of $\mathbf{x}_s$) is much larger does not matter much computationally if the *revised simplex method* is used. The one apparent flaw is that it would be tedious to identify all the $\mathbf{x}_{jk}^*$. Fortunately, it is not necessary to do this when using the revised simplex method. The procedure is outlined below.

The Algorithm Based on This Decomposition

Let $\mathbf{A}'$ be the matrix of constraint coefficients for this reformulation of the problem, and let $\mathbf{c}'$ be the vector of objective function coefficients. (The individual elements of $\mathbf{A}'$ and $\mathbf{c}'$ are determined only when they are needed.) As usual, let $\mathbf{B}$ be the current basis matrix, and let $\mathbf{c}_B$ be the corresponding vector of basic variable coefficients in the objective function.

For a portion of the work required for the optimality test and part 1 of the iterative step, the *revised simplex method* needs to find the *minimum* element of $(\mathbf{c}_B \mathbf{B}^{-1} \mathbf{A}' - \mathbf{c}')$, the vector of coefficients of the *original* variables (the ρ_{jk} in this case) in the *current* Eq. (0). Let $(z_{jk} - c_{jk})$ denote the element in this vector corresponding to ρ_{jk}. Let m_0 denote the number of elements of $\mathbf{b}_0$. Let $(\mathbf{B}^{-1})_{1;m_0}$ be the matrix consisting of the first m_0 columns of $\mathbf{B}^{-1}$, and let $(\mathbf{B}^{-1})_i$ be the vector consisting of the ith column of $\mathbf{B}^{-1}$. Then $(z_{jk} - c_{jk})$ reduces to

$$z_{jk} - c_{jk} = \mathbf{c}_B(\mathbf{B}^{-1})_{1;m_0}\mathbf{A}_j\mathbf{x}_{jk}^* + \mathbf{c}_B(\mathbf{B}^{-1})_{m_0+j} - \mathbf{c}_j\mathbf{x}_{jk}^*$$

$$= (\mathbf{c}_B(\mathbf{B}^{-1})_{1;m_0}\mathbf{A}_j - \mathbf{c}_j)\mathbf{x}_{jk}^* + \mathbf{c}_B(\mathbf{B}^{-1})_{m_0+j}.$$

Since $\mathbf{c}_B(\mathbf{B}^{-1})_{m_0+j}$ is independent of k, the *minimum* value of $(z_{jk} - c_{jk})$ over $k = 1, 2, \ldots, n_j$ can be found as follows. The $\mathbf{x}_{jk}^*$ are just the corner-point feasible solutions for the set of constraints, $\mathbf{x}_j \geq \mathbf{0}$ and $\mathbf{A}_{N+j}\mathbf{x}_j \leq \mathbf{b}_j$, and the simplex method identifies the corner-point feasible solution which minimizes (or maximizes) a given objective function. Therefore, solve the linear programming problem

$$\text{Minimize} \quad W_j = (\mathbf{c}_B(\mathbf{B}^{-1})_{1;m_0}\mathbf{A}_j - \mathbf{c}_j)\mathbf{x}_j + \mathbf{c}_B(\mathbf{B}^{-1})_{m_0+j},$$

subject to $\quad \mathbf{A}_{N+j}\mathbf{x}_j \leq \mathbf{b}_j \quad \text{and} \quad \mathbf{x}_j \geq \mathbf{0}.$

The optimal value of W_j (denoted by W_j^*) is the desired minimum value of $(z_{jk} - c_{jk})$ over k. Furthermore, the optimal solution for $\mathbf{x}_j$ is the corresponding $\mathbf{x}_{jk}^*$.

Therefore, the first step at each iteration requires solving N linear programming problems of the above type to find W_j^* for $j = 1, 2, \ldots, N$. In addition, the *current* Eq. (0) coefficients of the elements of $\mathbf{x}_s$ that are *nonbasic* variables would be found in the usual way as the elements of $\mathbf{c}_B(\mathbf{B}^{-1})_{1:m_0}$. If all these coefficients [the W_j^* and the elements of $\mathbf{c}_B(\mathbf{B}^{-1})_{1:m_0}$] are *nonnegative*, the current solution is *optimal* by the optimality test. Otherwise, the *minimum* of these coefficients is found, and the corresponding variable is selected as the new *entering basic variable*. If that variable is ρ_{jk}, then the solution to the linear programming problem involving W_j has identified $\mathbf{x}_{jk}^*$, so that the original constraint coefficients of ρ_{jk} are now identified. Hence, the revised simplex method can complete the iteration in the usual way.

Assuming that $\mathbf{x} = \mathbf{0}$ is feasible for the original problem, the *initialization step* would use the corresponding solution in the reformulated problem as the *initial basic feasible solution*. This involves selecting the initial *set of basic variables* (the elements of $\mathbf{x}_B$) to be the elements of $\mathbf{x}_s$ and the one variable ρ_{jk} for each subproblem j ($j = 1, 2, \ldots, N$) such that $\mathbf{x}_{jk}^* = \mathbf{0}$. Following the initialization step, the above procedure is repeated for a succession of iterations until an optimal solution is reached. The optimal values of the ρ_{jk} are then substituted into the equations for the $\mathbf{x}_j$ for the optimal solution to conform to the original form of the problem.

EXAMPLE: To illustrate this procedure, consider the problem

$$\text{Maximize} \quad Z = 4x_1 + 6x_2 + 8x_3 + 5x_4,$$

subject to

$$x_1 + 3x_2 + 2x_3 + 4x_4 \leq 20$$

$$2x_1 + 3x_2 + 6x_3 + 4x_4 \leq 25$$

$$x_1 + x_2 \qquad\qquad \leq 5$$

$$x_1 + 2x_2 \qquad\qquad \leq 8$$

$$4x_3 + 3x_4 \leq 12$$

and

$$x_j \geq 0, \quad \text{for } j = 1, 2, 3, 4.$$

Thus the $\mathbf{A}$ matrix is

$$\mathbf{A} = \begin{bmatrix} 1 & 3 & 2 & 4 \\ 2 & 3 & 6 & 4 \\ \hline 1 & 1 & 0 & 0 \\ 1 & 2 & 0 & 0 \\ \hline 0 & 0 & 4 & 3 \end{bmatrix},$$

so that $N = 2$ and

$$\mathbf{A}_1 = \begin{bmatrix} 1 & 3 \\ 2 & 3 \end{bmatrix}, \quad \mathbf{A}_2 = \begin{bmatrix} 2 & 4 \\ 6 & 4 \end{bmatrix}, \quad \mathbf{A}_3 = \begin{bmatrix} 1 & 1 \\ 1 & 2 \end{bmatrix}, \quad \mathbf{A}_4 = [4, 3].$$

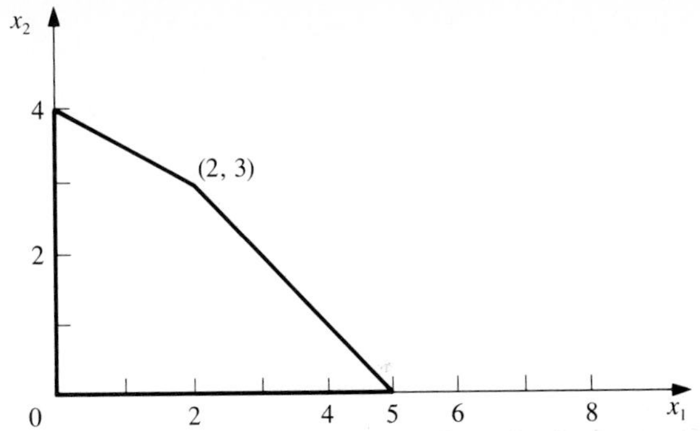

Figure 9.9 Subproblem 1 for example illustrating the decomposition principle.

In addition,

$$\mathbf{c}_1 = [4, 6], \qquad \mathbf{c}_2 = [8, 5],$$

$$\mathbf{x}_1 = \begin{bmatrix} x_1 \\ x_2 \end{bmatrix}, \qquad \mathbf{x}_2 = \begin{bmatrix} x_3 \\ x_4 \end{bmatrix}, \qquad \mathbf{b}_0 = \begin{bmatrix} 20 \\ 25 \end{bmatrix}, \qquad \mathbf{b}_1 = \begin{bmatrix} 5 \\ 8 \end{bmatrix}, \qquad \mathbf{b}_2 = [12].$$

To prepare for demonstrating how this problem would be solved, we shall first examine its two subproblems individually and then construct the reformulation of the overall problem. Thus *subproblem 1* is

$$\text{Maximize} \qquad Z_1 = [4, 6] \begin{bmatrix} x_1 \\ x_2 \end{bmatrix},$$

subject to
$$\begin{bmatrix} 1 & 1 \\ 1 & 2 \end{bmatrix} \begin{bmatrix} x_1 \\ x_2 \end{bmatrix} \le \begin{bmatrix} 5 \\ 8 \end{bmatrix} \qquad \text{and} \qquad \begin{bmatrix} x_1 \\ x_2 \end{bmatrix} \ge \begin{bmatrix} 0 \\ 0 \end{bmatrix},$$

so that its set of feasible solutions is as shown in Fig. 9.9.

It can be seen that this subproblem has four *extreme points* ($n_1 = 4$). One of these is the origin, considered the "first" of these extreme points, so

$$\mathbf{x}_{11}^* = \begin{bmatrix} 0 \\ 0 \end{bmatrix}, \qquad \mathbf{x}_{12}^* = \begin{bmatrix} 5 \\ 0 \end{bmatrix}, \qquad \mathbf{x}_{13}^* = \begin{bmatrix} 2 \\ 3 \end{bmatrix}, \qquad \mathbf{x}_{14}^* = \begin{bmatrix} 0 \\ 4 \end{bmatrix},$$

where $\rho_{11}, \rho_{12}, \rho_{13}, \rho_{14}$ are the respective weights on these points.

Similarly, *subproblem 2* is

$$\text{Maximize} \qquad Z_2 = [8, 5] \begin{bmatrix} x_3 \\ x_4 \end{bmatrix},$$

subject to
$$[4, 3] \begin{bmatrix} x_3 \\ x_4 \end{bmatrix} \le [12] \qquad \text{and} \qquad \begin{bmatrix} x_3 \\ x_4 \end{bmatrix} \ge \begin{bmatrix} 0 \\ 0 \end{bmatrix},$$

and its set of feasible solutions is shown in Fig. 9.10. Thus its three extreme points are

$$\mathbf{x}_{21}^* = \begin{bmatrix} 0 \\ 0 \end{bmatrix}, \qquad \mathbf{x}_{22}^* = \begin{bmatrix} 3 \\ 0 \end{bmatrix}, \qquad \mathbf{x}_{23}^* = \begin{bmatrix} 0 \\ 4 \end{bmatrix},$$

where $\rho_{21}, \rho_{22}, \rho_{23}$ are the respective weights on these points.

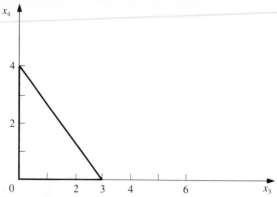

Figure 9.10 Subproblem 2 for example illustrating the decomposition principle.

By performing the $\mathbf{c}_j \mathbf{x}^*_{jk}$ vector multiplications and the $\mathbf{A}_j \mathbf{x}^*_{jk}$ matrix multiplications, the following reformulated version of the overall problem can be obtained:

$$\text{Maximize} \quad Z = 20\rho_{12} + 26\rho_{13} + 24\rho_{14} + 24\rho_{22} + 20\rho_{23},$$

subject to

$$5\rho_{12} + 11\rho_{13} + 12\rho_{14} + 6\rho_{22} + 16\rho_{23} + x_{s1} \qquad = 20$$

$$10\rho_{12} + 13\rho_{13} + 12\rho_{14} + 18\rho_{22} + 16\rho_{23} \qquad + x_{s2} = 25$$

$$\rho_{11} + \rho_{12} + \rho_{13} + \rho_{14} = 1$$

$$\rho_{21} + \rho_{22} + \rho_{23} = 1$$

and

$$\rho_{1k} \geq 0, \quad \text{for } k = 1, 2, 3, 4,$$

$$\rho_{2k} \geq 0, \quad \text{for } k = 1, 2, 3,$$

$$x_{si} \geq 0, \quad \text{for } i = 1, 2.$$

However, we should emphasize that the complete reformulation normally is *not* constructed *explicitly*; rather, just parts of it are generated as needed during the progress of the revised simplex method.

To begin solving this problem, the *initialization step* selects x_{s1}, x_{s2}, ρ_{11}, and ρ_{21} to be the initial basic variables, so that

$$\mathbf{x}_B = \begin{bmatrix} x_{s1} \\ x_{s2} \\ \rho_{11} \\ \rho_{21} \end{bmatrix}.$$

Therefore, since $\mathbf{A}_1 \mathbf{x}^*_{11} = 0$, $\mathbf{A}_2 \mathbf{x}^*_{21} = 0$, $\mathbf{c}_1 \mathbf{x}^*_{11} = 0$, and $\mathbf{c}_2 \mathbf{x}^*_{21} = 0$, then

$$\mathbf{B} = \begin{bmatrix} 1 & 0 & 0 & 0 \\ 0 & 1 & 0 & 0 \\ 0 & 0 & 1 & 0 \\ 0 & 0 & 0 & 1 \end{bmatrix} = \mathbf{B}^{-1}, \qquad \mathbf{x}_B = \mathbf{b}' = \begin{bmatrix} 20 \\ 25 \\ 1 \\ 1 \end{bmatrix}, \qquad \mathbf{c}_B = [0, 0, 0, 0]$$

for the initial basic feasible solution.

To begin testing for optimality, let $j = 1$, and solve the linear programming problem

$$\text{Minimize} \quad W_1 = (\mathbf{0} - \mathbf{c}_1)\mathbf{x}_1 + 0 = -4x_1 - 6x_2,$$

subject to $\qquad \mathbf{A}_3\mathbf{x}_1 \leq \mathbf{b}_1 \qquad$ and $\qquad \mathbf{x}_1 \leq \mathbf{0},$

so the feasible region is that shown in Fig. 9.9. Using Fig. 9.9 to solve graphically, the solution is

$$\mathbf{x}_1 = \begin{bmatrix} 2 \\ 3 \end{bmatrix} = \mathbf{x}_{13}^*,$$

so that $W_1^* = -26.$

Next, let $j = 2$, and solve the problem

$$\text{Minimize} \quad W_2 = (\mathbf{0} - \mathbf{c}_2)\mathbf{x}_2 + 0 = -8x_3 - 5x_4,$$

subject to $\qquad \mathbf{A}_4\mathbf{x}_2 \leq \mathbf{b}_2 \qquad$ and $\qquad \mathbf{x}_2 \geq \mathbf{0},$

so Fig. 9.10 shows this feasible region. Using Fig. 9.10, the solution is

$$\mathbf{x}_2 = \begin{bmatrix} 3 \\ 0 \end{bmatrix} = \mathbf{x}_{22}^*,$$

so $W_2^* = -24.$ Finally, since *none* of the *slack* variables are *nonbasic*, no more coefficients in the current Eq. (0) need to be calculated. It can now be concluded that because both $W_1^* < 0$ and $W_2^* < 0$, the current basic feasible solution is *not* optimal. Furthermore, since W_1^* is the smaller of these, ρ_{13} is the new *entering basic variable*.

For the revised simplex method to now determine the *leaving basic variable*, it is first necessary to calculate the column of $\mathbf{A}'$ giving the *original* coefficients of ρ_{13}. This column is

$$\mathbf{A}_k' = \begin{bmatrix} \mathbf{A}_1\mathbf{x}_{13}^* \\ 1 \\ 0 \end{bmatrix} = \begin{bmatrix} 11 \\ 13 \\ 1 \\ 0 \end{bmatrix}.$$

Proceeding in the usual way to calculate the *current* coefficients of ρ_{13} and the right-side column,

$$\mathbf{B}^{-1}\mathbf{A}_k' = \begin{bmatrix} 11 \\ 13 \\ 1 \\ 0 \end{bmatrix}, \qquad \mathbf{B}^{-1}\mathbf{b}' = \begin{bmatrix} 20 \\ 25 \\ 1 \\ 1 \end{bmatrix}.$$

Considering just the strictly positive coefficients, the *minimum ratio* of the right side to the coefficient is the $\frac{1}{1}$ in the third row, so that $r = 3$; that is, ρ_{11} is the new *leaving basic variable*. Thus the new values of $\mathbf{x}_B$ and $\mathbf{c}_B$ are

$$\mathbf{x}_B = \begin{bmatrix} x_{s1} \\ x_{s2} \\ \rho_{13} \\ \rho_{21} \end{bmatrix}, \qquad \mathbf{c}_B = [0, 0, 26, 0].$$

To find the new value of $\mathbf{B}^{-1}$, set

$$
\mathbf{E} = \begin{bmatrix} 1 & 0 & -11 & 0 \\ 0 & 1 & -13 & 0 \\ 0 & 0 & 1 & 0 \\ 0 & 0 & 0 & 1 \end{bmatrix},
$$

so

$$
\mathbf{B}^{-1}_{\text{new}} = \mathbf{E}\mathbf{B}^{-1}_{\text{old}} = \begin{bmatrix} 1 & 0 & -11 & 0 \\ 0 & 1 & -13 & 0 \\ 0 & 0 & 1 & 0 \\ 0 & 0 & 0 & 1 \end{bmatrix}.
$$

The stage is now set for again testing whether the current basic feasible solution is optimal. In this case

$$
W_1 = (\mathbf{0} - \mathbf{c}_1)\mathbf{x}_1 + 26 = -4x_1 - 6x_2 + 26,
$$

so the minimum feasible solution from Fig. 9.9 is again

$$
\mathbf{x}_1 = \begin{bmatrix} 2 \\ 3 \end{bmatrix} = \mathbf{x}^*_{13},
$$

with $W_1^* = 0$. Similarly,

$$
W_2 = (\mathbf{0} - \mathbf{c}_2)\mathbf{x}_2 + 0 = -8x_3 - 5x_4,
$$

so the minimizing solution from Fig. 9.10 is again

$$
\mathbf{x}_2 = \begin{bmatrix} 3 \\ 0 \end{bmatrix} = \mathbf{x}^*_{22},
$$

with $W_2^* = -24$. Finally, there are no nonbasic slack variables to be considered. Since $W_2^* < 0$, the current solution is not optimal, and ρ_{22} is the new *entering basic variable*.

Proceeding with the revised simplex method,

$$
\mathbf{A}'_k = \begin{bmatrix} \mathbf{A}_2\mathbf{x}^*_{22} \\ 0 \\ 1 \end{bmatrix} = \begin{bmatrix} 6 \\ 18 \\ 0 \\ 1 \end{bmatrix},
$$

so

$$
\mathbf{B}^{-1}\mathbf{A}'_k = \begin{bmatrix} 6 \\ 18 \\ 0 \\ 1 \end{bmatrix}, \qquad \mathbf{B}^{-1}\mathbf{b}' = \begin{bmatrix} 9 \\ 12 \\ 1 \\ 1 \end{bmatrix}.
$$

Therefore, the minimum positive ratio is $\frac{12}{18}$ from the second row, so $r = 2$; that is, x_{s2} is the new *leaving basic variable*. Thus

$$\mathbf{E} = \begin{bmatrix} 1 & -\frac{1}{3} & 0 & 0 \\ 0 & \frac{1}{18} & 0 & 0 \\ 0 & 0 & 1 & 0 \\ 0 & -\frac{1}{18} & 0 & 1 \end{bmatrix},$$

$$\mathbf{B}_{new}^{-1} = \mathbf{E}\mathbf{B}_{old}^{-1} = \begin{bmatrix} 1 & -\frac{1}{3} & -\frac{20}{3} & 0 \\ 0 & \frac{1}{18} & -\frac{13}{18} & 0 \\ 0 & 0 & 1 & 0 \\ 0 & -\frac{1}{18} & \frac{13}{18} & 1 \end{bmatrix}, \quad \mathbf{x}_B = \begin{bmatrix} x_{s1} \\ \rho_{22} \\ \rho_{13} \\ \rho_{21} \end{bmatrix},$$

and $\mathbf{c}_B = [0, 24, 26, 0]$.

Now test whether the new basic feasible solution is optimal. Since

$$W_1 = \left([0, 24, 26, 0] \begin{bmatrix} 1 & -\frac{1}{3} \\ 0 & \frac{1}{18} \\ 0 & 0 \\ 0 & -\frac{1}{18} \end{bmatrix} \begin{bmatrix} 1 & 3 \\ 2 & 3 \end{bmatrix} - [4, 6] \right) \begin{bmatrix} x_1 \\ x_2 \end{bmatrix} + [0, 24, 26, 0] \begin{bmatrix} -\frac{20}{3} \\ -\frac{13}{18} \\ 1 \\ \frac{13}{18} \end{bmatrix}$$

$$= \left([0, \tfrac{4}{3}] \begin{bmatrix} 1 & 3 \\ 2 & 3 \end{bmatrix} - [4, 6] \right) \begin{bmatrix} x_1 \\ x_2 \end{bmatrix} + \tfrac{26}{3}$$

$$= -\tfrac{4}{3}x_1 - 2x_2 + \tfrac{26}{3}.$$

Fig. 9.9 indicates that the minimum feasible solution is again

$$\mathbf{x}_1 = \begin{bmatrix} 2 \\ 3 \end{bmatrix} = \mathbf{x}_{13}^*,$$

so $W_1^* = \frac{2}{3}$. Similarly,

$$W_2 = \left([0, \tfrac{4}{3}] \begin{bmatrix} 2 & 4 \\ 6 & 4 \end{bmatrix} - [8, 5] \right) \begin{bmatrix} x_3 \\ x_4 \end{bmatrix} + 0$$

$$= 0x_3 + \tfrac{1}{3}x_4,$$

so the minimizing solution from Fig. 9.10 now is

$$\mathbf{x}_2 = \begin{bmatrix} 0 \\ 0 \end{bmatrix} = \mathbf{x}_{21}^*,$$

and $W_2^* = 0$. Finally, $\mathbf{c}_B(\mathbf{B}^{-1})_{1;m_0} = [-, \frac{4}{3}]$. Therefore, since $W_1^* \geq 0$, $W_2^* \geq 0$, and $\mathbf{c}_B(\mathbf{B}^{-1})_{1;m_0} \geq \mathbf{0}$, the current basic feasible solution is *optimal*. To identify this solution, set

$$\mathbf{x}_B = \begin{bmatrix} x_{s1} \\ \rho_{22} \\ \rho_{13} \\ \rho_{21} \end{bmatrix} = \mathbf{B}^{-1}\mathbf{b}' = \begin{bmatrix} 1 & -\frac{1}{3} & -\frac{20}{3} & 0 \\ 0 & \frac{1}{18} & -\frac{13}{18} & 0 \\ 0 & 0 & 1 & 0 \\ 0 & -\frac{1}{18} & \frac{13}{18} & 1 \end{bmatrix} \begin{bmatrix} 20 \\ 25 \\ 1 \\ 1 \end{bmatrix} = \begin{bmatrix} 5 \\ \frac{2}{3} \\ 1 \\ \frac{1}{3} \end{bmatrix},$$

so
$$\mathbf{x}_1 = \begin{bmatrix} x_1 \\ x_2 \end{bmatrix} = \sum_{k=1}^{4} \rho_{1k}\mathbf{x}_{1k}^{*} = \mathbf{x}_{12}^{*} = \begin{bmatrix} 2 \\ 3 \end{bmatrix},$$

$$\mathbf{x}_2 = \begin{bmatrix} x_3 \\ x_4 \end{bmatrix} = \sum_{k=1}^{3} \rho_{2k}\mathbf{x}_{2k}^{*} = \tfrac{1}{3}\begin{bmatrix} 0 \\ 0 \end{bmatrix} + \tfrac{2}{3}\begin{bmatrix} 3 \\ 0 \end{bmatrix} = \begin{bmatrix} 2 \\ 0 \end{bmatrix}.$$

Thus, an optimal solution for this problem is $x_1 = 2$, $x_2 = 3$, $x_3 = 2$, $x_4 = 0$, with $Z = 42$.

9.6 Conclusions

The *upper bound technique* provides a way of streamlining the simplex method for the common situation in which many or all of the variables have explicit upper bounds. It can greatly reduce the computational effort for large problems.

The *dual simplex method* and *parametric linear programming* are especially valuable for sensitivity analysis, although they also can be very useful in other contexts as well.

Mathematical-programming computer packages usually include all three of these procedures, and they are widely used. Because their basic structure is based largely upon the simplex method as presented in Chap. 4, they retain the exceptional computational efficiency to handle very large problems of the sizes described in Sec. 4.8.

Various other special-purpose algorithms also have been developed to exploit the special structure of particular types of linear programming problems (such as those discussed in Chap. 7). One of these is the algorithm based on the *decomposition principle* for exploiting the block angular structure that characterizes multidivisional problems.

Karmarkar's interior-point algorithm has been an exciting new development in linear programming. This algorithm and its variants hold much promise as a powerful new approach for efficiently solving some very large problems.

SELECTED REFERENCES

1. Bradley, Stephen P., Arnoldo C. Hax, and Thomas L. Magnanti: *Applied Mathematical Programming,* Addison-Wesley, Reading, Mass., 1977.

2. Dantzig, George B.: *Linear Programming and Extensions,* Princeton University Press, Princeton, N.J., 1963.

3. Hooker, J. N.: "Karmarkar's Linear Programming Algorithm," *Interfaces,* **16**:75–90, July–August 1986.

4. Lasdon, Leon S.: *Optimization Theory for Large Systems,* Macmillan, New York, 1970.

5. Murty, Katta G.: *Linear Programming,* 2d ed., Wiley, New York, 1983.

6. Orchard-Hays, William: *Advanced Linear-Programming Computing Techniques,* McGraw-Hill, New York, 1968.

7. Shapiro, Jeremy: *Mathematical Programming: Structures and Algorithms,* Wiley, New York, 1979.

8. Vanderbei, R. J.: "Affine-Scaling for Linear Programs with Free Variables," *Mathematical Programming,* **43**:31–44, 1989.

1. Use the *upper bound technique* to solve the Wyndor Glass Co. problem presented in Sec. 3.1.

2. Consider the following problem.

$$\text{Maximize} \quad Z = 2x_1 + x_2,$$

subject to

$$x_1 - x_2 \le 5$$
$$x_1 \quad \le 10$$
$$x_2 \le 10$$

and

$$x_1 \ge 0, \quad x_2 \ge 0.$$

(a) Solve this problem graphically.
(b) Use the *upper bound technique* to solve this problem.
(c) Trace graphically the path taken by the upper bound technique.

3.* Use the *upper bound technique* to solve the following problem.

$$\text{Maximize} \quad Z = x_1 + 3x_2 - 2x_3,$$

subject to

$$x_2 - 2x_3 \le 1$$
$$2x_1 + x_2 + 2x_3 \le 8$$
$$x_1 \le 1$$
$$x_2 \le 3$$
$$x_3 \le 2$$

and

$$x_1 \ge 0, \quad x_2 \ge 0, \quad x_3 \ge 0.$$

4. Use the *upper bound technique* to solve the following problem.

$$\text{Maximize} \quad Z = 2x_1 + 3x_2 - 2x_3 + 5x_4,$$

subject to

$$2x_1 + 2x_2 + x_3 + 2x_4 \le 5$$
$$x_1 + 2x_2 - 3x_3 + 4x_4 \le 5$$

and

$$0 \le x_j \le 1, \quad \text{for} \quad j = 1, 2, 3, 4.$$

5. Use the *upper bound technique* to solve the linear programming model given in Prob. 4(b), Chap. 6.

6. Consider the following problem.

$$\text{Maximize} \quad Z = -x_1 - x_2,$$

subject to

$$x_1 + x_2 \le 8$$
$$x_2 \ge 3$$
$$-x_1 + x_2 \le 2$$

and

$$x_1 \ge 0, \quad x_2 \ge 0.$$

(a) Solve this problem graphically.
(b) Use the *dual simplex method* to solve this problem.
(c) Trace graphically the path taken by the dual simplex method.

7. Use the *dual simplex method* to solve each of the following linear programming models:

    (*a*) Model given in Prob. 17, Chap. 4.

    (*b*) Model given in Prob. 26, Chap. 4.

8.* Use the *dual simplex method* to solve the following problem.

$$\text{Minimize} \quad Z = 5x_1 + 2x_2 + 4x_3,$$

subject to

$$3x_1 + x_2 + 2x_3 \geq 4$$
$$6x_1 + 3x_2 + 5x_3 \geq 10$$

and

$$x_1 \geq 0, \quad x_2 \geq 0, \quad x_3 \geq 0.$$

9. Use the *dual simplex method* to solve the following problem.

$$\text{Minimize} \quad Z = 7x_1 + 2x_2 + 5x_3 + 4x_4,$$

subject to

$$2x_1 + 4x_2 + 7x_3 + x_4 \geq 5$$
$$8x_1 + 4x_2 + 6x_3 + 4x_4 \geq 8$$
$$3x_1 + 8x_2 + x_3 + 4x_4 \geq 4$$

and

$$x_j \geq 0, \quad \text{for} \quad j = 1, 2, 3, 4.$$

10. Consider the following problem.

$$\text{Maximize} \quad Z = 3x_1 + 2x_2,$$

subject to

$$3x_1 + x_2 \leq 12$$
$$x_1 + x_2 \leq 6$$
$$5x_1 + 3x_2 \leq 27$$

and

$$x_1 \geq 0, \quad x_2 \geq 0.$$

    (*a*) Solve by the *original simplex method* (in tabular form). Identify the *complementary* basic solution for the dual problem obtained at each iteration.

    (*b*) Solve the *dual* of this problem by the *dual simplex method*. Compare the resulting sequence of basic solutions with the complementary basic solutions obtained in part (*a*).

11. Consider the example for case 1 of sensitivity analysis given in Sec. 6.7, where the initial simplex tableau of Table 4.8 is modified by changing b_2 from 12 to 24, thereby changing the respective entries in the right-side column of the *final* simplex tableau to 54, 6, 12, and -2. Starting from this revised final simplex tableau, use the *dual simplex method* to obtain the new optimal solution shown in Table 6.20. Show your work.

12.* Consider Prob. 35 (*a*) and (*b*), Chap. 6. Use the *dual simplex method* to obtain the new optimal solution for each of these two cases.

13. Use *both* the *upper bound technique* and the *dual simplex method* to solve the following problem.

$$\text{Minimize} \quad Z = 3x_1 + 4x_2 + 2x_3,$$

subject to

$$x_1 + x_2 \qquad \geq 15$$
$$x_2 + x_3 \geq 10$$

and

$$0 \leq x_1 \leq 25, \quad 0 \leq x_2 \leq 5, \quad 0 \leq x_3 \leq 15.$$

14. Use *both* the *upper bound technique* and the *dual simplex method* to solve the Nori & Leets Co. problem given in Sec. 3.4 for controlling air pollution.

15.* Consider the following problem.

$$\text{Maximize} \quad Z = 8x_1 + 24x_2,$$

subject to

$$x_1 + 2x_2 \leq 10$$

$$2x_1 + x_2 \leq 10$$

and

$$x_1 \geq 0, \quad x_2 \geq 0.$$

Suppose that Z represents profit and that it is possible to modify the objective function somewhat by an appropriate shifting of key personnel between the two activities. In particular, suppose that the unit profit of activity 1 can be increased above 8 (to a maximum of 18) at the expense of decreasing the unit profit of activity 2 below 24 by twice the amount. Thus Z can actually be represented as

$$Z(\theta) = (8 + \theta)x_1 + (24 - 2\theta)x_2,$$

where θ is also a decision variable such that $0 \leq \theta \leq 10$.

(a) Solve the original form of this problem graphically. Then extend this graphical procedure to solve the parametric extension of the problem; i.e., find the optimal solution and the optimal value of $Z(\theta)$ as a function of θ, for $0 \leq \theta \leq 10$.

(b) Find the optimal solution to the original form of the problem by the simplex method. Then use *parametric linear programming* to find the optimal solution and the optimal value of $Z(\theta)$ as a function of θ, for $0 \leq \theta \leq 10$. Plot $Z(\theta)$.

(c) Determine the optimal value of θ. Then indicate how this optimal value could have been identified directly by solving only two ordinary linear programming problems. (*Hint:* A convex function achieves its maximum at an endpoint.)

16. Use *parametric linear programming* to find the optimal solution of the following problem as a function of θ, for $0 \leq \theta \leq 20$.

$$\text{Maximize} \quad Z(\theta) = (20 + 4\theta)x_1 + (30 - 3\theta)x_2 + 5x_3,$$

subject to

$$3x_1 + 3x_2 + x_3 \leq 30$$

$$8x_1 + 6x_2 + 4x_3 \leq 75$$

$$6x_1 + x_2 + x_3 \leq 45$$

and

$$x_1 \geq 0, \quad x_2 \geq 0, \quad x_3 \geq 0.$$

17. Consider the following problem.

$$\text{Maximize} \quad Z(\theta) = (10 - \theta)x_1 + (12 + \theta)x_2 + (7 + 2\theta)x_3,$$

subject to

$$x_1 + 2x_2 + 2x_3 \leq 30$$

$$x_1 + x_2 + x_3 \leq 20$$

and

$$x_1 \geq 0, \quad x_2 \geq 0, \quad x_3 \geq 0.$$

(a) Use *parametric linear programming* to find the optimal solution for this problem as a function of θ, for $\theta \geq 0$.

(b) Construct the dual model for this problem. Then find the optimal solution for this dual problem as a function of θ, for $\theta \geq 0$, by the method described in the latter part of Sec. 9.3. Indicate graphically what this algebraic procedure is doing. Compare the basic solutions obtained with the complementary basic solutions obtained in part (a).

18. Consider Prob. 43, Chap. 6. Use *parametric linear programming* to find the optimal solution as a function of θ, for $\theta \geq 0$.

19. Consider Prob. 45(b), Chap. 6. Extend the *parametric linear programming* procedure for making systematic changes in the c_j parameters to consider also systematic changes in the a_{ij} parameters in order to find the optimal solution as a function of θ, for $0 \leq \theta \leq 1$.

20.* Use the *parametric linear programming* procedure for making systematic changes in the b_i parameters to find the optimal solution for the following problem as a function of θ, for $0 \leq \theta \leq 25$.

$$\text{Maximize} \qquad Z(\theta) = 2x_1 + 2x_2,$$

subject to

$$x_1 \qquad \leq 10 + 2\theta$$
$$x_1 + x_2 \leq 25 - \theta$$
$$x_2 \leq 10 + 2\theta$$

and

$$x_1 \geq 0, \qquad x_2 \geq 0.$$

Indicate graphically what this algebraic procedure is doing.

21. Use *parametric linear programming* to find the optimal solution for the following problem as a function of θ, for $0 \leq \theta \leq 30$.

$$\text{Maximize} \qquad Z(\theta) = 5x_1 + 6x_2 + 4x_3 + 7x_4,$$

subject to

$$3x_1 - 2x_2 + x_3 + 3x_4 \leq 135 - 2\theta$$
$$2x_1 + 4x_2 - x_3 + 2x_4 \leq 78 - \theta$$
$$x_1 + 2x_2 + x_3 + 2x_4 \leq 30 + \theta$$

and

$$x_j \geq 0, \qquad \text{for} \qquad j = 1, 2, 3, 4.$$

Then identify the value of θ that gives the largest optimal value of $Z(\theta)$.

22. Consider Prob. 36, Chap. 6. Use *parametric linear programming* to find the optimal solution as a function of θ over the following ranges of θ.
   (a) $0 \leq \theta \leq 20$.
   (b) $-20 \leq \theta \leq 0$. (*Hint:* Substitute $-\theta'$ for θ, and then increase θ' from zero.)

23. Follow the instructions of Prob. 22 for Prob. 40, Chap. 6.

24. Consider the $Z^*(\theta)$ function shown in Fig. 9.1 for *parametric linear programming* with systematic changes in the c_j parameters.
   (a) Explain why this function is *piecewise linear*.
   (b) Show that this function must be *convex*.

25. Consider the $Z^*(\theta)$ function shown in Fig. 9.2 for *parametric linear programming* with systematic changes in the b_i parameters.
   (a) Explain why this function is *piecewise linear*.
   (b) Show that this function must be *concave*.

26. Let

$$Z^* = \max \left\{ \sum_{j=1}^{n} c_j x_j \right\},$$

subject to

$$\sum_{j=1}^{n} a_{ij} x_j \leq b_i, \qquad \text{for} \qquad i = 1, 2, \dots, m,$$

and

$$x_j \geq 0, \qquad \text{for} \qquad j = 1, 2, \dots, n$$

(where the a_{ij}, b_i, and c_j are fixed constants), and let $(y_1^*, y_2^*, \dots, y_m^*)$ be the corresponding optimal dual solution. Then let

$$Z^{**} = \max \left\{ \sum_{j=1}^{n} c_j x_j \right\},$$

subject to

$$\sum_{j=1}^{n} a_{ij} x_j \le b_i + k_i, \qquad \text{for} \qquad i = 1, 2, , \ldots, m,$$

and

$$x_j \ge 0, \qquad \text{for} \qquad j = 1, 2, , \ldots, n,$$

where $k_1, k_2, \ldots, k_m$ are given constants. Show that

$$Z^{**} \le Z^* + \sum_{i=1}^{m} k_i y^*_i.$$

27. Reconsider the example used to illustrate the interior-point algorithm in Sec. 9.4. Suppose that $(x_1, x_2) = (1, 3)$ were used instead as the initial feasible trial solution. Perform two iterations starting from this solution.

28. Consider the following problem.

$$\text{Maximize} \qquad Z = 3x_1 + x_2,$$

subject to

$$x_1 + x_2 \le 4$$

and

$$x_1 \ge 0, \qquad x_2 \ge 0.$$

(a) Solve this problem graphically. Also identify all corner-point feasible solutions.
(b) Starting from the initial trial solution $(x_1, x_2) = (1, 1)$, perform four iterations of the interior-point algorithm presented in Sec. 9.4.
(c) Draw figures corresponding to Figs. 9.4, 9.5, 9.6, 9.7, and 9.8 for this problem. In each case, identify the basic (or corner-point) feasible solutions in the current coordinate system. (Trial solutions can be used to determine projected gradients.)

29. Consider the following problem.

$$\text{Maximize} \qquad Z = x_1 + 2x_2,$$

subject to

$$x_1 + x_2 = 8$$

and

$$x_1 \ge 0, \qquad x_2 \ge 0.$$

(a) Near the end of Sec. 9.4, there is a discussion of what the interior-point algorithm does on this problem when starting from the initial feasible trial solution $(x_1, x_2) = (4, 4)$. Verify the results presented there by performing two iterations.
(b) Use these results to predict what subsequent trial solutions would be if additional iterations were to be performed.
(c) Suppose that the stopping rule adopted for the algorithm in this application is that the algorithm stops when two successive trial solutions differ by no more than 0.01 in any component. Use your predictions from part (b) to predict the final trial solution and the total number of iterations required to get there. How close would this solution be to the optimal solution $(x_1, x_2) = (0, 8)$?

30. Consider the following problem.

$$\text{Maximize} \qquad Z = x_1 + x_2,$$

subject to

$$x_1 + 2x_2 \le 9$$

$$2x_1 + x_2 \le 9$$

and

$$x_1 \ge 0, \qquad x_2 \ge 0.$$

(a) Solve the problem graphically.
(b) Find the *gradient* of the objective function in the original x_1-x_2 coordinate system.

If you move from the origin in the direction of the gradient until you reach the boundary of the feasible region, where does it lead relative to the optimal solution?

(c) Starting from the initial trial solution $(x_1, x_2) = (1, 1)$, perform 10 iterations of the interior-point algorithm presented in Sec. 9.4.

(d) Repeat part (c) with $\alpha = 0.9$.

31. Consider the following problem.

$$\text{Maximize} \quad Z = 2x_1 + 5x_2 + 7x_3,$$

subject to

$$x_1 + 2x_2 + 3x_3 = 6$$

and

$$x_1 \geq 0, \qquad x_2 \geq 0, \qquad x_3 \geq 0.$$

(a) Graph the feasible region.

(b) Find the *gradient* of the objective function and then find the *projected gradient* onto the feasible region.

(c) Starting from the initial trial solution $(x_1, x_2, x_3) = (1, 1, 1)$, perform two iterations of the interior-point algorithm presented in Sec. 9.4.

(d) Perform eight additional iterations.

32. Starting from the initial trial solution $(x_1, x_2) = (2, 2)$, apply 15 iterations of the interior-point algorithm presented in Sec. 9.4 to the Wyndor Glass Co. problem presented in Sec. 3.1. Also draw a figure like Fig. 9.8 to show the trajectory of the algorithm in the original x_1-x_2 coordinate system.

33. Use the *decomposition principle* to solve the *Wyndor Glass Co.* problem presented in Sec. 3.1.

34. Consider the *multidivisional problem*

$$\text{Maximize} \quad Z = 10x_1 + 5x_2 + 8x_3 + 7x_4,$$

subject to

$$6x_1 + 5x_2 + 4x_3 + 6x_4 \leq 40$$

$$3x_1 + x_2 \qquad\qquad \leq 15$$

$$x_1 + x_2 \qquad\qquad \leq 10$$

$$x_3 + 2x_4 \leq 10$$

$$2x_3 + x_4 \leq 10$$

and

$$x_j \geq 0, \qquad \text{for} \quad j = 1, 2, 3, 4.$$

(a) Explicitly construct the complete *reformulated* version of this problem in terms of the ρ_{jk} decision variables that would be generated (as needed) and used by the *decomposition principle*.

(b) Use the *decomposition principle* to solve this problem.

35. Using the *decomposition principle, begin* solving the *Good Foods Corp.* multidivisional problem presented in Sec. 7.5 by executing the first *two* iterations.

10

Network Analysis, Including PERT-CPM

Networks arise in numerous settings and in a variety of guises. Transportation, electrical, and communication networks pervade our daily lives. Network representations also are widely used for problems in such diverse areas as production, distribution, project planning, facilities location, resource management, and financial planning—to name just a few examples. In fact, a network representation provides such a powerful visual and conceptual aid for portraying the relationships between the components of systems that it is used in virtually every field of scientific, social, and economic endeavor.

One of the most exciting developments in operations research in recent years has been the unusually rapid advance in both the methodology and application of network optimization models. A number of algorithmic breakthroughs in the 1970s and 1980s have had a major impact, as have ideas from computer science concerning data structures and efficient data manipulation. Consequently, algorithms and software now are available *and are being used* to solve huge problems on a routine basis that would have been completely intractable a couple of decades ago.

As one example of a recent application, an award-winning study (see Selected References 8 and 9) was conducted during the mid-1980s at Citgo Petroleum Corporation, which is devoted to petroleum refining and marketing operations and has annual sales of several billion dollars. When Citgo was acquired by Southland Corporation (best known for its 7-Eleven stores) in 1983, top management saw an urgent need for a modeling system to help Citgo overcome the pressures of volatile crude oil prices and a 30-fold increase in the costs of financing working capital. The operations research team developed an optimization-based decision support system using network methodology, and coupled this system with an on-line corporate database. The model takes in all aspects of the business, helping management decide everything from run levels at the various refineries to what prices to pay or charge. A network representation is essential because of the flow of goods through several stages: purchase of crude oil from various suppliers, shipping it to refineries, refining it into various products, and sending the products to distribution centers and product storage terminals for subsequent sale. The modeling system enabled the company to reduce its petroleum products inventory by $116 million. This has meant a savings in annual interest of $14 million, and improvements in coordination, pricing, and purchasing decisions worth another $2.5 million each year.

In this Citgo study, the model used for each product fits the model for the *minimum cost flow problem* presented in Sec. 10.6. Each product's model has about 3,000 equations (nodes) and 15,000 variables (arcs), which is of very modest size by today's standards for the application of network optimization models.

In this one chapter we shall be able only to scratch the surface of the current state of the art of network methodology. However, we shall introduce you to five important kinds of network problems and some basic ideas of how to solve them (without delving into issues of data structures, etc., that are so vital to successful large-scale implementations). Each of the first three problem types—the *shortest path problem,* the *minimum spanning tree problem,* and the *maximum flow problem*—has a very specific structure that arises frequently in applications.

The fourth type—the *minimum cost flow problem*—provides a unified approach to many other applications because of its far more general structure. In fact, this structure is so general that it includes as special cases both the shortest path problem and the maximum flow problem, as well as the transportation problem, the transshipment problem, and the assignment problem from Chap. 7. Like many network optimization models, the minimum cost flow problem can be formulated as a linear programming problem, and it can be solved extremely efficiently by a streamlined version of the simplex method called the *network simplex method.* (We shall not discuss even more general network problems that are more difficult to solve.)

The last problem type considered is *project planning and control* with PERT (Program Evaluation and Review Technique) and CPM (Critical Path Method). Although limited to this one area of application, PERT and CPM have proven to be invaluable tools there. In fact, since their development in the late 1950s, PERT and CPM have been (and probably continue to be) the most widely used kind of network technique in operations research.

The first section introduces a prototype example that will be used subsequently to illustrate the approach to the first three of these problems. Section 10.2 presents some basic terminology for networks. The next four sections deal with the first four

problems in turn. Section 10.7 then is devoted to the network simplex method, and **355**
Sec. 10.8 discusses the last problem type.

Network Analysis,
Including PERT-CPM

10.1 Prototype Example

SEERVADA PARK has recently been set aside for a limited amount of sightseeing
and backpack hiking. Cars are not allowed into the park, but there is a narrow, winding
road system for trams and for jeeps driven by the park rangers. This road system is
shown (without the curves) in Fig. 10.1, where location O is the entrance into the
park; other letters designate the locations of ranger stations (and other limited facilities). The numbers give the distances of these winding roads in miles.

The park contains a scenic wonder at station T. A small number of trams are
used to transport sightseers from the park entrance to station T and back.

The park management currently faces three problems. One is to determine which
route from the park entrance to station T has the *smallest total distance* for the operation of the trams. (This is an example of the *shortest path problem* to be discussed
in Sec. 10.3.)

A second problem is that telephone lines must be installed under the roads to
establish telephone communication among all the stations (including the park entrance). Because the installation is both expensive and disruptive to the natural environment, lines will be installed under just enough roads to provide some connection
between every pair of stations. The question is where the lines should be laid to
accomplish this with a *minimum* total number of miles of line installed. (This is an
example of the *minimum spanning tree problem* to be discussed in Sec. 10.4.)

The third problem is that more people want to take the tram ride from the park
entrance to station T than can be accommodated during the peak season. To avoid
unduly disturbing the ecology and wildlife of the region, a strict ration has been placed
on the number of tram trips that can be made on each of the roads per day. (These
limits differ for the different roads, as we shall describe in detail in Sec. 10.5.)
Therefore, during the peak season, various routes might be followed regardless of
distance to increase the number of tram trips that can be made each day. The question
is how to route the various trips to *maximize* the number of trips that can be made
per day without violating the limits on any individual road. (This is an example of
the *maximum flow problem* to be discussed in Sec. 10.5.)

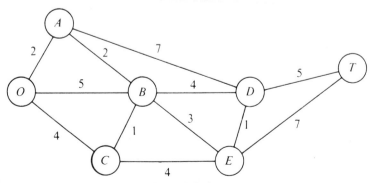

Figure 10.1 The road system for Seervada Park.

10.2 The Terminology of Networks

A relatively extensive terminology has been developed to describe the various kinds of networks and their components. Although we have avoided as much of this special vocabulary as we could, we still need to introduce a considerable number of terms for use throughout the chapter. We suggest that you read through this section once at the outset to understand the definitions, and then plan to return to refresh your memory as the terms are used in subsequent sections. To assist you, each term is highlighted in **boldface** at the point where it is defined.

A network consists of a set of *points* and a set of *lines* connecting certain pairs of the points. The points are called **nodes** (or vertices); e.g., the network in Fig. 10.1 has seven *nodes* designated by the seven circles. The lines are called **arcs** (or links or edges or branches); e.g., the network in Fig. 10.1 has 12 *arcs* corresponding to the 12 roads in the road system. Arcs are labeled by naming the nodes at either end; e.g., *AB* is the arc between nodes *A* and *B* in Fig. 10.1.

The arcs of a network may have a flow of some type through them, e.g., the flow of trams on the roads of Seervada Park in Sec. 10.1. Table 10.1 gives several examples of flow in typical networks. If flow through an arc is allowed in only one direction (e.g., a one-way street), the arc is said to be a **directed arc.** The direction is indicated by adding an arrowhead at the end of the line representing the arc. When labeling a directed arc by listing two nodes it connects, the *from* node always is given before the *to* node; e.g., an arc that is directed *from* node A *to* node B must be labeled as *AB* rather than *BA*. Alternatively, this arc may be labeled as $A \to B$.

If the flow through an arc is allowed in both directions (e.g., a two-way street), the arc is said to be an **undirected arc.** To help you distinguish between the two kinds of arcs, we shall frequently refer to *undirected* arcs by the suggestive alternative name of **links.**

A network that has only *directed* arcs is called a **directed network.** Similarly, if all of its arcs are *undirected,* the network is said to be an **undirected network.** A network with a mixture of *directed* and *undirected* arcs (or even all undirected arcs) can be converted into a *directed network,* if desired, by replacing each *undirected* arc by a pair of *directed* arcs in opposite directions.

When two nodes are not connected by an arc, a natural question is whether they are connected by a series of arcs. A **path** between two nodes is a *sequence of distinct arcs* connecting these nodes. For example, one of the paths connecting nodes *O* and *T* in Fig. 10.1 is the sequence of arcs *OB–BD–DT* ($O \to B \to D \to T$), or vice versa. When some or all of the arcs in the network are *directed arcs,* we then distinguish between *directed paths* and *undirected paths.* A **directed path** from node *i* to node *j*

Table 10.1 **Components of Typical Networks**

Nodes	Arcs	Flow
Intersections	Roads	Vehicles
Airports	Air lanes	Aircraft
Switching points	Wires, channels	Messages
Pumping stations	Pipes	Fluid
Work centers	Materials-handling routes	Jobs

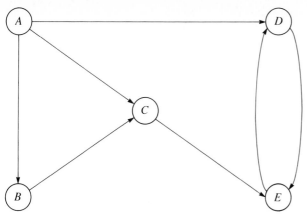

Figure 10.2 Example of a directed network.

is a sequence of connecting arcs whose direction (if any) is *toward* node *j,* so that flow from node *i* to node *j* along this path is feasible. An **undirected path** from node *i* to node *j* is a sequence of connecting arcs whose direction (if any) can be *either* toward or away from node *j.* (Notice that a *directed* path also satisfies the definition of an *undirected* path, but not vice versa.) Frequently, an undirected path will have some arcs directed toward node *j* but others directed away (i.e., toward node *i*). You will see in Secs. 10.5 and 10.6 that, perhaps surprisingly, *undirected* paths play a major role in the analysis of *directed* networks.

To illustrate these definitions, Fig. 10.2 shows a typical *directed network.* The sequence of arcs *AB–BC–CE* ($A \rightarrow B \rightarrow C \rightarrow E$) is a *directed path* from node *A* to node *E,* since flow toward node *E* along this entire path is feasible. On the other hand, *BC–AC–AD* ($B \rightarrow C \rightarrow A \rightarrow D$) is *not* a directed path from node *B* to node *D,* because the direction of arc *AC* is away from node *D* (on this path). However, $B \rightarrow C \rightarrow A \rightarrow D$ is an *undirected path* from node *B* to node *D.* As an example of the relevance of this undirected path, suppose that 2 units of flow from node *A* to node *C* had previously been assigned to arc *AC.* Given this previous assignment, it now is feasible to assign a smaller flow—say, 1 unit—to the entire undirected path $B \rightarrow C \rightarrow A \rightarrow D$ from node *B* to node *D,* because this involves *reducing* the flow on arc *AC* by 1 unit. Reducing a previously assigned flow in the "wrong direction" when adding a flow to an undirected path will prove to be a key concept in Secs. 10.5 and 10.6.

A path that begins and ends at the same node is called a **cycle.** In a *directed* network, a cycle is either a *directed* or an *undirected* cycle, depending on whether the path involved is a directed or an undirected path. (Since a *directed* path also is an *undirected* path, a *directed* cycle is an *undirected* cycle, but not vice versa in general.) In Fig. 10.2, for example, *DE–ED* is a *directed* cycle. By contrast, *AB–BC–AC* is *not* a *directed* cycle, because the direction of arc *AC* opposes the direction of arcs *AB* and *BC.* On the other hand, *AB–BC–AC* is an *undirected* cycle, because $A \rightarrow B \rightarrow C \rightarrow A$ *is* an *undirected* path. In the *undirected* network shown in Fig. 10.1, there are many cycles, e.g., *OA–AB–BC–CO.* However, note that the definition of *path* (a sequence of *distinct* arcs) rules out retracing one's steps in forming a cycle. For example, *OB–BO* in Fig. 10.1 does not qualify as a cycle, because *OB* and *BO* are two labels for the *same* arc (link). On the other hand, *DE–ED* is a (directed) cycle in Fig. 10.2, because *DE* and *ED* are distinct arcs.

Two nodes are said to be **connected** if the network contains at least one *undirected* path between them. (Note that the path does not need to be directed even if the network is directed.) A **connected network** is a network where every pair of nodes is connected. Thus the networks in Figs. 10.1 and 10.2 are both connected. However, the latter network would not be connected if arcs *AD* and *CE* were removed.

Consider a set of *n* nodes (e.g., the *n* = 5 nodes in Fig. 10.2) without any arcs. A "tree" can then be "grown" by adding one arc (or branch) at a time in a certain way. The first arc can go anywhere to connect some pair of nodes. Thereafter, each new arc should be between a node that already is connected to other nodes and a new node not previously connected to any other nodes. Adding an arc in this way avoids creating a cycle, and also ensures that the number of connected nodes is one greater than the number of arcs. Each new arc creates a larger **tree,** which is a *connected network* (for some subset of the *n* nodes) that contains *no undirected cycles.* Once the (*n* − 1)st arc has been added, the process stops because the resulting tree spans (connects) all *n* nodes. This tree is called a **spanning tree,** i.e., a *connected network* for all *n* nodes that contains *no undirected cycles.* Every spanning tree has exactly (*n* − 1) arcs, since this is the *minimum* number of arcs needed to have a connected network and the *maximum* number possible without having undirected cycles.

Figure 10.3 uses the five nodes and some of the arcs of Fig. 10.2 to illustrate this process of growing a tree one arc (branch) at a time until a spanning tree has

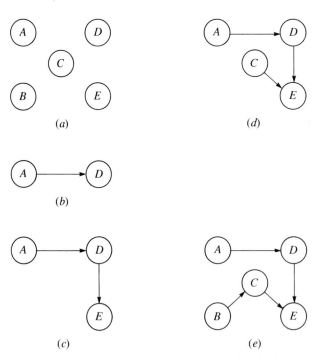

Figure 10.3 Example of growing a tree one arc at a time for the network of Fig. 10.2. (*a*) The nodes without arcs; (*b*) a tree with one arc; (*c*) a tree with two arcs; (*d*) a tree with three arcs; (*e*) a spanning tree.

the process, so Fig. 10.3 shows only one of many ways to construct a spanning tree in this case. Note, however, how each new added arc satisfies the conditions specified in the preceding paragraph. We shall discuss and illustrate spanning trees further in Sec. 10.4.

Spanning trees play a key role in the analysis of many networks. For example, they form the basis for the *minimum spanning tree problem* discussed in Sec. 10.4. Another prime example is that (feasible) spanning trees correspond to the *basic feasible solutions* for the network simplex method discussed in Sec. 10.6.

Finally, we shall need a little additional terminology about *flows* in networks. The maximum amount of flow (possibly infinity) that can be carried on a directed arc is referred to as the **arc capacity.** For nodes, a distinction is made among those that are net generators of flow, net absorbers of flow, or neither. A **supply node** (or source node or source) has the property that the flow *out* of the node exceeds the flow *into* the node. The reverse case is a **demand node** (or sink node or sink), where the flow *into* the node exceeds the flow *out* of the node. A **transshipment node** (or intermediate node) satisfies *conservation of flow,* so flow in equals flow out.

10.3 The Shortest Path Problem

Although several other versions of the shortest path problem (including some for directed networks) are mentioned at the end of the section, we shall focus on the following simple version. Consider an *undirected* and *connected* network with two special nodes called the *origin* and the *destination*. Associated with each of the *links* (undirected arcs) is a nonnegative *distance*. The objective is to find the *shortest path* (the path with the minimum total distance) from the origin to the destination.

A relatively straightforward algorithm is available for this problem. The essence of this procedure is that it fans out from the origin, successively identifying the shortest path to each of the nodes of the network in the ascending order of their (shortest) distances from the origin, thereby solving the problem when the destination node is reached. We shall first outline the method and then illustrate it by solving the shortest path problem encountered by the *Seervada Park* management in Sec. 10.1.

Algorithm for Shortest Path Problem

Objective of nth iteration: Find nth nearest node to origin. (To be repeated for $n = 1, 2, \ldots$, until nth nearest node is the destination.)

Input for nth iteration: $(n - 1)$ nearest nodes to origin (solved for at previous iterations), including their shortest path and distance from the origin. (These nodes, plus the origin, will be called *solved nodes;* the others are *unsolved nodes.*)

Candidates for nth nearest node: Each solved node that is directly connected by a link to one or more unsolved nodes provides *one* candidate—the unsolved node with the *shortest* connecting link. (Ties provide additional candidates.)

Calculation of nth nearest node: For each such solved node and its candidate, add the distance between them and the distance of the shortest path from the

Table 10.2 **Applying Algorithm for Shortest Path Problem to Seervada Park Problem**

n	Solved Nodes Directly Connected to Unsolved Nodes	Closest Connected Unsolved Node	Total Distance Involved	*n*th Nearest Node	Minimum Distance	Last Connection
1	*O*	*A*	2	*A*	2	*OA*
2, 3	*O* *A*	*C* *B*	4 2 + 2 = 4	*C* *B*	4 4	*OC* *AB*
4	*A* *B* *C*	*D* *E* *E*	2 + 7 = 9 4 + 3 = 7 4 + 4 = 8	*E*	7	*BE*
5	*A* *B* *E*	*D* *D* *D*	2 + 7 = 9 4 + 4 = 8 7 + 1 = 8	*D* *D*	8 8	*BD* *ED*
6	*D* *E*	*T* *T*	8 + 5 = 13 7 + 7 = 14	*T*	13	*DT*

origin to this solved node. The candidate with the smallest such total distance is the *n*th nearest node (ties provide additional solved nodes), and its shortest path is the one generating this distance.

EXAMPLE: The Seervada Park management needs to find the shortest path from the park entrance (node *O*) to the scenic wonder (node *T*) through the road system shown in Fig. 10.1. Applying the preceding algorithm to this problem yields the results shown in Table 10.2 (where the tie for the second nearest node allows skipping directly to seeking the fourth nearest node next). The first column (*n*) indicates the iteration count. The second column simply lists the *solved nodes* for beginning the current iteration after deleting the irrelevant ones (those not connected directly to any unsolved node). The third column then gives the *candidates* for the *n*th nearest node (the unsolved nodes with the *shortest* connecting link to a solved node). The fourth column calculates the distance of the shortest path from the origin to each of these candidates (namely, the distance to the solved node plus the link distance to the candidate). The candidate with the smallest such distance is the *n*th nearest node to the origin, as listed in the fifth column. The last two columns summarize the information for this *newest solved node* that is needed to proceed to subsequent iterations (namely, the distance of the shortest path from the origin to this node and the last link on this shortest path).

The shortest path *from the destination to the origin* can now be traced back through the last column of Table 10.2 as *either* $T \to D \to E \to B \to A \to O$ or $T \to D \to B \to A \to O$. Therefore, the two alternates for the shortest path *from the origin to the destination* have been identified as $O \to A \to B \to E \to D \to T$ and $O \to A \to B \to D \to T$, with a total distance of 13 miles on either path.

Other Applications

Before concluding this discussion of the shortest path problem, we need to emphasize one point. The problem thus far has been described in terms of minimizing the *distance* from the origin to the destination. However, in actuality the network problem being solved is finding which path connecting two specified nodes minimizes the sum of the *link values* on the path. There is no reason that these link values need to represent *distances,* even indirectly. For example, the links might correspond to *activities* of some kind, where the value associated with each link is the *cost* of that activity. The problem then would be to find which sequence of activities that accomplishes a specified objective minimizes the total *cost* involved. (See Prob. 2.) Another alternative is that the value associated with each link is the *time* required for that activity. The problem then would be to find which sequence of activities that accomplishes a specified objective minimizes the total *time* involved. (See Prob. 6.) Thus some of the most important applications of the shortest path problem have nothing to do with distances.

Many of these applications require finding the shortest *directed* path from the origin to the destination through a *directed* network. The algorithm already presented can be easily modified to deal just with directed paths at each iteration. In particular, when identifying candidates for the *n*th nearest node, only directed arcs *from* a solved node *to* an unsolved node would be considered.

Another version of the shortest path problem is to find the shortest paths from the origin to *all* of the other nodes of the network. Notice that the algorithm already solves for the shortest path to each node that is closer to the origin than the destination. Therefore, when *all* nodes are potential destinations, the only modification needed in the algorithm is that it does not stop until *all* of the nodes are solved nodes.

An even more general version of the shortest path problem is to find the shortest paths from *every* node to every other node. Another option is to drop the restriction that ''distances'' (arc values) be nonnegative. Constraints also can be imposed on the paths that can be followed. All of these variations occasionally arise in applications, and so have been studied by researchers.

The algorithms for a wide variety of combinatorial optimization problems, such as certain vehicle routing or network design problems, often call for the solution of a large number of shortest path problems as subroutines. Although we lack the space to pursue this topic further, this use may now be the most important kind of application of the shortest path problem.

10.4 The Minimum Spanning Tree Problem

The *minimum spanning tree problem* bears some similarities to the main version of the shortest path problem presented in the preceding section. In both cases, an *undirected* network is being considered, where the given information includes the nodes and the distances[1] between pairs of nodes. However, the crucial difference for the minimum spanning tree problem is that the *links* (undirected arcs) between the nodes are no longer specified. Thus, rather than finding a shortest path through a fully defined network, the problem involves *choosing* for the network the *links* that have the *shortest total length* while providing a path between each pair of nodes. The links need to be chosen in such a way that the resulting network forms a *tree* (as defined in Sec. 10.2) that spans (connects) all the given nodes. In short, the problem is to find the *spanning tree* with a minimum total length of the links.

Figure 10.4 illustrates this concept of a *spanning tree* for the Seervada Park problem (see Sec. 10.1). Thus Fig. 10.4*a* is *not* a *spanning* tree because the (*O, A, B, C*) nodes are not connected with the (*D, E, T*) nodes. It needs another link to make this connection. This network actually consists of *two* trees, one for each of

---

[1] Once again, ''distance'' instead can be cost, time, or some other quantity.

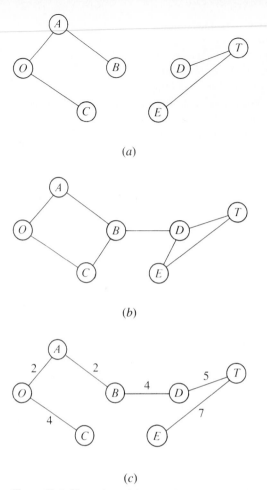

Figure 10.4 Illustrations of the spanning tree concept for the Seervada Park problem. (*a*) Not a spanning tree; (*b*) not a spanning tree; (*c*) a spanning tree.

these two sets of nodes. The links in Fig. 10.4*b* do *span* the network (i.e., the network is *connected* as defined in Sec. 10.2), but it is *not a tree* because there are two *cycles* (*O–A–B–C–O* and *D–T–E–D*). It has too many links. Because the Seervada Park problem has $n = 7$ nodes, Sec. 10.2 indicates that the network must have exactly $(n - 1) = 6$ links, with *no cycles*, to qualify as a spanning tree. This condition is achieved in Fig. 10.4*c*, so this network is a *feasible* solution (with a value of 24 miles for the total length of the links) for the minimum spanning tree problem. (You soon will see that this solution is not *optimal* because it is possible to construct a spanning tree with only 14 miles of links.)

This problem has a number of important practical applications. For example, it can sometimes be helpful in planning *transportation networks* that will not be used much, where the primary consideration is to provide *some* path between all pairs of nodes in the *most economical* way. (See Prob. 8.) The nodes would be the locations that require access to the other locations, the branches would be transportation lanes (highways, railroad tracks, air lanes, and so forth), and the "distances" (link values) would be the costs of providing the transportation lanes. In this context, the minimum spanning tree problem is to determine which transportation lanes would service all the locations with a minimum total cost. Other examples where a comparable decision arises include the planning of large-scale *communication networks* and *distribution networks*. Both represent important application areas.

The minimum spanning tree problem can be solved in a very straightforward way because it happens to be one of the few operations research problems where being *greedy* at each stage of the solution procedure still leads to an overall optimal solution at the end! Thus, beginning with any node, the first stage involves choosing the shortest possible link to another node, without worrying about the effect this choice would have on subsequent decisions. The second stage involves identifying the unconnected node that is closest to either of these connected nodes and then adding the corresponding link to the network. This process would be repeated, as per the following summary, until all the nodes have been connected. (Note that this is the same process already illustrated in Fig. 10.3 for constructing a spanning tree, but now with a specific rule for selecting each new link.) The resulting network is guaranteed to be a minimum spanning tree.

Algorithm for Minimum Spanning Tree Problem

1. Select any node arbitrarily, and then connect it (i.e., add a link) to the nearest distinct node.
2. Identify the unconnected node that is closest to a connected node, and then connect these two nodes (i.e., add a link between them). Repeat this step until all nodes have been connected.

   Tie Breaking: Ties for the nearest distinct node (step 1) or the closest unconnected node (step 2) may be broken arbitrarily and the algorithm must still yield an optimal solution. However, such ties are a signal that there may be (but need not be) multiple optimal solutions. All such optimal solutions can be identified by pursuing all ways of breaking ties to their conclusion.

The fastest way of executing this algorithm manually is the graphical approach illustrated as follows.

EXAMPLE: The Seervada Park management (see Sec. 10.1) needs to determine under which roads telephone lines should be installed to connect all stations with a minimum total length of line. Using the data given in Fig. 10.1, we outline the step-by-step solution of this problem next.

Nodes and distances for the problem are summarized below, where the thin lines now represent *potential* links.

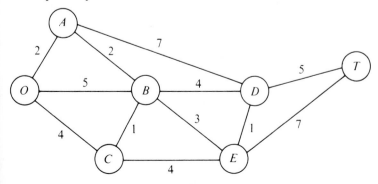

Arbitrarily select node O to start. The unconnected node closest to node O is node A. Connect node A to node O.

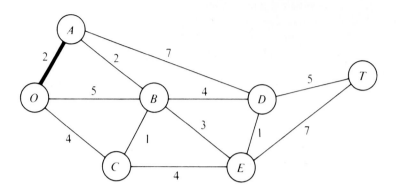

The unconnected node closest to either node O or A is node B (closest to A). Connect node B to node A.

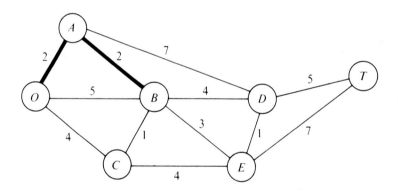

The unconnected node closest to node O, A, or B is node C (closest to B). Connect node C to node B.

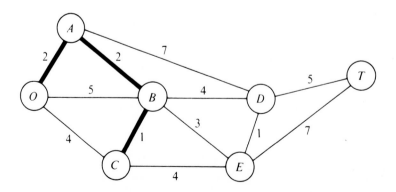

The unconnected node closest to node O, A, B, or C is node E (closest to B). Connect node E to node B.

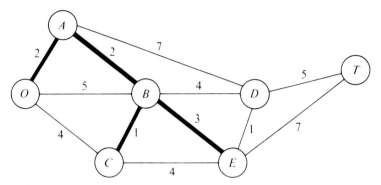

The unconnected node closest to node O, A, B, C, or E is node D (closest to E). Connect node D to node E.

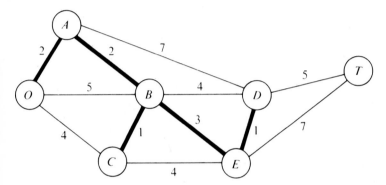

The only remaining unconnected node is node T. It is closest to node D. Connect node T to node D.

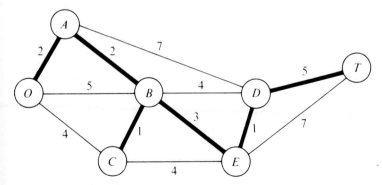

All nodes are now connected, so this solution to the problem is the desired (optimal) one. The total length of the links is 14 miles.

Although it may appear at first glance that the choice of the initial node will affect the resulting final solution (and its total link length) with this procedure, it really doesn't. We suggest you verify this fact for the example by reapplying the algorithm, starting with nodes other than node O.

The minimum spanning tree problem is the one problem we consider in this chapter that falls into the broad category of *network design*. In this category, the objective is to *design* the most appropriate network for the given application (frequently involving transportation systems) rather than analyzing an already designed network. Selected Reference 10 provides a survey of this important area.

10.5 The Maximum Flow Problem

Now recall that the third problem facing the Seervada Park management (see Sec. 10.1) during the peak season is to determine how to route the various tram trips from the park entrance (station O in Fig. 10.1) to the scenic wonder (station T) to maximize the number of trips per day. (Each tram will return by the same route it took on the outgoing trip, so the analysis focuses on outgoing trips only.) In order to avoid unduly disturbing the ecology and wildlife of the region, strict upper limits have been imposed on the number of outgoing trips allowed per day in each direction on each individual road. These limits are shown in Fig. 10.5, where the numbers next to each station and road give the limit for that road in the direction leading away from that station. For example, only *one* loaded trip per day is allowed from station A to station B, but one other also is allowed from station B to station A. Given the limits, one *feasible solution* is to send seven trams per day, with five using the route $O \rightarrow B \rightarrow E \rightarrow T$, one using $O \rightarrow B \rightarrow C \rightarrow E \rightarrow T$, and one using $O \rightarrow B \rightarrow C \rightarrow E \rightarrow D \rightarrow T$. However, because this solution blocks the use of any routes starting with $O \rightarrow C$ (because the $E \rightarrow T$ and $E \rightarrow D$ capacities are fully used), it is easy to find better feasible solutions. Many *combinations* of routes (and the number of trips to assign to each one) need to be considered to find the one(s) maximizing the number of trips made per day. This kind of problem is called a *maximum flow problem*.

Using the terminology introduced in Sec. 10.2, the maximum flow problem can be described formally as follows. Consider a *directed* and *connected* network where just one node is a *supply node,* one node is a *demand node,* and the rest are transshipment nodes. Given the arc capacities, the objective is to determine the feasible pattern of flows through the network that *maximizes the total flow* from the supply node to the demand node.

To formally fit the Seervada Park problem into this format with a directed network, each link in Fig. 10.5 with a 0 at one end would be replaced by a directed

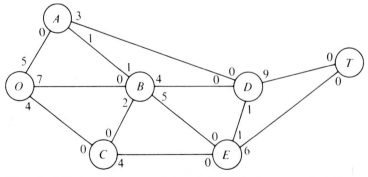

Figure 10.5 Limits on the number of trips per day for the Seervada Park problem.

arc in the direction of feasible flow. For example, the link between nodes O and A would be replaced by a directed arc from node O to node A with an arc capacity of 5. The two links with a 1 at either end (AB and DE) would be replaced by a *pair* of directed arcs in opposite directions, each with an arc capacity of 1. With these understandings, we shall continue to operate on the network as shown in Fig. 10.5.

Because the maximum flow problem can be formulated as a *linear programming problem* (see Prob. 11), it can be solved by the simplex method. However, an even more efficient *augmenting path algorithm* is available for solving this problem. This algorithm is based on two intuitive concepts, those of a *residual network* and of an *augmenting path*.

At the outset, the **residual network** differs from the original network only in that each directed arc ($i \rightarrow j$) lacking a directed arc in the opposite direction ($j \rightarrow i$) now has such an arc added with *zero* arc capacity. Subsequently, the arc capacities in the residual network (called *residual capacities*) are adjusted as follows. Each time some amount of flow Δ is added to arc $i \rightarrow j$ in the original network, the residual capacity of arc $i \rightarrow j$ is *decreased* by Δ but the residual capacity of arc $j \rightarrow i$ is *increased* by Δ. Thus the residual capacity represents the *unused* arc capacity in the original network *or* the amount of flow in the opposite direction in this network that can be *cancelled* (or a combination of both if the original network has arcs in both directions). Therefore, after assigning various flows to the original network, the residual network shows how much more can be done either by increasing flows further or by cancelling previously assigned flows.

An **augmenting path** is a directed path from the supply node to the demand node in the residual network such that *every* arc on this path has *strictly positive* residual capacity. The *minimum* of these residual capacities is called the *residual capacity of the augmenting path* because it represents the amount of flow that can feasibly be added to the entire path. Therefore, each augmenting path provides an opportunity to further augment the flow through the original network.

The augmenting path algorithm repeatedly selects some augmenting path and adds a flow equal to its residual capacity to that path in the original network. This process continues until there are no more augmenting paths, so the flow from the supply node to the demand node cannot be increased further. The key to ensuring that the final solution necessarily is optimal is the fact that augmenting paths can cancel some previously assigned flows in the original network, so an indiscriminate selection of paths for assigning flows cannot prevent the use of a better combination of flow assignments.

To summarize, each *iteration* of the algorithm consists of the following three steps.

Algorithm for Maximum Flow Problem[1]

1. Identify an augmenting path by finding some directed path from the supply node to the demand node in the residual network such that every arc on this path has strictly positive residual capacity. (If no augmenting path exists; the net flows already assigned constitute an optimal flow pattern.)
2. Identify the residual capacity c^* of this augmenting path by finding the *minimum* of the residual capacities of the arcs on this path. *Increase* the flow in this path by c^*.

---

[1] It is assumed that the arc capacities are either integers or rational numbers.

3. *Decrease* by $c*$ the residual capacity of each arc on this augmenting path. *Increase* by $c*$ the residual capacity of each arc in the opposite direction on this augmenting path. Return to step 1.

When performing step 1, there often will be a number of alternative augmenting paths from which to choose. Although the algorithmic strategy for making this selection is of some importance for the efficiency of large-scale implementations, we shall not delve into this relatively specialized topic. (Later in the section, we do describe a systematic procedure for finding some augmenting path.) Therefore, for the following example (and the problems at the end of the chapter), the selection is just made arbitrarily.

EXAMPLE: Applying this algorithm to the Seervada Park problem (see Fig. 10.5 for the original network) yields the results summarized next. For each iteration, the residual network is shown after completing all three steps, where a single line is used to represent the pair of directed arcs in opposite directions between each pair of nodes. The residual capacity of arc $i \rightarrow j$ is shown next to node i, whereas the residual capacity of arc $j \rightarrow i$ is shown next to node j. Using this format, the network shown in Fig. 10.5 actually is the residual network at the outset, before assigning any flows. After subsequent iterations, we show in **boldface** (next to nodes O and T) the total amount of flow achieved thus far.

Iteration 1: Referring to Fig. 10.5, one of several augmenting paths is $O \rightarrow B \rightarrow E \rightarrow T$, which has a residual capacity of min{7, 5, 6} = 5. Assigning a flow of 5 to this path, the resulting residual network is

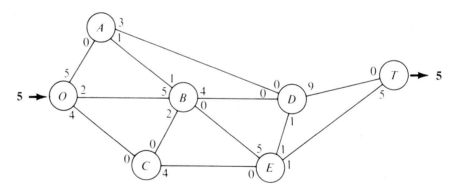

Iteration 2: Assign a flow of 3 to the augmenting path, $O \rightarrow A \rightarrow D \rightarrow T$. The resulting residual network is

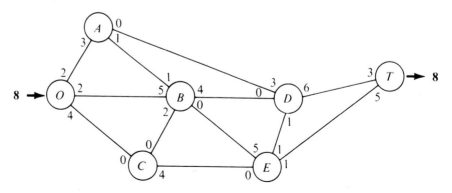

Iteration 3: Assign a flow of 1 to the augmenting path, $O \rightarrow A \rightarrow B \rightarrow D \rightarrow T$.

Iteration 4: Assign a flow of 2 to the augmenting path, $O \rightarrow B \rightarrow D \rightarrow T$. The resulting residual network is

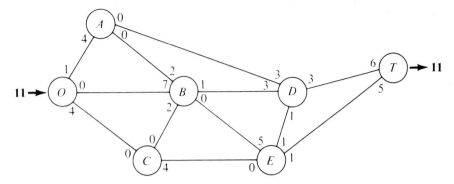

Iteration 5: Assign a flow of 1 to the augmenting path, $O \rightarrow C \rightarrow E \rightarrow D \rightarrow T$.

Iteration 6: Assign a flow of 1 to the augmenting path, $O \rightarrow C \rightarrow E \rightarrow T$. The resulting residual network is

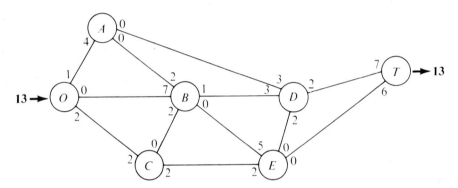

Iteration 7: Assign a flow of 1 to the augmenting path, $O \rightarrow C \rightarrow E \rightarrow B \rightarrow D \rightarrow T$. The resulting residual network is

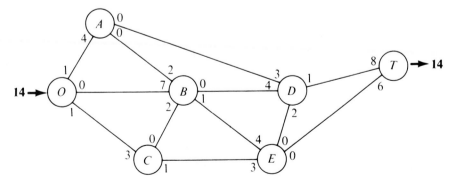

There are no more augmenting paths, so the current flow pattern is optimal.

The current flow pattern may be identified by either cumulating the flow assignments or by comparing the final residual capacities with the original arc capacities.

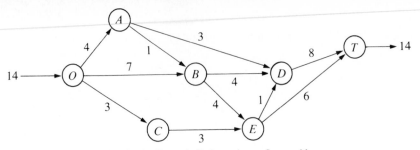

Figure 10.6 Optimal solution for the Seervada Park maximum flow problem.

If we use the latter method, there is flow along an arc if the final residual capacity is less than the original capacity. The magnitude of this flow equals the difference in these capacities. Applying this method by comparing the residual network obtained from the last iteration with Fig. 10.5 yields the optimal flow pattern shown in Fig. 10.6.

This example nicely illustrates the reason for supplementing an arc $i \rightarrow j$ in the original network by an arc $j \rightarrow i$ in the residual network and then increasing the residual capacity of the latter arc by c^* when a flow of c^* is assigned to arc $i \rightarrow j$. Without this refinement, the first six iterations would be unchanged. However, at that point it would appear that no augmenting paths remain (because the real unused arc capacity for $E \rightarrow B$ is *zero*). Therefore, the refinement permits adding the flow assignment of 1 for $O \rightarrow C \rightarrow E \rightarrow B \rightarrow D \rightarrow T$ in *iteration 7*. In effect, this additional flow assignment cancels out one unit of flow assigned at iteration 1 ($O \rightarrow B \rightarrow E \rightarrow T$) and replaces it by assignments of one unit of flow to *both* $O \rightarrow B \rightarrow D \rightarrow T$ and $O \rightarrow C \rightarrow E \rightarrow T$.

The most difficult part of this algorithm when *large* networks are involved is finding an augmenting path. This task may be simplified by the following systematic procedure. Begin by determining all nodes that can be reached from the supply node along a (single) arc with strictly positive residual capacity. Then, for each of these nodes that were reached, determine all *new* nodes (those not yet reached) that can be reached from this node along an arc with strictly positive residual capacity. Repeat this successively with the new nodes as they are reached. The result will be the identification of a tree of all the nodes that can be reached from the supply node along a path with strictly positive residual flow capacity. Hence this *fanning-out procedure*

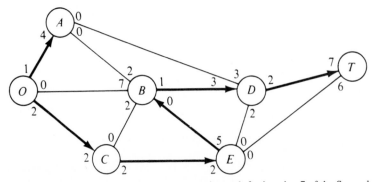

Figure 10.7 Procedure for finding an augmenting path for iteration 7 of the Seervada Park example.

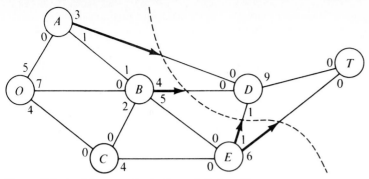

Figure 10.8 A minimum cut for the Seervada Park problem.

will always identify an augmenting path if one exists. The procedure is illustrated in Fig. 10.7 for the residual network that results from *iteration 6* in the preceding example.

Although the procedure illustrated in Fig. 10.7 is a relatively straightforward one, it would be helpful to be able to recognize when optimality has been reached without an exhaustive search for a nonexistent path. It is sometimes possible to recognize this event because of an important theorem of network theory known as the *max-flow min-cut theorem*. A **cut** may be defined as *any set of directed arcs containing at least one arc from every directed path from the supply node to the demand node*. The *cut value* is the *sum of the arc capacities of the arcs (in the specified direction) of the cut*. The **max-flow min-cut theorem** states that, for any network with a single supply node and demand node, the *maximum feasible flow* from the supply node to the demand node *equals* the *minimum cut value* for all of the cuts of the network. Thus, if we let F denote the amount of flow from the supply node to the demand node for any feasible flow pattern, the value of any cut provides an upper bound to F, and the smallest of the cut values is equal to the maximum value of F. Therefore, if a cut whose value equals the value of F currently attained by the solution procedure can be found in the original network, the current flow pattern must be *optimal*. Eventually, optimality has been attained whenever there exists a cut in the residual network whose value is zero.

To illustrate, consider the cut in the network of Fig. 10.5 that is indicated in Fig. 10.8. Notice that the value of the cut is $(3 + 4 + 1 + 6) = 14$, which was found to be the maximum value of F, so this cut is a *minimum cut*. Notice also that, in the residual network resulting from *iteration 7*, where $F = 14$, the corresponding cut has a value of *zero*. If this had been noticed, it would not have been necessary to search for additional augmenting paths.

10.6 The Minimum Cost Flow Problem

The *minimum cost flow problem* holds a central position among network optimization models, both because it encompasses such a broad class of applications and because it can be solved extremely efficiently. Like the *maximum flow problem*, it considers flow through a network with limited arc capacities. Like the *shortest path problem*, it considers a cost (or distance) for flow through an arc. Like the *transportation problem* or *assignment problem* of Chap. 7, it can consider multiple sources (supply nodes) and multiple destinations (demand nodes) for the flow, again with associated costs. Like the *transshipment problem* of Chap. 7, it also can consider various junction points (transshipment nodes) between the sources and destinations for this flow. In fact, all five of these previously studied problems are *special cases* of the minimum cost flow problem, as we will demonstrate shortly.

The reason that the minimum cost flow problem can be solved so efficiently is that it can be formulated as a linear programming problem, so it can be solved by a streamlined version of the simplex method called the *network simplex method*. We describe this algorithm in the next section.

Formulation

Consider a *directed* and *connected* network, where the n nodes include at least one supply node and at least one demand node. The decision variables are

$$x_{ij} = \text{flow through arc } i \rightarrow j,$$

and the given information includes

$$c_{ij} = \text{cost per unit flow through arc } i \rightarrow j,$$
$$u_{ij} = \text{arc capacity for arc } i \rightarrow j,$$
$$b_i = \text{net flow generated at node } i.$$

The value of b_i depends on the nature of node i, where

$$b_i > 0, \quad \text{if node } i \text{ is a supply node,}$$
$$b_i < 0, \quad \text{if node } i \text{ is a demand node,}$$
$$b_i = 0, \quad \text{if node } i \text{ is a transshipment node.}$$

The objective is to minimize the total cost of sending the available supply through the network to satisfy the given demand.

By using the convention that summations are taken only over existing arcs, the linear programming formulation of this problem is

$$\text{Minimize} \quad Z = \sum_{i=1}^{n} \sum_{j=1}^{n} c_{ij} x_{ij},$$

subject to

$$\sum_{j=1}^{n} x_{ij} - \sum_{j=1}^{n} x_{ji} = b_i, \quad \text{for each node } i,$$

and

$$0 \leq x_{ij} \leq u_{ij}, \quad \text{for each arc } i \rightarrow j.$$

The first summation in the node constraints represents the total flow *out* of node i, whereas the second summation represents the total flow *into* node i, so the difference is the net flow generated at this node.

In some applications, it is necessary to have a lower bound $L_{ij} > 0$ for the flow through each arc $i \rightarrow j$. When this occurs, use a translation of variables, $x'_{ij} = x_{ij} - L_{ij}$, with $(x'_{ij} + L_{ij})$ substituted for x_{ij} throughout the model, in order to convert the model back into the above format with nonnegativity constraints.

It is not guaranteed that the problem actually will possess *feasible* solutions, depending partially upon which arcs are present in the network and their arc capacities.

However, for a reasonably designed network, the main condition needed is the following.

>**Feasible solutions property:** A necessary condition for a minimum cost flow problem to have any feasible solutions is that

$$\sum_{i=1}^{n} b_i = 0;$$

>i.e., the total flow being generated at the supply nodes equals the total flow being absorbed at the demand nodes.

If the values of the b_i provided for some application violate this condition, the usual interpretation is that either the supplies or the demands (whichever is in excess) actually represent upper bounds rather than exact amounts. When this situation arose for the *transportation problem* in Sec. 7.1, either a *dummy destination* was added to receive the excess supply or a *dummy source* was added to send the excess demand. The analogous step now is that either a *dummy demand node* should be added to absorb the excess supply (with $c_{ij} = 0$ arcs added *from* every supply node *to* this node) or a *dummy supply node* should be added to generate the flow for the excess demand (with $c_{ij} = 0$ arcs added *from* this node *to* every demand node).

For many applications, the b_i and u_{ij} quantities will have *integer* values, and implementation will require that the flow quantities (the x_{ij}) also be *integer*. Fortunately, just as for the transportation problem, this outcome is guaranteed without explicitly imposing integer constraints on the variables because of the following property.

>**Integer solutions property:** For minimum cost flow problems where every b_i and u_{ij} has an integer value, all the basic variables in *every* basic feasible solution (including an optimal one) also have *integer* values.

An example of a minimum cost flow problem is shown in Fig. 10.9. This network is the same as in Fig. 10.2 except that now values of the b_i, c_{ij}, and u_{ij} have

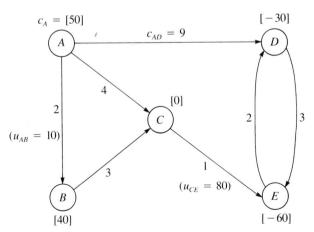

Figure 10.9 Example of a minimum cost flow problem.

been added. The b_i values are shown in square brackets by the nodes, so the *supply nodes* ($b_i > 0$) are A and B, the *demand nodes* ($b_i < 0$) are D and E, and the one *transshipment node* ($b_i = 0$) is C. The c_{ij} values are shown next to the arcs. In this example, all but two of the arcs have arc capacities exceeding the total flow generated (90), so $u_{ij} = \infty$ for all practical purposes. The two exceptions are arc $A \rightarrow B$, where $u_{AB} = 10$, and arc $C \rightarrow E$, which has $u_{CE} = 80$.

The linear programming model for this example is

$$\text{Minimize} \quad Z = 2x_{AB} + 4x_{AC} + 9x_{AD} + 3x_{BC} + x_{CE} + 3x_{DE} + 2x_{ED},$$

$$\text{subject to} \quad x_{AB} + x_{AC} + x_{AD} \qquad\qquad\qquad\qquad = 50$$

$$-x_{AB} \qquad\qquad + x_{BC} \qquad\qquad\qquad\qquad = 40$$

$$-x_{AC} \qquad - x_{BC} + x_{CE} \qquad\qquad\qquad = 0$$

$$-x_{AD} \qquad\qquad\qquad + x_{DE} - x_{ED} = -30$$

$$-x_{CE} - x_{DE} + x_{ED} = -60$$

and $\qquad\qquad x_{AB} \leq 10, \qquad x_{CE} \leq 80, \qquad$ and all $x_{ij} \geq 0$.

Now note the pattern of coefficients for each variable in the set of five link constraints. Each variable has exactly *two* nonzero coefficients, where one is $+1$ and the other is -1. This pattern recurs in *every* minimum cost flow problem, and it is this special structure that leads to the *integer solutions property*.

Another implication of this special structure is that (any) one of the link constraints is *redundant*. The reason is that summing all these constraint equations yields nothing but zeroes on both sides (assuming feasible solutions exist, so the b_i sum to zero), so the negative of any one of these equations equals the sum of the rest of the equations. With just ($n - 1$) nonredundant link constraints, these equations provide just ($n - 1$) basic variables for a basic feasible solution. In the next section, you will see that the network simplex method treats the $x_{ij} \leq u_{ij}$ constraints as mirror images of the nonnegativity constraints, so the *total* number of basic variables is ($n - 1$). This leads to a direct correspondence between the ($n - 1$) arcs of a *spanning tree* and the ($n - 1$) basic variables—but more about that story later.

We shall soon solve this example by the network simplex method. However, let us first see how the five special cases mentioned earlier fit into the network format of the minimum cost flow problem. For each case, we shall show how to formulate its *prototype example* in this more general way.

Special Cases

THE TRANSPORTATION PROBLEM: To formulate the transportation problem presented in Sec. 7.1 as a minimum cost flow problem, a *supply node* is provided for each *source*, as well as a *demand node* for each *destination*, but no transshipment nodes are included in the network. All of the arcs are directed from a supply node to a demand node, where distributing x_{ij} units from source i to destination j corresponds to a flow of x_{ij} through arc $i \rightarrow j$. The cost c_{ij} per unit distributed becomes the cost c_{ij} per unit of flow. Since the transportation problem does not impose upper bound constraints on individual x_{ij}, all of the $u_{ij} = \infty$.

Using this formulation for the P & T Co. transportation problem presented in Table 7.2 yields the network shown in Fig. 10.10.

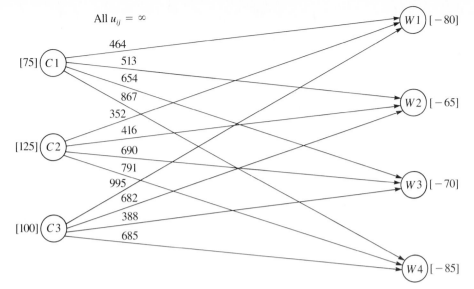

Figure 10.10 Formulation of the P & T Co. transportation problem as a minimum cost flow problem.

THE TRANSSHIPMENT PROBLEM: Recall that the transshipment problem presented in Sec. 7.3 is the generalization of the transportation problem where units being distributed from a source to a destination can first pass through intermediate points, which can be either *transshipment points* or other sources and destinations. Therefore, the formulation of the transshipment problem as a minimum cost flow problem is the same as for the transportation problem *except* that now a *transshipment node* is provided for each *transshipment point* and *arcs* are added for each feasible intermediate trip from one point (source, transshipment point, or destination) to another.

With these additions, this formulation actually includes all of the general features of the minimum cost flow problem except for not having (finite) arc capacities. For this reason, the minimum cost flow problem sometimes is called the *capacitated transshipment problem*.

Using this formulation for the P & T Co. transshipment problem presented in Table 7.24 yields the network shown in Fig. 10.11. Because *every* arc has a companion arc in the opposite direction between the same pair of nodes, we have simplified the depiction of this large network by using a single link with arrowheads at both ends to represent the pair of arcs. We also have deleted the values of the c_{ij}, but each one is shown in Table 7.24.

THE ASSIGNMENT PROBLEM: Since the assignment problem discussed in Sec. 7.4 is a special type of transportation problem, its formulation as a minimum cost flow problem fits into the format already demonstrated in Fig. 10.10. The additional specializations are that (1) the number of supply nodes equals the number of demand nodes, (2) $b_i = 1$ for each supply node, and (3) $b_i = -1$ for each demand node.

Figure 10.12 shows this formulation for the Job Shop Co. assignment problem presented in Table 7.27.

THE SHORTEST PATH PROBLEM: Now consider the main version of the shortest path problem presented in Sec. 10.3 (finding the shortest path from one origin to one

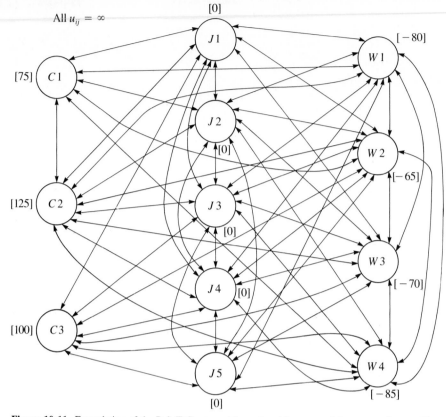

Figure 10.11 Formulation of the P & T Co. transshipment problem as a minimum cost flow problem.

destination through an *undirected* network). To formulate this problem as a minimum cost flow problem, one *supply node* with a supply of 1 is provided for the origin, one *demand node* with a demand of 1 is provided for the destination, and the rest of the nodes are *transshipment nodes*. Because the network for our shortest path problem is *undirected,* whereas the minimum cost flow problem is assumed to have a *directed* network, we replace each link by a pair of directed arcs in opposite directions (depicted by a single line with arrowheads at both ends). The only exceptions are that there is no need to bother with arcs *into* the supply node or *out of* the demand node. The distance between nodes i and j becomes the unit cost c_{ij} or c_{ji} for flow in either direction between these nodes. As with the preceding special cases, no arc capacities are imposed, so all $u_{ij} = \infty$.

Figure 10.13 depicts this formulation for the Seervada Park shortest path problem shown in Fig. 10.1, where the numbers next to the lines now represent the unit cost of flow in either direction.

THE MAXIMUM FLOW PROBLEM: The last special case we shall consider is the maximum flow problem described in Sec. 10.5. In this case a network already is provided with one supply node, one demand node, and various transshipment nodes, as well as the various arcs and arc capacities. Only three adjustments are needed to fit this problem into the format for the minimum cost flow problem. One is to set $c_{ij} = 0$ for all existing arcs to reflect the absence of costs in the maximum flow

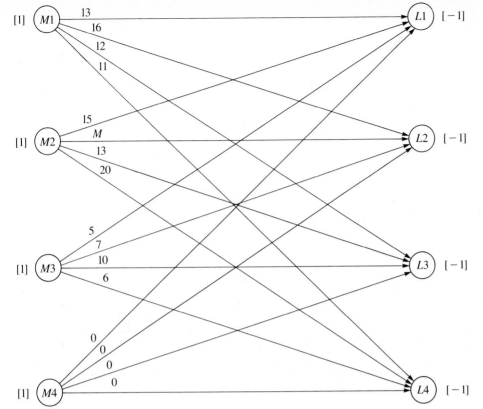

Figure 10.12 Formulation of the Job Shop Co. assignment problem as a minimum cost flow problem.

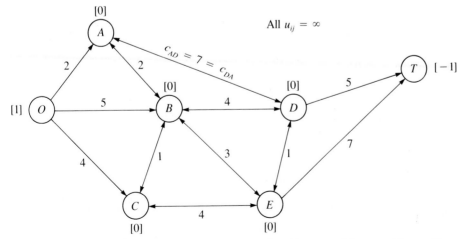

Figure 10.13 Formulation of the Seervada Park shortest path problem as a minimum cost flow problem.

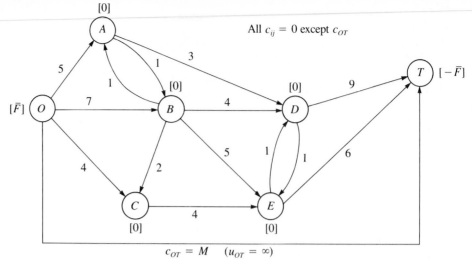

Figure 10.14 Formulation of the Seervada Park maximum flow problem as a minimum cost flow problem.

problem. A second is to select a quantity $\bar{F}$, which is a safe upper bound on the maximum feasible flow through the network, and then to assign a supply and a demand of $\bar{F}$ to the supply node and the demand node, respectively. The third is to add an arc going directly from the supply node to the demand node and to assign it an arbitrarily large unit cost of $c_{ij} = M$ as well as an unlimited arc capacity ($u_{ij} = \infty$). Because of this huge cost, the minimum cost flow problem will send the maximum feasible flow through the *other* arcs, which achieves the objective of the maximum flow problem.

Applying this formulation to the Seervada Park maximum flow problem shown in Fig. 10.5 yields the network given in Fig. 10.14.

FINAL COMMENTS: When each of these five problems was first presented, we described a special-purpose algorithm for solving it very efficiently. Therefore, it certainly is not necessary to reformulate these special cases to fit the format of the minimum cost flow problem in order to solve them. However, when a computer code is not readily available for the special-purpose algorithm, it is very reasonable to use the network simplex method instead. In fact, recent implementations of the network simplex method have become so powerful that it now provides an excellent alternative to the special-purpose algorithm. This is especially true for the (uncapacitated) trans-shipment problem and, to some extent, the transportation problem.

The fact that these five problems are special cases of the minimum cost flow problem is of interest for other reasons as well. One is that the underlying theory for the minimum cost flow problem and the network simplex method provides a unifying theory for all of these special cases. Another is that some of the many applications of the minimum cost flow problem include features of one or more of the special cases within them, so it is important to know how to reformulate these features into the broader framework of the general problem.

10.7 The Network Simplex Method

The network simplex method is a highly streamlined version of the simplex method for solving minimum cost flow problems. As such, it goes through the same basic steps at each iteration—finding the entering basic variable, determining the leaving basic variable, and solving for the new basic feasible solution—in order to move from the current basic feasible solution to a better adjacent one. However, it executes these steps in ways that exploit the special network structure of the problem without ever needing a simplex tableau.

You may note some similarities between the network simplex method and the *transportation simplex method* presented in Sec. 7.2. In fact, both are streamlined versions of the simplex method that provide alternative algorithms for solving transportation problems in similar ways. The network simplex method extends these ideas to solving other types of minimum cost flow problems as well.

In this section, we provide a somewhat abbreviated description of the network simplex method that focuses just on the main concepts. We omit certain details needed for a full computer implementation, including how to construct an initial basic feasible solution, or how to perform certain calculations (such as for finding the entering basic variable) in the most efficient manner. These details are provided in various more specialized textbooks, such as Selected References 1, 2, 6, 7, 11, 12, and 14.

Incorporating the Upper Bound Technique

The first concept is to incorporate the upper bound technique described in Sec. 9.1 to deal efficiently with the arc capacity constraints, $x_{ij} \leq u_{ij}$. Thus, rather than treating these constraints as *functional* constraints, they are handled just like *nonnegativity* constraints. Therefore, they are considered only when determining the *leaving basic variable*. In particular, as the entering basic variable is increased from zero, the leaving basic variable is the *first* basic variable that reaches *either* its lower bound (0) or its upper bound (u_{ij}). A nonbasic variable at its upper bound, $x_{ij} = u_{ij}$, is replaced by $x_{ij} = u_{ij} - y_{ij}$, so $y_{ij} = 0$ becomes the nonbasic variable. See Sec. 9.1 for further details.

In our current context, y_{ij} has an interesting network interpretation. Whenever y_{ij} becomes a basic variable with a strictly positive value ($\leq u_{ij}$), this value can be thought of as flow from node j to node i (so in the "wrong" direction through arc $i \to j$) that, in actuality, is *cancelling* that amount of the previously assigned flow ($x_{ij} = u_{ij}$) from node i to node j. Thus, when $x_{ij} = u_{ij}$ is replaced by $x_{ij} = u_{ij} - y_{ij}$, we also replace the *real* arc $i \to j$ by the *reverse* arc $j \to i$, where this new arc has arc capacity u_{ij} (the maximum amount of the $x_{ij} = u_{ij}$ flow that can be cancelled) and unit cost $- c_{ij}$ (since each unit of flow cancelled saves c_{ij}). To reflect the flow of $x_{ij} = u_{ij}$ through the deleted arc, we shift this amount of net flow generated from node i to node j by *decreasing* b_i by u_{ij} and *increasing* b_j by u_{ij}. Later, if y_{ij} becomes the leaving basic variable by reaching its upper bound, $y_{ij} = u_{ij}$ is replaced by $y_{ij} = u_{ij} - x_{ij}$ with $x_{ij} = 0$ as the new nonbasic variable, so the above process would be reversed (replace arc $j \to i$ by arc $i \to j$, etc.) back to the original configuration.

To illustrate this process, consider the minimum cost flow problem shown in Fig. 10.9. While the network simplex method is generating a sequence of basic

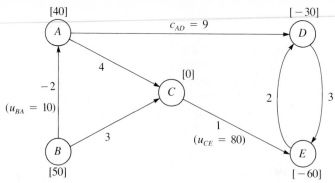

Figure 10.15 The adjusted network for the example when the upper bound technique leads to replacing $x_{AB} = 10$ by $x_{AB} = 10 - y_{AB}$.

feasible solutions, suppose that x_{AB} has become the leaving basic variable for some iteration by reaching its upper bound of 10. Consequently, $x_{AB} = 10$ is replaced by $x_{AB} = 10 - y_{AB}$, so $y_{AB} = 0$ becomes the new nonbasic variable. At the same time, we replace arc $A \rightarrow B$ by arc $B \rightarrow A$ (with y_{AB} as its flow quantity), and assign this new arc a capacity of 10 and a unit cost of -2. To take $x_{AB} = 10$ into account, we also decrease b_A from 50 to 40 and increase b_B from 40 to 50. The resulting adjusted network is shown in Fig. 10.15.

We shall soon illustrate the entire network simplex method with this same example, starting with $y_{AB} = 0$ ($x_{AB} = 10$) as a nonbasic variable and so using Fig. 10.15. A later iteration will show x_{CE} reaching its upper bound of 80 and so being replaced by $x_{CE} = 80 - y_{CE}$, etc., and then the next iteration has y_{AB} reaching its upper bound of 10. You will see that all of these operations are performed directly on the network, so we won't need to use the x_{ij} or y_{ij} labels for arc flows or even keep track of which arcs are *real* arcs and which are *reverse* arcs (except when recording the final solution).

Using the upper bound technique leaves the *node constraints* (flow out minus flow in $= b_i$) as the only functional constraints. Minimum cost flow problems tend to have far more arcs than nodes, so the resulting number of functional constraints generally is only a small fraction of what it would have been if the arc capacity constraints had been included. The computation time for the simplex method goes up relatively rapidly with the number of functional constraints but only slowly with the number of variables (or the number of bounding constraints on these variables). Therefore, incorporating the upper bound technique here tends to provide a tremendous saving in computation time.

However, this technique is not needed for *uncapacitated* minimum cost flow problems (including the first four special cases considered in the preceding section), where there are no arc capacity constraints.

Correspondence between Basic Feasible Solutions and Feasible Spanning Trees

The most important concept underlying the network simplex method is its network representation of *basic feasible solutions*. Recall from Sec. 10.6 that with n nodes, every basic feasible solution has $(n - 1)$ basic variables, where each basic variable x_{ij} represents the flow through arc $i \rightarrow j$. These $(n - 1)$ arcs are referred to as **basic**

arcs. (Similarly, the arcs corresponding to the *nonbasic* variables, $x_{ij} = 0$ or $y_{ij} = 0$, are called **nonbasic arcs**.)

A key property of basic arcs is that they never form undirected *cycles*. (This property prevents the resulting solution from being a weighted average of another pair of feasible solutions, which would violate one of the general properties of basic feasible solutions.) However, *any* set of $(n - 1)$ arcs that contains no undirected cycles forms a *spanning tree*. Therefore, any set of basic arcs forms a spanning tree.

Thus basic feasible solutions can be obtained by "solving" spanning trees, as summarized below.

A **spanning tree solution** is obtained as follows:

1. For the arcs *not* in the spanning tree (the nonbasic arcs), set the corresponding variables (x_{ij} or y_{ij}) equal to zero.
2. For the arcs that *are* in the spanning tree (the basic arcs), solve for the corresponding variables (x_{ij} or y_{ij}) in the system of linear equations provided by the node constraints.

(The network simplex method actually solves for the new basic feasible solution from the current one much more efficiently, without resolving this system of equations from scratch.) Note that this solution process does not consider either the nonnegativity constraints or the arc capacity constraints for the basic variables, so the resulting spanning tree solution may or may not be feasible with respect to these constraints— which leads to our next definition.

A **feasible spanning tree** is a spanning tree whose solution from the node constraints also satisfies all the other constraints ($0 \leq x_{ij} \leq u_{ij}$ or $0 \leq y_{ij} \leq u_{ij}$).

With these definitions, we now can summarize our key conclusion as follows:

The fundamental theorem for the network simplex method: Basic solutions are *spanning tree solutions* (and conversely). Basic feasible solutions are solutions for *feasible spanning trees* (and conversely).

To begin illustrating the application of this fundamental theorem, consider the network shown in Fig. 10.15 that results from replacing $x_{AB} = 10$ by $x_{AB} = 10 - y_{AB}$ for our example in Fig. 10.9. One spanning tree for this network is the one shown in Fig. 10.3*e*, where the arcs are $A \rightarrow D$, $D \rightarrow E$, $C \rightarrow E$, and $B \rightarrow C$. With these as the *basic arcs*, the process of finding the spanning tree solution is shown below. On the left is the set of node constraints given in Sec. 10.6 after substituting $(10 - y_{AB})$ for x_{AB}, where the *basic* variables are shown in **boldface**. On the right, starting at the top and moving down, is the sequence of steps for setting or calculating the values of the variables.

$$(1) \quad y_{AB} = 0, \; x_{AC} = 0, \; x_{ED} = 0$$

$$-y_{AB} + x_{AC} + \boldsymbol{x_{AD}} \qquad\qquad = \quad 40 \qquad x_{AD} = 40.$$

$$y_{AB} \qquad\qquad + \boldsymbol{x_{BC}} \qquad\qquad = \quad 50 \qquad x_{BC} = 50.$$

$$- x_{AC} \qquad - x_{BC} + \boldsymbol{x_{CE}} \qquad = \quad 0 \quad \text{so} \quad x_{CE} = 50.$$

$$- \boldsymbol{x_{AD}} \qquad\qquad + \boldsymbol{x_{DE}} - x_{ED} = -30 \quad \text{so} \quad x_{DE} = 10.$$

$$- x_{CE} - x_{DE} + x_{ED} = -60 \qquad \text{Redundant.}$$

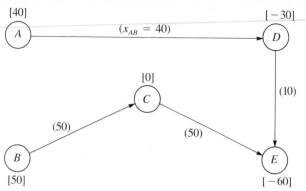

Figure 10.16 The initial feasible spanning tree and its solution for the example.

Since the values of all these basic variables satisfy the nonnegativity constraints and the one relevant arc capacity constraint ($x_{CE} \leq 80$), the spanning tree is a *feasible spanning tree,* so we have a *basic feasible solution.*

We shall use this solution as the *initial basic feasible solution* for demonstrating the network simplex method. Figure 10.16 shows its network representation, namely, the feasible spanning tree and its solution. Thus the numbers given next to the arcs now represent *flows* (values of the x_{ij}) rather than the unit costs c_{ij} previously given. (To help you distinguish, we shall always put parentheses around flows but not around costs.)

Selecting the Entering Basic Variable

To begin an iteration of the network simplex method, recall that the standard simplex method criterion for selecting the entering basic variable is to choose the nonbasic variable that, when increased from zero, will *improve Z at the fastest rate.* Now let us see how this is done without having a simplex tableau.

To illustrate, consider the nonbasic variable x_{AC} in our initial basic feasible solution, i.e., the nonbasic arc $A \rightarrow C$. Increasing x_{AC} from zero to some value θ means that the arc $A \rightarrow C$ with flow θ must be added to the network shown in Fig. 10.16. Adding a nonbasic arc to a spanning tree *always* creates a unique undirected *cycle,* where the cycle in this case is seen in Fig. 10.17 to be AC–CE–DE–AD. Figure

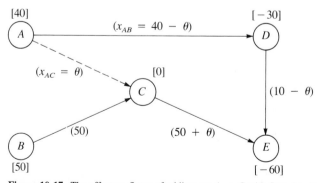

Figure 10.17 The effect on flows of adding arc $A \rightarrow C$ with flow θ to the initial feasible spanning tree.

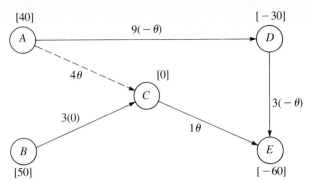

Figure 10.18 The incremental effect on costs of adding arc $A \to C$ with flow θ to the initial feasible spanning tree.

10.17 also shows the effect of adding the flow θ to arc $A \to C$ on the other flows in the network. Specifically, the flow is thereby *increased* by θ for other arcs that have the *same* direction as $A \to C$ in the cycle (arc $C \to E$), whereas the *net* flow is *decreased* by θ for other arcs whose direction is *opposite* to $A \to C$ in the cycle (arcs $D \to E$ and $A \to D$). In the latter case, the new flow is, in effect, *cancelling* a flow of θ in the opposite direction. Arcs not in the cycle ($B \to C$) are unaffected by the new flow. (Check these conclusions by noting the effect of the change in x_{AC} on the values of the other variables in the solution just derived for the initial feasible spanning tree.)

Now what is the incremental effect on Z (total flow cost) from adding the flow θ to arc $A \to C$, etc.? Figure 10.18 shows most of the answer by giving the unit cost times the change in the flow for each arc of Fig. 10.17. Therefore, the overall increment in Z is

$$\Delta Z = c_{AC}\theta + c_{CE}\theta + c_{DE}(-\theta) + c_{AD}(-\theta)$$

$$= 4\theta + 1\theta - 3\theta - 9\theta$$

$$= -7\theta.$$

Setting $\theta = 1$ then gives the *rate* of change of Z as x_{AC} is increased, namely,

$$\Delta Z = -7, \quad \text{when } \theta = 1.$$

Because the objective is to *minimize Z*, this large rate of decrease in Z by increasing x_{AC} is very desirable, so x_{AC} becomes a prime candidate to be the entering basic variable.

We now need to perform the same analysis for the other nonbasic variables before making the final selection of the entering basic variable. The only other nonbasic variables are y_{AB} and x_{ED}, corresponding to the two other *nonbasic arcs*, $B \to A$ and $E \to D$, in Fig. 10.15.

Figure 10.19 shows the incremental effect on costs of adding arc $B \to A$ with flow θ to the initial feasible spanning tree given in Fig. 10.16. Adding this arc creates the undirected cycle BA–AD–DE–CE–BC, so the flow increases by θ for arcs $A \to D$ and $D \to E$, but decreases by θ for the two arcs in the opposite direction on this cycle, $C \to E$ and $B \to C$. These flow increments, θ and $-\theta$, are the multiplicands

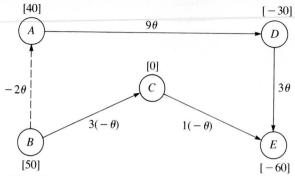

Figure 10.19 The incremental effect on costs of adding arc $B \rightarrow A$ with flow θ to the initial feasible spanning tree.

for the c_{ij} values in the figure. Therefore,

$$\Delta Z = -2\theta + 9\theta + 3\theta + 1(-\theta) + 3(-\theta) = 6\theta$$

$$= 6, \quad \text{when } \theta = 1.$$

The fact that Z *increases* rather than decreases when y_{AB} (flow through the reverse arc $B \rightarrow A$) is increased from zero rules out this variable as a candidate to be the entering basic variable. (Remember that increasing y_{AB} from zero really means decreasing x_{AB}, flow through the real arc $A \rightarrow B$, from its upper bound of 10.)

A similar result is obtained for the last nonbasic arc $E \rightarrow D$. Adding this arc with flow θ to the initial feasible spanning tree creates the undirected cycle ED–DE shown in Fig. 10.20, so the flow also increases by θ for arc $D \rightarrow E$, but no other arcs are affected. Therefore,

$$\Delta Z = 2\theta + 3\theta = 5\theta$$

$$= 5, \quad \text{when } \theta = 1,$$

so x_{ED} is ruled out as a candidate to be the entering basic variable.

To summarize,

$$\Delta Z = \begin{cases} -7, & \text{if } \Delta x_{AC} = 1 \\ 6, & \text{if } \Delta y_{AB} = 1 \\ 5, & \text{if } \Delta x_{ED} = 1 \end{cases}$$

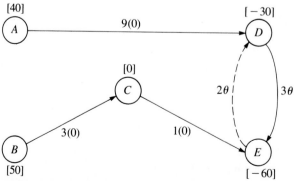

Figure 10.20 The incremental effect on costs of adding arc $E \rightarrow D$ with flow θ to the initial feasible spanning tree.

so the negative value for x_{AC} implies that x_{AC} becomes the entering basic variable for the first iteration. If there had been more than one nonbasic variable with a *negative* value of ΔZ, the one having the *largest* absolute value would have been chosen. (If there had been *no* nonbasic variables with a negative value of ΔZ, the current basic feasible solution would have been optimal.)

Rather than identifying undirected cycles, etc., the network simplex method actually obtains these ΔZ values by an algebraic procedure that is considerably more efficient (especially for large networks). The procedure is analogous to that used by the *transportation simplex method* (see Sec. 7.2) to solve for the u_i and v_j in order to obtain the value of $(c_{ij} - u_i - v_j)$ for each nonbasic variable x_{ij}. We shall not describe this procedure further, so you should just use the undirected cycles method when doing problems at the end of the chapter.

Finding the Leaving Basic Variable and the Next Basic Feasible Solution

After selecting the entering basic variable, only one more quick step is needed to simultaneously determine the leaving basic variable and solve for the next basic feasible solution. For the first iteration of the example, the key is Fig. 10.17. Since x_{AC} is the entering basic variable, the flow θ through arc $A \to C$ is to be increased from zero as far as possible until one of the basic variables reaches *either* its lower bound (0) or its upper bound (u_{ij}). For those arcs whose flow *increases* with θ in Fig. 10.17 (arcs $A \to C$ and $C \to E$), only the *upper* bounds $(u_{AC} = \infty$ and $u_{CE} = 80)$ need to be considered:

$$x_{AC} = \theta \leq \infty.$$

$$x_{CE} = 50 + \theta \leq 80, \qquad \text{so } \theta \leq 30.$$

For those arcs whose flow *decreases* with θ (arcs $D \to E$ and $A \to D$), only the *lower* bound of 0 needs to be considered:

$$x_{DE} = 10 - \theta \geq 0, \qquad \text{so } \theta \leq 10.$$

$$x_{AD} = 40 - \theta \geq 0, \qquad \text{so } \theta \leq 40.$$

Arcs whose flow is unchanged by θ (i.e., those not part of the undirected cycle), which is just arc $B \to C$ in Fig. 10.17, can be ignored since no bound will be reached as θ is increased.

For the five arcs in Fig. 10.17, the conclusion is that x_{DE} must be the leaving basic variable because it reaches a bound for the smallest value of θ (10). Setting $\theta = 10$ in this figure thereby yields the flows through the basic arcs in the next basic feasible solution:

$$x_{AC} = \theta = 10,$$

$$x_{CE} = 50 + \theta = 60,$$

$$x_{AD} = 40 - \theta = 30,$$

$$x_{BC} = 50.$$

The corresponding feasible spanning tree is shown in Fig. 10.21.

If the leaving basic variable had reached its *upper* bound, then the adjustments discussed for the *upper bound technique* would have been needed at this point (as

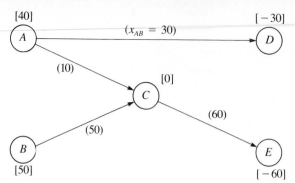

Figure 10.21 The second feasible spanning tree and its solution for the example.

you will see illustrated during the next two iterations). However, because it was the *lower* bound of 0 that was reached, nothing more needs to be done.

COMPLETING THE EXAMPLE: For the two remaining iterations needed to reach the optimal solution, the primary focus will be on some features of the upper bound technique they illustrate. The pattern for finding the entering basic variable, the leaving basic variable, and the next basic feasible solution will be very similar to that described for the first iteration, so we shall only summarize these steps briefly.

Iteration 2: Starting with the feasible spanning tree shown in Fig. 10.21, and referring back to Fig. 10.15 for the unit costs (c_{ij}), the calculations for selecting the entering basic variable are given in Table 10.3. The second column identifies the unique undirected cycle that is created by adding the nonbasic arc in the first column to this spanning tree, and the third column shows the incremental effect on costs because of the changes in flows on this cycle caused by adding a flow of $\theta = 1$ to the nonbasic arc. Arc $E \rightarrow D$ has the largest (in absolute terms) negative value of ΔZ, so x_{ED} is the entering basic variable.

We now make the flow θ through arc $E \rightarrow D$ as large as possible while satisfying the following flow bounds:

$$x_{ED} = \theta \leq u_{ED} = \infty, \qquad \text{so } \theta \leq \infty.$$

$$x_{AD} = 30 - \theta \geq 0, \qquad \text{so } \theta \leq 30.$$

$$x_{AC} = 10 + \theta \leq u_{AC} = \infty, \qquad \text{so } \theta \leq \infty.$$

$$x_{CE} = 60 + \theta \leq u_{CE} = 80, \qquad \text{so } \theta \leq 20. \qquad \leftarrow \text{Minimum}$$

Because x_{CE} imposes the smallest upper bound (20) on θ, x_{CE} becomes the leaving

Table 10.3 Calculations for Selecting the Entering Basic Variable for Iteration 2

Nonbasic Arc	Cycle Created	ΔZ when $\theta = 1$	
$B \rightarrow A$	BA–AC–BC	$-2 + 4 - 3 = -1$	
$D \rightarrow E$	DE–CE–AC–AD	$3 - 1 - 4 + 9 = 7$	
$E \rightarrow D$	ED–AD–AC–CE	$2 - 9 + 4 + 1 = -2$	←Minimum

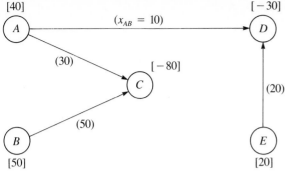

Figure 10.22 The third feasible spanning tree and its solution for the example.

basic variable. Setting $\theta = 20$ in the above expressions for x_{ED}, x_{AD}, and x_{AC} then yields the flow through the basic arcs for the next basic feasible solution (with $x_{BC} = 50$ unaffected by θ), as shown in Fig. 10.22.

What is of special interest here is that the leaving basic variable x_{CE} was obtained by the variable reaching its *upper* bound (80). Therefore, by using the upper bound technique, x_{CE} is replaced by $80 - y_{CE}$, where $y_{CE} = 0$ is the new nonbasic variable. At the same time, the original arc $C \to E$ with $c_{CE} = 1$ and $u_{CE} = 80$ is replaced by the reverse arc $E \to C$ with $c_{EC} = -1$ and $u_{EC} = 80$. The values of b_E and b_C also are adjusted by adding 80 to b_E and subtracting 80 from b_C. The resulting *adjusted* network is shown in Fig. 10.23, where the *nonbasic* arcs are shown as dashed lines, and the numbers by all the arcs are unit costs.

Iteration 3: Using Figs. 10.22 and 10.23 to initiate the next iteration, Table 10.4 shows the calculations that lead to selecting y_{AB} (reverse arc $B \to A$) as the entering basic variable. We then add as much flow θ through arc $B \to A$ as possible while satisfying the flow bounds below:

$$y_{AB} = \theta \leq u_{BA} = 10, \qquad \text{so } \theta \leq 10. \qquad \leftarrow \text{Minimum}$$

$$x_{AC} = 30 + \theta \leq u_{AC} = \infty, \qquad \text{so } \theta \leq \infty.$$

$$x_{BC} = 50 - \theta \geq 0, \qquad \text{so } \theta \leq 50.$$

The smallest upper bound (10) on θ is imposed by y_{AB}, so this variable becomes the leaving basic variable. Setting $\theta = 10$ in these expressions for x_{AC} and x_{BC} (along

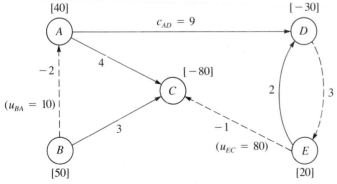

Figure 10.23 The adjusted network with unit costs at the completion of iteration 2.

Table 10.4 Calculations for Selecting the Entering Basic
Variable for Iteration 3

Nonbasic Arc	Cycle Created	ΔZ when $\theta = 1$	
$B \to A$	$BA\text{–}AC\text{–}BC$	$-2 + 4 - 3 = -1$	$\leftarrow$Minimum
$D \to E$	$DE\text{–}EC\text{–}AC\text{–}AD$	$3 - 1 - 4 + 9 = 7$	
$E \to C$	$EC\text{–}AC\text{–}AD\text{–}ED$	$-1 - 4 + 9 - 2 = 2$	

with the unchanged values of $x_{AC} = 10$ and $x_{ED} = 20$) then yields the next basic
feasible solution, as shown in Fig. 10.24.

As with iteration 2, the leaving basic variable (y_{AB}) was obtained here by the
variable reaching its *upper* bound. In addition, there are two other points of special
interest concerning this particular choice. One is that the *entering* basic variable y_{AB}
also became the *leaving* basic variable on the same iteration! This event occurs occasionally with the upper bound technique whenever increasing the entering basic
variable from zero causes *its* upper bound to be reached first before any of the other
basic variables reach a bound.

The other interesting point is that the arc $B \to A$ that now needs to be replaced
by a *reverse* arc $A \to B$ (because of the leaving basic variable reaching an *upper*
bound) already is a reverse arc! This is no problem, because the reverse arc for a
reverse arc is simply the original *real* arc. Therefore, the arc $B \to A$ (with $c_{BA} = -2$
and $u_{BA} = 10$) in Fig. 10.23 now is replaced by arc $A \to B$ (with $c_{AB} = 2$ and
$u_{AB} = 10$), which is the arc between nodes A and B in the original network shown
in Fig. 10.9, and a generated net flow of 10 is shifted from node B ($b_B = 50 \to 40$)
to node A ($b_A = 40 \to 50$). Simultaneously, the variable $y_{AB} = 10$ is replaced by
$10 - x_{AB}$, with $x_{AB} = 0$ as the new nonbasic variable.

The resulting adjusted network is shown in Fig. 10.25.

Passing the Optimality Test: At this point, the algorithm would attempt to use Figs.
10.24 and 10.25 to find the next entering basic variable with the usual calculations
shown in Table 10.5. However, *none* of the nonbasic arcs gives a *negative* value of
ΔZ, so an improvement in Z cannot be achieved by introducing flow through any of
them. This means that the current basic feasible solution shown in Fig. 10.24 has
passed the optimality test, so the algorithm stops.

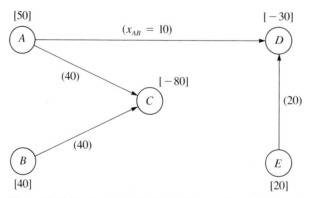

Figure 10.24 The fourth (and final) feasible spanning tree and its solution for the example.

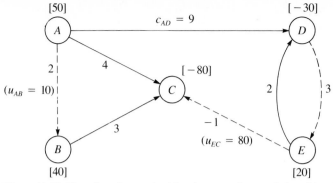

Figure 10.25 The adjusted network with unit costs at the completion of iteration 3.

To identify the flows through *real* arcs rather than *reverse* arcs for this optimal solution, the current adjusted network (Fig. 10.25) should be compared with the original network (Fig. 10.9). Note that each of the arcs has the same direction in the two networks with the one exception of the arc between nodes C and E. This means that the only *reverse* arc in Fig. 10.25 is arc $E \rightarrow C$, where its flow is given by the variable y_{CE}. Therefore, calculate $x_{CE} = u_{CE} - y_{CE} = 80 - y_{CE}$. Arc $E \rightarrow C$ happens to be a *nonbasic* arc, so $y_{CE} = 0$, so $x_{CE} = 80$ is the flow through the *real* arc $C \rightarrow E$. All the other flows through real arcs are the flows given in Fig. 10.24. Therefore, the optimal solution is the one shown in Fig. 10.26.

10.8 Project Planning and Control with PERT-CPM

The successful management of large-scale projects requires careful *planning, scheduling,* and *coordinating* of numerous interrelated activities. To aid in these tasks, formal procedures based on the use of *networks* and *network techniques* were developed beginning in the late 1950s. The most prominent of these procedures are PERT (Program Evaluation and Review Technique) and CPM (Critical Path Method), although there have been many variants under different names. As you will see later, there are a few important differences between these two procedures. However, in recent years the trend has been to merge the two approaches into what is usually referred to as a *PERT-type system.*

Although the original application of PERT-type systems was for evaluating a schedule for a research and development program, it is also used to measure and control progress on numerous other types of special projects. Examples of these project types include construction programs, programming of computers, preparation of bids

Table 10.5 **Calculations for the Optimality Test at the End of Iteration 3**

Nonbasic Arc	Cycle Created	ΔZ when $\theta = 1$
$A \rightarrow B$	$AB–BC–AC$	$2 + 3 - 4 = 1$
$D \rightarrow E$	$DE–EC–AC–AD$	$3 - 1 - 4 + 9 = 7$
$E \rightarrow C$	$EC–AC–AD–ED$	$-1 - 4 + 9 - 2 = 2$

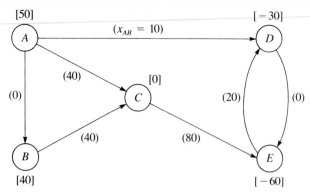

Figure 10.26 The optimal flow pattern in the original network for the example.

and proposals, maintenance planning, and the installation of computer systems. This kind of approach has even been applied to the production of movies, political campaigns, and complex surgery.

A PERT-type system is designed to *aid* in planning and control, so it may not involve much direct *optimization*. Sometimes one of the primary objectives is to determine the probability of meeting specified deadlines. It also identifies the activities that are most likely to be bottlenecks and, therefore, the places where the greatest effort should be made to stay on schedule. A third objective is to evaluate the effect of changes in the program. For example, it will evaluate the effect of a contemplated shift of resources from the less critical activities to the activities identified as probable bottlenecks. Other resource and performance trade-offs may also be evaluated. Another important use is to evaluate the effect of deviations from schedule.

All PERT-type systems use a **project network** to portray graphically the interrelationships among the elements of a project. This network representation of the project plan shows all the *precedence relationships* regarding the order in which tasks must be performed. This feature is illustrated by Fig. 10.27, which shows the initial project network for building a house. This network indicates that the excavation must be done before laying the foundation, and then the foundation must be completed before putting up the rough wall. Once the rough wall is up, three tasks (rough electrical work, rough exterior plumbing, and putting up the roof) can be done in parallel. Tracing through the network further then spells out the ordering of subsequent tasks.

In the terminology of PERT, each *arc* of the project network represents an **activity** that is one of the tasks required by the project. Each *node* represents an **event** that usually is defined as the point in time when all activities leading into that node are completed. The *arrowheads* indicate the sequences in which the events must be achieved. Furthermore, an event must precede the initiation of the activities leading out of that node. (In reality, it is often possible to overlap successive phases of a project, so the network may represent an approximate idealization of the project plan.)

The node toward which all activities lead is the event that corresponds to the completion of the currently planned project. The network may represent either the plan for the project from its inception or, if the project has already begun, the plan for the completion of the project. In the latter case, each node without incoming arcs represents either the event of continuing a current activity or the event of initiating a new activity that may begin at any time.

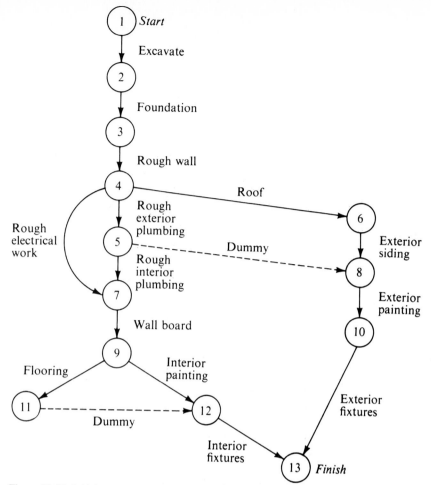

Figure 10.27 Initial *project network* for constructing a house.

Each arc plays a dual role of representing an activity and helping to show the precedence relationships between the various activities. Occasionally, an arc is needed to further define precedence relationships even when there is no *real* activity to be represented. In this case, a **dummy activity** requiring *zero* time is introduced, where the arc representing this fictional activity is shown as a dashed-line arrow that indicates a precedence relationship. To illustrate, consider arc $5 \rightarrow 8$ representing a dummy activity in Fig. 10.27. The sole purpose of this arc is to indicate that the rough exterior plumbing must be completed before the exterior painting can begin.

A common rule for constructing these project networks is that two nodes can be directly connected by *no more than one* arc. Dummy activities can also be used to avoid violating this rule when there are two or more concurrent activities, as illustrated by arc $11 \rightarrow 12$ in Fig. 10.27. The sole purpose of this arc is to indicate that the flooring must be completed before installing the interior fixtures, without having two arcs from node 9 to node 12.

After the network for a project has been developed, the next step is to estimate the *time* required for each of the activities. These estimates for the house-construction example of Fig. 10.27 are shown by the darker numbers (in units of *work days*) next to the arcs in Fig. 10.28. These times are used to calculate two basic quantities for *each event,* namely, its *earliest time* and its *latest time.*

The **earliest time** for an event is the (estimated) time at which the event will occur if the *preceding* activities are started *as early as possible.*

The *earliest times* are obtained by making a *forward pass* through the network, starting with the initial events and working forward in time toward the final events. For each event, a calculation is made of the time at which that event will occur if each immediately preceding event occurs at its earliest time and each intervening activity consumes exactly its estimated time. The initiation of the project should be labeled as time 0. This process is shown in Table 10.6 for the example considered in Figs. 10.27 and 10.28. The resulting earliest times are recorded in Fig. 10.28 as the *first* of the two numbers given by each node.

The **latest time** for an event is the (estimated) last time at which the event can occur *without delaying the completion of the project* beyond its earliest time.

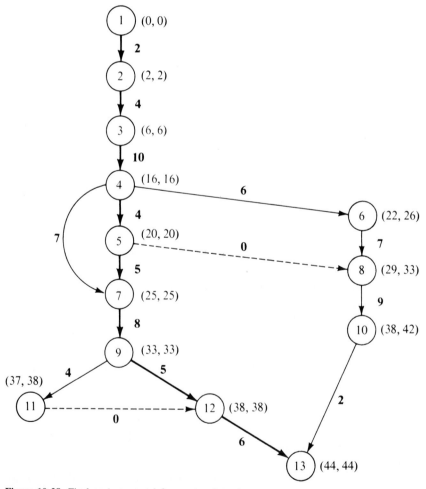

Figure 10.28 Final *project network* for constructing a house.

Table 10.6 **Calculation of Earliest Times
for House-Construction Example**

Event	Immediately Preceding Event	Earliest Time + Activity Time	Maximum = Earliest Time
1	—	—	0
2	1	0 + 2	2
3	2	2 + 4	6
4	3	6 + 10	16
5	4	16 + 4	20
6	4	16 + 6	22
7	4	16 + 7	25
	5	20 + 5	
8	5	20 + 0	29
	6	22 + 7	
9	7	25 + 8	33
10	8	29 + 9	38
11	9	33 + 4	37
12	9	33 + 5	38
	11	37 + 0	
13	10	38 + 2	44
	12	38 + 6	

In this case, the latest times are obtained successively for the events by making a *backward pass* through the network, starting with the final events and working backward in time toward the initial events. For each event, a calculation is made of the final time the event can occur in order for each immediately *following* event to occur at its latest time if each intervening activity consumes exactly its estimated time. This process is illustrated in Table 10.7, with 44 as the earliest time *and* latest time for the completion of the house-construction project. The resulting latest times are recorded in Fig. 10.28 as the *second* of the two numbers given by each node.

Let activity (i, j) denote the activity going from event i to event j in the project network.

The **slack for an event** is the *difference* between its latest and its earliest time. The **slack for an activity** (i, j) is the *difference* between [the latest time of event j] and [the earliest time of event i plus the estimated activity time].

Thus, assuming everything else remains on schedule, the *slack for an event* indicates how much delay in reaching the event can be tolerated without delaying the project's completion, and the *slack for an activity* indicates the same thing regarding a delay in the completion of that activity. The calculation of these slacks is illustrated in Table 10.8 for the house-construction project.

A **critical path** for a project is a path through the network such that the activities on this path have *zero slack*. (*All* activities and events having zero slack must lie on a critical path, but no others can.)

If we check the activities in Table 10.8 that have zero slack, we find that the house-construction example has one critical path, $1 \to 2 \to 3 \to 4 \to 5 \to 7 \to 9 \to 12 \to 13$, as shown in Fig. 10.28 by the dark arrows. Thus this sequence of critical activities must be kept strictly on schedule in order to avoid slippage in completing

Table 10.7 **Calculation of Latest Times for House-Construction Example**

Event	Immediately Following Event	Latest Time − Activity Time	Minimum = Latest Time
13	—	—	44
12	13	44 − 6	38
11	12	38 − 0	38
10	13	44 − 2	42
9	12	38 − 5	33
	11	38 − 4	
8	10	42 − 9	33
7	9	33 − 8	25
6	8	33 − 7	26
5	8	33 − 0	20
	7	25 − 5	
4	7	25 − 7	16
	6	26 − 6	
	5	20 − 4	
3	4	16 − 10	6
2	3	6 − 4	2
1	2	2 − 2	0

the project. Other projects may have more than one such critical path; e.g., note what would happen in Fig. 10.28 if the estimated time for activity (4, 6) were changed from 6 to 10.

It is interesting to observe in Table 10.8 that, whereas every event on the critical path (including events 4 and 7) necessarily has zero slack, activity (4, 7) does not because its estimated time is less than the sum of the estimated times for activities (4, 5) and (5, 7). Consequently, the latter activities are on the critical path but activity (4, 7) is not.

This information on earliest and latest times, slack, and the critical path is invaluable for the project manager. Among other things, it enables the manager to

Table 10.8 **Calculation of Slacks for House-Construction Example**

Event	Slack	Activity	Slack
1	0 − 0 = 0	(1, 2)	2 − (0 + 2) = 0
2	2 − 2 = 0	(2, 3)	6 − (2 + 4) = 0
3	6 − 6 = 0	(3, 4)	16 − (6 + 10) = 0
4	16 − 16 = 0	(4, 5)	20 − (16 + 4) = 0
5	20 − 20 = 0	(4, 6)	26 − (16 + 6) = 4
6	26 − 22 = 4	(4, 7)	25 − (16 + 7) = 2
7	25 − 25 = 0	(5, 7)	25 − (20 + 5) = 0
8	33 − 29 = 4	(6, 8)	33 − (22 + 7) = 4
9	33 − 33 = 0	(7, 9)	33 − (25 + 8) = 0
10	42 − 38 = 4	(8, 10)	42 − (29 + 9) = 4
11	38 − 37 = 1	(9, 11)	38 − (33 + 4) = 1
12	38 − 38 = 0	(9, 12)	38 − (33 + 5) = 0
13	44 − 44 = 0	(10, 13)	44 − (38 + 2) = 4
		(12, 13)	44 − (38 + 6) = 0

investigate the effect of possible improvements in the project plan, to determine where special effort should be expended to stay on schedule, and to assess the impact of schedule slippages.

The PERT Three-Estimate Approach

Thus far we have implicitly assumed that reasonably accurate estimates can be made of the time required for each *activity* of the project. In actuality, there frequently is considerable uncertainty about what the time will be; it really is a *random variable* having some probability distribution. The original version of PERT took this uncertainty into account by using *three* different types of estimates of the activity time to obtain basic information about its probability distribution. This information for all the activity times is then used to estimate the *probability* of completing the project by the scheduled date.

The three time estimates used by PERT for each activity are a *most likely* estimate, an *optimistic* estimate, and a *pessimistic* estimate. The **most likely estimate** (denoted by m) is intended to be the *most realistic* estimate of the time the activity might consume. Statistically speaking, it is an estimate of the *mode* (the highest point) of the probability distribution for the activity time. The **optimistic estimate** (denoted by a) is intended to be the *unlikely but possible time if everything goes well*. Statistically speaking, it is an estimate of essentially the *lower bound* of the probability distribution. The **pessimistic estimate** (denoted by b) is intended to be the *unlikely but possible time if everything goes badly*. Statistically speaking, it is an estimate of essentially the *upper bound* of the probability distribution. The intended location of these three estimates with respect to the probability distribution is shown in Fig. 10.29.

Two assumptions are made to convert m, a, and b into estimates of the *expected value* (t_e) and *variance* (σ^2) of the elapsed time required by the activity. One assumption is that σ, the standard deviation (square root of the variance), equals one-sixth the *range* of reasonably possible time requirements; that is,

$$\sigma^2 = [\tfrac{1}{6}(b - a)]^2$$

is the desired estimate of the variance. The rationale for this assumption is that the tails of many probability distributions (such as the normal distribution) are considered to lie about three standard deviations from the mean, so that there is a spread of about six standard deviations between the tails. For example, the control charts commonly used for statistical quality control are constructed so that the spread between the control limits is estimated to be six standard deviations.

To obtain the estimated *expected value* (t_e), we also need an assumption about the *form* of the probability distribution. This assumption is that the distribution is (at

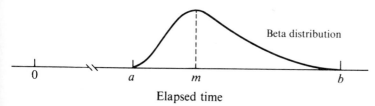

Figure 10.29 Model of the probability distribution of activity times for the PERT three-estimate approach: m = most likely estimate, a = optimistic estimate, and b = pessimistic estimate.

least approximately) a **beta distribution**. This type of distribution has the form shown in Fig. 10.29, which is a reasonable one for this purpose.

If we use the model illustrated in Fig. 10.29, the expected value of the activity time is approximately

$$t_e = \tfrac{1}{3}[2m + \tfrac{1}{2}(a + b)].$$

Notice that the midrange $(a + b)/2$ lies midway between a and $b,$ so that t_e is the weighted arithmetic mean of the mode and the midrange, the mode carrying two-thirds of the entire weight. Although the assumption of a beta distribution is an arbitrary one, it serves its purpose of locating the expected value with respect to $m,$ $a,$ and b in what seems to be a reasonable way.

After calculating the estimated *expected value* and *variance* for each of the activity times, we need three additional assumptions (or approximations) to enable us to calculate the probability of completing the project on schedule. One is that the activity times are *statistically independent*. A second is that the *critical path* (in terms of expected times) *always* requires a longer total elapsed time than any other path. The resulting implication is that the expected value and variance of *project time* are just the *sum* of the expected values and variances (respectively) of the times for the activities on the critical path.

The third assumption is that the project time has a *normal distribution*. The rationale for this assumption is that this time is the sum of many independent random variables, and the general version of the *central limit theorem* implies that the probability distribution of such a sum is approximately *normal* under a wide range of conditions. Given the mean and variance, it is then straightforward (see Table A5.1) to find the probability that this *normal* random variable (project time) will be less than the scheduled completion time.[1]

To illustrate, suppose that the house-construction project of Fig. 10.27 is scheduled to be completed after 50 working days and that *both* the expected value and variance of each activity time happen to equal the estimated time given for that activity in Fig. 10.28. Therefore, when we add these quantities (separately) over the critical path, both the expected value and variance of *project time* are 44, so its standard deviation is $\sqrt{44} \approx 6.63$. Thus the scheduled completion time is approximately 0.9 standard deviation above the *expected* project time. Table A5.1 then gives an approximate probability of $1 - 0.1841 \approx 0.82$ that this schedule will be met.

The CPM Method of Time-Cost Trade-Offs

The original versions of CPM and PERT differ in two important ways. First, CPM assumes that activity times are *deterministic* (i.e., they can be reliably predicted without significant uncertainty), so that the three-estimate approach just described is not needed. Second, rather than primarily emphasizing time (explicitly), CPM places equal emphasis on *time and cost*. This dual emphasis is achieved by constructing a **time-cost curve** for *each activity*, such as the one shown in Fig. 10.30. This curve plots the relationship between the budgeted *direct cost*[2] for the activity and its resulting

---

[1] The same procedure can also be used to find the probability that an *intermediate* event will be accomplished before a scheduled time.

[2] *Direct* cost includes the cost of the material, equipment, and direct labor required to perform the activity but *excludes* indirect project costs such as supervision and other customary overhead costs, interest charges, and so forth.

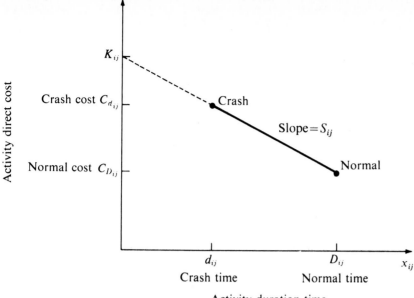

Figure 10.30 Time-cost curve for activity (i, j).

duration time. The plot normally is based on two points[1]: the *normal* and the *crash points*. The **normal point** gives the cost and time involved when the activity is performed in the *normal* way *without* any extra costs (overtime labor, special time-saving materials or equipment, etc.) being expended to speed up the activity. By contrast, the **crash point** gives the time and cost involved when the activity is performed on a *crash basis;* i.e., it is *fully expedited* with no cost spared to reduce the duration time as much as possible. As an approximation, it is then assumed that *all* intermediate *time-cost trade-offs* also are possible and that they lie on the *line segment* between these two points (see the solid line segment shown in Fig. 10.30). Thus the only estimates that need to be obtained from the project personnel for this activity are the cost and time for the two points.

The basic objective of CPM is to determine just *which* time-cost trade-off should be used for each activity to *meet the scheduled project completion time at a minimum cost*. One way of determining the optimal combination of time-cost trade-offs for all the activities is to use *linear programming*. To describe this approach, we need to introduce considerable notation, some of which is summarized in Fig. 10.30. Let

D_{ij} = *normal time* for activity (i, j).

C_{Dij} = *normal (direct) cost* for activity (i, j).

d_{ij} = *crash time* for activity (i, j).

C_{dij} = *crash (direct) cost* for activity (i, j).

The *decision variables* for the problem are the x_{ij}, where

x_{ij} = *duration time* for activity (i, j).

[1] More than two points can be used under certain circumstances.

Thus there is one decision variable x_{ij} for each activity, but there is none for those values of i and j that do not have a corresponding activity.

To express the direct cost for activity (i, j) as a (linear) function of x_{ij}, denote the *slope* of the line through the normal and crash points for activity (i, j) by

$$S_{ij} = \frac{C_{Dij} - C_{dij}}{D_{ij} - d_{ij}}.$$

Also define K_{ij} as the *intercept* with the *direct cost axis* of this line, as shown in Fig. 10.30. Therefore,

$$\text{Direct cost for activity } (i, j) = K_{ij} + S_{ij}x_{ij},$$

Consequently,

$$\text{Total direct cost for the project} = \sum_{(i,j)} (K_{ij} + S_{ij}x_{ij}),$$

where the summation is over *all* activities (i, j). We are now ready to state and formulate the problem mathematically.

The problem: For a given (maximum) project completion time T, choose the x_{ij} to *minimize total direct cost* for the project.

LINEAR PROGRAMMING FORMULATION: To take the project completion time into account, we need one more variable for each event in the linear programming formulation of the problem. This additional variable is

$$y_k = \text{(unknown) } earliest \text{ } time \text{ for event } k, \text{ which is a deterministic}$$
$$\text{function of the } x_{ij}.$$

Each y_k is an *auxiliary variable*, i.e., a variable that is introduced into the model as a convenience in the formulation rather than representing a decision. However, the simplex method treats auxiliary variables just like the regular decision variables (the x_{ij}).

To illustrate how the y_k are worked into the formulation, consider event 7 in Fig. 10.27. By definition, its earliest time is

$$y_7 = \max\{y_4 + x_{47}, y_5 + x_{57}\}.$$

In other words, y_7 is the *smallest* quantity such that *both* of the following constraints hold:

$$y_4 + x_{47} \le y_7$$

$$y_5 + x_{57} \le y_7,$$

so that these two constraints can be incorporated directly into the linear programming formulation (after bringing y_7 to the left-hand side for proper form). Furthermore, we shall soon describe why the optimal solution obtained by the simplex method for the overall model *automatically* will have y_7 at the *smallest* quantity that satisfies these constraints, so no further constraints are needed to incorporate the definition of y_7 into the model.

In the process of adding these constraints for all the events, *every* variable x_{ij}

will appear in exactly one constraint of this type,

$$y_i + x_{ij} \leq y_j,$$

which then is expressed in proper form as

$$y_i + x_{ij} - y_j \leq 0.$$

To continue the preparations for writing down the complete linear programming model, label

$$\text{Event } 1 = \text{project start}$$
$$\text{Event } n = \text{project completion},$$

so $\qquad\qquad y_1 = 0$

$$y_n = \text{(unknown) project completion time.}$$

Also note that ΣK_{ij} is just a fixed constant that can be dropped from the objective function, so that minimizing total direct cost for the project is *equivalent* (see Sec. 4.6) to *maximizing* $\Sigma (-S_{ij})x_{ij}$. Therefore, the linear programming problem is to find the x_{ij} (and the corresponding y_k) that

$$\text{Maximize} \qquad Z = \sum_{(i,j)} (-S_{ij})x_{ij},$$

subject to
$$\left.\begin{array}{l} x_{ij} \geq d_{ij} \\ x_{ij} \leq D_{ij} \\ y_i + x_{ij} - y_j \leq 0 \end{array}\right\} \text{ for all activities } (i, j)$$

$$y_n \leq T.$$

From a computational viewpoint, this formulation can be improved somewhat by replacing each x_{ij} by

$$x_{ij} = d_{ij} + x'_{ij}$$

throughout the model, so that the first set of functional constraints ($x_j \geq d_{ij}$) would be replaced by simple *nonnegativity constraints*

$$x'_{ij} \geq 0.$$

As a convenience we can also introduce nonnegativity constraints for the other variables,

$$y_k \geq 0,$$

although these variables already are forced to be nonnegative by setting $y_1 = 0$ because of the $x'_{ij} \geq 0$ and $y_j \geq y_i + d_{ij} + x'_{ij}$ constraints.

One interesting property of an optimal solution for this model is that (under ordinary circumstances) *every* path through the network will be a critical path requiring a time of T. The reason is that such a solution satisfies the $y_n \leq T$ constraint while avoiding the extra cost involved in shortening the time for any path.

The key to this formulation is the way that the y_k are introduced into the model through the $y_i + x_{ij} - y_j \leq 0$ constraints in order to provide *earliest times* for the respective events (given the values of the x_{ij} in the current basic feasible solution). Since earliest times must be obtained sequentially, all of these y_k are needed for the sole purpose of ultimately obtaining the correct value of y_n (for the current values of the x_{ij}), thereby enabling the $y_n \leq T$ constraint to be enforced. However, obtaining

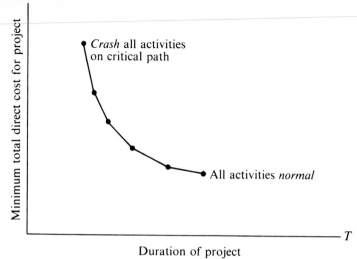

Figure 10.31 Time-cost curve for the overall project.

the correct value does require that the value of each y_j (including y_n) must be the *smallest* quantity that satisfies all the $y_i + x_{ij} \leq y_j$ constraints. Now let us briefly describe why (under ordinary circumstances) this property holds for an optimal solution.

Consider any solution for the x_{ij} variables such that every path through the network is a critical path requiring a time of T. If the values of the y_k variables satisfy the above property, then the y_k are true *earliest times* with $y_n = T$ exactly, and the overall solution for the x_{ij} and y_k satisfy all the constraints. However, if any particular y_i is made a little larger, this would create a chain reaction whereby some y_j would need to be made a little larger to still satisfy the $y_i + x_{ij} \leq y_j$ constraints, etc., until ultimately y_n must be made a little larger, thereby violating the $y_n \leq T$ constraint. The only way to avoid violating this constraint with the larger y_i is to make the duration times for some activities (subsequent to event i) a little *smaller*, thereby increasing the cost. Therefore, an optimal solution will avoid making any y_k larger than need be to satisfy the $y_i + x_{ij} \leq y_j$ constraints.

The problem as stated here assumes that a specified *deadline T* has been fixed (perhaps by contract) for the completion of the project. In fact, some projects do not have such a deadline, in which case it is not clear what value should be assigned to T in the linear programming formulation. In such situations, the decision on T actually is a question of what is the *best trade-off* between the *total cost* and the *total time* for the project.

The basic information we need to address this question is how the *minimum total direct cost* changes as T is changed in the preceding formulation, as illustrated in Fig. 10.31. This information can be obtained by using *parametric linear programming* (see Secs. 4.7, 6.7, and 9.3) to solve for the optimal solution *as a function of T* over its entire range.[1] Even more efficient procedures that exploit the special structure of the problem also are available for obtaining this information.[2]

---

[1] The *slope* of the time-cost curve changes at the points shown in Fig. 10.31 because the set of basic variables that give the optimal solution changes at these values of T. This fact is discussed further in a more general context in Sec. 9.3.

[2] See Selected Reference 13 for further information.

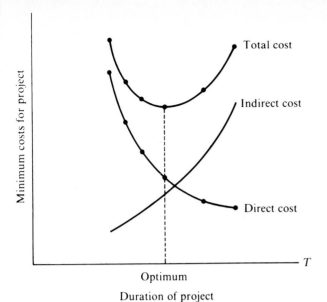

Figure 10.32 Minimum cost curves for the overall project.

Figure 10.31 provides a useful basis for a managerial decision on T (and the corresponding optimal solution for the x_{ij}) when the important effects of the project duration (other than direct costs) are largely intangible. However, when these other effects are primarily financial (indirect costs), it is appropriate to combine the minimum total direct cost curve of Fig. 10.31 with a curve of *minimum total indirect cost* (supervision, facilities, clerical, interest, contractual penalties) versus t, as shown in Fig. 10.32. The *sum* of these curves thereby gives the *minimum total project cost* curve for the various values of T. The *optimal* value of T is then the one that minimizes this total cost curve.

Choosing between PERT and CPM

The choice between the PERT *three-estimate approach* and the CPM *method of time-cost trade-offs* depends primarily upon the *type of project* and the *managerial objectives*. PERT is particularly appropriate when there is considerable uncertainty in predicting activity times and when it is important to effectively *control* the project schedule; for example, most *research and development* projects fall into this category. On the other hand, CPM is particularly appropriate when activity times can be predicted well (perhaps based on previous experience) but these times can be adjusted readily (e.g., by changing crew sizes), and when it is important to plan an appropriate trade-off between project *time* and *cost*. This latter type is typified by most *construction* and *maintenance* projects.

Actually, differences between *current* versions of PERT and CPM are not necessarily as pronounced as we have described them. Most versions of PERT now allow using only a *single* estimate (the most likely estimate) of each activity time and thus omit the probabilistic investigation. A version called *PERT/Cost* also considers *time-cost trade-offs* in a manner similar to CPM.

10.9 Conclusions

Networks of some type arise in a wide variety of contexts. Network representations are very useful for portraying the relationships and connections between the components of systems. Frequently, flow of some type must be sent through a network, so a decision needs to be made on the best way to do this. The kinds of network optimization models and algorithms introduced in this chapter provide a powerful tool for making such decisions.

The minimum cost flow problem plays a central role among these network optimization models, both because it is so broadly applicable and because it can be solved extremely efficiently by the network simplex method. Two of its special cases included in this chapter, the shortest path problem and the maximum flow problem, also are basic network optimization models, as are three additional special cases discussed in Chap. 7 (the transportation problem, the transshipment problem, and the assignment problem).

Whereas all of these models are concerned with optimizing the *operation* of an *existing* network, the minimum spanning tree problem is a prominent example of a model for optimizing the *design* of a *new* network.

This chapter has only scratched the surface of the current state of the art of network methodology. Because of their combinatorial nature, network problems often are extremely difficult to solve. However, great progress is being made in developing powerful modeling techniques and solution methodologies that are opening up new vistas for important applications. In fact, recent algorithmic advances are enabling us to solve successfully some complex network problems of enormous size.

The most widely used network technique has been the *PERT-type system* for project planning and control. It has been very valuable for organizing planning effort, testing alternative plans, revealing the overall dimensions and details of the project plan, establishing well-understood management responsibilities, and identifying realistic expectations for the project. It also lays the foundation for *anticipatory* management action against potential trouble spots during the course of the project. Although not a panacea, it has greatly aided project management on numerous occasions.

SELECTED REFERENCES

1. Bazaraa, Mokhtar S., and John J. Jarvis: *Linear Programming and Network Flows,* Wiley, New York, 1977.

2. Chachra, V., P. Ghare, and J. Moore: *Applications of Graph Theory Algorithms,* North-Holland, New York, 1979.

3. Elmaghraby, S. E.: *Activity Networks: Project Planning and Control by Network Models,* Wiley, New York, 1977.

4. Ford, L. R., Jr., and D. R. Fulkerson: *Flows in Networks,* Princeton University Press, Princeton, N.J., 1962.

5. Glover, Fred, and Darwin Klingman: ''Network Application in Industry and Government,'' *AIIE Transactions,* **9**:363–376, 1977.

6. Jensen, Paul A., and J. Wesley Barnes: *Network Flow Programming,* Wiley, New York, 1980.

7. Kennington, Jeff L., and Richard V. Helgason: *Algorithms for Network Programming,* Wiley-Interscience, Somerset, N.J., 1980.

8. Klingman, Darwin, Nancy Phillips, David Steiger, Ross Wirth, and Warren Young: "The Challenges and Success Factors in Implementing an Integrated Products Planning System for Citgo," *Interfaces,* **16**(3):1–19, May–June 1986.

9. Klingman, Darwin, Nancy Phillips, David Steiger, and Warren Young: "The Successful Deployment of Management Science throughout Citgo Petroleum Corporation," *Interfaces,* **17**(1):4–25, Jan.–Feb. 1987.

10. Magnanti, Thomas L., and Richard T. Wong: "Network Design and Transportation Planning: Models and Algorithms," *Transportation Science,* **18**:1–55, 1984.

11. Mandl, C.: *Applied Network Optimization,* Academic Press, Orlando, Fla., 1979.

12. Minieka, Edward: *Optimization Algorithms for Networks and Graphs,* Dekker, New York, 1978.

13. Moder, Joseph J., Cecil R. Phillips, and Edward W. Davis: *Project Management with CPM, PERT and Precedence Diagramming,* 3d ed., Van Nostrand Reinhold, New York, 1983.

14. Phillips, D., and A. Diaz: *Fundamentals of Network Analysis,* Prentice-Hall, Englewood Cliffs, N.J., 1981.

15. Sheffi, Yosef: *Urban Transportation Networks: Equilibrium Analysis with Mathematical Programming Methods,* Prentice-Hall, Englewood Cliffs, N.J., 1985.

16. Weist, Jerome D., and Ferdinand K. Levy: *Management Guide to PERT/CPM,* 2d ed., Prentice-Hall, Englewood Cliffs, N.J., 1977.

PROBLEMS

1. Consider the following directed network.

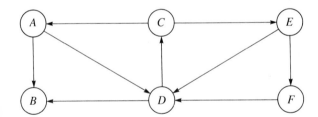

(a) Find a *directed path* from node *A* to node *F,* and then identify three other *undirected paths* from node *A* to node *F.*

(b) Find three *directed cycles.* Then identify an *undirected cycle* that includes every node.

(c) Identify a set of arcs that forms a *spanning tree.*

(d) Use the process illustrated in Fig. 10.3 to grow a tree one arc at a time until a spanning tree has been formed. Then repeat this process to obtain another spanning tree. [Do not duplicate the spanning tree identified in part (c).]

2. At a small but growing airport, the local airline company is purchasing a new tractor for a tractor-trailer train to bring luggage to and from the airplanes. A new mechanized luggage system will be installed in 3 years, so the tractor will not be needed after that. However, because it will receive heavy use, so that the running and maintenance costs will increase rapidly as it ages, it may still be more economical to replace the tractor after 1 or 2 years. The following table gives the total net discounted cost associated with purchasing a tractor (purchase price minus trade-in allowance, plus running and maintenance costs) at the end of year i and trading it in at the end of year j (where year 0 is now).

		j	
	1	2	3
0	8	18	31
i *1*		10	21
2			12

The problem is to determine at what times (if any) the tractor should be replaced to minimize the total cost for the tractors over the 3 years.

(*a*) Formulate this problem as a shortest path problem.

(*b*) Use the algorithm described in Sec. 10.3 to solve this shortest path problem.

3.* Use the algorithm described in Sec. 10.3 to find the *shortest path* through networks (*a*) and (*b*), where the numbers represent actual distances between the corresponding nodes.

(*a*)

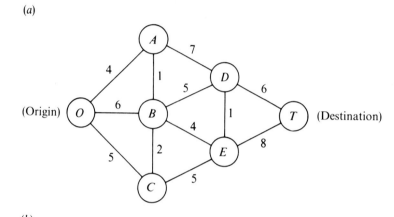

(*b*)

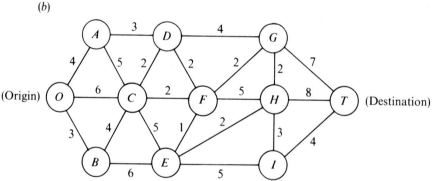

4. Reconsider the student's "car problem" described in Prob. 26 at the end of Chap. 7.

(*a*) Formulate the student's problem as a shortest path problem.

(*b*) Use the algorithm described in Sec. 10.3 to solve this shortest path problem.

5. Formulate the shortest path problem as a linear programming problem.

6. A company has learned that a competitor is planning to come out with a new kind of product with a great sales potential. This company has been working on a similar product, and research is nearly complete. It now wishes to rush the product out to meet the competition. There are four nonoverlapping phases left to be accomplished, including the remaining research that currently is being conducted at a normal pace. However, each phase can instead be conducted at a priority or crash level to expedite completion. The times required (in months) at these levels are

			Time	
Level	Remaining Research	Development	Design of Manufacturing System	Initiate Production and Distribution
Normal	5			
Priority	4	3	5	2
Crash	2	2	3	1

$30,000,000 is available for these four phases. The cost (in millions of dollars) at the different levels is

			Cost	
Level	Remaining Research	Development	Design of Manufacturing System	Initiate Production and Distribution
Normal	3			
Priority	6	6	9	3
Crash	9	9	12	6

The problem is to determine at which level to conduct each of the four phases to minimize the total time until the product can be marketed subject to the budget restriction.

(a) Formulate this problem as a shortest path problem.

(b) Use the algorithm described in Sec. 10.3 to solve this shortest path problem.

7.* Reconsider the networks shown in Prob. 1. Assume that the nodes and actual distances between nodes are as shown there (where unspecified distances between nodes are greater than any of the given distances), but assume that the arcs have not yet been specified. Use the algorithm described in Sec. 10.4 to find the *minimum spanning tree* for each of these networks.

8. A logging company will soon begin logging eight groves of trees in the same general area. Therefore, it must develop a system of dirt roads that makes each grove accessible from every other grove. The distance (in miles) between every pair of groves is

Distance between Pairs of Groves

Grove	1	2	3	4	5	6	7	8
1	—	1.3	2.1	0.9	0.7	1.8	2.0	1.5
2	1.3	—	0.9	1.8	1.2	2.6	2.3	1.1
3	2.1	0.9	—	2.6	1.7	2.5	1.9	1.0
4	0.9	1.8	2.6	—	0.7	1.6	1.5	0.9
5	0.7	1.2	1.7	0.7	—	0.9	1.1	0.8
6	1.8	2.6	2.5	1.6	0.9	—	0.6	1.0
7	2.0	2.3	1.9	1.5	1.1	0.6	—	0.5
8	1.5	1.1	1.0	0.9	0.8	1.0	0.5	—

The problem is to determine between which pairs of groves the roads should be constructed to connect all groves with a minimum total length of road.

(a) Describe how this problem fits the network description of the minimum spanning tree problem.

(b) Use the algorithm described in Sec. 10.4 to solve the problem.

9. A bank soon will be hooking up computer terminals at each of its branch offices to the computer at its main office using special phone lines with telecommunications devices. The phone line from a branch office need not be connected directly to the main office. It can be connected indirectly by being connected to another branch office that is connected (directly or indirectly) to the main office. The only requirement is that every branch office be connected by some route to the main office.

The charge for the special phone lines is directly proportional to the mileage involved, where the distance (in miles) between every pair of offices is

Distance between Pairs of Offices

	Main	B.1	B.2	B.3	B.4	B.5
Main office	—	190	70	115	270	160
Branch 1	190	—	100	240	215	50
Branch 2	70	100	—	140	120	220
Branch 3	115	240	140	—	175	80
Branch 4	270	215	120	175	—	310
Branch 5	160	50	220	80	310	—

The problem is to determine which pairs of offices should be directly connected by special phone lines in order to connect every branch office (directly or indirectly) to the main office at a minimum total cost.

   (a) Describe how this problem fits the network description of the minimum spanning tree problem.

   (b) Use the algorithm described in Sec. 10.4 to solve the problem.

   10.* For networks (a) and (b), use the augmenting path algorithm described in Sec. 10.5 to find the flow pattern giving the *maximum flow* from the supply node (the left-most node) to the demand node (the right-most node), given that the arc capacity from node i to node j is the number nearest node i along the link between these nodes.

(a)

(b)

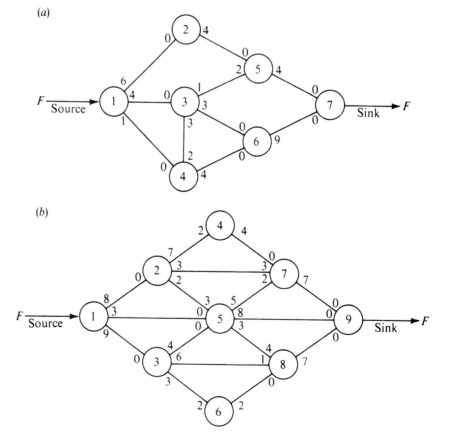

11. Formulate the maximum flow problem as a linear programming problem.

12. One track of the Eura Railroad system runs from the major industrial city of Faire-parc to the major port city of Portstown. This track is heavily used by both express passenger and freight trains. The passenger trains are carefully scheduled and have priority over the slow freight trains (this is a European railroad), so that the freight trains must pull over onto a siding whenever a passenger train is scheduled to pass them soon. It is now necessary to increase the freight service, so the problem is to schedule the freight trains so as to maximize the number that can be sent each day without interfering with the fixed schedule for passenger trains.

Consecutive freight trains must maintain a schedule differential of at least 0.1 hour, and this is the time unit used for scheduling them (so that the daily schedule indicates the status of each freight train at times 0.0, 0.1, 0.2, . . . , 23.9). There are S sidings between Faireparc and Portstown, where siding i is long enough to hold n_i freight trains ($i = 1, . . . , S$). It requires t_i time units (rounded up to an integer) for a freight train to travel from siding i to siding $i + 1$ (where t_0 is the time from the Faireparc station to siding 1 and t_s is the time from siding S to the Portstown station). A freight train is allowed to pass or leave siding i ($i = 0, 1, . . . , S$) at time j ($j = 0.0, 0.1, . . . , 23.9$) only if it would not be overtaken by a scheduled passenger train before reaching siding $i + 1$ (let $\delta_{ij} = 1$ if it would not be overtaken, and let $\delta_{ij} = 0$ if it would be). A freight train also is required to stop at a siding if there will not be room for it at all subsequent sidings that it would reach before being overtaken by a passenger train.

Formulate this problem as a maximum flow problem by identifying every node (including the supply node and the demand node) as well as every arc and its arc capacity for the network representation of the problem. (*Hint:* Use a different set of nodes for each of the 240 times.)

13. Consider the maximum flow problem shown below, where the supply node is node A, the demand node is node F, and the arc capacities are the numbers shown next to these directed arcs.

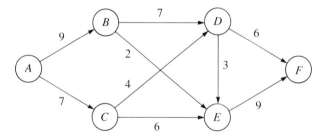

(a) Use the augmenting path algorithm described in Sec. 10.5 to solve this problem.
(b) Formulate the network representation of this problem as a minimum cost flow problem, including adding the arc $A \rightarrow F$. Use $\overline{F} = 20$.
(c) Obtain an initial basic feasible solution by solving the feasible spanning tree with basic arcs $A \rightarrow B$, $A \rightarrow C$, $A \rightarrow F$, $B \rightarrow D$, and $E \rightarrow F$, where two of the nonbasic arcs ($E \rightarrow C$ and $F \rightarrow D$) are *reverse* arcs.
(d) Use the network simplex method to solve this problem.

14. A company will be producing the same new product at two different factories, and then the product must be shipped to two warehouses. Factory 1 can send an unlimited amount by rail to Warehouse 1 only, whereas Factory 2 can send an unlimited amount by rail to Warehouse 2 only. However, independent truckers can be used to ship up to 50 units from each factory to a distribution center, from which up to 50 units can be shipped to each warehouse. The shipping cost per unit for each alternative is shown in the following table, along with the amounts to be produced at the factories and the amounts needed at the warehouses.

From \ To	Distr. Center	Warehouse 1	Warehouse 2	Output
Factory 1	3	7	—	80
Factory 2	4	—	9	70
D. Center		2	4	
Allocation		60	90	

Unit Shipping Cost (table heading)

(a) Formulate the network representation of this problem as a minimum cost flow problem.

(b) Formulate the linear programming model for this problem.

(c) Obtain an initial basic feasible solution by solving the feasible spanning tree that corresponds to using just the two rail lines plus Factory 1 shipping to Warehouse 2 via the distribution center.

(d) Use the network simplex method to solve this problem.

15. Reconsider Prob. 2.

(a) Now formulate this problem as a minimum cost flow problem by showing the appropriate network representation.

(b) Starting with the initial basic feasible solution that corresponds to replacing the tractor every year, use the network simplex method to solve this problem.

16. For the P & T Co. transportation problem given in Table 7.2, consider its network representation as a minimum cost flow problem presented in Fig. 10.10. Use the northwest corner rule to obtain an initial basic feasible solution from Table 7.2. Then use the network simplex method to solve this problem (and verify the optimal solution given in Sec. 7.1).

17. Consider the Metro Water District transportation problem presented in Table 7.12.

(a) Formulate the network representation of this problem as a minimum cost flow problem. (*Hint:* Arcs where flow is prohibited should be deleted.)

(b) Starting with the initial basic feasible solution given in Table 7.19, use the network simplex method to solve this problem. Compare the sequence of basic feasible solutions obtained with the sequence obtained by the transportation simplex method in Table 7.23.

18. Consider the transportation problem having the following cost and requirements table:

		Destination 1	Destination 2	Destination 3	Supply
Source	1	6	7	4	40
	2	5	8	6	60
Demand		30	40	30	

Formulate the network representation of this problem as a minimum cost flow problem. Use the northwest corner rule to obtain an initial basic feasible solution. Then use the network simplex method to solve the problem.

19. Consider the minimum cost flow problem shown below, where the b_i are given by the nodes, the c_{ij} are given by the arcs, and the *finite* u_{ij} are given in parentheses by the arcs.

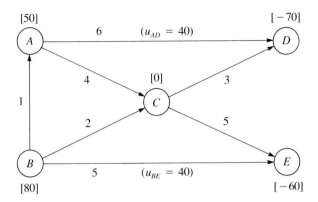

Obtain an initial basic feasible solution by solving the feasible spanning tree with basic arcs $A \rightarrow C$, $B \rightarrow A$, $C \rightarrow D$, and $C \rightarrow E$, where one of the nonbasic arcs $(D \rightarrow A)$ is a *reverse* arc. Then use the network simplex method to solve this problem.

20.* Consider the following project network. Assume that the time required (in weeks) for each activity is a predictable constant and that it is given by the number along the corresponding arc. Find the earliest time, latest time, and slack for each event, as well as the slack for each activity. Also identify the critical path.

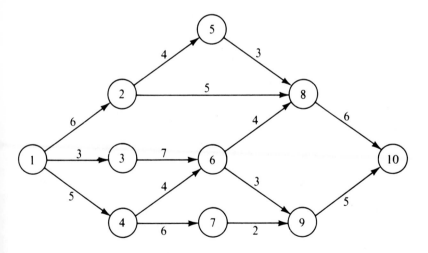

21. Consider the following project network. Assume that the time required (in days) for each activity is a predictable constant and that it is given by the number along the corresponding arc. Find the earliest time, latest time, and slack for each event, as well as the slack for each activity. Also identify the critical path.

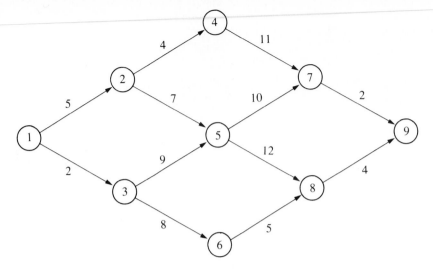

22. You and several friends are about to prepare a lasagne dinner. The tasks to be performed, their times (in minutes), and the precedence constraints are as follows:

Task No.	Task	Time	Tasks That Must Precede
1	Buy the mozzarella cheese*	30	
2	Slice the mozzarella	5	1
3	Beat 2 eggs	2	
4	Mix eggs and ricotta cheese	3	3
5	Cut up onions and mushrooms	7	
6	Cook the tomato sauce	25	5
7	Boil large quantity of water	15	
8	Boil the lasagne noodles	10	7
9	Drain the lasagne noodles	2	8
10	Assemble all the ingredients	10	9, 6, 4, 2
11	Preheat the oven	15	
12	Bake the lasagne	30	10, 11

* There is none in the refrigerator.

(a) Formulate this problem as a *PERT-type system* by drawing the project network. Use one event to represent the simultaneous initiation of the initial tasks. On one side of each arc, identify the number of the task being performed in parentheses, e.g., (Task 7). On the other side, show the times required.

(b) Find the earliest time, latest time, and slack for each event, as well as the slack for each activity. Also identify the critical path.

(c) Because of a phone call you were interrupted for 6 minutes when you should have been cutting the onions and mushrooms. By how much will the dinner be delayed? If you use your food processor, which reduces the cutting time from 7 minutes to 2 minutes, will the dinner still be delayed?

23.* Using the PERT three-estimate approach, the three estimates for one of the activities are as follows: optimistic estimate = 30 days, most likely estimate = 36 days, pessimistic estimate = 48 days. What are the resulting estimates of the *expected value* and *variance* of the time required by the activity?

24. Consider the following project network.

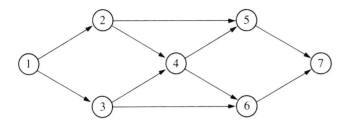

The PERT three-estimate approach has been used, and it has led to the following estimates of the *expected value* (in months) and *variance* of the time required for the respective activities:

	Activity Times	
Activity	Estimated Expected Value	Estimated Variance
$1 \rightarrow 2$	4	5
$1 \rightarrow 3$	6	10
$2 \rightarrow 4$	4	8
$2 \rightarrow 5$	8	12
$3 \rightarrow 4$	3	6
$3 \rightarrow 6$	7	14
$4 \rightarrow 5$	5	12
$4 \rightarrow 6$	3	5
$5 \rightarrow 7$	5	8
$6 \rightarrow 7$	5	7

The scheduled project completion time is 22 months after the start of the project.

(*a*) Using expected values, determine the *critical path* for the project.

(*b*) Using the procedure described in Sec. 10.8, find the approximate probabjlity that the project will be completed by the scheduled time.

(*c*) In addition to the critical path, there are five other paths through the network. For each of these other paths, find the approximate probability that the sum of the activity times along the path is not more than 22 months.

25. Consider the following project network.

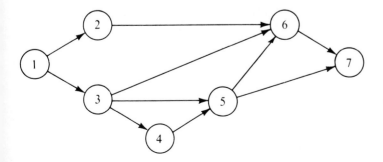

Using the PERT three-estimate approach, suppose that the usual three estimates for the time required (in weeks) for each of these activities are

Activity	Optimistic Estimate	Most Likely Estimate	Pessimistic Estimate
1 → 2	28	32	36
1 → 3	22	28	32
2 → 6	26	36	46
3 → 4	14	16	18
3 → 5	32	32	32
3 → 6	40	52	74
4 → 5	12	16	24
5 → 6	16	20	26
5 → 7	26	34	42
6 → 7	12	16	30

The project is ready to start now and the deadline for completing the project is 100 weeks hence.

(a) On the basis of the estimates just listed, calculate the *expected value* and *standard deviation* of the time required for each activity.

(b) Using expected times, determine the *critical path* for the project.

(c) Using the procedure described in Sec. 10.8, find the approximate probability that the project will be completed by the deadline.

26. Reconsider the project network shown in Prob. 25. Suppose that the CPM method of time-cost trade-offs is to be used to determine how to meet the project deadline (100 weeks hence) in the most economical way. Also suppose that the *crash time* and *normal time* for each of the activities correspond to the times shown in the *Optimistic Estimate* and *Pessimistic Estimate* columns of the table for Prob. 25 (except that activity 3 → 5 has a crash time of 28 and a normal time of 36) and that the difference between the *crash cost* and *normal cost* is 10 (in units of thousands of dollars) for *every* activity. Formulate the linear programming model for this problem.

27. Suppose that the scheduled completion time for the house-construction project described in Figs. 10.27 and 10.28 has been moved forward to 40. Therefore, the CPM method of time-cost trade-offs is to be used to determine how to accelerate the project to meet this deadline in the most economical way. The relevant data are:

Activity	Normal Time	Crash Time	Normal Cost, $	Crash Cost, $
1 → 2	2	1	1,800	2,300
2 → 3	4	2	3,200	3,600
3 → 4	10	7	6,200	7,300
4 → 5	4	3	4,100	4,900
4 → 6	6	4	2,600	3,000
4 → 7	7	5	2,100	2,400
5 → 7	5	3	1,800	2,200
6 → 8	7	4	9,000	9,600
7 → 9	8	6	4,300	4,600
8 → 10	9	6	2,000	2,500
9 → 11	4	3	1,600	1,800
9 → 12	5	3	2,500	3,000
10 → 13	2	1	1,000	1,500
12 → 13	6	3	3,300	4,000

Formulate the linear programming model for this problem.

11

Dynamic Programming

Dynamic programming is a useful mathematical technique for making a sequence of interrelated decisions. It provides a systematic procedure for determining the optimal combination of decisions.

In contrast to linear programming, there does not exist a standard mathematical formulation of "the" dynamic programming problem. Rather, dynamic programming is a general type of approach to problem solving, and the particular equations used must be developed to fit each individual situation. Therefore, a certain degree of ingenuity and insight into the general structure of dynamic programming problems is required to recognize when a problem can be solved by dynamic programming procedures and how it can be done. These abilities can best be developed by an exposure to a wide variety of dynamic programming applications and a study of the characteristics that are common to all these situations. A large number of illustrative examples are presented for this purpose.

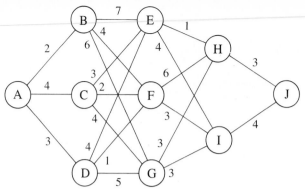

Figure 11.1 The road system and costs for the stagecoach problem.

11.1 Prototype Example

The STAGECOACH PROBLEM is a problem especially constructed[1] to illustrate the features and to introduce the terminology of dynamic programming. It concerns a mythical fortune seeker in Missouri who decided to go west to join the 49'er gold rush in California during the mid-nineteenth century. The journey would require traveling by stagecoach through unsettled country where there was serious danger of attack by marauders. Although his starting point and destination were fixed, he had considerable choice as to which **states** (or territories that subsequently became states) to travel through en route. The possible routes are shown in Fig. 11.1, where each state is represented by a lettered circle. Thus four **stages** (*stage*coach runs) were required to travel from his point of embarkation in state A (Missouri) to his destination in state J (California).

This fortune seeker was a prudent man who was quite concerned about his safety. After some thought, he came up with a rather clever way of determining the safest route. Life insurance policies were offered to stagecoach passengers. Because the cost of the **policy** for taking any given stagecoach run was based on a careful evaluation of the safety of that run, the safest route should be the one with the cheapest total life insurance policy.

The cost for the standard policy on the stagecoach run from state i to state j, which will be denoted by c_{ij}, is

	B	C	D			E	F	G			H	I			J
A	2	4	3		B	7	4	6		E	1	4		H	3
					C	3	2	4		F	6	3		I	4
					D	4	1	5		G	3	3			

These costs are also shown in Fig. 11.1.

[1] This problem was developed by Professor Harvey M. Wagner while he was at Stanford University.

We shall now focus on the question of which route minimizes the total cost of the policy.

415

Dynamic Programming

Solving the Problem

First note that the shortsighted approach of selecting the cheapest run offered by each successive stage need not yield an overall optimal decision. Following this strategy would give the route $A \rightarrow B \rightarrow F \rightarrow I \rightarrow J$, at a total cost of 13. However, sacrificing a little on one stage may permit greater savings thereafter. For example, $A \rightarrow D \rightarrow F$ is cheaper overall than $A \rightarrow B \rightarrow F$.

One possible approach to solving this problem is to use trial and error.[1] However, the number of possible routes is large (18) and having to calculate the total cost for each route is not an appealing task.

Fortunately, dynamic programming provides a solution with much less effort than exhaustive enumeration. (The computational savings are enormous for larger versions of this problem.) Dynamic programming starts with a small portion of the original problem and finds the optimal solution for this smaller problem. It then gradually enlarges the problem, finding the current optimal solution from the preceding one, until the original problem is solved in its entirety.

For the stagecoach problem, we start with the smaller problem where the fortune seeker has nearly completed his journey and has only one more stage (stagecoach run) to go. The obvious optimal solution for this smaller problem is to go from his current state (whatever it is) to his ultimate destination (state J). At each subsequent iteration, the problem is enlarged by increasing by one the number of stages left to go to complete the journey. For this enlarged problem, the optimal solution for where to go next from each possible state can be found relatively easily from the results obtained at the preceding iteration. The details involved in implementing this approach follow.

FORMULATION: Let the decision variables x_n ($n = 1, 2, 3, 4$) be the immediate destination on stage n (the nth stagecoach run to be taken). Thus the route selected is $A \rightarrow x_1 \rightarrow x_2 \rightarrow x_3 \rightarrow x_4$, where $x_4 = J$.

Let $f_n(s, x_n)$ be the total cost of the best overall *policy* for the *remaining* stages, given that the fortune seeker is in state s, ready to start stage n, and selects x_n as the immediate destination. Given s and n, let x_n^* denote the value of x_n that minimizes $f_n(s, x_n)$, and let $f_n^*(s)$ be the corresponding minimum value of $f_n(s, x_n)$. Thus

$$f_n^*(s) = \min_{x_n} f_n(s, x_n) = f_n(s, x_n^*),$$

where

$f_n(s, x_n) = $ immediate cost (stage n) + minimum future cost (stages $n + 1$ onward)

$$= c_{sx_n} + f_{n+1}^*(x_n).$$

---

[1] This problem also can be formulated as a *shortest path problem* (see Sec. 10.3), where *costs* here play the role of *distances* in the shortest path problem. The solution procedure presented in Sec. 10.3 actually uses the philosophy of dynamic programming. However, because the present problem has a fixed number of stages, the dynamic programming approach presented here is even better.

The value of c_{sx_n} is given by the preceding tables for c_{ij} by setting $i = s$ (the current state) and $j = x_n$ (the immediate destination). Because the ultimate destination (state J) is reached at the end of stage 4, $f_5^*(J) = 0$.

The objective is to find $f_1^*(A)$ and the corresponding route. Dynamic programming finds it by successively finding $f_4^*(s)$, $f_3^*(s)$, $f_2^*(s)$ for each of the possible states s and then using $f_2^*(s)$ to solve for $f_1^*(A)$.[1]

SOLUTION PROCEDURE: When the fortune seeker has only one more stage to go ($n = 4$), his route thereafter is determined entirely by his current state s (either H or I) and his final destination, $x_4 = J$, so the route for this final stagecoach run is $s \rightarrow J$. Therefore, since $f_4^*(s) = f_4(s, J) = c_{s,J}$, the immediate solution to the $n = 4$ problem is

$n = 4$:

s	$f_4^*(s)$	x_4^*
H	3	J
I	4	J

When the fortune seeker has two more stages to go ($n = 3$), the solution procedure requires a few calculations. For example, suppose that the fortune seeker is in state F. Then, as depicted below, he must next go to either state H or I at an immediate cost of $c_{F,H} = 6$ or $c_{F,I} = 3$, respectively. If he chooses state H, the minimum additional cost after he reaches there is given in the preceding table as $f_4^*(H) = 3$, as shown next to the H box in the diagram. Therefore, the total cost for this decision would be $6 + 3 = 9$. If he chooses state I instead, the total cost is $3 + 4 = 7$, which is smaller. Therefore, the optimal choice would be this latter one, $x_3^* = I$, because it gives the minimum cost, $f_3^*(F) = 7$.

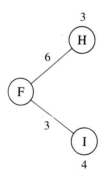

Similar calculations need to be made when starting from the other two possible states, $s = E$ and $s = G$, with two stages to go. Try it, proceeding both graphically

---

[1] Because this procedure involves moving *backward* stage by stage, some writers also count n backward to denote the number of *remaining stages* to the destination. We use the more natural *forward counting* for greater simplicity.

(Fig. 11.1) and algebraically [combining c_{ij} and $f_4^*(s)$ values], to verify the following complete results for the $n = 3$ problem.

	x_3	$f_3(s,x_3) = c_{sx_3} + f_4^*(x_3)$			
$n = 3$:	s	H	I	$f_3^*(s)$	x_3^*
	E	4	8	4	H
	F	9	7	7	I
	G	6	7	6	H

The solution for the three-stage problem ($n = 2$) is obtained in a similar fashion. In this case, $f_2(s,x_2) = c_{sx_2} + f_3^*(x_2)$. For example, suppose that the fortune seeker is in state C, as depicted below.

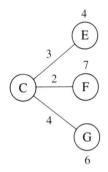

He must next go to state E, F, or G at an immediate cost of $c_{C,E} = 3$, $c_{C,F} = 2$, or $c_{C,G} = 4$, respectively. After getting there, the minimum additional cost for stage 3 to the end is given by the $n = 3$ table as $f_3^*(E) = 4$, $f_3^*(F) = 7$, or $f_3^*(G) = 6$, respectively, as shown next to the E, F, and G states in the above diagram. The resulting calculations for the three alternatives are summarized below.

$x_2 = E$: $\qquad f_2(C, E) = c_{C,E} + f_3^*(E) = 3 + 4 = 7.$

$x_2 = F$: $\qquad f_2(C, F) = c_{C,F} + f_3^*(F) = 2 + 7 = 9.$

$x_2 = G$: $\qquad f_2(C, G) = c_{C,G} + f_3^*(G) = 4 + 6 = 10.$

The minimum of these three numbers is 7, so the minimum total cost from state C to the end is $f_2^*(C) = 7$ and the immediate destination should be $x_2^* = E$.

Making similar calculations when starting from state B or D (try it) yields the following results for the $n = 2$ problem:

	x_2	$f_2(s, x_2) = c_{sx_2} + f_3^*(x_2)$				
$n = 2$:	s	E	F	G	$f_2^*(s)$	x_2^*
	B	11	11	12	11	E or F
	C	7	9	10	7	E
	D	8	8	11	8	E or F

Moving to the four-stage problem ($n = 1$), the calculations are similar to those just shown for the three-stage problem ($n = 2$), except now there is just *one* possible starting state, $s = $ A, as depicted below.

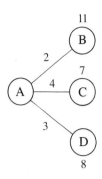

These calculations are summarized next for the three alternatives for the immediate destination:

$$x_1 = \text{B:} \qquad f_1(\text{A, B}) = c_{A,B} + f_2^*(\text{B}) = 2 + 11 = 13.$$

$$x_1 = \text{C:} \qquad f_1(\text{A, C}) = c_{A,C} + f_2^*(\text{C}) = 4 + 7 = 11.$$

$$x_1 = \text{D:} \qquad f_1(\text{A, D}) = c_{A,D} + f_2^*(\text{D}) = 3 + 8 = 11.$$

Since 11 is the minimum, $f_1^*(\text{A}) = 11$ and $x_1^* = $ C or D, as shown in the following table.

	x_1	$f_1(s, x_1) = c_{sx_1} + f_2^*(x_1)$				
$n = 1$:	s	B	C	D	$f_1^*(s)$	x_1^*
	A	13	11	11	11	C or D

The optimal solution for the entire problem can now be identified. Results for the $n = 1$ problem indicate that the fortune seeker should go initially to either state C or state D. Suppose that he chooses $x_1^* = $ C. For $n = 2$, the result for $s = $ C is $x_2^* = $ E. This result leads to the $n = 3$ problem, which gives $x_3^* = $ H for $s = $ E, and the $n = 4$ problem yields $x_4^* = $ J for $s = $ H. Hence one optimal route is A $\rightarrow$ C $\rightarrow$ E $\rightarrow$ H $\rightarrow$ J. Choosing $x_1^* = $ D leads to the other two optimal routes, A $\rightarrow$ D $\rightarrow$ E $\rightarrow$ H $\rightarrow$ J and A $\rightarrow$ D $\rightarrow$ F $\rightarrow$ I $\rightarrow$ J. They all yield a total cost of $f_1^*(\text{A}) = 11$.

You will see in the next section that the *special terms* describing the particular context of this problem—*stage, state, policy*—actually are part of the *general terminology* of dynamic programming with an analogous interpretation in other contexts.

11.2 Characteristics of Dynamic Programming Problems

The stagecoach problem is a literal prototype of dynamic programming problems. In fact, this example was purposely designed to provide a literal physical interpretation of the rather abstract structure of such problems. Therefore, one way to recognize a situation that can be formulated as a dynamic programming problem is to notice that its basic structure is analogous to that of the stagecoach problem.

These basic features that characterize dynamic programming problems are presented and discussed here.

1. The problem can be divided into **stages**, with a **policy decision** required at each stage.

The stagecoach problem was literally divided into its four *stages* (stagecoaches) that correspond to the four legs of the journey. The policy decision at each stage was which life insurance *policy* to choose (i.e., which destination to select for the next stagecoach ride). Similarly, other dynamic programming problems require making a *sequence of interrelated decisions,* where each decision corresponds to one stage of the problem.

2. Each stage has a number of **states** associated with it.

The states associated with each stage in the stagecoach problem were the states (or territories) in which the fortune seeker could be located when embarking on that particular leg of the journey. In general, the states are the various *possible conditions* in which the system might be at that stage of the problem. The number of states may be either *finite* (as in the stagecoach problem) or *infinite* (as in some subsequent examples).

3. The effect of the policy decision at each stage is to *transform the current state into a state associated with the next stage* (possibly according to a probability distribution).

The fortune seeker's decision as to his next destination led him from his current state to the next state on his journey. This procedure suggests that dynamic programming problems can be interpreted in terms of the *networks* described in Chap. 10. Each *node* would correspond to a *state*. The network would consist of columns of nodes, with each *column* corresponding to a *stage,* so that the flow from a node can go only to a node in the next column to the right. The links from a node to nodes in the next column correspond to the possible policy decisions on which state to go to next. The value assigned to each link usually can be interpreted as the *immediate contribution* to the objective function from making that policy decision. In most cases, the objective corresponds to finding either the *shortest* or the *longest route* through the network.

4. The solution procedure is designed to find an **optimal policy** for the overall problem, i.e., a prescription of the optimal policy decision at each stage for *each* of the possible states.

For the stagecoach problem, the solution procedure constructed a table for each stage (n) that prescribes the optimal decision (x_n^*) for *each* possible state (s). Thus, in addition to identifying three *optimal solutions* (optimal routes) for the overall problem, the results also show the fortune seeker how he should proceed if he gets detoured to a state that is not on an optimal route. For any problem, dynamic programming provides this kind of *policy* prescription of what to do under every possible circumstance (which is why the actual decision made upon reaching a particular state at a given stage is referred to as a *policy* decision). Providing this additional information beyond simply specifying an optimal solution (optimal sequence of decisions) can be helpful in a variety of ways, including sensitivity analysis.

5. Given the current state, an *optimal policy for the remaining stages* is *independent* of the policy adopted in *previous stages*. (This is the **principle of optimality** for dynamic programming.)

Given the state in which the fortune seeker is currently located, the optimal life insurance policy (and its associated route) from this point onward is independent of how he got there. For dynamic programming problems in general, knowledge of the current state of the system conveys all the information about its previous behavior necessary for determining the optimal policy henceforth. (This property is the *Markovian property*.) Any problem lacking this property cannot be formulated as a dynamic programming problem.

6. The solution procedure begins by finding the *optimal policy for the last stage*.

The optimal policy for the last stage prescribes the optimal policy decision for *each* of the possible states at that stage. The solution of this one-stage problem is usually trivial, as it was for the stagecoach problem.

7. A **recursive relationship** that identifies the optimal policy for stage n, given the optimal policy for stage $(n + 1)$, is available.

For the stagecoach problem, this recursive relationship was

$$f_n^*(s) = \min_{x_n} \{c_{sx_n} + f_{n+1}^*(x_n)\}.$$

Therefore, finding the *optimal policy decision* when starting in state s at stage n requires finding the minimizing value of x_n. The corresponding minimum cost is achieved by using this value of x_n and then following the optimal policy when starting in state x_n at stage $(n + 1)$.

The precise form of the recursive relationship differs somewhat among dynamic programming problems. However, notation analogous to that introduced in the preceding section will continue to be used here, as summarized below.

N = number of stages.
n = label for current stage ($n = 1, 2, \ldots, N$).
s_n = current *state* for stage n.
x_n = decision variable for stage n.
x_n^* = optimal value of x_n (given s_n).
$f_n(s_n, x_n)$ = contribution of stages $n, n + 1, \ldots, N$ to the objective function if the system starts in state s_n at stage n, the immediate decision is x_n, and optimal decisions are made thereafter.
$f_n^*(s_n) = f_n(s_n, x_n^*)$.

The recursive relationship will always be of the form

$$f_n^*(s_n) = \max_{x_n} \{f_n(s_n, x_n)\} \quad \text{or} \quad f_n^*(s_n) = \min_{x_n} \{f_n(s_n, x_n)\},$$

where $f_n(s_n, x_n)$ would be written in terms of s_n, x_n, $f_{n+1}^*(s_{n+1})$ and probably some measure of the first-stage effectiveness (or ineffectiveness) of x_n.

The recursive relationship is given its name because it keeps *recurring* as we move backward stage by stage. When the current stage number n is decreased by one, the new $f_n^*(s_n)$ function is derived by using the $f_{n+1}^*(s_{n+1})$ function that was just derived during the preceding iteration, and then this process keeps repeating. This property is emphasized in our next (and final) characteristic of dynamic programming.

8. When we use this recursive relationship, the solution procedure moves *backward* stage by stage—each time finding the optimal policy for that stage—until it finds the optimal policy starting at the *initial* stage.

This backward movement was demonstrated by the stagecoach problem, where the optimal policy was found successively beginning in each state at stages 4, 3, 2, and 1, respectively.[1] For all dynamic programming problems, a table such as the following one would be obtained for each stage ($n = N, N - 1, \ldots, 1$).

s_n \\ x_n	$f_n(s_n, x_n)$	$f_n^*(s_n)$	x_n^*

When this table is finally obtained for the initial stage ($n = 1$), the problem of interest is solved. Because the initial state is known, the initial decision is specified by x_1^* in this table. The optimal value of the other decision variables is then specified by the other tables in turn according to the state of the system that results from the preceding decisions.

11.3 Deterministic Dynamic Programming

This section further elaborates upon the dynamic programming approach to *deterministic* problems, where the *state* at the *next stage* is *completely determined* by the *state* and *policy decision* at the *current stage*. The *probabilistic* case, where there is a probability distribution for what the next state will be, is discussed in the next section.

Deterministic dynamic programming can be described diagrammatically as shown in Fig. 11.2. Thus at stage n the process will be in some state s_n. Making policy decision x_n then moves the process to some state s_{n+1} at stage $(n + 1)$. The contribution *thereafter* to the objective function under an optimal policy has been previously calculated to be $f_{n+1}^*(s_{n+1})$. The policy decision x_n also makes some contribution to the objective function. Combining these two quantities in an appropriate way provides $f_n(s_n, x_n)$, the contribution of stages n onward to the objective function. Optimizing with respect to x_n then gives $f_n^*(s_n) = f_n(s_n, x_n^*)$. After finding x_n^* and $f_n^*(s_n)$ for each possible value of s_n, the solution procedure is ready to move back one stage.

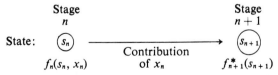

Figure 11.2 The basic structure for deterministic dynamic programming.

---

[1] Actually, for this problem the solution procedure can move *either* backward or forward. However, for many problems (especially when the stages correspond to *time periods*), the solution procedure *must* move backward.

One way of categorizing deterministic dynamic programming problems is by the *form of the objective function*. For example, the objective might be to *minimize* the *sum* of the contributions from the individual stages (as for the stagecoach problem), or to *maximize* such a sum, or to minimize a *product* of such terms, and so on. Another categorization is in terms of the nature of the *set of states* for the respective stages. In particular, the states s_n might be representable by a *discrete* state variable (as for the stagecoach problem), or by a *continuous* state variable, or perhaps a state *vector* (more than one variable) is required.

Several examples are presented to illustrate these various possibilities. More important, they illustrate that these apparently major differences are actually quite inconsequential (except in terms of computational difficulty) because the underlying basic structure shown in Fig. 11.2 always remains the same.

The first new example arises in a much different context from the stagecoach problem, but it has the same *mathematical formulation* except that the objective is to *maximize* rather than minimize a sum.

Example 2—Distributing Medical Teams to Countries

The WORLD HEALTH COUNCIL is devoted to improving health care in the underdeveloped countries of the world. It now has five *medical teams* available to allocate among three such countries to improve their medical care, health education, and training programs. Therefore, the council needs to determine how many teams (if any) to allocate to each of these countries to maximize the total effectiveness of the five teams. The teams must be kept intact, so the number allocated to each country must be integer.

The measure of performance being used is *additional person-years of life*. (For a particular country, this measure equals the country's *increased life expectancy* in years times its population.) Table 11.1 gives the estimated additional person-years of life (in multiples of 1,000) for each country for each possible allocation of medical teams.

Which allocation maximizes the measure of performance?

FORMULATION: This problem requires making three *interrelated decisions,* namely, how many medical teams to allocate to each of the three countries. Therefore, even though there is no fixed sequence, these three countries can be considered as the three

Table 11.1 **Data for the World Health Council Problem**

No. of Medical Teams	Thousands of Additional Person-Years of Life		
	Country		
	1	2	3
0	0	0	0
1	45	20	50
2	70	45	70
3	90	75	80
4	105	110	100
5	120	150	130

stages in a dynamic programming formulation. The decision variables x_n ($n = 1, 2, 3$) would be the number of teams to allocate to stage (country) n.

The identification of the *states* may not be readily apparent. To determine the states, we ask questions such as the following. What is it that changes from one stage to the next? Given that the decisions have been made at the previous stages, how can the status of the situation at the current stage be described? What information about the current state of affairs is necessary to determine the optimal policy hereafter? On these bases, an appropriate choice for the "state of the system" is

s_n = number of medical teams still available for allocation to the remaining countries ($n, \ldots, 3$).

Thus, at stage 1 (country 1), where all three countries remain under consideration for allocations, $s_1 = 5$. However, at stage 2 or 3 (country 2 or 3), s_n is just 5 minus the number of teams allocated at preceding stages. With the dynamic programming procedure of solving backward stage by stage, when we are solving at stage 2 or 3, we shall not yet have solved for the allocations at the preceding stages. Therefore, we shall consider every possible state we could be in at stage 2 or 3, namely, $s_n = 0$, 1, 2, 3, 4, or 5.

Let $p_i(x_i)$ be the measure of performance from allocating x_i medical teams to country i, as given in Table 11.1. Thus the objective is to choose x_1, x_2, x_3 so as to

$$\text{Maximize} \quad \sum_{i=1}^{3} p_i(x_i),$$

subject to

$$\sum_{i=1}^{3} x_i = 5,$$

and   the x_i are nonnegative integers.

Using the notation presented in Sec. 11.2, $f_n(s_n, x_n)$ is then

$$f_n(s_n, x_n) = p_n(x_n) + \text{maximum} \sum_{i=n+1}^{3} p_i(x_i),$$

where the maximum is taken over $x_{n+1}, \ldots, x_3$ such that

$$\sum_{i=n}^{3} x_i = s_n$$

and the x_i are nonnegative integers, for $n = 1, 2, 3$. In addition,

$$f_n^*(s_n) = \max_{x_n = 0, 1, \ldots, s_n} f_n(s_n, x_n)$$

Therefore, $f_n(s_n, x_n) = p_n(x_n) + f_{n+1}^*(s_n - x_n)$

(with f_4^* defined to be *zero*). These basic relationships are summarized in Fig. 11.3.

Consequently, the *recursive relationship* relating the f_1^*, f_2^*, and f_3^* functions for this problem is

$$f_n^*(s_n) = \max_{x_n = 0, 1, \ldots, s_n} \{p_n(x_n) + f_{n+1}^*(s_n - x_n)\}, \quad \text{for } n = 1, 2.$$

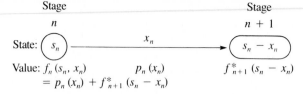

Value: $f_n(s_n, x_n)$ $p_n(x_n)$ $f^*_{n+1}(s_n - x_n)$
$\qquad = p_n(x_n) + f^*_{n+1}(s_n - x_n)$

Figure 11.3 The basic structure for the World Health Council problem.

For the last stage ($n = 3$),

$$f^*_3(s_3) = \max_{x_3=0,1,\ldots,s_3} p_3(x_3).$$

The resulting dynamic programming calculations are given next.

SOLUTION PROCEDURE: Beginning with the last stage ($n = 3$), we note that the values of $p_3(x_3)$ are given in the last column of Table 11.1, and that these values keep increasing as we move down the column. Therefore, with s_3 medical teams still available for allocation to country 3, the maximum of $p_3(x_3)$ is automatically achieved by allocating all s_3 teams, so $x^*_3 = s_3$ and $f^*_3(s_3) = p_3(s_3)$ as shown in the following table.

$n = 3$:

s_3	$f^*_3(s_3)$	x^*_3
0	0	0
1	50	1
2	70	2
3	80	3
4	100	4
5	130	5

We now move backward to start from the next-to-last stage ($n = 2$). Here, finding x^*_2 requires calculating and comparing $f_2(s_2, x_2)$ for the alternative values of x_2, namely, $x_2 = 0, 1, \ldots, s_2$. To illustrate, we depict this situation when $s_2 = 2$ graphically:

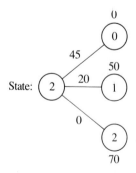

This diagram corresponds to Fig. 11.3 except that all three possible states at stage 3 are shown. Thus, if $x_2 = 0$, the resulting state at stage 3 will be $s_2 - x_2 = 2 - 0 = 2$, whereas $x_2 = 1$ leads to state 1 and $x_2 = 2$ leads to state 0. The corresponding values of $p_2(x_2)$ from the country 2 column of Table 11.1 are shown along the links, and the values of $f_3^*(s_2 - x_2)$ from the $n = 3$ table are given next to the stage 3 nodes. The required calculations for this case of $s_2 = 2$ are summarized below.

$x_2 = 0$: $f_2(2, 0) = p_2(0) + f_3^*(2) = \;\; 0 + 70 = 70.$

$x_2 = 1$: $f_2(2, 1) = p_2(1) + f_3^*(1) = 20 + 50 = 70.$

$x_2 = 2$: $f_2(2, 2) = p_2(2) + f_3^*(0) = 45 + \;\; 0 = 45.$

Because the objective is *maximization*, $x_2^* = 0$ or 1 with $f_2^*(2) = 70$.

Proceeding in a similar way with the other possible values of s_2 (try it) yields the following table.

	x_2	$f_2(s_2, x_2) = p_2(x_2) + f_3^*(s_2 - x_2)$							
$n = 2$:	s_2	0	1	2	3	4	5	$f_2^*(s_2)$	x_2^*
	0	0						0	0
	1	50	20					50	0
	2	70	70	45				70	0 or 1
	3	80	90	95	75			95	2
	4	100	100	115	125	110		125	3
	5	130	120	125	145	160	150	160	4

We now are ready to move backward to solve the original problem where we are starting from stage 1 ($n = 1$). In this case, the only state to be considered is the starting state of $s_1 = 5$, as depicted below

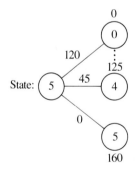

Since allocating x_1 medical teams to country 1 leads to a state of $(5 - x_1)$ at stage 2, a choice of $x_1 = 0$ leads to the bottom node on the right, $x_1 = 1$ leads to the next node up, and so forth up to the top node with $x_1 = 5$. The corresponding $p_1(x_1)$ values from Table 11.1 are shown next to the links. The numbers next to the nodes are obtained from the $f_2^*(s_2)$ column of the $n = 2$ table. As with $n = 2$, the calculation needed for each alternative value of the decision variable involves adding the corresponding link value and node value, as summarized below.

$x_1 = 0$: $f_1(5, 0) = p_1(0) + f_2^*(5) = \;\; 0 + 160 = 160.$

$x_1 = 1$: $f_1(5, 1) = p_1(1) + f_2^*(4) = 45 + 125 = 170.$

     $\vdots$

$x_1 = 5$: $f_1(5, 5) = p_1(5) + f_2^*(0) = 120 + \;\; 0 = 120.$

The similar calculations for $x_1 = 2, 3, 4$ (try it) verify that $x_1^* = 1$ with $f_1^*(5) = 170$, as shown in the following table.

$n = 1:$	x_1	$f_1(s_1, x_1) = p_1(x_1) + f_2^*(s_1 - x_1)$							
	s_1	0	1	2	3	4	5	$f_1^*(s_1)$	x_1^*
	5	160	170	165	160	155	120	170	1

Thus the optimal solution has $x_1^* = 1$, which makes $s_2 = 5 - 1 = 4$, so $x_2^* = 3$, which makes $s_3 = 4 - 3 = 1$, so $x_3^* = 1$. Since $f_1^*(5) = 170$, this $(1, 3, 1)$ allocation of medical teams to the three countries will yield an estimated total of 170,000 *additional person-years of life,* which is at least 5,000 more than for any other allocation.

A Prevalent Problem Type—The Distribution of Effort Problem

The preceding example illustrates a particularly common type of dynamic programming problem called the *distribution of effort problem.* For this type of problem, there is just one kind of *resource* that is to be allocated to a number of *activities.* The objective is to determine how to *distribute the effort* (the resource) among the activities most effectively. For the World Health Council example, the resource involved is the *medical teams,* and the three activities are the health care work in the three *countries.*

ASSUMPTIONS: This interpretation of allocating resources to activities should ring a bell for you, because it is the typical interpretation for *linear programming* problems given at the beginning of Chap. 3. However, there also are some key differences between the distribution of effort problem and linear programming that help illuminate the general distinctions between dynamic programming and other areas of mathematical programming.

One key difference is that the distribution of effort problem involves only *one resource* (one functional constraint), whereas linear programming can deal with hundreds or even thousands of resources. (In principle, dynamic programming can handle slightly more than one resource, as we shall illustrate in Example 5 by solving the *three-resource* Wyndor Glass Co. problem, but it quickly becomes very inefficient when the number of resources is increased.)

On the other hand, the distribution of effort problem is far more general than linear programming in other ways. Consider the four assumptions of linear programming presented in Sec. 3.3—proportionality, additivity, divisibility, and certainty. *Proportionality* is routinely violated by nearly all dynamic programming problems, including distribution of effort problems (e.g., Table 11.1 violates proportionality). *Divisibility* also is often violated, as in Example 2, where the decision variables must be integer. In fact, dynamic programming calculations become more complex when divisibility does hold (as in Examples 4 and 5). Although we shall consider the distribution of effort problem only under the assumption of *certainty,* this is not necessary, and many other dynamic programming problems violate this assumption as well (as described in Sec. 11.4).

Of the four assumptions of linear programming, the *only* one needed by the distribution of effort problem (or other dynamic programming problems) is *additivity* (or its analog for functions involving a *product* of terms). This assumption is needed to satisfy the *principle of optimality* for dynamic programming (characteristic 5 in Sec. 11.2).

FORMULATION: Because they always involve allocating one kind of resource to a number of activities, distribution of effort problems always have the following dynamic programming formulation (where the ordering of the activities is arbitrary):

Stage n = activity n ($n = 1, 2, \ldots, N$).

$\quad x_n$ = amount of the resource allocated to activity n.

State s_n = amount of the resource still available for allocation to the remaining activities ($n, \ldots, N$).

Therefore, when starting at stage n in state s_n, the choice of x_n always results in the next state at stage ($n + 1$) being $s_{n+1} = (s_n - x_n)$, as depicted below:

Stage: n $n + 1$

State: s_n $s_n - x_n$

Note how the structure of this diagram corresponds to the one shown in Fig. 11.3 for the World Health Council example of a distribution of effort problem. What will differ from one such example to the next is the *rest* of what is shown in Fig. 11.3, namely, the relationship between $f_n(s_n, x_n)$ and $f_{n+1}^*(s_n - x_n)$, and then the resulting *recursive relationship* between the f_n^* and f_{n+1}^* functions. These relationships depend on the particular objective function for the overall problem.

The *structure* of the next example is similar to the one for the World Health Council because it too is a *distribution of effort problem*. However, its *recursive relationship* differs in that its objective is to *minimize* a *product* of terms for the respective stages.

At first glance this example may appear *not* to be a *deterministic* dynamic programming problem because probabilities are involved. However, it does indeed fit our definition because the state at the next stage is completely determined by the state and policy decision at the current stage.

Example 3—Distributing Scientists to Research Teams

A government space project is conducting research on a certain engineering problem that must be solved before people can fly safely to Mars. Three research teams are currently trying three different approaches for solving this problem. The estimate has been made that, under present circumstances, the probability that the respective teams—call them 1, 2, and 3—will not succeed is 0.40, 0.60, and 0.80, respectively. Thus the current probability that all three teams will fail is $(0.40)(0.60)(0.80) = 0.192$. Because the objective is to minimize the probability of failure, two more top scientists have been assigned to the project.

Table 11.2 gives the estimated probability that the respective teams will fail when 0, 1, or 2 additional scientists are added to that team. Only integer numbers of scientists are considered because each new scientist will need to devote full attention

Table 11.2 **Data for the Government Space Project Problem**

Number of New Scientists	Probability of Failure		
	Team		
	1	2	3
0	0.40	0.60	0.80
1	0.20	0.40	0.50
2	0.15	0.20	0.30

to one team. The problem is to determine how to allocate the two additional scientists to minimize the probability that all three teams will fail.

FORMULATION: Because both Examples 2 and 3 are *distribution of effort problems*, their underlying structure is actually very similar. In this case, scientists replace medical teams as the kind of resource involved, and research teams replace countries as the activities. Therefore, instead of medical teams being allocated to countries, scientists are being allocated to research teams. The only basic difference between the two problems is in their objective functions.

With so few scientists and teams involved, this problem could be solved very easily by a process of exhaustive enumeration. However, the dynamic programming solution is presented for illustrative purposes.

In this case, *stage n* ($n = 1, 2, 3$) corresponds to research team n, and the *state* s_n is the number of new scientists *still available* for allocation to the remaining teams. The decision variables x_n ($n = 1, 2, 3$) are the number of additional scientists allocated to team n.

Let $p_i(x_i)$ denote the probability of failure for team i if it is assigned x_i additional scientists, as given by Table 11.2. Letting Π denote multiplication, the government's objective is to choose x_1, x_2, x_3 so as to

$$\text{Minimize} \quad \prod_{i=1}^{3} p_i(x_i) = p_1(x_1)p_2(x_2)p_3(x_3),$$

subject to

$$\sum_{i=1}^{3} x_i = 2$$

and   the x_i are nonnegative integers.

Consequently, $f_n(s_n, x_n)$ for this problem is

$$f_n(s_n, x_n) = p_n(x_n) \cdot \text{minimum} \prod_{i=n+1}^{3} p_i(x_i),$$

where the minimum is taken over $x_{n+1}, \ldots, x_3$ such that

$$\sum_{i=n}^{3} x_i = s_n$$

and   the x_i are nonnegative integers,

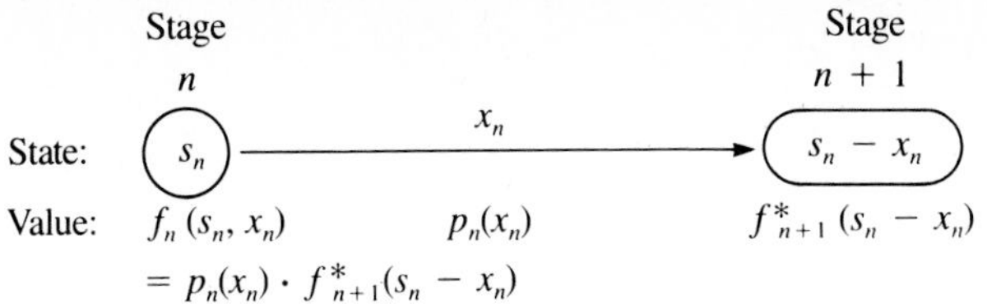

Figure 11.4 The basic structure for the government space project problem.

for $n = 1, 2, 3$. Thus

$$f_n^*(s_n) = \min_{x_n \leq s_n} f_n(s_n, x_n).$$

Hence

$$f_n(s_n, x_n) = p_n(x_n) \cdot f_{n+1}^*(s_n - x_n)$$

(with f_4^* defined to be *one*). Figure 11.4 summarizes these basic relationships.

Thus the *recursive relationship* relating the f_1^*, f_2^*, and f_3^* functions in this case is

$$f_n^*(s_n) = \min_{x_n \leq s_n} \{p_n(x_n) \cdot f_{n+1}^*(s_n - x_n)\}, \qquad \text{for } n = 1, 2,$$

and, when $n = 3$,

$$f_3^*(s_3) = \min_{x_3 \leq s_3} p_3(x_3).$$

SOLUTION PROCEDURE: The resulting dynamic programming calculations are:

$n = 3$:

s_3	$f_3^*(s_3)$	x_3^*
0	0.80	0
1	0.50	1
2	0.30	2

$n = 2$:

s_2 \ x_2	$f_2(s_2, x_2) = p_2(x_2) \cdot f_3^*(s_2 - x_2)$ 0	1	2	$f_2^*(s_2)$	x_2^*
0	0.48			0.48	0
1	0.30	0.32		0.30	0
2	0.18	0.20	0.16	0.16	2

$n = 1$:

s_1 \ x_1	$f_1(s_1, x_1) = p_1(x_1) \cdot f_2^*(s_1 - x_1)$ 0	1	2	$f_1^*(s_1)$	x_1^*
2	0.064	0.060	0.072	0.060	1

Therefore, the optimal solution must have $x_1^* = 1$, which makes $s_2 = 2 - 1 = 1$, so that $x_2^* = 0$, which makes $s_3 = 1 - 0 = 1$, so that $x_3^* = 1$. Thus teams 1 and 3 should each receive one additional scientist. The new probability that all three teams will fail would then be 0.060.

All the examples thus far have had a *discrete* state variable s_n at each stage. Furthermore, they all have been *reversible* in the sense that the solution procedure actually could have moved *either* backward or forward stage by stage. (The latter alternative amounts to renumbering the stages in reverse order and then applying the procedure in the standard way.) This reversibility is a general characteristic of *distribution of effort problems* such as Examples 2 and 3, since the activities (stages) can be ordered in any desired manner.

The next example is different in both respects. Rather than being restricted to *integer* values, its state variable s_n at stage n is a *continuous* variable that can take on *any* value over certain intervals. Since s_n now has an infinite number of values, it is no longer possible to consider each of its feasible values individually. Rather, the solution for $f_n^*(s_n)$ and x_n^* must be expressed as *functions* of s_n. Furthermore, this example is *not* reversible because its stages correspond to *time periods,* so the solution procedure *must* proceed backward.

Example 4—Scheduling Employment Levels

The workload for the LOCAL JOB SHOP is subject to considerable seasonal fluctuation. However, machine operators are difficult to hire and costly to train, so the manager is reluctant to lay off workers during the slack seasons. He is likewise reluctant to maintain his peak season payroll when it is not required. Furthermore, he is definitely opposed to overtime work on a regular basis. Since all work is done to custom orders, it is not possible to build up inventories during slack seasons. Therefore, the manager is in a dilemma as to what his policy should be regarding employment levels.

The following estimates are given for the minimum employment requirements during the four seasons of the year for the foreseeable future:

Season	Spring	Summer	Autumn	Winter	Spring
Requirements	255	220	240	200	255

Employment will not be permitted to fall below these levels. Any employment above these levels is wasted at an approximate cost of \$2,000/person/season. It is estimated that the hiring and firing costs are such that the total cost of changing the level of employment from one season to the next is \$200 times the square of the difference in employment levels. Fractional levels of employment are possible because of a few part-time employees, and the cost data also apply on a fractional basis.

FORMULATION: On the basis of the data available, it is not worthwhile to have the employment level go above the peak season requirements of 255. Therefore, spring employment should be at 255, and the problem is reduced to finding the employment level for the other three seasons.

For a dynamic programming formulation, the seasons should be the *stages.* There are actually an indefinite number of stages because the problem extends into the indefinite future. However, each year begins an identical cycle, and because spring employment is known, it is possible to consider only one cycle of four seasons ending with the spring season, as summarized below.

Stage 1 = summer,
Stage 2 = autumn,
Stage 3 = winter,
Stage 4 = spring.
x_n = employment level for stage n (n = 1, 2, 3, 4).
(x_4 = 255.)

It is necessary that the spring season be the last stage because the optimal value of the decision variable for each state at the last stage must be either known or obtainable without considering other stages. For every other season, the solution for the optimal employment level must consider the effect on costs in the following season.

Let

$$r_n = \text{minimum employment requirement for stage } n,$$

where these requirements were given earlier as r_1 = 220, r_2 = 240, r_3 = 200, and r_4 = 255. Thus the only feasible values for x_n are

$$r_n \leq x_n \leq 255.$$

Referring to the cost data given in the problem statement,

$$\text{Cost for stage } n = 200(x_n - x_{n-1})^2 + 2,000(x_n - r_n).$$

Note that the cost at the current stage depends only upon the current decision x_n and the employment in the preceding season x_{n-1}. Thus the preceding employment level is all the information about the current state of affairs that we need to determine the optimal policy henceforth. Therefore, the *state* s_n for stage n is

$$\text{State } s_n = x_{n-1}.$$

When n = 1, $s_1 = x_0 = x_4$ = 255.

For your ease of reference while working through the problem, a summary of the above data is given in Table 11.3.

The objective for the problem is to choose x_1, x_2, x_3 (with x_4 = 255) so as to

$$\text{Minimize} \quad \sum_{i=1}^{4} [200(x_i - x_{i-1})^2 + 2,000(x_i - r_i)],$$

subject to $\qquad r_i \leq x_i \leq 255, \qquad \text{for } i = 1, 2, 3, 4.$

Table 11.3 Data for the Local Job Shop Problem

n	r_n	Feasible x_n	Possible $s_n = x_{n-1}$	Cost
1	220	$220 \leq x_1 \leq 255$	s_1 = 255	$200(x_1 - 255)^2 + 2,000(x_1 - 220)$
2	240	$240 \leq x_2 \leq 255$	$220 \leq s_2 \leq 255$	$200(x_2 - x_1)^2 + 2,000(x_2 - 240)$
3	200	$200 \leq x_3 \leq 255$	$240 \leq s_3 \leq 255$	$200(x_3 - x_2)^2 + 2,000(x_3 - 200)$
4	255	x_4 = 255	$200 \leq s_4 \leq 255$	$200(255 - x_3)^2$

Thus for stage n onward ($n = 1, 2, 3, 4$),

$$f_n(s_n, x_n) = 200(x_n - s_n)^2 + 2,000(x_n - r_n)$$

$$+ \text{ minimum} \sum_{r_i \le x_i \le 255}^{4} [200(x_i - x_{i-1})^2 + 2,000(x_i - r_i)],$$

because $s_n = x_{n-1}$. Also,

$$f_n^*(s_n) = \min_{r_n \le x_n \le 255} f_n(s_n, x_n).$$

Hence

$$f_n(s_n, x_n) = 200(x_n - s_n)^2 + 2,000(x_n - r_n) + f_{n+1}^*(x_n)$$

(with f_5^* defined to be *zero* because costs after stage 4 are irrelevant to the analysis). A summary of these basic relationships is given in Fig. 11.5.

Consequently, the *recursive relationship* relating the f_n^* functions is

$$f_n^*(s_n) = \min_{r_n \le x_n \le 255} \{200(x_n - s_n)^2 + 2,000(x_n - r_n) + f_{n+1}^*(x_n)\}.$$

The dynamic programming approach uses this relationship to identify successively these functions—$f_4^*(s_4)$, $f_3^*(s_3)$, $f_2^*(s_2)$, $f_1^*(255)$—and the corresponding minimizing x_n.

SOLUTION PROCEDURE

Stage 4: Beginning at the *last* stage ($n = 4$), we already know that $x_4^* = 255$, so the necessary results are

$n = 4$:	s_4	$f_4^*(s_4)$	x_4^*
	$200 \le s_4 \le 255$	$200(255 - s_4)^2$	255

Stage 3: For the problem consisting of just the last *two* stages ($n = 3$), the recursive relationship reduces to

$$f_3^*(s_3) = \min_{200 \le x_3 \le 255} \{200(x_3 - s_3)^2 + 2,000(x_3 - 200) + f_4^*(x_3)\}$$

$$= \min_{200 \le x_3 \le 255} \{200(x_3 - s_3)^2 + 2,000(x_3 - 200) + 200(255 - x_3)^2\},$$

where the possible values of s_3 are $240 \le s_3 \le 255$.

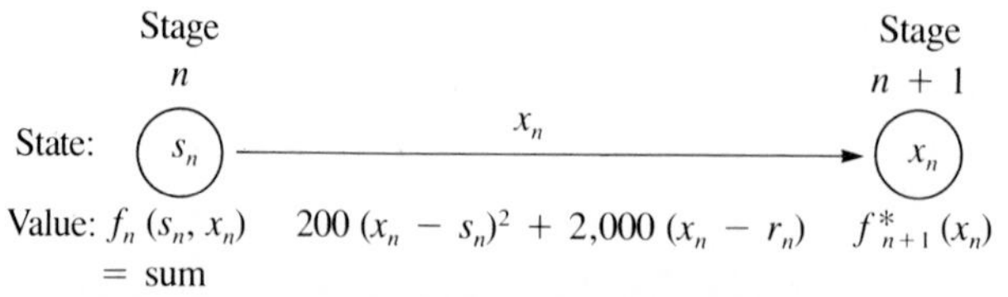

Stage
n

State: (s_n) $\xrightarrow{\quad x_n \quad}$ (x_n)

Stage
$n + 1$

Value: $f_n(s_n, x_n)$ $\quad 200(x_n - s_n)^2 + 2,000(x_n - r_n) \quad$ $f_{n+1}^*(x_n)$
$= $ sum

Figure 11.5 The basic structure for the Local Job Shop problem.

One way to solve for the value of x_3 that minimizes $f_3(s_3, x_3)$ for any particular value of s_3 is the graphical approach illustrated in Fig. 11.6.

However, a faster way is to use *calculus*. We want to solve for the minimizing x_3 in terms of s_3 by considering s_3 to have some fixed (but unknown) value. Therefore, equate to zero the first (partial) derivative of $f_3(s_3, x_3)$ with respect to x_3,

$$\frac{\partial}{\partial x_3} f_3(s_3, x_3) = 400(x_3 - s_3) + 2{,}000 - 400(255 - x_3)$$

$$= 400(2x_3 - s_3 - 250)$$

$$= 0,$$

which yields

$$x_3^* = \frac{s_3 + 250}{2}.$$

Because the second derivative is positive, and because this solution lies in the feasible interval for x_3 ($200 \le x_3 \le 255$) for all possible s_3 ($240 \le s_3 \le 255$), it is indeed the desired minimum.

Note a key difference between the nature of this solution and those obtained for the preceding examples where there were only a few possible states to consider. We now have an *infinite* number of possible states ($240 \le s_3 \le 255$), so it is no longer feasible to solve separately for x_3^* for each possible value of s_3. Therefore, we instead have solved for x_3^* as a *function* of the unknown s_3.

Using

$$f_3^*(s_3) = f_3(s_3, x_3^*) = 200\left(\frac{s_3 + 250}{2} - s_3\right)^2 + 200\left(255 - \frac{s_3 + 250}{2}\right)^2$$

$$+ 2{,}000\left(\frac{s_3 + 250}{2} - 200\right)$$

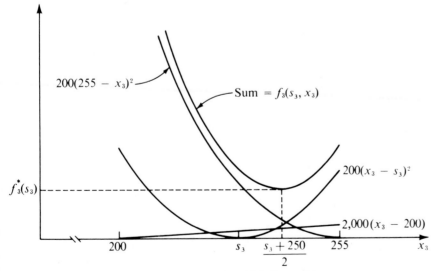

Figure 11.6 Graphical solution for $f_3^*(s_3)$ for the Local Job Shop problem.

and reducing this expression algebraically completes the required results for the two-stage problem summarized as follows.

$n = 3$:	s_3	$f_3^*(s_3)$	x_3^*
	$240 \le s_3 \le 255$	$50(250 - s_3)^2 + 50(260 - s_3)^2 + 1{,}000(s_3 - 150)$	$\dfrac{s_3 + 250}{2}$

Stage 2: The three-stage ($n = 2$) and four-stage problems ($n = 1$) are solved in a similar fashion. Thus for $n = 2$,

$$f_2(s_2, x_2) = 200(x_2 - s_2)^2 + 2{,}000(x_2 - r_2) + f_3^*(x_2)$$

$$= 200(x_2 - s_2)^2 + 2{,}000(x_2 - 240)$$

$$+ 50(250 - x_2)^2 + 50(260 - x_2)^2 + 1{,}000(x_2 - 150).$$

The possible values of s_2 are $220 \le s_2 \le 255$ and the feasible region for x_2 is $240 \le x_2 \le 255$. The problem is to find the minimizing value of x_2 in this region, so that

$$f_2^*(s_2) = \min_{240 \le x_2 \le 255} f_2(s_2, x_2).$$

Setting to zero the partial derivative with respect to x_2,

$$\frac{\partial}{\partial x_2} f_2(s_2, x_2) = 400(x_2 - s_2) + 2{,}000 - 100(250 - x_2) - 100(260 - x_2) + 1{,}000$$

$$= 200(3x_2 - 2s_2 - 240)$$

$$= 0,$$

yields

$$x_2 = \frac{2s_2 + 240}{3}.$$

Because

$$\frac{\partial^2}{\partial x_2^2} f_2(s_2, x_2) = 600 > 0,$$

this value of x_2 is the desired minimizing value *if it is feasible* ($240 \le x_2 \le 255$). Over the possible s_2 ($220 \le s_2 \le 255$), this solution actually is feasible *only* if $240 \le s_2 \le 255$.

Therefore, we still need to solve for the *feasible* value of x_2 that minimizes $f_2(s_2, x_2)$ when $220 \le s_2 < 240$. The key to analyzing the behavior of $f_2(s_2, x_2)$ over the feasible region for x_2 again is the partial derivative of $f_2(s_2, x_2)$. When $s_2 < 240$,

$$\frac{\partial}{\partial x_2} f_2(s_2, x_2) > 0, \quad \text{for } 240 \le x_2 \le 255,$$

so that $x_2 = 240$ is the desired minimizing value.

The next step is to plug these values of x_2 into $f_2(s_2, x_2)$ to obtain $f_2^*(s_2)$ for $s_2 \ge 240$ and $s_2 < 240$. After considerable algebraic manipulation, the following results are obtained.

s_2	$f_2^*(s_2)$	x_2^*
$220 \leq s_2 \leq 240$	$200(240 - s_2)^2 + 115{,}000$	240
$240 \leq s_2 \leq 255$	$\dfrac{200}{9} [2(250 - s_2)^2 + (265 - s_2)^2 + 30(3s_2 - 575)]$	$\dfrac{2s_2 + 240}{3}$

Stage 1: For the four-stage problem ($n = 1$),

$$f_1(s_1, x_1) = 200(x_1 - s_1)^2 + 2{,}000(x_1 - r_1) + f_2^*(x_1).$$

Because $r_1 = 220$, the feasible region for x_1 is $220 \leq x_1 \leq 255$. The expression for $f_2^*(x_1)$ will differ in the two portions, $220 \leq x_1 \leq 240$ and $240 \leq x_1 \leq 255$, of this region. Therefore,

$$f_1(s_1, x_1) = \begin{cases} 200(x_1 - s_1)^2 + 2{,}000(x_1 - 220) + 200(240 - x_1)^2 + 115{,}000, \\ \qquad\qquad\qquad\qquad\qquad\qquad\qquad \text{if } 220 \leq x_1 \leq 240 \\[2mm] 200(x_1 - s_1)^2 + 2{,}000(x_1 - 220) + \dfrac{200}{9} [2(250 - x_1)^2 \\[2mm] \qquad + (265 - x_1)^2 + 30(3x_1 - 575)], \text{ if } 240 \leq x_1 \leq 255. \end{cases}$$

Considering first the case where $220 \leq x_1 \leq 240$,

$$\frac{\partial}{\partial x_1} f_1(s_1, x_1) = 400(x_1 - s_1) + 2{,}000 - 400(240 - x_1)$$

$$= 400(2x_1 - s_1 - 235).$$

It is known that $s_1 = 255$ (spring employment), so that

$$\frac{\partial}{\partial x_1} f_1(s_1, x_1) = 800(x_1 - 245) < 0$$

for all $x_1 \leq 240$. Therefore, $x_1 = 240$ is the minimizing value of $f_1(s_1, x_1)$ over the region, $220 \leq x_1 \leq 240$.

When $240 \leq x_1 \leq 255$,

$$\frac{\partial}{\partial x_1} f_1(s_1, x_1) = 400(x_1 - s_1) + 2{,}000 - \frac{200}{9} [4(250 - x_1) + 2(265 - x_1) - 90]$$

$$= \frac{400}{3} (4x_1 - 3s_1 - 225).$$

Because

$$\frac{\partial^2}{\partial x_1^2} f_1(s_1, x_1) > 0, \qquad \text{for all } x_1,$$

set

$$\frac{\partial}{\partial x_1} f_1(s_1, x_1) = 0,$$

which yields

$$x_1 = \frac{3s_1 + 225}{4}.$$

Because $s_1 = 255$, it follows that $x_1 = 247.5$ minimizes $f_1(s_1, x_1)$ over the region, $240 \leq x_1 \leq 255$.

Note that this region, $240 \leq x_1 \leq 255$, includes $x_1 = 240$, so that $f_1(s_1, 240) > f_1(s_1, 247.5)$. In the next-to-last paragraph, we found that $x_1 = 240$ minimizes $f_1(s_1, x_1)$ over the region $220 \leq x_1 \leq 240$. Consequently, we now can conclude that $x_1 = 247.5$ also minimizes $f_1(s_1, x_1)$ over the *entire* feasible region, $220 \leq x_1 \leq 255$.

Our final calculation is to find $f_1^*(s_1)$ for $s_1 = 255$ by plugging $x_1 = 247.5$ into the expression for $f_1(255, x_1)$ that holds for $240 \leq x_1 \leq 255$. Hence

$$f_1^*(255) = 200(247.5 - 255)^2 + 2,000(247.5 - 220)$$

$$+ \frac{200}{9}[2(250 - 247.5)^2 + (265 - 247.5)^2 + 30(742.5 - 575)]$$

$$= 185,000.$$

These results are summarized as follows:

$n = 1$:	s_1	$f_1^*(s_1)$	x_1^*
	255	185,000	247.5

Therefore, tracing back through the tables for $n = 2$, $n = 3$, and $n = 4$, respectively, and setting $s_n = x_{n-1}^*$ each time, the resulting optimal solution is $x_1^* = 247.5$, $x_2^* = 245$, $x_3^* = 247.5$, $x_4^* = 255$, with a total estimated cost per cycle of \$185,000.

To conclude our illustrations of deterministic dynamic programming, we give one example that requires *more than one* variable to describe the state at each stage.

Example 5—Wyndor Glass Company Problem

Consider the following linear programming problem:

$$\text{Maximize} \quad Z = 3x_1 + 5x_2,$$

subject to

$$x_1 \qquad\qquad \leq 4$$

$$2x_2 \leq 12$$

$$3x_1 + 2x_2 \leq 18$$

and

$$x_1 \geq 0, \qquad x_2 \geq 0.$$

(You might recognize this model as the prototype example for linear programming in Chap. 3.) One way of solving small linear (or nonlinear) programming problems like this one is by dynamic programming, which is illustrated below.

FORMULATION: This problem requires making two interrelated decisions, namely, the level of activity 1, x_1, and the level of activity 2, x_2. Therefore, these two activities can be interpreted as the two *stages* in a dynamic programming formulation. Although they can be taken in either order, let stage n = activity n ($n = 1, 2$). Thus x_n is the decision variable at stage n.

What are the *states?* In other words, given that the decision had been made at prior stages (if any), what information is needed about the current state of affairs before the decision can be made at stage n? Reflection might suggest that the required information is the *amount of slack* left in the functional constraints. Interpret the right-hand side of these constraints (4, 12, and 18) as the total available amount of resources 1, 2, and 3, respectively (as described in Sec. 3.1). Then the state s_n can be defined as

> State s_n = amount of the respective resources still available for
> allocation to the remaining activities ($n, \ldots, 2$).

(Note that the definition of the state is analogous to that for *distribution of effort* problems, including Examples 2 and 3, except that there are now three resources to be allocated instead of just one.) Thus

$$s_n = (R_1, R_2, R_3),$$

where R_i is the amount of resource i remaining to be allocated ($i = 1, 2, 3$). Therefore,

$$s_1 = (4, 12, 18),$$

$$s_2 = (4 - x_1, 12, 18 - 3x_1).$$

However, when we begin by solving for stage 2, we shall not yet know the value of x_1, and so use $s_2 = (R_1, R_2, R_3)$ at that point.

Therefore, in contrast to the preceding examples, this problem has *three* state variables (i.e., a *state vector* with three components) at each stage rather than one. From a theoretical standpoint, this difference is not particularly serious. It only means that, instead of considering all possible values of the one state variable, we must consider all possible *combinations* of values of the several state variables. However, from the standpoint of computational efficiency, this difference tends to be a very serious complication. Because the number of combinations, in general, can be as large as the *product* of the number of possible values of the respective variables, the number of required calculations tends to "blow up" rapidly when additional state variables are introduced. This phenomenon has been given the apt name of the **curse of dimensionality**.

Each of the three state variables is *continuous*. Therefore, rather than consider each possible combination of values separately, we must use the approach introduced in Example 4 of solving for the required information as a *function* of the state of the system.

Despite these complications, this problem is small enough that it can still be solved without great difficulty. To solve it, we need to introduce the usual dynamic programming notation. Thus,

$$f_2(R_1, R_2, R_3, x_2) = \text{contribution of activity 2 to } Z \text{ if the system starts in state}$$
$$(R_1, R_2, R_3) \text{ at stage 2 and the decision is } x_2$$

$$= 5x_2,$$

$f_1(4, 12, 18, x_1)$ = contribution of activities 1 and 2 to Z if the system starts in state $(4, 12, 18)$ at stage 1, the immediate decision is x_1, and then an optimal decision is made at stage 2.

$$= 3x_1 + \underset{\substack{x_2 \leq 12 \\ 2x_2 \leq 18 - 3x_1 \\ x_2 \geq 0}}{\text{maximum}} \{5x_2\}.$$

Similarly, for $n = 1, 2,$

$$f_n^*(R_1, R_2, R_3) = \max_{x_n} f_n(R_1, R_2, R_3, x_n),$$

where this maximum is taken over the *feasible* values of x_n. Consequently, using the relevant portions of the constraints of the problem,

(1) $\qquad f_2^*(R_1, R_2, R_3) = \underset{\substack{2x_2 \leq R_2 \\ 2x_2 \leq R_3 \\ x_2 \geq 0}}{\text{maximum}} \{5x_2\},$

(2) $\qquad f_1(4, 12, 18, x_1) = 3x_1 + f_2^*(4 - x_1, 12, 18 - 3x_1),$

(3) $\qquad f_1^*(4, 12, 18) = \underset{\substack{x_1 \leq 4 \\ 3x_1 \leq 18 \\ x_1 \geq 0}}{\text{maximum}} \{3x_1 + f_2^*(4 - x_1, 12, 18 - 3x_1)\}.$

Equation (1) will be used to solve the stage 2 problem. Equation (2) shows the basic dynamic programming structure for this problem, as also depicted in Fig. 11.7. Equation (3) gives the *recursive relationship* between f_1^* and f_2^* that will be used to solve the stage 1 problem.

SOLUTION PROCEDURE

Stage 2: To solve at the last stage $(n = 2)$, Eq. (1) indicates that x_2^* must be the *largest* value of x_2 that *simultaneously* satisfies $2x_2 \leq R_2$, $2x_2 \leq R_3$, and $x_2 \geq 0$. Assuming that $R_2 \geq 0$ and $R_3 \geq 0$, so that feasible solutions exist, this largest value is the smaller of $R_2/2$ and $R_3/2$. Thus the solution is

$n = 2$:	(R_1, R_2, R_3)	$f_2^*(R_1, R_2, R_3)$	x_2^*
	$R_2 \geq 0, R_3 \geq 0$	$5 \min\left\{\dfrac{R_2}{2}, \dfrac{R_3}{2}\right\}$	$\min\left\{\dfrac{R_2}{2}, \dfrac{R_3}{2}\right\}$

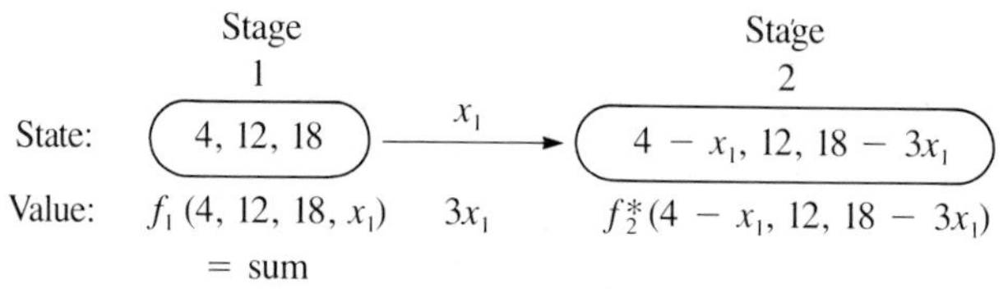

State: $\boxed{4, 12, 18} \xrightarrow{\quad x_1 \quad} \boxed{4 - x_1, 12, 18 - 3x_1}$

Stage 1   Stage 2

Value: $f_1(4, 12, 18, x_1) \qquad 3x_1 \qquad f_2^*(4 - x_1, 12, 18 - 3x_1)$

$= $ sum

Figure 11.7 The basic structure for the Wyndor Glass Co. linear programming problem.

Stage 1: To solve the two-stage problem ($n = 1$), we plug the solution just obtained for $f_2^*(R_1, R_2, R_3)$ into Eq. (3). For stage 2,

$$(R_1, R_2, R_3) = (4 - x_1, 12, 18 - 3x_1),$$

so that

$$f_2^*(4 - x_1, 12, 18 - 3x_1) = 5 \min \left\{ \frac{R_2}{2}, \frac{R_3}{2} \right\} = 5 \min \left\{ \frac{12}{2}, \frac{18 - 3x_1}{2} \right\}$$

is the specific solution plugged into Eq. (3). After combining its constraints on x_1, Eq. (3) then becomes

$$f_1^*(4, 12, 18) = \max_{0 \le x_1 \le 4} \left\{ 3x_1 + 5 \min \left\{ \frac{12}{2}, \frac{18 - 3x_1}{2} \right\} \right\}.$$

Over the feasible interval, $0 \le x_1 \le 4$, notice that

$$\min \left\{ \frac{12}{2}, \frac{18 - 3x_1}{2} \right\} = \begin{cases} 6, & \text{if } 0 \le x_1 \le 2 \\ 9 - \dfrac{3}{2}x_1, & \text{if } 2 \le x_1 \le 4, \end{cases}$$

so that

$$3x_1 + 5 \min \left\{ \frac{12}{2}, \frac{18 - 3x_1}{2} \right\} = \begin{cases} 3x_1 + 30, & \text{if } 0 \le x_1 \le 2 \\ 45 - \dfrac{9}{2}x_1, & \text{if } 2 \le x_1 \le 4. \end{cases}$$

Because both

$$\max_{0 \le x_1 \le 2} \{3x_1 + 30\} \quad \text{and} \quad \max_{2 \le x_1 \le 4} \left\{ 45 - \frac{9}{2}x_1 \right\}$$

achieve their maximum at $x_1 = 2$, it follows that $x_1^* = 2$, and that this maximum is 36, as given in the following table.

$n = 1$:	(R_1, R_2, R_3)	$f_1^*(R_1, R_2, R_3)$	x_1^*
	$(4, 12, 18)$	36	2

Because $x_1^* = 2$ leads to

$$R_1 = 4 - 2 = 2, \qquad R_2 = 12, \qquad R_3 = 18 - 3(2) = 12$$

for stage 2, the $n = 2$ table yields $x_2^* = 6$. Consequently, $x_1^* = 2$, $x_2^* = 6$ is the optimal solution for this problem (as originally found in Sec. 3.1), and the $n = 1$ table shows that the resulting value of Z is 36.

11.4 Probabilistic Dynamic Programming

Probabilistic dynamic programming differs from deterministic dynamic programming in that the state at the next stage is *not* completely determined by the state and policy decision at the current stage. Rather, there is a *probability distribution* for what the next state will be. However, this probability distribution still is completely determined by the state and policy decision at the current stage. The resulting basic structure for probabilistic dynamic programming is described diagrammatically in Fig. 11.8.

For purposes of this diagram, we let S denote the number of possible states at stage $n + 1$ and label these states on the right-side as $1, 2, \ldots, S$. The system goes to state i with probability p_i ($i = 1, 2, \ldots, S$) given the state s_n and decision x_n at stage n. If the system goes to state i, C_i is the contribution of stage n to the objective function.

When Fig. 11.8 is expanded to include all the possible states and decisions at all the stages, it is sometimes referred to as a **decision tree**. If the decision tree is not too large, it provides a useful way of summarizing the various possibilities that may occur.

Because of the probabilistic structure, the relationship between $f_n(s_n, x_n)$ and the $f^*_{n+1}(s_{n+1})$ necessarily is somewhat more complicated than for deterministic dynamic programming. The precise form of this relationship will depend upon the form of the overall objective function.

To illustrate, suppose that the objective is to *minimize* the *expected sum* of the contributions from the individual stages. In this case, $f_n(s_n, x_n)$ would represent the minimum expected sum from stage n onward, *given* that the state and policy decision at stage n are s_n and x_n, respectively. Consequently,

$$f_n(s_n, x_n) = \sum_{i=1}^{S} p_i[C_i + f^*_{n+1}(i)],$$

with

$$f^*_{n+1}(i) = \min_{x_{n+1}} f_{n+1}(i, x_{n+1}),$$

where this minimization is taken over the *feasible* values of x_{n+1}.

Example 6 has this same form. Example 7 will illustrate another form.

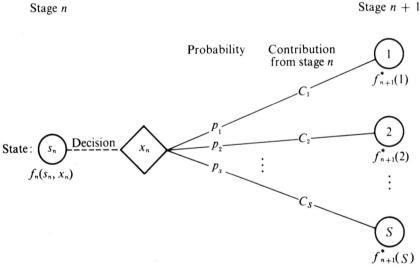

Figure 11.8 The basic structure for probabilistic dynamic programming.

Example 6—Determining Reject Allowances

441

Dynamic Programming

The HIT-AND-MISS MANUFACTURING COMPANY has received an order to supply one item of a particular type. However, the customer has specified such stringent quality requirements that the manufacturer may have to produce more than one item to obtain an item that is acceptable. The number of *extra* items produced in a production run is called the *reject allowance*. Including a reject allowance is common practice when producing for a custom order, and it seems advisable in this case.

The manufacturer estimates that each item of this type that is produced will be *acceptable* with probability $\frac{1}{2}$ and *defective* (without possibility for rework) with probability $\frac{1}{2}$. Thus the number of acceptable items produced in a lot of size L will have a *binomial distribution;* that is, the probability of producing *zero* acceptable items in such a lot is $(\frac{1}{2})^L$.

Marginal production costs for this product are estimated to be $100 per item (even if defective), and excess items are worthless. In addition, a setup cost of $300 must be incurred whenever the production process is set up for this product, and a completely new setup at this same cost is required for each subsequent production run if a lengthy inspection procedure reveals that a completed lot has not yielded an acceptable item. The manufacturer has time to make no more than three production runs. If an acceptable item has not been obtained by the end of the third production run, the cost to the manufacturer in lost sales income and in penalty costs would be $1,600.

The objective is to determine the policy regarding the lot size (1 + reject allowance) for the required production run(s) that minimizes total expected cost for the manufacturer.

FORMULATION: A dynamic programming formulation for this problem is

$$\text{Stage } n = \text{production run } n \ (n = 1, 2, 3),$$
$$x_n = \text{lot size for stage } n,$$
$$\text{State } s_n = \text{number of acceptable items still needed (one or zero)}$$
$$\text{when beginning stage } n.$$

Thus, at stage 1, the state $s_1 = 1$. If at least one acceptable item is obtained subsequently, the state changes to $s_n = 0$, after which no additional costs need to be incurred.

Because of the stated objective for the problem,

$$f_n(s_n, x_n) = \text{total expected cost for stages } n, \ldots ,3 \text{ if the system starts in}$$
$$\text{state } s_n \text{ at stage } n, \text{ the immediate decision is } x_n, \text{ and optimal}$$
$$\text{decisions are made thereafter,}$$
$$f_n^*(s_n) = \min_{x_n=0,1,\ldots} f_n(s_n, x_n),$$

where $f_n^*(0) = 0$. Using $100 as the unit of money, the contribution to cost from stage n is $(K + x_n)$ regardless of the next state, where

$$K = \begin{cases} 0, & \text{if } x_n = 0 \\ 3, & \text{if } x_n > 0. \end{cases}$$

Therefore, for $s_n = 1$,

$$f_n(1, x_n) = K + x_n + (\tfrac{1}{2})^{x_n} f^*_{n+1}(1) + [1 - (\tfrac{1}{2})^{x_n}] f^*_{n+1}(0)$$

$$= K + x_n + (\tfrac{1}{2})^{x_n} f^*_{n+1}(1)$$

[where $f^*_4(1)$ is defined to be 16, the terminal cost if no acceptable items have been obtained]. A summary of these basic relationships is given in Fig. 11.9.

Consequently, the *recursive relationship* for the dynamic programming calculations is

$$f^*_n(1) = \min_{x_n = 0,1,\ldots} \{K + x_n + (\tfrac{1}{2})^{x_n} f^*_{n+1}(1)\}$$

for $n = 1, 2, 3$.

SOLUTION PROCEDURE: The calculations using this recursive relationship are summarized as follows.

$n = 3$:

x_3 / s_3	$f_3(1, x_3) = K + x_3 + 16(\tfrac{1}{2})^{x_3}$						$f^*_3(s_3)$	x^*_3
	0	1	2	3	4	5		
0	0						0	0
1	16	12	9	8	8	$8\tfrac{1}{2}$	8	3 or 4

$n = 2$:

x_2 / s_2	$f_2(1, x_2) = K + x_2 + (\tfrac{1}{2})^{x_2} f^*_3(1)$					$f^*_2(s_2)$	x^*_2
	0	1	2	3	4		
0	0					0	0
1	8	8	7	7	$7\tfrac{1}{2}$	7	2 or 3

$n = 1$:

x_1 / s_1	$f_1(1, x_1) = K + x_1 + (\tfrac{1}{2})^{x_1} f^*_2(1)$					$f^*_1(s_1)$	x^*_1
	0	1	2	3	4		
1	7	$7\tfrac{1}{2}$	$6\tfrac{3}{4}$	$6\tfrac{7}{8}$	$7\tfrac{7}{16}$	$6\tfrac{3}{4}$	2

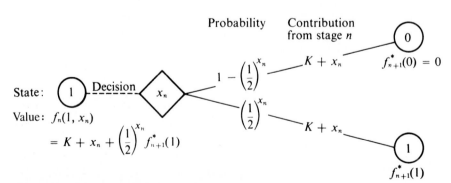

Figure 11.9 The basic structure for the Hit-and-Miss Manufacturing Co. problem.

Thus the optimal policy is to produce *two* items on the first production run; if none is acceptable, then produce either *two* or *three* items on the second production run; if none is acceptable, then produce either *three* or *four* items on the third production run. The *total expected cost* for this policy is $675.

Example 7—Winning in Las Vegas

An enterprising young statistician believes that she has developed a system for winning a popular Las Vegas game. Her colleagues do not believe that her system works, so they have made a large bet with her that, starting with three chips, she will not have five chips after three plays of the game. Each play of the game involves betting any desired number of available chips and then either winning or losing this number of chips. The statistician believes that her system will give her a probability of $\frac{2}{3}$ of winning a given play of the game.

Assuming the statistician is correct, we now will use dynamic programming to determine her optimal policy regarding how many chips to bet (if any) at each of the three plays of the game. The decision at each play should take into account the results of earlier plays. The objective is to maximize the probability of winning her bet with her colleagues.

FORMULATION: The dynamic programming formulation for this problem is

$$\text{Stage } n = n\text{th play of the game } (n = 1, 2, 3),$$
$$x_n = \text{number of chips to bet at stage } n,$$
$$\text{State } s_n = \text{number of chips in hand to begin stage } n.$$

This definition of the state is chosen because it provides the needed information about the current situation for making an optimal decision on how many chips to bet next.

Because the objective is to maximize the probability that the statistician will win her bet, the objective function to be maximized at each stage must be the probability of finishing the three plays with at least five chips. Therefore,

$$f_n(s_n, x_n) = \text{probability of finishing the three plays with at least five chips,}$$
given that the statistician starts stage n in state s_n, makes the immediate decision x_n, and makes optimal decisions thereafter,
$$f_n^*(s_n) = \max_{x_n = 0, 1, \ldots, s_n} f_n(s_n, x_n).$$

The expression for $f_n(s_n, x_n)$ must reflect the fact that it may still be possible to accumulate five chips eventually even if the statistician should lose the next play. If she should lose, the state at the next stage would be $(s_n - x_n)$, and the probability of finishing with at least five chips would then be $f_{n+1}^*(s_n - x_n)$. If she should win the next play instead, the state would become $(s_n + x_n)$, and the corresponding probability would be $f_{n+1}^*(s_n + x_n)$. Because the assumed probability of winning a given play is $\frac{2}{3}$, it now follows that

$$f_n(s_n, x_n) = \tfrac{1}{3} f_{n+1}^*(s_n - x_n) + \tfrac{2}{3} f_{n+1}^*(s_n + x_n)$$

[where $f_4^*(s_4)$ is defined to be *zero* for $s_4 < 5$ and 1 for $s_4 \geq 5$]. Thus there is no direct contribution to the objective function from stage n in addition to the effect of then being in the next state. These basic relationships are summarized in Fig. 11.10.

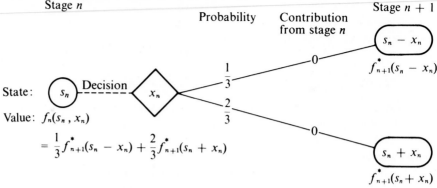

Figure 11.10 The basic structure for the Las Vegas problem.

Therefore, the *recursive relationship* for this problem is

$$f_n^*(s_n) = \max_{x_n = 0, 1, \ldots, s_n} \{\tfrac{1}{3} f_{n+1}^*(s_n - x_n) + \tfrac{2}{3} f_{n+1}^*(s_n + x_n)\},$$

for $n = 1, 2, 3$, with $f_4^*(s_4)$ as just defined.

SOLUTION PROCEDURE: This recursive relationship leads to the following computational results.

$n = 3$:

s_3	$f_3^*(s_3)$	x_3^*
0	0	—
1	0	—
2	0	—
3	$\tfrac{2}{3}$	2 (or more)
4	$\tfrac{2}{3}$	1 (or more)
≥ 5	1	0 (or $\leq s_3 - 5$)

$n = 2$:

	x_2	$f_2(s_2, x_2) = \tfrac{1}{3} f_3^*(s_2 - x_2) + \tfrac{2}{3} f_3^*(s_2 + x_2)$						
s_2		0	1	2	3	4	$f_2^*(s_2)$	x_2^*
0		0					0	—
1		0	0				0	—
2		0	$\tfrac{4}{9}$	$\tfrac{4}{9}$			$\tfrac{4}{9}$	1 or 2
3		$\tfrac{2}{3}$	$\tfrac{4}{9}$	$\tfrac{2}{3}$	$\tfrac{2}{3}$		$\tfrac{2}{3}$	0, 2, or 3
4		$\tfrac{2}{3}$	$\tfrac{8}{9}$	$\tfrac{2}{3}$	$\tfrac{2}{3}$	$\tfrac{2}{3}$	$\tfrac{8}{9}$	1
≥ 5		1					1	0 (or $\leq s_2 - 5$)

$n = 1$:

	x_1	$f_1(s_1, x_1) = \tfrac{1}{3} f_2^*(s_1 - x_1) + \tfrac{2}{3} f_2^*(s_1 + x_1)$					
s_1		0	1	2	3	$f_1^*(s_1)$	x_1^*
3		$\tfrac{2}{3}$	$\tfrac{20}{27}$	$\tfrac{2}{3}$	$\tfrac{2}{3}$	$\tfrac{20}{27}$	1

$$x_1^* = 1 \begin{cases} \text{if win, } x_2^* = 1 \begin{cases} \text{if win, } x_3^* = 0 \\ \text{if lose, } x_3^* = 2 \text{ or } 3. \end{cases} \\ \\ \text{if lose, } x_2^* = 1 \text{ or } 2 \begin{cases} \text{if win, } x_3^* = \begin{cases} 2 \text{ or } 3 \text{ (for } x_2^* = 1) \\ 1, 2, 3, \text{ or } 4 \text{ (for } x_2^* = 2) \end{cases} \\ \text{if lose, bet is lost.} \end{cases} \end{cases}$$

This policy gives the statistician a probability of $\frac{20}{27}$ of winning her bet with her colleagues.

11.5 Conclusions

Dynamic programming is a very useful technique for making a *sequence of interrelated decisions*. It requires formulating an appropriate *recursive relationship* for each individual problem. However, it provides a great computational savings over using exhaustive enumeration to find the best combination of decisions, especially for large problems. For example, if a problem has 10 stages with 10 states and 10 possible decisions at each stage, then exhaustive enumeration must consider up to 10^{10} combinations, whereas dynamic programming need make no more than 10^3 calculations (10 for each state at each stage).

This chapter has considered only dynamic programming with a *finite* number of stages. We have not included *Markovian decision processes*, which provide a general kind of model for probabilistic dynamic programming where the stages continue to recur indefinitely.

SELECTED REFERENCES

1. Bellman, Richard, and Stuart Dreyfus: *Applied Dynamic Programming*, Princeton University Press, Princeton, N.J., 1962.

2. Cooper, Leon L., and Mary W. Cooper: *Introduction to Dynamic Programming*, Pergamon Press, Elmsford, N.Y., 1981.

3. Denardo, Eric V.: *Dynamic Programming Theory and Applications*, Prentice-Hall, Englewood Cliffs, N.J., 1982.

4. Dreyfus, Stuart E., and Averill M. Law: *The Art and Theory of Dynamic Programming*, Academic Press, New York, 1977.

5. Howard, Ronald A.: "Dynamic Programming," *Management Science*, **12**:317–345, 1966.

PROBLEMS

1. Consider the following network, where each number along a link represents the actual distance between the pair of nodes connected by that link.

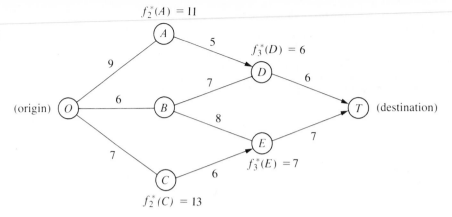

The objective is to find the shortest route from the origin to the destination.

(a) What are the stages and states for the dynamic programming formulation of this problem?

(b) Use dynamic programming to solve this problem. However, instead of using the usual tables, show your work graphically. In particular, redraw the preceding network, where the answers already are given for $f_n^*(s_n)$ for four of the nodes; then solve for and fill in $f_2^*(B)$ and $f_1^*(O)$. Draw an arrowhead that shows the optimal link to take out of each of the latter two nodes. Finally, identify the optimal route by following the arrows from node O onward to node T.

(c) Use dynamic programming to solve this problem by constructing the usual tables for $n = 3$, $n = 2$, and $n = 1$.

(d) Use the shortest-route algorithm presented in Sec. 10.3 to solve this problem. Compare and contrast this approach with the one in parts (b) and (c).

2. The sales manager for a publisher of college textbooks has six traveling salespeople to assign to three different regions of the country. She has decided that each region should be assigned at least one salesperson and that each individual salesperson should be restricted to one of the regions, but now she wants to determine how many salespeople should be assigned to the respective regions in order to maximize sales.

The following table gives the estimated increase in sales (in appropriate units) in each region if it were allocated various numbers of salespeople:

| Number | Region | | |
of Salespeople	1	2	3
1	35	21	28
2	48	42	41
3	70	56	63
4	89	70	75

(a) Use dynamic programming to solve this problem. Instead of using the usual tables, show your work graphically by constructing and filling in a network such as the one shown for Prob. 1. Proceed as in part (b) of Prob. 1 by solving for $f_n^*(s_n)$ for each node (except the terminal node) and writing its value by the node. Draw an arrowhead to show the optimal link (or links in the case of a tie) to take out of each node. Finally, identify the resulting optimal route (or routes) through the network and the corresponding optimal solution (or solutions).

(b) Use dynamic programming to solve this problem by constructing the usual tables for $n = 3$, $n = 2$, and $n = 1$.

3.* The owner of a chain of three grocery stores has purchased five crates of fresh strawberries. The estimated probability distribution of potential sales of the strawberries before spoilage differs among the three stores. Therefore, the owner wants to know how he should allocate the five crates to the three stores to maximize expected profit.

For administrative reasons, the owner does not wish to split crates between stores. However, he is willing to distribute zero crates to any of his stores.

The following table gives the estimated expected profit at each store when it is allocated various numbers of crates:

Number of Crates	Stores		
	1	2	3
0	0	0	0
1	5	6	4
2	9	11	9
3	14	15	13
4	17	19	18
5	21	22	20

Use dynamic programming to determine how many of the five crates should be assigned to each of the three stores to maximize the total expected profit.

4. A college student has 7 days remaining before final examinations begin in her four courses, and she wants to allocate this study time as effectively as possible. She needs at least 1 day on each course, and she likes to concentrate on just one course each day, so she wants to allocate 1, 2, 3, or 4 days to each course. Having recently taken an operations research course, she decides to use dynamic programming to make these allocations to maximize the total grade points to be obtained from the four courses. She estimates that the alternative allocations for each course would yield the number of grade points shown in the following table:

	Estimated Grade Points			
Number of Study Days	Course			
	1	2	3	4
1	3	5	2	6
2	5	5	4	7
3	6	6	7	9
4	7	9	8	9

Solve this problem by dynamic programming.

5. A company is planning its advertising strategy for next year for its three major products. Since the three products are quite different, each advertising effort will focus on a single product. In units of millions of dollars, a total of 6 is available for advertising next year, where the advertising expenditure for each product must be an integer greater than or equal to 1. The vice-president for marketing has established the objective, namely, determine how much to spend on each product in order to maximize total sales. The following table gives the estimated increase in sales (in appropriate units) for the different advertising expenditures:

Advertising Expenditure	Product		
	1	2	3
1	7	4	6
2	10	8	9
3	14	11	13
4	17	14	15

Use dynamic programming to solve this problem.

6. A political campaign is entering its final stage, and polls indicate a very close election. One of the candidates has enough funds left to purchase TV time for a total of five prime-time commercials on TV stations located in four different areas. Based on polling information, an estimate has been made of the number of additional votes that can be won in the different broadcasting areas depending upon the number of commercials run. These estimates are given in the following table in units of thousands of votes:

Number of Commercials	Area			
	1	2	3	4
0	0	0	0	0
1	4	6	5	3
2	7	8	9	7
3	9	10	11	12
4	12	11	10	14
5	15	12	9	16

Use dynamic programming to determine how the five commercials should be distributed among the four areas in order to maximize the estimated number of votes won.

7. A county chairwoman of a certain political party is making plans for an upcoming presidential election. She has received the services of six volunteer workers for precinct work, and she wants to assign them to four precincts in such a way as to maximize their effectiveness. She feels that it would be inefficient to assign a worker to more than one precinct, but she is willing to assign no workers to any one of the precincts if they can accomplish more in other precincts.

The following table gives the estimated increase in the number of votes for the party's candidate in each precinct if it were allocated various numbers of workers:

Number of Workers	Precinct			
	1	2	3	4
0	0	0	0	0
1	4	7	5	6
2	9	11	10	11
3	15	16	15	14
4	18	18	18	16
5	22	20	21	17
6	24	21	22	18

This problem has several optimal solutions for how many of the six workers should be assigned to each of the four precincts to maximize the total estimated increase in the plurality of the party's candidate. Use dynamic programming to find all of them so the chairwoman can make the final selection based on other factors.

8. Use dynamic programming to solve the Northern Airplane Co. production scheduling problem presented in Sec. 7.1 (see Table 7.7). Assume that production quantities must be integer multiples of 5.

9. Reconsider the Build-Em-Fast Co. problem described in Prob. 21 of Chap. 7 (also see Prob. 51 of Chap. 14). Use dynamic programming to solve this problem.

10.* A company will soon be introducing a new product into a very competitive market and is currently planning its marketing strategy. The decision has been made to introduce the product in three phases. Phase 1 will feature making a special introductory offer of the product to the public at a greatly reduced price to attract first-time buyers. Phase 2 will involve an intensive advertising campaign to persuade these first-time buyers to continue purchasing the product at a regular price. It is known that another company will be introducing a new competitive product at about the time phase 2 will end. Therefore, phase 3 will involve a follow-up advertising and promotion campaign to try to keep the regular purchasers from switching to the competitive product.

A total of \$4 million has been budgeted for this marketing campaign. The problem now is to determine how to allocate this money most effectively to the three phases. Let m denote the initial share of the market (expressed as a percentage) attained in phase 1, f_2 the fraction of this market share that is retained in phase 2, and f_3 the fraction of the remaining market share that is retained in phase 3. Given the following data, use dynamic programming to determine how to allocate the \$4 million to maximize the *final share* of the market for the new product, i.e., to maximize mf_2f_3.

(a) Assume that the money must be spent in integer multiples of \$1 million in each phase, where the minimum permissible multiple is 1 for phase 1 and 0 for phases 2 and 3. The following table gives the estimated effect of expenditures in each phase:

Millions of Dollars Expended	Effect on Market Share		
	m	f_2	f_3
0	—	0.2	0.3
1	20	0.4	0.5
2	30	0.5	0.6
3	40	0.6	0.7
4	50	—	—

(b) Now assume that *any* amount within the total budget can be spent in each phase, where the estimated effect of spending an amount x_i (in units of *millions* of dollars) in phase i ($i = 1, 2, 3$) is

$$m = 10x_1 - x_1^2$$

$$f_2 = 0.40 + 0.10x_2$$

$$f_3 = 0.60 + 0.07x_3.$$

[*Hint:* After solving for the $f_2^*(s)$ and $f_3^*(s)$ functions analytically, solve for x_1^* graphically.]

11. Consider an electronic system consisting of four components, each of which must function for the system to function. The reliability of the system can be improved by installing several parallel units in one or more of the components. The following table gives the probability that the respective components will function if they consist of one, two, or three parallel units:

Number of Parallel Units	Probability of Functioning			
	Component 1	Component 2	Component 3	Component 4
1	0.5	0.6	0.7	0.5
2	0.6	0.7	0.8	0.7
3	0.8	0.8	0.9	0.9

The probability that the system will function is the product of the probabilities that the respective components will function.

The cost (in hundreds of dollars) of installing one, two, or three parallel units in the respective components is given by the following table:

Number of Parallel Units	Cost			
	Component 1	Component 2	Component 3	Component 4
1	1	2	1	2
2	2	4	3	3
3	3	5	4	4

Because of budget limitations, a maximum of $1,000 can be expended.

Use dynamic programming to determine how many parallel units should be installed in each of the four components to maximize the probability that the system will function.

12. Consider the following integer nonlinear programming problem.

$$\text{Maximize} \quad Z = x_1 x_2^2 x_3^3,$$

subject to

$$x_1 + 2x_2 + 3x_3 \le 10,$$

$$x_1 \ge 1, \qquad x_2 \ge 1, \qquad x_3 \ge 1,$$

and

$$x_1, x_2, x_3 \text{ are integers.}$$

Use dynamic programming to solve this problem.

13.* Consider the following nonlinear programming problem.

$$\text{Maximize} \quad Z = 36x_1 + 9x_1^2 - 6x_1^3 + 36x_2 - 3x_2^3,$$

subject to

$$x_1 + x_2 \le 3$$

and

$$x_1 \ge 0, \qquad x_2 \ge 0.$$

Use dynamic programming to solve this problem.

14. Resolve the *Local Job Shop* employment scheduling problem (Example 4) when the total cost of changing the level of employment from one season to the next is changed to $100 times the square of the difference in employment levels.

15. Consider the following nonlinear programming problem.

$$\text{Maximize} \quad Z = x_1^2 x_2,$$

subject to

$$x_1^2 + x_2 \le 2.$$

(Note that there are no nonnegativity constraints.) Use dynamic programming to solve this problem.

16. Consider the following nonlinear programming problem.

$$\text{Maximize} \quad Z = x_1^3 + 4x_2^2 + 16x_3,$$

subject to

$$x_1 x_2 x_3 = 4$$

and

$$x_1 \ge 1, \qquad x_2 \ge 1, \qquad x_3 \ge 1.$$

(*a*) Solve by dynamic programming when, in addition to the given constraints, all three variables also are required to be *integer*.

(*b*) Use dynamic programming to solve the problem as given (continuous variables).

17. Consider the following nonlinear programming problem.

$$\text{Maximize} \quad Z = x_1(1 - x_2)x_3,$$

subject to

$$x_1 - x_2 + x_3 \leq 1$$

and

$$x_1 \geq 0, \quad x_2 \geq 0, \quad x_3 \geq 0.$$

Use dynamic programming to solve this problem.

18. Consider the following linear programming problem.

$$\text{Maximize} \quad Z = 15x_1 + 10x_2,$$

subject to

$$x_1 + 2x_2 \leq 6$$

$$3x_1 + x_2 \leq 8$$

and

$$x_1 \geq 0, \quad x_2 \geq 0.$$

Use dynamic programming to solve this problem.

19. Consider the following nonlinear programming problem.

$$\text{Maximize} \quad Z = 5x_1 + x_2,$$

subject to

$$2x_1^2 + x_2 \leq 13$$

$$x_1^2 + x_2 \leq 9$$

and

$$x_1 \geq 0, \quad x_2 \geq 0.$$

Use dynamic programming to solve this problem.

20. Consider the following "fixed-charge" problem.

$$\text{Maximize} \quad Z = 3x_1 + 7x_2 + 6f(x_3),$$

subject to

$$x_1 + 3x_2 + 2x_3 \leq 6$$

$$x_1 + x_2 \qquad \leq 5$$

and

$$x_1 \geq 0, \quad x_2 \geq 0, \quad x_3 \geq 0,$$

where

$$f(x_3) = \begin{cases} 0, & \text{if } x_3 = 0 \\ -1 + x_3, & \text{if } x_3 > 0. \end{cases}$$

Use dynamic programming to solve this problem.

21. Consider the *Food and Agriculture Organization* example presented in Sec. 8.3. Suppose that large additional amounts of equipment and money become available, so that the constraint on experts, $x_1 + 2x_2 \leq 10$, is the *only* relevant functional constraint. After deleting the other two functional constraints, use dynamic programming to solve directly the first model presented for this example (ignore the equivalent linear programming model) under the following alternative assumptions.

(a) Assume that x_1 and x_2 are required to be integer.

(b) Assume that the divisibility assumption (see Sec. 3.3) holds, so that the only restrictions on x_1 and x_2 are the one functional constraint and the nonnegativity constraints.

22. A backgammon player will be playing three consecutive matches with friends tonight. For each match, he will have the opportunity to place an even bet that he will win; the amount bet can be *any* quantity of his choice between zero and the amount of money he still has left after the bets on the preceding matches. For each match, the probability is $\frac{1}{2}$ that he will win the match and thus win the amount bet, whereas the probability is $\frac{1}{2}$ that he will lose the match and thus lose the amount bet. He will begin with $75, and his goal is to have $100 at the end. (Because these are friendly matches, he does not want to end up with more than $100.) Therefore, he wants to find the optimal betting policy (including all ties) that maximizes the probability that he will have exactly $100 after the three matches.

Use dynamic programming to solve this problem.

23. Imagine that you have $5,000 to invest and that you will have an opportunity to invest that amount in either of two investments (A or B) at the beginning of each of the next 3 years. Both investments have uncertain returns. For investment A you will either lose your money entirely or (with higher probability) get back $10,000 (a profit of $5,000) at the end of the year. For investment B you will either get back just your $5,000 or (with low probability) $10,000 at the end of the year. The probabilities for these events are

Investment	Amount Returned ($)	Probability
A	0	0.3
	10,000	0.7
B	5,000	0.9
	10,000	0.1

You are allowed to make only (at most) *one* investment each year, and can invest only $5,000 each time. (Any additional money accumulated is left idle.)

(*a*) Use dynamic programming to find the investment policy that maximizes the *expected amount of money* you will have after the 3 years.

(*b*) Use dynamic programming to find the investment policy that maximizes the *probability* that you will have at least $10,000 after the 3 years.

24.* Suppose that the situation for the Hit-and-Miss Manufacturing Co. problem (Example 6) has changed somewhat. After a more careful analysis, you now estimate that each item produced will be *acceptable* with probability $\frac{2}{3}$, rather than $\frac{1}{2}$, so that the probability of producing *zero* acceptable items in a lot of size L is $(\frac{1}{3})^L$. Furthermore, there now is only enough time available to make *two* production runs. Use dynamic programming to determine the new optimal policy for this problem.

25. Reconsider Example 7. Suppose that the bet is changed to: ''Starting with two chips, she will not have five chips after *five* plays of the game.'' By referring to the previous computational results, make additional calculations to determine what the new optimal policy is for the enterprising young statistician.

26. The Profit & Gambit Co. has a major product that has been losing money recently because of declining sales. In fact, during the current quarter of the year, sales will be 4 million units below the *break-even point*. Because the *marginal* revenue for each unit sold exceeds the *marginal* cost by $5, this amounts to a loss of $20 million for the quarter. Therefore, management must take action quickly to rectify this situation. Two alternative courses of action are being considered. One is to abandon the product immediately, incurring a cost of $20 million for shutting down. The other alternative is to undertake an intensive advertising campaign to increase sales and then abandon the product (at the cost of $20 million) only if the campaign is not sufficiently successful. Tentative plans for this advertising campaign have been developed and analyzed. It would extend over the next three quarters (subject to early cancellation), and the cost would be $30 million in each of the three quarters. It is estimated that the increase in sales would be approximately 3 million units in the first quarter, another 2 million units in the second quarter, and another 1 million units in the third quarter. However, because of a number of unpredictable market variables, there is considerable uncertainty as to what impact the advertising actually would have, and careful analysis indicates that the estimates for each quarter could turn out to be off by as much as 2 million units in either direction. (To quantify this uncertainty, assume that the additional increases in sales in the three quarters are independent random variables having a uniform distribution with a range from 1 to 5 million, from 0 to 4 million, and from −1 to 3 million, respectively.) If the actual increases are too small, the advertising campaign can be discontinued and the product abandoned at the end of either of the next two quarters.

If the intensive advertising campaign were to be initiated and continued to its completion, it is estimated that the sales for some time thereafter would continue to be at about the same level as in the third (last) quarter of the campaign. Therefore, if the sales in that quarter still are below the break-even point, the product would be abandoned. Otherwise, it is estimated that the expected discounted profit thereafter would be $40 for each unit sold over the break-even point in the third quarter.

Use dynamic programming to determine the optimal policy maximizing expected profit.

12

Game Theory

Life is full of conflict and competition. Numerous examples involving adversaries in conflict include parlor games, military battles, political campaigns, advertising and marketing campaigns by competing business firms, and so forth. A basic feature in many of these situations is that the final outcome depends primarily upon the combination of strategies selected by the adversaries. *Game theory* is a mathematical theory that deals with the general features of competitive situations like these in a formal, abstract way. It places particular emphasis on the decision-making processes of the adversaries.

As briefly surveyed in Sec. 12.6, research on game theory continues to delve into rather complicated types of competitive situations. However, the focus in this chapter is on the simplest case, called **two-person, zero-sum games.** As the name implies, these games involve only two adversaries or *players* (who may be armies, teams, firms, and so on). They are called *zero-sum* games because one player wins whatever the other one loses, so that the sum of their net winnings is zero.

After introducing the basic model for two-person, zero-sum games in Sec. 12.1, the following four sections describe and illustrate different approaches to solving such games. We then conclude the chapter by mentioning some other kinds of competitive situations that are dealt with by other branches of game theory.

12.1 The Formulation of Two-Person, Zero-Sum Games

To illustrate the basic characteristics of two-person, zero-sum games, consider the game called *Odds and Evens*. This game consists simply of each player simultaneously showing either one finger or two fingers. If the number of fingers matches, so that the total number for both players is even, then the player taking Evens (say, player I) wins the bet (say $1) from the player taking Odds (player II). If the number does not match, player I would pay $1 to player II. Thus each player has two *strategies:* to show either one finger or two fingers. The resulting payoff to player I in dollars is shown in the *payoff table* given in Table 12.1.

In general, a two-person game is characterized by

1. The strategies of player I.
2. The strategies of player II.
3. The payoff table.

Before the game begins, each player knows the strategies he or she has available, the ones the opponent has available, and the payoff table. The actual play of the game consists of the players simultaneously choosing a strategy without knowing the opponent's choice.

A *strategy* may involve only a simple action, as when showing a certain number of fingers in the Odds and Evens game. On the other hand, in more complicated games involving a series of moves, a **strategy** is a *predetermined rule that specifies completely how one intends to respond to each possible circumstance at each stage of the game.* For example, a strategy for one side in *chess* would indicate how to make the next move for *every* possible position on the board, so the total number of possible strategies would be astronomical. Applications of game theory normally involve far less complicated competitive situations than chess, but the strategies involved can be fairly complex.

The **payoff table** shows the gain (positive or negative) for player I that would result from each combination of strategies for the two players. It is given only for player I because the table for player II is just the negative of this one, due to the zero-sum nature of the game.

The entries in the payoff table may be in any units desired, such as dollars, provided that they accurately represent the *utility* to player I of the corresponding outcome. However, utility is not necessarily proportional to the amount of money (or

Table 12.1 **Payoff Table for the Odds and Evens Game**

		II 1	2
I	1	1	−1
	2	−1	1

any other commodity) when large quantities are involved. For example, $2 million (after taxes) is probably worth much less than twice as much as $1 million to a poor person. In other words, given the choice between (1) a 50 percent chance of receiving $2 million rather than nothing and (2) being sure of getting $1 million, such an individual probably would much prefer the latter. On the other hand, the outcome corresponding to an entry of 2 in a payoff table should be "worth twice as much" to player I as the outcome corresponding to an entry of 1. Thus, given the choice, he or she should be indifferent between a 50 percent chance of receiving the former outcome (rather than nothing) and definitely receiving the latter outcome instead.

A primary objective of game theory is the development of *rational criteria* for selecting a strategy. Two key assumptions are made:

1. *Both* players are *rational*.
2. *Both* players choose their strategies solely to *promote their own welfare* (no compassion for the opponent).

We shall develop the standard game theory criteria for choosing strategies by means of illustrative examples. In particular, the next section presents a prototype example that illustrates the formulation of a two-person, zero-sum game and its solution in some simple situations. A more complicated variation of this game is then carried into Sec. 12.3 to develop a more general criterion. Sections 12.4 and 12.5 describe a *graphical procedure* and a *linear programming formulation* for solving such games.

12.2 Solving Simple Games—A Prototype Example

Two politicians are running against each other for the U.S. Senate. Campaign plans must now be made for the final 2 days, which are expected to be crucial because of the closeness of the race. Therefore, both politicians want to spend these days campaigning in two key cities: *Bigtown* and *Megalopolis*. To avoid wasting campaign time, they plan to travel at night and spend either one full day in each city or two full days in just one of the cities. However, since the necessary arrangements must be made in advance, neither politician will learn his (or her)[1] opponent's campaign schedule until after he has finalized his own. Therefore each politician has asked his campaign manager in each of these cities to assess what the impact would be (in terms of votes won or lost) from the various possible combinations of days spent there by himself and by his opponent. He then wishes to use this information to choose his best strategy on how to use these 2 days.

FORMULATION: To formulate this problem as a two-person, zero-sum game, we must identify the two *players* (obviously the two politicians), the *strategies* for each player, and the *payoff table*.

As the problem has been stated, each player has the following three strategies:

---

[1] We use only *his* or *her* in our examples and problems for ease of reading; we do not mean to imply that only men or women are engaged in the various activities.

Strategy 1 = spend 1 day in each city.
Strategy 2 = spend both days in Bigtown.
Strategy 3 = spend both days in Megalopolis.

By contrast, the strategies would have been more complicated ones in a different situation where each politician would learn where his opponent will spend his first day before he finalizes his own plans for his second day. In that case, a typical strategy would be: Spend the first day in Bigtown; if the opponent also spends the first day in Bigtown, then spend the second day in Bigtown; however, if the opponent spends the first day in Megalopolis, then spend the second day in Megalopolis. There would be eight such strategies, one for each combination of the two first-day choices, the opponent's two first-day choices, and the two second-day choices.

Each entry in the payoff table for player I represents the *utility* to player I (or the negative utility to player II) of the outcome resulting from the corresponding strategies used by the two players. From the politician's viewpoint, the objective is to *win votes,* and each additional vote (before learning the outcome of the election) is of equal value to him. Therefore, the appropriate entries for the payoff table are the *total net votes won* from the opponent (i.e., the sum of the net vote changes in the two cities) resulting from these 2 days of campaigning. This formulation is summarized in Table 12.2.

However, we should also point out that this payoff table would *not* be appropriate if additional information were available to the politicians. In particular, if they knew exactly how the populace was planning to vote 2 days before the election, the only significance of the data prescribed by Table 12.2 would be to indicate which politician would win the election with each combination of strategies. Because the ultimate goal is to win the election, and because the size of the plurality is relatively inconsequential, the utility entries in the table then should be some positive constant (say, +1) when politician I would win and −1 when he would lose. Even if only a *probability* of winning can be determined for each combination of strategies, the appropriate entries would be *the probability of winning minus the probability of losing* because they then would represent *expected* utilities. However, sufficiently accurate data to make such determinations usually are not available.

Using the form given in Table 12.2, three alternative sets of data for the payoff table are given here to illustrate how to solve three different kinds of games.

Variation 1: Given that Table 12.3 is the payoff table for the two politicians (players), which strategy should each of them select?

Table 12.2 Formulation of Payoff Table for the Political Campaign Problem

	Strategy	Total Net Votes Won by Politician I (in Units of 1,000 Votes)		
		Politician II		
	Strategy	1	2	3
	1			
Politician I	*2*			
	3			

Table 12.3 **Payoff Table for Variation 1 of the Political Campaign Problem**

		II		
		1	2	3
	1	1	2	4
I	2	1	0	5
	3	0	1	−1

This situation is a rather special one, where the answer can be obtained just by applying the concept of **dominated strategies** to rule out a succession of inferior strategies until only one choice remains. Specifically, a strategy can be eliminated from further consideration if it is *dominated* by another strategy, i.e., if there is another strategy that is *always at least as good* regardless of what the opponent does.

At the outset, Table 12.3 includes *no* dominated strategies for player II. However, for player I, strategy 3 is dominated by strategy 1 because the latter has larger payoffs ($1 \geq 0$, $2 \geq 1$, $4 \geq -1$) regardless of what player II does. Eliminating strategy 3 from further consideration yields the following reduced payoff table:

	1	2	3
1	1	2	4
2	1	0	5

Because both players are assumed to be rational, player II also can deduce that player I has only these two strategies remaining under consideration. Therefore, player II now *does* have a dominated strategy—strategy 3, which is dominated by both strategies 1 and 2 because they always have smaller losses (payoffs to player I) in this reduced payoff table (for strategy 1: $1 \leq 4$, $1 \leq 5$; for strategy 2: $2 \leq 4$, $0 \leq 5$). Eliminating this strategy yields

	1	2
1	1	2
2	1	0

At this point, strategy 2 for player I becomes dominated by strategy 1 because the latter is better in column 2 ($2 \geq 0$) and equally good in column 1 ($1 \geq 1$). Eliminating the dominated strategy leads to

	1	2
1	1	2

where strategy 2 for player II is dominated by strategy 1 ($1 \leq 2$). Consequently, both players should select their strategy 1. Player I then will receive a payoff of 1 from player II (i.e., politician I will gain 1,000 votes from politician II).

In general, the payoff to player I when both players play optimally is referred to as the **value of the game**. A game that has a value of *zero* is said to be a **fair game**. Since this particular game has a value of 1, it is *not* a fair game.

The concept of a *dominated strategy* is a very useful one for reducing the size of the payoff table that needs to be considered and, in unusual cases like this one, actually identifying the optimal solution for the game. However, most games require another approach to at least finish solving, as illustrated by the next two variations of the example.

Variation 2: Now suppose that the current data give Table 12.4 as the payoff table for the politicians (players). This game does not have dominated strategies, so it is not obvious what the players should do. What line of reasoning does game theory say they should use?

Consider player I. By selecting strategy 1, he could win 6 or he could lose as much as 3. However, because player II is rational and thus will protect himself from large payoffs to I, it seems probable that playing strategy 1 would result in a loss to player I. Similarly, by selecting strategy 3, player I could win 5, but more probably his rational opponent would avoid this loss and instead administer him a loss, which could be as large as 4. On the other hand, if player I selects strategy 2, he is guaranteed not to lose anything, and he could even win something. Therefore, because it provides the *best guarantee,* strategy 2 seems to be a "rational" choice for player I against his rational opponent.

Now consider player II. He could lose as much as 5 or 6 by using strategy 1 or 3, but is guaranteed at least breaking even with strategy 2. Therefore, using the same reasoning of seeking the *best guarantee* against a rational opponent, his apparent choice is strategy 2.

If both players choose their strategy 2, the result is that both break even. Thus, in this case, neither player improves upon his *best guarantee,* but both also are forcing the opponent into the same position. Even when the opponent deduces a player's strategy, the opponent cannot exploit this information to improve his position. Stalemate.

The end product of this line of reasoning is that each player should play in such a way as to *minimize his maximum losses* whenever the resulting choice of strategy cannot be exploited by the opponent to then improve his position. This so-called **minimax criterion** is a standard criterion proposed by game theory for selecting a strategy. In terms of the payoff table, it implies that *player I should select the strategy whose minimum payoff is largest,* whereas *player II should choose the one whose maximum payoff to player I is the smallest.* This criterion is illustrated in Table 12.4, where strategy 2 is identified as the "maximin" strategy for player I, and strategy 2 is the minimax strategy for player II. The resulting payoff of 0 is the *value of the game,* so this is a *fair game.*

Table 12.4 Payoff Table for Variation 2 of the Political Campaign Problem

		II			
		1	2	3	Minimum
I	*1*	−3	−2	6	−3
	2	2	0	2	0 ← Maximin value
	3	5	−2	−4	−4
	Maximum:	5	0	6	

Minimax value (↑ under the 0 column)

Notice the interesting fact that the same entry in this payoff table yields both the maximin and minimax values. The reason is that this entry is both the minimum in its row and the maximum in its column. The position of any such entry is called a **saddle point**.

The fact that this game possesses a saddle point was actually crucial in determining how it should be played. Because of the saddle point, neither player can take advantage of the opponent's strategy to improve his own position. In particular, when player II predicts or learns that player I is using strategy 2, player II would only increase his losses if he were to change from his original plan of using his strategy 2. Similarly, player I would only worsen his position if he were to change his plan. Thus neither player has any motive to consider changing strategies, either to take advantage of his opponent or to prevent the opponent from taking advantage of him. Therefore, since this is a **stable solution**, players I and II should exclusively use their maximin and minimax strategies, respectively.

As the next variation illustrates, some games do not possess a saddle point, in which case a more complicated analysis is required.

Variation 3: Late developments in the campaign result in the *final* payoff table for the two politicians (players) given by Table 12.5. How should this game be played?

Suppose that both players attempt to apply the *minimax criterion* in the same way as in variation 2. Player I can guarantee that he will lose no more than 2 by playing strategy 1. Similarly, player II can guarantee that he will lose no more than 2 by playing strategy 3.

However, notice that the maximin value (-2) and the minimax value (2) do not coincide in this case. The result is that there is *no saddle point*.

What are the resulting consequences if both players plan to use the strategies just derived? It can be seen that player I would win 2 from player II, which would make player II unhappy. Because player II is rational and can therefore foresee this outcome, he would then conclude that he can do much better, actually winning 2 rather than losing 2, by playing strategy 2 instead. Because player I is also rational, he would anticipate this switch and conclude that he can improve considerably, from -2 to 4, by changing to strategy 2. Realizing this, player II would then consider switching back to strategy 3 to convert a loss of 4 to a gain of 3. This possibility of a switch would cause player I to consider again using strategy 1, after which the whole cycle would start over again.

In short, the originally suggested solution (player I to play strategy 1 and player II to play strategy 3) is an **unstable solution**, so it is necessary to develop a more satisfactory solution. But what kind of solution should it be?

Table 12.5 **Payoff Table for Variation 3 of the Political Campaign Problem**

		II			
		1	2	3	Minimum
I	*1*	0	-2	2	-2 ← Maximin value
	2	5	4	-3	-3
	3	2	3	-4	-4
	Maximum:	5	4	2	

↑
Minimax value

The key fact seems to be that whenever one player's strategy is predictable, the opponent here can take great advantage of this information to improve his position. Therefore, an essential feature of a rational plan for playing a game such as this one is that neither player should be able to deduce which strategy the other will use. Hence, rather than applying some known criterion for determining a single strategy that will definitely be used, it is necessary to choose among alternative acceptable strategies on some kind of random basis. By doing this, neither player knows in advance which of his own strategies will be used, let alone what his opponent will do.

This suggests, in very general terms, the kind of approach that is required for games lacking a saddle point. The next section discusses this approach more fully. Given this foundation, we turn our attention to procedures for finding an optimal way of playing such games. This particular variation of the political campaign problem will continue to be used to illustrate these ideas as they are developed.

12.3 Games with Mixed Strategies

Whenever a game does not possess a saddle point, game theory advises each player to assign a probability distribution over his or her set of strategies. To express this mathematically, let

x_i = probability that player I will use strategy i ($i = 1, 2, \ldots, m$),

y_j = probability that player II will use strategy j ($j = 1, 2, \ldots, n$),

where m and n are the respective numbers of available strategies. Thus player I would specify his plan for playing the game by assigning values to $x_1, x_2, \ldots, x_m$. Because these values are probabilities, they would need to be nonnegative and add up to 1. Similarly, the plan for player II would be described by the values he assigns to his decision variables $y_1, y_2, \ldots, y_n$. These plans $(x_1, x_2, \ldots, x_m)$ and $(y_1, y_2, \ldots, y_n)$ are usually referred to as **mixed strategies**, and the original strategies would then be called *pure strategies*.

When actually playing the game, it is necessary for each player to use one of his pure strategies. However, this pure strategy would be chosen by using some random device to obtain a random observation from the probability distribution specified by the mixed strategy, where this observation would indicate which particular pure strategy to use.

To illustrate, suppose that players I and II in *variation 3* of the political campaign problem (see Table 12.5) select the mixed strategies $(x_1, x_2, x_3) = (\frac{1}{2}, \frac{1}{2}, 0)$ and $(y_1, y_2, y_3) = (0, \frac{1}{2}, \frac{1}{2})$, respectively. This selection would say that player I is giving an equal chance (probability of $\frac{1}{2}$) to choosing either (pure) strategy 1 or 2, but he is discarding strategy 3 entirely. Similarly, player II is randomly choosing between his last two pure strategies. To play the game, each player could then flip a coin to determine which of his two acceptable pure strategies he will actually use.

Although no completely satisfactory measure of performance is available for evaluating mixed strategies, a very useful one is the *expected payoff*. Applying the probability theory definition of expected value, this quantity is

$$\text{Expected payoff} = \sum_{i=1}^{m} \sum_{j=1}^{n} p_{ij} x_i y_j,$$

where p_{ij} is the payoff if player I uses pure strategy i and player II uses pure strategy j. In the example of mixed strategies just given, there are four possible payoffs (-2, 2, 4, -3), each occurring with a probability of $\frac{1}{4}$, so the expected payoff is $\frac{1}{4}(-2 + 2 + 4 - 3) = \frac{1}{4}$. Thus this measure of performance does not disclose anything about the risks involved in playing the game, but it does indicate what the average payoff will tend to be if the game is played many times.

Using this measure, game theory extends the concept of the minimax criterion to games that lack a saddle point and thus need mixed strategies. In this context, the **minimax criterion** says that a given player should select the mixed strategy that *minimizes* the *maximum expected loss* to himself. Equivalently, when focusing on payoffs (player I) rather than losses (player II), this criterion says to *maximin* instead, i.e., *maximize* the *minimum expected payoff* to the player. By the *minimum expected payoff* we mean the smallest possible expected payoff that can result from any mixed strategy with which the opponent can counter. Thus the mixed strategy for player I that is *optimal* according to this criterion is the one that provides the *guarantee* (minimum expected payoff) that is *best* (maximal). (The value of this best guarantee is the *maximin value,* denoted by $\underline{v}$.) Similarly, the *optimal* strategy for player II is the one that provides the *best guarantee,* where best now means *minimal* and guarantee refers to the *maximum expected loss* that can be administered by any of the opponent's mixed strategies. (This best guarantee is the *minimax value,* denoted by $\bar{v}$.)

Recall that when only pure strategies were used, games not having a saddle point turned out to be *unstable* (no stable solutions). The reason was essentially that $\underline{v} < \bar{v}$, so that the players would want to change their strategies to improve their positions. Similarly, for games with mixed strategies, it is necessary that $\underline{v} = \bar{v}$ for the optimal solution to be *stable*. Fortunately, according to the *minimax theorem* of game theory this condition always holds for such games.

> **Minimax theorem:** If mixed strategies are allowed, the pair of mixed strategies that is optimal according to the minimax criterion provides a *stable solution* with $\underline{v} = \bar{v} = v$ (the value of the game), so that neither player can do better by unilaterally changing his strategy.

One proof of this theorem is included in Sec. 12.5.

Although the concept of mixed strategies becomes quite intuitive if the game is played *repeatedly,* it requires some interpretation when the game is to be played just *once.* In this case, using a mixed strategy still involves selecting and using *one* pure strategy (randomly selected from the specified probability distribution), so it might seem more sensible to ignore this randomization process and just choose the one "best" pure strategy to be used. However, we have already illustrated for variation 3 in the preceding section that a player must *not* allow his opponent to deduce what his strategy will be (i.e., the solution procedure under the rules of game theory must not *definitely* identify which pure strategy will be used when the game is unstable). Furthermore, even if the opponent is able to use only his knowledge of the tendencies of the first player to deduce *probabilities* (for the pure strategy chosen) that are different from those for the *optimal* mixed strategy, then he still can take advantage of this knowledge to reduce the expected payoff to the first player. Therefore, the only way to *guarantee* attaining the optimal *expected* payoff v is to *randomly* select the pure strategy to be used from the probability distribution for the optimal mixed strategy.

Now we must show how to find the optimal mixed strategy for each player.

463

Game Theory

There are several methods of doing this. One is a graphical procedure that may be used whenever one of the players has only two (undominated) pure strategies; this approach is described in the next section. When larger games are involved, the usual method is to transform the problem into a linear programming problem that would then be solved by the simplex method on a computer; Sec. 12.5 discusses this approach.

12.4 Graphical Solution Procedure

Consider any game with mixed strategies such that, after eliminating dominated strategies, one of the players has only two pure strategies. To be specific, let this player be player I. Because his mixed strategies are (x_1, x_2) and $x_2 = 1 - x_1$, it is necessary for him to solve only for the optimal value of x_1. However, it is straightforward to plot the expected payoff as a function of x_1 for each of his opponent's pure strategies. This graph can then be used to identify the point that maximizes the minimum expected payoff. The opponent's minimax mixed strategy can also be identified from the graph.

To illustrate this procedure, consider *variation 3* of the political campaign problem (see Table 12.5). Notice that the third pure strategy for player I is dominated by his second, so the payoff table can be reduced to the form given in Table 12.6. Therefore, for each of the pure strategies available to player II, the expected payoff for player I will be

(y_1, y_2, y_3)	Expected Payoff
$(1, 0, 0)$	$0x_1 + 5(1 - x_1) = 5 - 5x_1$
$(0, 1, 0)$	$-2x_1 + 4(1 - x_1) = 4 - 6x_1$
$(0, 0, 1)$	$2x_1 - 3(1 - x_1) = -3 + 5x_1$

Now plot these expected payoff lines on a graph, as shown in Fig. 12.1. For any given value of x_1 and of (y_1, y_2, y_3), the expected payoff will be the appropriate weighted average of the corresponding points on these three lines. In particular,

$$\text{Expected payoff} = y_1(5 - 5x_1) + y_2(4 - 6x_1) + y_3(-3 + 5x_1).$$

Table 12.6 Reduced Payoff Table for Variation 3 of the Political Campaign Problem

			II	
	Probability	y_1	y_2	y_3
Probability	Pure Strategy	1	2	3
x_1	1	0	-2	2
I $1 - x_1$	2	5	4	-3

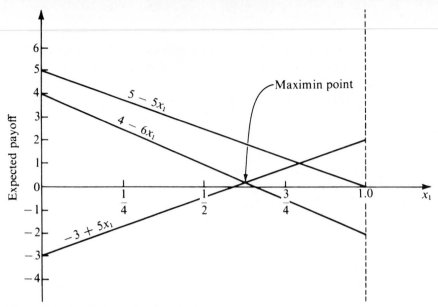

Figure 12.1 Graphical procedure for solving games.

Thus, given x_1, the minimum expected payoff is given by the corresponding point on the "bottom" line. According to the minimax (or maximin) criterion, player I should select the value of x_1 giving the *largest* minimum expected payoff, so that

$$\underline{v} = v = \max_{0 \le x_1 \le 1} \{\min(-3 + 5x_1, 4 - 6x_1)\}.$$

Therefore, the optimal value of x_1 is the one at the intersection of the two lines $(-3 + 5x_1)$ and $(4 - 6x_1)$. Solving algebraically,

$$-3 + 5x_1 = 4 - 6x_1,$$

so that $x_1 = \frac{7}{11}$; thus $(x_1, x_2) = (\frac{7}{11}, \frac{4}{11})$ is the *optimal mixed strategy* for player I, and

$$\underline{v} = v = -3 + 5(\tfrac{7}{11}) = \tfrac{2}{11}$$

is the value of the game.

To find the corresponding optimal mixed strategy for player II, we now reason as follows. According to the definition of the *minimax value* $\bar{v}$ and the minimax theorem, the expected payoff resulting from the optimal strategy $(y_1, y_2, y_3) = (y_1^*, y_2^*, y_3^*)$ will satisfy the condition

$$y_1^*(5 - 5x_1) + y_2^*(4 - 6x_1) + y_3^*(-3 + 5x_1) \le \bar{v} = v = \tfrac{2}{11}$$

for all values of x_1 ($0 \le x_1 \le 1$). Furthermore, when player I is playing optimally (that is, $x_1 = \frac{7}{11}$), this inequality will be an equality, so that

$$\tfrac{20}{11}y_1^* + \tfrac{2}{11}y_2^* + \tfrac{2}{11}y_3^* = v = \tfrac{2}{11}.$$

Because (y_1, y_2, y_3) is a probability distribution, it is also known that

$$y_1^* + y_2^* + y_3^* = 1.$$

Therefore, $y_1^* = 0$ because $y_1^* > 0$ would violate the next-to-last equation; i.e., the expected payoff on the graph at $x_1 = \frac{7}{11}$ would be above the maximin point. (In general, any line that does not pass through the maximin point must be given a zero weight to avoid increasing the expected payoff above this point.)

Hence

$$y_2^*(4 - 6x_1) + y_3^*(-3 + 5x_1) \begin{cases} \leq \frac{2}{11} & \text{for } 0 \leq x_1 \leq 1 \\ = \frac{2}{11} & \text{for } x_1 = \frac{7}{11}. \end{cases}$$

But y_2^* and y_3^* are numbers, so the left-hand side is the equation of a straight line, which is a fixed weighted average of the two "bottom" lines on the graph. Because the ordinate of this line must equal $\frac{2}{11}$ at $x_1 = \frac{7}{11}$, and because it must never exceed $\frac{2}{11}$, the line necessarily is horizontal. (This conclusion is always true unless the optimal value of x_1 is either zero or 1, in which case player II also should use a single pure strategy.) Therefore,

$$y_2^*(4 - 6x_1) + y_3^*(-3 + 5x_1) = \frac{2}{11}, \qquad \text{for } 0 \leq x_1 \leq 1.$$

Hence, to solve for y_2^* and y_3^*, select two values of x_1 (say, zero and 1), and solve the resulting two simultaneous equations. Thus

$$4y_2^* - 3y_3^* = \frac{2}{11},$$

$$-2y_2^* + 2y_3^* = \frac{2}{11},$$

so that $y_3^* = \frac{6}{11}$ and $y_2^* = \frac{5}{11}$. Therefore, the *optimal mixed strategy* for player II is $(y_1, y_2, y_3) = (0, \frac{5}{11}, \frac{6}{11})$.

If, in another problem, there should happen to be more than two lines passing through the maximin point, so that more than two of the y_j^* can be greater than zero, this condition would imply that there are many ties for the optimal mixed strategy for player II. One such strategy can then be identified by setting all but two of these y_j^* equal to zero and solving for the remaining two in the manner just described. For the remaining two, the associated lines must have positive slope in one case and negative slope in the other.

Although this graphical procedure has been illustrated for only one particular problem, essentially the same reasoning can be used to solve any game with mixed strategies that has only two undominated pure strategies for one of the players.

12.5 Solving by Linear Programming

Any game with mixed strategies can be solved by transforming the problem into a linear programming problem. As you will see, this transformation requires little more than applying the minimax theorem and using the definitions of the maximin value $\underline{v}$ and minimax value $\bar{v}$.

First, consider how to find the optimal mixed strategy for player I. As indicated in Sec. 12.3,

$$\text{Expected payoff} = \sum_{i=1}^{m} \sum_{j=1}^{n} p_{ij} x_i y_j$$

and the strategy $(x_1, x_2, \ldots, x_m)$ is optimal if

$$\sum_{i=1}^{m} \sum_{j=1}^{n} p_{ij} x_i y_j \geq \underline{v} = v$$

for all opposing strategies $(y_1, y_2, \ldots, y_n)$. Thus this inequality will need to hold, for example, for each of the pure strategies of player II, i.e., for each of the strategies $(y_1, y_2, \ldots, y_n)$ where one $y_j = 1$ and the rest equal zero. Substituting these values into the inequality yields

$$\sum_{i=1}^{m} p_{ij} x_i \geq v, \qquad \text{for } j = 1, 2, \ldots, n,$$

so that the inequality *implies* this set of n inequalities. Furthermore, this set of n inequalities *implies* the original inequality (rewritten),

$$\sum_{j=1}^{n} y_j \left(\sum_{i=1}^{m} p_{ij} x_i \right) \geq \sum_{j=1}^{n} y_j v = v,$$

since

$$\sum_{j=1}^{n} y_j = 1.$$

Because the implication goes in both directions, it follows that imposing this set of n linear inequalities is *equivalent* to requiring the original inequality to hold for all strategies $(y_1, y_2, \ldots, y_n)$. But these n inequalities are legitimate linear programming constraints, as are the additional constraints

$$x_1 + x_2 + \cdots + x_m = 1$$

$$x_i \geq 0, \qquad \text{for } i = 1, 2, \ldots, m$$

that are required to ensure that the x_i are probabilities. Therefore, any solution $(x_1, x_2, \ldots, x_m)$ that satisfies this entire set of linear programming constraints is the desired optimal mixed strategy.

Consequently, the problem of finding an optimal mixed strategy has been reduced to finding a *feasible solution* for a *linear programming* problem, which can be done as described in Chap. 4. The two remaining difficulties are (1) v is unknown and (2) the linear programming problem has no objective function. Fortunately, both these difficulties can be resolved at one stroke by replacing the unknown constant v by the variable x_{m+1} and then *maximizing* x_{m+1}, so that x_{m+1} automatically will equal v (by definition) at the *optimal* solution for the linear programming problem!

To summarize, player I would find his optimal mixed strategy by using the simplex method to solve the *linear programming* problem:

$$\text{Minimize} \qquad (-x_{m+1}),$$

subject to

$$p_{11} x_1 + p_{21} x_2 + \cdots + p_{m1} x_m - x_{m+1} \geq 0$$

$$p_{12} x_1 + p_{22} x_2 + \cdots + p_{m2} x_m - x_{m+1} \geq 0$$

$$\vdots$$

$$p_{1n} x_1 + p_{2n} x_2 + \cdots + p_{mn} x_m - x_{m+1} \geq 0$$

$$-(x_1 + x_2 + \cdots + x_m) = -1$$

and $$x_i \geq 0, \qquad \text{for } i = 1, 2, \ldots, m.$$

(The objective function and equality constraint have been rewritten here in an equivalent way for later convenience.) Note that x_{m+1} is not restricted to be nonnegative, whereas the simplex method can be applied only after *all* the variables have non-negativity constraints. However, this matter can be easily rectified, as will be discussed shortly.

Now consider player II. He could find his optimal mixed strategy by rewriting the payoff table as the payoff to himself rather than to player I and then by proceeding exactly as just described. However, it is enlightening to summarize his formulation in terms of the original payoff table. By proceeding in a way that is completely analogous to the one just described, player II would conclude that his optimal mixed strategy is given by the optimal solution to the *linear programming* problem:

$$\text{Maximize} \qquad (-y_{n+1}),$$

subject to

$$p_{11}y_1 + p_{12}y_2 + \cdots + p_{1n}y_n - y_{n+1} \leq 0$$

$$p_{21}y_1 + p_{22}y_2 + \cdots + p_{2n}y_n - y_{n+1} \leq 0$$

$$\vdots$$

$$p_{m1}y_1 + p_{m2}y_2 + \cdots + p_{mn}y_n - y_{n+1} \leq 0$$

$$-(y_1 + y_2 + \cdots + y_n) = -1$$

and

$$y_j \geq 0, \qquad \text{for } j = 1, 2, \ldots, n.$$

Notice the key fact that this linear programming problem and the one given for player I are *dual* to each other in the sense described in Secs. 6.1 and 6.4. (In particular, this problem is in the form given for the primal problem, and the one for player I is the corresponding dual problem.) This fact has several important implications. One implication is that the optimal mixed strategies for both players can be found by solving only one of the linear programming problems because the optimal dual solution is an automatic by-product of the simplex method calculations to find the optimal primal solution. A second implication is that this brings all *duality theory* (described in Chap. 6) to bear upon the interpretation and analysis of games.

A related implication is that this provides a simple proof of the minimax theorem. Let x_{m+1}^* and y_{n+1}^* denote the value of x_{m+1} and y_{n+1} in the optimal solution of the respective linear programming problems. It is known from the *strong duality property* given in Sec. 6.1 that $-x_{m+1}^* = -y_{n+1}^*$, so that $x_{m+1}^* = y_{n+1}^*$. However, it is evident from the definition of $\underline{v}$ and $\overline{v}$ that $\underline{v} = x_{m+1}^*$ and $\overline{v} = y_{n+1}^*$, so it follows that $\underline{v} = \overline{v}$ as claimed by the minimax theorem.

The objective functions of these two linear programming problems have been written as minimize $(-x_{m+1})$ and maximize $(-y_{n+1})$ simply to demonstrate that the two problems are *dual* to each other. Hereafter, these objective functions will be written in the more natural *equivalent* forms, maximize x_{m+1} and minimize y_{n+1}. For the same reason, the negative sign now should be deleted from both sides of the equality constraint in each problem.

One remaining loose end needs to be tied up, namely, what to do about x_{m+1} and y_{n+1} being unrestricted in sign in the linear programming formulations. If it is clear that $v \geq 0$ so that the optimal values of x_{m+1} and y_{n+1} are nonnegative, then it is safe to introduce nonnegativity constraints for these variables for purposes of applying the simplex method. However, if $v < 0$, then an adjustment needs to be made. One possibility is to use the approach described in Sec. 4.6 for replacing a variable without a nonnegativity constraint by the difference of two nonnegative variables. Another is to reverse players I and II so that the payoff table would be rewritten as the payoff to the original player II, which would make the corresponding value of v positive. A third, and the most commonly used, procedure is to add a sufficiently large fixed constant to all the entries in the payoff table so that the new value of the game will be positive. (For example, setting this constant equal to the absolute value of the largest negative entry will suffice.) Because this same constant is added to every entry, this adjustment cannot alter the optimal mixed strategies in any way, so they can now be obtained in the usual manner. The indicated value of the game would be increased by the amount of the constant, but this value can be readjusted after the solution has been obtained.

To illustrate this linear programming approach, consider again variation 3 of the political campaign problem after eliminating dominated strategy 3 for player I (see Table 12.6). Because there are some negative entries in the reduced payoff table, it is unclear at the outset whether the *value* of the game v is *nonnegative* (it turns out to be). For the moment, let us assume that $v \geq 0$ and proceed without making any of the adjustments discussed in the preceding paragraph.

To write out the linear programming model for player I for this example, note that p_{ij} in the general model is the entry in row i and column j of Table 12.6, for $i = 1, 2$ and $j = 1, 2, 3$. Using maximize x_{m+1} instead of the equivalent minimize $(-x_{m+1})$, with $m = 2$ and $n = 3$, the resulting model is

$$\text{Maximize} \quad x_3,$$

subject to

$$5x_2 - x_3 \geq 0$$

$$-2x_1 + 4x_2 - x_3 \geq 0$$

$$2x_1 - 3x_2 - x_3 \geq 0$$

$$x_1 + x_2 = 1$$

and

$$x_1 \geq 0, \quad x_2 \geq 0, \quad x_3 \geq 0.$$

Applying the simplex method to this linear programming problem yields $x_1^* = \frac{7}{11}, x_2^* = \frac{4}{11}, x_3^* = \frac{2}{11}$ as the optimal solution. (See Probs. 16 and 17.) Consequently, the optimal *mixed* strategy for player I according to the minimax criterion is $(x_1, x_2) = (\frac{7}{11}, \frac{4}{11})$, and the *value* of the game is $v = x_3^* = \frac{2}{11}$. The simplex method also yields the optimal solution to the *dual* (given next) of this problem, namely, $y_1^* = 0, y_2^* = \frac{5}{11}, y_3^* = \frac{6}{11}, y_4^* = \frac{2}{11}$, so the optimal *mixed* strategy for player II is $(y_1, y_2, y_3) = (0, \frac{5}{11}, \frac{6}{11})$.

The *dual* of the preceding problem is just the linear programming model for player II (the one with variables $y_1, y_2, \ldots, y_n, y_{n+1}$) shown earlier in this section. Plugging in the values of p_{ij} from Table 12.6, this model (in minimization form) is

$$\text{Minimize} \quad y_4,$$

subject to

$$-2y_2 + 2y_3 - y_4 \leq 0$$

$$5y_1 + 4y_2 - 3y_3 - y_4 \leq 0$$

$$y_1 + y_2 + y_3 = 1$$

Applying the simplex method directly to this model yields the optimal solution: $y_1^* = 0$, $y_2^* = \frac{5}{11}$, $y_3^* = \frac{6}{11}$, $y_4^* = \frac{2}{11}$ (as well as the optimal *dual* solution, $x_1^* = \frac{7}{11}$, $x_2^* = \frac{4}{11}$, $x_3^* = \frac{2}{11}$). Thus the optimal *mixed* strategy for player II is $(y_1, y_2, y_3) = (0, \frac{5}{11}, \frac{6}{11})$, and the *value* of the game is again seen to be $v = y_4^* = \frac{2}{11}$.

Because we already had found the optimal *mixed* strategy for player II while dealing with the first model, we did not have to solve the second one. In general, you always can find the optimal mixed strategies for *both* players by choosing just one of the models (either one) and then using the simplex method to solve for *both* the optimal solution and the optimal *dual* solution.

Both of these linear programming models assumed that $v \geq 0$. If this assumption were violated, what would happen is that both models would have *no feasible solutions,* so the simplex method would stop quickly with this message. To avoid this risk, we could have added a positive constant, say 3 (the absolute value of the largest negative entry), to all of the entries in Table 12.6. This then would increase by 3 all of the coefficients of x_1, x_2, y_1, y_2, and y_3 in the inequality constraints of the two models. (See Prob. 13.)

12.6 Extensions

Although this chapter has considered only *two-person, zero-sum games* with a *finite* number of pure strategies, game theory extends far beyond this kind of game. In fact, extensive research has been done on a number of more complicated types of games, including the ones summarized in this section.

One such type is the *n-person game,* where more than two players may participate in the game. This generalization is particularly important because, in many kinds of competitive situations, there frequently are more than two competitors involved. This may occur, for example, in competition among business firms, in international diplomacy, and so forth. Unfortunately, the existing theory for such games is less satisfactory than it is for two-person games.

Another generalization is the *nonzero-sum game,* where the sum of the payoffs to the players need not be zero (or any other fixed constant). This case reflects the fact that many competitive situations include noncompetitive aspects that contribute to the mutual advantage or mutual disadvantage of the players. For example, the advertising strategies of competing companies can affect not only how they will split the market but also the total size of the market for their competing products.

Because mutual gain is possible, nonzero-sum games are further classified in terms of the degree to which the players are permitted to cooperate. At one extreme is the *noncooperative game,* where there is no preplay communication between the players. At the other extreme is the *cooperative game,* where preplay discussions and binding agreements are permitted. For example, competitive situations involving trade regulations between countries, or collective bargaining between labor and management, might be formulated as cooperative games. When there are more than two players, cooperative games also allow some or all of the players to form coalitions.

Still another extension is to the class of *infinite games,* where the players have an infinite number of pure strategies available to them. These games are designed for the kind of situation where the strategy to be selected can be represented by a *continuous* decision variable. For example, this decision variable might be the time at which to take a certain action, or the proportion of one's resources to allocate to a certain activity, in a competitive situation. Much research has been concentrated on such games in recent years.

However, the analysis required in these extensions beyond the two-person, zero-sum, finite game is relatively complex and will not be pursued further here.

12.7 Conclusions

The general problem of how to make decisions in a competitive environment is a very common and important one. The fundamental contribution of game theory is that it provides a basic conceptual framework for formulating and analyzing such problems in simple situations. However, there is a considerable gap between what the theory can handle and the complexity of most competitive situations arising in practice. Therefore, the conceptual tools of game theory usually play just a supplementary role in dealing with these situations.

Because of the importance of the general problem, research is continuing with some success to extend the theory to more complex situations.

SELECTED REFERENCES

1. Begley, Sharon, with David Grant: ''The Games Scholars Play,'' *Newsweek,* Sept. 6, 1982, p. 72.
2. Davis, M.: *Game Theory: An Introduction,* Basic Books, New York, 1983.
3. Kaplan, E. L.: *Mathematical Programming and Games,* Wiley, New York, 1982.
4. Luce, R. Duncan, and Howard Raiffa: *Games and Decisions,* Wiley, New York, 1957.
5. May, Francis B.: *Introduction to Games of Strategy,* Allyn & Bacon, Boston, 1970.
6. Moulin, Herve: *Game Theory for the Social Sciences,* 2d ed., New York University Press, New York, 1986.
7. Owen, Guillermo: *Game Theory,* 2d ed., Academic Press, New York, 1982.
8. Shubik, Martin: *Game Theory in the Social Sciences,* Vols. I (1982) and II (1987), MIT Press, Cambridge, Mass.
9. Szep, J., and F. Forgo: *Introduction to the Theory of Games,* D. Reidel, Boston, 1985.

PROBLEMS

1. * For each of the following payoff tables, determine the optimal strategy for each player by successively eliminating dominated strategies. (Indicate the order in which you eliminated strategies.)

(a)

		II		
		1	2	3
	1	−3	1	2
I	2	1	2	1
	3	1	0	−2

(b)

		II		
		1	2	3
	1	1	2	0
I	2	2	−3	−2
	3	0	3	−1

2. Consider the game having the following payoff table.

		II			
		1	2	3	4
I	1	2	-3	-1	1
	2	-1	1	-2	2
	3	-1	2	-1	3

Determine the optimal strategy for each player by successively eliminating dominated strategies. Give a list of the dominated strategies (and the corresponding dominating strategies) in the order in which you were able to eliminate them.

3. Find the saddle point for the game having the following payoff table.

		II		
		1	2	3
I	1	1	-1	1
	2	-2	0	3
	3	3	1	2

4. Find the saddle point for the game having the following payoff table.

		II			
		1	2	3	4
I	1	3	-3	-2	-4
	2	-4	-2	-1	1
	3	1	-1	2	0

5. Two companies share the bulk of the market for a particular kind of product. Each is now planning its new marketing plans for the next year in an attempt to wrest some sales away from the other company. (The total sales for the product are relatively fixed, so one company can only increase its sales by winning them away from the other.) Each company is considering three possibilities: (1) better packaging of the product, (2) increased advertising, and (3) a slight reduction in price. The costs of the three alternatives are quite comparable and sufficiently large that each company will select just one. The estimated effect of each combination of alternatives on the *increased percentage of the sales* for company I is

		II		
		1	2	3
I	1	2	3	1
	2	1	4	0
	3	3	-2	-1

Each company must make its selection before learning the decision of the other company.

(*a*) Without eliminating dominated strategies, use the minimax (or maximin) criterion to determine the best strategy for each side.

(*b*) Now identify and eliminate dominated strategies as far as possible. Make a list of the dominated strategies showing the order in which you were able to eliminate them. Then show the resulting reduced payoff table with no remaining dominated strategies.

6. The labor union and management of a particular company have been negotiating a new labor contract. However, negotiations have now come to an impasse, with management making a "final" offer of a wage increase of $1.10/hour and the union making a "final" demand of a $1.60/hour increase. Therefore, both sides have agreed to have an impartial arbitrator set the wage increase somewhere between $1.10/hour and $1.60/hour (inclusively).

The arbitrator has asked each side to submit to her a confidential proposal for a fair and economically reasonable wage increase (rounded to the nearest dime). From past experience, both sides know that this arbitrator normally accepts the proposal of the side that gives the most from its "final" figure. If neither side changes its final figure, or if they both give in the same amount, then the arbitrator normally compromises halfway between ($1.35 in this case). Each side now needs to determine what wage increase to propose for its own maximum advantage.

(a) Formulate this problem as a *two-person, zero-sum game*.
(b) Use the concept of dominated strategies to determine the best strategy for each side.
(c) Without eliminating dominated strategies, use the minimax criterion to determine the best strategy for each side.

7.* Two politicians soon will be starting their campaigns against each other for a certain political office. Each must now select the main issue he will emphasize as the theme of his campaign. Each has three advantageous issues from which to choose, but the relative effectiveness of each one would depend upon the issue chosen by his opponent. In particular, the estimated increase in the vote for politician I (expressed as a percentage of the total vote) resulting from each combination of issues is

		Issue for Politician II		
		1	2	3
Issue for politician I	1	7	−1	3
	2	1	0	2
	3	−5	−3	1

However, because considerable staff work is required to research and formulate the issue chosen, each politician must make his own choice before learning his opponent's choice. Which issue should he choose?

For each of the situations described here, formulate this problem as a *two-person, zero-sum game,* and then determine which issue should be chosen by each politician according to the specified criterion.

(a) The current preferences of the voters are very uncertain, so each additional percent of votes won by one of the politicians has the same value to him. Use the *minimax* criterion.
(b) A reliable poll has found that the percentage of the voters currently preferring politician I (before the issues have been raised) lies between 45 and 50 percent. (Assume a uniform distribution over this range.) Use the concept of *dominated strategies,* beginning with the strategies for politician I.
(c) Suppose that the percentage described in part (b) actually were 45 percent. Should politician I use the minimax criterion? Explain. Which issue would you recommend? Why?

8. Two manufacturers currently are competing for sales in two different but equally profitable product lines. In both cases the sales volume for manufacturer II is three times as large as that for manufacturer I. Because of a recent technological breakthrough, both manu-facturers will be making a major improvement in both products. However, they are uncertain as to what development and marketing strategy they should follow.

If both product improvements are developed simultaneously, either manufacturer can have them ready for sale in 12 months. Another alternative is to have a "crash program" to develop only one product first to try to get it marketed ahead of the competition. By doing this, manufacturer II could have one product ready for sale in 9 months, whereas manufacturer I would require 10 months (because of previous commitments for its production facilities). For either manufacturer, the second product could then be ready for sale in an additional 9 months.

For either product line, if both manufacturers market their improved models simultane-ously, it is estimated that manufacturer I would increase its share of the total future sales of this product by 8 percent of the total (from 25 to 33 percent). Similarly, manufacturer I would increase its share by 20, 30, and 40 percent of the total if it markets the product sooner than manufacturer II by 2, 6, and 8 months, respectively. On the other hand, manufacturer I would lose 4, 10, 12, and 14 percent of the total if manufacturer II markets it sooner by 1, 3, 7, and 10 months, respectively.

Formulate this problem as a two-person, zero-sum game, and then determine which strategy the respective manufacturers should use according to the minimax criterion.

9. Consider the following parlor game to be played between two players. Each player begins with three chips: one red, one white, and one blue. Each chip can be used only once.

To begin, each player selects one of his chips and places it on the table, concealed. Both players then uncover the chips and determine the payoff to the winning player. In particular, if both players play the same kind of chip, it is a draw; otherwise, the following table indicates the winner and how much he receives from the other player. Next, each player selects one of his two remaining chips and repeats the procedure, resulting in another payoff according to the following table. Finally, each player plays his one remaining chip, resulting in the third and final payoff.

Winning Chip	Payoff
Red beats White	$50
White beats Blue	$40
Blue beats Red	$30
Matching colors	0

Formulate this problem as a two-person, zero-sum game by identifying the form of the strategies and payoffs.

10. Consider the game having the following payoff table.

		II	
		1	2
I	1	3	−2
	2	−1	2

Use the graphical procedure described in Sec. 12.4 to determine the value of the game and the optimal *mixed* strategy for each player according to the minimax criterion. Check your answer for player II by constructing *his* payoff table and applying the graphical procedure directly to this table.

11.* For each of the following payoff tables, use the graphical procedure described in Sec. 12.4 to determine the value of the game and the optimal *mixed* strategy for each player according to the minimax criterion:

(a)

	II		
I	1	2	3
1	4	3	1
2	0	1	2

(b)

	II		
I	1	2	3
1	1	−1	3
2	0	4	1
3	3	−2	5
4	−3	6	−2

12. Consider the following parlor game between two players. It begins when a referee flips a coin, notes whether it comes up heads or tails, and then shows this result to player I only. Player I may then either (1) pass and thereby pay $5 to player II, or (2) he may bet. If player I passes, the game is terminated. However, if he bets, the game continues, in which case player II may then either (1) pass and thereby pay $5 to player I, or (2) he may call. If player II calls, the referee then shows him the coin; if it came up heads, player II pays $10 to player I; if it came up tails, player II receives $10 from player I.

(a) Give the pure strategies for each player. (*Hint:* Player I will have four pure strategies, each one specifying how he would respond to each of the two results the referee can show him; player II will have two pure strategies, each one specifying how he will respond if player I bets.)

(b) Develop the payoff table for this game, using expected values for the entries when necessary. Determine whether it has a saddle point or not.

(c) Use the graphical procedure described in Sec. 12.4 to determine the optimal *mixed* strategy for each player according to the minimax criterion. Also give the corresponding value of the game.

13. Referring to the last paragraph of Sec. 12.5, suppose that 3 were added to all of the entries of Table 12.6 in order to ensure that the corresponding linear programming models for both players have feasible solutions with $x_3 \geq 0$ and $y_4 \geq 0$. Write out these two models. Based on the information given in Sec. 12.5, what are the optimal solutions for these two models? What is the relationship between x_3^* and y_4^*? What is the relationship between the *value* of the original game v and the values of x_3^* and y_4^*?

14.* Consider the game having the following payoff table.

	II			
I	1	2	3	4
1	5	0	3	1
2	2	4	3	2
3	3	2	0	4

Use the approach described in Sec. 12.5 to formulate the problem of finding the optimal *mixed* strategies according to the minimax criterion as a *linear programming* problem.

15. For each of the following payoff tables, transform the problem of finding the minimax *mixed* strategies into an equivalent *linear programming* problem.

(a)

	II		
I	1	2	3
1	4	2	−3
2	−1	0	3
3	2	3	−2

(b)

	II				
I	1	2	3	4	5
1	1	−3	2	−2	1
2	2	3	0	3	−2
3	0	4	−1	−3	2
4	−4	0	−2	2	−1

16. Consider variation 3 of the political campaign problem (see Table 12.6). Refer to the resulting *linear programming* model for player I given near the end of Sec. 12.5. Ignoring the objective function variable (x_3), plot the *feasible region* for x_1 and x_2 graphically (as described in Sec. 3.1). (*Hint:* This feasible region consists of a single line segment.) Next, write an algebraic expression for the maximizing value of x_3 for any point in this feasible region. Finally, use this expression to demonstrate that the optimal solution must, in fact, be the one given in Sec. 12.5.

17. Consider the *linear programming* model for player I given near the end of Sec. 12.5 for variation 3 of the political campaign problem (see Table 12.6). Verify the optimal *mixed* strategies for both players given in Sec. 12.5 by applying the automatic routine for the simplex method in your OR COURSEWARE to this model to find *both* its optimal solution and its optimal *dual* solution.

18. The A. J. Swim Team soon will have an important swim meet with the G. N. Swim Team. Each team has a star swimmer (John and Mark, respectively) who can swim very well in the 100-yard butterfly, backstroke, and breaststroke events. However, the rules prevent them from being used in more than *two* of these events. Therefore, their coaches now need to decide how to use them to maximum advantage.

Each team will enter three swimmers per event (the maximum allowed). For each event, the following table gives the best time previously achieved by John and Mark as well as the best time for each of the other swimmers who will definitely enter that event. (Whichever event John or Mark does not swim, his team's third entry for that event will be slower than the two shown in the table.)

	A. J. Swim Team			G. N. Swim Team		
	Entry			*Entry*		
	1	2	John	Mark	1	2
Fly	1:01.6	59.1	57.5	58.4	1:03.2	59.8
Back	1:06.8	1:05.6	1:03.3	1:02.6	1:04.9	1:04.1
Breast	1:13.9	1:12.5	1:04.7	1:06.1	1:15.3	1:11.8

The points awarded are 5 points for first place, 3 for second place, 1 for third place, and none for lower places. Both coaches believe that all swimmers will essentially equal their best times in this meet. Thus John and Mark each will definitely be entered in two of these three events.

(*a*) The coaches must submit all their entries before the meet without knowing the entries for the other team, and no changes are permitted later. The outcome of the meet is very uncertain, so each additional point has equal value for the coaches. Formulate this problem as a *two-person, zero-sum* game. Eliminate dominated strategies, and then use the graphical procedure described in Sec. 12.4 to find the optimal *mixed* strategy for each team according to the minimax criterion.

(*b*) The situation and assignment are the same as in part (*a*), except that both coaches now believe that the A. J. Swim Team will win the swim meet if it can win 13 or more points in these three events, but will lose with less than 13 points. [Compare the resulting optimal mixed strategies with those obtained in part (*a*).]

(*c*) Now suppose that the coaches submit their entries during the meet one event at a time. When submitting his entries for an event, the coach does not know who will be swimming that event for the other team, but he does know who has swum *preceding* events. The three key events just discussed are swum in the order listed in the table. Once again, the A. J. Swim Team needs 13 points in these events to win the swim meet. Formulate this problem as a *two-person, zero-sum* game. Then use the concept of dominated strategies to determine the best strategy for the G. N. team that actually "guarantees" it will win under the assumptions being made.

(*d*) The situation is the same as in part (*c*). However, assume now that the coach for the G. N. team does not know about game theory and so may, in fact, choose any of his available strategies that have Mark swimming two events. Use the concept of dominated strategies to determine the best strategies from which the coach for the A. J. team should choose. If this coach knows that the other coach has a tendency to enter Mark in the butterfly and the backstroke more often than in the breaststroke, which strategy should he choose?

19. Consider the general $m \times n$, two-person, zero-sum game. Let p_{ij} denote the payoff to player I if he plays his strategy i ($i = 1, \ldots, m$) and player II plays his strategy j ($j = 1, \ldots, n$). Strategy 1 (say) for player I is said to be *weakly dominated* by strategy 2 (say) if $p_{1j} \leq p_{2j}$ for $j = 1, \ldots, n$ and $p_{1j} = p_{2j}$ for one or more values of j.

(*a*) Assume that the payoff table possesses one or more *saddle points*, so that the players have corresponding optimal pure strategies under the minimax criterion. Prove that eliminating *weakly dominated* strategies from the payoff table cannot eliminate all these saddle points and cannot produce any new ones.

(*b*) Assume that the payoff table does not possess any saddle points, so that the optimal strategies under the minimax criterion are *mixed* strategies. Prove that eliminating weakly dominated pure strategies from the payoff table cannot eliminate all optimal mixed strategies and cannot produce any new ones.

20. Briefly describe what you feel are the advantages and disadvantages of the minimax criterion.

13

Integer Programming

In Part 2 you saw several examples of the numerous diverse applications of *linear programming*. However, one key limitation that prevents many more applications is the assumption of *divisibility* (see Sec. 3.3), which requires that *noninteger* values be permissible for decision variables. In many practical problems, the decision variables actually make sense only if they have *integer* values. For example, it is often necessary to assign people, machines, and vehicles to activities in integer quantities. If requiring integer values is the only way in which a problem deviates from a linear programming formulation, then it is an **integer programming (IP)** problem. (The more complete name is *integer linear programming,* but the adjective *linear* normally is dropped except when contrasting this problem to the more esoteric *integer nonlinear programming* problem, which is beyond the scope of this book.)

The mathematical model for integer programming is simply the linear programming model (see Sec. 3.2) with the one additional restriction that the variables must have integer values. If only *some* of the variables are required to have integer values (so the divisibility assumption holds for the rest), this model is referred to as **mixed**

integer programming (MIP). When distinguishing the all-integer problem from this mixed case, we call the former *pure* integer programming.

For example, the Wyndor Glass Co. problem presented in Sec. 3.1 actually would have been an IP problem if the two decision variables, x_1 and x_2, had represented the total number of units to be produced of products 1 and 2, respectively, instead of the production rates. Because both products (glass doors and wood-framed windows) necessarily come in whole units, x_1 and x_2 would have to be restricted to integer values.

There have been numerous such applications of integer programming that involve a direct extension of linear programming where the divisibility assumption must be dropped. However, another area of application may be of even greater importance, namely, problems involving a number of interrelated "yes-or-no decisions." In such decisions, the only two possible choices are *yes* or *no*. For example, should we undertake a particular fixed project? Should we make a particular fixed investment? Should we locate a facility in a particular site?

With just two choices, we can represent such decisions by decision variables that are restricted to just two values, say zero and one. Thus the jth yes-or-no decision would be represented by, say, x_j, such that

$$x_j = \begin{cases} 1, & \text{if decision } j \text{ is yes} \\ 0, & \text{if decision } j \text{ is no.} \end{cases}$$

Such variables are called **binary variables** (or 0–1 variables). Consequently, IP problems that contain only binary variables sometimes are called **binary integer programming (BIP)** problems (or 0–1 integer programming problems).

Section 13.1 presents a miniature version of a typical *BIP* problem. Additional formulation possibilities with binary variables are discussed in Sec. 13.2. The remaining sections then deal with ways to solve *IP* problems, including both *BIP* and *MIP* problems.

13.1 Prototype Example

The CALIFORNIA MANUFACTURING COMPANY is considering expansion by building a new factory in either Los Angeles or San Francisco, or perhaps even in both cities. It also is considering building at most one new warehouse, but the choice of the location is restricted to a city where a new factory is being built. The *net present value* (total profitability considering the time value of money) of each of these alternatives is shown in the fourth column of Table 13.1. The last column gives the capital required for the respective investments, where the total capital available is $10,000,000. The objective is to find the feasible combination of alternatives that maximizes the total net present value.

Although this problem is small enough that it can be solved very quickly by inspection (build factories in both cities but no warehouse), let us formulate the IP model for illustrative purposes. All the decision variables have the *binary* form,

$$x_j = \begin{cases} 1, & \text{if decision } j \text{ is yes} \\ 0, & \text{if decision } j \text{ is no} \end{cases} \quad (j = 1, 2, 3, 4).$$

Table 13.1 Data for California Manufacturing Co. Example

Decision Number	Yes-or-No Question	Decision Variable	Net Present Value	Capital Required
1	Build factory in L.A.?	x_1	$9 million	$6 million
2	Build factory in S.F.?	x_2	$5 million	$3 million
3	Build warehouse in L.A.?	x_3	$6 million	$5 million
4	Build warehouse in S.F.?	x_4	$4 million	$2 million

Capital available: $10 million

Because the last two decisions represent *mutually exclusive alternatives* (the company wants *at most* one new warehouse), we need the constraint

$$x_3 + x_4 \leq 1.$$

Furthermore, decisions 3 and 4 are *contingent decisions,* because they are contingent on decisions 1 and 2, respectively (the company would consider building a warehouse in a city only if a new factory also is going there). This contingency is taken into account by the constraints

$$x_3 - x_1 \leq 0$$

$$x_4 - x_2 \leq 0,$$

which force $x_3 = 0$ if $x_1 = 0$ and $x_4 = 0$ if $x_2 = 0$. Therefore, the complete BIP model is

$$\text{Maximize} \quad Z = 9x_1 + 5x_2 + 6x_3 + 4x_4,$$

subject to

$$6x_1 + 3x_2 + 5x_3 + 2x_4 \leq 10$$

$$x_3 + x_4 \leq 1$$

$$-x_1 \qquad\quad + x_3 \qquad\quad \leq 0$$

$$-x_2 \qquad\quad + x_4 \leq 0$$

$$x_j \leq 1$$

$$x_j \geq 0$$

and

$$x_j \text{ is an integer}, \quad \text{for } j = 1, 2, 3, 4$$

Equivalently, the last three lines of this model can be replaced by the single restriction

$$x_j \text{ is binary}, \quad \text{for } j = 1, 2, 3, 4.$$

Except for its small size, this example is typical of many real applications of integer programming where the basic decisions to be made are of the yes-or-no type. Like the second pair of decisions for this example, groups of yes-or-no decisions often constitute groups of **mutually exclusive alternatives** such that *only one* decision in the group can be yes. Each group requires a constraint that the sum of the corresponding binary variables must be $= 1$ (if *exactly one* decision in the group must be yes) or ≤ 1 (if *at most one* decision in the group can be yes). Occasionally, decisions of the yes-or-no type are **contingent decisions**; i.e., decisions depend upon previous decisions. In particular, one decision is said to be *contingent* on another decision if it is allowed to be yes *only if* the other is yes. This situation occurs when the contingent decision involves a follow-up action that would become irrelevant, or even impossible, if the other decision is no. The form that the resulting constraint takes always is that illustrated by the third and fourth constraints in the example.

13.2 Some Other Formulation Possibilities with Binary Variables

You have just seen a prototype example where the *basic decisions* of the problem are of the *yes-or-no type,* so that *binary variables* are introduced to represent these decisions. In addition, binary variables also can be very useful in other ways for formulating difficult problems in a tractable manner. In particular, these variables sometimes enable us to take a problem whose natural formulation is intractable and *reformulate* it as a pure or mixed IP problem.

This kind of situation arises when the original formulation of the problem fits either an IP or a linear programming format *except* for certain minor disparities involving combinatorial relationships in the model. By expressing these combinatorial relationships in terms of questions that must be answered yes or no, *auxiliary* binary variables can be introduced into the model to represent these yes-or-no decisions. Introducing these variables reduces the problem to an MIP problem (or a *pure* IP problem if all of the original variables also are required to have integer values).

Some cases that can be handled by this approach are discussed next, where the x_j denote the *original* variables of the problem (they may be either continuous or integer variables), and the y_i denote the *auxiliary* binary variables that are introduced for the reformulation.

Either-Or Constraints

Consider the important case where a choice can be made between two constraints, so that *only one* must hold. For example, there may be a choice as to which of two resources to use for a certain purpose, so that it is necessary for only one of the two resource availability constraints to hold mathematically. To illustrate the approach to such situations, suppose that one of the requirements in the overall problem is that

$$\text{Either} \quad 3x_1 + 2x_2 \leq 18$$

$$\text{or} \quad x_1 + 4x_2 \leq 16.$$

This requirement must be reformulated to fit it into the linear programming format where *all* specified constraints must hold. Let M be an *extremely* large positive number. Then this requirement can be rewritten as

$$\text{Either} \left\{ \begin{array}{l} 3x_1 + 2x_2 \leq 18 \\ \text{and} \quad\quad x_1 + 4x_2 \leq 16 + M \end{array} \right.$$

$$\text{or} \left\{ \begin{array}{l} 3x_1 + 2x_2 \leq 18 + M \\ \text{and} \quad\quad x_1 + 4x_2 \leq 16, \end{array} \right.$$

because adding M to the right-hand side of such constraints has the effect of eliminating them, because they would be satisfied automatically by any solutions that satisfy the other constraints of the problem. (This formulation assumes that the set of feasible solutions for the overall problem is a bounded set and that M is large enough so that it will not eliminate any feasible solutions.) This formulation is equivalent to the set of constraints

$$3x_1 + 2x_2 \leq 18 + yM$$

$$x_1 + 4x_2 \leq 16 + (1 - y)M.$$

Because the *auxiliary variable* y must be either zero or 1, this formulation guarantees that one of the original constraints must hold while the other is, in effect, eliminated. This new set of constraints would then be appended to the other constraints in the overall model to give a pure or mixed IP problem (depending upon whether the x_j are integer or continuous variables).

This approach is related directly to our earlier discussion about expressing combinatorial relationships in terms of questions that must be answered yes or no. The combinatorial relationship involved concerns the combination of the *other* constraints of the model with the *first* of the two *alternative* constraints and then with the *second*. Which of these two combinations of constraints is *better* (in terms of the value of the objective function that then can be achieved)? To rephrase this question in yes-or-no terms, we ask two complementary questions:

1. Should $x_1 + 4x_2 \leq 16$ be selected as the constraint that must hold?
2. Should $3x_1 + 2x_2 \leq 18$ be selected as the constraint that must hold?

Because exactly one of these questions is to be answered affirmatively, we let the binary terms, y and $(1 - y)$, respectively, represent these yes-or-no decisions, so that $y + (1 - y) = 1$ (*one* yes) automatically. If instead separate binary variables, y_1 and y_2, had been used to represent these yes-or-no decisions, then an additional constraint, $y_1 + y_2 = 1$, would have been needed to make them mutually exclusive.

A formal presentation of this approach is given next for a more general case.

K Out of N Constraints Must Hold

Consider the case where the overall model includes a set of N possible constraints such that only some K of these constraints *must* hold. (Assume that $K < N$.) Part of the optimization process is to choose *which combination* of K constraints permits the objective function to reach its best possible value. The $(N - K)$ constraints *not* chosen are, in effect, eliminated from the problem, although feasible solutions might coincidentally still satisfy some of them.

This case is a direct generalization of the preceding case, which had $K = 1$ and $N = 2$. Denote the N possible constraints by

$$f_1(x_1, x_2, \ldots, x_n) \leq d_1$$

$$f_2(x_1, x_2, \ldots, x_n) \leq d_2$$

$$\vdots$$

$$f_N(x_1, x_2, \ldots, x_n) \leq d_N.$$

Then, applying the same logic as for the preceding case, we find that an equivalent formulation of the requirement that some K of these constraints *must* hold is

$$f_1(x_1, x_2, \ldots, x_n) \le d_1 + My_1$$

$$f_2(x_1, x_2, \ldots, x_n) \le d_2 + My_2$$

$$\vdots$$

$$f_N(x_1, x_2, \ldots, x_n) \le d_N + My_N$$

$$\sum_{i=1}^{N} y_i = N - K,$$

and y_i is binary,   for $i = 1, 2, \ldots, N$,

where M is an extremely large positive number. Because the constraints on the y_i guarantee that K of these variables will equal zero and those remaining will equal 1, K of the original constraints will be unchanged and the rest will, in effect, be eliminated. The choice of *which* K of these constraints should be retained is made by applying the appropriate algorithm to the overall problem so it finds an optimal solution for *all* of the variables simultaneously.

Functions with N Possible Values

Consider the situation where a given function is required to take on any one of N given values. Denote this requirement by

$$f(x_1, x_2, \ldots, x_n) = d_1, \quad \text{or} \quad d_2, \ldots, \quad \text{or} \quad d_N.$$

One special case is where this function is

$$f(x_1, x_2, \ldots, x_n) = \sum_{j=1}^{n} a_j x_j,$$

as on the left-hand side of a linear programming constraint. Another special case is where $f(x_1, x_2, \ldots, x_n) = x_j$ for a given value of j, so the requirement becomes that x_j must take on any one of N given values.

The equivalent IP formulation of this requirement is the following:

$$f(x_1, x_2, \ldots, x_n) = \sum_{i=1}^{N} d_i y_i$$

$$\sum_{i=1}^{N} y_i = 1$$

and y_i is binary,   for $i = 1, 2, \ldots, N$,

so this new set of constraints would replace this requirement in the statement of the overall problem. This set of constraints provides an *equivalent* formulation because exactly one y_i must equal 1 and the others must equal zero, so exactly one d_i is being chosen as the value of the function. In this case, there are N yes-or-no questions being asked, namely, should d_i be the value chosen ($i = 1, 2, \ldots, N$)? Because the y_i respectively represent these *yes-or-no decisions*, the second constraint makes them *mutually exclusive alternatives.*

To illustrate how this case can arise, reconsider the Wyndor Glass Co. problem presented in Sec. 3.1. Eighteen percent of the total production capacity of Plant 3 currently is unused and available for the two new products *or* for certain future products that will be ready for production soon. In order to leave any remaining capacity in usable blocks for these future products, management now wants to impose the restriction that the amount of capacity used by the two current new products must be 6 percent *or* 12 percent *or* 18 percent. Thus the third constraint of the original model ($3x_1 + 2x_2 \leq 18$) now becomes

$$3x_1 + 2x_2 = 6 \quad \text{or} \quad 12 \quad \text{or} \quad 18.$$

In the preceding notation, $N = 3$ with $d_1 = 6$, $d_2 = 12$, and $d_3 = 18$. Consequently, management's new requirement should be formulated as follows:

$$3x_1 + 2x_2 = 6y_1 + 12y_2 + 18y_3$$

$$y_1 + y_2 + y_3 = 1$$

and $\qquad\qquad\qquad y_1, y_2, y_3 \text{ are binary.}$

The overall model for this new version of the problem then consists of the original model (see Sec. 3.1) plus this new set of constraints that replaces the original third constraint. This replacement yields a very tractable MIP formulation.

The Fixed-Charge Problem

It is quite common to incur a fixed charge or setup cost when undertaking an activity. For example, such a charge occurs when a production run to produce a small batch of a particular product is undertaken and the required production facilities must be set up to initiate the run. In such cases the total cost of the activity is the sum of a variable cost related to the level of the activity and the setup cost required to initiate the activity. Frequently the variable cost will be at least roughly proportional to the level of the activity. If it is, the *total cost* of the activity (say, activity j) can be represented by a function of the form

$$f_j(x_j) = \begin{cases} k_j + c_j x_j, & \text{if } x_j > 0 \\ 0, & \text{if } x_j = 0 \end{cases}$$

where x_j denotes the level of activity j ($x_j \geq 0$), k_j denotes the setup cost, and c_j denotes the cost for each incremental unit. Were it not for the setup cost k_j, this cost structure would suggest the possibility of a *linear programming* formulation to determine the optimal levels of the competing activities. Fortunately, even with the k_j, MIP can still be used.

To formulate the overall model, suppose that there are n activities, each with the preceding cost structure (with $k_j \geq 0$ in every case and $k_j > 0$ for some $j = 1, 2, \ldots, n$), and that the problem is to

$$\text{Minimize} \quad Z = f_1(x_1) + f_2(x_2) + \cdots + f_n(x_n),$$

subject to $\qquad\qquad$ given linear programming constraints.

To convert this problem into an MIP format, we begin by posing n questions that must be answered yes or no; namely, for each $j = 1, 2, \ldots, n$, should activity j be undertaken ($x_j > 0$)? Each of these *yes-or-no decisions* is then represented by an auxiliary *binary variable* y_j, so that

$$Z = \sum_{j=1}^{n} (c_j x_j + k_j y_j),$$

where
$$y_j = \begin{cases} 1, & \text{if } x_j > 0 \\ 0, & \text{if } x_j = 0. \end{cases}$$

Therefore, the y_j can be viewed as *contingent decisions* similar to (but not identical to) the type considered in Sec. 13.1. Let M be an extremely large positive number that exceeds the maximum feasible value of any x_j ($j = 1, 2, \ldots, n$). Then the constraints,

$$x_j \leq M y_j, \qquad \text{for } j = 1, 2, \ldots, n,$$

will ensure that $y_j = 1$ rather than zero whenever $x_j > 0$. The one difficulty remaining is that these constraints leave y_j free to be either zero or 1 when $x_j = 0$. Fortunately, this difficulty is automatically resolved because of the nature of the objective function. The case where $k_j = 0$ can be ignored because y_j can then be deleted from the formulation. So we consider the only other case, namely, where $k_j > 0$. When $x_j = 0$, so that the constraints permit a choice between $y_j = 0$ and $y_j = 1$, $y_j = 0$ must yield a smaller value of Z than $y_j = 1$. Therefore, because the objective is to minimize Z, an algorithm yielding an optimal solution would always choose $y_j = 0$ when $x_j = 0$.

To summarize, the MIP formulation of the fixed-charge problem is

$$\text{Minimize} \qquad Z = \sum_{j=1}^{n} (c_j x_j + k_j y_j),$$

subject to   the original constraints, plus

$$x_j - M y_j \leq 0$$

and y_j is binary,   for $j = 1, 2, \ldots, n$.

If the x_j also had been restricted to be integer, then this would be a *pure* IP problem.

To illustrate this approach, look again at Sec. 3.4 at the air pollution problem faced by the *Nori & Leets Co.* The first of the abatement methods considered—increasing the height of the smokestacks—actually would involve a substantial *fixed charge* to get ready for *any* increase in addition to a variable cost that would be roughly proportional to the amount of increase. After conversion to the equivalent annual costs used in the formulation, this fixed charge would be \$2,000,000 each for the blast furnaces and the open-hearth furnaces, whereas the variable costs are those identified in Table 3.14. Thus, in the preceding notation, $k_1 = 2$, $k_2 = 2$, $c_1 = 8$, and $c_2 = 10$. Because the other abatement methods do not involve any fixed charges, $k_j = 0$ for $j = 3, 4, 5, 6$. Consequently, the new MIP formulation of this problem is

$$\text{Minimize} \qquad Z = 8x_1 + 10x_2 + 7x_3 + 6x_4 + 11x_5 + 9x_6 + 2y_1 + 2y_2,$$

subject to   the constraints given in Sec. 3.4, plus

$$x_1 - M y_1 \leq 0$$

$$x_2 - M y_2 \leq 0,$$

and y_1, y_2 are binary.

Suppose that you have a pure IP problem where most of the variables are *binary* variables, but the presence of a few *general* integer variables prevents solving the problem by one of the very efficient BIP algorithms now available. A nice way to circumvent this difficulty is to use the *binary representation* for each of these general integer variables. Specifically, if the bounds on an integer variable x are

$$0 \leq x \leq u, \qquad \text{where } 2^N \leq u < 2^{N+1},$$

then its **binary representation** is

$$x = \sum_{i=0}^{N} 2^i y_i,$$

where the y_i variables are (auxiliary) binary variables. Substituting this binary representation for each of the general integer variables (with a different set of auxiliary binary variables for each) thereby reduces the entire problem to a BIP model.

For example, suppose that an IP problem has just two general integer variables, x_1 and x_2, along with many binary variables, and that the functional constraints include

$$x_1 \qquad \leq 5$$

$$2x_1 + 3x_2 \leq 30.$$

Using $u = 5$ for x_1 and $u = 10$ for x_2, their binary representations become

$$x_1 = y_0 + 2y_1 + 4y_2$$

$$x_2 = y_3 + 2y_4 + 4y_5 + 8y_6.$$

After substituting these expressions for the respective variables throughout all the functional constraints and the objective function, the two functional constraints noted above become

$$y_0 + 2y_1 + 4y_2 \qquad\qquad\qquad \leq 5$$

$$2y_0 + 4y_1 + 8y_2 + 3y_3 + 6y_4 + 12y_5 + 24y_6 \leq 30.$$

Note that each feasible value of x_1 corresponds to one of the feasible values of the vector (y_0, y_1, y_2), and similarly for x_2 and (y_3, y_4, y_5, y_6).

For an IP problem where *all* the variables are (bounded) general integer variables, it would be possible to use this same technique to reduce the problem to a BIP model. However, this is not advisable for most cases because of the explosion in the number of variables involved. Applying a good IP algorithm to the original IP model generally should be more efficient than applying a good BIP algorithm to the much larger BIP model.

In general terms, for *all* the formulation possibilities with auxiliary binary variables discussed in this section, we need to strike the same note of caution. This approach sometimes requires adding a relatively large number of such variables, which can make the model *computationally infeasible*. In fact, as the next section explains, you may even be in trouble with less than a hundred binary variables.

13.3 Some Perspectives on Solving Integer Programming Problems

It may seem that IP problems should be relatively easy to solve. After all, *linear programming* problems can be solved extremely efficiently, and the only difference is that IP problems have far fewer solutions to be considered. In fact, *pure* IP problems with a bounded feasible region are guaranteed to have just a *finite* number of feasible solutions.

Unfortunately, there are two fallacies in this line of reasoning. One is that having a finite number of feasible solutions ensures that the problem is readily solvable. Finite numbers can be astronomically large. For example, consider the simple case of BIP problems. With n variables, there are 2^n solutions to be considered (where some of these solutions can subsequently be discarded because they violate the functional constraints). Thus, each time n is increased by *one*, the number of solutions is *doubled*. This pattern is referred to as the **exponential growth** of the difficulty of the problem. With $n = 10$, there are more than a *thousand* solutions (1,024); with $n = 20$, there are more than a *million*; with $n = 30$, there are more than a *billion*; and so forth. Therefore, even the fastest computers are incapable of performing *exhaustive enumeration* (checking each solution for feasibility and, if it is feasible, calculating the value of the objective value) for BIP problems with more than a few dozen variables, let alone for *general* IP problems with the same number of integer variables. Sophisticated algorithms, such as those described in subsequent sections, can do somewhat better. In fact, Sec. 13.6 discusses how recently developed algorithms have successfully solved certain *vastly* larger BIP problems (up to 2,756 variables). Nevertheless, because of *exponential growth,* even the best algorithms cannot be guaranteed to solve every relatively small problem (less than a hundred binary or integer variables).

The second fallacy is that removing some feasible solutions (the noninteger ones) from a linear programming problem will make it easier to solve. To the contrary, it is only because all of these feasible solutions are there that the guarantee can be given (see Sec. 5.1) that there will be a corner-point feasible solution (basic feasible solution) that is optimal for the overall problem. *This* guarantee is the key to the remarkable efficiency of the simplex method. As a result, linear programming problems generally are *much* easier to solve than IP problems.

Consequently, most successful algorithms for integer programming incorporate the simplex method (or dual simplex method) as much as they can by relating portions of the IP problem under consideration to the corresponding linear programming problem (i.e., the same problem except that the integer restriction is deleted). For any given IP problem, this corresponding linear programming problem commonly is referred to as its **LP-relaxation**. The algorithms presented in the next two sections illustrate how a sequence of LP-relaxations for portions of an IP problem can be used to solve the overall IP problem effectively.

There is one special situation where solving an IP problem is no more difficult than solving its LP-relaxation once by the simplex method, namely, when the optimal solution to the latter problem turns out to satisfy the integer restriction of the IP problem. When this situation occurs, this solution *must* be optimal for the IP problem as well, because it is the best solution among all the feasible solutions for the LP-relaxation, which includes all the feasible solutions for the IP problem. Therefore, it is common for an IP algorithm to begin by applying the simplex method to the LP-relaxation to check whether this fortuitous outcome has occurred.

Although it generally is quite fortuitous indeed for the optimal solution to the LP-relaxation to be integer as well, there actually exist several *special types* of IP problems for which this outcome is *guaranteed*. You already have seen the most prominent of these special types in Chaps. 7 and 10, namely, the *minimum cost flow problem* (with integer parameters) and its special cases (the *transportation problem,* the *transshipment problem,* the *assignment problem,* the *shortest path problem,* and the *maximum flow problem*). The reason that this guarantee can be given for these types of problems is that they possess a certain *special structure* (e.g., see Table 7.6) that ensures that every basic feasible solution is integer, as stated in the *integer solutions property* given in Secs. 7.1 and 10.6. Consequently, these special types of IP problems can be treated like linear programming problems (which is why three of them appear in Chap. 7), because they can be solved completely by a streamlined version of the simplex method.

Although this much simplification is somewhat unusual, in practice IP problems frequently have *some* special structure that can be exploited to simplify the problem. Sometimes, very large versions of these problems can be solved successfully. *Special-purpose algorithms* designed specifically to exploit certain kinds of special structures are becoming increasingly important in integer programming.

Thus the two primary determinants of *computational difficulty* for an IP problem are (1) the *number of integer variables* and (2) the *structure* of the problem. This situation is in contrast to *linear programming,* where the number of (functional) constraints is much more important than the number of variables. In integer programming, the number of constraints is of *some* importance (especially if LP-relaxations are being solved), but it is strictly secondary to the other two factors. In fact, there occasionally are cases where *increasing* the number of constraints *decreases* the computation time because the number of feasible solutions has been reduced. For MIP problems, it is the number of *integer* variables rather than the *total* number of variables that is important, because the *continuous* variables have almost no effect on computational effort.

Because IP problems are, in general, much more difficult to solve than linear programming problems, it sometimes is tempting to use the approximate procedure of simply applying the simplex method to the LP-relaxation and then *rounding* the noninteger values to integers in the resulting solution. This approach may be adequate for some applications, especially if the values of the variables are quite large so that rounding creates relatively little error. However, you should beware of two pitfalls involved in this approach.

One pitfall is that the optimal linear programming solution is *not necessarily feasible* after it is rounded. Often it is difficult to see in which way the rounding should be done to retain feasibility. It may even be necessary to change the value of some variables by one or more units after rounding. To illustrate, suppose that some of the constraints are

$$-x_1 + x_2 \leq 3\tfrac{1}{2}$$

$$x_1 + x_2 \leq 16\tfrac{1}{2}$$

and that the simplex method has identified the optimal solution for the LP-relaxation as $x_1 = 6\tfrac{1}{2}$, $x_2 = 10$. Notice that it is impossible to round x_1 to 6 or 7 (or any other integer) and retain feasibility. Feasibility can be retained only by also changing the integer value of x_2. It is easy to imagine how such difficulties can be compounded when there are tens or hundreds of constraints and variables.

Even if the optimal solution for the LP-relaxation is rounded successfully, there remains another pitfall. There is no guarantee that this rounded solution will be the optimal integer solution. In fact, it may even be far from optimal in terms of the value of the objective function. This fact is illustrated by the following problem:

$$\text{Maximize} \quad Z = x_1 + 5x_2,$$

subject to

$$x_1 + 10x_2 \leq 20$$

$$x_1 \qquad \leq 2$$

and

$$x_1 \geq 0, \qquad x_2 \geq 0$$

$$x_1, x_2 \text{ are integers.}$$

Because there are only two decision variables, this problem can be depicted graphically as shown in Fig. 13.1. Either the graph or the simplex method may be used to find that the optimal solution for the LP-relaxation is $x_1 = 2$, $x_2 = \frac{9}{5}$, with $Z = 11$. If a graphical solution were not available (which it would not be with more decision variables), then the variable with the noninteger value $x_2 = \frac{9}{5}$ would normally be rounded in the feasible direction to $x_2 = 1$. The resulting integer solution is $x_1 = 2$, $x_2 = 1$, which yields $Z = 7$. Notice that this solution is far from the optimal solution $(x_1, x_2) = (0, 2)$, where $Z = 10$.

Because of these two pitfalls, a better approach for dealing with IP problems that are too large to be solved exactly is to use one of the available *heuristic algorithms*. These algorithms are extremely efficient for large problems, but they are not guaranteed to find an optimal solution. However, they do tend to be considerably more effective than the rounding approach just discussed in finding very good feasible solutions.

For IP problems that are small enough to be solved to optimality, a considerable number of algorithms now are available. Until recently, no IP algorithms possessed computational efficiency that was even remotely comparable to the *simplex method* (except on special types of problems). Therefore, developing IP algorithms has continued to be an active area of research. Fortunately, some exciting algorithmic advances were made during the mid- and late 1980s, and additional progress can be

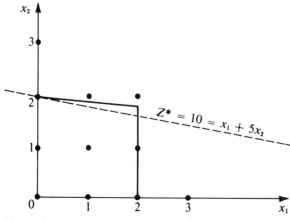

Figure 13.1 Illustrative integer programming problem.

anticipated during the next decade. These recent advances are discussed further in

Sec. 13.6.

489

Integer Programming

The most popular mode for IP algorithms is to use the *branch-and-bound technique* and related ideas to *implicitly enumerate* the feasible integer solutions, and we shall focus on this approach. The next section presents the branch-and-bound technique in a general context, and illustrates it with a basic branch-and-bound algorithm for BIP problems. Section 13.5 presents another algorithm of the same type for general MIP problems.

13.4 The Branch-and-Bound Technique and Its Application to Binary Integer Programming

Because any bounded *pure* IP problem has only a finite number of feasible solutions, it is natural to consider using some kind of *enumeration procedure* for finding an optimal solution. Unfortunately, as we discussed in the preceding section, this finite number can be, and usually is, very large. Therefore, it is imperative that any enumeration procedure be cleverly structured so that only a tiny fraction of the feasible solutions actually need be examined. For example, dynamic programming (see Chap. 11) provides one such kind of procedure for many problems having a finite number of feasible solutions (although it is not particularly efficient for most IP problems). Another such approach is provided by the *branch-and-bound technique*. This technique and variations of it have been applied with some success to a variety of operations research problems, but it is especially well known for its application to IP problems.

The basic concept underlying the branch-and-bound technique is to *divide and conquer*. Since the original "large" problem is too difficult to be solved directly, it is *divided* into smaller and smaller subproblems until these subproblems can be *conquered*. The dividing (*branching*) is done by partitioning the entire set of feasible solutions into smaller and smaller subsets. The conquering (*fathoming*) is done partially by *bounding* how good the best solution in the subset can be, and then discarding the subset if its bound indicates that it cannot possibly contain an optimal solution for the original problem.

We shall now describe in turn these three basic steps—branching, bounding, and fathoming—and illustrate them by applying a branch-and-bound algorithm to the prototype example (the California Manufacturing Co. problem) presented in Sec. 13.1.

Branching

When dealing with binary variables, the most straightforward way to partition the set of feasible solutions into subsets is to fix the value of one of the variables (say, x_1) at $x_1 = 0$ for one subset and at $x_1 = 1$ for the other subset. Doing this for the prototype example divides the whole problem into the two smaller subproblems shown below.

Subproblem 1: ($x_1 = 0$)

$$\text{Maximize} \quad Z = 5x_2 + 6x_3 + 4x_4,$$

subject to
$$3x_2 + 5x_3 + 2x_4 \le 10$$
$$x_3 + x_4 \le 1$$
$$x_3 \le 0$$
$$-x_2 + x_4 \le 0$$

and $\quad x_j$ is binary, $\quad$ for $j = 2, 3, 4$.

Subproblem 2: $(x_1 = 1)$

$$\text{Maximize} \quad Z = 9 + 5x_2 + 6x_3 + 4x_4,$$

subject to
$$3x_2 + 5x_3 + 2x_4 \le 4$$
$$x_3 + x_4 \le 1$$
$$x_3 \le 1$$
$$-x_2 + x_4 \le 0$$

and $\quad x_j$ is binary, $\quad$ for $j = 2, 3, 4$.

Figure 13.2 portrays this dividing (branching) into subproblems by a *tree* (defined in Sec. 10.2) with *branches* (arcs) from the *All* node (corresponding to the whole problem having *All* feasible solutions) to the two nodes corresponding to the two subproblems. This tree, which will continue "growing branches" iteration by iteration, is referred to as the **solution tree** (or enumeration tree) for the algorithm. The variable used to do this branching at any iteration by assigning values to the variable (as with x_1 above) is called the **branching variable**.

Later in the section you will see that one of these subproblems can be conquered (fathomed) immediately, whereas the other subproblem will need to be divided further into smaller subproblems by setting $x_2 = 0$ or $x_2 = 1$, etc.

For other IP problems where the integer variables have more than two possible values, the branching can still be done by setting the branching variable at its respective individual values, thereby creating more than two new subproblems. However, a good alternate approach is to specify a *range* of values (e.g., $x_j \le 2$ or $x_j \ge 3$) for the branching variable for each new subproblem. This is the approach used for the algorithm presented in Sec. 13.5.

Variable: $\quad x_1$

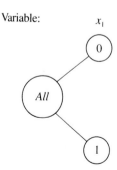

Figure 13.2 The solution tree created by the branching for the first iteration of the BIP branch-and-bound algorithm for the example.

For each of these subproblems, we now need to obtain a *bound* on how good its best feasible solution can be. The standard way of doing this is to quickly solve a simpler *relaxation* of the subproblem. In most cases, a **relaxation** of a problem is obtained simply by *deleting* ("relaxing") one set of constraints that had made the problem difficult to solve. For IP problems, the most troublesome constraints are those requiring the respective variables to be integer. Therefore, the most widely used relaxation is the *LP-relaxation* that deletes this set of constraints.

To illustrate for the example, consider first the whole problem given in Sec. 13.1. Its LP-relaxation is obtained by deleting the last line of the model (x_j is an integer, for $j = 1, 2, 3, 4$), but retaining the $x_j \leq 1$ and $x_j \geq 0$ constraints. Using the simplex method to quickly solve this LP-relaxation yields its optimal solution,

$$(x_1, x_2, x_3, x_4) = (\tfrac{5}{6}, 1, 0, 1), \quad \text{with } Z = 16\tfrac{1}{2}.$$

Therefore, $Z \leq 16\tfrac{1}{2}$ for all feasible solutions for the original BIP problem (since these solutions are a subset of the feasible solutions for the LP-relaxation). In fact, as summarized below, this *bound* of $16\tfrac{1}{2}$ can be rounded down to 16, because all coefficients in the objective function are integer, so all integer solutions must have an integer value for Z.

$$\text{Bound for whole problem:} \quad Z \leq 16.$$

Now let us obtain the bounds for the two subproblems in the same way. Their LP-relaxations are obtained from the models in the preceding subsection by replacing the constraints, x_j is binary for $j = 2, 3, 4$, by $0 \leq x_j \leq 1$ for $j = 2, 3, 4$. Applying the simplex method then yields their optimal solutions (plus the fixed value of x_1) shown below.

LP-relaxation of Subproblem 1: $(x_1, x_2, x_3, x_4) = (0, 1, 0, 1)$, with $Z = 9$.
LP-relaxation of Subproblem 2: $(x_1, x_2, x_3, x_4) = (1, \tfrac{4}{5}, 0, \tfrac{4}{5})$, with $Z = 16\tfrac{1}{5}$.

The resulting *bounds* for the subproblems then are

Bound for Subproblem 1: $Z \leq 9$,
Bound for Subproblem 2: $Z \leq 16$.

Figure 13.3 summarizes these results, where the numbers given just below the nodes are the *bounds,* and below each bound is the optimal solution obtained for the LP-relaxation.

Fathoming

A subproblem can be conquered (fathomed), and thereby dismissed from further consideration, in the three ways described below.

One way is illustrated by the results for Subproblem 1 given by the $x_1 = 0$ node in Fig. 13.3. Note that the (unique) optimal solution for its LP-relaxation, $(x_1, x_2, x_3, x_4) = (0, 1, 0, 1)$, is an *integer* solution. Therefore, this solution must also be the optimal solution for Subproblem 1 itself. This solution should be stored as the first **incumbent** (the best feasible solution found so far) for the whole problem, along with its value of Z. This value is denoted by

$$Z^* = \text{value of } Z \text{ for the current incumbent,}$$

Variable: x_1

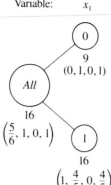

Figure 13.3 The results of bounding for the first iteration of the BIP branch-and-bound algorithm for the example.

so $Z^* = 9$ at this point. Having stored this solution, there is no reason to consider Subproblem 1 any further by branching from the $x_1 = 0$ node, etc. Doing so could only lead to other feasible solutions that are inferior to the incumbent, and we have no interest in such solutions. Because it has been solved, we *fathom* (dismiss) Subproblem 1 now.

The above results suggest a second key fathoming test. Since $Z^* = 9$, there is no reason to consider further any subproblem whose *bound* ≤ 9, since such a subproblem cannot have a feasible solution better than the *incumbent*. Stated more generally, a subproblem is fathomed whenever its

$$\text{Bound} \leq Z^*.$$

This outcome does not occur in the current iteration of the example because Subproblem 2 has a bound of 16 that is larger than 9. However, it might occur later for **descendants** of this subproblem (new smaller subproblems created by branching on this subproblem, and then perhaps branching further through subsequent "generations"). Furthermore, as new incumbents with larger values of Z^* are found, it will become easier to fathom in this way.

The third way of fathoming is quite straightforward. If the simplex method finds that a subproblem's LP-relaxation has *no feasible solutions,* then the subproblem itself must have *no feasible solutions,* so it can be dismissed (fathomed).

In all three cases, we are conducting our search for an optimal solution by retaining for further investigation only those subproblems that could possibly have a feasible solution better than the current incumbent.

Summary of Fathoming Tests

A subproblem is *fathomed* (dismissed from further consideration) if
   Test 1: Its bound $\leq Z^*$,
   or
   Test 2: Its LP-relaxation has no feasible solutions,
   or
   Test 3: The optimal solution for its LP-relaxation is *integer*. (If this solution is better than the incumbent, it becomes the new incumbent, and test 1 is reapplied to all unfathomed subproblems with the new larger Z^*.)

Variable: x_1

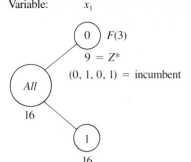

Figure 13.4 The solution tree after the first iteration of the BIP branch-and-bound algorithm for the example.

Figure 13.4 summarizes the results of applying these three tests to Subproblems 1 and 2 by showing the current *solution tree*. Only Subproblem 1 has been fathomed, by test 3, as indicated by the $F(3)$ next to the $x_1 = 0$ node. The resulting incumbent also is identified below this node.

The subsequent iterations will illustrate successful applications of all three tests. However, before continuing the example, let us summarize the algorithm being applied to this BIP problem. (This algorithm assumes that all coefficients in the objective function are integer, and that the ordering of the variables for branching is $x_1, x_2, \ldots, x_n$.)

Summary of BIP Branch-and-Bound Algorithm

Initialization Step: Set $Z^* = -\infty$. Apply the bounding step, fathoming step, and optimality test described below to the whole problem. If not fathomed, classify this problem as the one remaining "subproblem" for performing the first full iteration below.

Steps for Each Iteration:

1. *Branching:* Among the *remaining* (unfathomed) subproblems, select the one that was created *most recently*. (Break ties according to which has the *larger bound*.) Branch from the node for this subproblem to create two new subproblems by fixing the next variable (the branching variable) at either 0 or 1.

2. *Bounding:* For each new subproblem, obtain its *bound* by applying the simplex method to its LP-relaxation and rounding down the value of Z for the resulting optimal solution.

3. *Fathoming:* For each new subproblem, apply the three fathoming tests summarized above, and discard those subproblems that are fathomed by any of the tests.

Optimality Test: Stop when there are *no remaining* subproblems; the current *incumbent* is optimal.[1] Otherwise, return to perform another iteration.

The branching step for this algorithm warrants a comment as to why the subproblem to branch from is selected in this way. One option not used would have been

---

[1] If there is no incumbent, the conclusion is that the problem has no feasible solutions.

always to select the remaining subproblem with the *best bound,* because this sub-problem would be the most promising one to contain an optimal solution for the whole problem. The reason for using instead the option of selecting the *most recently created* subproblem is that *LP-relaxations* are being solved in the bounding step. Rather than starting the simplex method from scratch each time, each LP-relaxation generally is solved by *reoptimization* in large-scale implementations of this algorithm. This reop-timization involves revising the final simplex tableau from the preceding LP-relaxation as needed because of the few differences in the model (just as for sensitivity analysis), and then applying a few iterations of perhaps the dual simplex method. This reopti-mization tends to be *much* faster than starting from scratch, *provided* the preceding and current models are closely related. The models will tend to be closely related under the branching rule used, but *not* when you are skipping around in the solution tree by selecting the subproblem with the best bound.

Completing the Example

The pattern for the remaining iterations will be quite similar to that for the first iteration described above except for the ways in which fathoming occurs. Therefore, we shall summarize the branching and bounding steps fairly briefly and then focus on the fathoming step.

ITERATION 2: The only remaining subproblem corresponds to the $x_1 = 1$ node in Fig. 13.4, so we shall branch from this node to create the two new subproblems given below.

Subproblem 3: $(x_1 = 1, x_2 = 0)$

$$\text{Maximize} \quad Z = 9 + 6x_3 + 4x_4,$$

subject to

$$5x_3 + 2x_4 \le 4$$
$$x_3 + x_4 \le 1$$
$$x_3 \quad\ \le 1$$
$$x_4 \le 0$$

and

$$x_j \text{ is binary}, \quad \text{for } j = 3, 4.$$

Subproblem 4: $(x_1 = 1, x_2 = 1)$

$$\text{Maximize} \quad Z = 14 + 6x_3 + 4x_4,$$

subject to

$$5x_3 + 2x_4 \le 1$$
$$x_3 + x_4 \le 1$$
$$x_3 \quad\ \le 1$$
$$x_4 \le 0$$

and

$$x_j \text{ is binary}, \quad \text{for } j = 3, 4.$$

The LP-relaxations of these subproblems are obtained by replacing the con-straints, x_j is binary for $j = 3, 4$, by $0 \le x_j \le 1$ for $j = 3, 4$. Their optimal solutions (plus the fixed values of x_1 and x_2) are

LP-relaxation of Subproblem 3: $(x_1, x_2, x_3, x_4) = (1, 0, \frac{4}{5}, 0)$, with $Z = 13\frac{4}{5}$,
LP-relaxation of Subproblem 4: $(x_1, x_2, x_3, x_4) = (1, 1, 0, \frac{1}{2})$, with $Z = 16$.

The resulting bounds for the subproblems are

Bound for Subproblem 3: $Z \leq 13$,
Bound for Subproblem 4: $Z \leq 16$.

Note that both of these bounds are larger than $Z^* = 9$, so fathoming test 1 fails in both cases. Test 2 also fails, since both LP-relaxations have feasible solutions (as indicated by the existence of an optimal solution). Alas, test 3 fails as well, because both optimal solutions include variables with noninteger values.

Figure 13.5 shows the resulting solution tree at this point. The lack of an F to the right of either new node indicates that both remain unfathomed.

ITERATION 3: So far, the algorithm has created four subproblems. Subproblem 1 has been fathomed, and Subproblem 2 has been replaced by (separated into) Subproblems 3 and 4, but these latter two remain under consideration. Because they were created simultaneously, but Subproblem 4 ($x_1 = 1$, $x_2 = 1$) has the larger *bound* ($16 > 13$), the next branching is done from the $(x_1, x_2) = (1, 1)$ node in the solution tree, which creates the following new subproblems.

Subproblem 5: ($x_1 = 1, x_2 = 1, x_3 = 0$)

$$\text{Maximize} \quad Z = 14 + 4x_4,$$

subject to

$$2x_4 \leq 1$$

$$x_4 \leq 1 \quad \text{(twice)}$$

and

$$x_4 \text{ is binary.}$$

Subproblem 6: ($x_1 = 1, x_2 = 1, x_3 = 1$)

$$\text{Maximize} \quad Z = 20 + 4x_4,$$

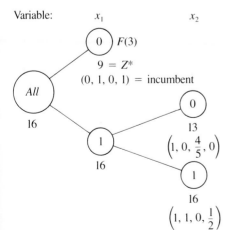

Figure 13.5 The solution tree after iteration 2 of the BIP branch-and-bound algorithm for the example.

$$\text{subject to} \qquad 2x_4 \leq -4$$

$$x_4 \leq 0$$

$$x_4 \leq 1$$

and $\qquad x_4$ is binary.

Forming their LP-relaxations by replacing x_4 is binary by $0 \leq x_4 \leq 1$, the following results are obtained:

LP-relaxation of Subproblem 5: $(x_1, x_2, x_3, x_4) = (1, 1, 0, \frac{1}{2})$, with $Z = 16$.
LP-relaxation of Subproblem 6: No feasible solutions.
Bound for Subproblem 5: $Z \leq 16$.

Note how the constraints, $2x_4 \leq -4$ and $x_4 \geq 0$, in the LP-relaxation of Subproblem 6 prevent any feasible solutions. Therefore, this subproblem is fathomed by test 2. However, Subproblem 5 fails this test, as well as test 1 ($16 > 9$) and test 3 ($x_4 = \frac{1}{2}$ is not integer), so it remains under consideration.

We now have the solution tree shown in Fig. 13.6.

ITERATION 4: The subproblems corresponding to nodes $(1, 0)$ and $(1, 1, 0)$ in Fig. 13.6 remain under consideration, but the latter node was created more recently, so it is selected for branching from next. Since the resulting branching variable x_4 is the *last* variable, fixing its value at either 0 or 1 actually creates a *single solution* rather than subproblems requiring fuller investigation. These single solutions are

$$x_4 = 0: (x_1, x_2, x_3, x_4) = (1, 1, 0, 0) \text{ is feasible, with } Z = 14,$$
$$x_4 = 1: (x_1, x_2, x_3, x_4) = (1, 1, 0, 1) \text{ is infeasible.}$$

Formally applying the fathoming tests, the first solution passes test 3 and the second passes test 2. Furthermore, this feasible first solution is better than the incumbent ($14 > 9$), so it becomes the new incumbent, with $Z^* = 14$.

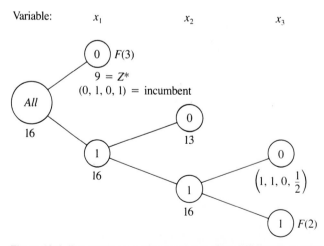

Figure 13.6 The solution tree after iteration 3 of the BIP branch-and-bound algorithm for the example.

Because a new incumbent has been found, we now reapply fathoming test 1 with the new larger value of $Z*$ to the only remaining subproblem, the one at node $(1, 0)$.

$$\textit{Subproblem 3:} \quad \text{Bound} = 13 \leq Z* = 14.$$

Therefore, this subproblem now is fathomed.

We now have the solution tree shown in Fig. 13.7. Note that there are *no remaining* (unfathomed) subproblems. Consequently, the *optimality test* indicates that the current *incumbent,*

$$(x_1, x_2, x_3, x_4) = (1, 1, 0, 0),$$

is optimal, so we are done.

Other Options with the Branch-and-Bound Technique

This section has illustrated the branch-and-bound technique by describing a basic branch-and-bound algorithm for solving BIP problems. However, the general framework of the branch-and-bound technique provides a great deal of flexibility in how to design a specific algorithm for any given type of problem such as BIP. There are many options available, and constructing an efficient algorithm requires tailoring the specific design to fit the specific structure of the problem type.

Every branch-and-bound algorithm has the same three basic steps of *branching, bounding,* and *fathoming.* The flexibility is in how these steps are performed.

Branching always involves *selecting* one remaining subproblem and *dividing* it into smaller subproblems. The flexibility here is in the rules for selecting and dividing. Our BIP algorithm selected the *most recently created* subproblem, because this is very efficient for *reoptimizing* each LP-relaxation from the preceding one. Selecting the subproblem with the *best bound* is the other most popular rule, because it tends to lead more quickly to better incumbents and so more fathoming. Combinations of the two rules also can be used. The *dividing* typically (but not always) is done by choosing

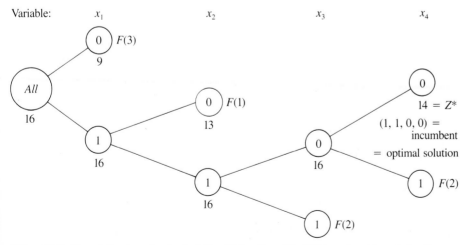

Figure 13.7 The solution tree after the final (fourth) iteration of the BIP branch-and-bound algorithm for the example.

a *branching variable* and assigning it either individual values (e.g., our BIP algorithm) or ranges of values (e.g., the algorithm in the next section). More sophisticated algorithms generally use a rule for strategically choosing a branching variable that should tend to lead to early fathoming.

Bounding usually is done by solving a *relaxation*. However, there are a variety of ways to form relaxations. For example, consider the **Lagrangian relaxation**, where the entire set of functional constraints, $\mathbf{Ax} \leq \mathbf{b}$ (in matrix notation), is *deleted* (except possibly for any ''convenient'' constraints), and then the objective function,

$$\text{Maximize} \quad Z = \mathbf{cx},$$

is replaced by

$$\text{Maximize} \quad Z_R = \mathbf{cx} - \boldsymbol{\lambda}(\mathbf{Ax} - \mathbf{b}),$$

where the fixed vector $\boldsymbol{\lambda} \geq \mathbf{0}$. If $\mathbf{x}^*$ is an optimal solution for the original problem, its $Z \leq Z_R$, so solving the Lagrangian relaxation for the optimal value of Z_R provides a valid *bound*. If $\boldsymbol{\lambda}$ is chosen well, this bound tends to be a reasonably tight one (at least comparable to the bound from the LP-relaxation). Without any functional constraints, this relaxation also can be solved extremely quickly. The drawbacks are that fathoming tests 2 and 3 (revised) are not as powerful as for the LP-relaxation. However, the two relaxations occasionally are used together to good advantage.

In general terms, the two features sought in choosing a relaxation are that it can be solved relatively quickly and that it provides a relatively tight bound. Neither alone is adequate. The LP-relaxation is popular because it provides an excellent trade-off between these two factors.

One option occasionally employed is to use a quickly solved relaxation and then, if fathoming is not achieved, to tighten the relaxation in some way to obtain a somewhat tighter bound.

Fathoming generally is done pretty much as described for the BIP algorithm. The three fathoming criteria can be stated in more general terms as follows.

Summary of Fathoming Criteria

A subproblem is *fathomed* if an analysis of its *relaxation* reveals that

Criterion 1: Feasible solutions of the subproblem must have $Z \leq Z^*$, or

Criterion 2: The subproblem has no feasible solution, or

Criterion 3: An optimal solution of the subproblem has been found.

Just as for the BIP algorithm, the first two criteria usually are applied by solving the relaxation to obtain a bound for the subproblem, and then checking whether this bound is $\leq Z^*$ (test 1) or whether the relaxation has no feasible solutions (test 2). If the relaxation differs from the subproblem *only* by the deletion (or loosening) of some constraints, then the third criterion usually is applied by checking whether the optimal solution for the relaxation is *feasible* for the subproblem, in which case it must be *optimal* for the subproblem. For other relaxations (such as the Lagrangian relaxation), additional analysis is required to determine whether the optimal solution for the relaxation is also optimal for the subproblem.

If the original problem involves *minimization* rather than maximization, two options are available. One is to convert to maximization in the usual way (see Sec. 4.6). The other is to convert the branch-and-bound algorithm directly into minimization form, where the most important adjustment is to change the direction of the

$$\text{Is the subproblem's bound} \leq Z^*?$$

to

$$\text{Is the subproblem's bound} \geq Z^*?$$

So far, we have described how to use the branch-and-bound technique to find only *one* optimal solution. However, in the case of ties for the optimal solution, it is sometimes desirable to identify *all* these optimal solutions so that the final choice among them can be made on the basis of intangible factors not incorporated into the mathematical model. To find them all, you need to make only a few slight alterations in the procedure. First, change the weak inequality for fathoming test 1 (Is the subproblem's bound $\leq Z^*?$) to a strict inequality (Is the subproblem's bound $< Z^*?$), so that fathoming will not occur if the subproblem can have a feasible solution *equal* to the incumbent. Second, if fathoming test 3 passes and the optimal solution for the subproblem has $Z = Z^*$, then store this solution as *another* (tied) incumbent. Third, if test 3 provides a new incumbent (tied or otherwise), then check whether the optimal solution obtained for the *relaxation* is *unique*. If it is not, then identify the other optimal solutions for the relaxation and check whether they are optimal for the subproblem as well, in which case they also become incumbents. Finally, when the *optimality test* finds that there are *no remaining* (unfathomed) subsets, *all* the current *incumbents* will be the *optimal* solutions.

Finally, it should be noted that rather than finding an optimal solution, the branch-and-bound technique can also be used to find a *nearly optimal* solution, generally with much less computational effort. For some applications, a solution is "good enough" if its Z is "close enough" to the value of Z for an optimal solution (call it Z^{**}). "Close enough" can be defined as either

$$Z \geq Z^{**} - K \quad \text{or} \quad Z \geq (1 - \alpha)Z^{**}$$

for a specified (positive) constant K or α. For example, if $\alpha = 0.05$, then the solution is required to be within 5 percent of optimal. To find a solution that is "close enough" to being optimal, only one change is needed in the usual branch-and-bound procedure. This change is to replace the usual fathoming test 1 for a subproblem,

$$\text{Bound} \leq Z^*?$$

by either

$$\text{Bound} - K \leq Z^*?$$

or

$$(1 - \alpha) \text{ bound} \leq Z^*?$$

and then perform this test *after* test 3 (so that a feasible solution found with $Z > Z^*$ is still kept as the new incumbent). The reason this weaker test 1 suffices is that regardless of how close Z for the subproblem's (unknown) optimal solution is to the subproblem's bound, the incumbent is still "close enough" to this solution (if the new inequality holds) that the subproblem does not need to be considered further. When there are no remaining subproblems, the current incumbent will be the desired *nearly optimal* solution. However, it is much easier to fathom with this new fathoming test (in either form), so the algorithm should run much faster. For a large problem, this acceleration may make the difference between finishing with a solution guaranteed to be close to optimal and never terminating.

13.5 A Branch-and-Bound Algorithm for Mixed Integer Programming

We shall now consider the general MIP problem, where *some* of the variables (say, I of them) are restricted to integer values (but not necessarily just 0 and 1), but the rest are ordinary continuous variables. For notational convenience, we shall order the variables so that the first I variables are the *integer-restricted* variables. Therefore, the general form of the problem being considered is

$$\text{Maximize} \quad Z = \sum_{j=1}^{n} c_j x_j,$$

subject to
$$\sum_{j=1}^{n} a_{ij} x_j \leq b_i, \quad \text{for } i = 1, 2, \ldots, m,$$

and
$$x_j \geq 0, \quad \text{for } j = 1, 2, \ldots, n,$$

$$x_j \text{ is integer}, \quad \text{for } j = 1, 2, \ldots, I \quad (I \leq n).$$

(When $I = n$, this problem becomes the *pure* IP problem.)

We shall describe a basic branch-and-bound algorithm for solving this problem that, with a variety of refinements, has provided the standard approach to MIP. The structure of this algorithm was first developed by R. J. Dakin,[1] based on a pioneering branch-and-bound algorithm by A. H. Land and A. G. Doig.[2]

This algorithm is quite similar in structure to the BIP algorithm presented in the preceding section. Solving *LP-relaxations* again provides the basis for both the *bounding* and *fathoming* steps. In fact, only four changes are needed in the BIP algorithm to deal with the generalizations from *binary* to *general* integer variables and from *pure* IP to *mixed* IP.

One change involves the choice of the *branching variable*. Before, the *next* variable in the natural ordering—$x_1, x_2, \ldots, x_n$—was chosen automatically. Now, the only variables considered are the *integer-restricted* variables that have a *noninteger* value in the optimal solution for the LP-relaxation of the current subproblem. Our rule for choosing among these variables is to select the *first* one in the natural ordering. (Production codes generally use a more sophisticated rule.)

The second change involves the values assigned to the branching variable for creating the new smaller subproblems. Before, the *binary* variable was fixed at 0 and 1, respectively, for the two new subproblems. Now, the *general* integer-restricted variable could have a very large number of possible integer values, and it would be inefficient to create *and* analyze *many* subproblems by fixing the variable at its individual integer values. Therefore, what is done instead is to create just *two* new subproblems (as before) by specifying two *ranges* of values for the variable.

To spell out how this is done, let x_j be the current branching variable, and let x_j^* be its (noninteger) value in the optimal solution for the LP-relaxation of the current subproblem. Using a square bracket to denote

$$[x_j^*] = \text{greatest integer less-than-or-equal-to } x_j^*,$$

[1] Dakin, R. J.: "A Tree Search Algorithm for Mixed Integer Programming Problems," *Computer Journal*, **8**(3):250–255, 1965.

[2] Land, A. H., and A. G. Doig: "An Automatic Method of Solving Discrete Programming Problems," *Econometrica*, **28**:497–520, 1960.

$$x_j \le [x_j^*] \quad \text{and} \quad x_j \ge [x_j^*] + 1,$$

respectively. Each inequality becomes an *additional constraint* for that new subproblem. For example, if $x_j^* = 3\frac{1}{2}$, then

$$x_j \le 3 \quad \text{and} \quad x_j \ge 4$$

are the respective additional constraints for the new subproblem.

When the two changes to the BIP algorithm described above are combined, an interesting phenomenon of a *recurring branching variable* can occur. To illustrate, let $j = 1$ in the above example where $x_j^* = 3\frac{1}{2}$, and consider the new subproblem where $x_1 \le 3$. When the LP-relaxation of a descendant of this subproblem is solved, suppose that $x_1^* = 1\frac{1}{4}$. Then x_1 *recurs* as the branching variable, and the two new subproblems created have the additional constraint, $x_1 \le 1$ and $x_1 \ge 2$, respectively (as well as the previous additional constraint, $x_1 \le 3$). Later, when the LP-relaxation for a descendant of, say, the $x_1 \le 1$ subproblem is solved, suppose that $x_1^* = \frac{3}{4}$. Then x_1 *recurs* again as the branching variable, and the two new subproblems created have $x_1 = 0$ (because of the new $x_1 \le 0$ constraint and the nonnegativity constraint on x_1) and $x_1 = 1$ (because of the new $x_1 \ge 1$ constraint and the previous $x_1 \le 1$ constraint).

The third change involves the *bounding step*. Before, with a *pure* IP problem and integer coefficients in the objective function, the value of Z for the optimal solution for the subproblem's LP-relaxation was *rounded down* to obtain the bound, because any feasible solution for the subproblem must have an *integer* Z. Now, with some of the variables *not* integer-restricted, the bound is the value of Z *without* rounding down.

The fourth (and final) change to the BIP algorithm to obtain our MIP algorithm involves fathoming test 3. Before, with a *pure* IP problem, the test was that the optimal solution for the subproblem's LP-relaxation is *integer*, since this ensures that the solution is feasible, and therefore optimal, for the subproblem. Now, with a *mixed* IP problem, the test requires only that the *integer-restricted* variables be *integer* in the optimal solution for the subproblem's LP-relaxation, because this suffices to ensure that the solution is feasible, and therefore optimal, for the subproblem.

Incorporating these four changes into the summary presented in the preceding section for the BIP algorithm yields the following summary for the new algorithm for MIP.

Summary of MIP Branch-and-Bound Algorithm

Initialization Step: Set $Z^* = -\infty$. Apply the bounding step, fathoming step, and optimality test described below to the whole problem. If not fathomed, classify this problem as the one remaining "subproblem" for performing the first full iteration below.

Steps for Each Iteration:

1. *Branching:* Among the *remaining* (unfathomed) subproblems, select the one that was created *most recently*. (Break ties according to which has the *larger bound*.) Among the *integer-restricted* variables that have a *noninteger* value in the optimal solution for the LP-relaxation of the subproblem, choose the *first one* in the natural ordering of the variables to be the *branching variable*. Let x_j be this variable, and x_j^* its value in this solution. Branch from the

node for the subproblem to create two new subproblems by adding the respective constraints, $x_j \leq [x_j^*]$ and $x_j \geq [x_j^*] + 1$.

2. *Bounding:* For each new subproblem, obtain its *bound* by applying the simplex method (or the dual simplex method when reoptimizing) to its LP-relaxation and using the value of Z for the resulting optimal solution.

3. *Fathoming:* For each new subproblem, apply the three fathoming tests given below, and discard those subproblems that are fathomed by any of the tests.

> *Test 1:* Its bound $\leq Z^*$, where Z^* is the value of Z for the current *incumbent.*
>
> *Test 2:* Its LP-relaxation has no feasible solutions.
>
> *Test 3:* The optimal solution for its LP-relaxation has *integer* values for the *integer-restricted* variables. (If this solution is better than the incumbent, it becomes the new incumbent and test 1 is reapplied to all unfathomed subproblems with the new larger Z^*.)

Optimality Test: Stop when there are *no remaining* subproblems; the current *incumbent* is optimal.[1] Otherwise, return to perform another iteration.

Example

We will now illustrate this algorithm by applying it to the following MIP problem:

$$\text{Maximize} \quad Z = 4x_1 - 2x_2 + 7x_3 - x_4,$$

subject to

$$x_1 \quad\quad + 5x_3 \quad\quad \leq 10$$

$$x_1 + x_2 - x_3 \quad\quad \leq 1$$

$$6x_1 - 5x_2 \quad\quad\quad \leq 0$$

$$-x_1 \quad\quad + 2x_3 - 2x_4 \leq 3$$

and

$$x_j \geq 0, \quad \text{for } j = 1, 2, 3, 4$$

$$x_j \text{ is an integer}, \quad \text{for } j = 1, 2, 3.$$

Note that the number of integer-restricted variables is $I = 3$, so x_4 is the only continuous variable.

INITIALIZATION STEP: After setting $Z^* = -\infty$, we form the LP-relaxation of this problem by *deleting* the set of constraints, x_j is an integer for $j = 1, 2, 3$. Applying the simplex method to this LP-relaxation yields its optimal solution below.

LP-relaxation of whole problem: $(x_1, x_2, x_3, x_4) = (\frac{5}{4}, \frac{3}{2}, \frac{7}{4}, 0)$, with $Z = 14\frac{1}{4}$.

Because it has *feasible* solutions and this optimal solution has *noninteger* values for its integer-restricted variables, the whole problem is not fathomed, so the algorithm continues with the first full iteration below.

ITERATION 1: In this optimal solution for the LP-relaxation, the *first* integer-restricted variable that has a noninteger value is $x_1 = \frac{5}{4}$, so x_1 becomes the branching

---

[1] If there is no incumbent, the conclusion is that the problem has no feasible solutions.

variable. Branching from the *All* node (*All* feasible solutions) with this branching variable then creates the following two subproblems:

Subproblem 1: Original problem plus additional constraint:

$$x_1 \leq 1.$$

Subproblem 2: Original problem plus additional constraint:

$$x_1 \geq 2.$$

Deleting the set of integer constraints again, and solving the resulting LP-relaxations of these two subproblems yields the following results.

> LP-relaxation of Subproblem 1: $(x_1, x_2, x_3, x_4) = (1, \frac{6}{5}, \frac{9}{5}, 0)$, with $Z = 14\frac{1}{5}$.
> Bound for Subproblem 1: $Z \leq 14\frac{1}{5}$.
> LP-relaxation of Subproblem 2: No feasible solutions.

This outcome for Subproblem 2 means that it is *fathomed* by test 2. However, just as for the whole problem, Subproblem 1 fails all fathoming tests.

These results are summarized in the *solution tree* shown in Fig. 13.8.

ITERATION 2: With only one remaining subproblem, corresponding to the $x_1 \leq 1$ node in Fig. 13.8, the next branching is from this node. Examining its LP-relaxation's optimal solution given below, this node reveals that the *branching variable* is x_2, because $x_2 = \frac{6}{5}$ is the first integer-restricted variable that has a noninteger value. Adding one of the constraints, $x_2 \leq 1$ or $x_2 \geq 2$, then creates the following two new subproblems.

Subproblem 3: Original problem plus additional constraints:

$$x_1 \leq 1$$

$$x_2 \leq 1.$$

Subproblem 4: Original problem plus additional constraints:

$$x_1 \leq 1$$

$$x_2 \geq 2.$$

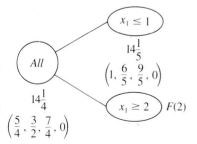

Figure 13.8 The solution tree after the first iteration of the MIP branch-and-bound algorithm for the example.

Solving their LP-relaxations gives the following results.

LP-relaxation of Subproblem 3: $(x_1, x_2, x_3, x_4) = (\frac{5}{6}, 1, \frac{11}{6}, 0)$, with $Z = 14\frac{1}{6}$.
Bound for Subproblem 3: $Z \leq 14\frac{1}{6}$.
LP-relaxation of Subproblem 4: $(x_1, x_2, x_3, x_4) = (\frac{5}{6}, 2, \frac{11}{6}, 0)$, with $Z = 12\frac{1}{6}$.
Bound for Subproblem 4: $Z \leq 12\frac{1}{6}$.

Because both solutions exist (feasible solutions) and have noninteger values for integer-restricted variables, neither subproblem is fathomed. (Test 1 still isn't operational, since $Z^* = -\infty$ until the first incumbent is found.)

The *solution tree* at this point is given in Fig. 13.9.

ITERATION 3: With two remaining subproblems (3 and 4) that were created simultaneously, the one with the *larger bound* (Subproblem 3, with $14\frac{1}{6} > 12\frac{1}{6}$) is selected for the next branching. Because $x_1 = \frac{5}{6}$ has a noninteger value in the optimal solution for this subproblem's LP-relaxation, x_1 becomes the *branching variable*. (Note that x_1 now is a *recurring* branching variable, since it also was chosen at iteration 1.) This leads to the following new subproblems.

Subproblem 5: Original problem plus additional constraints:

$$x_1 \leq 1$$

$$x_2 \leq 1$$

$$x_1 \leq 0 \qquad \text{(so } x_1 = 0\text{)}.$$

Subproblem 6: Original problem plus additional constraints:

$$x_1 \leq 1$$

$$x_2 \leq 1$$

$$x_1 \geq 1 \qquad \text{(so } x_1 = 1\text{)}.$$

The results from solving their LP-relaxations are given below.

LP-relaxation of Subproblem 5: $(x_1, x_2, x_3, x_4) = (0, 0, 2, \frac{1}{2})$, with $Z = 13\frac{1}{2}$.
Bound for Subproblem 5: $Z \leq 13\frac{1}{2}$.
LP-relaxation of Subproblem 6: No feasible solutions.

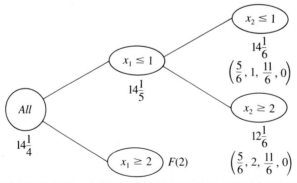

Figure 13.9 The solution tree after the second iteration of the MIP branch-and-bound algorithm for the example.

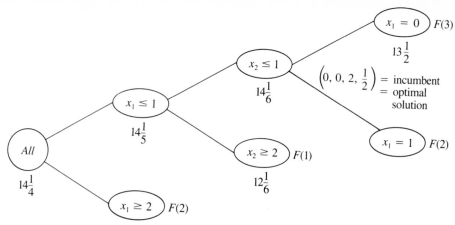

Figure 13.10 The solution tree after the final (third) iteration of the MIP branch-and-bound algorithm for the example.

Subproblem 6 is immediately fathomed by test 2. However, note that Subproblem 5 also can be fathomed. Test 3 passes because the optimal solution for its LP-relaxation has *integer* values ($x_1 = 0$, $x_2 = 0$, $x_3 = 2$) for all three *integer-restricted* variables. (It does not matter that $x_4 = \frac{1}{2}$, since x_4 is not integer-restricted.) This *feasible* solution for the original problem becomes our first incumbent:

$$\text{Incumbent} = (0, 0, 2, \tfrac{1}{2}), \text{ with } Z^* = 13\tfrac{1}{2}.$$

Using this Z^* to reapply fathoming test 1 to the only other subproblem (Subproblem 4) is successful, because its *bound* of $12\frac{1}{6}$ is $\leq Z^*$.

This iteration has succeeded in fathoming subproblems in all three possible ways. Furthermore, there now are *no remaining* subproblems, so the current *incumbent* is optimal.

$$\text{Optimal solution} = (0, 0, 2, \tfrac{1}{2}), \text{ with } Z = 13\tfrac{1}{2}.$$

These results are summarized by the final *solution tree* given in Fig. 13.10.

13.6 Recent Developments

Integer programming has been an especially exciting area of operations research in very recent years because of the dramatic progress being made in its solution methodology.

To place this progress into perspective, consider the historical background. One big breakthrough had come in the 1960s and early 1970s with the development and refinement of the branch-and-bound approach. But then the state of the art seemed to hit a plateau. Relatively small problems (well under a hundred variables) could be solved very efficiently, but even a modest increase in problem size might cause an explosion in computation time beyond feasible limits. Little progress was being made in overcoming this *exponential growth* in computation time as the problem size was increased. Many important problems arising in practice could not be solved.

Then came the next breakthrough in the mid-1980s, as reported largely in three papers published in 1983, 1985, and 1987. (See Selected References 1, 4, and 13.)

In the 1983 paper, Harlan Crowder, Ellis Johnson, and Manfred Padberg presented a new algorithmic approach to solving *pure* BIP problems that had successfully solved problems with no apparent special structure having up to 2,756 variables! This paper won the Lanchester Prize awarded by the Operations Research Society of America for the most notable publication in operations research during 1983. In the 1985 paper, Ellis Johnson, Michael Kostreva, and Uwe Suhl further refined this algorithmic approach.

However, both of these papers were limited to *pure* BIP. For IP problems arising in practice, it is quite common for all the integer-restricted variables to be *binary,* but a large proportion of these problems are *mixed* BIP problems. What was critically needed was a way of extending this same kind of algorithmic approach to *mixed* BIP. This came in the 1987 paper by Tony Van Roy and Laurence Wolsey of Belgium. Once again, problems of *very* substantial size (up to nearly a thousand binary variables and a larger number of continuous variables) were being solved successfully. And once again, this paper won a very prestigious award, the Orchard-Hays Prize given triannually by the Mathematical Programming Society.

We do need to add one note of caution. It is not yet clear just how *consistently* this algorithmic approach can successfully solve a wide variety of problems of this kind of very substantial size. The very large pure BIP problems solved had *sparse* **A** matrices; i.e., the percentage of coefficients in the functional constraints that were *nonzeroes* was quite small (perhaps less than 5 percent). In fact, the approach depends heavily upon this sparsity. (Fortunately, this kind of sparsity is typical in large practical problems.) Furthermore, there are other important factors besides sparsity and size that affect just how difficult a given IP problem will be to solve. It appears that IP formulations of fairly substantial size should still be approached with considerable caution.

On the other hand, each new algorithmic breakthrough in operations research always generates a flurry of new research activity to try to develop and refine the new approach further. We will undoubtedly see some further fruits of intensified research activity in integer programming over the next decade. Perhaps through this research the gap in efficiency between *integer* programming and *linear* programming algorithms can be further closed.

Although it would be beyond the scope and level of this book to describe the new algorithmic approach fully, we will now give a very brief overview. (You are encouraged to read Selected References 1, 4, and 13 for further information.)

The approach uses a combination of three kinds of techniques: (1) *automatic problem preprocessing,* (2) the *generation of cutting planes,* and (3) clever *branch-and-bound* techniques. You already are familiar with branch-and-bound techniques, and we will not elaborate further on the more advanced versions incorporated here. A conceptual introduction to the other two kinds of techniques is given below.

Automatic problem preprocessing involves a "computer inspection" of the user-supplied formulation of the IP problem in order to spot reformulations that make the problem quicker to solve without eliminating any feasible solutions. Some of the ideas are very simple, as we will now illustrate with *pure binary* variables. It may be possible to *fix* a variable at either 0 or 1 because of some constraint, e.g.,

$$3x_1 + 2x_2 \leq 2 \quad \Rightarrow \quad x_1 = 0$$

$$3x_3 - 2x_4 \leq -1 \quad \Rightarrow \quad x_3 = 0 \quad and \quad x_4 = 1,$$

so the variable can then be deleted from the model (after substituting its fixed value).
A functional constraint may be *redundant* because of the binary constraints, e.g.,

$$3x_1 + 2x_2 \leq 6,$$

and so can be deleted. It may be possible to *tighten* a functional constraint by reducing some of its coefficients because of the binary constraints, e.g.,

$$4x_1 + 5x_2 + x_3 \geq 2 \quad \Rightarrow \quad 2x_1 + 2x_2 + x_3 \geq 2.$$

Doing so has the major advantage of *tightening* the LP-relaxation (eliminating some of its feasible solutions) *without* eliminating any feasible solutions for the BIP problem. Another technique is to prepare a "trigger" that will fix the value of one or more variables after another variable has been set at a certain value during the course of the algorithm. For example, for the group of variables representing a set of *mutually exclusive alternatives,* as soon as one of the variables is set equal to 1, the rest can be fixed at 0. Such a trigger can also be prepared for constraints representing *contingent decisions* and other constraints with a similar form.

Using the computer to implement these and similar ideas *automatically* has proven helpful in accelerating an algorithm. However, an even more important technique is the *generation of cutting planes*. A **cutting plane** for an IP problem is a new constraint that *eliminates* some *feasible* solutions for the original LP-relaxation, *including* its optimal solution, *without* eliminating any feasible solutions for the IP problem. The purpose of adding new constraints with this property is to *tighten* the LP-relaxation, thereby tightening the bound obtained in the bounding step of the branch-and-bound technique. Tightening the LP-relaxation also increases the chance that its optimal solution will be feasible (and so optimal) for the IP problem, thereby improving fathoming test 3.

To illustrate a cutting plane, consider the California Manufacturing Co. pure BIP problem presented in Sec. 13.1 and used to illustrate the BIP branch-and-bound algorithm in Sec. 13.4. The optimal solution for its LP-relaxation is given in Fig. 13.3 as $(x_1, x_2, x_3, x_4) = (\frac{5}{6}, 1, 0, 1)$. One of the functional constraints is

$$6x_1 + 3x_2 + 5x_3 + 2x_4 \leq 10.$$

Now note that the binary constraints and this constraint together imply that

$$x_1 + x_2 \quad + x_4 \leq 2.$$

This new constraint is a *cutting plane*. It "cuts off" the optimal solution for the LP-relaxation $(\frac{5}{6}, 1, 0, 1)$, but it does not eliminate any feasible *integer* solutions. Adding just this one cutting plane to the original model would improve the performance of the BIP branch-and-bound algorithm in Sec. 13.4 (see Fig. 13.7) in two ways. First, the optimal solution for the new (tighter) LP-relaxation would be $(1, 1, \frac{1}{5}, 0)$, with $Z = 15\frac{1}{5}$, so the bounds for the *All*, $x_1 = 1$, and $x_2 = 1$ nodes now would be 15 instead of 16. Second, one less iteration would be needed because the optimal solution for the LP-relaxation at the $x_3 = 0$ node now would be $(1, 1, 0, 0)$, which provides a new *incumbent* with $Z^* = 14$. Therefore, on the *third* iteration (see Fig. 13.6), this node would be fathomed by test 3 and the $x_2 = 0$ node would be fathomed by test 1, thereby revealing that this incumbent is the optimal solution for the original BIP problem.

The new algorithmic approach presented in Selected References 1, 4, and 13 involves generating *many* cutting planes in a similar manner before then applying clever branch-and-bound techniques. The results of including the cutting planes can be quite dramatic in tightening the LP-relaxations. For example, for the test problem with 2,756 binary variables considered in the 1983 paper, 326 cutting planes were generated. The result was that the *gap* between Z for the optimal solution for the LP-relaxation of the whole BIP problem and Z for this problem's optimal solution was reduced by 98 percent. Similar results were obtained on about half of the problems.

Ironically, the very first algorithms developed for integer programming, including Ralph Gomory's celebrated algorithm announced in 1958, were based on cutting planes (generated in a different way), but this approach proved to be unsatisfactory in practice (except for special classes of problems). However, these algorithms relied *solely* on cutting planes. We now know that judiciously *combining* cutting planes and branch-and-bound techniques (along with automatic problem preprocessing) provides a powerful algorithmic approach for solving large-scale BIP problems.

13.7 Conclusions

IP problems arise frequently because some or all of the decision variables must be restricted to integer values. There also are many applications involving yes-or-no decisions (including combinatorial relationships expressible in terms of such decisions) that can be represented by binary (0–1) variables. These problems are more difficult than they would be without the integer restriction, so the algorithms available for integer programming are generally much less efficient than the simplex method. The most important determinants of computation time are the *number of integer variables* and the *structure* of the problem. For a fixed number of integer variables, BIP problems generally are much easier to solve than problems with general integer variables, but adding continuous variables (MIP) may not increase computation time substantially. For special types of BIP problems containing a special structure that can be exploited by a *special-purpose algorithm,* it may be possible to solve very large problems (well over a thousand binary variables) routinely. Other much smaller problems without such special structure may not be solvable.

Computer codes for IP algorithms now are commonly available in mathematical programming software packages. These algorithms usually are based on the *branch-and-bound* technique and variations thereof.

It appears that a new era in IP solution methodology has now been ushered in by a series of landmark papers in the mid-1980s. The new algorithmic approach involves combining automatic problem preprocessing, the generation of cutting planes, and clever branch-and-bound techniques. Research in this area is continuing.

IP problems arising in practice sometimes are so large that they cannot be solved by even the latest algorithms. In these cases, it is common to simply apply the simplex method to the LP-relaxation and then round the optimal solution to a feasible integer solution. However, this approach is sometimes quite unsatisfactory because it may be difficult (or impossible) to find a feasible integer solution in this way, and the solution found may be far from optimal. This is especially true when dealing with binary variables or even general integer variables with small values.

To circumvent these difficulties with rounding, considerable progress has been made in developing *efficient heuristic algorithms*. Even with very large IP problems, these algorithms generally will quickly find very good feasible solutions that are not necessarily optimal but usually are better than those that can be found by simple rounding.

In recent years, there has been considerable investigation into the development of algorithms for integer *nonlinear* programming, and this area continues to be a very active area of research.

SELECTED REFERENCES

1. Crowder, H., E. L. Johnson, and M. Padberg: "Solving Large-Scale Zero-One Linear Programming Problems," *Operations Research,* **31**:803–834, 1983.

2. Garfinkel, Robert S., and George L. Nemhauser: *Integer Programming,* Wiley, New York, 1972.

3. Geoffrion, A. M., and R. E. Marsten: "Integer Programming Algorithms: A Framework and State-of-the-Art Survey," *Management Science,* **18**:465–491, 1972.

4. Johnson, E. L., M. M. Kostreva, and U. H. Suhl: "Solving 0–1 Integer Programming Problems Arising from Large Scale Planning Models," *Operations Research,* **33**:803–819, 1985.

5. Lawler, E. L., J. K. Lenstra, A. H. G. Rinnooy Kan, and D. B. Shmoys (eds.): *The Traveling Salesman Problem: A Guided Tour of Combinatorial Optimization,* Wiley, New York, 1985.

6. Lawler, E. L., and D. E. Wood: "Branch-and-Bound Methods: A Survey," *Operations Research,* **14**:699–719, 1966.

7. Mitten, L. G.: "Branch-and-Bound Methods: General Formulation and Properties," *Operations Research,* **18**:24–34, 1970.

8. Nemhauser, George L., and Lawrence A. Wolsey: *Integer and Combinatorial Optimization,* Wiley, New York, 1988.

9. Parker, R. Gary, and Ronald L. Rardin: *Discrete Optimization,* Academic Press, San Diego, 1988.

10. Salkin, Harvey M.: *Integer Programming,* Addison-Wesley, Reading, Mass., 1975.

11. Schriver, Alexander: *Theory of Linear and Integer Programming,* Wiley, New York, 1986.

12. Taha, Hamdy A.: *Integer Programming: Theory, Applications, and Computations,* Academic Press, New York, 1975.

13. Van Roy, T. J., and L. A. Wolsey: "Solving Mixed 0–1 Programs by Automatic Reformulation," *Operations Research,* **35**:45–57, 1987.

14. Williams, H. P.: *Model Building in Mathematical Programming,* 2d ed., Wiley, New York, 1985.

15. Zionts, Stanley: *Linear and Integer Programming,* Prentice-Hall, Englewood Cliffs, N.J., 1974.

PROBLEMS

1. A young couple, Eve and Steven, want to divide their main household chores (marketing, cooking, dishwashing, and laundering) between them so that each has two tasks but the total time they spend on household duties is kept to a minimum. Their efficiencies on these tasks differ, where the time each would need to perform the task is given by the following table:

	Hours Per Week Needed			
	Marketing	Cooking	Dishwashing	Laundering
Eve	4.5	7.8	3.6	2.9
Steven	4.9	7.2	4.3	3.1

(a) Formulate a BIP model for this problem.

(b) Use the BIP branch-and-bound algorithm presented in Sec. 13.4 to solve this problem. (*Hint:* Reduce the model to four binary variables before solving.)

2. The board of directors of the *General Wheels Co.* is considering seven large capital investments. These investments differ in the estimated long-run profit (net present value) they will generate, as well as in the amount of capital required, as shown by the following table (in units of millions of dollars):

	Investment Opportunity						
	1	2	3	4	5	6	7
Estimated profit	17	10	15	19	7	13	9
Capital required	43	28	34	48	17	32	23

The total amount of capital available for these investments is $100,000,000. Investment opportunities 1 and 2 are mutually exclusive, and so are 3 and 4. Furthermore, neither 3 nor 4 can be undertaken unless either 1 or 2 is undertaken. There are no such restrictions on investment opportunities 5, 6, and 7. The objective is to select the combination of capital investments that will maximize the total estimated long-run profit (net present value).

(a) Formulate a BIP model for this problem.

(b) Use the BIP branch-and-bound algorithm presented in Sec. 13.4 to solve this problem. (*Hint:* Reduce the problem to solving seven three-variable problems.)

3. Reconsider Prob. 3, Chap. 7. Now suppose that trucks (and their drivers) need to be hired to do the hauling, and each truck can only be used to haul gravel from a single pit to a single site. In addition to the hauling and gravel costs specified previously, there now is a fixed cost of 5 associated with hiring each truck. A truck can haul 5 tons, but it is not required to go full. For each combination of pit and site, there now are two decisions to be made: (1) the number of trucks to be used, and (2) the amount of gravel to be hauled. Formulate an MIP model for this problem.

4. Reconsider Prob. 30, Chap. 7. Formulate a BIP model for this problem.

5. Consider the following special type of *shortest path problem* (see Sec. 10.3) where the nodes are in columns and the only routes considered always move forward one column at a time.

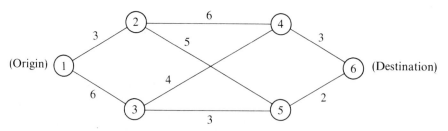

The numbers along the links represent distances, and the objective is to find the shortest route from the origin to the destination.

This problem also can be formulated as a BIP model involving both mutually exclusive alternatives and contingent decisions. Formulate this model.

6. Consider the following project network for a PERT-type system as described in Sec. 10.8.

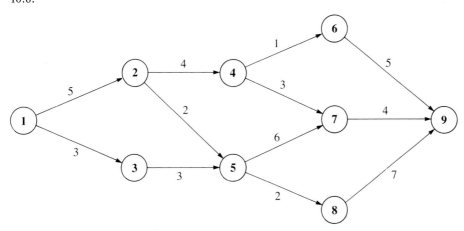

Formulate a BIP model for the problem of finding a *critical path* for this project network. (*Hint:* Do not try to apply the definition of "critical path" literally. Instead, use a resulting property of a critical path to obtain the objective function.)

7. A new planned community is being developed, and one of the decisions to be made is where to locate the two fire stations that have been allocated to the community. For planning purposes, the community has been divided into five tracts, with no more than one fire station to be located in any given tract. Each station is to respond to *all* of the fires that occur in the tract in which it is located as well as in the other tracts that are assigned to this station. Thus the decisions to be made consist of (1) the tracts to receive a fire station and (2) the assignment of each of the other tracts to one of the fire stations. The objective is to minimize the overall average of the *response times* to fires.

The following table gives the average response time (in minutes) to a fire in each tract (the columns) if that tract is served by a station in a given tract (the rows). The bottom row gives the forecasted average number of fires that will occur in each of the tracts each day.

		Response Times				
		Fire in Tract				
		1	2	3	4	5
Assigned station located in tract	1	5	12	30	20	15
	2	20	4	15	10	25
	3	15	20	6	15	12
	4	25	15	25	4	10
	5	10	25	15	12	5
Frequency of emergencies		2	1	3	1	3

Formulate a complete BIP model for this problem. Identify any constraints that correspond to *mutually exclusive alternatives* or *contingent decisions*.

8. Reconsider the Middletown racial balance study presented in Sec. 8.5. Suppose that the school board changes the current policy by prohibiting the splitting of a tract among different schools, so that an entire tract must be assigned to the same school. However, they will continue to require that the fraction of students in a school who are white (or who are black) must be between $\frac{1}{3}$ and $\frac{2}{3}$. Formulate a BIP model for this problem under this new policy.

9. Suppose that a state sends R persons to the U.S. House of Representatives. There are D counties in the state $(D > R)$, and the state legislature wants to group these counties into R distinct electoral districts, each of which sends a delegate to Congress. The total population of the state is P, and the legislature wants to form districts whose population approximates $p = P/R$. Suppose that the appropriate legislative committee studying the electoral districting problem generates a long list of N *candidates* to be districts $(N > R)$. Each of these candidates contains contiguous counties and a total population p_j $(j = 1, 2, \ldots, N)$ that is acceptably close to p. Define $c_j = |p_j - p|$. Each county i $(i = 1, 2, \ldots, D)$ is included in at least one candidate and typically will be included in a considerable number of candidates (in order to provide many feasible ways of selecting a set of R candidates that includes each county exactly once). Define

$$a_{ij} = \begin{cases} 1, & \text{if county } i \text{ is included in candidate } j \\ 0, & \text{if not.} \end{cases}$$

Given the values of the c_j and the a_{ij}, the objective is to select R of these N possible districts such that each county is contained in a single district and such that the largest of the associated c_j is as small as possible.

Formulate a BIP model for this problem.

10.* An airline company is considering the purchase of new long-, medium-, and short-range jet passenger airplanes. The purchase price would be $33,500,000 for each long-range plane, $25,000,000 for each medium-range plane, and $17,500,000 for each short-range plane. The board of directors has authorized a maximum commitment of $750,000,000 for these purchases. Regardless of which airplanes are purchased, air travel of all distances is expected to be sufficiently large enough so that these planes would be utilized at essentially maximum capacity. It is estimated that the net annual profit (after subtracting capital recovery costs) would be $2,100,000 per long-range plane, $1,500,000 per medium-range plane, and $1,150,000 per short-range plane.

It is predicted that enough trained pilots will be available to the company to crew 30 new airplanes. If only short-range planes were purchased, the maintenance facilities would be able to handle 40 new planes. However, each medium-range plane is equivalent to $1\frac{1}{3}$ short-range planes, and each long-range plane is equivalent to $1\frac{2}{3}$ short-range planes in terms of their use of the maintenance facilities.

The information given here was obtained by a preliminary analysis of the problem. A more detailed analysis will be conducted subsequently. However, using the preceding data as a first approximation, management wishes to know how many planes of each type should be purchased to maximize profit.

(*a*) Formulate the IP model for this problem.

(*b*) Use the MIP branch-and-bound algorithm presented in Sec. 13.5 to solve this problem.

(*c*) Use a binary representation of the variables to reformulate the IP model in part (*a*) as a BIP problem.

11. An American professor will be spending a short sabbatical leave at the University of Iceland. She wishes to bring all needed items with her on the airplane. After collecting together the professional items that she must have, she finds that airline regulations on space and weight for checked luggage will severely limit the clothes she can take. (She plans to carry on a warm coat, and then purchase a warm Icelandic sweater upon arriving in Iceland.) Clothes under consideration for checked luggage include 3 skirts, 3 slacks, 4 tops, and 3 dresses. The professor wants to maximize the number of outfits she will have in Iceland (including the special dress she will wear on the airplane). Each dress constitutes an outfit. Other outfits consist of a combination of a top and either a skirt or slacks. However, certain combinations are not fashionable and so will not qualify as an outfit.

In the following table, the combinations that will make an outfit are marked with an x.

					Top	
		1	2	3	4	Icelandic Sweater
	1	x	x			x
Skirt	2	x			x	
	3		x	x	x	x
	1	x		x		
Slacks	2	x	x		x	x
	3			x	x	x

The weight (in grams) and volume (in cubic centimeters) of each item is shown in the following table:

		Weight	Volume
	1	600	5,000
Skirt	2	450	3,500
	3	700	3,000
	1	600	3,500
Slacks	2	550	6,000
	3	500	4,000
	1	350	4,000
Top	2	300	3,500
	3	300	3,000
	4	450	5,000
	1	600	6,000
Dress	2	700	5,000
	3	800	4,000
Total allowed		4,000	32,000

Formulate a BIP model to choose which items of clothing to take. (*Hint:* After using binary decision variables to represent the individual items, introduce *auxiliary* binary variables to represent outfits involving combinations of items. Then use constraints and the objective function to ensure that these auxiliary variables have the correct values given the values of the decision variables.)

12. The Research and Development Division of a company has been developing four possible new product lines. Management must now make a decision as to which of these four products actually will be produced and at what levels. Therefore, they have asked the Operations Research Department to formulate a mathematical programming model to find the most profitable product mix.

A substantial cost is associated with beginning the production of any product, as given in the first row of the following table. The marginal net revenue from each unit produced is given in the second row of the table.

	Product			
	1	2	3	4
Startup cost	$50,000	$40,000	$70,000	$60,000
Marginal revenue	$70	$60	$90	$80

Let the continuous decision variables x_1, x_2, x_3, and x_4 be the production levels of products 1, 2, 3, and 4, respectively. Management has imposed the following policy constraints on these variables:

1. No more than two of the products can be produced.
2. Either product 3 or 4 can be produced only if either product 1 or 2 is produced.
3. Either $5x_1 + 3x_2 + 6x_3 + 4x_4 \leq 6000$
   or $\quad 4x_1 + 6x_2 + 3x_3 + 5x_4 \leq 6000$.

Introduce auxiliary binary variables to formulate an MIP model for this problem.

13. Consider the following mathematical model.

$$\text{Minimize} \quad Z = f_1(x_1) + f_2(x_2),$$

subject to the restrictions

1. Either $x_1 \geq 3$ or $x_2 \geq 3$.
2. At least one of the following inequalities holds:

$$2x_1 + x_2 \geq 7$$

$$x_1 + x_2 \geq 5$$

$$x_1 + 2x_2 \geq 7.$$

3. $|x_1 - x_2| = 0$, or 3, or 6.
4. $x_1 \geq 0$, $x_2 \geq 0$,

where
$$f_1(x_1) = \begin{cases} 7 + 5x_1, & \text{if } x_1 > 0 \\ 0, & \text{if } x_1 = 0, \end{cases}$$

$$f_2(x_2) = \begin{cases} 5 + 6x_2, & \text{if } x_2 > 0 \\ 0, & \text{if } x_2 = 0. \end{cases}$$

Formulate this problem as an MIP problem.

14. Consider the following mathematical model.

$$\text{Maximize} \quad Z = 3x_1 + 2f(x_2) + 2x_3 + 3g(x_4),$$

subject to the restrictions

1. $2x_1 - x_2 + x_3 + 3x_4 \leq 15$.
2. At least one of the following two inequalities holds:

$$x_1 + x_2 + x_3 + x_4 \leq 4$$

$$3x_1 - x_2 - x_3 + x_4 \leq 3.$$

3. At least two of the following four inequalities hold:

$$5x_1 + 3x_2 + 3x_3 - x_4 \leq 10$$

$$2x_1 + 5x_2 - x_3 + 3x_4 \leq 10$$

$$-x_1 + 3x_2 + 5x_3 + 3x_4 \leq 10$$

$$3x_1 - x_2 + 3x_3 + 5x_4 \leq 10.$$

4. $x_3 = 1$, or 2, or 3.
5. $x_j \geq 0$ $(j = 1, 2, 3, 4)$,

where
$$f(x_2) = \begin{cases} -5 + 3x_2, & \text{if } x_2 > 0 \\ 0, & \text{if } x_2 = 0, \end{cases}$$

and
$$g(x_4) = \begin{cases} -3 + 5x_4, & \text{if } x_4 > 0 \\ 0, & \text{if } x_4 = 0. \end{cases}$$

Formulate this problem as an MIP problem.

15. Consider the two-variable IP model discussed in Sec. 13.3 and illustrated in Fig. 13.1.

(a) Use a binary representation of the variables to reformulate this model as a BIP problem.

(b) Use the BIP branch-and-bound algorithm presented in Sec. 13.4 to solve this problem.

(c) Use the MIP branch-and-bound algorithm presented in Sec. 13.5 to solve the *original* model.

16. Consider the following IP problem.

$$\text{Maximize} \quad Z = 5x_1 + x_2,$$

subject to

$$-x_1 + 2x_2 \leq 4$$

$$x_1 - x_2 \leq 1$$

$$4x_1 + x_2 \leq 12$$

and

$$x_1 \geq 0, \quad x_2 \geq 0$$

$$x_1, x_2 \text{ are integers.}$$

(a) Solve this problem graphically.

(b) Solve the LP-relaxation graphically. Round this solution to the *nearest* integer solution and check whether it is feasible. Then enumerate *all* of the rounded solutions (rounding each noninteger value *either* up or down), check them for feasibility, and calculate Z for those that are feasible. Are any of these feasible rounded solutions optimal for the IP problem?

(c) Use the MIP branch-and-bound algorithm presented in Sec. 13.5 to solve this problem. For each subproblem, solve its LP-relaxation *graphically*.

17. Consider the following IP problem.

$$\text{Maximize} \quad Z = 220x_1 + 80x_2,$$

subject to

$$-x_1 + 2x_2 \leq 4$$

$$5x_1 + 2x_2 \leq 16$$

$$2x_1 - x_2 \leq 4.$$

and

$$x_1 \geq 0, \quad x_2 \geq 0$$

$$x_1, x_2 \text{ are integers.}$$

(a) Solve this problem graphically.

(b) Solve the LP-relaxation graphically. Round this solution to the *nearest* integer solution and check whether it is feasible. Then enumerate *all* of the rounded solutions (rounding each noninteger value *either* up or down), check them for feasibility, and calculate Z for those that are feasible. Are any of these feasible rounded solutions optimal for the IP problem?

(c) Use the MIP branch-and-bound algorithm presented in Sec. 13.5 to solve this problem. For each subproblem, solve its LP-relaxation *graphically*.

18.* Consider the assignment problem with the following cost table:

		Assignment				
		1	2	3	4	5
	1	39	65	69	66	57
	2	64	84	24	92	22
Assignee	*3*	49	50	61	31	45
	4	48	45	55	23	50
	5	59	34	30	34	18

(a) Design a *branch-and-bound algorithm* for solving such assignment problems by specifying how the branching, bounding, and fathoming steps would be performed. (*Hint:* For the assignees not yet assigned for the current subproblem, form the relaxation by deleting the constraints that each of these assignees must perform exactly one assignment.)

(b) Use this algorithm to solve this problem.

19. Five jobs need to be done on a certain machine. However, the setup time for each job depends upon which job immediately preceded it, as shown by the following table:

		Setup Time				
		Job				
		1	2	3	4	5
	None	4	5	8	9	4
	1	—	7	12	10	9
Immediately	*2*	6	—	10	14	11
preceding job	*3*	10	11	—	12	10
	4	7	8	15	—	7
	5	12	9	8	16	—

The objective is to schedule the *sequence* of jobs that minimizes the sum of the resulting setup times.

(a) Design a *branch-and-bound algorithm* for sequencing problems of this type by specifying how the branch, bound, and fathoming steps would be performed.

(b) Use this algorithm to solve this problem.

20.* Consider the following BIP problem.

Maximize $Z = 80x_1 + 60x_2 + 40x_3 + 20x_4 - (7x_1 + 5x_2 + 3x_3 + 2x_4)^2$,

subject to x_j is binary, for $j = 1, 2, 3, 4$.

Given the value of the first k variables $(x_1, \ldots, x_k)$, where $k = 0, 1, 2,$ or 3, an upper bound on the value of Z that can be achieved by the corresponding feasible solutions is

$$\sum_{j=1}^{k} c_j x_j - \left(\sum_{j=1}^{k} d_j x_j \right)^2 + \sum_{j=k+1}^{4} \max \left\{ 0, c_j - \left[\left(\sum_{i=1}^{k} d_i x_i + d_j \right)^2 - \left(\sum_{i=1}^{k} d_i x_i \right)^2 \right] \right\},$$

where $c_1 = 80, c_2 = 60, c_3 = 40, c_4 = 20, d_1 = 7, d_2 = 5, d_3 = 3, d_4 = 2$. Use this bound to solve the problem by the *branch-and-bound technique*.

21. Use the BIP branch-and-bound algorithm presented in Sec. 13.4 to solve the following problem.

Minimize $Z = 5x_1 + 6x_2 + 7x_3 + 8x_4 + 9x_5$,

subject to
$$3x_1 - x_2 + x_3 + x_4 - 2x_5 \geq 2$$
$$x_1 + 3x_2 - x_3 - 2x_4 + x_5 \geq 0$$
$$-x_1 - x_2 + 3x_3 + x_4 + x_5 \geq 1$$

and
$$x_j \text{ is binary,} \quad \text{for } j = 1, 2, \ldots, 5.$$

22.* Use the BIP branch-and-bound algorithm presented in Sec. 13.4 to solve the following problem.

$$\text{Maximize} \quad Z = 2x_1 - x_2 + 5x_3 - 3x_4 + 4x_5,$$

subject to
$$3x_1 - 2x_2 + 7x_3 - 5x_4 + 4x_5 \leq 6$$
$$x_1 - x_2 + 2x_3 - 4x_4 + 2x_5 \leq 0$$

and
$$x_j \text{ is binary,} \quad \text{for } j = 1, 2, \ldots, 5.$$

23. Use the BIP branch-and-bound algorithm presented in Sec. 13.4 to solve the following problem.

$$\text{Maximize} \quad Z = 5x_1 + 5x_2 + 8x_3 - 2x_4 - 4x_5,$$

subject to
$$-3x_1 + 6x_2 - 7x_3 + 9x_4 + 9x_5 \geq 10$$
$$x_1 + 2x_2 \quad - x_4 - 3x_5 \leq 0$$

and
$$x_j \text{ is binary,} \quad \text{for } j = 1, 2, \ldots, 5.$$

24. Consider the following IP problem.

$$\text{Maximize} \quad Z = -3x_1 + 5x_2,$$

subject to
$$5x_1 - 7x_2 \geq 3$$

and
$$x_j \leq 3$$
$$x_j \geq 0$$

$$x_j \text{ is integer,} \quad \text{for } j = 1, 2.$$

(a) Use the MIP branch-and-bound algorithm presented in Sec. 13.5 to solve this problem. For each subproblem, solve its LP-relaxation *graphically*.
(b) Use the binary representation for integer variables to reformulate this problem as a BIP problem.
(c) Use the BIP branch-and-bound algorithm presented in Sec. 13.4 to solve the problem as formulated in part (b).

25. Use the MIP branch-and-bound algorithm presented in Sec. 13.5 to solve the following problem. (For each subproblem, solve its LP-relaxation *graphically*.)

$$\text{Minimize} \quad Z = 2x_1 + 3x_2,$$

subject to
$$x_1 + x_2 \geq 3$$
$$x_1 + 3x_2 \geq 6$$

and
$$x_1 \geq 0, \quad x_2 \geq 0$$

$$x_1, x_2 \text{ are integers.}$$

26. Use the MIP branch-and-bound algorithm presented in Sec. 13.5 to solve the following MIP problem.

$$\text{Maximize} \quad Z = 5x_1 + 4x_2 + 4x_3 + 2x_4,$$

subject to

$$x_1 + 3x_2 + 2x_3 + x_4 \le 10$$

$$5x_1 + x_2 + 3x_3 + 2x_4 \le 15$$

$$x_1 + x_2 + x_3 + x_4 \le 6$$

and

$$x_j \ge 0, \qquad \text{for } j = 1, 2, 3, 4$$

$$x_j \text{ is integer}, \qquad \text{for } j = 1, 2, 3.$$

27. Use the MIP branch-and-bound algorithm presented in Sec. 13.5 to solve the following MIP problem.

$$\text{Maximize} \qquad Z = 3x_1 + 4x_2 + 2x_3 + x_4 + 2x_5,$$

subject to

$$2x_1 - x_2 + x_3 + x_4 + x_5 \le 3$$

$$-x_1 + 3x_2 + x_3 - x_4 - 2x_5 \le 2$$

$$2x_1 + x_2 - x_3 + x_4 + 3x_5 \le 1$$

and

$$x_j \ge 0, \qquad \text{for } j = 1, 2, 3, 4, 5$$

$$x_j \text{ is binary}, \qquad \text{for } j = 1, 2, 3.$$

28. Use the MIP branch-and-bound algorithm presented in Sec. 13.5 to solve the following MIP problem.

$$\text{Minimize} \qquad Z = 5x_1 + x_2 + x_3 + 2x_4 + 3x_5,$$

subject to

$$x_2 - 5x_3 + x_4 + 2x_5 \ge -2$$

$$5x_1 - x_2 \qquad\qquad + x_5 \ge 7$$

$$x_1 + x_2 + 6x_3 + x_4 \qquad \ge 4$$

and

$$x_j \ge 0, \qquad \text{for } j = 1, 2, 3, 4, 5$$

$$x_j \text{ is integer}, \qquad \text{for } j = 1, 2, 3.$$

29. The optimal solution for the LP-relaxation of a certain four-variable pure BIP problem is $(x_1, x_2, x_3, x_4) = (\frac{3}{4}, 1, \frac{3}{4}, \frac{1}{4})$. One of several functional constraints for this problem is

$$3x_1 + 5x_2 + 4x_3 + 8x_4 \le 10.$$

Use this constraint to generate two valid *cutting planes* for the problem.

14

Nonlinear Programming

The fundamental role of *linear programming* in operations research is accurately reflected by the fact that it is the focus of *seven* chapters of this book, and it is used in several other chapters. A key assumption of linear programming is that *all its functions* (objective function and constraint functions) *are linear*. Although this assumption essentially holds for numerous practical problems, it frequently does not hold. In fact, many economists have found that some degree of nonlinearity is the rule and not the exception in economic planning problems.[1] Therefore it often is necessary to deal directly with *nonlinear programming problems,* so we turn our attention to this important area.

In one general form,[2] the *nonlinear programming problem* is to find $\mathbf{x} = (x_1, x_2, \ldots, x_n)$ so as to

$$\text{Maximize} \quad f(\mathbf{x}),$$

[1] For example, see Baumol, W. J., and R. C. Bushnell: "Error Produced by Linearization in Mathematical Programming," *Econometrica,* **35**:447–471, 1967.

[2] The other *legitimate forms* correspond to those for *linear programming* listed in Sec. 3.2. Section 4.6 describes how to convert these other forms into the form given here.

519

subject to $\qquad g_i(\mathbf{x}) \leq b_i, \qquad$ for $i = 1, 2, \ldots, m$

and $\qquad\qquad\qquad\qquad \mathbf{x} \geq \mathbf{0},$

where $f(\mathbf{x})$ and the $g_i(\mathbf{x})$ are given functions of the n decision variables.[1]

No algorithm that will solve *every* specific problem fitting this format is available. However, substantial progress has been made for some important special cases of this problem by making various assumptions about these functions, and research is continuing very actively. This area is a large one, and we do not have space to survey it completely. However, we do present a few sample applications and then introduce some of the basic ideas for solving certain important types of nonlinear programming problems.

Both Appendixes 1 and 2 provide useful background for this chapter, and we recommend that you review these appendixes as you study the next few sections.

14.1 Sample Applications

The following examples illustrate a few of the many important types of problems to which nonlinear programming has been applied.

The Product Mix Problem with Price Elasticity

In *product mix* problems, such as the Wyndor Glass Co. problem of Sec. 3.1, the goal is to determine the optimal mix of production levels for a firm's products, given limitations on the resources needed to produce those products, in order to maximize the firm's total profit. In some cases, there is a fixed unit profit associated with each of the products, so the resulting objective function will be linear. However, in many product mix problems, certain factors introduce *nonlinearities* into the objective function. For example, a large manufacturer may encounter *price elasticity,* whereby the amount of a product that can be sold has an inverse relationship to the price charged. Thus the *price-demand curve* might look like the one shown in Fig. 14.1, where $p(x)$

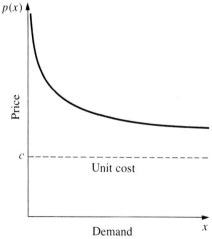

Figure 14.1 Price-demand curve.

---

[1] For simplicity, we assume throughout the chapter that *all* of these functions are *differentiable* everywhere.

is the price required in order to be able to sell x units. If the unit cost for producing the product is fixed at c (see the dashed line in Fig. 14.1), the firm's profit for producing and selling x units is given by the nonlinear function,

$$P(x) = x\, p(x) - cx,$$

as plotted in Fig. 14.2. If *each* of the firm's n products has a similar profit function, say $P_j(x_j)$ for producing and selling x_j units of product j ($j = 1, 2, \ldots, n$), then the overall objective function is

$$f(\mathbf{x}) = \sum_{j=1}^{n} P_j(x_j),$$

a sum of nonlinear functions.

Another reason that nonlinearities can arise in the objective function is due to the fact that the *marginal cost* of producing another unit of a given product varies with the production level. For example, the marginal cost may decrease when the production level is increased because of a *learning curve effect* (more efficient production with more experience). On the other hand, it may increase instead because special measures such as overtime or more expensive production facilities may be needed to increase production further.

Nonlinearities also may arise in the $g_i(\mathbf{x})$ constraint functions in a similar fashion. For example, if there is a budget constraint on total production cost, the cost function will be nonlinear if the marginal cost of production varies as just described. For constraints on the other kinds of resources, $g_i(\mathbf{x})$ will be nonlinear whenever the use of the corresponding resource is not strictly proportional to the production levels of the respective products.

The Transportation Problem with Volume Discounts on Shipping Costs

As illustrated by the P & T Company example in Sec. 7.1, a typical application of the *transportation problem* is to determine an optimal plan for shipping goods from various sources to various destinations, given supply and demand constraints, in order to minimize total shipping cost. It was assumed in Chap. 7 that the *cost per unit*

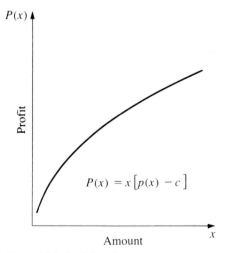

Figure 14.2 Profit function.

shipped from a given source to a given destination is *fixed,* regardless of the amount shipped. In actuality, this cost may not be fixed. *Volume discounts* sometimes are available for large shipments, so that the *marginal cost* of shipping one more unit might follow a pattern like the one shown in Fig. 14.3. The resulting cost of shipping *x* units then is given by a *nonlinear* function $C(x)$, which is a *piecewise linear function* with slope equal to the marginal cost, like the one shown in Fig. 14.4. Consequently, if *each* combination of source and destination has a similar shipping cost function, so that the cost of shipping x_{ij} units from source i ($i = 1, 2, \ldots, m$) to destination j ($j = 1, 2, \ldots, n$) is given by a nonlinear function $C_{ij}(x_{ij})$, then the overall objective function to be *minimized* is

$$f(\mathbf{x}) = \sum_{i=1}^{m} \sum_{j=1}^{n} C_{ij}(x_{ij}).$$

Even with this nonlinear objective function, the constraints normally are still the special linear constraints that fit the transportation problem model in Sec. 7.1.

Portfolio Selection with Risky Securities

It now is common practice for professional managers of large stock portfolios to use computer models based partially on nonlinear programming to guide them. Because investors are concerned about both the expected return (gain) and the risk associated with their investments, nonlinear programming is used to determine a portfolio that, under certain assumptions, provides an optimal trade-off between these two factors.

A nonlinear programming model can be formulated for this problem as follows. Suppose that *n* stocks (securities) are being considered for inclusion in the portfolio, and let the decision variables x_j ($j = 1, 2, \ldots, n$) be the number of shares of stock *j* to be included. Let μ_j and σ_{jj} be the (estimated) *mean* and *variance* of the return on each share of stock *j,* where σ_{jj} measures the risk of this stock. For $i = 1, 2, \ldots, n$

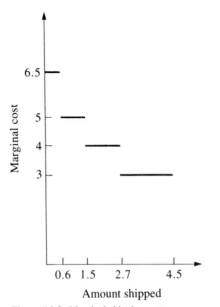

Figure 14.3 Marginal shipping cost.

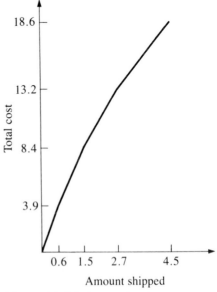

Figure 14.4 Shipping cost function.

$(i \neq j)$, let σ_{ij} be the *covariance* of the return on one share each of stock i and stock j. (Because it would be difficult to estimate all of the σ_{ij}, the usual approach is to make certain assumptions about market behavior that enable us to calculate σ_{ij} directly from σ_{ii} and σ_{jj}.) Then the expected value $R(\mathbf{x})$ and the variance $V(\mathbf{x})$ of the total return from the entire portfolio are

$$R(\mathbf{x}) = \sum_{j=1}^{n} \mu_j x_j,$$

$$V(\mathbf{x}) = \sum_{i=1}^{n} \sum_{j=1}^{n} \sigma_{ij} x_i x_j,$$

where $V(\mathbf{x})$ measures the risk associated with the portfolio. The device used to consider the trade-off between these two factors is to combine them together in the objective function to be maximized,

$$f(\mathbf{x}) = R(\mathbf{x}) - \beta V(\mathbf{x}),$$

where the parameter β is a nonnegative constant that reflects the investor's desired trade-off between expected return and risk. Thus choosing $\beta = 0$ implies that risk should be ignored completely, whereas choosing a large value for β places a heavy weight on minimizing risk [by maximizing the negative of $V(\mathbf{x})$].

The complete nonlinear programming model might be

$$\text{Maximize} \quad f(\mathbf{x}) = \sum_{j=1}^{n} \mu_j x_j - \beta \sum_{i=1}^{n} \sum_{j=1}^{n} \sigma_{ij} x_i x_j,$$

subject to

$$\sum_{j=1}^{n} P_j x_j \leq B$$

and

$$x_j \geq 0, \quad \text{for } j = 1, 2, \ldots, n,$$

where P_j is the price for each share of stock j and B is the amount of money budgeted for the portfolio. Under certain assumptions about the investor's *utility function* (measuring the relative value to the investor of different total returns), it can be shown that an optimal solution for this nonlinear programming problem maximizes the investor's *expected utility*.[1]

One drawback of the preceding formulation is that, because $R(\mathbf{x})$ and $V(\mathbf{x})$ are somewhat incommensurable, it is relatively difficult to choose an appropriate value for β. Therefore, rather than stopping with one choice of β, it is common to use a *parametric* (nonlinear) programming approach to generate the optimal solution as a function of β over a wide range of values of β. The next step is to examine the values of $R(\mathbf{x})$ and $V(\mathbf{x})$ for these solutions that are optimal for some value of β, and then choose the solution that seems to give the best trade-off between these two quantities. This procedure often is referred to as generating the solutions on the *efficient frontier* of the two-dimensional graph of $(R(\mathbf{x}), V(\mathbf{x}))$ points for feasible $\mathbf{x}$. The reason is that the $(R(\mathbf{x}), V(\mathbf{x}))$ point for an optimal $\mathbf{x}$ (for some β) lies on the *frontier* (boundary) of the feasible points. Furthermore, each optimal $\mathbf{x}$ is *efficient* in the sense that no other feasible solution is at least equally good with one measure (R or V) and strictly better with the other measure (smaller V or larger R).

---

[1] See Selected Reference 2, pp. 21–22, for further details.

14.2 Graphical Illustration of Nonlinear Programming Problems

When a nonlinear programming problem has just one or two variables, it can be represented graphically much like the Wyndor Glass Co. example for linear programming in Sec. 3.1. Because such a graphical representation gives considerable insight into the properties of optimal solutions for linear and nonlinear programming, let us look at a few examples. In order to highlight these differences, we shall use some *nonlinear* variations of the Wyndor Glass Co. problem.

Figure 14.5 shows what happens to this problem if the only changes in the model shown in Sec. 3.1 are that both the second and third functional constraints are replaced by the single nonlinear constraint, $9x_1^2 + 5x_2^2 \leq 216$. Compare Fig. 14.5 with Fig. 3.3. The optimal solution still happens to be $(x_1, x_2) = (2, 6)$. Furthermore, it still lies on the boundary of the feasible region. However, it is *not* a corner-point feasible solution. The optimal solution could have been a corner-point feasible solution with a different objective function (check $Z = 3x_1 + x_2$), but the fact that it need not be one means that we no longer have the tremendous simplification used in linear programming of limiting the search for an optimal solution to just the corner-point feasible solutions.

Now suppose that the linear constraints of Sec. 3.1 are kept unchanged, but the objective function is made nonlinear. For example, if

$$Z = 126x_1 - 9x_1^2 + 182x_2 - 13x_2^2,$$

then the graphical representation in Fig. 14.6 indicates that the optimal solution is $x_1 = \frac{8}{3}, x_2 = 5$, which again lies on the boundary of the feasible region. (The value

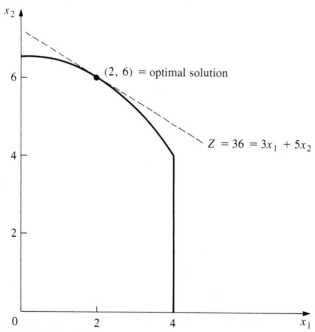

Figure 14.5 Wyndor Glass Co. example with a nonlinear constraint.

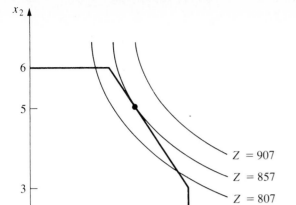

Figure 14.6 Wyndor Glass Co. example with a nonlinear objective function.

of Z for this optimal solution is $Z = 857$, so Fig. 14.6 depicts the fact that the locus of all points with $Z = 857$ intersects the feasible region at just this one point, whereas the locus of points with any larger Z does not intersect the feasible region at all.) On the other hand, if

$$Z = 54x_1 - 9x_1^2 + 78x_2 - 13x_2^2,$$

then the optimal solution turns out to be $(x_1, x_2) = (3, 3)$, which lies *inside* the boundary of the feasible region. (You can check that this solution is optimal by using calculus to derive it as the unconstrained global maximum; because it also satisfies the constraints, it must be optimal for the constrained problem.) Therefore, a general algorithm for solving similar problems needs to consider *all* solutions in the feasible region, not just those on the boundary.

Another complication that arises in nonlinear programming is that a *local* maximum need not be a *global* maximum (the overall optimal solution). For example, consider the function of a single variable plotted in Fig. 14.7. Over the interval $0 \leq x \leq 5$, this function has three local maxima—$x = 0$, $x = 2$, $x = 4$—but only one of these, $x = 4$, is a *global maximum*. (Similarly, there are local minima at $x = 1$, 3, and 5, but only $x = 5$ is a *global minimum*.)

Nonlinear programming algorithms generally are unable to distinguish between a *local* maximum and a *global* maximum (except by finding another *better* local maximum). Therefore, it becomes crucial to know the conditions under which any local maximum is *guaranteed* to be a global maximum over the feasible region. You may recall from calculus that when we maximize an ordinary (doubly differentiable) function of a single variable $f(x)$ without any constraints, this guarantee can be given when

$$\frac{d^2f}{dx^2} \leq 0 \qquad \text{for all } x.$$

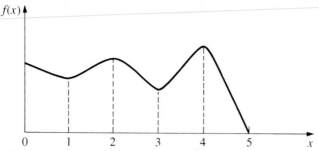

Figure 14.7 A function with several local maxima.

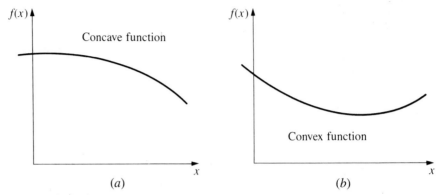

Figure 14.8 Examples of (*a*) *concave* function, (*b*) *convex* function.

Such a function that is always "curving downward" (or not curving at all) is called a **concave** function.[1] Similarly, if $\le$ is replaced by $\ge$, so that the function is always "curving upward" (or not curving at all), it is called a **convex** function.[2] (Thus a *linear* function is both concave and convex.) See Fig. 14.8 for examples. Then note that Fig. 14.7 illustrates a function that is *neither* concave nor convex because it alternates between curving upward and curving downward.

Functions of multiple variables also can be characterized as *concave* or *convex* if they always curve downward or curve upward. For example, consider a function consisting of a sum of terms. If *each* term is *concave* (we can check to see if it is from its second derivative when the term involves just one of the variables), then the function is concave. Similarly, the function is *convex* if *each* term is *convex*. These intuitive definitions are restated in precise terms, along with further elaboration on these concepts, in Appendix 1.

If a nonlinear programming problem has no constraints, the objective function being *concave* guarantees that a local maximum is a *global maximum*. (Similarly, the objective function being *convex* ensures that a local minimum is a *global minimum*.) If there are constraints, then one more condition will provide this guarantee—namely, that the *feasible region* is a **convex set**. As discussed in Appendix 1, a convex set is simply a set of points such that, for each pair of points in the collection, the entire line segment joining these two points is also in the collection. Thus the feasible region

---

[1] *Concave* functions sometimes are referred to as *concave downward*.

[2] *Convex* functions sometimes are referred to as *concave upward*.

for the Wyndor Glass Co. problem in Fig. 3.3 (or the feasible region for any other linear programming problem) is a convex set. Similarly, the feasible region in Fig. 14.5 is a convex set, which occurs whenever all of the $g_i(\mathbf{x})$ [for the constraints $g_i(\mathbf{x}) \leq b_i$] are *convex*. Therefore, to guarantee that a local maximum is a global maximum for a nonlinear programming problem with constraints $g_i(\mathbf{x}) \leq b_i$ ($i = 1, 2, \ldots , m$) and $\mathbf{x} \geq \mathbf{0}$, the objective function $f(\mathbf{x})$ must be *concave* and each $g_i(\mathbf{x})$ must be *convex*.

14.3 Types of Nonlinear Programming Problems

Nonlinear programming problems come in many different shapes and forms. Unlike the simplex method for linear programming, no single algorithm that will solve all of these different types of problems exists. Instead, algorithms have been developed for various individual *classes* (special types) of nonlinear programming problems. The most important classes are introduced next, and then the subsequent sections describe how some problems of these types can be solved.

Unconstrained Optimization

Unconstrained optimization problems have *no* constraints, so the objective is simply

$$\text{Maximize} \quad f(\mathbf{x})$$

over *all* values of $\mathbf{x} = (x_1, x_2, \ldots , x_n)$. As reviewed in Appendix 2, the *necessary* condition that a particular solution $\mathbf{x} = \mathbf{x}^*$ be optimal when $f(\mathbf{x})$ is a differentiable function is

$$\frac{\partial f}{\partial x_j} = 0 \quad \text{at } \mathbf{x} = \mathbf{x}^*, \quad \text{for } j = 1, 2, \ldots , n.$$

When $f(\mathbf{x})$ is *concave*, this condition also is *sufficient*, so then solving for $\mathbf{x}^*$ reduces to solving the system of n equations obtained by setting the n partial derivatives equal to zero. Unfortunately, for *nonlinear* functions $f(\mathbf{x})$, these equations often are going to be *nonlinear* as well, in which case you are unlikely to be able to solve analytically for their simultaneous solution. What then? Sections 14.4 and 14.5 describe *algorithmic search procedures* for finding $\mathbf{x}^*$, first for $n = 1$ and then for $n > 1$. These procedures also play an important role in solving many of the problem types described next, where there are constraints. The reason is that many algorithms for *constrained* problems are designed so that they can focus on an *unconstrained* version of the problem during a portion of each iteration.

When a variable x_j does have a nonnegativity constraint, $x_j \geq 0$, the preceding necessary and (perhaps) sufficient condition changes slightly to

$$\frac{\partial f}{\partial x_j} \begin{cases} \leq 0 & \text{at } \mathbf{x} = \mathbf{x}^*, \quad \text{if } x_j^* = 0 \\ = 0 & \text{at } \mathbf{x} = \mathbf{x}^*, \quad \text{if } x_j^* > 0 \end{cases}$$

for each such j. A problem that has some such nonnegativity constraints but no functional constraints is one special case ($m = 0$) of the next class of problems.

Linearly constrained optimization problems are characterized by constraints that completely fit linear programming, so that *all* of the $g_i(\mathbf{x})$ constraint functions are *linear*, but the objective function $f(\mathbf{x})$ is *nonlinear*. The problem is considerably simplified by having just one nonlinear function to take into account, along with a linear programming feasible region. A number of special algorithms based upon *extending* the simplex method to consider the nonlinear objective function have been developed.

One important special case, which we consider next, is *quadratic programming*.

Quadratic Programming

Quadratic programming problems again have linear constraints, but now the objective function $f(\mathbf{x})$ must be *quadratic*. Thus, the only difference between them and a linear programming problem is that some of the terms in the objective function involve the *square* of a variable or the *product* of two variables.

Many algorithms have been developed for this case under the additional assumption that $f(\mathbf{x})$ is *concave*. Section 14.7 presents an algorithm that involves a direct extension of the simplex method.

Quadratic programming is very important, partially because such formulations arise naturally in many applications. For example, the problem of portfolio selection with risky securities described in Sec. 14.1 fits into this format. However, another major reason for its importance is that a common approach to solving general linearly constrained optimization problems is to solve a sequence of quadratic programming approximations.

Convex Programming

Convex programming covers a broad class of problems that actually encompasses as special cases all of the preceding types when $f(\mathbf{x})$ is concave. The assumptions are:

> **1.** $f(\mathbf{x})$ is concave,
> **2.** Each $g_i(\mathbf{x})$ is convex.

As discussed at the end of Sec. 14.2, these assumptions are enough to ensure that a *local maximum* is a *global maximum*. You will see in Sec. 14.6 that the necessary and sufficient conditions for such an optimal solution are a natural generalization of the conditions just given for *unconstrained optimization* and its extension to include *nonnegativity constraints*. Section 14.9 then describes algorithmic approaches to solving convex programming problems.

Separable Programming

Separable programming is a special case of *convex programming*, where the one additional assumption is:

> **3.** All of the $f(\mathbf{x})$ and $g_i(\mathbf{x})$ functions are *separable functions*.

A **separable function** is simply a function where *each term* involves just a *single variable*, so that the function is *separable* into a sum of functions of individual variables. For example, if $f(\mathbf{x})$ is a *separable function*, it can be expressed as

$$f(\mathbf{x}) = \sum_{j=1}^{n} f_j(x_j),$$

where each $f_j(x_j)$ function includes only the terms involving just x_j. In the terminology of linear programming (see Sec. 3.3), separable programming problems satisfy the assumption of *additivity* but not the assumption of *proportionality* (for nonlinear functions).

It is important to distinguish these problems from other convex programming problems, because any separable programming problem can be closely approximated by a *linear programming* problem so that the extremely efficient simplex method can be used. This approach is described in Sec. 14.8. (For simplicity, we focus there on the *linearly constrained* case where the special approach is needed only on the objective function.)

Nonconvex Programming

Nonconvex programming encompasses all nonlinear programming problems that do not satisfy the assumptions of convex programming. Now, even if you are successful in finding a *local maximum*, there is no assurance that it also will be a *global maximum*. Therefore, there is no algorithm that will guarantee finding an optimal solution for all such problems. However, there do exist some algorithms that are relatively well suited for finding *local maxima*, especially when the forms of the nonlinear functions do not deviate too strongly from those assumed for convex programming. One such algorithm is presented in Sec. 14.10.

However, certain specific types of nonconvex programming problems can be solved without great difficulty by special methods. Two especially important such types are discussed briefly next.

Geometric Programming

When we apply nonlinear programming to engineering design problems, the objective function and the constraint functions frequently take the form

$$g(\mathbf{x}) = \sum_{i=1}^{N} c_i P_i(\mathbf{x}),$$

where
$$P_i(\mathbf{x}) = x_1^{a_{i1}} x_2^{a_{i2}} \ldots x_n^{a_{in}}, \quad \text{for } i = 1, 2, \ldots, N.$$

In such cases, the c_i and a_{ij} typically represent physical constants and the x_j are design variables. These functions generally are neither convex nor concave, so the techniques of convex programming cannot be applied directly to these *geometric programming problems*. However, there is one important case where the problem can be *transformed* into an *equivalent* convex programming problem. This case is where *all* of the c_i coefficients in each function are strictly positive, so that the functions are *generalized positive polynomials* (now called **posynomials**), and the objective function is to be minimized. The *equivalent* convex programming problem with decision variables $y_1, y_2, \ldots, y_n$ is then obtained by setting

$$x_j = e^{y_j}, \quad \text{for } j = 1, 2, \ldots, n$$

throughout the original model, so now a convex programming algorithm can be applied. Alternative solution procedures also have been developed for solving these *posynomial programming problems,* as well as for geometric programming problems of other types.[1]

Fractional Programming

Suppose that the objective function is in the form of a *fraction,* i.e., the *ratio* of two functions,

$$\text{Maximize} \quad f(\mathbf{x}) = \frac{f_1(\mathbf{x})}{f_2(\mathbf{x})}.$$

Such *fractional programming* problems arise, for example, when maximizing the ratio of *output* to *man-hours expended* (productivity), or *profit* to *capital expended* (rate of return), or *expected value* to *standard deviation* of some measure of performance for an investment portfolio (return/risk). Some special solution procedures have been developed for certain forms of $f_1(\mathbf{x})$ and $f_2(\mathbf{x})$.[2]

When it can be done, the most straightforward approach to solving a fractional programming problem is to *transform* it into an *equivalent* problem of a standard type for which effective solution procedures already are available. To illustrate, suppose that $f(\mathbf{x})$ is of the *linear fractional programming* form

$$f(\mathbf{x}) = \frac{\mathbf{cx} + c_0}{\mathbf{dx} + d_0},$$

where $\mathbf{c}$ and $\mathbf{d}$ are row vectors, $\mathbf{x}$ is a column vector, and c_0 and d_0 are scalars. Also assume that the constraint functions $g_i(\mathbf{x})$ are linear, so that the constraints in matrix form are $\mathbf{Ax} \leq \mathbf{b}$ and $\mathbf{x} \geq \mathbf{0}$.

Under mild additional assumptions, we can transform the problem into an equivalent *linear programming problem* by letting

$$\mathbf{y} = \frac{\mathbf{x}}{\mathbf{dx} + d_0} \quad \text{and} \quad t = \frac{1}{\mathbf{dx} + d_0},$$

so that $\mathbf{x} = \mathbf{y}/t$. This result yields

$$\text{Maximize} \quad Z = \mathbf{cy} + c_0 t,$$

subject to

$$\mathbf{Ay} - \mathbf{b}t \leq \mathbf{0}$$

$$\mathbf{dy} + d_0 t = 1,$$

and

$$\mathbf{y} \geq \mathbf{0}, \quad t \geq 0,$$

---

[1] Duffin, Richard J., Elmur L. Peterson, and Clarence M. Zehner: *Geometric Programming,* Wiley, New York, 1967; Beightler, Charles, and Donald T. Phillips: *Applied Geometric Programming,* Wiley, New York, 1976.

[2] The pioneering work on fractional programming was done by Charnes, A., and W. W. Cooper: "Programming with Linear Fractional Functionals," *Naval Research Logistics Quarterly,* **9**:181–186, 1962. Also see Schaible, Siegfried: "A Survey of Fractional Programming," in Schaible, Siegfried, and William T. Ziemba (eds.), *Generalized Concavity in Optimization and Economics,* Academic Press, New York, 1981, pp. 417–440.

which can be solved by the simplex method. More generally, the same kind of transformation can be used to convert a fractional programming problem with concave $f_1(\mathbf{x})$, convex $f_2(\mathbf{x})$, and convex $g_i(\mathbf{x})$ into an equivalent convex programming problem.

The Complementarity Problem

When we deal with quadratic programming in Sec. 14.7, you will see one example of how solving certain nonlinear programming problems can be reduced to solving the complementarity problem. Given variables $w_1, w_2, \ldots, w_p$ and $z_1, z_2, \ldots, z_p$, the **complementarity problem** is to find a *feasible* solution for the set of constraints,

$$\mathbf{w} = F(\mathbf{z}), \qquad \mathbf{w} \geq \mathbf{0}, \qquad \mathbf{z} \geq \mathbf{0},$$

that also satisfies the **complementarity constraint**,

$$\mathbf{w}^T \mathbf{z} = 0.$$

Here, $\mathbf{w}$ and $\mathbf{z}$ are column vectors, F is a given vector-valued function, and the superscript T denotes *transpose* (see Appendix 3). The problem has no objective function, so technically it is not a full-fledged nonlinear programming problem. It is called the **complementarity problem** because of the complementary relationships that *either*

$$w_i = 0 \qquad \text{or} \qquad z_i = 0 \qquad \text{(or both)} \qquad \text{for each } i = 1, 2, \ldots, p.$$

An important special case is the **linear complementarity problem**, where

$$F(\mathbf{z}) = \mathbf{q} + \mathbf{Mz},$$

where $\mathbf{q}$ is a given column vector and $\mathbf{M}$ is a given $p \times p$ matrix. Efficient algorithms have been developed for solving this problem under suitable assumptions about the properties of the matrix $\mathbf{M}$.[1] One type involves pivoting from one basic feasible solution to the next, much like the simplex method for linear programming.

In addition to having applications in nonlinear programming, complementarity problems have applications in game theory, economic equilibrium problems, and engineering equilibrium problems.

14.4 One-Variable Unconstrained Optimization

We now begin discussing how to solve some of the types of problems just described by considering the simplest case—*unconstrained optimization* with just a single variable x ($n = 1$), where the differentiable function $f(x)$ to be maximized is *concave*.[2] Thus the *necessary and sufficient condition* for a particular solution $x = x^*$ to be optimal (a *global maximum*) is

$$\frac{df}{dx} = 0 \qquad \text{at } x = x^*,$$

---

[1] See Cottle, R. W., and G. B. Dantzig: "Complementary Pivot Theory of Mathematical Programming," *Linear Algebra and Its Applications,* **1**:103–125, 1966, and Murty, K. G.: *Linear and Combinatorial Programming,* Wiley, New York, 1976, chap. 16.

[2] See the beginning of Appendix 2 for a review of the corresponding case when $f(x)$ is not concave.

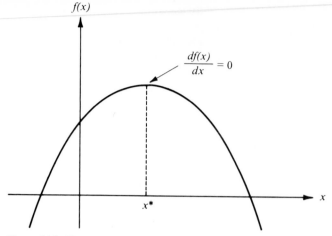

Figure 14.9 The one-variable unconstrained programming problem when the function is concave.

as depicted in Fig. 14.9. If this equation can be solved directly for x^*, you are done. However, if $f(x)$ is not a particularly simple function, so the derivative is not just a linear or quadratic function, you may not be able to solve the equation *analytically*. If not, the *one-dimensional search procedure* provides a straightforward way of solving the problem *numerically*.

The One-Dimensional Search Procedure

Like other *search procedures* in nonlinear programming, the *one-dimensional* search procedure finds a sequence of *trial solutions* that leads toward an optimal solution. At each iteration, you begin at the current trial solution to conduct a systematic search that culminates by identifying a new *improved* (hopefully substantially improved) trial solution.

The idea behind the one-dimensional search procedure is a very intuitive one, namely, that whether the *slope* (derivative) is positive or negative at a *trial solution* definitely indicates whether this solution needs to be made larger or smaller to move toward an optimal solution. Thus, if the derivative evaluated at a particular value of x is *positive*, then x^* must be larger than this x (see Fig. 14.9), so this x becomes a *lower bound* on the trial solutions that need to be considered thereafter. Conversely, if the derivative is *negative*, then x^* must be *smaller* than this x, so x would become an *upper bound*. Therefore, after both types of bounds have been identified, each new trial solution selected between the current bounds provides a new *tighter* bound of one type, thereby narrowing the search further. As long as a reasonable rule is used to select each trial solution in this way, the resulting *sequence* of trial solutions must *converge* to x^*. In practice, this means continuing the sequence until the distance between the bounds is sufficiently small that the next trial solution must be within a prespecified *error tolerance* of x^*.

This entire process is summarized next, using the notation

$$x' = \text{current trial solution,}$$

$$\underline{x} = \text{current lower bound on } x^*,$$

$$\bar{x} = \text{current upper bound on } x^*,$$

$$\varepsilon = \text{error tolerance for } x^*.$$

Although there are several reasonable rules for selecting each new trial solution, the one used in the following procedure is the **midpoint rule** (traditionally called the *Bolzano search plan*), which says simply to select the *midpoint* between the two current bounds.

Summary of One-Dimensional Search Procedure

Initialization Step: Select ε. Find an initial $\underline{x}$ and $\bar{x}$ by inspection (or by respectively finding any value of x at which the derivative is positive and then negative). Select an initial trial solution,

$$x' = \frac{\underline{x} + \bar{x}}{2}.$$

Iterative Step:

1. Evaluate $\dfrac{df(x)}{dx}$ at $x = x'$.

2. If $\dfrac{df(x)}{dx} \geq 0$, reset $\underline{x} = x'$.

3. If $\dfrac{df(x)}{dx} \leq 0$, reset $\bar{x} = x'$.

4. Select a new $x' = \dfrac{\underline{x} + \bar{x}}{2}$.

Stopping Rule: If $(\bar{x} - \underline{x}) \leq 2\varepsilon$, so the new x' must be within ε of x^*, stop. Otherwise, return to the iterative step.

We shall now illustrate this procedure by applying it to the following example.

EXAMPLE: Suppose that the function to be maximized is

$$f(x) = 12x - 3x^4 - 2x^6,$$

as plotted in Fig. 14.10. Its first two derivatives are

$$\frac{df(x)}{dx} = 12(1 - x^3 - x^5),$$

$$\frac{d^2f(x)}{dx^2} = -12(3x^2 + 5x^4).$$

Because the second derivative is nonpositive everywhere, $f(x)$ is a concave function, so the one-dimensional search procedure can be applied safely to find its global maximum. A quick inspection of this function (without even constructing its graph as shown in Fig. 14.10) indicates that $f(x)$ is positive for small positive values of x, but it is negative for $x < 0$ or $x > 2$. Therefore, $\underline{x} = 0$ and $\bar{x} = 2$ can be used as the initial bounds, with their midpoint, $x' = 1$, as the initial trial solution. Let 0.01 be the error tolerance for x^* in the stopping rule, so the final $(\bar{x} - \underline{x}) \leq 0.02$ with the final x' at the midpoint. Applying the one-dimensional search procedure then yields

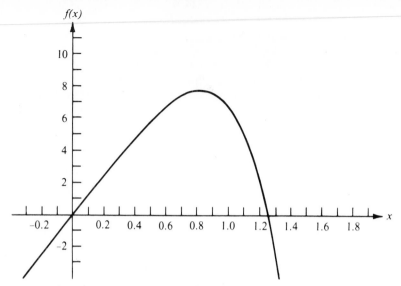

Figure 14.10 Example for one-dimensional search procedure.

Table 14.1 **Application of One-Dimensional Search Procedure to Example**

Iteration	$\dfrac{df(x)}{dx}$	$\underline{x}$	$\bar{x}$	New x'	$f(x')$
0		0	2	1	7.0000
1	−12	0	1	0.5	5.7812
2	+10.12	0.5	1	0.75	7.6948
3	+4.09	0.75	1	0.875	7.8439
4	−2.19	0.75	0.875	0.8125	7.8672
5	+1.31	0.8125	0.875	0.84375	7.8829
6	−0.34	0.8125	0.84375	0.828125	7.8815
7	+0.51	0.828125	0.84375	0.8359375	7.8839
Stop					

the sequence of results shown in Table 14.1. [This table includes both the function and derivative values for your information, where the derivative is evaluated at the trial solution generated at the *preceding* iteration. However, note that the algorithm actually doesn't need to calculate $f(x')$ at all and that it only needs to calculate the derivative far enough to determine its sign.] The conclusion is that

$$x^* \approx 0.836,$$

$$0.828125 < x^* < 0.84375.$$

14.5 Multivariable Unconstrained Optimization

Now consider the problem of maximizing a *concave* function $f(\mathbf{x})$ of *multiple* variables, $\mathbf{x} = (x_1, x_2, \ldots, x_n)$, when there are no constraints on the feasible values. Suppose again that the *necessary and sufficient condition* for optimality, given by the system of equations obtained by setting the respective partial derivatives equal to zero (see Sec. 14.3), cannot be solved *analytically*, so that a *numerical* search procedure must be used. How can the preceding *one-dimensional* search procedure be extended to this *multidimensional* problem?

In Sec. 14.4, the value of the *ordinary* derivative was used to select one of just *two* possible directions (increase x or decrease x) in which to move from the current trial solution to the next one. The goal was to reach a point eventually where this derivative is (essentially) *zero*. Now, there are *innumerable* possible directions in which to move; they correspond to the possible *proportional rates* at which the respective variables can be changed. The goal is to reach a point eventually where all of the partial derivatives are (essentially) *zero*. Therefore, extending the one-dimensional search procedure requires using the values of the *partial* derivatives to select the specific direction in which to move. This selection involves using the *gradient* of the objective function, as described next.

Because the objective function $f(\mathbf{x})$ is assumed to be differentiable, it possesses a *gradient* denoted by $\nabla f(\mathbf{x})$ at each point $\mathbf{x}$. In particular, the **gradient** at a specific point $\mathbf{x} = \mathbf{x}'$ is the *vector* whose elements are the respective *partial derivatives* evaluated at $\mathbf{x} = \mathbf{x}'$, so that

$$\nabla f(\mathbf{x}') = \left(\frac{\partial f}{\partial x_1}, \frac{\partial f}{\partial x_2}, \ldots, \frac{\partial f}{\partial x_n} \right) \quad \text{at } \mathbf{x} = \mathbf{x}'.$$

The significance of the gradient is that the (infinitesimal) change in $\mathbf{x}$ that *maximizes* the rate at which $f(\mathbf{x})$ increases is the change that is *proportional* to $\nabla f(\mathbf{x})$. To express this idea geometrically, the "direction" of the gradient, $\nabla f(\mathbf{x}')$, is interpreted as the *direction* of the directed line segment (arrow) from the origin $(0, 0, \ldots, 0)$ to the point $(\partial f/\partial x_1, \partial f/\partial x_2, \ldots, \partial f/\partial x_n)$, where $\partial f/\partial x_j$ is evaluated at $x_j = x_j'$. Therefore, it may be said that the rate at which $f(\mathbf{x})$ increases is maximized if (infinitesimal) changes in $\mathbf{x}$ are in the *direction* of the gradient $\nabla f(\mathbf{x})$. Because the objective is to find the feasible solution *maximizing* $f(\mathbf{x})$, it would seem expedient to attempt to move in the direction of the gradient as much as possible.

The Gradient Search Procedure

Because the current problem has *no* constraints, this interpretation of the gradient suggests that an efficient *search procedure* should keep moving in the direction of the gradient until it (essentially) reaches an optimal solution $\mathbf{x}^*$, where $\nabla f(\mathbf{x}^*) = \mathbf{0}$. However, it normally would not be practical to change $\mathbf{x}$ *continuously* in the direction of $\nabla f(\mathbf{x})$, because this series of changes would require continuously *reevaluating* the $\partial f/\partial x_j$ and changing the direction of the path. Therefore, a better approach is to keep moving in a *fixed* direction from the current trial solution, not stopping until $f(\mathbf{x})$ stops increasing. This stopping point would be the next trial solution, so the gradient then would be recalculated to determine the new direction in which to move. With this approach, each *iteration* involves changing the *current* trial solution $\mathbf{x}'$ as follows:

$$\text{Reset} \quad \mathbf{x}' = \mathbf{x}' + t^* \, \nabla f(\mathbf{x}'),$$

where t^* is the positive value of t that *maximizes* $f(\mathbf{x}' + t \, \nabla f(\mathbf{x}'))$; that is,

$$f(\mathbf{x}' + t^* \, \nabla f(\mathbf{x}')) = \max_{t \geq 0} f(\mathbf{x}' + t \, \nabla f(\mathbf{x}')).$$

[Note that $f(\mathbf{x}' + t\,\nabla f(\mathbf{x}'))$ is simply $f(\mathbf{x})$ where

$$x_j = x_j' + t\left(\frac{\partial f}{\partial x_j}\right)_{\mathbf{x}=\mathbf{x}'}, \qquad \text{for } j = 1, 2, \ldots, n,$$

and that these expressions for the x_j involve only constants and t, so $f(\mathbf{x})$ becomes a function of just the single variable t.] The iterations of this gradient search procedure continue until $\nabla f(\mathbf{x}) = \mathbf{0}$ within a small tolerance ε, that is, until

$$\left|\frac{\partial f}{\partial x_j}\right| \le \varepsilon \qquad \text{for } all \ j = 1, 2, \ldots, n.$$

An analogy may help to clarify this procedure. Suppose that you need to climb to the top of a hill. You are near-sighted, so you can't see the top of the hill in order to walk directly in that direction. However, when you stand still, you can see the ground around your feet well enough to determine the direction in which the hill is sloping upward most sharply. You are able to walk in a straight line. While walking, you also are able to tell when you stop climbing (zero slope in your direction). Assuming that the hill is *concave*, you now can use the *gradient search procedure* for climbing to the top efficiently. This problem is a *two-variable* problem, where (x_1, x_2) represents the coordinates (ignoring height) of your current location. The function $f(x_1, x_2)$ gives the height of the hill at (x_1, x_2). You start each iteration at your current location (current trial solution) by determining the direction [in the (x_1, x_2) coordinate system] in which the hill is sloping upward most sharply (the direction of the gradient) at this point. You then begin walking in this fixed direction and continue as long as you still are climbing. You eventually stop at a new trial location (solution) when the hill becomes level in your direction, at which point you prepare to do another iteration in another direction. You continue these iterations, following a zigzag path up the hill, until you reach a trial location where the slope is essentially zero in all locations. Under the assumption that the hill [$f(x_1, x_2)$] is concave, you must then be essentially at the top of the hill.

The most difficult part of the gradient search procedure usually is to find t^*, the value of t that maximizes f in the direction of the gradient, at each iteration. Because $\mathbf{x}$ and $\nabla f(\mathbf{x})$ have fixed values for the maximization, and because $f(\mathbf{x})$ is concave, this problem should be viewed as maximizing a *concave* function of a *single variable* t. Therefore, it can be solved by the *one-dimensional search procedure* of Sec. 14.4 (where the initial lower bound on t must be nonnegative because of the $t \ge 0$ constraint). Alternatively, if f is a simple function, it may be possible to obtain an analytical solution by setting the derivative with respect to t equal to zero and solving.

Summary of Gradient Search Procedure

Initialization Step: Select ε and any initial trial solution $\mathbf{x}'$. Go first to the stopping rule.

Iterative Step: 1. Express $f(\mathbf{x}' + t\,\nabla f(\mathbf{x}'))$ as a function of t by setting

$$x_j = x_j' + t\left(\frac{\partial f}{\partial x_j}\right)_{\mathbf{x}=\mathbf{x}'}, \qquad \text{for } j = 1, 2, \ldots, n,$$

and then substituting these expressions into $f(\mathbf{x})$.

2. Use the one-dimensional search procedure (or calculus) to find $t = t^*$ that maximizes $f(\mathbf{x}' + t \nabla f(\mathbf{x}'))$ over $t \geq 0$.
3. Reset $\mathbf{x}' = \mathbf{x}' + t^* \nabla f(\mathbf{x}')$. Then go to the stopping rule.

Stopping Rule: Evaluate $\nabla f(\mathbf{x}')$ at $\mathbf{x} = \mathbf{x}'$. Check if

$$\left| \frac{\partial f}{\partial x_j} \right| \leq \varepsilon \qquad \text{for all } j = 1, 2, \ldots, n.$$

If so, stop with the current $\mathbf{x}'$ as the desired approximation of an optimal solution $\mathbf{x}^*$. Otherwise, go to the iterative step.

Now let us illustrate this procedure.

EXAMPLE: Consider the following two-variable problem.

$$\text{Maximize} \qquad f(\mathbf{x}) = 2x_1 x_2 + 2x_2 - x_1^2 - 2x_2^2.$$

Thus

$$\frac{\partial f}{\partial x_1} = 2x_2 - 2x_1,$$

$$\frac{\partial f}{\partial x_2} = 2x_1 + 2 - 4x_2.$$

We also can verify (see Appendix 1) that $f(\mathbf{x})$ is *concave*. To begin the gradient search procedure, suppose that $\mathbf{x} = (0, 0)$ is selected as the initial trial solution. Because the respective partial derivatives are 0 and 2 at this point, the gradient is

$$\nabla f(0, 0) = (0, 2).$$

Therefore, to begin the first iteration, set

$$x_1 = 0 + t(0) = 0,$$

$$x_2 = 0 + t(2) = 2t,$$

and then substitute these expressions into $f(\mathbf{x})$ to obtain

$$f(\mathbf{x}' + t \nabla f(\mathbf{x}')) = f(0, 2t)$$

$$= 2(0)(2t) + 2(2t) - (0)^2 - 2(2t)^2$$

$$= 4t - 8t^2.$$

Because

$$f(0, 2t^*) = \max_{t \geq 0} f(0, 2t) = \max_{t \geq 0} \{4t - 8t^2\},$$

and

$$\frac{d}{dt} (4t - 8t^2) = 4 - 16t = 0,$$

it follows that

$$t^* = \tfrac{1}{4},$$

so

Reset $\qquad \mathbf{x}' = (0, 0) + \tfrac{1}{4}(0, 2) = (0, \tfrac{1}{2}).$

For this new trial solution, the gradient is

$$\nabla f(0, \tfrac{1}{2}) = (1, 0).$$

Thus for the second iteration, set

$$\mathbf{x} = (0, \tfrac{1}{2}) + t(1, 0) = (t, \tfrac{1}{2}),$$

so

$$f(\mathbf{x}' + t\,\nabla f(\mathbf{x}')) = f(0 + t, \tfrac{1}{2} + 0t) = f(t, \tfrac{1}{2})$$

$$= (2t)(\tfrac{1}{2}) + 2(\tfrac{1}{2}) - t^2 - 2(\tfrac{1}{2})^2$$

$$= t - t^2 + \tfrac{1}{2}.$$

Because

$$f(t^*, \tfrac{1}{2}) = \max_{t \geq 0} f(t, \tfrac{1}{2}) = \max_{t \geq 0} \{t - t^2 + \tfrac{1}{2}\},$$

$$\frac{d}{dt}(t - t^2 + \tfrac{1}{2}) = 1 - 2t = 0,$$

then

$$t^* = \tfrac{1}{2},$$

so   Reset $\mathbf{x}' = (0, \tfrac{1}{2}) + \tfrac{1}{2}(1, 0) = (\tfrac{1}{2}, \tfrac{1}{2}).$

A nice way of organizing this work is to write out a table such as Table 14.2, which summarizes the preceding two iterations. At each iteration, the second column shows the current trial solution, and the last column shows the eventual new trial solution, which then is carried down into the second column for the next iteration. The fourth column gives the expressions for the x_j in terms of t that need to be substituted into $f(\mathbf{x})$ to give the fifth column.

Continuing in this fashion, the subsequent trial solutions would be $(\tfrac{1}{2}, \tfrac{3}{4})$, $(\tfrac{3}{4}, \tfrac{3}{4})$, $(\tfrac{3}{4}, \tfrac{7}{8})$, $(\tfrac{7}{8}, \tfrac{7}{8})$, . . . , as shown in Fig. 14.11. Because these points are converging to $\mathbf{x}^* = (1, 1)$, this solution is the optimal solution, as verified by the fact that

$$\nabla f(1, 1) = (0, 0).$$

However, because this converging sequence of trial solutions never *reaches* its limit, the procedure actually will stop somewhere (depending on ε) slightly below $(1, 1)$ as its final approximation of $\mathbf{x}^*$.

As Fig. 14.11 suggests, the gradient search procedure *zigzags* to the optimal solution rather than moving in a straight line. Some modifications of the procedure have been developed that *accelerate* movement toward the optimum by taking this zigzag behavior into account.

If $f(\mathbf{x})$ were *not* a *concave* function, the gradient search procedure still would converge to a *local* maximum. The only change in the description of the procedure for this case is that t^* now would correspond to the *first local maximum* of $f(\mathbf{x}' + t\,\nabla f(\mathbf{x}'))$ as t is increased from zero.

If the objective were to *minimize* $f(\mathbf{x})$ instead, one change in the procedure would be to move in the *opposite* direction of the gradient at each iteration. In other words, the rule for obtaining the next point now would be

Reset $\mathbf{x}' = \mathbf{x}' - t^*\,\nabla f(\mathbf{x}').$

Table 14.2 Application of Gradient Search Procedure to Example

Iteration	$\mathbf{x}'$	$\nabla f(\mathbf{x}')$	$\mathbf{x}' + t\,\nabla f(\mathbf{x}')$	$f(\mathbf{x}' + t\,\nabla f(\mathbf{x}'))$	t^*	$\mathbf{x}' + t^*\,\nabla f(\mathbf{x}')$
1	$(0, 0)$	$(0, 2)$	$(0, 2t)$	$4t - 8t^2$	$\tfrac{1}{4}$	$(0, \tfrac{1}{2})$
2	$(0, \tfrac{1}{2})$	$(1, 0)$	$(t, \tfrac{1}{2})$	$t - t^2 + \tfrac{1}{2}$	$\tfrac{1}{2}$	$(\tfrac{1}{2}, \tfrac{1}{2})$

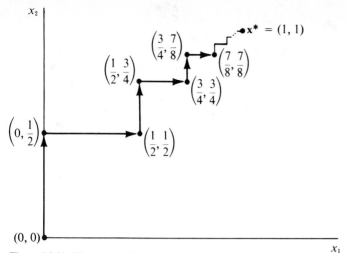

Figure 14.11 Illustration of the gradient search procedure.

The only other change is that t^* now would be the nonnegative value of t that *minimizes* $f(\mathbf{x}' - t\,\nabla f(\mathbf{x}'))$; that is,

$$f(\mathbf{x}' - t^*\,\nabla f(\mathbf{x}')) = \min_{t \geq 0} f(\mathbf{x}' - t\,\nabla f(\mathbf{x}')).$$

14.6 The Karush-Kuhn-Tucker (KKT) Conditions for Constrained Optimization

We now focus on the question of how to recognize an *optimal solution* for a nonlinear programming problem (with differentiable functions). What are the necessary and (perhaps) sufficient conditions that such a solution must satisfy?

In the preceding sections we already have noted these conditions for *unconstrained optimization*, as summarized in the first two lines of Table 14.3. Early in

Table 14.3 **Necessary and Sufficient Conditions for Optimality**

Problem	Necessary Conditions for Optimality	Also Sufficient if
One-variable unconstrained	$\dfrac{df}{dx} = 0$	$f(\mathbf{x})$ concave
Multivariable unconstrained	$\dfrac{\partial f}{\partial x_j} = 0 \quad (j = 1, 2, \ldots, n)$	$f(\mathbf{x})$ concave
Constrained, nonnegativity constraints only	$\dfrac{\partial f}{\partial x_j} \leq 0$ $(j = 1, 2, \ldots, n)$ $x_j \dfrac{\partial f}{\partial x_j} = 0$	$f(\mathbf{x})$ concave
General constrained problem	Karush-Kuhn-Tucker conditions	$f(\mathbf{x})$ concave $g_i(\mathbf{x})$ convex $(i = 1, 2, \ldots, m)$

Sec. 14.3 we also gave these conditions for the slight *extension* of unconstrained optimization where the *only* constraints are nonnegativity constraints. These conditions are shown in the third line of Table 14.3 in another *equivalent* form that is suggestive of their generalization for *general* constrained optimization. As indicated in the last line of the table, the conditions for the general case are called the **Karush-Kuhn-Tucker conditions** (or **KKT conditions**), because they were derived independently by Karush[1] and by Kuhn and Tucker.[2] Their basic result is embodied in the following theorem.

THEOREM: Assume that $f(\mathbf{x})$, $g_1(\mathbf{x})$, $g_2(\mathbf{x})$, . . . , $g_m(\mathbf{x})$ are *differentiable* functions satisfying certain regularity conditions.[3] Then

$$\mathbf{x}^* = (x_1^*, x_2^*, \ldots, x_n^*)$$

can be an *optimal solution* for the nonlinear programming problem only if there exist m numbers, $u_1, u_2, \ldots, u_m$, such that *all* of the *following conditions* are satisfied:

$$\left. \begin{aligned} &\textbf{1. } \frac{\partial f}{\partial x_j} - \sum_{i=1}^{m} u_i \frac{\partial g_i}{\partial x_j} \leq 0 \\ &\textbf{2. } x_j^* \left(\frac{\partial f}{\partial x_j} - \sum_{i=1}^{m} u_i \frac{\partial g_i}{\partial x_j} \right) = 0 \end{aligned} \right\} \quad \text{at } \mathbf{x} = \mathbf{x}^*, \quad \text{for } j = 1, 2, \ldots, n.$$

$$\left. \begin{aligned} &\textbf{3. } g_i(\mathbf{x}^*) - b_i \leq 0 \\ &\textbf{4. } u_i[g_i(\mathbf{x}^*) - b_i] = 0 \end{aligned} \right\} \quad \text{for } i = 1, 2, \ldots, m.$$

5. $x_j^* \geq 0,$ for $j = 1, 2, \ldots, n.$

6. $u_i \geq 0,$ for $i = 1, 2, \ldots, m.$

In the preceding *KKT conditions*, the u_i correspond to the *dual variables* of linear programming (we expand on this correspondence at the end of the section), and they have a comparable economic interpretation. (However, the u_i actually arose in the mathematical derivation as Lagrange multipliers.) Conditions 3 and 5 do nothing more than ensure the feasibility of the solution. The other conditions eliminate most of the feasible solutions as possible candidates to be an optimal solution. However, it should be noted that satisfying these conditions does not guarantee that the solution is optimal. As summarized in the last column of Table 14.3, certain additional *convexity* assumptions are needed to obtain this guarantee. These assumptions are spelled out in the following extension of the theorem.

COROLLARY: Assume that $f(\mathbf{x})$ is a *concave function* and that $g_1(\mathbf{x})$, $g_2(\mathbf{x})$, . . . , $g_m(\mathbf{x})$ are *convex functions* (i.e., this problem is a *convex programming* problem), where all of these functions satisfy the regularity conditions. Then $\mathbf{x}^* =$

[1] Karush, W.: "Minima of Functions of Several Variables with Inequalities as Side Conditions," M.S. Thesis, Department of Mathematics, University of Chicago, 1939.

[2] Kuhn, H. W., and A. W. Tucker: "Nonlinear Programming," in Jerzy Neyman (ed.), *Proceedings of the Second Berkeley Symposium,* University of California Press, Berkeley, 1951, pp. 481–492.

[3] Ibid., p. 483.

$(x_1^*, x_2^*, \ldots, x_n^*)$ is an *optimal solution* if and only if *all* the conditions of the theorem are satisfied.

EXAMPLE: To illustrate the formulation and application of the *KKT conditions*, we consider the following two-variable nonlinear programming problem.

$$\text{Maximize} \qquad f(\mathbf{x}) = \ln(x_1 + 1) + x_2,$$

subject to $\qquad\qquad\qquad 2x_1 + x_2 \leq 3$

and $\qquad\qquad\qquad x_1 \geq 0, \qquad x_2 \geq 0,$

where ln denotes *natural logarithm*. Thus $m = 1$ and $g_1(\mathbf{x}) = 2x_1 + x_2$, so $g_1(\mathbf{x})$ is *convex*. Furthermore, it can be easily verified (see Appendix 1) that $f(\mathbf{x})$ is *concave*. Hence the corollary applies, so any solution that satisfies the KKT conditions will definitely be an optimal solution. These conditions are

1(*a*). $\dfrac{1}{x_1 + 1} - 2u_1 \leq 0.$

2(*a*). $x_1 \left(\dfrac{1}{x_1 + 1} - 2u_1 \right) = 0.$

1(*b*). $1 - u_1 \leq 0.$
2(*b*). $x_2(1 - u_1) = 0.$
   3. $2x_1 + x_2 \leq 3.$
   4. $u_1(2x_1 + x_2 - 3) = 0.$
   5. $x_1 \geq 0, x_2 \geq 0.$
   6. $u_1 \geq 0.$

The steps in solving the KKT conditions for this particular example are outlined below.

1. $u_1 \geq 1$, from condition 1(*b*).
   $x_1 \geq 0$, from condition 5.

2. Therefore, $\dfrac{1}{x_1 + 1} - 2u_1 < 0.$

3. Therefore, $x_1 = 0$, from condition 2(*a*).
4. $u_1 \neq 0$ implies that $2x_1 + x_2 - 3 = 0$, from condition 4.
5. Steps 3 and 4 imply that $x_2 = 3$.
6. $x_2 \neq 0$ implies that $u_1 = 1$, from condition 2(*b*).
7. No conditions are violated by $x_1 = 0, x_2 = 3, u_1 = 1$.

Therefore, there exists a number $u_1 = 1$ such that $x_1 = 0, x_2 = 3$, and $u_1 = 1$ satisfy all the conditions. Consequently, $\mathbf{x}^* = (0, 3)$ is an optimal solution for this problem.

The particular progression of steps needed to solve the KKT conditions will differ from one problem to the next. When the logic is not apparent, it is sometimes helpful to consider separately the different cases where each x_j and u_i is specified to be either $= 0$ or > 0, and then trying each case until one leads to a solution. In the example, there are eight such cases corresponding to the eight combinations of $x_1 = 0$ versus $x_1 > 0$, $x_2 = 0$ versus $x_2 > 0$, and $u_1 = 0$ versus $u_1 > 0$. Each case leads to a simpler statement and analysis of the conditions. To illustrate, consider first the case shown next, where $x_1 = 0, x_2 = 0$, and $u_1 = 0$.

1(a). $\dfrac{1}{0 + 1} - 0 \leq 0.$ Contradiction.

1(b). $1 - 0 \leq 0.$ Contradiction.

3. $0 + 0 \leq 3.$

(All the other conditions are redundant.)

As listed below, the other three cases where $u_1 = 0$ also give immediate contradictions in a similar way, so no solution is available.

Case $(x_1 = 0, x_2 > 0, u_1 = 0)$: Contradicts conditions 1(a), 1(b), and 2(b).
Case $(x_1 > 0, x_2 = 0, u_1 = 0)$: Contradicts conditions 1(a), 2(a), and 1(b).
Case $(x_1 > 0, x_2 > 0, u_1 = 0)$: Contradicts conditions 1(a), 2(a), 1(b), and 2(b).

The case $(x_1 > 0, x_2 > 0, u_1 > 0)$ enables deleting these nonzero multipliers from conditions 2(a), 2(b), and (4), which then enables deleting conditions 1(a), 1(b), and (3) as redundant, as summarized below.

KKT Conditions for the Case $(x_1 > 0, x_2 > 0, u_1 > 0)$

2(a). $\dfrac{1}{x_1 + 1} - 2u_1 = 0.$

2(b). $1 - u_1 = 0.$

4. $2x_1 + x_2 - 3 = 0.$

(All the other conditions are redundant.)

Therefore, $u_1 = 1$, so $x_1 = -\frac{1}{2}$, which contradicts $x_1 > 0$.

Now suppose that the case $(x_1 = 0, x_2 > 0, u_1 > 0)$ is tried next.

KKT Conditions for the Case $(x_1 = 0, x_2 > 0, u_1 > 0)$

1(a). $\dfrac{1}{0 + 1} - 2u_1 \leq 0.$

2(b). $1 - u_1 = 0.$

4. $0 + x_2 - 3 = 0.$

(All the other conditions are redundant.)

Therefore, $x_1 = 0, x_2 = 3, u_1 = 1$.
Having found a solution, no additional cases need be considered.

For problems more complicated than this example, it may be difficult, if not essentially impossible, to derive an optimal solution *directly* from the KKT conditions. Nevertheless, these conditions still provide valuable clues as to the identity of an optimal solution, and they also permit us to check whether a proposed solution may be optimal.

There also are many valuable *indirect* applications of the KKT conditions. One of these applications arises in the *duality theory* that has been developed for nonlinear programming to parallel the duality theory for linear programming presented in Chap. 6. In particular, for any given constrained maximization problem (call it the *primal problem*), the KKT conditions can be used to define a closely associated dual problem that is a constrained minimization problem. The variables in the dual problem consist of both the Lagrange multipliers, u_i $(i = 1, 2, \ldots, m)$, and the primal variables, x_j

$(j = 1, 2, \ldots, n)$.[1] In the special case where the primal problem is a linear programming problem, the x_j variables drop out of the dual problem and it becomes the familiar dual problem of linear programming (where the u_i variables here correspond to the y_i variables in Chap. 6). When the primal problem is a convex programming problem, it is possible to establish between the primal problem and the dual problem relationships that are similar to those for linear programming. For example, the *strong duality property* of Sec. 6.1, which states that the optimal objective function values of the two problems are equal, also holds here. Furthermore, the values of the u_i variables in an optimal solution for the dual problem can again be interpreted as *shadow prices* (see Secs. 4.7 and 6.2); i.e., they give the rate at which the optimal objective function value for the primal problem could be increased by (slightly) increasing the right-hand side of the corresponding constraint. Because duality theory for nonlinear programming is a relatively advanced topic, the interested reader is referred elsewhere for further information.[2]

You will see another indirect application of the KKT conditions in the next section.

14.7 Quadratic Programming

As indicated in Sec. 14.3, the *quadratic programming* problem differs from the linear programming problem only in that the *objective function* also includes x_j^2 and $x_i x_j$ $(i \neq j)$ terms. Thus, if we use matrix notation like that introduced at the beginning of Sec. 5.2, the problem is to find $\mathbf{x}$ so as to

$$\text{Maximize} \quad f(\mathbf{x}) = \mathbf{c}\mathbf{x} - \tfrac{1}{2}\mathbf{x}^T\mathbf{Q}\mathbf{x},$$

subject to
$$\mathbf{A}\mathbf{x} \leq \mathbf{b} \quad \text{and} \quad \mathbf{x} \geq \mathbf{0},$$

where $\mathbf{c}$ is a row vector, $\mathbf{x}$ and $\mathbf{b}$ are column vectors, $\mathbf{Q}$ and $\mathbf{A}$ are matrices, and the superscript T denotes *transpose* (see Appendix 3). The q_{ij} (elements of Q) are given constants such that $q_{ij} = q_{ji}$ (which is the reason for the factor of $\tfrac{1}{2}$ in the objective function). By performing the indicated vector and matrix multiplications, the objective function then is expressed in terms of these q_{ij}, the c_j (elements of $\mathbf{c}$), and the variables as follows:

$$f(\mathbf{x}) = \mathbf{c}\mathbf{x} - \tfrac{1}{2}\mathbf{x}^T\,\mathbf{Q}\mathbf{x} = \sum_{j=1}^{n} c_j x_j - \tfrac{1}{2}\sum_{i=1}^{n}\sum_{j=1}^{n} q_{ij} x_i x_j \,.$$

If $i = j$ in this double summation, then $x_i x_j = x_j^2$, so $-\tfrac{1}{2}q_{jj}$ is the coefficient of x_j^2. If $i \neq j$, then $-\tfrac{1}{2}(q_{ij}x_i x_j + q_{ji}x_j x_i) = -q_{ij}\,x_i x_j$, so $-q_{ij}$ is the total coefficient for the product of x_i and x_j.

To illustrate this notation, consider the following example of a quadratic programming problem.

---

[1] For details on this formulation, see Mangasarian, Olvi T.: *Nonlinear Programming*, McGraw-Hill, New York, 1969, chap 8. For a unified survey of various approaches to duality in nonlinear programming, see Geoffrion, A. M.: "Duality in Nonlinear Programming: A Simplified Applications-Oriented Development," *SIAM Review*, **13**:1–37, 1971.

[2] Ibid.

Maximize $f(x_1, x_2) = 15x_1 + 30x_2 + 4x_1x_2 - 2x_1^2 - 4x_2^2,$

subject to $x_1 + 2x_2 \leq 30$

and $x_1 \geq 0, \qquad x_2 \geq 0.$

In this case,

$$\mathbf{c} = [15 \quad 30], \qquad \mathbf{x} = \begin{bmatrix} x_1 \\ x_2 \end{bmatrix}, \qquad \mathbf{Q} = \begin{bmatrix} 4 & -4 \\ -4 & 8 \end{bmatrix},$$

$$\mathbf{A} = [1 \quad 2], \qquad \mathbf{b} = [30].$$

Note that

$$-\tfrac{1}{2}\mathbf{x}^T\mathbf{Q}\mathbf{x} = -\tfrac{1}{2}[x_1 \quad x_2]\begin{bmatrix} 4 & -4 \\ -4 & 8 \end{bmatrix}\begin{bmatrix} x_1 \\ x_2 \end{bmatrix} = 4x_1x_2 - 2x_1^2 - 4x_2^2.$$

Several algorithms have been developed for the special case of the quadratic programming problem where the objective function is a *concave* function. (A way to verify that the objective function is concave is to verify the equivalent condition that

$$\mathbf{x}^T\mathbf{Q}\mathbf{x} \geq 0$$

for all $\mathbf{x}$, that is, $\mathbf{Q}$ is a *positive semidefinite* matrix.) We shall describe one[1] of these algorithms (the *modified simplex method*) that has been quite popular because it requires using only the *simplex method* with a slight modification. The key to this approach is to construct the *KKT conditions* from the preceding section, and then to reexpress these conditions in a convenient form that closely resembles linear programming. Therefore, before describing the algorithm, we shall develop this convenient form.

The KKT Conditions for Quadratic Programming

For concreteness, let us first consider the above example. Starting with the form given in the preceding section, its KKT conditions are the following.

1(a). $15 + 4x_2 - 4x_1 - u_1 \leq 0.$
2(a). $x_1(15 + 4x_2 - 4x_1 - u_1) = 0.$
1(b). $30 + 4x_1 - 8x_2 - 2u_1 \leq 0.$
2(b). $x_2(30 + 4x_1 - 8x_2 - 2u_1) = 0.$
   3. $x_1 + 2x_2 - 30 \leq 0.$
   4. $u_1(x_1 + 2x_2 - 30) = 0.$
   5. $x_1 \geq 0, \quad x_2 \geq 0.$
   6. $u_1 \geq 0.$

To begin reexpressing these conditions in a more convenient form, we move the constants in conditions 1(a), 1(b), and 3 to the right-hand side, and then introduce nonnegative *slack variables* (denoted by y_1, y_2, and v_1, respectively) to convert these inequalities to equations.

[1] Wolfe, Philip: "The Simplex Method for Quadratic Programming," *Econometrics*, **27**:382–398, 1959. This paper develops both a short form and a long form of the algorithm. We present a version of the *short form*, which assumes further that *either* $\mathbf{c} = \mathbf{0}$ *or* the objective function is *strictly* concave.

1(a). $-4x_1 + 4x_2 - u_1 + y_1 \qquad\qquad = -15$
1(b). $4x_1 - 8x_2 - 2u_1 \qquad + y_2 \qquad = -30$
3. $x_1 + 2x_2 \qquad\qquad + v_1 = \quad 30$

Having introduced y_1, note that condition 2(a) can now be reexpressed as simply requiring that either x_1 or 0 or $y_1 = 0$; that is,

2(a). $x_1 y_1 = 0.$

In just the same way, conditions 2(b) and 4 can be replaced by

2(b). $x_2 y_2 = 0,$
4. $u_1 v_1 = 0.$

For each of these three pairs—(x_1, y_1), (x_2, y_2), (u_1, v_1)—the two variables are called **complementary variables**, because only one of the two variables can be nonzero. Since all six variables are required to be nonnegative, these new forms of conditions 2(a), 2(b), and 4 can be combined into one constraint,

$$x_1 y_1 + x_2 y_2 + u_1 v_1 = 0,$$

called the **complementarity constraint**.

After multiplying through the equations for conditions 1(a) and 1(b) by (-1) to obtain nonnegative right-hand sides, we now have the desired convenient form for the entire set of conditions shown below.

$$4x_1 - 4x_2 + u_1 - y_1 \qquad\qquad = 15$$

$$-4x_1 + 8x_2 + 2u_1 \qquad - y_2 \qquad = 30$$

$$x_1 + 2x_2 \qquad\qquad + v_1 = 30$$

$$x_1 \geq 0, \quad x_2 \geq 0, \quad u_1 \geq 0, \quad y_1 \geq 0, \quad y_2 \geq 0, \quad v_1 \geq 0$$

$$x_1 y_1 + x_2 y_2 + u_1 v_1 = 0$$

This form is particularly convenient because, except for the complementarity constraint, these conditions are *linear programming constraints*.

For *any* quadratic programming problem, its KKT conditions can be reduced to this same convenient form containing just linear programming constraints plus one complementarity constraint. Using matrix notation again, this general form is

$$\mathbf{Qx} + \mathbf{A}^T\mathbf{u} - \mathbf{y} = \mathbf{c}^T,$$

$$\mathbf{Ax} + \mathbf{v} = \mathbf{b},$$

$$\mathbf{x} \geq \mathbf{0}, \quad \mathbf{u} \geq \mathbf{0}, \quad \mathbf{y} \geq \mathbf{0}, \quad \mathbf{v} \geq \mathbf{0},$$

$$\mathbf{x}^T\mathbf{y} + \mathbf{u}^T\mathbf{v} = \mathbf{0},$$

where the elements of the column vector $\mathbf{u}$ are the u_i of the preceding section, and the elements of the column vectors $\mathbf{y}$ and $\mathbf{v}$ are slack variables.

Because the objective function of the original problem is assumed to be concave and because the constraint functions are linear and therefore convex, the corollary to the theorem of Sec. 14.6 applies. Thus $\mathbf{x}$ is *optimal* if, and only if, there exist values of $\mathbf{y}$, $\mathbf{u}$, and $\mathbf{v}$ such that all four vectors together satisfy all these conditions. The original problem is thereby reduced to the equivalent problem of finding a *feasible solution* to these *constraints*.

It is of interest to note that this equivalent problem is one example of the *linear complementarity problem* introduced in Sec. 14.3 (see Prob. 13), and that a key constraint for the linear complementarity problem is its *complementarity constraint*.

The Modified Simplex Method

The *modified simplex method* exploits the key fact that, with the exception of the complementarity constraint, the KKT conditions in the convenient form obtained above are nothing more than *linear programming constraints*. Furthermore, the complementarity constraint simply implies that it is not permissible for *both* complementary variables of any pair to be (nondegenerate) *basic variables* (the only variables > 0) when considering (nondegenerate) basic feasible solutions. Therefore, the problem reduces to finding an *initial basic feasible solution* to any linear programming problem that has these constraints, subject to this additional restriction on the identity of the basic variables. (This initial basic feasible solution may be the only feasible solution in this case.)

As we discussed in Sec. 4.6, finding such an initial basic feasible solution is relatively straightforward. In the simple case where $c^T \leq 0$ (unlikely) and $b \geq 0$, the initial basic variables are the elements of y and v (multiply through the first set of equations by -1), so that the desired solution is $x = 0$, $u = 0$, $y = -c^T$, $v = b$. Otherwise, you need to revise the problem by introducing an *artificial variable* into each of the equations where $c_j > 0$ (add the variable on the left) or $b_i < 0$ (subtract the variable on the left and then multiply through by -1) in order to use these artificial variables (call them z_1, z_2, and so on) as initial basic variables for the revised problem. (Note that this choice of initial basic variables satisfies the *complementarity constraint*, because as nonbasic variables $x = 0$ and $u = 0$ automatically.)

Next, use *phase 1* of the *two-phase method* (see Sec. 4.6) to find a basic feasible solution for the real problem; i.e., apply the simplex method (with one modification) to the following linear programming problem.

$$\text{Minimize} \quad Z = \sum_j z_j,$$

subject to the linear programming constraints obtained from the KKT conditions, but with these artificial variables included.

The one modification in the simplex method is the following change in the procedure for selecting an *entering basic variable*.

RESTRICTED ENTRY RULE: When choosing an entering basic variable, exclude from consideration any nonbasic variable whose *complementary variable* already is a basic variable; the choice should be made from among the *other* nonbasic variables according to the usual criterion for the simplex method.

This rule keeps the complementarity constraint satisfied throughout the course of the algorithm. When an optimal solution

$$x^*, u^*, y^*, v^*, z_1 = 0, \ldots, z_n = 0$$

is obtained for the *phase 1* problem, x^* is the desired optimal solution for the original quadratic programming problem. *Phase 2* of the two-phase method is not needed.

We shall now illustrate this approach on the example given at the beginning of the section. As can be verified from the results in Appendix 1 [see Prob. 43(a)], $f(x_1, x_2)$ is *strictly concave;* that is,

$$Q = \begin{bmatrix} 4 & -4 \\ -4 & 8 \end{bmatrix}$$

is *positive definite,* so the algorithm can be applied.

The starting point for solving this example is its KKT conditions in the convenient form obtained earlier in the section. After introducing the needed artificial variables, the linear programming problem to be addressed *explicitly* by the *modified simplex method* then is

$$\text{Minimize} \quad Z = z_1 + z_2,$$

subject to

$$4x_1 - 4x_2 + u_1 - y_1 \qquad + z_1 \qquad = 15$$
$$-4x_1 + 8x_2 + 2u_1 \qquad - y_2 \qquad + z_2 = 30$$
$$x_1 + 2x_2 \qquad + v_1 \qquad = 30$$

and

$$x_1 \geq 0, \quad x_2 \geq 0, \quad u_1 \geq 0, \quad y_1 \geq 0, \quad y_2 \geq 0, \quad v_1 \geq 0, \quad z_1 \geq 0, \quad z_2 \geq 0.$$

The additional *complementarity constraint,*

$$x_1 y_1 + x_2 y_2 + u_1 v_1 = 0,$$

is not included explicitly, because the algorithm *automatically* enforces this constraint because of the *restricted entry rule.* In particular, for each of the three pairs of *complementary variables*—(x_1, y_1), (x_2, y_2), (u_1, v_1)—whenever one of the two variables already is a basic variable, the other variable is *excluded* as a candidate to be the entering basic variable. Remember that the only *nonzero* variables are basic variables. Because the initial set of basic variables for the linear programming problem— z_1, z_2, v_1—gives an initial basic feasible solution that satisfies the complementarity constraint, there is no way that this constraint can be violated by any subsequent basic feasible solution.

Table 14.4 shows the results of applying the *modified simplex method* to this problem. The first simplex tableau exhibits the initial system of equations *after* converting from minimizing Z to maximizing $(-Z)$ *and* algebraically eliminating the initial basic variables from Eq. (0), just as was done for the *radiation therapy example* in Sec. 4.6. The three iterations proceed just as for the regular simplex method, *except* for eliminating certain candidates to be the entering basic variable because of the *restricted entry rule.* In the first tableau, u_1 is eliminated as a candidate because its *complementary variable* (v_1) already is a basic variable (but x_2 would have been chosen anyway because $-4 < -3$). In the second tableau, both u_1 and y_2 are eliminated as candidates (because v_1 and x_2 are basic variables), so x_1 automatically is chosen as the only candidate with a negative coefficient in row 0 (whereas the *regular* simplex method would have permitted choosing *either* x_1 or u_1 because they are tied for having the largest negative coefficient). In the third tableau, both y_1 and y_2 are eliminated (because x_1 and x_2 are basic variables). However, u_1 is *not* eliminated because v_1 no longer is a basic variable, so u_1 is chosen as the entering basic variable in the usual way.

Table 14.4 Application of Modified Simplex Method to Quadratic Programming Example

Iteration	Basic Variable	Eq. No.	Z	x_1	x_2	u_1	y_1	y_2	v_1	z_1	z_2	Right Side
0	Z	0	-1	0	-4	-3	1	1	0	0	0	-45
	z_1	1	0	4	-4	1	-1	0	0	1	0	15
	z_2	2	0	-4	8	2	0	-1	0	0	1	30
	v_1	3	0	1	2	0	0	0	1	0	0	30
1	Z	0	-1	-2	0	-2	1	$\frac{1}{2}$	0	0	$\frac{1}{2}$	-30
	z_1	1	0	2	0	2	-1	$-\frac{1}{2}$	0	1	$\frac{1}{2}$	30
	x_2	2	0	$-\frac{1}{2}$	1	$\frac{1}{4}$	0	$-\frac{1}{8}$	0	0	$\frac{1}{8}$	$3\frac{3}{4}$
	v_1	3	0	2	0	$-\frac{1}{2}$	0	$\frac{1}{4}$	1	0	$-\frac{1}{4}$	$22\frac{1}{2}$
2	Z	0	-1	0	0	$-\frac{5}{2}$	1	$\frac{3}{4}$	1	0	$\frac{1}{4}$	$-7\frac{1}{2}$
	z_1	1	0	0	0	$\frac{5}{2}$	-1	$-\frac{3}{4}$	-1	1	$\frac{3}{4}$	$7\frac{1}{2}$
	x_2	2	0	0	1	$\frac{1}{8}$	0	$-\frac{1}{16}$	$\frac{1}{4}$	0	$\frac{1}{16}$	$9\frac{3}{8}$
	x_1	3	0	1	0	$-\frac{1}{4}$	0	$\frac{1}{8}$	$\frac{1}{2}$	0	$-\frac{1}{8}$	$11\frac{1}{4}$
3	Z	0	-1	0	0	0	0	0	0	1	1	0
	u_1	1	0	0	0	1	$-\frac{2}{5}$	$-\frac{3}{10}$	$-\frac{2}{5}$	$\frac{2}{5}$	$\frac{3}{10}$	3
	x_2	2	0	0	1	0	$\frac{1}{20}$	$-\frac{1}{40}$	$\frac{3}{10}$	$-\frac{1}{20}$	$\frac{1}{40}$	9
	x_1	3	0	1	0	0	$-\frac{1}{10}$	$\frac{1}{20}$	$\frac{2}{5}$	$\frac{1}{10}$	$-\frac{1}{20}$	12

The resulting optimal solution for this *phase 1* problem is $x_1 = 12$, $x_2 = 9$, $u_1 = 3$, with the rest of the variables zero. [Problem 43(c) asks you to verify that this solution is optimal by showing that $x_1 = 12$, $x_2 = 9$, $u_1 = 3$ satisfy the KKT conditions for the original problem when they are written in the form given in Sec. 14.6.] Therefore, the optimal solution for the quadratic programming problem (which includes only the x_1 and x_2 variables) is $(x_1, x_2) = (12, 9)$.

14.8 Separable Programming

The preceding section showed how one class of nonlinear programming problems can be solved by an extension of the *simplex method*. We now consider another class, called *separable programming,* that actually can be solved by the simplex method itself, because any such problem can be approximated as closely as desired by a *linear programming* problem with a larger number of variables.

As indicated in Sec. 14.3, separable programming assumes that the objective function $f(\mathbf{x})$ is *concave,* that *each* of the constraint functions $g_i(\mathbf{x})$ is *convex,* and that *all* of these functions are *separable functions* (functions where *each* term involves just a *single* variable). However, in order to simplify the discussion, we focus here on the special case where the convex and separable $g_i(\mathbf{x})$ are, in fact, *linear functions,* just as for linear programming. Thus only the objective function requires special treatment.

Under the preceding assumptions, the objective function can be expressed as a *sum* of *concave* functions of individual variables,

$$f(\mathbf{x}) = \sum_{j=1}^{n} f_j(x_j),$$

so that each $f_j(x_j)$ has a shape such as the one shown in Fig. 14.12 (either case) over the feasible range of values of x_j.[1] Because $f(\mathbf{x})$ represents the measure of performance

---

[1] $f(\mathbf{x})$ is concave if and only if *every* $f_j(x_j)$ is concave.

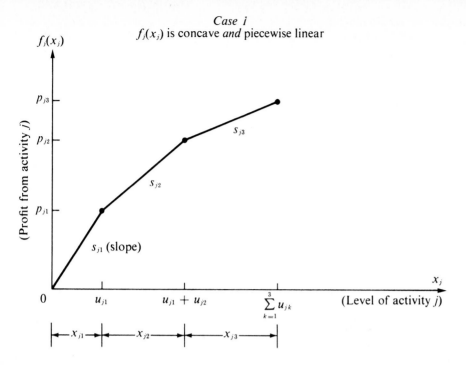

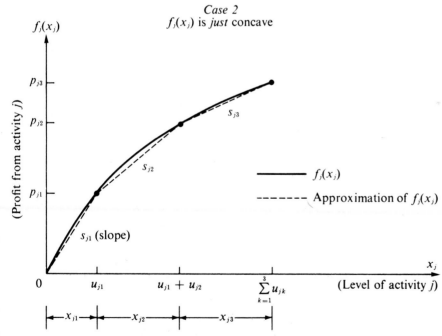

Figure 14.12 Shape of profit curves for separable programming.

(say *profit*) for all of the activities together, $f_j(x_j)$ represents the *contribution to profit* from activity j when it is conducted at the level x_j. The condition of $f(\mathbf{x})$ being *separable* simply implies *additivity* (see Sec. 3.3); i.e., there are no interactions between the activities (no cross-product terms) that affect total profit beyond their independent contributions. The assumption that each $f_j(x_j)$ is concave says that the *marginal profitability* (slope of the profit curve) either stays the same or decreases (*never* increases) as x_j is increased.

Concave profit curves occur quite frequently. For example, it may be possible to sell a limited amount of some product at a certain price, and then a further amount at a lower price, and perhaps finally a further amount at a still lower price. Similarly, it may be necessary to purchase raw materials from increasingly expensive sources. Another common situation is where a more expensive production process must be used (e.g., overtime rather than regular-time work) to increase the production rate beyond a certain point.

These kinds of situations can lead to either type of profit curve shown in Fig. 14.12. In case 1, the slope decreases only at certain *breakpoints,* so that $f_j(x_j)$ is a *piecewise linear function* (a sequence of connected line segments). For case 2, the slope may decrease continuously as x_j increases, so that $f_j(x_j)$ is a general concave function. Any such function can be approximated as closely as desired by a piecewise linear function, and this kind of approximation is used as needed for separable programming problems. (Figure 14.12 shows an approximating function that consists of just three line segments, but the approximation can be made even better just by introducing additional breakpoints.) This approximation is very convenient because a piecewise linear function of a single variable can be rewritten as a *linear function* of several variables, with one special restriction on the values of these variables, as described next.

Reformulation as a Linear Programming Problem

The key to rewriting a piecewise linear function as a linear function is to use a separate variable for each line segment. To illustrate, consider the piecewise linear function $f_j(x_j)$ shown in Fig. 14.12, case 1 (or its approximation for case 2), which has three line segments over the feasible range of values of x_j. Introduce the three new variables—x_{j1}, x_{j2}, x_{j3}—and set

$$x_j = x_{j1} + x_{j2} + x_{j3},$$

where $\qquad 0 \leq x_{j1} \leq u_{j1}, \qquad 0 \leq x_{j2} \leq u_{j2}, \qquad 0 \leq x_{j3} \leq u_{j3}.$

Then use the slopes—s_{j1}, s_{j2}, s_{j3}—to rewrite $f_j(x_j)$ as

$$f_j(x_j) = s_{j1}x_{j1} + s_{j2}x_{j2} + s_{j3}x_{j3},$$

with the *special restriction* that

$$x_{j2} = 0 \qquad \text{whenever} \qquad x_{j1} < u_{j1},$$

$$x_{j3} = 0 \qquad \text{whenever} \qquad x_{j2} < u_{j2}.$$

To see why this special restriction is required, suppose that $x_j = 1$, where $u_{jk} > 1$ ($k = 1, 2, 3$), so that $f_j(1) = s_{j1}$. Note that

$$x_{j1} + x_{j2} + x_{j3} = 1$$

permits

$$x_{j1} = 1, \quad x_{j2} = 0, \quad x_{j3} = 0 \implies f_j(1) = s_{j1},$$

$$x_{j1} = 0, \quad x_{j2} = 1, \quad x_{j3} = 0 \implies f_j(1) = s_{j2},$$

$$x_{j1} = 0, \quad x_{j2} = 0, \quad x_{j3} = 1 \implies f_j(1) = s_{j3},$$

and so on, where

$$s_{j1} > s_{j2} > s_{j3}.$$

However, the special restriction permits only the first possibility, which is the only one giving the correct value for $f_j(1)$.

Unfortunately, the *special restriction* does not fit into the required format for linear programming constraints, so *some* piecewise linear functions cannot be rewritten in a linear programming format. However, our $f_j(x_j)$ are assumed to be concave, so $s_{j1} > s_{j2} > \ldots$, so that an algorithm for maximizing $f(\mathbf{x})$ *automatically* gives the highest priority to using x_{j1} when (in effect) increasing x_j from zero, the next highest priority to using x_{j2}, and so on, without even including the special restriction explicitly in the model. This observation leads to the following key property.

KEY PROPERTY OF SEPARABLE PROGRAMMING: When $f(\mathbf{x})$ and the $g_i(\mathbf{x})$ satisfy the assumptions of separable programming, and when the resulting piecewise linear functions are rewritten as linear functions, deleting the *special restriction* gives a *linear programming model* whose optimal solution automatically satisfies the special restriction.

We shall elaborate further on the logic behind this key property later in this section in the context of a specific example. [Also see Prob. 54(a).]

To write down the complete linear programming model using the above notation, let n_j be the number of line segments in $f_j(x_j)$ (or the piecewise linear function approximating it), so that

$$x_j = \sum_{k=1}^{n_j} x_{jk}$$

would be substituted throughout the original model and

$$f_j(x_j) = \sum_{k=1}^{n_j} s_{jk} x_{jk}$$

would be substituted into the objective function for $j = 1, 2, \ldots, n.$[1] The resulting model is

$$\text{Maximize} \quad Z = \sum_{j=1}^{n} \left(\sum_{k=1}^{n_j} s_{jk} x_{jk} \right),$$

subject to

$$\sum_{j=1}^{n} a_{ij} \left(\sum_{k=1}^{n_j} x_{jk} \right) \leq b_i, \quad \text{for } i = 1, 2, \ldots, m$$

$$x_{jk} \leq u_{jk}, \quad \text{for } k = 1, 2, \ldots, n_j \quad \text{and} \quad j = 1, 2, \ldots, n,$$

and $\quad x_{jk} \geq 0, \quad \text{for } k = 1, 2, \ldots, n_j \quad \text{and} \quad j = 1, 2, \ldots, n.$

[1] If one or more of the $f_j(x_j)$ already are *linear functions*, $f_j(x_j) = c_j x_j$, then $n_j = 1$ so neither of these substitutions would be made for the j.

(The $\sum_{k=1}^{n_j} x_{jk} \geq 0$ constraints are deleted because they are ensured by the $x_{jk} \geq 0$ constraints.) If some original variable x_j has no upper bound, then $u_{jn_j} = \infty$, so the constraint involving this quantity would be deleted.

The most efficient way of solving this model is to use the streamlined version of the simplex method for dealing with *upper bound constraints* mentioned at the end of Sec. 7.5 (and described in Sec. 9.1). After obtaining an optimal solution for this model, you then would calculate

$$x_j = \sum_{k=1}^{n_j} x_{jk},$$

for $j = 1, 2, \ldots, n$ in order to identify an optimal solution for the original separable programming program (or its piecewise linear approximation).

Example

The Wyndor Glass Co. (see Sec. 3.1) has received a special order for handcrafted goods to be made in Plants 1 and 2 throughout the next 4 months. Filling this order will require borrowing certain employees from the work crews for the regular products, so the remaining workers would need to work overtime to utilize the full production capacity of the plant's machinery and equipment for these regular products. In particular, for the two new regular products discussed in Sec. 3.1, overtime would be required to utilize the last 25 percent of the production capacity available in Plant 1 for product 1, and for the last 50 percent of the capacity available in Plant 2 for product 2. The additional cost of using overtime work would reduce the profit for each unit involved from $3 to $2 for product 1, and from $5 to $1 for product 2, giving the *profit curves* of Fig. 14.13, both of which fit the form for case 1 of Fig. 14.12.

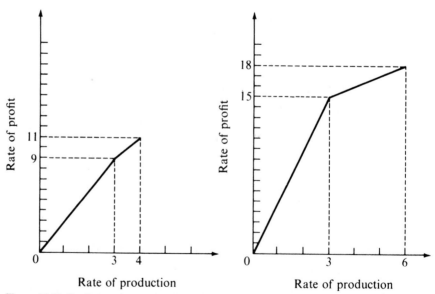

Figure 14.13 Profit data during the next 4 months for the Wyndor Glass Co.

Management has decided to go ahead and use overtime work rather than hire additional workers during this temporary situation. However, it does insist that the work crew for each product be fully utilized on regular time before any overtime is used. Furthermore, it feels that the current production rates ($x_1 = 2$ for product 1 and $x_2 = 6$ for product 2) should be changed temporarily if this would improve overall profitability. Therefore, it has instructed the OR Department to review products 1 and 2 again to determine the most profitable product mix during the next 4 months.

FORMULATION: At first glance it may appear straightforward to modify the Wyndor Glass Co. linear programming model in Sec. 3.1 to fit this new situation. In particular, let the production rate for product 1 be $x_1 = x_{1R} + x_{1O}$, where x_{1R} is the production rate achieved on regular time and x_{1O} is the incremental production rate from using overtime. Define $x_2 = x_{2R} + x_{2O}$ in the same way for product 2. Thus $n = 2$, $n_1 = 2$, and $n_2 = 2$ in the preceding general model. The new linear programming problem is to determine the values of x_{1R}, x_{1O}, x_{2R}, x_{2O} so as to

$$\text{Maximize} \quad Z = 3x_{1R} + 2x_{1O} + 5x_{2R} + x_{2O},$$

subject to

$$
\begin{aligned}
x_{1R} & \leq 3 \\
x_{1O} & \leq 1 \\
2x_{2R} & \leq 6 \\
2x_{2O} & \leq 6 \\
3(x_{1R} + x_{1O}) + 2(x_{2R} + x_{2O}) & \leq 18
\end{aligned}
$$

and

$$x_{1R} \geq 0, \qquad x_{1O} \geq 0, \qquad x_{2R} \geq 0, \qquad x_{2O} \geq 0.$$

However, there is one important factor that is not taken into account explicitly in this formulation. Specifically, there is nothing in the model that requires all available regular time for a product to be fully utilized before any overtime is used for that product. In other words, it may be feasible to have $x_{1O} > 0$ even when $x_{1R} < 3$ and to have $x_{2O} > 0$ even when $x_{2R} < 3$. Such solutions would not, however, be acceptable to management. (Prohibiting such solutions is the *special restriction* discussed earlier in this section.)

Now we come to the *key property of separable programming*. Even though the model does not take this factor into account explicitly, the model does take it into account implicitly! Despite the model having excess "feasible" solutions that actually are unacceptable, any *optimal* solution for the model is *guaranteed* to be a legitimate one that does not replace any available regular-time work with overtime work. (The reasoning here is analogous to that for the *Big M method* discussed in Sec. 4.6, where excess feasible but *nonoptimal* solutions also were allowed in the model as a matter of convenience.) Therefore, the simplex method can be safely applied to this model to find the most profitable acceptable product mix. The reason is twofold. First, the two decision variables for each product *always* appear together as a *sum*, ($x_{1R} + x_{1O}$) or ($x_{2R} + x_{2O}$), in *each* functional constraint (one in this case) other than the upper bound constraints on individual variables. Therefore, it *always* is possible to convert an unacceptable feasible solution to an acceptable one having the same total production rates, $x_1 = x_{1R} + x_{1O}$ and $x_2 = x_{2R} + x_{2O}$, merely by replacing overtime production by regular-time production as much as possible. Second, overtime production is less profitable than regular-time production (i.e., the *slope* of each profit curve in Fig. 14.13 is a monotonically *decreasing* function of the rate of production), so converting an unacceptable feasible solution to an acceptable one in this way *must* increase the total rate of profit Z. Consequently, any feasible solution that uses overtime production for a product when regular-time production is still available *cannot* be optimal with respect to the model.

For example, consider the unacceptable feasible solution $x_{1R} = 1$, $x_{1O} = 1$, $x_{2R} = 1$, $x_{2O} = 3$, which yields a total rate of profit $Z = 13$. The acceptable way of achieving the same total production rates $x_1 = 2$ and $x_2 = 4$ is $x_{1R} = 2$, $x_{1O} = 0$, $x_{2R} = 3$, $x_{2O} = 1$. This latter solution is still feasible, but it also increases Z by $(3 - 2)(1) + (5 - 1)(2) = 9$.

Similarly, the optimal solution for this model turns out to be $x_{1R} = 3$, $x_{1O} = 1$, $x_{2R} = 3$, $x_{2O} = 0$, which is an acceptable feasible solution.

Notice that most of the functional constraints in the model are *upper bound constraints*, i.e., constraints that simply specify the maximum value allowed for an individual variable. When a computer code is available for the special streamlined version of the simplex method for dealing with such constraints (see Secs. 7.5 and 9.1), it provides a very efficient way of solving even extremely large problems of this type.

Extensions

Thus far we have focused on the special case of separable programming where the only nonlinear function is the objective function $f(\mathbf{x})$. Now consider briefly the general case where the constraint functions $g_i(\mathbf{x})$ need not be linear, but are convex and separable, so that each $g_i(\mathbf{x})$ can be expressed as a sum of functions of individual variables,

$$g_i(\mathbf{x}) = \sum_{j=1}^{n} g_{ij}(x_j),$$

where each $g_{ij}(x_j)$ is a *convex* function. Once again, each of these new functions may be approximated as closely as desired by a *piecewise linear* function (if it is not already in that form). The one new restriction is that for each variable x_j ($j = 1, 2, \ldots, n$), all of the piecewise linear approximations of the functions of this variable $[f_j(x_j), g_{1j}(x_j), \ldots, g_{mj}(x_j)]$ must have the *same* breakpoints so that the same new variables $(x_{j1}, x_{j2}, \ldots, x_{jn_j})$ can be used for all of these piecewise linear functions. This formulation leads to a linear programming model just like the one given for the special case except that for each i and j, the x_{jk} variables now have different coefficients in constraint i [where these coefficients are the corresponding slopes of the piecewise linear function approximating $g_{ij}(x_j)$]. Because the $g_{ij}(x_j)$ are required to be convex, essentially the same logic as before implies that the *key property of separable programming* still must hold. [See Prob. 54(b).]

One drawback of approximating functions by piecewise linear functions as described in this section is that achieving a close approximation requires a large number of line segments (variables), whereas such a fine grid for the breakpoints is needed only in the immediate neighborhood of an optimal solution. Therefore, more sophisticated approaches that use a succession of *two-segment* piecewise linear functions have been developed[1] to obtain *successively closer approximations* within this immediate neighborhood. This kind of approach tends to be both *faster* and *more accurate* in closely approximating an optimal solution.

14.9 Convex Programming

We already have discussed some special cases of convex programming in Secs. 14.4 and 14.5 (unconstrained problems), 14.7 (quadratic objective function with linear constraints), and 14.8 (separable functions). You also have seen some theory for the general case (necessary and sufficient conditions for optimality) in Sec. 14.6. In this section, we briefly discuss some of the types of approaches used to solve the general convex programming problem [where the objective function $f(\mathbf{x})$ to be maximized is concave and the $g_i(\mathbf{x})$ constraint functions are convex] and then present one example of an algorithm for convex programming.

There is no single standard algorithm that always is used to solve convex programming problems. Many different algorithms have been developed, each with its own advantages and disadvantages, and research continues to be active in this area. Roughly speaking, most of these algorithms fall into one of the following three categories.

One category is **gradient algorithms,** where the *gradient search procedure* of Sec. 14.5 is modified in some way to keep the search path from penetrating any constraint boundary. For example, one popular gradient method is the *generalized reduced gradient* (GRG) method.[2]

The second category—**sequential unconstrained algorithms**—includes *penalty function* and *barrier function* methods. These algorithms convert the original constrained optimization problem into a sequence of *unconstrained optimization* problems whose optimal solutions converge to the optimal solution for the original problem. Each of these unconstrained optimization problems can be solved by the *gradient search procedure* of Sec. 14.5. This conversion is accomplished by incorporating the constraints into a *penalty function* (or *barrier function*) that is subtracted from the objective function in order to impose large penalties for violating constraints (or even being near constraint boundaries). You will see one example of this category of algorithms in the next section.

A third category—**sequential-approximation algorithms**—includes *linear-approximation* and *quadratic-approximation* methods. These algorithms replace the nonlinear objective function by a succession of linear or quadratic approximations. For linearly constrained optimization problems, these approximations allow repeated application of linear or quadratic programming algorithms. This work is accompanied by other analysis that yields a sequence of solutions that converges to an optimal solution for the original problem. Although these algorithms are particularly suitable for linearly constrained optimization problems, some of them also can be extended to

---

[1] Meyer, R. R.: "Two-Segment Separable Programming," *Management Science,* **25**:385–395, 1979.

[2] Lasdon, L. S., and A. D. Warren: "Generalized Reduced Gradient Software for Linearly and Nonlinearly Constrained Problems," in H. G. Greenberg (ed.), *Design and Implementation of Optimization Software,* Sijthoff and Noordhoff, Alphem aan den Rijn, The Netherlands, 1978.

problems with nonlinear constraint functions by the use of appropriate linear approximations.

As one example of a *sequential-approximation* algorithm, we present here the **Frank-Wolfe algorithm**[1] for the case of *linearly constrained* convex programming (so the constraints are $\mathbf{Ax} \leq \mathbf{b}$, $\mathbf{x} \geq \mathbf{0}$ in matrix form). This procedure is particularly straightforward; it combines *linear* approximations of the objective function (enabling us to use the simplex method) with the one-dimensional search procedure of Sec. 14.4.

A Sequential-Linear-Approximation Algorithm (Frank-Wolfe)

Given a feasible trial solution $\mathbf{x}'$, the linear approximation used for the objective function $f(\mathbf{x})$ is the first-order Taylor's series expansion of $f(\mathbf{x})$ around $\mathbf{x} = \mathbf{x}'$, namely,

$$f(\mathbf{x}) \approx f(\mathbf{x}') + \sum_{j=1}^{n} \frac{\partial f(\mathbf{x}')}{\partial x_j} (x_j - x_j') = f(\mathbf{x}') + \nabla f(\mathbf{x}')(\mathbf{x} - \mathbf{x}').$$

Because $f(\mathbf{x}')$ and $\nabla f(\mathbf{x}')\mathbf{x}'$ have fixed values, they can be dropped to give an equivalent linear objective function,

$$g(\mathbf{x}) = \nabla f(\mathbf{x}')\mathbf{x}.$$

The simplex method (or the graphical procedure if $n = 2$) then is applied to the resulting *linear programming* problem to find *its* optimal solution $\mathbf{x}_{\text{LP}}$. Note that the linear objective function necessarily increases steadily as one moves along the line segment from $\mathbf{x}'$ to $\mathbf{x}_{\text{LP}}$ (which is on the boundary of the feasible region). However, the linear approximation may not be a particularly close one for $\mathbf{x}$ far from $\mathbf{x}'$, so the *nonlinear* objective function may not continue to increase all the way from $\mathbf{x}'$ to $\mathbf{x}_{\text{LP}}$. Therefore, rather than just accepting $\mathbf{x}_{\text{LP}}$ as the next trial solution, we choose the point that maximizes the nonlinear objective function along this line segment. This point may be found by conducting the *one-dimensional search procedure* of Sec. 14.4, where the one variable for purposes of this search is the fraction t of the total distance from $\mathbf{x}'$ to $\mathbf{x}_{\text{LP}}$. This point then becomes the new trial solution for initiating the next iteration of the algorithm, as just described. The sequence of trial solutions generated by repeated iterations converges to an optimal solution for the original problem, so the algorithm stops as soon as the successive trial solutions are close enough together to have essentially reached this optimal solution.

Summary of Frank-Wolfe Algorithm

Initialization Step: Find a feasible initial trial solution $\mathbf{x}^{(0)}$, e.g., by applying linear programming procedures to find an initial basic feasible solution. Set $k = 1$.

Iterative Step
 Part 1: For $j = 1, 2, \ldots, n$, evaluate

$$\frac{\partial f(\mathbf{x})}{\partial x_j} \quad \text{at } \mathbf{x} = \mathbf{x}^{(k-1)} \quad \text{and set} \quad c_j = \frac{\partial f(\mathbf{x})}{\partial x_j}.$$

---

[1] Frank, M., and P. Wolfe: "An Algorithm for Quadratic Programming," *Naval Research Logistics Quarterly,* **3**:95–110, 1956. Although originally designed for quadratic programming, this algorithm is easily adapted to the case of a general concave objective function considered here.

Part 2: Find an optimal solution $\mathbf{x}_{LP}^{(k)}$ to the following linear programming problem.

$$\text{Maximize} \quad g(\mathbf{x}) = \sum_{j=1}^{n} c_j x_j,$$

subject to $\quad \mathbf{Ax} \le \mathbf{b} \quad$ and $\quad \mathbf{x} \ge \mathbf{0}.$

Part 3: For the variable t $(0 \le t \le 1)$, set

$$h(t) = f(\mathbf{x}) \quad \text{for } \mathbf{x} = \mathbf{x}^{(k-1)} + t(\mathbf{x}_{LP}^{(k)} - \mathbf{x}^{(k-1)}).$$

Use some procedure such as the one-dimensional search procedure (see Sec. 14.4) to maximize $h(t)$ over $0 \le t \le 1$, and set $\mathbf{x}^{(k)}$ equal to the corresponding $\mathbf{x}$. Go to the stopping rule.

Stopping Rule: If $\mathbf{x}^{(k-1)}$ and $\mathbf{x}^{(k)}$ are sufficiently close, stop and use $\mathbf{x}^{(k)}$ (or some extrapolation of $\mathbf{x}^{(0)}, \mathbf{x}^{(1)}, \ldots, \mathbf{x}^{(k-1)}, \mathbf{x}^{(k)}$) as your estimate of an optimal solution. Otherwise, reset $k = k + 1$ and return to the iterative step.

Now let us illustrate this procedure.

EXAMPLE: Consider the following linearly constrained convex programming problem.

$$\text{Maximize} \quad f(\mathbf{x}) = 5x_1 - x_1^2 + 8x_2 - 2x_2^2,$$

subject to $\quad 3x_1 + 2x_2 \le 6$

and $\quad x_1 \ge 0, \quad x_2 \ge 0.$

Note that

$$\frac{\partial f}{\partial x_1} = 5 - 2x_1, \quad \frac{\partial f}{\partial x_2} = 8 - 4x_2,$$

so that the *unconstrained* maximum, $\mathbf{x} = (\frac{5}{2}, 2)$, violates the functional constraint. Thus more work is needed to find the *constrained* maximum.

Because $\mathbf{x} = (0, 0)$ is clearly feasible (and corresponds to the initial basic feasible solution for the linear programming constraints), let us choose it as the initial trial solution $\mathbf{x}^{(0)}$ for the Frank-Wolfe algorithm. Plugging $x_1 = 0$ and $x_2 = 0$ into the expressions for the partial derivatives gives $c_1 = 5$ and $c_2 = 8$, so that $g(\mathbf{x}) = 5x_1 + 8x_2$ is the initial linear approximation of the objective function. Graphically, solving this linear programming problem (see Fig. 14.14a) yields $\mathbf{x}_{LP}^{(1)} = (0, 3)$. For part 3 of the iterative step, the points on the line segment between $(0, 0)$ and $(0, 3)$ shown in Fig. 14.14a are expressed by

$$(x_1, x_2) = (0, 0) + t[(0, 3) - (0, 0)] \quad \text{for } 0 \le t \le 1$$

$$= (0, 3t) \quad \text{for } 0 \le t \le 1,$$

as shown in the sixth column of Table 14.5. This expression then gives

$$h(t) = f(0, 3t) = 8(3t) - 2(3t)^2$$

$$= 24t - 18t^2,$$

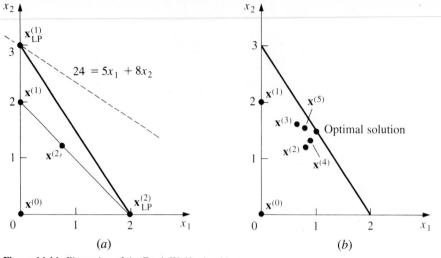

Figure 14.14 Illustration of the Frank-Wolfe algorithm.

so that the value $t = t^*$ that maximizes $h(t)$ over $0 \leq t \leq 1$ may be obtained in this case by setting

$$\frac{dh(t)}{dt} = 24 - 36t = 0,$$

so that $t^* = \frac{2}{3}$. This result yields the next trial solution,

$$\mathbf{x}^{(1)} = (0, 0) + \tfrac{2}{3}[(0, 3) - (0, 0)]$$

$$= (0, 2),$$

which completes the first iteration.

To sketch the calculations that lead to the results in the second row of Table 14.5, note that $\mathbf{x}^{(1)} = (0, 2)$ gives

$$c_1 = 5 - 2(0) = 5,$$

$$c_2 = 8 - 4(2) = 0.$$

For the objective function, $g(\mathbf{x}) = 5x_1$, graphically solving the problem over the feasible region in Fig. 14.14a gives $\mathbf{x}_{LP}^{(2)} = (2, 0)$. Therefore, the expression for the line segment between $\mathbf{x}^{(1)}$ and $\mathbf{x}_{LP}^{(2)}$ (see Fig. 14.14a) is

$$\mathbf{x} = (0, 2) + t[(2, 0) - (0, 2)]$$

$$= (2t, 2 - 2t),$$

Table 14.5 Application of Frank-Wolfe Algorithm to Example

k	$\mathbf{x}^{(k-1)}$	c_1	c_2	$\mathbf{x}_{LP}^{(k)}$	x for $h(t)$	$h(t)$	t^*	$\mathbf{x}^{(k)}$
1	(0, 0)	5	8	(0, 3)	(0, 3t)	$24t - 18t^2$	$\frac{2}{3}$	(0, 2)
2	(0, 2)	5	0	(2, 0)	(2t, 2 − 2t)	$8 + 10t - 12t^2$	$\frac{5}{12}$	$(\frac{5}{6}, \frac{7}{6})$

$$h(t) = f(2t, 2 - 2t)$$

$$= 5(2t) - (2t)^2 + 8(2 - 2t) - 2(2 - 2t)^2$$

$$= 8 + 10t - 12t^2.$$

Setting

$$\frac{dh(t)}{dt} = 10 - 24t = 0$$

yields $t^* = \frac{5}{12}$. Hence

$$\mathbf{x}^{(2)} = (0, 2) + \tfrac{5}{12}[(2, 0) - (0, 2)]$$

$$= (\tfrac{5}{6}, \tfrac{7}{6}).$$

You can see in Fig. 14.14*b* how the trial solutions keep alternating between two trajectories that appear to intersect at approximately the point $\mathbf{x} \doteq (1, \tfrac{3}{2})$. This point is, in fact, the optimal solution, as can be verified by applying the *KKT conditions* from Sec. 14.6.

This example illustrates a common feature of the Frank-Wolfe algorithm, namely, that the trial solutions alternate between two (or more) trajectories. When they alternate in this way, we can extrapolate the trajectories to their approximate point of intersection to estimate an optimal solution. This estimate tends to be better than using the last trial solution generated. The reason is that the trial solutions tend to converge rather slowly toward an optimal solution, so the last trial solution may still be quite far from optimal.

In conclusion, we should emphasize that the Frank-Wolfe algorithm is just one example of sequential-approximation algorithms. Many of these algorithms use *quadratic* instead of *linear* approximations at each iteration because quadratic approximations provide a considerably closer fit to the original problem and thus enable the sequence of solutions to converge considerably more rapidly toward an optimal solution than was the case in Fig. 14.14*b*. For this reason, even though sequential-linear-approximation methods such as the Frank-Wolfe algorithm are relatively straightforward to use, *sequential-quadratic-approximation methods*[1] now are generally preferred in actual applications. Popular among these are the so-called *quasi-Newton* (or *variable metric*) methods, which compute a quadratic approximation to the curvature of a nonlinear function without explicitly calculating second (partial) derivatives. (For linearly constrained optimization problems, this nonlinear function is just the objective function; whereas with nonlinear constraints, it is the Lagrangian function described in Appendix 2.) Some quasi-Newton algorithms don't even explicitly form and solve an approximating quadratic programming problem at each iteration, but instead incorporate some of the basic ingredients of *gradient algorithms*.

For further information about the state of the art in convex programming algorithms, see Selected References 6 and 7.

14.10 Nonconvex Programming

The assumptions of convex programming are very convenient ones, because they ensure that any *local maximum* also is a *global maximum*. Unfortunately, the nonlinear programming problems that arise in practice frequently only come fairly close to

[1] For a survey of these methods, see Powell, M. J. D.: "Variable Metric Methods for Constrained Optimization," in Bachem, A., M. Grotschel, and B. Korte (eds.), *Mathematical Programming: The State of the Art*, Springer-Verlag, Berlin, Heidelberg, New York, and Tokyo, 1983, pp. 288–311.

satisfying these assumptions, but they have some relatively minor disparities. What kind of approach can be used to deal with such *nonconvex programming* problems?

A common approach is to apply an algorithmic *search procedure* that will stop when it finds a *local maximum* and then to restart it a number of times from a variety of initial trial solutions in order to find as many distinct local maxima as possible. The best of these local maxima is then chosen for implementation. Normally, the search procedure is one that has been designed to find a global maximum when all of the assumptions of convex programming hold, but it also can operate to find a local maximum when they do not.

One such search procedure that has been widely used since its development in the 1960s is the *sequential unconstrained minimization technique* (or *SUMT* for short).[1] There actually are two main versions of SUMT, one of which is as an *exterior-point* algorithm that deals with *infeasible* solutions while using a *penalty function* to force convergence to the feasible region. We shall describe the other version, which is as an *interior-point* algorithm that deals directly with *feasible* solutions while using a *barrier function* to force staying inside the feasible region. Although SUMT was originally presented as a *minimization* technique, we shall convert it to a maximization technique in order to be consistent with the rest of the chapter. Therefore, we continue to assume that the problem is in the form given at the beginning of the chapter, and that all of the functions are differentiable.

Sequential Unconstrained Minimization Technique (SUMT)

As the name implies, SUMT replaces the original problem by a *sequence* of *unconstrained* optimization problems whose solutions *converge* to a solution (local maximum) of the original problem. This approach is very attractive because unconstrained optimization problems are much easier to solve (see the *gradient search procedure* in Sec. 14.5) than those with constraints. Each of the unconstrained problems in this sequence involves choosing a (successively smaller) strictly positive value of a scalar r and then solving for $\mathbf{x}$ so as to

$$\text{Maximize} \quad P(\mathbf{x}; r) = f(\mathbf{x}) - rB(\mathbf{x}).$$

$B(\mathbf{x})$ is a **barrier function** that has the following properties (for $\mathbf{x}$ that are feasible for the original problem):

1. $B(\mathbf{x})$ is *small* when $\mathbf{x}$ is *far* from the boundary of the feasible region,
2. $B(\mathbf{x})$ is *large* when $\mathbf{x}$ is *close* to the boundary of the feasible region,
3. $B(\mathbf{x}) \rightarrow \infty$ as the distance from the (nearest) boundary of the feasible region $\rightarrow 0$.

Thus, by starting the search procedure with a *feasible* initial trial solution and then attempting to increase $P(\mathbf{x}; r)$, $B(\mathbf{x})$ provides a *barrier* that prevents the search from ever crossing (or even reaching) the boundary of the feasible region for the original problem.

The most common choice of $B(\mathbf{x})$ is

$$B(\mathbf{x}) = \sum_{i=1}^{m} \frac{1}{b_i - g_i(\mathbf{x})} + \sum_{j=1}^{n} \frac{1}{x_j}.$$

[1] Fiacco, Anthony V., and Garth P. McCormick: *Nonlinear Programming: Sequential Unconstrained Minimization Techniques*, Wiley, New York, 1968.

For feasible values of **x**, note that the denominator of each term is proportional to the distance of **x** from the constraint boundary for the corresponding functional or non-negativity constraint. Consequently, *each* term is a *boundary repulsion term* that has all of the preceding three properties with respect to this particular constraint boundary. Another attractive feature of this $B(\mathbf{x})$ is that, when all the assumptions of *convex programming* are satisfied, $P(\mathbf{x}; r)$ is a *concave* function.

Because $B(\mathbf{x})$ keeps the search away from the boundary of the feasible region, you probably are asking the very legitimate question: What happens if the desired solution lies there? This concern is the reason that SUMT involves solving a *sequence* of these unconstrained optimization problems for successively smaller values of r approaching zero (where the final trial solution from each one becomes the initial trial solution for the next). For example, each new r might be obtained from the preceding one by multiplying by a constant θ ($0 < \theta < 1$), where a typical value is $\theta = 0.01$. As r approaches zero, $P(\mathbf{x}; r)$ approaches $f(\mathbf{x})$, so the corresponding local maximum of $P(\mathbf{x}; r)$ converges to a local maximum of the original problem. Therefore, it is necessary to solve only enough unconstrained optimization problems to permit extrapolating their solutions to this limiting solution.

How many are enough to permit this extrapolation? When the original problem satisfies the assumptions of *convex programming,* useful information is available to guide us in this decision. In particular, if $\bar{x}$ is a global maximum of $P(\mathbf{x}; r)$, then

$$f(\bar{\mathbf{x}}) \le f(\mathbf{x}^*) \le f(\bar{\mathbf{x}}) + rB(\bar{\mathbf{x}}),$$

where $\mathbf{x}^*$ is the (unknown) *optimal* solution for the original problem. Thus, $rB(\bar{\mathbf{x}})$ is the *maximum error* (in the value of the objective function) that can result by using $\bar{\mathbf{x}}$ to approximate $\mathbf{x}^*$, and extrapolating beyond $\bar{\mathbf{x}}$ to increase $f(\mathbf{x})$ further decreases this error. If an *error tolerance* is established in advance, then you can stop as soon as $rB(\bar{\mathbf{x}})$ is less than this quantity.

Unfortunately, no such guarantee for the maximum error can be given for *non-convex programming* problems. However, $rB(\bar{\mathbf{x}})$ still is *likely* to exceed the actual error when $\bar{\mathbf{x}}$ and $\mathbf{x}^*$ now are corresponding *local maxima of $P(\mathbf{x}; r)$ and the original problem,* respectively.

Summary of SUMT

Initialization Step: Identify a *feasible* initial trial solution $\mathbf{x}^{(0)}$ that is not on the boundary of the feasible region. Set $k = 1$ and choose appropriate strictly positive values for the initial r and for $\theta < 1$ (say, $r = 1$ and $\theta = 0.01$).[1]

Iterative Step: Starting from $\mathbf{x}^{(k-1)}$, apply the *gradient search procedure* described in Sec. 14.5 (or some similar method) to find a *local maximum,* $\mathbf{x}^{(k)}$, of

$$P(\mathbf{x}; r) = f(\mathbf{x}) - r \left[\sum_{i=1}^{m} \frac{1}{b_i - g_i(\mathbf{x})} + \sum_{j=1}^{n} \frac{1}{x_j} \right].$$

Stopping Rule: If the change from $\mathbf{x}^{(k-1)}$ to $\mathbf{x}^{(k)}$ is negligible, stop and use $\mathbf{x}^{(k)}$ (or an extrapolation of $\mathbf{x}^{(0)}, \mathbf{x}^{(1)}, \ldots, \mathbf{x}^{(k-1)}, \mathbf{x}^{(k)}$) as your estimate of a *local maximum* of the original problem. Otherwise, reset $k = k + 1$ and $r = \theta r$ and return to the iterative step.

---

[1] A reasonable criterion for choosing the initial r is one that makes $rB(\mathbf{x})$ about the same order of magnitude as $f(\mathbf{x})$ for feasible solutions **x** that are not particularly close to the boundary.

When the assumptions of *convex programming* are not satisfied, this algorithm should be repeated a number of times by starting from a variety of feasible initial trial solutions. The best of the *local maxima* thereby obtained for the original problem should be used as the best available approximation of a *global maximum*.

Finally, note that SUMT also can be easily extended to deal with *equality constraints*, $g_i(\mathbf{x}) = b_i$. One standard way is as follows. For each equality constraint,

$$\frac{-[b_i - g_i(\mathbf{x})]^2}{\sqrt{r}} \qquad \text{replaces} \qquad \frac{-r}{b_i - g_i(\mathbf{x})}$$

in the expression for $P(\mathbf{x}; r)$ given under Summary of SUMT, and then the same procedure is used. The numerator, $-[b_i - g_i(\mathbf{x})]^2$, imposes a large penalty for deviating substantially from satisfying the equality constraint, and then the denominator tremendously increases this penalty as r is decreased to a tiny amount, thereby forcing the sequence of trial solutions to converge toward a point that satisfies the constraint.

SUMT has been widely used because of its simplicity and versatility. However, numerical analysts have found that it is relatively prone to *numerical instability,* so considerable caution is advised. For further information on this issue, as well as similar analyses for alternative algorithms, see Selected Reference 7.

EXAMPLE: To illustrate *SUMT,* consider the following two-variable problem.

$$\text{Maximize} \qquad f(\mathbf{x}) = x_1 x_2,$$

subject to

$$x_1^2 + x_2 \leq 3$$

and

$$x_1 \geq 0, \qquad x_2 \geq 0.$$

Even though $g_1(\mathbf{x}) = x_1^2 + x_2$ is convex (because each term is convex), this problem is a *nonconvex programming* problem because $f(\mathbf{x}) = x_1 x_2$ is *not* concave (see Appendix 1).

For the initialization step, $(x_1, x_2) = (1, 1)$ is one obvious feasible solution that is not on the boundary of the feasible region, so we can set $\mathbf{x}^{(0)} = (1, 1)$. Reasonable choices for r and θ are $r = 1$ and $\theta = 0.01$.

For the iterative step,

$$P(\mathbf{x}; r) = x_1 x_2 - r \left(\frac{1}{3 - x_1^2 - x_2} + \frac{1}{x_1} + \frac{1}{x_2} \right).$$

With $r = 1$, applying the *gradient search procedure* starting from $(1, 1)$ to

Table 14.6 **Illustration of SUMT**

k	r	$x_1^{(k)}$	$x_2^{(k)}$
0		1	1
1	1	0.90	1.36
2	10^{-2}	0.983	1.933
3	10^{-4}	0.998	1.994
		$\downarrow$	$\downarrow$
		1	2

maximize this expression eventually leads to $\mathbf{x}^{(1)} = (0.90, 1.36)$. Resetting $r = 0.01$ and restarting the gradient search procedure from $(0.90, 1.36)$ then leads to $\mathbf{x}^{(2)} = (0.983, 1.933)$. One more iteration with $r = 0.01(0.01) = 0.0001$ leads from $\mathbf{x}^{(2)}$ to $\mathbf{x}^{(3)} = (0.998, 1.994)$. This sequence of points, summarized in Table 14.6, quite clearly is converging to $(1, 2)$. Applying the *KKT conditions* to this solution verifies that it does indeed satisfy the necessary condition for optimality. Graphical analysis demonstrates that $(x_1, x_2) = (1, 2)$ is, in fact, a *global maximum*. [See Prob. 60(*b*).]

For this problem, there are no local maxima other than $(x_1, x_2) = (1, 2)$,[1] so reapplying SUMT from various feasible initial trial solutions always leads to this same solution.

14.11 Conclusions

Practical optimization problems frequently involve *nonlinear* behavior that must be taken into account. It is sometimes possible to *reformulate* these nonlinearities to fit into a linear programming format, as can be done for *separable programming* problems. However, it is frequently necessary to use a *nonlinear programming* formulation.

In contrast to the case of the simplex method for linear programming, there is no efficient all-purpose algorithm that can be used to solve *all* nonlinear programming problems. In fact, some of these problems cannot be solved in a very satisfactory manner by *any* method. However, considerable progress has been made for some important classes of problems, including *quadratic programming, convex programming,* and certain special types of *nonconvex programming*. A variety of algorithms that frequently perform well are available for these cases. Some of these algorithms incorporate highly efficient procedures for *unconstrained optimization* for a portion of each iteration, and some use a succession of linear or quadratic approximations to the original problem.

There has been a strong emphasis in recent years on developing high-quality, reliable *software packages* for general use in applying the best of these algorithms on mainframe computers. (See Selected References 6 and 7.) For example, several powerful software packages such as MINOS (discussed in Sec. 4.9) have been developed in the Systems Optimization Laboratory at Stanford University. These packages are widely used elsewhere for solving many of the types of problems discussed in this chapter (as well as linear programming problems). The steady improvements being made in both algorithmic techniques and software now are bringing some rather large problems into the range of computational feasibility.

With the current rapid growth in the use and power of personal computers, good progress is being made in nonlinear programming software development for microcomputers. For example, the GAMS/MINOS package (a combination of two well-known mainframe programs) now is available for use on IBM personal computers. Another prominent package called GINO (see Selected Reference 8) was developed specifically for microcomputers.

Research in nonlinear programming remains very active.

---

[1] The technical reason is that $f(\mathbf{x})$ is a (strictly) *quasiconcave* function that shares the property of concave functions that a local maximum always is a global maximum. For further information, see Avriel, Mordecai, W. Erwin Diewert, Siegfried Schaible, and Israel Zang: *Generalized Concavity,* Plenum, New York, 1985.

Nonlinear
Programming

1. Avriel, Mordecai: *Nonlinear Programming: Analysis and Methods,* Prentice-Hall, Engle-wood Cliffs, N.J., 1976.

2. Bazaraa, Mokhtar S., and C. M. Shetty: *Nonlinear Programming: Theory and Algorithms,* Wiley, New York, 1979.

3. Beightler, Charles S., Don T. Phillips, and Douglas J. Wilde: *Foundations of Optimization,* 2d ed., Prentice-Hall, Englewood Cliffs, N.J., 1979.

4. Bracken, Jerome, and Garth P. McCormick: *Selected Applications of Nonlinear Program-ming,* Wiley, New York, 1968.

5. Fletcher, R.: *Practical Methods of Optimization,* 2d ed., Wiley, New York, 1988.

6. Gill, Philip E., Walter Murray, Michael A. Saunders, and Margaret H. Wright: "Trends in Nonlinear Programming Software," *European Journal of Operational Research,* **17**:141–149, 1984.

7. Gill, Philip E., Walter Murray, and Margaret H. Wright: *Practical Optimization,* Academic Press, London, 1981.

8. Liebman, J., L. Schrage, L. Lasdon, and A. Waren: *Applications of Modeling and Opti-mization with GINO,* Scientific Press, Palo Alto, Calif., 1985.

9. Luenberger, David G.: *Introduction to Linear and Nonlinear Programming,* 2d ed., Addison-Wesley, Reading, Mass., 1984.

10. McCormick, G. P.: *Nonlinear Programming: Theory, Algorithms and Applications,* Wiley, New York, 1983.

11. Reklaitis, G. V., A. Ravindran, and K. M. Ragsdell: *Engineering Optimization: Methods and Applications,* Wiley, New York, 1983.

12. Schittkowski, Klaus: *Nonlinear Programming Codes,* Lecture Notes in Economics and Mathematical Systems No. 183, Springer-Verlag, Berlin, Heidelberg, New York, 1980.

13. Shapiro, Jeremy F.: *Mathematical Programming: Structures and Algorithms,* Wiley, New York, 1979, chaps. 5–7.

14. Zangwill, Willard I.: *Nonlinear Programming: A Unified Approach,* Prentice-Hall, Engle-wood Cliffs, N.J., 1969.

PROBLEMS

1. Consider the *product mix* problem described in Chap. 3, Prob. 2. Suppose that this manufacturing firm actually encounters *price elasticity* in selling the three products, so that the profits would be different from those stated in Chap. 3. In particular, suppose that the unit costs for producing products 1, 2, and 3 are $25, $10, and $15, respectively, and that the prices required (in dollars) in order to be able to sell x_1, x_2, and x_3 units are $(35 + 100x_1^{-1/3})$, $(15 + 40x_2^{-1/4})$, and $(20 + 50x_3^{-1/2})$, respectively.

    (a) Formulate a *linearly constrained optimization* model for the problem of determining how many units of each product the firm should produce to maximize profit.

    (b) Verify that this problem is a *convex programming* problem.

    (c) Starting from the initial trial solution $(x_1, x_2, x_3) = (10, 10, 10)$, apply two itera-tions of the *Frank-Wolfe algorithm.*

2. For the P & T Co. problem described in Sec. 7.1, suppose that there is a 10 percent discount in the shipping cost for all truckloads *beyond* the first 40 for each combination of cannery and warehouse. Show that this situation leads to a *nonconvex programming* problem.

3. Consider the following example of the *portfolio selection* problem with risky secu-rities described in Sec. 14.1. Just two stocks are being considered for inclusion in the portfolio.

The estimated mean and variance of the return on each share of stock 1 are 5 and 4, respectively, whereas the corresponding quantities for stock 2 are 10 and 100, respectively. The covariance of the return on one share each of the two stocks is 5. The price per share is 20 for stock 1 and 30 for stock 2, where the total amount budgeted for the portfolio is 50.

    (a) Without assigning a specific numerical value to β, formulate the *quadratic programming* model for this problem.

    (b) Verify that the model in part (a) is a *convex programming* problem by using the test in Appendix 1 to show that the objective function is *concave*.

    (c) Starting from the initial trial solution $(x_1, x_2) = (0, 0)$, apply five iterations of the *Frank-Wolfe algorithm* to this problem with $\beta = 0.1$.

    (d) Repeat part (c) for $\beta = 0.02$ (low aversion to risk) and $\beta = 0.5$ (high aversion to risk). Also solve by inspection the extreme cases of $\beta = 0$ (no aversion to risk) and $\beta = \infty$ (complete aversion to risk), where the latter case means to minimize $V(\mathbf{x})$. Characterize the solutions for the five values of β in terms of the relative concentration on the conservative investment (stock 1) or the risky investment (stock 2).

4. Consider the variation of the Wyndor Glass Co. example represented in Fig. 14.5, where the second and third functional constraints of the original problem (see Sec. 3.1) have been replaced by $9x_1^2 + 5x_2^2 \leq 216$.

    (a) Demonstrate that $(x_1, x_2) = (2, 6)$ with $Z = 36$ is indeed optimal by showing that the objective function line, $36 = 3x_1 + 5x_2$, is *tangent* to this constraint boundary at $(2, 6)$. (*Hint:* Express x_2 in terms of x_1 on this boundary, and then differentiate this expression with respect to x_1 to find the slope of the boundary.)

    (b) Starting from the initial trial solution $(x_1, x_2) = (2, 3)$, apply *SUMT* to this problem, with $r = 1, 10^{-2}, 10^{-4}$.

5. Consider the variation of the Wyndor Glass Co. problem represented in Fig. 14.6, where the original objective function (see Sec. 3.1) has been replaced by $Z = 126x_1 - 9x_1^2 + 182x_2 - 13x_2^2$. Demonstrate that $(x_1, x_2) = (\frac{8}{3}, 5)$ with $Z = 857$ is indeed optimal by showing that the ellipse, $857 - 126x_1 - 9x_1^2 + 182x_2 - 13x_2^2$, is *tangent* to the constraint boundary, $3x_1 + 2x_2 = 18$, at $(\frac{8}{3}, 5)$. (*Hint:* Solve for x_2 in terms of x_1 for the ellipse, and then differentiate this expression with respect to x_1 to find the slope of the ellipse.)

6. Consider the following function.

$$f(x) = 48x - 60x^2 + x^3.$$

    (a) Use the first and second derivatives to find the *local maxima* and *local minima* of $f(x)$.

    (b) Use the first and second derivatives to show that $f(x)$ has neither a *global maximum* nor a *global minimum* because it is unbounded in both directions.

7. For each of the following functions, show whether it is convex, concave, or neither.

    (a) $f(x) = 10x - x^2$.

    (b) $f(x) = x^4 + 6x^2 + 12x$.

    (c) $f(x) = 2x^3 - 3x^2$.

    (d) $f(x) = x^4 + x^2$.

    (e) $f(x) = x^3 + x^4$.

8. * For each of the following functions, use the test given in Appendix 1 to determine whether it is convex, concave, or neither.

    (a) $f(\mathbf{x}) = x_1 x_2 - x_1^2 - x_2^2$.

    (b) $f(\mathbf{x}) = 3x_1 + 2x_1^2 + 4x_2 + x_2^2 - 2x_1 x_2$.

    (c) $f(\mathbf{x}) = x_1^2 + 3x_1 x_2 + 2x_2^2$.

    (d) $f(\mathbf{x}) = 20x_1 + 10x_2$.

    (e) $f(\mathbf{x}) = x_1 x_2$.

9. Consider the following function.

$$f(\mathbf{x}) = 5x_1 + 2x_2^2 + x_3^2 - 3x_3x_4 + 4x_4^2 + 2x_5^4 + x_5^2 + 3x_5x_6 + 6x_6^2 + 3x_6x_7 + x_7^2.$$

Show that $f(\mathbf{x})$ is *convex* by expressing it as a *sum* of functions of one or two variables and then showing (see Appendix 1) that all of these functions are convex.

10. Consider the following nonlinear programming problem.

$$\text{Maximize} \quad f(\mathbf{x}) = x_1 + x_2,$$

subject to

$$x_1^2 + x_2^2 \le 1$$

and

$$x_1 \ge 0, \quad x_2 \ge 0.$$

(a) Verify that this is a *convex programming* problem.
(b) Solve this problem graphically.
(c) Use the *KKT conditions* to verify that the solution you obtained in part (b) is optimal.

11. Consider the following *constrained optimization problem.*

$$\text{Maximize} \quad f(x) = -6x + 3x^2 - 2x^3,$$

subject to

$$x \ge 0.$$

Use just the first and second derivatives of $f(x)$ to derive an optimal solution.

12. Consider the following nonlinear programming problem.

$$\text{Minimize} \quad Z = x_1^4 + 2x_1^2 + 2x_1x_2 + 4x_2^2,$$

subject to

$$2x_1 + x_2 \ge 10$$

$$x_1 + 2x_2 \ge 10$$

and

$$x_1 \ge 0, \quad x_2 \ge 0.$$

(a) Of the special types of nonlinear programming problems described in Sec. 14.3, to which type or types can this particular problem be fitted? Justify your answer.
(b) What are the *KKT conditions* for this problem? Use these conditions to determine whether $(x_1, x_2) = (0, 10)$ can be optimal.
(c) If *SUMT* were to be applied directly to this problem, what would be the unconstrained function $P(\mathbf{x}; r)$ to be *minimized* at each iteration?
(d) Setting $r = 100$ and using $(x_1, x_2) = (5, 5)$ as the initial trial solution, apply the *gradient search procedure* with $\epsilon = 10$ to minimize the function $P(\mathbf{x}; r)$ you obtained in part (c).
(e) Now suppose that the problem were changed slightly by replacing the nonnegativity constraints by $x_1 \ge 1$, $x_2 \ge 1$. Convert this new problem into an *equivalent* problem that has just two functional constraints, two variables, and two nonnegativity constraints.

13. Consider the expression given in Sec. 14.7 for the *KKT conditions* for the quadratic programming problem. Show that the problem of finding a *feasible solution* for these conditions is a *linear complementarity problem,* as introduced in Sec. 14.3, by identifying $\mathbf{w}$, $\mathbf{z}$, $\mathbf{q}$, and $\mathbf{M}$ in terms of the vectors and matrices in Sec. 14.7.

14. Consider the following *geometric programming* problem.

$$\text{Minimize} \quad f(\mathbf{x}) = 2x_1^{-2}x_2^{-1} + x_1^{-1}x_2^{-2},$$

subject to

$$4x_1x_2 + x_1^2x_2^2 \le 12$$

and $$x_1 \geq 0, \qquad x_2 \geq 0.$$

(a) Transform this problem into an equivalent *convex programming* problem.
(b) Use the test given in Appendix 1 to verify that the model formulated in part (a) is indeed a convex programming problem.

15. Consider the following *linear fractional programming* problem.

$$\text{Maximize} \qquad f(\mathbf{x}) = \frac{10x_1 + 20x_2 + 10}{3x_1 + 4x_2 + 20},$$

subject to $$x_1 + 3x_2 \leq 50$$

$$3x_1 + 2x_2 \leq 80$$

and $$x_1 \geq 0, \qquad x_2 \geq 0.$$

(a) Transform this problem into an equivalent *linear programming* problem.
(b) Use the simplex method (the *automatic* routine in your OR COURSEWARE) to solve the model formulated in part (a). What is the resulting optimal solution for the original problem?

16.* Use the *one-dimensional search procedure* to solve (approximately) the following problem.

$$\text{Maximize} \qquad f(x) = x^3 + 2x - 2x^2 - 0.25x^4.$$

Use an error tolerance $\varepsilon = 0.04$ and initial bounds $\underline{x} = 0, \bar{x} = 2.4$.

17. Use the *one-dimensional search procedure* with an error tolerance $\varepsilon = 0.04$ and with the following initial bounds to solve (approximately) each of the following problems.
(a) Maximize $\quad f(x) = 6x - x^2$, with $\underline{x} = 0, \bar{x} = 4.8$.
(b) Minimize $\quad f(x) = 6x + 7x^2 + 4x^3 + x^4$, with $\underline{x} = -4, \bar{x} = 1$.

18. Use the *one-dimensional search procedure* to solve (approximately) the following problem.

$$\text{Maximize} \qquad f(x) = 48x^5 + 42x^3 + 3.5x - 16x^6 - 61x^4 - 16.5x^2.$$

Use an error tolerance $\varepsilon = 0.08$ and initial bounds $\underline{x} = -1, \bar{x} = 4$.

19. Use the *one-dimensional search procedure* to solve (approximately) the following problem.

$$\text{Maximize} \qquad f(x) = x^3 + 30x - x^6 - 2x^4 - 3x^2.$$

Use an error tolerance $\varepsilon = 0.07$ and find appropriate initial bounds by inspection.

20. Consider the following convex programming problem.

$$\text{Minimize} \qquad Z = x^4 + x^2 - 4x,$$

subject to $$x \leq 2$$

and $$x \geq 0.$$

(a) Use one simple calculation *just* to check whether the optimal solution lies in the interval $0 \leq x \leq 1$ or the interval $1 \leq x \leq 2$. (Do *not* actually solve for the optimal solution in order to determine in which interval it must lie.) Explain your logic.
(b) Use the *one-dimensional search procedure* with initial bounds $\underline{x} = 0, \bar{x} = 2$ and with an error tolerance $\varepsilon = 0.02$ to solve (approximately) this problem.
(c) Use the *KKT conditions* to derive the optimal solution.

21. Consider the problem of maximizing a differentiable function $f(x)$ of a single unconstrained variable x. Let $\underline{x}_0$ and $\bar{x}_0$, respectively, be a valid lower bound and upper bound on the same global maximum (if one exists). Prove the following general properties of the *one-dimensional search procedure* (as presented in Sec. 14.4) for attempting to solve such a problem.

(a) Given $\underline{x}_0$, $\bar{x}_0$, and $\varepsilon = 0$, the sequence of trial solutions selected by the *midpoint rule* must *converge* to a limiting solution. [*Hint:* First show that $\lim_{n\to\infty}(\bar{x}_n - \underline{x}_n) = 0$, where $\bar{x}_n$ and $\underline{x}_n$ are the upper and lower bounds identified at iteration n.]

(b) If $f(x)$ is *concave* {so that $[df(x)/dx]$ is a monotone decreasing function of x}, then the limiting solution in part (a) must be a global maximum.

(c) If $f(x)$ is not concave everywhere, but would be concave if its domain were restricted to the interval between $\underline{x}_0$ and $\bar{x}_0$, then the limiting solution in part (a) must be a global maximum.

(d) If $f(x)$ is not concave even over the interval between $\underline{x}_0$ and $\bar{x}_0$, then the limiting solution in part (a) need not be a global maximum. (Prove this by graphically constructing a counterexample.)

(e) If $[df(x)/dx] < 0$ for all x, then no $\underline{x}_0$ exists. If $[df(x)/dx] > 0$ for all x, then no $\bar{x}_0$ exists. In either case, $f(x)$ does not possess a global maximum.

(f) If $f(x)$ is concave and $\lim_{x\to-\infty} [df(x)/dx] < 0$, then no $\underline{x}_0$ exists. If $f(x)$ is concave and $\lim_{x\to\infty} [df(x)/dx] > 0$, then no $\bar{x}_0$ exists. In either case, $f(x)$ does not possess a global maximum.

22. Consider the following *unconstrained optimization* problem.

$$\text{Maximize} \qquad f(\mathbf{x}) = 2x_1x_2 + x_2 - x_1^2 - 2x_2^2.$$

(a) Starting from the initial trial solution $(x_1, x_2) = (1, 1)$, apply the *gradient search procedure* with $\varepsilon = 0.25$ to obtain an approximate solution.

(b) Solve the system of linear equations obtained by setting $\nabla f(\mathbf{x}) = \mathbf{0}$ to obtain the exact solution.

(c) Referring to Fig. 14.11 as a sample for a similar problem, draw the path of trial solutions you obtained in part (a). Then show the apparent *continuation* of this path with your best guess for the next three trial solutions [based on the pattern in part (a) and in Fig. 14.11]. Also show the exact solution from part (b) toward which this sequence of trial solutions is converging.

23. Repeat the three parts of Prob. 22 (except with $\varepsilon = 0.5$) for the following *unconstrained optimization* problem.

$$\text{Maximize} \qquad f(\mathbf{x}) = 2x_1x_2 - 2x_1^2 - x_2^2.$$

24. Starting from the initial trial solution $(x_1, x_2) = (1, 1)$, do two iterations of the *gradient search procedure* to begin solving the following problem.

$$\text{Maximize} \qquad f(\mathbf{x}) = 4x_1x_2 - 2x_1^2 - 3x_2^2.$$

Then solve $\nabla f(\mathbf{x}) = \mathbf{0}$ directly to obtain the exact solution.

25.* Starting from the initial trial solution $(x_1, x_2) = (0, 0)$, use the *gradient search procedure* with $\varepsilon = 0.3$ to obtain an approximate solution for the following problem.

$$\text{Maximize} \qquad f(\mathbf{x}) = 8x_1 - x_1^2 - 12x_2 - 2x_2^2 + 2x_1x_2.$$

Then solve $\nabla f(\mathbf{x}) = \mathbf{0}$ directly to obtain the exact solution.

26. Starting from the initial trial solution $(x_1, x_2) = (0, 0)$, do two iterations of the *gradient search procedure* to begin solving the following problem.

$$\text{Maximize} \qquad f(\mathbf{x}) = 6x_1 + 2x_1x_2 - 2x_2 - 2x_1^2 - x_2^2.$$

Then solve $\nabla f(\mathbf{x}) = \mathbf{0}$ directly to obtain the exact solution.

27. Starting from the initial trial solution $(x_1, x_2) = (0, 0)$, apply *two* iterations of the *gradient search procedure* to the following problem.

$$\text{Maximize} \qquad f(\mathbf{x}) = 4x_1 + 2x_2 + x_1^2 - x_1^4 - 2x_1x_2 - x_2^2.$$

For each of these iterations, approximately solve for t^* by applying *two* iterations of the *one-dimensional search procedure* with initial bounds $\underline{t} = 0$, $\bar{t} = 1$.

28. Starting from the initial trial solution $(x_1, x_2, x_3) = (1, 1, 1)$, use the *gradient search procedure* with $\varepsilon = 0.05$ to solve (approximately) the following problem.

$$\text{Maximize} \qquad f(\mathbf{x}) = 3x_1x_2 + 3x_2x_3 - x_1^2 - 6x_2^2 - x_3^2.$$

29.* Starting from the initial trial solution $(x_1, x_2) = (0, 0)$, use the *gradient search procedure* with $\varepsilon = 1$ to solve (approximately) each of the following problems.

(a) Maximize $\quad f(\mathbf{x}) = x_1x_2 + 3x_2 - x_1^2 - x_2^2.$

(b) Minimize $\quad f(\mathbf{x}) = x_1^2x_2^2 + 2x_1^2 + 2x_2^2 - 4x_1 + 4x_2.$

30. Consider the following *linearly constrained optimization* problem.

$$\text{Maximize} \qquad f(\mathbf{x}) = \ln(x_1 + x_2),$$

subject to

$$x_1 + 2x_2 \le 5$$

and

$$x_1 \ge 0, \qquad x_2 \ge 0,$$

where ln denotes *natural logarithm*.

(a) Verify that this problem is a *convex programming* problem.

(b) Use the *KKT conditions* to derive an optimal solution.

(c) Use intuitive reasoning to demonstrate that the solution obtained in part (b) is indeed optimal. [*Hint:* Note that $\ln(x_1 + x_2)$ is a monotone strictly increasing function of $(x_1 + x_2)$.]

(d) Starting from the initial trial solution $(x_1, x_2) = (1, 1)$, use one iteration of the *Frank-Wolfe algorithm* to obtain exactly the same solution you found in part (b), and then use a second iteration to verify that it is an optimal solution (because it is replicated exactly). Explain why exactly the same results would be obtained on these two iterations with any other initial trial solution except $(0, 0)$. What complication arises with $(0, 0)$?

31. Consider the following *linearly constrained optimization* problem.

$$\text{Maximize} \qquad f(\mathbf{x}) = \ln(x_1 + 1) - x_2^2,$$

subject to

$$x_1 + 2x_2 \le 3$$

and

$$x_1 \ge 0, \qquad x_2 \ge 0,$$

where ln denotes *natural logarithm*.

(a) Verify that this problem is a *convex programming* problem.

(b) Use the *KKT conditions* to derive an optimal solution.

(c) Use intuitive reasoning to demonstrate that the solution obtained in part (b) is indeed optimal.

(d) Starting from the initial trial solution $(x_1, x_2) = (0, 0)$, use one iteration of the *Frank-Wolfe algorithm* to obtain exactly the same solution you found in part (b), and then use a second iteration to verify that it is an optimal solution (because it is replicated exactly).

32. Consider the following *convex programming* problem.

$$\text{Maximize} \qquad f(\mathbf{x}) = 10x_1 - 2x_1^2 - x_1^3 + 8x_2 - x_2^2,$$

subject to

$$x_1 + x_2 \le 2$$

and $\qquad x_1 \geq 0, \qquad x_2 \geq 0.$

(a) Use the *KKT conditions* to demonstrate that $(x_1, x_2) = (1, 1)$ is *not* an optimal solution.

(b) Use the *KKT conditions* to derive an optimal solution.

33.* Consider the nonlinear programming problem given in Chap. 11, Prob. 13. Determine whether $(x_1, x_2) = (1, 2)$ can be optimal by applying the *KKT conditions*.

34. Consider the following *convex programming* problem.

$$\text{Maximize} \qquad f(\mathbf{x}) = 24x_1 - x_1^2 + 10x_2 - x_2^2,$$

subject to $\qquad\qquad\qquad\qquad\qquad x_1 \leq 8$

$$x_2 \leq 7$$

and $\qquad\qquad\qquad\qquad\qquad x_1 \geq 0, \qquad x_2 \geq 0.$

(a) Use the *KKT conditions* for this problem to derive an optimal solution.

(b) Decompose this problem into two separate constrained optimization problems involving just x_1 and just x_2, respectively. For each of these two problems, plot the objective function over the feasible region in order to *demonstrate* that the value of x_1 or x_2 derived in part (a) is indeed optimal. Then *prove* that this value is optimal by using just the first and second derivatives of the objective function and the constraints for the respective problems.

35. Consider the following nonlinear programming problem.

$$\text{Maximize} \qquad f(\mathbf{x}) = \frac{x_1}{x_2 + 1},$$

subject to $\qquad\qquad\qquad\qquad\qquad x_1 - x_2 \leq 2$

and $\qquad\qquad\qquad\qquad\qquad x_1 \geq 0, \qquad x_2 \geq 0.$

(a) Use the *KKT conditions* to demonstrate that $(x_1, x_2) = (4, 2)$ is *not* optimal.

(b) Derive a solution that does satisfy the *KKT conditions*.

(c) Show that this problem is *not* a *convex programming* problem.

(d) Despite the conclusion in part (c), use *intuitive* reasoning to show that the solution obtained in part (b) is, in fact, optimal. [The theoretical reason is that $f(\mathbf{x})$ is *pseudoconcave*.]

(e) Use the fact that this problem is a *linear fractional programming* problem to transform it into an equivalent *linear programming* problem. Solve the latter problem and thereby identify the optimal solution for the original problem. (*Hint:* Use the equality constraint in the linear programming problem to substitute one of the variables out of the model, and then solve the model graphically.)

36.* Use the *KKT conditions* to derive an optimal solution for each of the following problems.

(a) $\qquad\qquad\qquad$ Maximize $\qquad f(\mathbf{x}) = x_1 + 2x_2 - x_2^3,$

$\qquad$ subject to $\qquad\qquad\qquad x_1 + x_2 \leq 1$

$\qquad$ and $\qquad\qquad\qquad\qquad x_1 \geq 0, \qquad x_2 \geq 0.$

(b) $\qquad\qquad\qquad$ Maximize $\qquad f(\mathbf{x}) = 20x_1 + 10x_2,$

$\qquad$ subject to $\qquad\qquad\qquad x_1^2 + x_2^2 \leq 1$

$$x_1 + 2x_2 \leq 2$$

$\qquad$ and $\qquad\qquad\qquad\qquad x_1 \geq 0, \qquad x_2 \geq 0.$

37. What are the *KKT conditions* for nonlinear programming problems of the following form?

$$\text{Minimize} \quad f(\mathbf{x}),$$

subject to
$$g_i(\mathbf{x}) \geq b_i, \quad \text{for } i = 1, 2, \ldots, m$$

and
$$\mathbf{x} \geq \mathbf{0}.$$

(*Hint:* Convert this form into our standard form assumed in this chapter by using the techniques presented in Sec. 4.6, and then applying the *KKT conditions* as given in Sec. 14.6.)

38. Consider the following nonlinear programming problem.

$$\text{Minimize} \quad Z = 2x_1^2 + x_2^2,$$

subject to
$$x_1 + x_2 = 10$$

and
$$x_1 \geq 0, \quad x_2 \geq 0.$$

(*a*) Of the special types of nonlinear programming problems described in Sec. 14.3, to which type or types can this particular problem be fitted? Justify your answer. (*Hint:* First convert this problem to an *equivalent* nonlinear programming problem that fits the form given in the second paragraph of the chapter, with $m = 2$ and $n = 2$.)

(*b*) Obtain the *KKT conditions* for this problem.

(*c*) Use the *KKT conditions* to derive an optimal solution.

39. Consider the following *linearly constrained programming* problem.

$$\text{Minimize} \quad f(\mathbf{x}) = x_1^3 + 4x_2^2 + 16x_3,$$

subject to
$$x_1 + x_2 + x_3 = 5$$

and
$$x_1 \geq 1, \quad x_2 \geq 1, \quad x_3 \geq 1.$$

(*a*) Convert this problem to an *equivalent* nonlinear programming problem that fits the form given at the beginning of the chapter (second paragraph), with $m = 2$ and $n = 3$.

(*b*) Use the form obtained in part (*a*) to construct the *KKT conditions* for this problem.

(*c*) Use the KKT conditions to check whether $(x_1, x_2, x_3) = (2, 1, 2)$ is optimal.

(*d*) Starting from the initial trial solution $(x_1, x_2, x_3) = (1, 2, 2)$, apply two iterations of the Frank-Wolfe algorithm.

40. Consider the following *linearly constrained convex programming* problem.

$$\text{Minimize} \quad Z = x_1^2 - 6x_1 + x_2^3 - 3x_2,$$

subject to
$$x_1 + x_2 \leq 1$$

and
$$x_1 \geq 0, \quad x_2 \geq 0.$$

(*a*) Obtain the *KKT conditions* for this problem.

(*b*) Use the *KKT conditions* to check whether $(x_1, x_2) = (\frac{1}{2}, \frac{1}{2})$ is an optimal solution.

(*c*) Use the *KKT conditions* to derive an optimal solution.

(*d*) Starting from the initial trial solution $(x_1, x_2) = (0, 0)$, use one iteration of the *Frank-Wolfe algorithm* to obtain exactly the same solution you found in part (*c*), and then use a second iteration to verify that it is an optimal solution (because it is replicated exactly). Explain why exactly the same results would be obtained on these two iterations with any other trial solution as well.

41. Consider the following *linearly constrained convex programming* problem.

$$\text{Maximize} \quad f(\mathbf{x}) = 8x_1 - x_1^2 + 2x_2 + x_3,$$

subject to $\qquad x_1 + 3x_2 + 2x_3 \le 12$

and $\qquad\qquad x_1 \ge 0, \qquad x_2 \ge 0, \qquad x_3 \ge 0.$

(a) Use the *KKT conditions* to demonstrate that $(x_1, x_2, x_3) = (2, 2, 2)$ is *not* an optimal solution.

(b) Use the *KKT conditions* to derive an optimal solution. (*Hint:* Do some preliminary intuitive analysis to determine the most promising case regarding which variables are nonzero and which are zero.)

(c) Starting from the initial trial solution $(x_1, x_2, x_3) = (0, 0, 0)$, apply three iterations of the *Frank-Wolfe algorithm*.

42. Use the *KKT conditions* to determine whether $(x_1, x_2, x_3) = (1, 1, 1)$ can be optimal for the following problem.

$$\text{Minimize} \qquad Z = 2x_1 + x_2^3 + x_3^2,$$

subject to $\qquad\qquad x_1^2 + 2x_2^2 + x_3^2 \ge 4$

and $\qquad\qquad x_1 \ge 0, \qquad x_2 \ge 0, \qquad x_3 \ge 0.$

43. Consider the *quadratic programming* example presented in Sec. 14.7.

(a) Use the test given in Appendix 1 to show that the objective function is *strictly concave*.

(b) Verify that the objective function is *strictly concave* by demonstrating that $\mathbf{Q}$ is a *positive definite* matrix; that is, $\mathbf{x}^T\mathbf{Q}\mathbf{x} > \mathbf{0}$ for all $\mathbf{x} \ne \mathbf{0}$. (*Hint:* Reduce $\mathbf{x}^T\mathbf{Q}\mathbf{x}$ to a sum of squares.)

(c) Show that $x_1 = 12$, $x_2 = 9$, and $u_1 = 3$ satisfy the *KKT conditions* when they are written in the form given in Sec. 14.6.

(d) Starting from the initial trial solution $(x_1, x_2) = (5, 5)$, apply three iterations of the *Frank-Wolfe algorithm*.

44.* Consider the following *quadratic programming* problem.

$$\text{Maximize} \qquad f(\mathbf{x}) = 8x_1 - x_1^2 + 4x_2 - x_2^2,$$

subject to $\qquad\qquad x_1 + x_2 \le 2$

and $\qquad\qquad x_1 \ge 0, \qquad x_2 \ge 0.$

(a) Use the *KKT conditions* to derive an optimal solution.

(b) Now suppose that this problem is to be solved by the *modified simplex method*. Formulate the *linear programming* problem that is to be addressed *explicitly*, and then identify the additional *complementarity constraint* that is enforced automatically by the algorithm.

(c) Apply the *modified simplex method* to the problem as formulated in part (b).

45. Consider the following *quadratic programming* problem.

$$\text{Maximize} \qquad f(\mathbf{x}) = 20x_1 - 20x_1^2 + 50x_2 - 5x_2^2 + 20x_1x_2,$$

subject to $\qquad\qquad x_1 + x_2 \le 6$

$$x_1 + 4x_2 \le 18$$

and $\qquad\qquad x_1 \ge 0, \qquad x_2 \ge 0.$

Suppose that this problem is to be solved by the *modified simplex method.*

(a) Formulate the *linear programming* problem that is to be addressed *explicitly*, and then identify the additional *complementarity constraint* that is enforced automatically by the algorithm.

(b) Apply the *modified simplex method* to the problem as formulated in part (a).

46. Consider the following *quadratic programming* problem.

$$\text{Maximize} \quad f(\mathbf{x}) = 2x_1 + 3x_2 - x_1^2 - x_2^2,$$

subject to

$$x_1 + x_2 \le 2$$

and

$$x_1 \ge 0, \quad x_2 \ge 0.$$

(a) Starting from the initial trial solution $(x_1, x_2) = (0, 0)$, use the *Frank-Wolfe algorithm* (six iterations) to solve the problem (approximately).

(b) Show graphically how the sequence of trial solutions obtained in part (a) can be extrapolated to obtain a closer approximation of an optimal solution. What is your resulting estimate of this solution?

(c) Use the *KKT conditions* to derive an optimal solution directly.

(d) Now suppose that this problem is to be solved by the *modified simplex method*. Formulate the *linear programming* problem that is to be addressed *explicitly*, and then identify the additional *complementarity constraint* that is enforced automatically by the algorithm.

(e) Without applying the *modified simplex method*, show that the solution derived in part (c) is indeed optimal ($Z = 0$) for the equivalent problem formulated in part (d).

(f) Apply the *modified simplex method* to the problem as formulated in part (d).

47. Repeat parts (a) (three iterations only), (c), (d), and (e) of Prob. 46 for the first *quadratic programming* variation of the *Wyndor Glass Co.* problem presented in Sec. 14.2 (see Fig. 14.6). That is:

$$\text{Maximize} \quad Z = 126x_1 - 9x_1^2 + 182x_2 - 13x_2^2,$$

subject to the linear constraints given in Sec. 3.1.

48. Consider the following *quadratic programming* problem.

$$\text{Minimize} \quad f(x_1, x_2) = (x_1 - 1)^2 + (x_2 - 2)^2 - 3(x_1 + x_2),$$

subject to

$$4x_1 + x_2 \le 20$$

$$x_1 + 4x_2 \le 20$$

and

$$x_1 \ge 0, \quad x_2 \ge 0.$$

(a) Obtain the *KKT conditions* for this problem in the form given in Sec. 14.6. (*Hint:* These conditions assume that the objective function is to be maximized.)

(b) You are given the information that the optimal solution does *not* lie on the boundary of the feasible region. Use this information to derive the optimal solution from the KKT conditions.

(c) Now suppose that this problem is to be solved by the *modified simplex method*. Formulate the *linear programming* problem that is to be addressed *explicitly*, and then identify the additional *complementarity constraint* that is enforced automatically by the algorithm.

(d) Apply the *modified simplex method* to the problem as formulated in part (c).

(e) Use the *separable programming* formulation presented in Sec. 14.8 to formulate an approximate linear programming model for this problem. Use $x_1, x_2 = 0, 2.5, 5$ as the breakpoints of the piecewise linear functions.

(f) Use the simplex method (the *automatic* routine in your OR COURSEWARE) to solve the model formulated in part (e). Then reexpress this solution in terms of the *original* variables of the problem.

49. A certain corporation is planning to produce and market three different products. Let x_1, x_2, and x_3 denote the number of units of the three respective products to be produced. The preliminary estimates of their potential profitability are as follows.

For the first 15 units produced of product 1, the unit profit would be approximately $36. The unit profit would only be $3 for any additional units of product 1. For the first 20 units produced of product 2, the unit profit is estimated at $24. The unit profit would be $12 for each of the next 20 units and $9 for any additional units. For the first 10 units of product 3, the unit profit would be $45. The unit profit would be $30 for each of the next 5 units and $18 for any additional units.

Certain limitations on the use of needed resources impose the following constraints on the production of the three products:

$$x_1 + x_2 + x_3 \leq 60$$

$$3x_1 + 2x_2 \leq 200$$

$$x_1 + 2x_3 \leq 70.$$

Management wants to know what values of x_1, x_2, and x_3 should be chosen to maximize total profit.

(a) Use the *separable programming* technique presented in Sec. 14.8 to formulate a linear programming model for this problem.

(b) Now suppose that there is an additional constraint that the profit from products 1 and 2 must total at least $900. Use the technique presented in the Extensions subsection of Sec. 14.8 to add this constraint to the model formulated in part (a).

50.* Consider the following *convex programming* problem.

$$\text{Maximize} \quad f(\mathbf{x}) = 4x_1 + 6x_2 - x_1^3 - 2x_2^2,$$

subject to

$$x_1 + 3x_2 \leq 8$$

$$5x_1 + 2x_2 \leq 14$$

and

$$x_1 \geq 0, \quad x_2 \geq 0.$$

(a) Verify that $(x_1, x_2) = \left(\frac{2}{\sqrt{3}}, \frac{3}{2}\right)$ is an optimal solution by applying the *KKT conditions*.

(b) Use the *separable programming* technique presented in Sec. 14.8 to formulate an *approximate* linear programming model for this problem. Use $x_1, x_2 = 0, 1.5, 3$ as the breakpoints of the piecewise linear functions.

(c) Use the simplex method (the *automatic* routine in your OR COURSEWARE) to solve the approximate model formulated in part (b). Verify that the optimal solution satisfies the *special restriction* for the model. Compare this solution with the exact optimal solution for the original problem [see part (a)].

51. Reconsider the production scheduling problem of the Build-Em-Fast Company described in Prob. 21 of Chap. 7. The *special restriction* for such a situation is that overtime should not be used in any particular period unless regular time in that period is completely used up. Explain why the logic of *separable programming* implies that this restriction will be satisfied automatically by any optimal solution for the transportation problem formulation of the problem.

52. Consider the following *linearly constrained convex programming* problem.

$$\text{Maximize} \quad f(\mathbf{x}) = 32x_1 + 50x_2 - 10x_2^2 + x_2^3 - x_1^4 - x_2^4,$$

subject to

$$3x_1 + x_2 \leq 11$$

$$2x_1 + 5x_2 \leq 16$$

and
$$x_1 \geq 0, \qquad x_2 \geq 0.$$

(a) Use the *separable programming* technique presented in Sec. 14.8 to formulate an *approximate* linear programming model for this problem. Use $x_1 = 0, 1, 2, 3$ and $x_2 = 0, 1, 2, 3$ as the breakpoints of the piecewise linear functions.

(b) Use the *KKT conditions* to determine whether $(x_1, x_2) = (2, 2)$ can be optimal for this original problem (not the approximate model).

(c) Starting from the initial trial solution $(x_1, x_2) = (0, 0)$, use the *Frank-Wolfe algorithm* (four iterations) to solve the original problem (approximately).

(d) Ignore the constraints and solve the resulting two *one-variable unconstrained optimization* problems. Use calculus to solve the problem involving x_1 and use the *one-dimensional search procedure* with $\varepsilon = 0.1$ and initial bounds 0 and 4 to solve the problem involving x_2. Show that the resulting solution for (x_1, x_2) satisfies all of the constraints, so it is actually optimal for the original problem.

53. Suppose that the *separable programming* technique has been applied to a certain problem (the "original problem") to convert it into the following equivalent linear programming problem.

$$\text{Maximize} \quad Z = 5x_{11} + 4x_{12} + 2x_{13} + 4x_{21} + x_{22},$$

subject to

$$3x_{11} + 3x_{12} + 3x_{13} + 2x_{21} + 2x_{22} \leq 25$$

$$2x_{11} + 2x_{12} + 2x_{13} - x_{21} - x_{22} \leq 10$$

and

$$0 \leq x_{11} \leq 2$$

$$0 \leq x_{12} \leq 3$$

$$0 \leq x_{13}$$

$$0 \leq x_{21} \leq 3$$

$$0 \leq x_{22} \leq 1.$$

What was the mathematical model for the original problem? (You may define the objective function either algebraically or graphically, but express the constraints algebraically.)

54. For each of the following cases, *prove* that the *key property of separable programming* given in Sec. 14.8 must hold. [*Hint:* Assume that there exists an optimal solution that violates this property, and then contradict this assumption by showing that there exists a better feasible solution.]

(a) The special case of separable programming where all of the $g_i(\mathbf{x})$ are linear functions.

(b) The general case of separable programming where all of the functions are nonlinear functions of the designated form. [*Hint:* Think of the functional constraints as constraints on resources, where $g_{ij}(x_j)$ represents the amount of resource i used by running activity j at level x_j, and then use what the convexity assumption implies about the slopes of the approximating piecewise linear function.]

55. The *MFG Company* produces a certain subassembly in each of two separate plants. These subassemblies are then brought to a third nearby plant where they are used in the production of a certain product. The peak season of demand for this product is approaching, so in order to maintain the production rate within a desired range, it is necessary to use temporarily some overtime in making the subassemblies. The cost per subassembly on regular time (RT) and on overtime (OT) is shown in the following table for both plants, along with the maximum number of subassemblies that can be produced on RT and on OT each day.

	Unit Cost		Capacity	
	RT	OT	RT	OT
Plant 1	$15	$25	2,000	1,000
Plant 2	$16	$24	1,000	500

Let x_1 and x_2 denote the total number of subassemblies produced per day at plants 1 and 2, respectively. Suppose that the objective is to maximize $Z = x_1 + x_2$, subject to the constraint that the total daily cost should not exceed $60,000. Note that the mathematical programming formulation of this problem (with x_1 and x_2 as decision variables) has the same form as the main case of the *separable programming* model described in Sec. 14.8, except that the separable functions appear in a constraint function rather than the objective function. However, if it is allowable to use OT even when the RT capacity at that plant is not fully used, the same approach can be used to reformulate the problem as a linear programming problem.

(a) Formulate this linear programming problem.
(b) Explain why the logic of *separable programming* also applies here to guarantee that an optimal solution for the model formulated in part (a) never uses OT unless the RT capacity at that plant has been fully used.

56. Consider the following nonlinear programming problem (first considered in Chap. 11, Prob. 19).

$$\text{Maximize} \quad Z = 5x_1 + x_2,$$

subject to

$$2x_1^2 + x_2 \leq 13$$

$$x_1^2 + x_2 \leq 9$$

and

$$x_1 \geq 0, \quad x_2 \geq 0.$$

(a) Show that this problem is a *convex programming* problem.
(b) Use the *separable programming* technique discussed at the end of Sec. 14.8 to formulate an approximate linear programming model for this problem. Use the integers as the breakpoints of the piecewise linear function.
(c) Use the simplex method (the *automatic* routine in your OR COURSEWARE) to solve the model formulated in part (b). Then reexpress this solution in terms of the *original* variables of the problem.

57. Consider the following *linearly constrained convex programming* problem.

$$\text{Maximize} \quad f(\mathbf{x}) = 3x_1x_2 + 40x_1 + 30x_2 - 4x_1^2 - x_1^4 - 3x_2^2 - x_2^4,$$

subject to

$$4x_1 + 3x_2 \leq 12$$

$$x_1 + 2x_2 \leq 4$$

and

$$x_1 \geq 0, \quad x_2 \geq 0.$$

Starting from the initial trial solution $(x_1, x_2) = (0, 0)$, apply two iterations of the *Frank-Wolfe* algorithm.

58.* Consider the following *linearly constrained convex programming* problem.

$$\text{Maximize} \quad f(\mathbf{x}) = 3x_1 + 4x_2 - x_1^3 - x_2^2,$$

subject to

$$x_1 + x_2 \leq 1$$

and

$$x_1 \geq 0, \quad x_2 \geq 0.$$

(a) Starting from the initial trial solution $(x_1, x_2) = (\frac{1}{4}, \frac{1}{4})$, apply three iterations of the *Frank-Wolfe* algorithm.

(b) Use the *KKT conditions* to check whether the solution obtained in part (*a*) is, in fact, optimal.

(c) Starting from the initial trial solution $(x_1, x_2) = (\frac{1}{4}, \frac{1}{4})$, apply *SUMT*. Use the *gradient search procedure* to obtain the maximizing solution of $P(\mathbf{x}; r)$ at each iteration, with $r = 1, 10^{-2}, 10^{-4}$.

59. Consider the following *linearly constrained convex programming* problem.

$$\text{Maximize} \quad f(\mathbf{x}) = 4x_1 - x_1^4 + 2x_2 - x_2^2,$$

subject to

$$4x_1 + 2x_2 \le 5$$

and

$$x_1 \ge 0, \quad x_2 \ge 0.$$

(a) Starting from the initial trial solution $(x_1, x_2) = (\frac{1}{2}, \frac{1}{2})$, apply four iterations of the *Frank-Wolfe algorithm*.

(b) Show graphically how the sequence of trial solutions obtained in part (*a*) can be extrapolated to obtain a closer approximation of an optimal solution. What is your resulting estimate of this solution?

(c) Use the *KKT conditions* to check whether the solution you obtained in part (*b*) is, in fact, optimal. If not, use these conditions to derive the exact optimal solution.

(d) Starting from the initial trial solution $(x_1, x_2) = (\frac{1}{2}, \frac{1}{2})$, apply *SUMT*. Use the *gradient search procedure* to obtain the maximizing solution of $P(\mathbf{x}; r)$ at each iteration, with $r = 1, 10^{-2}, 10^{-4}, 10^{-6}$.

60. Consider the example for applying SUMT given in Sec. 14.10.

(a) Show that $(x_1, x_2) = (1, 2)$ satisfies the *KKT conditions*.

(b) Display the feasible region graphically, and then plot the locus of points, $x_1 x_2 = 2$, to demonstrate that $(x_1, x_2) = (1, 2)$ with $f(1, 2) = 2$ is, in fact, a *global maximum*.

61.* Use SUMT to solve the following *convex programming* problem.

$$\text{Maximize} \quad f(\mathbf{x}) = -2x_1 - (x_2 - 3)^2,$$

subject to

$$x_1 \ge 3$$

$$x_2 \ge 3.$$

Derive the maximizing solution of $P(\mathbf{x}; r)$ analytically, and use $r = 1, 10^{-2}, 10^{-4}, 10^{-6}$.

62. Use SUMT to solve the following *convex programming* problem.

$$\text{Minimize} \quad f(\mathbf{x}) = \frac{(x_1 + 1)^3}{3} + x_2,$$

subject to

$$x_1 \ge 1$$

$$x_2 \ge 0.$$

Derive the minimizing solution of $P(\mathbf{x}; r)$ analytically, and use $r = 1, 10^{-2}, 10^{-4}, 10^{-6}$.

63. Use SUMT to solve the following *convex programming* problem.

$$\text{Maximize} \quad f(\mathbf{x}) = x_1 x_2 - x_1 - x_1^2 - x_2 - x_2^2,$$

subject to

$$x_2 \ge 0.$$

Use the *gradient search procedure* to obtain the maximizing solution of $P(\mathbf{x}; r)$ at each iteration, with $r = 1, 10^{-2}, 10^{-4}$. Begin with the initial trial solution $(x_1, x_2) = (1, 1)$.

64. Apply SUMT to Prob. 46. Use the *gradient search procedure* to obtain the maximizing solution of $P(\mathbf{x}; r)$ at each iteration, with $r = 1, 10^{-2}, 10^{-4}$. Begin with the initial trial solution $(x_1, x_2) = (\frac{1}{2}, \frac{1}{2})$.

65. Apply SUMT to Prob. 47. Use the *gradient search procedure* to obtain the maximizing solution of $P(\mathbf{x}; r)$ at each iteration, with $r = 10^2, 1, 10^{-2}, 10^{-4}$. Begin with the initial trial solution $(x_1, x_2) = (2, 3)$.

66. Consider the following *nonconvex programming* problem.

$$\text{Maximize} \quad f(x) = 1{,}000x - 400x^2 + 40x^3 - x^4,$$

subject to

$$x^2 + x \le 500$$

and

$$x \ge 0.$$

(a) Identify the feasible values for x. Obtain general expressions for the first three derivatives of $f(x)$. Use this information to help you draw a rough sketch of $f(x)$ over the feasible region for x. Without calculating their values, mark the points on your graph that correspond to *local maxima* and *local minima*.

(b) Use the *one-dimensional search procedure* with $\varepsilon = 0.05$ to find each of the local maxima. Use your sketch from part (a) to identify appropriate initial bounds for each of these searches. Which of the local maxima is a *global maximum*?

(c) Use SUMT with $r = 10^4, 10^2, 1$ (and with $\varepsilon = 25$ for the gradient search procedure) to find each of the local maxima. Use $x = 3$ and $x = 15$ as the initial trial solutions for these searches. To find the maximizing solution of $P(\mathbf{x}; r)$ each time, use the *one-dimensional search procedure* as described in part (b). Which of the local maxima is a *global maximum*?

67. Consider the following *nonconvex programming* problem.

$$\text{Maximize} \quad f(\mathbf{x}) = 3x_1 x_2 - 2x_1^2 - x_2^2,$$

subject to

$$x_1^2 + 2x_2^2 \le 4$$

$$2x_1 - x_2 \le 3$$

$$x_1 x_2^2 + x_1^2 x_2 = 2$$

and

$$x_1 \ge 0, \qquad x_2 \ge 0.$$

(a) If SUMT were to be applied to this problem, what would be the unconstrained function $P(\mathbf{x}; r)$ to be maximized at each iteration?

(b) Starting from the initial trial solution $(x_1, x_2) = (1, 1)$, apply SUMT. Use the *gradient search procedure* to obtain the maximizing solution of $P(\mathbf{x}; r)$ at each iteration, with $r = 1, 10^{-2}, 10^{-4}$.

68. Reconsider the convex programming problem with an *equality constraint* given in Prob. 39.

(a) If SUMT were to be applied to this problem, what would be the unconstrained function $P(\mathbf{x}; r)$ to be *minimized* at each iteration?

(b) Starting from the initial trial solution $(x_1, x_2, x_3) = (\frac{3}{2}, \frac{3}{2}, 2)$, apply SUMT with $r = 10^{-2}, 10^{-4}, 10^{-6}, 10^{-8}$.

69. Consider the following *nonconvex programming* problem.

$$\text{Minimize} \quad f(\mathbf{x}) = \sin 3x_1 + \cos 3x_2 + \sin(x_1 + x_2),$$

subject to

$$x_1^2 - 10x_2 \ge -1$$

$$10x_1 + x_2^2 \le 100$$

and

$$x_1 \ge 0, \qquad x_2 \ge 0.$$

(a) If SUMT were to be applied to this problem, what would be the unconstrained function $P(\mathbf{x}; r)$ to be *minimized* at each iteration?

(b) Describe how SUMT should be applied to attempt to obtain a *global minimum*. (Do not actually solve.)

15

Inventory Theory

15.1 Introduction

Keeping an inventory (stock of goods) for future sale or use is very common in business. Retail firms, wholesalers, manufacturing companies—and even blood banks—generally have a stock of goods on hand. How does such a facility decide upon its "inventory policy"; i.e., when and how much does it replenish? In a small firm the manager may keep track of inventory and make these decisions. However, since this may not be feasible even in small firms, many companies have saved large sums of money by using "scientific inventory management." In particular, they:

1. Formulate a mathematical model describing the behavior of the inventory system.
2. Derive an optimal inventory policy with respect to this model.
3. Frequently use a computer to maintain a record of the inventory levels and to signal when and how much to replenish.

There are several basic considerations involved in determining an inventory policy that must be reflected in the mathematical inventory model; these are illustrated in the following examples.

EXAMPLE 1: A television manufacturing company produces its own speakers, which are used in the production of its television sets. The television sets are assembled on a continuous production line at a rate of 8,000 per month. The speakers are produced in batches because they do not warrant setting up a continuous production line, and relatively large quantities can be produced in a short time. The company is interested in determining when and how many to produce. Several costs must be considered:

1. Each time a batch is produced, a setup cost of $12,000 is incurred. This cost includes the cost of "tooling up," administrative costs, record keeping, and so forth. Note that the existence of this cost argues for producing speakers in large batches.
2. The production of speakers in large batches leads to a large inventory. The estimated cost of keeping a speaker in stock is 30 cents/month. This cost includes the cost of capital tied up, storage space, insurance, taxes, protection, and so on. The existence of a storage or holding cost argues for producing small batches.
3. The production cost of a single speaker (excluding the setup cost) is $10 and can be assumed to be a unit cost independent of the batch size produced. (In general, however, the unit production cost need not be constant and may decrease with batch size.)
4. Company policy prohibits deliberately planning for shortages of any of its components. However, a shortage of speakers occasionally crops up, and it has been estimated that each speaker that is not available when required costs $1.10/month. This cost includes the cost of installing speakers after the television set is fully assembled, storage space, delayed revenue, record keeping, and so forth.

EXAMPLE 2: A wholesale distributor of bicycles is having trouble with shortages of the most popular inexpensive 10-speed model and is currently reviewing the inventory policy for this model. The distributor purchases this model bicycle from the manufacturer monthly and then supplies them to various bicycle shops in the western United States. Upon request from shops, the distributor wholesales bicycles to the individual shops in its region. The distributor has analyzed his costs and has determined that the following are important:

1. The shortage cost, i.e., the cost of not having a bicycle on hand when needed. Most models are easily reordered from the manufacturer, and stores usually accept a delay in delivery. Still, although shortages are permissible, the distributor feels that he incurs a loss, which he estimates to be $15 per bicycle. This cost represents an evaluation of the cost of the loss of goodwill, additional clerical costs incurred, and the cost of the delay in revenue received. On a very few competitive (in price) models, stores do not accept a delay, which results in lost sales. In this case, the cost of lost revenue must be included in the shortage cost.
2. The holding cost, i.e., the cost of maintaining an inventory, is $1 per bicycle remaining at the end of the month. This cost represents the costs of capital tied up, warehouse space, insurance, taxes, and so on.

3. The ordering cost, i.e., the cost of placing an order plus the cost of the bicycle, consists of two components: The paperwork involved in placing an order is estimated as $200, and the actual cost of a bicycle is $35.

These two examples indicate that there exists a trade-off between the costs involved; the next section discusses the basic cost components of inventory models.

15.2 Components of Inventory Models

Because inventory policies obviously affect profitability, the choice among policies depends upon their relative profitability. Some of the costs that determine this profitability are (1) the costs of ordering or manufacturing, (2) holding or storage costs, (3) unsatisfied demand or shortage penalty costs, (4) revenues, (5) salvage costs, and (6) discount rates. [Costs (1) to (3) were encountered in Examples 1 and 2.]

The cost of ordering or manufacturing an amount z can be represented by a function $c(z)$. The simplest form of this function is one that is directly proportional to the amount ordered, that is, $c \cdot z$, where c represents the unit price paid. Another common assumption is that $c(z)$ is composed of two parts: a term that is directly proportional to the amount ordered and a term that is constant K for z positive and zero for $z = 0$. For this case, if z is positive, the ordering, or production cost, is given by $K + c \cdot z$. The constant K is often referred to as the setup cost and generally includes the administrative cost of ordering, the preliminary labor, and other expenses of starting a production run. There are other assumptions that can be made about this ordering function, but this chapter is restricted to the two cases just described. In Example 1, the speakers are manufactured, and the setup cost for the production run is $12,000. Furthermore, each speaker costs $10, so that the *production* cost is given by

$$c(z) = 12,000 + 10z, \quad \text{for } z > 0.$$

In Example 2, the distributor orders bicycles from the manufacturer, and the *ordering* cost is given by

$$c(z) = 200 + 35z, \quad \text{for } z > 0.$$

The holding or storage costs represent the costs associated with the storage of the inventory until it is sold or used. They may include the cost of capital tied up, space, insurance, protection, and taxes attributed to storage. These costs may be a function of the maximum quantity held during a period, the average amount held, or the cumulated excess of supply over the amount required (demand). The latter viewpoint is usually taken in this chapter. In the bicycle example, the holding cost was $1 per bicycle remaining at the end of the month. This cost can be interpreted as the interest lost in keeping capital tied up in an ''unnecessary'' bicycle for a month, cost of extra storage space, insurance, and so forth.

The unsatisfied demand or shortage penalty cost is incurred when the amount of the commodity required (demand) exceeds the available stock. This cost depends upon the structure of the model. One such case occurs when the demand exceeds the available inventory, and (1) it is met by a priority shipment, or (2) it is not met at all. In (1) the penalty cost can be viewed as the entire cost of the priority shipment that is used to meet the excess demand. In (2), the situation where the unsatisfied demand is lost, the penalty cost can be viewed as the loss in revenue. Either situation is known as ''no backlogging of unsatisfied demand.'' The scenario of the bicycle example implies that there exist a few competitive (in price) bicycle models where unsatisfied demand is lost, thereby resulting in lost revenue, and hence the example is one where unsatisfied demand is not backlogged. The second case of demand not being fulfilled out of stock assumes that it is satisfied when the commodity next becomes available. The penalty cost can be interpreted as the loss of customers' goodwill, their subsequent reluctance to do business with the firm, the cost of delayed revenue, and extra record keeping. This case is known as ''backlogging of unsatisfied demand.'' The speaker example calls for backlogging of unsatisfied demand. If a shortage occurs, the final assembly of the television set awaits the production of the next batch of speakers. Usually the unsatisfied demand cost is a function of the excess of demand over supply.

The revenue cost may or may not be included in the model. If it is assumed that both the price and the demand for the product are not under the control of the company, the revenue from sales is independent of the firm's inventory policy and may be neglected. However, if revenue is neglected in the model, the *loss in revenue* must then be included in the unsatisfied demand penalty cost whenever the firm cannot meet the demand and the sale is lost. For example, in the case of the 10-speed bicycles suppose that the distributor is offered very favorable terms on the purchase of a model of a name-brand bicycle whose production is to be discontinued. Stores have been informed that no reorders are possible. The distributor sells each bicycle for $45, and it may be assumed that unsatisfied demand is lost. If we adopt the criterion of maximizing net income, we must include revenue in the model. Indeed, net income is equal to total revenue minus the costs incurred (ordering, inventory holding, and unsatisfied demand). The total revenue component is simply the sales price of a bicycle ($45) times the demand *minus* the sales price times the unsatisfied demand whenever a shortage occurs. The former is independent of the inventory policy, and, hence, can be neglected, whereas the latter is just the lost revenue when a shortage exists. This latter term behaves just like the description of the unsatisfied demand cost, and, hence, the two (the loss of goodwill and the lost revenue) can be combined (added) and considered to be the resultant *unsatisfied demand cost*. The unsatisfied demand cost will be so interpreted throughout this chapter. Thus, in the bicycle example, the unsatisfied demand is simply $45 times the unsatisfied demand whenever a shortage exists.[1]

Because the revenue times the demand term is independent of the inventory policy chosen, its omission from the net income expression results in terms that are the negative of total cost. Hence, maximizing net income in this model is equivalent to minimizing total cost.

The discussion about the interpretation of the unsatisfied demand cost assumed that the unsatisfied demand was lost. For the case where this unsatisfied demand is met by a priority shipment, the same principles prevail. The total revenue component becomes the sales price of a bicycle ($45) times the demand *minus* the unit cost of the priority shipment times the unsatisfied demand whenever a shortage occurs. If the West Coast distributor is forced to meet the unsatisfied demand by purchasing bicycles from the Midwest distributor at the same cost that bicycles are sold to the retail outlets ($45) plus an air freight charge of, say, $2, then the appropriate unsatisfied demand cost is $47 per bicycle. Of course, any costs associated with loss of goodwill, if present, would be added to this amount.

---

[1] In general, the cost of the loss of goodwill, assumed to be negligible in the bicycle example, must be added to the revenue to determine the complete *unsatisfied demand cost*.

The salvage value of an item is the value of a leftover item at the termination of the inventory period. If the inventory policy is carried on for an indefinite number of periods, and if there is no obsolescence, there are no leftover items. What is left over at the end of one period is the amount available at the beginning of the next period. On the other hand, if the policy is to be carried out for only one period, the salvage value represents the disposal value of the item to the firm, say, the selling price. The negative of the salvage value is called the salvage cost. If there is a cost associated with the disposal of an item, the salvage cost may be positive. Because the storage costs generally are assumed to be a function of excess of supply over demand, the salvage costs can be combined with this cost and hence they are usually neglected in this chapter.

Finally, the discount rate takes into account the time value of money. When a firm ties up capital in inventory, it is prevented from using this money for alternative purposes. For example, it could invest this money in secure investments, say, government bonds, and have a return on investment a year hence of, say, 7 percent. Thus a dollar invested today would be worth $1.07 a year hence, or alternatively, a dollar profit a year hence is equivalent to $\alpha = 1/\$1.07$ today. The quantity α is known as the discount factor. Thus, in considering the profitability of an inventory policy, the profit or costs a year hence should be multiplied by α; 2 years hence, by α^2; and so on.

Of course, the convention of choosing a discount factor α that is based upon the current value of a dollar delivered *one year* hence is arbitrary, and any time period could have been used, for example, one month. It is also evident that in problems having short time horizons, α may be assumed to be 1 (and thereby neglected) because the current value of a dollar delivered during this short time horizon does not change very much. However, in problems having long time horizons, the discount factor must be included.

In using quantitative techniques to seek optimal inventory policies, we use the criterion of minimizing the total (expected) discounted cost. Under the assumptions that the price and demand for the product are not under the control of the company and that the lost or delayed revenue is included in the shortage penalty cost, minimizing cost is equivalent to maximizing net income. Another criterion to be considered, although it is nonquantitative but nevertheless important in practice, is that the resultant inventory policy be simple; i.e., the rule for indicating *when to order* and *how much to order* must be able to be easily described. Most of the policies considered possess this property.

Inventory models are usually classified according to whether the demand for a period is known (deterministic demand) or whether it is a random variable having a known probability distribution (nondeterministic or random demand). The production of batches of speakers is an example of deterministic demand because it is assumed that they are used in television assemblies at a rate of 8,000 per month. The bicycle shops' purchases of bicycles from the distributor is an example of random demand. This classification is frequently coupled with whether or not there exist time lags in the delivery of the items ordered or produced. In both the speaker and the bicycle examples, there was an implication that the items appeared immediately after an order was placed. In fact, the production of speakers may require some time, and similarly, the delivery of bicycles may not be instantaneous, so that time lags may have to be incorporated into the inventory model.

Another possible classification relates to the way the inventory is reviewed, either continuously or periodically. In continuous review, an order is placed as soon as the stock level falls below the prescribed reorder point, whereas in the periodic review case, the inventory level is checked at discrete intervals, e.g., at the end of each week, and ordering decisions are made only at these times even if the inventory level dips below the reorder point during the preceding period. The production of speakers is an example of a continuous review, whereas the bicycle problem is an example of periodic review. Incidentally, in practice, a periodic review policy can be used to approximate a continuous review policy by making the time interval sufficiently small.

This chapter is concerned with inventory problems where the actual demand is assumed to be known. Several models are considered, including the well-known economic lot-size formulation.

15.3 Continuous Review—Uniform Demand

The most common inventory problem faced by manufacturers, retailers, and wholesalers is concerned with the case where stock levels are depleted with time and then are replenished by the arrival of new items. A simple model representing this situation is given by the economic lot-size model. Items are assumed to be withdrawn continuously at a known constant rate denoted by a; that is, a units are required per unit time, say, per month. It is further assumed that items are produced (or ordered) in equal numbers, Q at a time, and all Q items arrive simultaneously when desired (fixed delivery lags will be considered later). The only costs to be considered are the setup cost K, charged at the time of the production (or ordering), a production cost (or purchase cost) of c dollars per item, and an inventory holding cost of h dollars per item per unit of time. The inventory problem is to determine how often to make a production run and what size it should be so that the cost per unit of time is a minimum. This is a continuous review inventory policy. We shall first assume that shortages are not allowed, and then we shall relax this assumption. The example of the production of speakers in television sets satisfies this model.

Shortages Not Permitted

A cycle can be viewed as the time between production runs. Thus, if 24,000 speakers are produced at each production run and are used at the rate of 8,000 per month, then the cycle length is $24{,}000/8{,}000 = 3$ months. In general, the cycle length is Q/a. Figure 15.1 illustrates how the inventory level varies over time.

The cost per unit time is obtained as follows: The production cost per cycle is given by

$$\begin{cases} 0, & \text{if } Q = 0 \\ K + cQ, & \text{if } Q > 0. \end{cases}$$

The holding cost per cycle is easily obtained. The average inventory level during a cycle is $(Q + 0)/2 = Q/2$ items per unit of time, and the corresponding cost is $hQ/2$ per unit of time. Because the cycle length is Q/a, the holding cost per cycle is given by

$$\frac{hQ^2}{2a}.$$

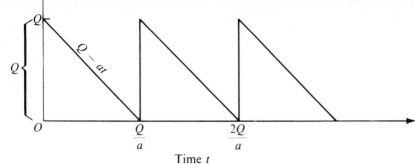

Figure 15.1 Diagram of inventory level as a function of time—no shortages permitted.

Therefore, the total cost per cycle is

$$K + cQ + \frac{hQ^2}{2a},$$

and the total cost per unit of time is

$$T = \frac{K + cQ + hQ^2/2a}{Q/a} = \frac{aK}{Q} + ac + \frac{hQ}{2}.$$

It is evident that the value of Q, say Q^*, that minimizes T is found from $dT/dQ = 0$. $dT/dQ = -aK/Q^2 + h/2 = 0$, so that

$$Q^* = \sqrt{\frac{2aK}{h}}$$

(because $d^2T/dQ^2 > 0$), which is the well-known economic lot-size result. Similarly, the time it takes to withdraw this optimum value of Q^*, say, t^*, is given by

$$t^* = \frac{Q^*}{a} = \sqrt{\frac{2K}{ah}}.$$

These results will now be applied to the speaker example. The appropriate parameters are

$$K = 12{,}000$$
$$h = 0.30$$
$$a = 8{,}000,$$

so that

$$Q^* = \sqrt{\frac{(2)(8{,}000)(12{,}000)}{0.30}} = 25{,}298$$

and

$$t^* = \frac{25{,}298}{8{,}000} = 3.2 \text{ months.}$$

Hence the production line is to be set up every 3.2 months and produce 25,298 speakers. Incidentally, the cost curve is rather flat near this optimal value, so that any production between 20,000 and 30,000 speakers is acceptable; this fact can be seen in Fig. 15.3.

It may be profitable to permit shortages to occur because the cycle length can be increased with a resultant saving in setup costs. However, this benefit may be offset by the cost that is incurred when shortages occur, and hence a detailed analysis is required.

If shortages are allowed and are priced out at a cost of p dollars for each unit of demand unfilled for one unit of time, results similar to the no-shortage case can be obtained. Denote by S the stock on hand at the beginning of a cycle. The problem is summarized in Fig. 15.2.

The cost per unit time is obtained as follows: The production cost per cycle is given by

$$\begin{cases} 0, & \text{if } Q = 0 \\ K + cQ, & \text{if } Q > 0. \end{cases}$$

The holding cost per cycle is easily obtained. Note that the inventory level is positive for a time of S/a. The average inventory level *during this time* is $(S + 0)/2 = S/2$ items per unit of time, and the corresponding cost is $hS/2$ per unit of time. Hence the total holding cost incurred over the time the inventory level is positive is the holding cost per cycle, which is given by

$$\frac{hS}{2} \frac{S}{a} = \frac{hS^2}{2a}.$$

Similarly, shortages occur for a time $(Q - S)/a$. The average amount of shortages *during this time* is $[0 + (Q - S)]/2 = (Q - S)/2$ items per unit of time, and the corresponding cost is $p(Q - S)/2$ per unit of time. Hence the total shortage cost incurred over the time shortages exist is the shortage cost per cycle, which is given by

$$\frac{p(Q - S)}{2} \frac{(Q - S)}{a} = \frac{p(Q - S)^2}{2a}.$$

Therefore, the total cost per cycle is

$$K + cQ + \frac{hS^2}{2a} + \frac{p(Q - S)^2}{2a},$$

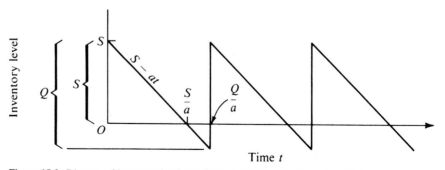

Figure 15.2 Diagram of inventory level as a function of time—shortages permitted.

and the total cost per unit of time is

$$T = \frac{K + cQ + hS^2/2a + p(Q - S)^2/2a}{Q/a}$$

$$= \frac{aK}{Q} + ac + \frac{hS^2}{2Q} + \frac{p(Q - S)^2}{2Q}.$$

In this model there are two decision variables (S and Q), so the optimum values (S^* and Q^*) are found by setting the partial derivatives $\partial T/\partial S$ and $\partial T/\partial Q$ equal to zero. Thus

$$\frac{\partial T}{\partial S} = \frac{hS}{Q} - \frac{p(Q - S)}{Q} = 0.$$

$$\frac{\partial T}{\partial Q} = -\frac{aK}{Q^2} - \frac{hS^2}{2Q^2} + \frac{p(Q - S)}{Q} - \frac{p(Q - S)^2}{2Q^2} = 0.$$

Solving these equations simultaneously leads to

$$S^* = \sqrt{\frac{2aK}{h}} \sqrt{\frac{p}{p + h}}, \qquad Q^* = \sqrt{\frac{2aK}{h}} \sqrt{\frac{p + h}{p}}.$$

The optimal period length t^* is given by

$$t^* = \frac{Q^*}{a} = \sqrt{\frac{2K}{ah}} \sqrt{\frac{p + h}{p}}.$$

The maximum shortage is expressed as

$$Q^* - S^* = \sqrt{\frac{2aK}{p}} \sqrt{\frac{h}{p + h}}.$$

Further, from Fig. 15.2, the fraction of time that no shortage exists is given by

$$\frac{S^*/a}{Q^*/a} = \frac{p}{p + h},$$

which is independent of K.

If shortages are permitted in the speaker example, the cost is estimated as $p = \$1.10$ per speaker. Again,

$$K = 12{,}000$$

$$h = 0.30$$

$$a = 8{,}000,$$

so that
$$S^* = \sqrt{\frac{(2)(8{,}000)(12{,}000)}{0.30}} \sqrt{\frac{1.1}{1.1 + 0.3}} = 22{,}424,$$

$$Q^* = \sqrt{\frac{(2)(8{,}000)(12{,}000)}{0.30}} \sqrt{\frac{1.1 + 0.3}{1.1}} = 28{,}540,$$

and
$$t^* = \frac{28{,}540}{8{,}000} = 3.6 \text{ months.}$$

Hence, when shortages are permitted, the production line is to be set up every 3.6 months to produce 28,540 speakers. A shortage of 6,116 speakers is permitted. Note that Q^* and t^* are not very different from the no-shortage case.

Quantity Discounts, Shortages Not Permitted

The models considered have assumed that the unit cost of an item is the same, independent of the quantity produced. In fact, this assumption resulted in the optimal solutions being independent of this unit cost. Suppose, however, that there exist cost breaks; i.e., the unit cost varies with the quantity ordered. For example, suppose the unit cost of producing a speaker is $c_1 = \$11$ if less than 10,000 speakers are produced, $c_2 = \$10$ if production falls between 10,000 and 80,000 speakers, and $c_3 = \$9.50$ if production exceeds 80,000 speakers. What is the optimal policy? The solution to this specific problem will reveal the general method.

From the results of the previously considered economic lot-size model (shortages not permitted), the total cost per unit time if the production cost is c_j is given by

$$T_j = \frac{aK}{Q} + ac_j + \frac{hQ}{2}, \qquad \text{for } j = 1, 2, 3.$$

A plot of T_j versus Q is shown in Fig. 15.3.

The feasible values of Q are shown by the solid lines, and it is only these regions that must be investigated. For each curve, the value of Q that minimizes T_j is easily found by the methods used in the previously considered economic lot-size model. For $K = 12,000$, $h = 0.30$, and $a = 8,000$, this value is

$$\sqrt{\frac{(2)(8,000)(12,000)}{0.30}} = 25,298.$$

This number is a feasible value for the cost function T_2. Because it is evident that for fixed Q, $T_j < T_{j-1}$ for all j, T_1 can be eliminated from further consideration. However, T_3 cannot be immediately discarded. Its minimum feasible value (which occurs at $Q = 80,000$) must be compared to T_2 evaluated at 25,298 (which is \$87,589). Because

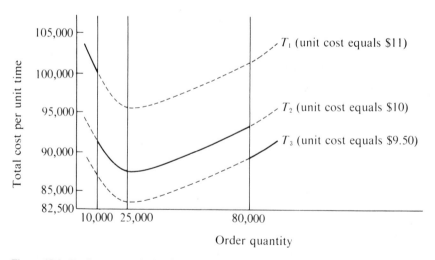

Figure 15.3 Total cost per unit time for speaker example with quantity discounts.

T_3 evaluated at 80,000 equals $89,200, it is better to produce in quantities of 25,298, and thus this quantity is the optimal value for this set of quantity discounts. If the quantity discount led to a cost of $9 (instead of $9.50) when production exceeded 80,000, then T_3 evaluated at 80,000 would equal 85,200, and the optimal production quantity would become 80,000.

Although this analysis concerned a very specific problem, its extension to a general problem is evident. Furthermore, a similar analysis can be made for other types of quantity discounts, such as incremental quantity discounts, where a cost c_0 is incurred for the first q_0 items, c_1 for the next q_1 items, and so on.

Remarks: Several remarks can be made about economic lot-size models.

1. If it is assumed that the production (or purchase) cost of an item is constant throughout time, it does not appear in the optimal solution. This result is evident because no matter what policy is used, the same quantity is required, and hence this cost is fixed.

2. It was previously assumed that Q, the number of units produced, was constant from cycle to cycle. A little reflection will reveal that this assumption is really a result rather than an assumption.

3. These models can be viewed as a special case of an (s, S) policy. An (s, S) policy is usually used in the context of a periodic review policy, where at review time an order is placed to bring the inventory level up to S if the current inventory is less than or equal to s. Otherwise, no order is placed.[1] The symbol S then denotes the reorder level, and s denotes the reorder point. In the economic lot-size models, s denotes the inventory level when items are ordered, so that when shortages are not permitted the reorder point s is zero, and when shortages are permitted, s is equal to the negative of the maximum shortage; that is,

$$s = -\sqrt{\frac{2aK}{p}}\sqrt{\frac{h}{p+h}}.$$

Furthermore, because the economic lot-size models are continuously reviewed, when the inventory level equals s an order is placed, bringing the inventory up to the reorder level S. Hence, for economic lot-size models the (s, S) policy can be described as follows: When the inventory level reaches the reorder point s, place an order to bring the inventory level up to reorder level S; that is, order $Q = S - s$.

4. It is evident from the analysis presented that the reorder point will never be positive. A policy that calls for $s > 0$ cannot be optimal because it is dominated by a policy that calls for ordering the same Q, but only when the reorder point reaches zero. It is dominated in that the latter policy has the same setup and purchase costs but a uniformly smaller holding cost.

5. A known fixed delivery lag is easily accommodated. Denote by λ the lead time between the placing and receiving of the order. It is assumed that λ is constant over time and independent of the size of the order. It is evident that if it is desired to have the order arrive the moment the inventory level reaches s, then the order must

---

[1] Usually (s, S) policies are described as ordering when the inventory level is less than s rather than when the inventory level is less than or equal to s. However, the costs are often the same if an optimal policy is followed.

be placed λ periods earlier. Thus the reorder point is simply

$$s + \lambda a,$$

where s is determined for the no-lag situation.[1]

15.4 Periodic Review—A General Model for Production Planning

The last section explored the economic lot-size model. The results were dependent upon the assumption of a constant demand rate. When this assumption is relaxed, i.e., when the amounts required from period to period are allowed to vary, the square-root formula no longer ensures a minimum cost solution.

Consider the following model, due to Wagner and Whitin.[2] As before, the only costs to be considered are the setup cost K, charged at the beginning of the period; a production cost (or purchase cost) of c dollars per item; and an inventory holding cost of h dollars per item, which is charged (arbitrarily) at the end of the time period. The choice of charging the inventory at the end of the period, and hence as a function of the excess of the supply over the requirement, is somewhat different from the holding charge incurred in the economic lot-size models. In the latter case, the average cost per unit of time was charged. Clearly, different policies can result from alternative ways of dealing with holding costs. In addition, $r_i = 1, 2, \ldots, n$ represents the requirements at time i, and it is assumed that these requirements must be met. Initially there is no stock on hand. For a horizon of n periods, the inventory problem is to determine how much should be produced at the beginning of each time period (assumed to be instantaneous) to minimize the total cost incurred over the n periods.

The model can be illustrated by the following variation of the speaker example.

EXAMPLE: A market survey conducted by the television manufacturer has indicated that the demand for television sets is seasonal rather than uniform. In particular, sales of 30,000 sets is forecast for the Christmas season (October to December), 20,000 for the winter slack season (January to March), 30,000 for the "new model" season (April to June), and 20,000 for the summer season (July to September). Because of the necessity of meeting the increased demand during the peak seasons, the television set production line was revamped. This revised production line enabled the company to introduce new equipment as well as to redesign some of its components, including the speakers. Hence the setup cost for speaker production is now $20,000, but the unit cost is down to $1. Furthermore, the holding cost of a speaker has been reduced to 20 cents per (3-month) period. Finally, the labor and equipment costs are such that the speakers must be produced in increments of 10,000. It is assumed that production of the television sets is completed and ready for shipment in the 3-month period prior to the season in which it is required. Thus the 30,000 sets required for the Christmas season are to be assembled during the July-to-September period. The speaker is the last component added to the television set, and it is easily installed. Furthermore, large quantities of speakers can be produced in a very short time, so their production

---

[1] This result holds only when $\lambda a < Q$.

[2] Wagner, H. M., and T. M. Whitin: "Dynamic Version of the Economic Lot-Size Model," *Management Science*, **5**(1):89–96, 1958.

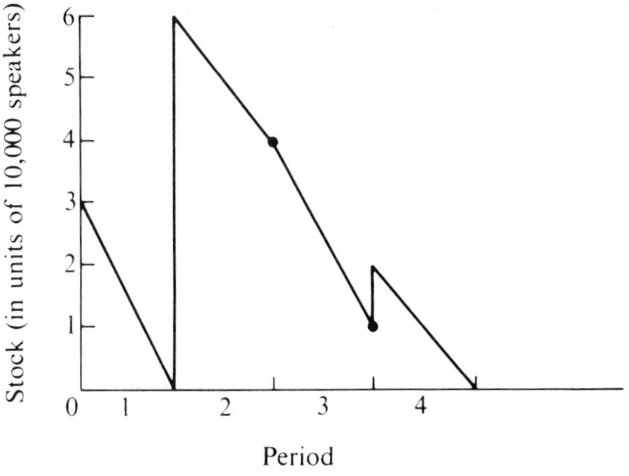

Figure 15.4 Production schedule that satisfies requirements.

and subsequent installation can be viewed as instantaneous. The problem is to determine how many to produce in each period while satisfying the requirements and minimizing the total cost. A solution, though not an optimal one, is given in Fig. 15.4; this policy calls for producing 30,000 speakers at the beginning of the first period (Christmas season), 60,000 speakers at the beginning of the second period, and 10,000 speakers at the beginning of the fourth period.

One approach to the solution of this model is to enumerate, for each of the 2^{n-1} combinations of either producing or not producing in each period, the possible quantities that can be produced. This approach is rather cumbersome, even for moderate-sized n. Hence a more efficient method is desirable. In particular, the method of dynamic programming introduced in Chap. 8 will be applied, followed by the introduction of an algorithm (in the ensuing section) that exploits the structure. Finally, a mathematical programming solution will be presented.

Periodic Review—Production Planning: Dynamic Programming Solution

Following the notation introduced for dynamic programming (Chap. 8) and interpreting the variables in the inventory context, the ith stage corresponds to the ith period; the state corresponds to the inventory entering period i and will be denoted by x_i; and the decision variable corresponds to the quantity produced (or ordered) at the beginning of period i and will be denoted by z_i.[1]

Let $B_i(x_i, z_i)$ denote the costs incurred in period i, given the entering inventory and the quantity produced. Recall that the only costs considered are the setup cost K, charged at the beginning of the period, a production cost (or purchase cost) of c dollars per item, and an inventory holding cost of h dollars per item charged at the end of the time period. The requirement for period i is r_i. Then $B_i(x_i, z_i)$ is given by

$$B_i(x_i, z_i) = \begin{cases} K + cz_i + h(x_i + z_i - r_i), & \text{if } z_i > 0 \\ h(x_i - r_i), & \text{if } z_i = 0. \end{cases}$$

---

[1] For the notation in this chapter to be consistent with that used in inventory theory in general, it may differ somewhat from the notation introduced in Chap. 8.

Denote by $C_i(x_i, z_i)$ the total cost of the best overall policy from the beginning of period i to the end of the planning horizon, given that the inventory level entering period i is x_i and z_i is chosen to be produced; and let $C_i^*(x_i)$ denote the corresponding minimum value of $C_i(x_i, z_i)$, subject to the constraints that $z_i \geq 0$ and that the requirements for the periods are met. Therefore,

$$C_i^*(x_i) = \min_{\substack{z_i \geq 0 \\ z_i \geq r_i - x_i}} [B_i(x_i, z_i) + C_{i+1}^*(x_i + z_i - r_i)],$$

for $i = 1, 2, \ldots, n$, and $C_{n+1}^*(\cdot)$ is defined to be zero. In the speaker example, there are four periods to be considered. (All calculations have been reduced in scale by a factor of 10,000.) The first iteration corresponds to $i = 4$, that is, a description of the optimal policy at the beginning of period 4. For this case,

$$C_4^*(x_4) = \min_{\substack{z_4 \geq 0 \\ z_4 \geq 2 - x_4}} [B_4(x_4, z_4)],$$

so that the immediate solution to the fourth-period problem is as follows:

x_4	z_4	$C_4^*(x_4)$	z_4^*
0	2	4	2
1	1	3	1
2	0	0	0

The second iteration requires finding the optimal policy from the beginning of period 3 to the end of period 4. For this case,

$$C_3^*(x_3) = \min_{\substack{z_3 \geq 0 \\ z_3 \geq 3 - x_3}} [B_3(x_3, z_3) + C_4^*(x_3 + z_3 - 3)],$$

so that the solution to the problem beginning at period 3 is as follows:

	$B_3(x_3, z_3) + C_4^*(x_3 + z_3 - 3)$							
x_3 \ z_3	0	1	2	3	4	5	$C_3^*(x_3)$	z_3^*
0	—	—	—	9.0	9.2	7.4	7.4	5
1	—	—	8	8.2	6.4	—	6.4	4
2	—	7	7.2	5.4	—	—	5.4	3
3	4.0	6.2	4.4	—	—	—	4.0	0
4	3.2	3.4	—	—	—	—	3.2	0
5	0.4	—	—	—	—	—	0.4	0

A typical entry can be verified. Let $x_3 = 2$ and $z_3 = 3$. Then 3 units are produced, given an initial inventory of 2. The cost of production is then $2 + 1(3) = 5$, and the inventory holding cost is $\frac{1}{5}(2 + 3 - 3) = 0.4$. The entering inventory in period 4 is then 2, so that $C_4^*(2) = 0$. Hence the total cost is given by $5 + 0.4 = 5.4$, which is the value given.

The third iteration requires finding the optimal policy from the beginning of period 2 to the end of period 4. For this case,

$$C_2^*(x_2) = \underset{\substack{z_2 \geq 0 \\ z_2 \geq 2-x_2}}{\text{minimum}} [B_2(x_2, z_2) + C_3^*(x_2 + z_2 - 2)],$$

so that the solution to the problem beginning at period 2 is as follows:

x_2	z_2									
				$B_2(x_2, z_2) + C_3^*(x_2 + z_2 - 2)$						
	0	1	2	3	4	5	6	7	$C_2^*(x_2)$	z_2^*
0	—	—	11.4	11.6	11.8	11.6	12.0	10.4	10.4	7
1	—	10.4	10.6	10.8	10.6	11.0	9.4	—	9.4	6
2	7.4	9.6	9.8	9.6	10.0	8.4	—	—	7.4	0
3	6.6	8.8	8.6	9.0	7.4	—	—	—	6.6	0
4	5.8	7.6	8.0	6.4	—	—	—	—	5.8	0
5	4.6	7.0	5.4	—	—	—	—	—	4.6	0
6	4.0	4.4	—	—	—	—	—	—	4.0	0
7	1.4	—	—	—	—	—	—	—	1.4	0

Finally, the last iteration requires finding the optimal policy from the beginning of period 1 to the end of period 4. For this case,

$$C_1^*(0) = \underset{z_1 \geq 3}{\text{minimum}} [B_1(0, z_1) + C_2^*(z_1 - 3)],$$

so that the solution to the production planning problem is as follows:

x_1	z_1												
					$B_1(0, z_1) + C_2^*(z_1 - 3)$								
	0	1	2	3	4	5	6	7	8	9	10	$C_1^*(x_1)$	z_1^*
0	—	—	—	15.4	15.6	14.8	15.2	15.6	15.6	16.2	14.8	14.8	5 or 10

Therefore, the optimal production schedule is to produce all the speakers at the beginning of the first period, or produce 50,000 speakers (5 units) at the beginning of the first period and 50,000 speakers (5 units) at the beginning of the third period. The minimum cost is $148,000.

It should be noted that the unit cost c is irrelevant to the problem because over all the time periods, all policies use the same number of items at the same total cost. Hence this cost could have been neglected, and the same optimal policies would have been obtained. Different costs in different periods are easily handled by this method of solution. For example, during the peak production periods (when 30,000 speakers are produced), the workers are fully engaged in the assembly of television sets, so that the speakers must be produced during overtime hours. Hence the unit cost is increased. This increase would be reflected in the calculations of $C_3^*(x_3)$ and $C_1^*(x_1)$.

Periodic Review—Production Planning: An Algorithm

In the previous subsection, dynamic programming was applied to solve the production planning model. Alternatively, by streamlining the dynamic programming approach, we shall develop an algorithm that exploits the structure of the model. Initially the production planning model we consider will be the same as that presented in the previous subsection, i.e., arbitrary demand requirement, a fixed setup cost, and linear production and holding costs.

The following result characterizes an optimal policy:

For an arbitrary demand requirement, a fixed setup cost, and linear production and holding costs, there is an optimum policy that produces only when the inventory level is zero.

To show why this result is true, choose any policy. Consider the time period that begins when production is made with zero stock level and ends with the first time production is made when the stock is not at zero level. For the policy given in Fig. 15.4, this interval starts at the beginning of period 2 and ends at the beginning of period 4, when one unit is produced. This time period is also shown by the solid lines in Fig. 15.5.

Consider the alternative policy, which implies production of 50,000 speakers at the beginning of period 2 and production of 20,000 speakers at the beginning of period 4. This policy is shown by the dashed lines in Fig. 15.5. This policy, B, dominates policy A in that the total cost is smaller. The setup and the production costs for both policies are the same. It is evident that the holding cost for B is smaller than that for A because there is always less stock on hand at the end of a period. Therefore, B is better than A, so that A cannot be optimal.

This characterization of optimal policies can be used to determine which policies are not optimal. In addition, because it implies that the amount produced at

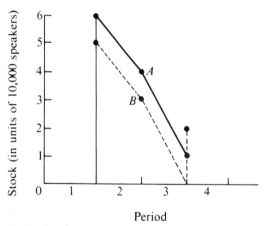

Figure 15.5 Production schedules.

$\cdots + r_n$, it can be exploited to obtain an efficient algorithm.

Suppose an optimal policy is presented. Consider the time from the initial production at the beginning of the first period to the first time the inventory level is again zero. The total cost of the subsequent periods must be a minimum for this reduced problem because the overall policy is optimal. Therefore, in the context of the speaker example, if the inventory level is zero for the first time (after the initial production) at the end of the second period, and if an optimal policy is being followed, all that remains to be done is to determine the optimal policy for the last two periods (periods 3 and 4) that have a requirement of 30,000 speakers for period 3 and 20,000 speakers for period 4.

Let C_i denote the total cost of the best overall policy from the beginning of period *i*, when no stock is available, to the end of the planning horizon, $i = 1, 2, \ldots, n$. A recursive relationship for C_i is given by

$$C_i = \underset{j=i,i+1,\ldots,n}{\text{minimum}} [C_{j+1} + K + c(r_i + r_{i+1} + \cdots + r_j)$$

$$+ h(r_{i+1} + 2r_{i+2} + 3r_{i+3} + \cdots + (j - i)r_j)],$$

where *j* can be viewed as an index that denotes the (end of the) period when the inventory reaches a zero level for the first time after the production at the beginning of period *i*. The cost C_{n+1} is 0, $c(r_i + r_{i+1} + \cdots + r_j)$ is the cost of the production from period *i* until the inventory level next reaches zero, and the quantity $h(\ )$ is the total holding cost of the inventory that results from the production from period *i* and remaining until the inventory level next reaches zero. This latter cost is charged at the end of every period as a function of the excess, if any, over the requirement.

The solution of this algorithm is much simpler than the dynamic programming approach. As in dynamic programming, $C_n, C_{n-1}, \ldots, C_2$ must be found before C_1 is obtained. However, the number of calculations is much smaller, and the number of possible production values is greatly reduced.

EXAMPLE: Returning to the speaker example, first consider the case of finding C_4, the cost of the optimal policy from the beginning of period 4 to the end of the planning horizon:

$$C_4 = C_5 + 2 + 1(2) = 0 + 2 + 2 = 4.0.$$

To find C_3 we must consider two cases; i.e., the first time after period 3 when the inventory reaches a zero level can occur at (1) the end of the third period or (2) the end of the fourth period. In the recursive relationship *j* may range over period 3 or 4, resulting in the costs $C_3^{(3)}$ or $C_3^{(4)}$, respectively. The policy associated with $C_3^{(3)}$ calls for producing only for period 3 and then following the optimal policy for period 4, whereas the policy associated with $C_3^{(4)}$ calls for producing for periods 3 and 4. The cost C_3 is then the minimum of $C_3^{(3)}$ and $C_3^{(4)}$. These cases are reflected by the policies given in Fig. 15.6.

$$C_3^{(3)} = C_4 + 2 + 1(3) = 4 + 2 + 3 = 9$$

and $\quad C_3^{(4)} = C_5 + 2 + 1(3 + 2) + \tfrac{1}{5}(2) = 0 + 2 + 5 + 0.4 = 7.4.$

Hence $\qquad\qquad C_3 = \min(7.4, 9.0) = 7.4.$

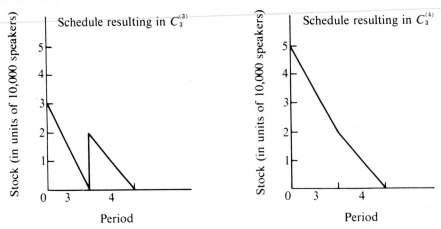

Figure 15.6 Alternative production schedules when production is required at the beginning of period 3.

To find C_2 we must consider three cases; i.e., the first time after period 2 when the inventory reaches a zero level can occur at (1) the end of the second period, (2) the end of the third period, or (3) the end of the fourth period. In the recursive relationship j may range over period 2, 3, or 4, resulting in costs $C_2^{(2)}$, $C_2^{(3)}$, or $C_2^{(4)}$, respectively. The cost C_2 is then the minimum of $C_2^{(2)}$, $C_2^{(3)}$, and $C_2^{(4)}$.

$$C_2^{(2)} = C_3 + 2 + 1(2) \qquad = 7.4 + 2 + 2 = 11.4,$$

$$C_2^{(3)} = C_4 + 2 + 1(2 + 3) + \tfrac{1}{5}(3) = 4 + 2 + 5 + 0.6 = 11.6,$$

and

$$C_2^{(4)} = C_5 + 2 + 1(2 + 3 + 2) + \tfrac{1}{5}[3 + 2(2)] = 0 + 2 + 7 + 1.4 = 10.4.$$

Hence $\qquad\qquad C_2 = \min(11.4, 11.6, 10.4) = 10.4.$

Finally, to find C_1 we must consider four cases; i.e., the first time after period 1 when the inventory reaches zero can occur at (1) the end of the first period, (2) the end of the second period, (3) the end of the third period, or (4) the end of the fourth period. In the recursive relationship j may range over period 1, 2, 3, or 4, resulting in the costs $C_1^{(1)}$, $C_1^{(2)}$, $C_1^{(3)}$, $C_1^{(4)}$. The cost C_1 is then the minimum of $C_1^{(1)}$, $C_1^{(2)}$, $C_1^{(3)}$, and $C_1^{(4)}$.

$$C_1^{(1)} = C_2 + 2 + 1(3) \qquad\qquad = 10.4 + 2 + 3 = 15.4,$$

$$C_1^{(2)} = C_3 + 2 + 1(3 + 2) + \tfrac{1}{5}(2) \qquad = 7.4 + 2 + 5 + 0.4 = 14.8,$$

$$C_1^{(3)} = C_4 + 2 + 1(3 + 2 + 3) + \tfrac{1}{5}[2 + 2(3)] = 4 + 2 + 8 + 1.6 = 15.6,$$

and $\qquad C_1^{(4)} = C_5 + 2 + 1(3 + 2 + 3 + 2) + \tfrac{1}{5}[2 + 2(3) + 3(2)]$

$$= 0 + 2 + 10 + 2.8 = 14.8.$$

Hence, $\qquad\qquad C_1 = \min(15.4, 14.8, 15.6, 14.8) = 14.8,$

so that the optimal production schedule is to produce all the speakers at the beginning of the first period or to produce 50,000 speakers at the beginning of the first period and 50,000 speakers at the beginning of the third period (the same solution as obtained previously). Note that the 14.8 comes from $C_1^{(4)}$ and $C_1^{(2)}$. The policy associated with $C_1^{(4)}$ calls for producing items for periods 1, 2, 3, 4 at period 1 whereas $C_1^{(2)}$ calls for producing items for periods 1 and 2 at period 1, and then following an optimal policy from period 3 on. Such a policy has already been shown to call for production for periods 3 and 4 together.

Again it should be noted that the unit cost c is irrelevant to the problem because over all the time periods, all policies use the same number of items at the same total cost. Hence this cost could have been neglected, and the same optimal policies would have been obtained.

The characterization of the optimal policy, and the subsequent algorithm for finding an optimal policy, depended upon the assumption that the holding and production costs were linear. This constraint can be relaxed to include *concave production and holding costs*. In fact, any increasing function of the holding cost will serve as an alternative condition if the production cost is linear. However, these alternative conditions require a modification in the algorithm for finding an optimal policy. If the production cost function is denoted by $c[\cdot]$ and holding cost function $h[\cdot]$, the recursive relationship for C_i becomes

$$C_i = \underset{j=i,i+1,\ldots,n}{\text{minimum}} \{C_{j+1} + K + c[r_i + r_{i+1} + \cdots + r_j]$$

$$+ \ h[r_{i+1} + r_{i+2} + r_{i+3} + \cdots + r_j]$$

$$+ \ h[r_{i+2} + r_{i+3} + \cdots + r_j] + \cdots + h[r_j]\},$$

where $C_{n+1} = 0$.[1]

A natural extension of the production planning model permits shortages to occur. This extension has been studied by Zangwill[2] and differs from the production planning model studied in that shortage costs are incurred for each unit of demand unfilled for one unit of time. Zangwill characterizes the form of the optimal policy and gives an efficient recursive relationship for finding the optimal policy. These results apply when the production, holding, and shortage costs, per period, are concave functions (thereby including the case of linear costs).

Other results have been obtained for this production planning model under the assumption that the production, holding, and shortage costs are convex. Convex production costs arise, for example, when there are several sources of limited production at different unit costs in a period. If one uses these sources up to capacity in order of ascending unit cost (as is optimal), the resulting production cost is convex in the total amount produced. Such an assumption about the production cost precludes the use of the setup charge.

Periodic Review—Production Planning: Integer Programming Formulation

The final technique for solving the production planning model is to formulate it as an integer programming problem. Instead of presenting the general solution, we shall again solve the speaker production example.

---

[1] In the expression for C_i there is no holding cost when $j = i$.

[2] Zangwill, W. I.: "A Deterministic Multi-Period Production Scheduling Model with Backlogging," *Management Science,* **13**(1):105–119, 1966.

Again, let z_i denote the quantity produced at the beginning of period i.[1] The costs to be considered are

$$\text{Production costs} = \text{setup costs} + c(z_1 + z_2 + z_3 + z_4)$$

and

$$\text{Holding costs} = h(z_1 - r_1) + h(z_1 + z_2 - r_1 - r_2)$$
$$+ h(z_1 + z_2 + z_3 - r_1 - r_2 - r_3)$$
$$= 3h(z_1 - r_1) + 2h(z_2 - r_2) + h(z_3 - r_3)$$
$$= 0.20[3(z_1 - 3) + 2(z_2 - 2) + (z_3 - 3)].$$

Hence the problem to be solved is to minimize the sum of the production and holding costs; that is,

Minimize $W = \text{setup cost} + 1(z_1 + z_2 + z_3 + z_4)$

$$+ 0.20[3(z_1 - 3) + 2(z_2 - 2) + (z_3 - 3)],$$

subject to

$$z_1 \leq r_1 + r_2 + r_3 + r_4 = 10$$
$$z_2 \leq r_2 + r_3 + r_4 = 7$$
$$z_3 \leq r_3 + r_4 = 5$$
$$z_4 \leq r_4 = 2$$
$$z_1 \geq r_1 = 3$$
$$z_1 + z_2 \geq r_1 + r_2 = 5$$
$$z_1 + z_2 + z_3 \geq r_1 + r_2 + r_3 = 8$$
$$z_1 + z_2 + z_3 + z_4 = r_1 + r_2 + r_3 + r_4 = 10$$

and $z_1 \geq 0,$ $z_2 \geq 0,$ $z_3 \geq 0,$ $z_4 \geq 0,$ and integer valued.

Unfortunately, this model is not an integer linear programming model because the objective function is not linear as a result of the "setup costs" term that appears in the objective function. However, this problem can be resolved as follows: Define variables v_1, v_2, v_3, and v_4 such that

$$v_1 \leq 1, \qquad v_2 \leq 1, \qquad v_3 \leq 1, \qquad v_4 \leq 1,$$
$$v_1 \geq 0, \qquad v_2 \geq 0, \qquad v_3 \geq 0, \qquad v_4 \geq 0,$$

and v_1, v_2, v_3, and v_4 are integers (v_i is then 0 or 1).

Add the following constraints to the problem:

$$z_1 \leq \left(\sum_{i=1}^{4} r_i\right) v_1 = 10v_1$$

$$z_2 \leq \left(\sum_{i=2}^{4} r_i\right) v_2 = 7v_2$$

[1] Again, all values are expressed in units of 10,000 speakers.

$$z_3 \le \left(\sum_{i=3}^{4} r_i \right) v_3 = 5v_3$$

$$z_4 \le r_4 v_4 = 2v_4.$$

Thus if $v_i = 0$, then $z_i = 0$, and if $v_i = 1$, then z_i is less than or equal to the maximum production in period i. The objective function can now be written as a linear function subject to linear constraints; that is,

Minimize $\quad W = K(v_1 + v_2 + v_3 + v_4) + 1(z_1 + z_2 + z_3 + z_4)$

$$+ 0.20[3(z_1 - 3) + 2(z_2 - 2) + (z_3 - 3)],$$

subject to

$$z_1 \le 10v_1$$

$$z_2 \le 7v_2$$

$$z_3 \le 5v_3$$

$$z_4 \le 2v_4$$

$$z_1 \ge 3$$

$$z_1 + z_2 \ge 5$$

$$z_1 + z_2 + z_3 \ge 8$$

$$z_1 + z_2 + z_3 + z_4 = 10$$

and
$$\begin{cases} z_1 \ge 0, & z_2 \ge 0, & z_3 \ge 0, & z_4 \ge 0, & \text{and integer valued,} \\ v_1 \ge 0, & v_2 \ge 0, & v_3 \ge 0, & v_4 \ge 0, \\ v_1 \le 1, & v_2 \le 1, & v_3 \le 1, & v_4 \le 1, & \text{and integer valued.} \end{cases}$$

The solution to this problem may be obtained by using an integer programming algorithm; it yields the same solution as obtained previously, i.e., produce 100,000 speakers at the beginning of the first period ($z_1 = 10$, $v_1 = 1$, and all other z_i and v_i equal zero), or produce 50,000 speakers at the beginning of the first period and 50,000 speakers at the beginning of the third period ($z_1 = z_3 = 5$, $v_1 = v_3 = 1$, and all other z_i and v_i equal zero).

15.5 Conclusions

The inventory models presented here are rather simplified, but they serve the purpose of introducing the general nature of inventory models. Furthermore, they are sufficiently accurate representations of many actual inventory situations so that they frequently are useful in practice. For example, the economic lot-size formulas have been particularly widely used, although they are sometimes modified to include some type of stochastic demand. Nevertheless, many inventory situations possess complications that must still be taken into account, e.g., interaction between products. Several complex models have been formulated in an attempt to fit such situations, but they still leave a wide gap between practice and theory. Continued growth is occurring in the computerization of inventory data processing, along with accompanying growth in scientific inventory management.

Inventory Theory

1. Arrow, K. J., S. Karlin, and H. Scarf: *Studies in the Mathematical Theory of Inventory and Production,* Stanford Univ. Press. Stanford, Calif., 1958.

2. Brown, R. G.: *Decision Rules for Inventory Management,* Holt, Rinehart and Winston, New York, 1967.

3. Buchan, J., and E. Koenigsberg: *Scientific Inventory Management,* Prentice-Hall, Englewood Cliffs, N.J., 1963.

4. Buffa, E. S.: *Modern Production/Operations Management,* 6th ed., Wiley, New York, 1980.

5. Buffa, E. S., and J. G. Miller: *Production-Inventory Systems: Planning and Control,* 3d ed., Richard D. Irwin, Homewood, Ill., 1979.

6. Hadley, G., and T. Whitin: *Analysis of Inventory Systems,* Prentice-Hall, Englewood Cliffs, N.J., 1963.

7. Hax, A., and D. Candea: *Production and Inventory Management,* Prentice-Hall, Englewood Cliffs, N.J., 1984.

8. Johnson, L. A., and D. C. Montgomery: *Operations Research in Production Planning, Scheduling and Inventory Control,* Wiley, New York, 1974.

9. Peterson, R., and E. A. Silver: *Decision Systems for Inventory Management and Production Planning,* Wiley, New York, 1979.

10. Starr, M., and D. Miller: *Inventory Control: Theory and Practice,* Prentice-Hall, Englewood Cliffs, N.J., 1962.

PROBLEMS

1.* Suppose that the demand for a product is 30 units per month, and the items are withdrawn uniformly. The setup cost each time a production run is made is $15. The production cost is $1 per item, and the inventory holding cost is $0.30 per item per month.

    (*a*) Assuming shortages are not allowed, determine how often to make a production run and what size it should be.

    (*b*) If shortages cost $3 per item per month, determine how often to make a production run and what size it should be.

2. The demand for a product is 600 units per week, and the items are withdrawn uniformly. The items are ordered, and the setup cost is $25. The unit cost of each item is $3, and the inventory holding cost is $0.05 per item per week.

    (*a*) Assuming shortages are not allowed, determine how often to order and what size the order should be.

    (*b*) If shortages cost $2 per item per week, determine how often to order and what size the order should be.

3. Solve Prob. 2 with shortages permitted and assume a delivery lag of 1 week.

4.* Solve the economic lot-size model problem presented in Sec. 15.3 when shortages are permitted but when the cost is $5 per speaker.

5. A taxi company uses gasoline at the rate of 8,500 gallons/month. The gasoline costs $1.05/gallon, with a setup cost of $1,000. The inventory holding cost is 1 cent/gallon/month.

    (*a*) Assuming shortages are not allowed, determine how often and how much to order.

    (*b*) If shortages cost 50 cents/gallon/month, determine how often and how much to order.

6. Solve Prob. 5(*a*) by assuming that the cost of gasoline drops to $1.00/gallon if at least 50,000 gallons are purchased.

7. Solve Prob. 5(*a*) if the cost of gasoline is $1.20/gallon for the first 20,000 gallons purchased, $1.10 for the next 20,000 gallons, and $1.00/gallon thereafter.

8. Consider the economic lot-size problem with shortages permitted, as presented in Sec. 15.3. Suppose, however, that it has been determined that $S/Q = 0.8$. Derive the expression for the optimal value of Q.

9. In the economic lot-size model, suppose the stock is replenished uniformly (rather than instantaneously) at the rate of b items per unit time until the lot size Q is fulfilled. Withdrawals from the inventory are made at the rate of a items per unit time, where $a < b$. Replenishments and withdrawals of the inventory are made simultaneously. For example, if Q is 60, b is 3 per day, and a is 2 per day, then 3 units of stock arrive each day until day 20. Units are withdrawn at the rate of 2 per day for 30 days, at which time 3 units of stock arrive each day for 20 days, etc. The diagram of inventory versus time is given below.

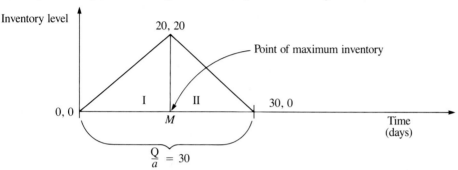

(*a*) Find the total cost per unit time in terms of the setup cost K, production number Q, unit cost c, holding cost h, withdrawal rate a, and replenishment rate b. (*Hint:* In determining the average inventory level, divide the cycle into the two intervals I and II shown in the diagram. Then find the average inventory level in each interval.)

(*b*) Determine the economic lot size.

10. Suppose the requirement for the next 5 months is given by $r_1 = 2$, $r_2 = 4$, $r_3 = 2$, $r_4 = 2$, and $r_5 = 3$. Items are ordered; the setup cost is $4, the purchase cost is $1, and the holding cost is $0.30. Determine the optimal production schedule that satisfies the monthly requirements. Use dynamic programming.

11. Solve Prob. 10 by assuming that the production costs are given by $3 (1 + \log_e X)$, where X is the amount produced in a month.

12. Solve Prob. 10 by using the algorithm presented in Sec. 15.4.

13. Solve Prob. 11 by using the algorithm presented in Sec. 15.4.

14. Formulate Prob. 10 as an integer programming problem.

15.* Solve the production planning model for the production of speakers when the requirements are increased by 1 unit in each period.

16. Solve the production planning model for the production of speakers when the unit costs during the first and third periods are increased to $1.40. Use dynamic programming.

17. Develop an algorithm to solve the production planning model that uses forward induction. (*Hint:* If an optimal policy is being followed and if the time of the *last* production is known for this optimal policy, the total cost of the previous periods must be a minimum for this reduced problem because the overall policy is optimal.)

18.* Consider a situation where a particular product is produced and placed in in-process inventory until it is needed in a subsequent production process. The number of units required in each of the next 3 months, as well as the setup cost and the regular-time unit production cost that would be incurred in each month, are

Month	Requirements	Setup Cost ($)	Regular-Time Unit Cost ($)
1	1	5	8
2	3	10	10
3	2	5	9

There currently is 1 unit in inventory, and we want to have 2 units in inventory at the end of the 3 months. A maximum of 3 units can be produced on regular-time production in each month, although one additional unit can be produced on overtime at a cost that is $2 larger than the regular-time unit production cost. The cost of storage is $2 per unit for each extra month that it is stored.

Use dynamic programming to determine how many units should be produced in each month to minimize total cost.

19. Consider a situation where a particular product is produced and placed in in-process inventory until it is needed in a subsequent production process. The number of units required in each of the next two months, as well as the setup cost, holding cost (charged as a function of excess of supply over requirement and charged at the end of the period), and regular-time unit production cost, are as follows:

Month	Requirement	Setup Cost ($)	Holding Cost ($)	Unit Cost ($)
1	3	5	0.30	9
2	4	5	0.30	9

Determine the optimal production schedule that satisfies the monthly requirements. Use the algorithm presented in Sec. 15.4.

Appendixes

Appendix 1

Convexity

The concept of *convexity* is frequently used in operations research work. Therefore, we introduce the properties of convex (or concave) functions and convex sets.

Definition: A *function* of a single variable, $f(x)$, is a **convex function** if, for each pair of values of x, say, x' and x'',

$$f(\lambda x'' + (1 - \lambda)x') \leq \lambda f(x'') + (1 - \lambda)f(x')$$

for all values of λ such that $0 \leq \lambda \leq 1$. It is a **strictly convex function** if $\leq$ can be replaced by $<$. It is a **concave function** (or a **strictly concave function**) if this statement holds when $\leq$ is replaced by $\geq$ (or by $>$).

This definition has an enlightening geometric interpretation. Consider the graph of the function $f(x)$ drawn as a function of x. Then $[x', f(x')]$ and $[x'', f(x'')]$ are two points on the graph of $f(x)$, and $[\lambda x'' + (1 - \lambda)x', \lambda f(x'') + (1 - \lambda)f(x')]$ represents the various points on the line segment between these two points when $0 \leq \lambda \leq 1$. Thus the original inequality in the definition indicates that this line segment lies entirely above or on the graph of the function. Therefore, $f(x)$ is convex if, for each pair of points on the graph of $f(x)$, the line

605

segment joining these two points lies entirely above or on the graph of $f(x)$. In other words, $f(x)$ is convex if it is "always bending upward." (This condition is sometimes referred to as "concave upward," as opposed to "concave downward" for a concave function.) To be more precise, if $f(x)$ possesses a second derivative everywhere, then $f(x)$ is convex if and only if $d^2f(x)/dx^2 \geq 0$ for all values of x [for which $f(x)$ is defined]. Similarly, $f(x)$ is strictly convex when $d^2f(x)/dx^2 > 0$, concave when $d^2f(x)/dx^2 \leq 0$, and strictly concave when $d^2f(x)/dx^2 < 0$. Some examples are given in Figs. A1.1 to A1.4.

The concept of a convex function also generalizes to functions of more than one variable. Thus if $f(x)$ is replaced by $f(x_1, x_2, \ldots, x_n)$, the definition just given still applies if x is replaced everywhere by $(x_1, x_2, \ldots, x_n)$. Similarly, the corresponding geometric interpretation is still valid after generalizing the concepts of *points* and *line segments*. Thus, just as a particular value of (x, y) is interpreted as a point in two-dimensional space, each possible value of $(x_1, x_2, \ldots, x_m)$ may be thought of as a point in m-dimensional (Euclidean) space. By letting $m = n + 1$, the points on the graph of $f(x_1, x_2, \ldots, x_n)$ become the possible values of $[x_1, x_2, \ldots, x_n, f(x_1, x_2, \ldots, x_n)]$. Another point, $(x_1, x_2, \ldots, x_n, x_{n+1})$, is said to lie above, on, or below the graph of $f(x_1, x_2, \ldots, x_n)$, according to whether x_{n+1} is larger, equal to, or smaller than $f(x_1, x_2, \ldots, x_n)$, respectively.

Definition: The **line segment** joining any two points $(x_1', x_2', \ldots, x_m')$ and $(x_1'', x_2'', \ldots, x_m'')$ is the collection of points.

$$(x_1, x_2, \ldots, x_m) = [\lambda x_1'' + (1 - \lambda)x_1', \lambda x_2'' + (1 - \lambda)x_2', \ldots, \lambda x_m'' + (1 - \lambda)x_m']$$

such that $0 \leq \lambda \leq 1$.

Thus a line segment in m-dimensional space is a direct generalization of a line segment in two-dimensional space. For example, if

$$(x_1', x_2') = (2, 6), \qquad (x_1'', x_2'') = (3, 4),$$

then the line segment joining them is the collection of points,

$$(x_1, x_2) = [3\lambda + 2(1 - \lambda), 4\lambda + 6(1 - \lambda)]$$

where $0 \leq \lambda \leq 1$.

Definition: $f(x_1, x_2, \ldots, x_n)$ is a **convex function** if, for each pair of points on the graph of $f(x_1, x_2, \ldots, x_n)$, the line segment joining these two points lies entirely above or on the graph of $f(x_1, x_2, \ldots, x_n)$. It is a **strictly convex function** if this line segment actually lies entirely above this graph except at the endpoints of the line segment. **Concave functions** and **strictly concave functions** are defined in exactly the same way, except that *above* is replaced by *below*.

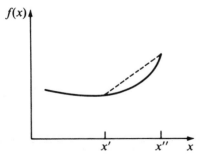

Figure A1.1 A convex function.

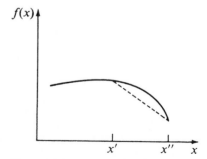

Figure A1.2 A concave function.

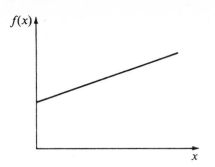

Figure A1.3 A function that is both convex and concave.

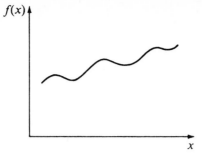

Figure A1.4 A function that is neither convex nor concave.

Just as the second derivative can be used (when it exists everywhere) to check whether a function of a single variable is convex, so second partial derivatives can be used to check functions of several variables, although in a more complicated way. For example, if there are two variables, then $f(x_1, x_2)$ is convex if and only if

(1)
$$\frac{\partial^2 f(x_1, x_2)}{\partial x_1^2} \frac{\partial^2 f(x_1, x_2)}{\partial x_2^2} - \left[\frac{\partial^2 f(x_1, x_2)}{\partial x_1 \, \partial x_2} \right]^2 \geq 0,$$

(2)
$$\frac{\partial^2 f(x_1, x_2)}{\partial x_1^2} \geq 0,$$

and

(3)
$$\frac{\partial^2 f(x_1, x_2)}{\partial x_2^2} \geq 0,$$

for all possible values of (x_1, x_2), assuming that these partial derivatives exist everywhere. It is strictly convex if $\geq$ can be replaced by $>$ in all three conditions [but now condition (3) is superfluous and can be omitted because it is implied by the other two conditions], whereas $f(x_1, x_2)$ is concave if $\geq$ can be replaced by $\leq$ in conditions (2) and (3). When there are more than two variables, the conditions for convexity are a generalization of the ones just shown. In mathematical terminology, $f(x_1, x_2, \ldots, x_n)$ is convex if and only if its $n \times n$ Hessian matrix is positive semidefinite for all possible values of $(x_1, x_2, \ldots, x_n)$.

Thus far convexity has been treated as a general property of a function. However, many nonconvex functions do satisfy the conditions for convexity over certain intervals for the respective variables. Therefore, it is meaningful to talk about a function being convex over a certain region. For example, a function is said to be convex within a neighborhood of a specified point if its second derivative or partial derivatives satisfy the conditions for convexity at that point. This concept is useful in Appendix 2.

Finally, two particularly important properties of convex functions should be mentioned. First, if $f(x_1, x_2, \ldots, x_n)$ is a convex function, then $g(x_1, x_2, \ldots, x_n) = -f(x_1, x_2, \ldots, x_n)$ is a concave function, and vice versa. Second, the sum of convex functions is a convex function. To illustrate,

$$f_1(x_1) = x_1^4 + 2x_1^2 - 5x_1$$

and

$$f_2(x_1, x_2) = x_1^2 + 2x_1x_2 + x_2^2$$

are both convex functions, as you can verify by calculating their second derivatives. Therefore, the sum of these functions,

$$f(x_1, x_2) = x_1^4 + 3x_1^2 - 5x_1 + 2x_1x_2 + x_2^2,$$

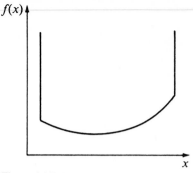

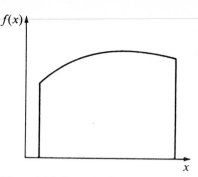

Figure A1.5 Example of a convex set determined by a convex function.

Figure A1.6 Example of a convex set determined by a concave function.

is a convex function, whereas its negative,

$$g(x_1, x_2) = -x_1^4 - 3x_1^2 + 5x_1 - 2x_1x_2 - x_2^2,$$

is a concave function.

The concept of a convex function leads quite naturally to the related concept of a **convex set**. Thus, if $f(x_1, x_2, \ldots, x_n)$ is a convex function, then the collection of points that lie above or on the graph of $f(x_1, x_2, \ldots, x_n)$ forms a convex set. Similarly, the collection of points that lie below or on the graph of a concave function is a convex set. These cases are illustrated in Figs. A1.5 and A1.6 for the case of a single independent variable. Furthermore, convex sets have the important property that, for any given group of convex sets, the collection of points that lie in all of them (i.e., the intersection of these convex sets) is also a convex set. Therefore, the collection of points that lie both above or on a convex function and below or on a concave function is a convex set, as illustrated in Fig. A1.7. Thus convex sets may be viewed intuitively as a collection of points whose bottom boundary is a convex function and whose top boundary is a concave function. To be a bit more precise, a convex set may be defined as follows:

> **Definition:** A **convex set** is a collection of points such that, for each pair of points in the collection, the entire line segment joining these two points is also in the collection.

The distinction between nonconvex sets and convex sets is illustrated in Figs. A1.8 and A1.9. Thus the set of points shown in Fig. A1.8 is not a convex set because there exist many pairs of these points, for example, $(1, 2)$ and $(2, 1)$, such that the line segment between them

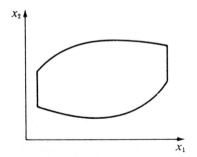

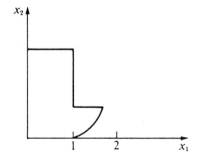

Figure A1.7 Example of a convex set determined by both convex and concave functions.

Figure A1.8 Example of a set that is not convex.

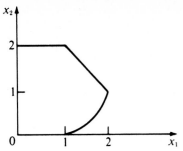

Figure A1.9 Example of a convex set.

does not lie entirely within the set. This is not the case for the set in Fig. A1.9, which is convex.

In conclusion, the useful concept of an *extreme point* of a convex set needs to be introduced.

> **Definition:** An **extreme point** of a convex set is a point in the set that does not lie on any line segment that joins two other points in the set.

Thus the extreme points of the convex set in Fig. A1.9 are $(0, 0)$, $(0, 2)$, $(1, 2)$, $(2, 1)$, $(1, 0)$, and all the infinite number of points on the boundary between $(2, 1)$ and $(1, 0)$. If this particular boundary were a line segment instead, then the set would have only the five listed extreme points.

Appendix 2

Classical Optimization Methods

This appendix reviews the classical methods of calculus for finding a solution that maximizes or minimizes (1) a function of a single variable, (2) a function of several variables, and (3) a function of several variables subject to constraints on the values of these variables. It is assumed that the functions considered possess continuous first and second derivatives and partial derivatives everywhere. Some of the concepts discussed next have been introduced briefly in Secs. 14.2 and 14.3.

Consider a function of a single variable, such as that shown in Fig. A2.1. A necessary condition for a particular solution, $x = x^*$, to be either a minimum or a maximum is that

$$\frac{df(x)}{dx} = 0 \qquad \text{at } x = x^*.$$

Thus in Fig. A2.1 there are five solutions satisfying these conditions. To obtain more information about these five so-called *critical points*, it is necessary to examine the second derivative. Thus, if

$$\frac{d^2f(x)}{dx^2} > 0 \qquad \text{at } x = x^*,$$

610

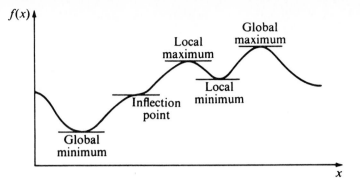

Figure A2.1 A function having several maxima and minima.

then x^* must be at least a *local minimum* [that is, $f(x^*) \leq f(x)$ for all x sufficiently close to x^*]. Using the language introduced in Appendix 1, we can say that x^* must be a local minimum if $f(x)$ is *strictly convex* within a neighborhood of x^*. Similarly, a sufficient condition for x^* to be a *local maximum* (given that it satisfies the necessary condition) is that $f(x)$ is *strictly concave* within a neighborhood of x^* (that is, the second derivative is *negative* at x^*). If the second derivative is zero, the issue is not resolved (the point may even be an *inflection point*), and it is necessary to examine higher derivatives.

To find a *global minimum* [i.e., a solution x^* such that $f(x^*) \leq f(x)$ for all x], it is necessary to compare the local minima and identify the one that yields the smallest value of $f(x)$. If this value is less than $f(x)$ as $x \rightarrow -\infty$ and as $x \rightarrow +\infty$ (or at the endpoints of the function, if it is defined only over a finite interval), then this point is a global minimum. Such a point is shown in Fig. A2.1, along with the global maximum, which is identified in an analogous way.

However, if $f(x)$ is known to be either a convex or concave function (see Appendix 1 for a description of such functions), the analysis becomes much simpler. In particular, if $f(x)$ is a *convex* function, such as the one shown in Fig. A1.1, then any solution x^* such that

$$\frac{df(x)}{dx} = 0 \quad \text{at } x = x^*$$

is known automatically to be a *global minimum*. In other words, this condition is not only a *necessary* but a *sufficient* condition for a global minimum of a convex function. If this function actually is strictly convex, then this solution must be the only global minimum. (However, if the function is either always decreasing or always increasing, so the derivative is nonzero for all values of x, then there will be no global minimum at a finite value of x.) Otherwise, there could be a tie for the global minimum over a single interval where the derivative is zero. Similarly, if $f(x)$ is a *concave* function, then having

$$\frac{df(x)}{dx} = 0 \quad \text{at } x = x^*$$

becomes both a *necessary* and *sufficient* condition for x^* to be a *global maximum*.

The analysis for an unconstrained function of several variables $f(\mathbf{x})$, where $\mathbf{x} = (x_1, x_2, \ldots, x_n)$, is similar. Thus a *necessary* condition for a solution $\mathbf{x} = \mathbf{x}^*$ to be either a minimum or a maximum is that

$$\frac{\partial f(\mathbf{x})}{\partial x_j} = 0 \quad \text{at } \mathbf{x} = \mathbf{x}^*, \quad \text{for } j = 1, 2, \ldots, n.$$

After identifying the critical points that satisfy this condition, each such point is then classified as a local minimum or maximum if the function is *strictly convex* or *strictly concave,* respectively, within a neighborhood of the point. (Additional analysis is required if the function is neither.) The *global minimum* and *maximum* would be found by comparing the local minima and maxima and then checking the value of the function as some of the variables approach $-\infty$ or $+\infty$. However, if the function is known to be *convex* or *concave*, then a critical point must be a *global minimum* or a *global maximum*, respectively.

Now consider the problem of finding the *minimum* or *maximum* of the function $f(\mathbf{x})$, subject to the restriction that $\mathbf{x}$ must satisfy all the equations

$$g_1(\mathbf{x}) = b_1$$

$$g_2(\mathbf{x}) = b_2$$

$$\vdots$$

$$g_m(\mathbf{x}) = b_m,$$

where $m < n$. For example, if $n = 2$ and $m = 1$, the problem might be

Maximize $\qquad f(x_1, x_2) = x_1^2 + 2x_2,$

subject to $\qquad g(x_1, x_2) = x_1^2 + x_2^2 = 1.$

In this case (x_1, x_2) is restricted to be on the circle of radius 1, whose center is at the origin, so that the goal is to find the point on this circle that yields the largest value of $f(x_1, x_2)$. This example is soon solved after a general approach to the problem is outlined.

A classical method of dealing with this problem is the **method of Lagrange multipliers**. This procedure begins by formulating the **Lagrangian function**,

$$h(\mathbf{x}, \boldsymbol{\lambda}) = f(\mathbf{x}) - \sum_{i=1}^{m} \lambda_i[g_i(\mathbf{x}) - b_i],$$

where the new variables $\boldsymbol{\lambda} = (\lambda_1, \lambda_2, \ldots, \lambda_m)$ are called *Lagrange multipliers*. Notice the key fact that for the *feasible* values of $\mathbf{x}$,

$$g_i(\mathbf{x}) - b_i = 0, \qquad \text{for all } i,$$

so $h(\mathbf{x}, \boldsymbol{\lambda}) = f(\mathbf{x})$. Therefore, it can be shown that if $(\mathbf{x}, \boldsymbol{\lambda}) = (\mathbf{x}^*, \boldsymbol{\lambda}^*)$ is a *local* or *global minimum* or *maximum* for the unconstrained function $h(\mathbf{x}, \boldsymbol{\lambda})$, then $\mathbf{x}^*$ is a corresponding *critical point* for the original problem. As a result, the method now reduces to analyzing $h(\mathbf{x}, \boldsymbol{\lambda})$ by the procedure just described for unconstrained functions. Thus the $(n + m)$ partial derivatives would be set equal to zero; that is,

$$\frac{\partial h}{\partial x_j} = \frac{\partial f}{\partial x_j} - \sum_{i=1}^{m} \lambda_i \frac{\partial g_i}{\partial x_j} = 0, \qquad \text{for } j = 1, 2, \ldots, n,$$

$$\frac{\partial h}{\partial \lambda_i} = -g_i(\mathbf{x}) + b_i = 0, \qquad \text{for } i = 1, 2, \ldots, m,$$

and then the critical points would be obtained by solving these equations for $(\mathbf{x}, \boldsymbol{\lambda})$. Notice that the last m equations are equivalent to the constraints in the original problem, so only feasible solutions are considered. After further analysis to identify the *global minimum* or *maximum* of $h(\cdot)$, the resulting value of $\mathbf{x}$ is then the desired solution to the original problem.

It should be pointed out that from a practical computational viewpoint, the method of Lagrange multipliers is not a particularly powerful procedure. It is often essentially impossible to solve the equations to obtain the critical points. Furthermore, even when they can be obtained, the number of critical points may be so large (often infinite) that it is impractical to attempt to identify a global minimum or maximum. However, for certain types of small problems, this method can sometimes be used successfully. To illustrate, consider the example introduced earlier. In this case,

$$h(x_1, x_2) = x_1^2 + 2x_2 - \lambda[x_1^2 + x_2^2 - 1],$$

so that

$$\frac{\partial h}{\partial x_1} = 2x_1 - 2\lambda x_1 = 0,$$

$$\frac{\partial h}{\partial x_2} = 2 - 2\lambda x_2 = 0,$$

$$\frac{\partial h}{\partial \lambda} = -[x_1^2 + x_2^2 - 1] = 0.$$

The first equation implies that either $\lambda = 1$ or $x_1 = 0$. If $\lambda = 1$, then the other two equations imply that $x_2 = 1$ and $x_1 = 0$. If $x_1 = 0$, then the third equation implies that $x_2 = \pm 1$. Therefore, the two critical points for the original problem are $(x_1, x_2) = (0, 1)$ and $(0, -1)$. Thus it is apparent that these points are the *global maximum* and *minimum*, respectively.

In presenting the classical optimization methods just described, we have assumed that you are already familiar with derivatives and how to obtain them. However, there is a special case of importance in operations research work that warrants additional explanation, namely, the derivative of an integral. In particular, consider how to find the derivative of the function

$$F(y) = \int_{g(y)}^{h(y)} f(x, y)\, dx,$$

where $g(y)$ and $h(y)$ are the limits of integration expressed as functions of y. To begin, suppose that these limits of integration are constants, so that $g(y) = a$ and $h(y) = b$, respectively. For this special case, it can be shown that, given the regularity conditions assumed at the beginning of this appendix, the derivative is simply

$$\frac{d}{dy} \int_a^b f(x, y)\, dx = \int_a^b \frac{\partial f(x, y)}{\partial y}\, dx.$$

For example, if $f(x, y) = e^{-xy}$, $a = 0$, and $b = \infty$, then

$$\frac{d}{dy} \int_0^\infty e^{-xy}\, dx = \int_0^\infty (-x)e^{-xy}\, dx = -\frac{1}{y^2}$$

at any positive value of y. Thus the intuitive procedure of interchanging the order of differentiation and integration is valid for this case. However, finding the derivative becomes a little more complicated than this when the limits of integration are functions. In particular,

$$\frac{d}{dy} \int_{g(y)}^{h(y)} f(x, y)\, dx = \int_{g(y)}^{h(y)} \frac{\partial f(x, y)}{\partial y}\, dx + f(h(y), y)\frac{dh(y)}{dy} - f(g(y), y)\frac{dg(y)}{dy},$$

where $f(h(y), y)$ is obtained by writing out $f(x, y)$ and then replacing x by $h(y)$ wherever it appears, and similarly for $f(g(y), y)$. To illustrate, if $f(x, y) = x^2y^3$, $g(y) = y$, and $h(y) = 2y$, then

$$\frac{d}{dy} \int_y^{2y} x^2y^3\, dx = \int_y^{2y} 3x^2y^2\, dx + (2y)^2y^3(2) - y^2y^3(1) = 14y^5$$

at any positive value of y.

Appendix 3

Matrices and Matrix Operations

A **matrix** is defined to be a *rectangular array of numbers*. For example,

$$\mathbf{A} = \begin{bmatrix} 2 & 5 \\ 3 & 0 \\ 1 & 1 \end{bmatrix}$$

is a 3 × 2 matrix (where 3 × 2 denotes "3 by 2") because it is a rectangular array of numbers with three rows and two columns. (Matrices are denoted in this book by **boldface capital letters**.) The numbers in the rectangular array are called the **elements** of the matrix. For example,

$$\mathbf{B} = \begin{bmatrix} 1 & 2.4 & 0 & \sqrt{3} \\ -4 & 2 & -1 & 15 \end{bmatrix}$$

is a 2 × 4 matrix whose elements are 1, 2.4, 0, $\sqrt{3}$, -4, 2, -1, and 15. Thus, in more general terms,

$$\mathbf{A} = \begin{bmatrix} a_{11} & a_{12} & \cdots & a_{1n} \\ a_{21} & a_{22} & \cdots & a_{2n} \\ \vdots & \vdots & & \vdots \\ a_{m1} & a_{m2} & & a_{mn} \end{bmatrix} = \|a_{ij}\|$$

is an $m \times n$ matrix, where $a_{11}, \ldots, a_{mn}$ represent the numbers that are the elements of this matrix; $\|a_{ij}\|$ is shorthand notation for identifying the matrix whose element in row i and column j is a_{ij} for every $i = 1, 2, \ldots, m$ and $j = 1, 2, \ldots, n$.

Because matrices do not possess a numerical value, they cannot be added, multiplied, and so on as if they were individual numbers. However, it is sometimes desirable to perform certain manipulations on arrays of numbers. Therefore, rules have been developed for performing operations on matrices that are analogous to arithmetic operations. To describe these, let $\mathbf{A} = \|a_{ij}\|$ and $\mathbf{B} = \|b_{ij}\|$ be two matrices having the same number of rows and the same number of columns. Then $\mathbf{A}$ and $\mathbf{B}$ are said to be equal ($\mathbf{A} = \mathbf{B}$) if and only if *all* of the corresponding elements are equal ($a_{ij} = b_{ij}$ for all i and j). The operation of multiplying a matrix by a number (denote this number by k) is performed by multiplying each element of the matrix by k, so that

$$k\mathbf{A} = \|ka_{ij}\|.$$

For example,

$$3 \begin{bmatrix} 1 & \frac{1}{3} & 2 \\ 5 & 0 & -3 \end{bmatrix} = \begin{bmatrix} 3 & 1 & 6 \\ 15 & 0 & -9 \end{bmatrix}.$$

To add $\mathbf{A}$ and $\mathbf{B}$, simply add the corresponding elements, so that

$$\mathbf{A} + \mathbf{B} = \|a_{ij} + b_{ij}\|.$$

To illustrate,

$$\begin{bmatrix} 5 & 3 \\ 1 & 6 \end{bmatrix} + \begin{bmatrix} 2 & 0 \\ 3 & 1 \end{bmatrix} = \begin{bmatrix} 7 & 3 \\ 4 & 7 \end{bmatrix}.$$

Similarly, subtraction is done as

$$\mathbf{A} - \mathbf{B} = \mathbf{A} + (-1)\mathbf{B},$$

so that

$$\mathbf{A} - \mathbf{B} = \|a_{ij} - b_{ij}\|.$$

For example,

$$\begin{bmatrix} 5 & 3 \\ 1 & 6 \end{bmatrix} - \begin{bmatrix} 2 & 0 \\ 3 & 1 \end{bmatrix} = \begin{bmatrix} 3 & 3 \\ -2 & 5 \end{bmatrix}.$$

Note that, with the exception of multiplication by a number, all the preceding operations are defined only when the two matrices involved are of the same size. However, all of them are straightforward because they involve performing only the same comparison or arithmetic operation on the corresponding elements of the matrices.

There exists one additional elementary operation that has not been defined, **matrix multiplication**, but it is considerably more complicated. To find the element in row i, column j of the matrix resulting from multiplying $\mathbf{A}$ times $\mathbf{B}$, it is necessary to multiply each element in row i of $\mathbf{A}$ by the corresponding element in column j of $\mathbf{B}$ and then to add these products. Therefore, the matrix multiplication is defined if and only if the number of columns of $\mathbf{A}$ equals the number of rows of $\mathbf{B}$, because this condition is required if we are to perform the specified element-by-element multiplication. Thus, if $\mathbf{A}$ is an $m \times n$ matrix and $\mathbf{B}$ is an $n \times r$ matrix, then their product is

$$\mathbf{AB} = \left\| \sum_{k=1}^{n} a_{ik}b_{kj} \right\|.$$

To illustrate,
$$\begin{bmatrix} 1 & 2 \\ 4 & 0 \\ 2 & 3 \end{bmatrix} \begin{bmatrix} 3 & 1 \\ 2 & 5 \end{bmatrix} = \begin{bmatrix} 1(3) + 2(2) & 1(1) + 2(5) \\ 4(3) + 0(2) & 4(1) + 0(5) \\ 2(3) + 3(2) & 2(1) + 3(5) \end{bmatrix} = \begin{bmatrix} 7 & 11 \\ 12 & 4 \\ 12 & 17 \end{bmatrix}.$$

On the other hand, if one attempts to multiply these matrices in the reverse order, the resulting product,

$$\begin{bmatrix} 3 & 1 \\ 2 & 5 \end{bmatrix} \begin{bmatrix} 1 & 2 \\ 4 & 0 \\ 2 & 3 \end{bmatrix},$$

is not even defined. Even when both **AB** and **BA** are defined,

$$\mathbf{AB} \neq \mathbf{BA}$$

in general. Thus *matrix multiplication* should be viewed as a specially designed operation whose properties are quite different from those of *arithmetic multiplication*. To understand why this special definition was adopted, consider the following system of equations:

$$2x_1 - x_2 + 5x_3 + x_4 = 20$$

$$x_1 + 5x_2 + 4x_3 + 5x_4 = 30$$

$$3x_1 + x_2 - 6x_3 + 2x_4 = 20.$$

Rather than writing these equations out as shown here, they can be written much more concisely in matrix form as

$$\mathbf{Ax} = \mathbf{b},$$

where
$$\mathbf{A} = \begin{bmatrix} 2 & -1 & 5 & 1 \\ 1 & 5 & 4 & 5 \\ 3 & 1 & -6 & 2 \end{bmatrix}, \qquad \mathbf{x} = \begin{bmatrix} x_1 \\ x_2 \\ x_3 \\ x_4 \end{bmatrix}, \qquad \mathbf{b} = \begin{bmatrix} 20 \\ 30 \\ 20 \end{bmatrix}.$$

It is this kind of multiplication for which matrix multiplication is designed.

Carefully note that *matrix division* is not defined.

Although the matrix operations described here do not possess certain of the properties of arithmetic operations, they do satisfy the following laws:

$$\mathbf{A} + \mathbf{B} = \mathbf{B} + \mathbf{A},$$

$$(\mathbf{A} + \mathbf{B}) + \mathbf{C} = \mathbf{A} + (\mathbf{B} + \mathbf{C}),$$

$$\mathbf{A}(\mathbf{B} + \mathbf{C}) = \mathbf{AB} + \mathbf{AC},$$

$$\mathbf{A}(\mathbf{BC}) = (\mathbf{AB})\mathbf{C},$$

when the relative sizes of these matrices are such that the indicated operations are defined.

Another type of matrix operation, which has no arithmetic analog, is the **transpose operation**. This operation involves nothing more than interchanging the rows and columns of the matrix, which is frequently useful for performing the multiplication operation in the desired way. Thus, for any matrix $\mathbf{A} = \|a_{ij}\|$, its transpose $\mathbf{A}^{\mathrm{T}}$ is

$$\mathbf{A}^{\mathrm{T}} = \|a_{ji}\|.$$

For example, if
$$\mathbf{A} = \begin{bmatrix} 2 & 5 \\ 1 & 3 \\ 4 & 0 \end{bmatrix},$$

then
$$\mathbf{A}^{\mathrm{T}} = \begin{bmatrix} 2 & 1 & 4 \\ 5 & 3 & 0 \end{bmatrix}.$$

Zero and 1 are numbers that play a special role in arithmetic. There also exist special matrices that play a similar role in matrix theory. In particular, the matrix that is analogous to 1 is the **identity matrix I**, which is a square matrix whose elements are zeroes except for ones along the main diagonal. Thus

$$I = \begin{bmatrix} 1 & 0 & 0 & \cdots & 0 \\ 0 & 1 & 0 & \cdots & 0 \\ 0 & 0 & 1 & \cdots & 0 \\ \vdots & \vdots & \vdots & & \vdots \\ 0 & 0 & 0 & \cdots & 1 \end{bmatrix}.$$

The number of rows or columns of **I** can be specified as desired. The analogy of **I** to 1 follows from the fact that for any matrix **A**,

$$IA = A = AI,$$

where **I** is assigned the appropriate number of rows and columns in each case for the multiplication operation to be defined. Similarly, the matrix that is analogous to zero is the so-called **null matrix 0**, which is a matrix of any size whose elements are *all zeroes*. Thus

$$0 = \begin{bmatrix} 0 & 0 & \cdots & 0 \\ 0 & 0 & \cdots & 0 \\ \vdots & \vdots & & \vdots \\ 0 & 0 & \cdots & 0 \end{bmatrix}.$$

Therefore, for any matrix **A**,

$$A + 0 = A, \qquad A - A = 0, \qquad \text{and} \qquad 0A = 0 = A0,$$

where **0** is the appropriate size in each case for the operations to be defined.

On certain occasions, it is useful to partition a matrix into several smaller matrices called **submatrices**. For example, one possible way of partitioning a 3×4 matrix would be

$$A = \begin{bmatrix} a_{11} & a_{12} & a_{13} & a_{14} \\ a_{21} & a_{22} & a_{23} & a_{24} \\ a_{31} & a_{32} & a_{33} & a_{34} \end{bmatrix} = \begin{bmatrix} a_{11} & A_{12} \\ A_{21} & A_{22} \end{bmatrix},$$

where $\quad A_{12} = [a_{12}, \; a_{13}, \; a_{14}], \qquad A_{21} = \begin{bmatrix} a_{21} \\ a_{31} \end{bmatrix}, \qquad A_{22} = \begin{bmatrix} a_{22} & a_{23} & a_{24} \\ a_{32} & a_{33} & a_{34} \end{bmatrix}$

all are submatrices. Rather than perform operations element by element on such partitioned matrices, we can instead do them in terms of the submatrices, provided the partitionings are such that the operations are defined. For example, if **B** is a partitioned 4×1 matrix such that

$$B = \begin{bmatrix} b_1 \\ b_2 \\ b_3 \\ b_4 \end{bmatrix} = \begin{bmatrix} b_1 \\ B_2 \end{bmatrix},$$

then

$$AB = \begin{bmatrix} a_{11}b_1 + A_{12}B_2 \\ A_{21}b_1 + A_{22}B_2 \end{bmatrix}.$$

A special kind of matrix that plays an important role in matrix theory is the kind that has either a *single row* or a *single column*. Such matrices are often referred to as **vectors**. Thus

$$x = [x_1, x_2, \ldots, x_n]$$

is a **row vector**, and

$$\mathbf{x} = \begin{bmatrix} x_1 \\ x_2 \\ \vdots \\ x_n \end{bmatrix}$$

is a **column vector**. (Vectors are denoted in this book by **boldface lowercase letters**.) These vectors also are sometimes called *n-vectors* to indicate that they have *n* elements. For example,

$$\mathbf{x} = [1, 4, -2, \tfrac{1}{3}, 7]$$

is a 5-vector. A **null vector 0** is either a row vector or a column vector whose elements are *all zeroes,* i.e.,

$$\mathbf{0} = [0, 0, \ldots, 0], \qquad \mathbf{0} = \begin{bmatrix} 0 \\ 0 \\ \vdots \\ 0 \end{bmatrix}$$

(Although the same symbol **0** is used for either kind of *null vector,* as well as for a *null matrix,* the context normally will identify which it is.)

One reason vectors play an important role in matrix theory is that any $m \times n$ matrix can be partitioned into either *m* row vectors or *n* column vectors, and important properties of the matrix can be analyzed in terms of these vectors. To amplify, consider a set of *n*-vectors, $\mathbf{x}_1, \mathbf{x}_2, \ldots, \mathbf{x}_m$, of the same type (i.e., they are either all row vectors or all column vectors).

Definition: A set of vectors $\mathbf{x}_1, \mathbf{x}_2, \ldots, \mathbf{x}_m$ is said to be **linearly dependent** if there exist *m* numbers (denoted by $c_1, c_2, \ldots, c_m$), some of which are not zero, such that

$$c_1\mathbf{x}_1 + c_2\mathbf{x}_2 + \cdots + c_m\mathbf{x}_m = \mathbf{0}.$$

Otherwise, the set is said to be **linearly independent**.

To illustrate, if $m = 3$ and

$$\mathbf{x}_1 = [1, 1, 1]$$

$$\mathbf{x}_2 = [0, 1, 1]$$

$$\mathbf{x}_3 = [2, 5, 5],$$

then
$$2\mathbf{x}_1 + 3\mathbf{x}_2 - \mathbf{x}_3 = \mathbf{0},$$

so that
$$\mathbf{x}_3 = 2\mathbf{x}_1 + 3\mathbf{x}_2.$$

Thus $\mathbf{x}_1, \mathbf{x}_2, \mathbf{x}_3$ would be linearly dependent because one of them is a linear combination of the others. However, if $\mathbf{x}_3$ were changed to

$$\mathbf{x}_3 = [2, 5, 6]$$

instead, then $\mathbf{x}_1, \mathbf{x}_2, \mathbf{x}_3$ would be linearly independent.

Definition: The **rank** of a *set* of vectors is the largest number of *linearly independent vectors* that can be chosen from the set.

Continuing the preceding example, the rank of the set of vectors $\mathbf{x}_1, \mathbf{x}_2, \mathbf{x}_3$ was 2, but it became 3 after changing $\mathbf{x}_3$.

Definition: A **basis** for a *set* of vectors is a *collection* of linearly independent vectors taken from the set such that every vector in the set is a linear combination of the vectors in the collection (i.e., every vector in the set equals the sum of certain multiples of the vectors in the collection).

To illustrate, x_1 and x_2 constituted a basis for x_1, x_2, x_3 in the preceding example before x_3 was changed.

Theorem A3.1: A *collection* of r linearly independent vectors chosen from a set of vectors is a *basis* for the set if and only if the set has rank r.

Given the preceding results regarding vectors, it is now possible to present certain important concepts regarding matrices.

Definition: The **row rank** of a matrix is the *rank* of its set of *row vectors*. The **column rank** of a matrix is the *rank* of its *column vectors*.

For example, if the matrix **A** is

$$\mathbf{A} = \begin{bmatrix} 1 & 1 & 1 \\ 0 & 1 & 1 \\ 2 & 5 & 5 \end{bmatrix},$$

then its row rank was shown to be 2. Note that the column rank of **A** is also 2. This fact is no coincidence, as the following general theorem indicates.

Theorem A3.2: The *row rank* and *column rank* of a matrix are *equal*.

Thus it is only necessary to speak of *the rank* of a matrix.

The final concept to be discussed is that of the **inverse of a matrix**. For any nonzero number k, there exists a reciprocal or inverse, $k^{-1} = 1/k$, such that

$$kk^{-1} = k^{-1}k = 1.$$

Is there an analogous concept that is valid in matrix theory? In other words, for a given matrix **A** other than the null matrix, does there exist a matrix $\mathbf{A}^{-1}$ such that

$$\mathbf{A}\mathbf{A}^{-1} = \mathbf{A}^{-1}\mathbf{A} = \mathbf{I}?$$

If **A** is not a square matrix (i.e., if the number of rows and columns of **A** differ), the answer is *never*, because these matrix products would necessarily have a different number of rows for the multiplication to be defined (so that the equality operation would not be defined). However, if **A** is square, then the answer is *under certain circumstances,* as indicated in Theorem A3.3.

Definition: A matrix is called **nonsingular** if its rank equals both the number of rows and the number of columns. Otherwise, it is called **singular**.

Thus only square matrices can be *nonsingular*. A useful way of testing for nonsingularity is provided by the fact that a square matrix is nonsingular if and only if *its determinant is nonzero*.

Theorem A3.3: (*a*) If **A** is *nonsingular,* there is a unique nonsingular matrix $\mathbf{A}^{-1}$, called the **inverse** of **A**, such that $\mathbf{A}\mathbf{A}^{-1} = \mathbf{I} = \mathbf{A}^{-1}\mathbf{A}$.
(*b*) If **A** is *nonsingular* and **B** is a matrix for which either $\mathbf{A}\mathbf{B} = \mathbf{I}$ or $\mathbf{B}\mathbf{A} = \mathbf{I}$, then $\mathbf{B} = \mathbf{A}^{-1}$.
(*c*) *Only nonsingular* matrices have *inverses*.

To illustrate, consider the matrix

$$A = \begin{bmatrix} 5 & -4 \\ 1 & -1 \end{bmatrix}.$$

Notice that the rank of **A** is 2, so it is *nonsingular*. Therefore, **A** must have an *inverse*, which happens to be

$$A^{-1} = \begin{bmatrix} 1 & -4 \\ 1 & -5 \end{bmatrix}.$$

Hence,

$$AA^{-1} = \begin{bmatrix} 5 & -4 \\ 1 & -1 \end{bmatrix}\begin{bmatrix} 1 & -4 \\ 1 & -5 \end{bmatrix} = \begin{bmatrix} 1 & 0 \\ 0 & 1 \end{bmatrix},$$

and

$$A^{-1}A = \begin{bmatrix} 1 & -4 \\ 1 & -5 \end{bmatrix}\begin{bmatrix} 5 & -4 \\ 1 & -1 \end{bmatrix} = \begin{bmatrix} 1 & 0 \\ 0 & 1 \end{bmatrix}.$$

Appendix 4

Simultaneous Linear Equations

Consider the system of simultaneous linear equations

$$a_{11}x_1 + a_{12}x_2 + \cdots + a_{1n}x_n = b_1,$$

$$a_{21}x_1 + a_{22}x_2 + \cdots + a_{2n}x_n = b_2,$$

$$\vdots$$

$$a_{m1}x_1 + a_{m2}x_2 + \cdots + a_{mn}x_n = b_m.$$

It is commonly assumed that this system has a solution, and a unique solution, if and only if $m = n$. However, this assumption is an oversimplification. It raises the questions: Under what conditions will these equations have a simultaneous solution? Given that they do, when will there be only one such solution? If there is a unique solution, how can it be identified in a systematic way? These questions are the ones we explore in this appendix. The discussion of the first two questions assumes that you are familiar with the basic information about matrices in Appendix 3.

621

The preceding system of equations can also be written in matrix form as

$$\mathbf{Ax} = \mathbf{b},$$

where

$$\mathbf{A} = \begin{bmatrix} a_{11} & a_{12} & \cdots & a_{1n} \\ a_{21} & a_{22} & \cdots & a_{2n} \\ & & \vdots & \\ a_{m1} & a_{m2} & \cdots & a_{mn} \end{bmatrix}, \quad \mathbf{x} = \begin{bmatrix} x_1 \\ x_2 \\ \vdots \\ x_n \end{bmatrix}, \quad \mathbf{b} = \begin{bmatrix} b_1 \\ b_2 \\ \vdots \\ b_m \end{bmatrix}.$$

The first two questions can be answered immediately in terms of the properties of these matrices. First, the system of equations possesses at least one solution if and only if the *rank* of $\mathbf{A}$ equals the *rank* of $[\mathbf{A}, \mathbf{b}]$. (Notice that equality is guaranteed if the *rank* of $\mathbf{A}$ equals m.) This result follows immediately from the definitions of *rank* and *linear independence* given in Appendix 3, because if the *rank* of $[\mathbf{A}, \mathbf{b}]$ exceeds the *rank* of $\mathbf{A}$ by 1 (the only other possibility), then $\mathbf{b}$ is *linearly independent* of the column vectors of $\mathbf{A}$ (that is, $\mathbf{b}$ cannot equal any linear combination $\mathbf{Ax}$ of these vectors).

Second, given that these ranks are equal, there are then two possibilities. If the *rank* of $\mathbf{A}$ is n (its maximum possible value), then the system of equations will possess exactly *one solution*. [This result follows from Theorem A3.1, the definition of a *basis*, and part (*b*) of Theorem A3.3.] If the *rank* of $\mathbf{A}$ is *less* than n, then there will exist an *infinite number of solutions*. (This result follows from the fact that for any *basis* of the column vectors of $\mathbf{A}$, the x_j corresponding to column vectors not in this basis can be assigned any value, and there will still exist a solution for the other variables as before.)

Finally, it should be noted that if $\mathbf{A}$ and $[\mathbf{A}, \mathbf{b}]$ have a *common rank* r such that $r < m$, then $(m - r)$ of the equations must be linear combinations of the other ones, so that these $(m - r)$ *redundant* equations can be deleted without affecting the solution(s). It then follows from the preceding results that this system of equations (with or without the redundant equations) possesses at least one solution, where the number of solutions is *one* if $r = n$ or *infinite* if $r < n$.

Now consider how to find a solution to the system of equations. Assume for the moment that $m = n$ and $\mathbf{A}$ is nonsingular, so that a unique solution exists. This solution can be obtained by the **Gauss-Jordan method of elimination** (commonly called **Gaussian elimination**), which proceeds as follows. To begin, eliminate the first variable from all but one (say, the first) of the equations by adding an appropriate multiple (positive or negative) of this equation to each of the others. (For convenience, this one equation would be divided by the coefficient of this variable, so that the final value of this coefficient is 1.) Next, proceed in the same way to eliminate the second variable from all equations except one new one (say, the second). Then repeat this procedure for the third variable, the fourth variable, and so on, until each of the n variables remains in only one of the equations and each of the n equations contains exactly one of these variables. The desired solution can then be read from the equations directly.

To illustrate the *Gauss-Jordan method of elimination,* we consider the following system of linear equations:

(1) $$x_1 - x_2 + 4x_3 = 10$$

(2) $$-x_1 + 3x_2 \qquad = 10$$

(3) $$2x_2 + 5x_3 = 22.$$

The method begins by eliminating x_1 from all but the first equation. This first step is executed simply by adding Eq. (1) to Eq. (2), which yields

(1) $$x_1 - x_2 + 4x_3 = 10$$

(2) $$2x_2 + 4x_3 = 20$$

(3) $$2x_2 + 5x_3 = 22.$$

The next step is to eliminate x_2 from all but the second equation. Begin this step by dividing Eq. (2) by 2, so that x_2 will have a coefficient of $+1$, as follows:

(1) $$x_1 - x_2 + 4x_3 = 10$$

(2) $$x_2 + 2x_3 = 10$$

(3) $$2x_2 + 5x_3 = 22.$$

Then add Eq. (2) to Eq. (1), and subtract two times Eq. (2) from Eq. (3), which yields

(1) $$x_1 + 6x_3 = 20$$

(2) $$x_2 + 2x_3 = 10$$

(3) $$x_3 = 2.$$

The final step is to eliminate x_3 from all but the third equation. This step requires subtracting six times Eq. (3) from Eq. (1) and subtracting two times Eq. (3) from Eq. (2), which yields

(1) $$x_1 = 8$$

(2) $$x_2 = 6$$

(3) $$x_3 = 2.$$

Thus the desired solution is $(x_1, x_2, x_3) = (8, 6, 2)$, and the procedure is completed.

Now consider briefly what happens if the Gauss-Jordan method of elimination is applied when $m \neq n$ and/or $\mathbf{A}$ is singular. As we discussed earlier, there are three possible cases to consider. First, if the rank of $[\mathbf{A}, \mathbf{b}]$ exceeds the rank of $\mathbf{A}$ by 1, then *no solution* to the system of equations will exist. In this case, the Gauss-Jordan method obtains an equation where the left-hand side has vanished (i.e., all the coefficients of the variables are zero), whereas the right-hand side is nonzero. This signpost indicates that no solution exists, so there is no reason to proceed further.

The second case is where both of these ranks are equal to n, so that a *unique solution* exists. This case implies that $m \geq n$. If $m = n$, then the previous assumptions must hold and no difficulty arises. Therefore, suppose that $m > n$, so that there are $(m - n)$ redundant equations. In this case, all these redundant equations are eliminated (i.e., both the left-hand and right-hand sides would become zero) during the process of executing the Gauss-Jordan method, so the unique solution is identified just as it was before.

The final case is where both the ranks are equal to r, where $r < n$, so that the system of equations possesses an *infinite number of solutions*. In this case, at the completion of the Gauss-Jordan method, each of the r variables remains in only one of the equations, and each of the r equations (any additional equations have vanished) contains exactly one of these variables. However, each of the other $(n - r)$ variables either vanishes or remains in some of the equations. Therefore, any solution obtained by assigning arbitrary values to the $(n - r)$ variables, and then identifying the respective values of the r variables from the single final equation in which each one appears, is a solution to the system of simultaneous equations. Equivalently, the transfer of these $(n - r)$ variables to the right-hand side of the equations (either before or after the method is executed) identifies the solutions for the r variables as a function of these extra variables.

Appendix 5

Table for the Normal Distribution

Table A5.1 **Areas Under the Normal Curve from K_α to ∞**

$$P\{\text{normal} \geq K_\alpha\} = \int_{K_\alpha}^{\infty} \frac{1}{\sqrt{2\pi}} e^{-x^2/2}\, dx = \alpha$$

K_α	.00	.01	.02	.03	.04	.05	.06	.07	.08	.09
0.0	.5000	.4960	.4920	.4880	.4840	.4801	.4761	.4721	.4681	.4641
0.1	.4602	.4562	.4522	.4483	.4443	.4404	.4364	.4325	.4286	.4247
0.2	.4207	.4168	.4129	.4090	.4052	.4013	.3974	.3936	.3897	.3859
0.3	.3821	.3783	.3745	.3707	.3669	.3632	.3594	.3557	.3520	.3483
0.4	.3446	.3409	.3372	.3336	.3300	.3264	.3228	.3192	.3156	.3121
0.5	.3085	.3050	.3015	.2981	.2946	.2912	.2877	.2843	.2810	.2776
0.6	.2743	.2709	.2676	.2643	.2611	.2578	.2546	.2514	.2483	.2451
0.7	.2420	.2389	.2358	.2327	.2296	.2266	.2236	.2206	.2177	.2148
0.8	.2119	.2090	.2061	.2033	.2005	.1977	.1949	.1922	.1894	.1867
0.9	.1841	.1814	.1788	.1762	.1736	.1711	.1685	.1660	.1635	.1611
1.0	.1587	.1562	.1539	.1515	.1492	.1469	.1446	.1423	.1401	.1379
1.1	.1357	.1335	.1314	.1292	.1271	.1251	.1230	.1210	.1190	.1170
1.2	.1151	.1131	.1112	.1093	.1075	.1056	.1038	.1020	.1003	.0985
1.3	.0968	.0951	.0934	.0918	.0901	.0885	.0869	.0853	.0838	.0823
1.4	.0808	.0793	.0778	.0764	.0749	.0735	.0721	.0708	.0694	.0681
1.5	.0668	.0655	.0643	.0630	.0618	.0606	.0594	.0582	.0571	.0559
1.6	.0548	.0537	.0526	.0516	.0505	.0495	.0485	.0475	.0465	.0455
1.7	.0446	.0436	.0427	.0418	.0409	.0401	.0392	.0384	.0375	.0367
1.8	.0359	.0351	.0344	.0336	.0329	.0322	.0314	.0307	.0301	.0294
1.9	.0287	.0281	.0274	.0268	.0262	.0256	.0250	.0244	.0239	.0233
2.0	.0228	.0222	.0217	.0212	.0207	.0202	.0197	.0192	.0188	.0183
2.1	.0179	.0174	.0170	.0166	.0162	.0158	.0154	.0150	.0146	.0143
2.2	.0139	.0136	.0132	.0129	.0125	.0122	.0119	.0116	.0113	.0110
2.3	.0107	.0104	.0102	.00990	.00964	.00939	.00914	.00889	.00866	.00842
2.4	.00820	.00798	.00776	.00755	.00734	.00714	.00695	.00676	.00657	.00639
2.5	.00621	.00604	.00587	.00570	.00554	.00539	.00523	.00508	.00494	.00480
2.6	.00466	.00453	.00440	.00427	.00415	.00402	.00391	.00379	.00368	.00357
2.7	.00347	.00336	.00326	.00317	.00307	.00298	.00289	.00280	.00272	.00264
2.8	.00256	.00248	.00240	.00233	.00226	.00219	.00212	.00205	.00199	.00193
2.9	.00187	.00181	.00175	.00169	.00164	.00159	.00154	.00149	.00144	.00139

K_α	.0	.1	.2	.3	.4	.5	.6	.7	.8	.9
3	.00135	$.0^3968$	$.0^3687$	$.0^3483$	$.0^3337$	$.0^3233$	$.0^3159$	$.0^3108$	$.0^4723$	$.0^4481$
4	$.0^4317$	$.0^4207$	$.0^4133$	$.0^5854$	$.0^5541$	$.0^5340$	$.0^5211$	$.0^5130$	$.0^6793$	$.0^6479$
5	$.0^6287$	$.0^6170$	$.0^7996$	$.0^7579$	$.0^7333$	$.0^7190$	$.0^7107$	$.0^8599$	$.0^8332$	$.0^8182$
6	$.0^9987$	$.0^9530$	$.0^9282$	$.0^9149$	$.0^{10}777$	$.0^{10}402$	$.0^{10}206$	$.0^{10}104$	$.0^{11}523$	$.0^{11}260$

Source: Croxton, F. E.: *Tables of Areas in Two Tails and in One Tail of the Normal Curve.* Copyright 1949 by Prentice-Hall, Inc., Englewood Cliffs, N.J.

Answers to Selected Problems

Chapter 3

1. (b)

$$\text{Maximize} \quad Z = 4{,}500x_1 + 4{,}500x_2,$$

subject to

$$x_1 \qquad\qquad\;\; \leq \quad 1$$

$$x_2 \leq \quad 1$$

$$5{,}000x_1 + 4{,}000x_2 \leq 6{,}000$$

$$400x_1 + \quad 500x_2 \leq \quad 600$$

and

$$x_1 \geq 0, \qquad x_2 \geq 0.$$

3. $(x_1, x_2) = (13, 5); Z = 31.$

Chapter 4

4. $(x_1, x_2, x_3) = (0, 10, 6\frac{2}{3}); Z = 70.$

16. $(x_1, x_2) = (2, 1); Z = 7.$

626

21. $(x_1, x_2) = (-\frac{8}{7}, \frac{18}{7}); Z = \frac{80}{7}.$

26. $(x_1, x_2, x_3) = (\frac{4}{5}, \frac{9}{5}, 0),$ with $Z = 7.$

Chapter 5

1. (a) $(x_1, x_2) = (2, 2)$ is optimal. Other corner-point feasible solutions are $(0, 0), (3, 0),$ and $(0, 3).$

10. $(x_1, x_2, x_3) = (0, \frac{5}{2}, \frac{5}{2})$ is optimal.

14. $(x_1, x_2, x_3, x_4, x_5) = (0, 5, 0, \frac{5}{2}, 0)$ with $Z = 50$ is optimal.

27. (a) Right side is $Z = 8, x_2 = 14, x_6 = 5, x_3 = 11.$
(b) $x_1 = 0, 2x_1 - 2x_2 + 3x_3 = 5, x_1 + x_2 - x_3 = 3.$

Chapter 6

2. (a) $\qquad$ Minimize $\quad y_0 = 15y_1 + 12y_2 + 45y_3,$

$\quad$ subject to $\qquad\qquad -y_1 + y_2 + 5y_3 \geq 10$

$$2y_1 + y_2 + 3y_3 \geq 20$$

$\quad$ and $\qquad\qquad y_1 \geq 0, \qquad y_2 \geq 0, \qquad y_3 \geq 0.$

7. (c)

Complementary Basic Solutions					
Primal Problem				**Dual Problem**	
Basic Solution	Feasible?	$Z = y_0$	Feasible?	Basic Solution	
$(0, 0, 20, 10)$	Yes	0	No	$(0, 0, -6, -8)$	
$(4, 0, 0, 6)$	Yes	24	No	$(1\frac{1}{5}, 0, 0, -5\frac{3}{5})$	
$(0, 5, 10, 0)$	Yes	40	No	$(0, 4, -2, 0)$	
$(2\frac{1}{2}, 3\frac{3}{4}, 0, 0)$	Yes and optimal	45	Yes and optimal	$(\frac{1}{2}, 3\frac{1}{2}, 0, 0)$	
$(10, 0, -30, 0)$	No	60	Yes	$(0, 6, 0, 4)$	
$(0, 10, 0, -10)$	No	80	Yes	$(4, 0, 14, 0)$	

18. $\qquad\qquad\qquad$ Maximize $\quad y_0 = 8y_1 + 6y_2,$

$\quad$ subject to $\qquad\qquad y_1 + 3y_2 \leq 2$

$$4y_1 + 2y_2 \leq 3$$

$$2y_1 \qquad\quad \leq 1$$

$\quad$ and $\qquad\qquad y_1 \geq 0, \qquad y_2 \geq 0.$

21. (a) $\qquad\qquad$ Minimize $\quad y_0 = 30y_1 + 20y_2 + 25y_3,$

$\quad$ subject to $\qquad\qquad y_2 - 3y_3 = -1$

$$3y_1 - y_2 + y_3 = 2$$

$$y_1 - 4y_2 + 2y_3 = 1$$

$\quad$ and $\qquad\qquad y_1 \geq 0, \qquad y_2 \geq 0, \qquad y_3 \geq 0.$

24. (d) Not optimal, since $2y_1 + 3y_2 \geq 3$ is violated for $y_1^* = \frac{1}{5}, y_2^* = \frac{3}{5}.$
(f) Not optimal, since $3y_1 + 2y_2 \geq 2$ is violated for $y_1^* = \frac{1}{5}, y_2^* = \frac{3}{5}.$

35.

Part	New Basic Solution $(x_1, x_2, x_3, x_4, x_5)$	Feasible?	Optimal?
(a)	$(0, 30, 0, 0, -30)$	No	No
(b)	$(0, 20, 0, 0, -10)$	No	No
(c)	$(0, 10, 0, 0, 60)$	Yes	Yes
(d)	$(0, 20, 0, 0, 10)$	Yes	Yes
(e)	$(0, 20, 0, 0, 10)$	Yes	Yes
(f)	$(0, 10, 0, 0, 40)$	Yes	No
(g)	$(0, 20, 0, 0, 10)$	Yes	Yes
(h)	$(0, 20, 0, 0, 10, x_6 = -10)$	No	No
(i)	$(0, 20, 0, 0, 0)$	Yes	Yes

36. $-10 \le \theta \le \frac{10}{9}$.

Chapter 7

1. Let x_{ij} be the shipment from plant i to distribution center j. Then $x_{13} = 2$, $x_{14} = 10$, $x_{22} = 9$, $x_{23} = 8$, $x_{31} = 10$, $x_{32} = 1$; cost = \$20,200.

5. (Answer in millions of acres) England → 70 oats; France → 110 wheat; Spain → 15 wheat, 60 barley, 5 oats.

12. (a) $x_{11} = 3$, $x_{12} = 2$, $x_{22} = 1$, $x_{23} = 1$, $x_{33} = 1$, $x_{34} = 2$; four iterations to reach optimality.

(b) and (c) $x_{11} = 3$, $x_{12} = 0$, $x_{13} = 0$, $x_{14} = 2$, $x_{23} = 2$, $x_{32} = 3$; already optimal.

19. $x_{11} = 10$, $x_{12} = 15$, $x_{22} = 0$, $x_{23} = 5$, $x_{25} = 30$, $x_{33} = 20$, $x_{34} = 10$, $x_{44} = 10$; cost = 77.30. Also have other tied optimal solutions.

24. $x_{14} = 20$, $x_{16} = 50$, $x_{23} = 10$, $x_{24} = 10$, $x_{25} = 60$, $x_{37} = 60$.

30. Back → David, breast → Tony, butterfly → Chris, freestyle → Carl; time = 126.2.

39.

Master Problem	Subproblem 1	Subproblem 2
$3x_1 + 2x_2 \le 18$	$x_1 \le 4$	$2x_2 \le 12$

Chapter 8

7. $(x_1, x_2) = (10, 5)$ is optimal.

Chapter 9

3. $(x_1, x_2, x_3) = (1, 3, 1)$ with $Z = 8$ is optimal.

8. $(x_1, x_2, x_3) = (\frac{2}{3}, 2, 0)$ with $Z = \frac{22}{3}$ is optimal.

12.

Part	New Optimal Solution	Value of Z
(a)	$(x_1, x_2, x_3, x_4, x_5) = (0, 0, 9, 3, 0)$	117
(b)	$(x_1, x_2, x_3, x_4, x_5) = (0, 5, 5, 0, 0)$	90

15. (b)

Range of θ	Optimal Solution	$Z(\theta)$
$0 \le \theta \le 2$	$(x_1, x_2) = (0, 5)$	$120 - 10\theta$
$2 \le \theta \le 8$	$(x_1, x_2) = (\frac{10}{3}, \frac{10}{3})$	$\dfrac{320 - 10\theta}{3}$
$8 \le \theta$	$(x_1, x_2) = (5, 0)$	$40 + 5\theta$

20.

Range of θ	Optimal Solution		$Z(\theta)$
	x_1	x_2	
$0 \le \theta \le 1$	$10 + 2\theta$	$10 + 2\theta$	$30 + 6\theta$
$1 \le \theta \le 5$	$10 + 2\theta$	$15 - 3\theta$	$35 + \theta$
$5 \le \theta \le 25$	$25 - \theta$	0	$50 - 2\theta$

Chapter 10

3. (a) $O \to A \to B \to D \to T$ or $O \to A \to B \to E \to D \to T$, with length $= 16$.

7. (a) $\{(O, A); (A, B); (B, C); (B, E); (E, D); (D, T)\}$, with length $= 18$.

10. (a)

Arc	(1, 2)	(1, 3)	(1, 4)	(2, 5)	(3, 4)	(3, 5)	(3, 6)	(4, 6)	(5, 7)	(6, 7)
Flow	4	4	1	4	1	0	3	2	4	5

20.

Event	1	2	3	4	5	6	7	8	9	10
Earliest time	0	6	3	5	10	10	11	14	13	20
Latest time	0	7	3	6	11	10	13	14	15	20
Slack	0	1	0	1	1	0	2	0	2	0

Critical path: $1 \to 3 \to 6 \to 8 \to 10$.

23. $t_e = 37, \sigma^2 = 9$.

Chapter 11

3.

	Store		
	1	2	3
Allocations	1	2	2
	3	2	0

10.

Phase	(a)	(b)
1	$2 M$	$2.945 M$
2	$1 M$	$1.055 M$
3	$1 M$	0
Market share	6%	6.302%

13. $x_1 = -2 + \sqrt{13} \approx 1.6056$, $x_2 = 5 - \sqrt{13} \approx 1.3944$; $Z = 98.233$.

24. Produce 2 on first production run; if none acceptable, produce 2 on second run.
Expected cost $= \$573$.

Chapter 12

1. (a) Player I: strategy 2; player II: strategy 1.

7. (a) Politician I: issue 2; politician II: issue 2.
   (b) Politician I: issue 1; politician II: issue 2.
   (c) Minimax criterion says politician I can use any issue, but issue 1 offers politician I
   the only chance of winning if politician II is not "smart."

11. (a) $(x_1, x_2) = (\frac{2}{5}, \frac{3}{5})$; $(y_1, y_2, y_3) = (\frac{1}{5}, 0, \frac{4}{5})$; $v = \frac{8}{5}$.

14.

$$\text{Minimize} \quad -x_4,$$

subject to

$$5x_1 + 2x_2 + 3x_3 - x_4 \geq 0$$
$$4x_2 + 2x_3 - x_4 \geq 0$$
$$3x_1 + 3x_2 \qquad\quad - x_4 \geq 0$$
$$x_1 + 2x_2 + 4x_3 - x_4 \geq 0$$
$$-x_1 - x_2 - x_3 \qquad\quad = -1$$

and

$$x_1 \geq 0, \quad x_2 \geq 0, \quad x_3 \geq 0, \quad x_4 \geq 0.$$

Chapter 13

10. (*b*) (long, medium, short) = (14, 0, 16), with profit of $9,560,000.

18.

Assignment	1	2	3	4	5
Assignee	1	3	2	4	5

20. $(x_1, x_2, x_3, x_4) = (0, 1, 1, 0)$, with $Z = 36$.
22. $(x_1, x_2, x_3, x_4, x_5) = (0, 0, 1, 1, 1)$, with $Z = 6$.

Chapter 14

8. (*a*) Concave.
16. Approximate solution $= 1.0125$.
25. Exact solution is $(x_1, x_2) = (2, -2)$.
29. (*a*) Approximate solution is $(x_1, x_2) = (0.75, 1.875)$.
33. $(x_1, x_2) = (1, 2)$ cannot be optimal.
36. (*a*) $(x_1, x_2) = (1 - 3^{-1/2}, 3^{-1/2})$.
44. (*a*) $(x_1, x_2) = (2, 0)$ is optimal.

(*b*)

$$\text{Minimize} \quad Z = Z_1 + Z_2,$$

subject to

$$2x_1 \qquad\quad + u_1 - y_1 \qquad\qquad + z_1 \qquad\quad = 8$$
$$2x_2 + u_1 \qquad\quad - y_2 \qquad\qquad + z_2 = 4$$
$$x_1 + x_2 \qquad\qquad\quad + v_1 \qquad\qquad = 2$$
$$x_1 \geq 0, \quad x_2 \geq 0, \quad u_1 \geq 0, \quad y_1 \geq 0, \quad y_2 \geq 0, \quad v_1 \geq 0, \quad z_1 \geq 0, \quad z_2 \geq 0.$$

(*c*) $(x_1, x_2, u_1, y_1, y_2, v_1, z_1, z_2) = (2, 0, 4, 0, 0, 0, 0, 0)$ is optimal.

50. (*b*)

$$\text{Maximize} \quad Z = 2.625x_{11} - 17.625x_{12} + 4.5x_{21} - 4.5x_{22},$$

subject to

$$x_{11} + x_{12} + 3x_{21} + 3x_{22} \leq 8$$
$$5x_{11} + 5x_{12} + 2x_{21} + 2x_{22} \leq 14$$

and

$$0 \leq x_{ij} \leq 1.5, \quad \text{for } i = 1, 2 \text{ and } j = 1, 2.$$

58. (*a*) $(x_1, x_2) = (\tfrac{1}{3}, \tfrac{2}{3})$.

61.

$$(x_1, x_2) = \left[3 + \left(\frac{r}{2}\right)^{1/3}, \ 3 + \left(\frac{r}{2}\right)^{1/2} \right] \quad \text{maximizes } P(\mathbf{x}; r), \text{ so that } (x_1, x_2) = (3, 3) \text{ is}$$

optimal.

1. (*a*) $t = 1.83$, $Q = 54.77$.
   (*b*) $t = 1.91$, $Q = 57.45$, $S = 52.22$.
4. $t = 3.26$, $Q = 26{,}046$, $S = 24{,}572$.
15. Produce 7 units in period 1 and 7 units in period 3.
18. Produce 3 units in period 1 and 4 units in period 3.

Indexes

Author Index

Subject Index

639

Instructions for Using
OR COURSEWARE

IBM Compatible Computers

1. Insert diskette in drive A.
2. Make drive A the current drive.
3. Type the name of the program (for example, LinProg) and press <ENTER>.
4. Programs may be copied to and run from a hard disk.

Macintosh Computers

1. The diskette does not include a system folder, so you must provide your own.
2. Insert the diskette and double click on the program icon (for example, LinProg).
3. If the computer you are using does not have arrow keys, use the following substitutes:

```
                    UP
                    {
      LEFT [               ] RIGHT
                    }
                  DOWN
```

3. Programs may be copied to and run from a hard disk.